S0-BMG-581

Physiology
and Biochemistry
of the Domestic Fowl

Physiology and Biochemistry of the Domestic Fowl

Volume 1

Edited by

D. J. BELL
Department of Physiology,
University Medical School,
Edinburgh, Scotland

and

B. M. FREEMAN
Houghton Poultry Research Station,
Houghton, Huntingdon,
England

1971

ACADEMIC PRESS · London · New York

ACADEMIC PRESS INC. (LONDON) LTD
24–28 Oval Road,
Camden Town,
NW1 7DD

U.S. Edition published by
ACADEMIC PRESS INC.
111 Fifth Avenue,
New York, New York 10003

Library of Congress Catalog Card Number: 75–170757

International Standard Book Number: 0–12–085001–X

PRINTED IN GREAT BRITAIN BY
WILLIAM CLOWES & SONS, LIMITED
LONDON, BECCLES AND COLCHESTER

Contributors

E. F. ANNISON, *Unilever Research Laboratory, Colworth House, Sharnbrook, Bedford, England*

K. N. BOORMAN, *Department of Applied Biochemistry and Nutrition, School of Agriculture, University of Nottingham, Sutton Bonington, Loughborough, LE12 5RD, England*

W. O. BROWN, *Department of Agricultural Chemistry, The Queen's University of Belfast, Elmwood Avenue, Belfast, BT9 6BB, Northern Ireland*

E. J. BUTLER, *Department of Physiology and Biochemistry, Houghton Poultry Research Station, Houghton, Huntingdon, PE17 2DA, England*

M. E. COATES, *National Institute for Research in Dairying, University of Reading, Shinfield, Reading, RG2 9AT, England*

F. J. CUNNINGHAM, *Department of Physiology and Biochemistry, University of Reading, Whiteknights, Reading, RG6 2AJ, England*

C. G. DACKE, *Department of Physiology and Biochemistry, University of Reading, Whiteknights, Reading, RG6 2AJ, England*

I. R. FALCONER, *Department of Applied Biochemistry and Nutrition, School of Agriculture, University of Nottingham, Sutton Bonington, Loughborough, LE12 5RD, England*

B. M. FREEMAN, *Department of Physiology and Biochemistry, Houghton Poultry Research Station, Houghton, Huntingdon, PE17 2DA, England*

R. FULLER, *National Institute for Research in Dairying, University of Reading, Shinfield, Reading, RG2 9AT, England*

C. N. HALES,[1] *Department of Biochemistry, University of Cambridge, Tennis Court Road, Cambridge, England*

K. J. HILL, *Unilever Research Laboratory, Colworth House, Sharnbrook, Bedford, England*

E. W. HORTON, *Department of Pharmacology, University of Edinburgh, 1 George Square, Edinburgh, EH8 9JZ, Scotland*

D. A. HUDSON,[2] *Department of Physiology, University of Sheffield, S10 2TN, England*

D. J. JAYNE-WILLIAMS, *National Institute for Research in Dairying, University of Reading, Shinfield, Reading, RG2 9AT, England*

M. G. M. JUKES, *Department of Physiology, Royal Veterinary College, Royal College Street, London, N.W.1, England*

[1] Present address: The Welsh National School of Medicine, Dept. of Chemical Pathology, The Royal Infirmary, Cardiff, Wales.

[2] Present address: Dunn Nutritional Laboratories, Milton Road, Cambridge, CB4 1XJ, England.

A. S. KING, *Department of Veterinary Anatomy, University of Liverpool, Brownlow Hill and Crown Street, Liverpool, L69 3BX, England*

D. R. LANGSLOW, *Department of Physiology and Biochemistry, Houghton Poultry Research Station, Houghton, Huntingdon, PE17 2DA, England*

R. J. LEVIN, *Department of Physiology, University of Sheffield, Sheffield, S10 2TN, England*

D. LEWIS, *Department of Applied Biochemistry and Nutrition, School of Agriculture, University of Nottingham, Sutton Bonington, Loughborough, LE12 5RD, England*

V. MOLONY, *Unit of Comparative Neurobiology, University of Liverpool, Brownlow Hill and Crown Street, Liverpool, L69 3BX, England*

J. PEARCE, *Department of Agricultural Chemistry, The Queen's University of Belfast, Elmwood Avenue, Belfast, BT9 6BB, Northern Ireland*

W. G. SILLER, *Anatomy Section, Agricultural Research Council's Poultry Research Centre, West Mains Road, Edinburgh, EH9 3JS, Scotland*

K. SIMKISS, *Department of Zoology and Comparative Physiology, Queen Mary College, University of London, Mile End Road, London E.1, England*

D. H. SMYTH, *Department of Physiology, University of Sheffield, Sheffield, S10 2TN, England*

A. STOCKELL HARTREE, *Department of Biochemistry, University of Cambridge, Tennis Court Road, Cambridge, England*

A. H. SYKES, *Wye College, University of London, Wye, Ashford, Kent, England*

T. G. TAYLOR, *Department of Physiology and Biochemistry, University of Southampton, Southampton, SO9 5NH, England*

J. W. WELLS, *Reproduction Section, Agricultural Research Council's Poultry Research Centre, West Mains Road, Edinburgh, EH9 3JS, Scotland*

P. A. L. WIGHT, *Anatomy Section, Agricultural Research Council's Poultry Research Centre, West Mains Road, Edinburgh, EH9 3JS, Scotland*

Preface

To those uninitiated into the recent activities in avian science the familiar domesticated fowl often engenders indifference or even contempt. This is regrettable because this bird possesses many fascinating and occasionally unique scientific features which deserve, and indeed nowadays receive, serious attention. We hope that these volumes may improve the appreciation of what has been discovered about this valuable bird.

For more than 5,000 years man has used the domesticated fowl as a source of protein. Its growth potential, its reproductive capacity and its efficiency as a converter of vegetable protein have been heavily exploited so that in many countries the domestic fowl is now second only to the dairy cow as a source of human food.

In multi-authored works some overlap and repetition is often unavoidable; this work is no exception. For such we make no apology; indeed we believe that editing more drastic than we have attempted would have produced a less readable and less valuable text. In preparing their contributions the authors have, for the most part, been able to survey the literature published to the end of 1970.

To conclude, both editors record their indebtedness to their wives for their forbearance during the many months of editing and to the staff of the Academic Press for their help in bringing the book to press.

D. J. BELL
B. M. FREEMAN

JULY 1971

Contents

3 Absorption from the Alimentary Tract

D. A. HUDSON, R. J. LEVIN and D. H. SMYTH

13 Protein Metabolism

K. N. BOORMAN and D. LEWIS

14 The Role of Vitamins in Metabolic Processes

M. E. COATES

15 The Role of Trace Elements in Metabolic Processes
E. J. BUTLER

16 The Pituitary Gland
A. STOCKELL HARTREE and F. J. CUNNINGHAM

Contents of Volume 2

Contents of Volume 3

1 The Structure of the Alimentary Tract

K. J. HILL

Unilever Research Laboratory, Colworth House, Sharnbrook, Bedford, England

I. Introduction

There is a considerable literature on the structure, development and histology of the avian digestive tract (see Bradley, 1960; Farner, 1960; Romanoff, 1960; Calhoun, 1961) and the characteristic features resulting from adaptation to flight and to wide variations in diet have been amply documented. Information on the functional significance of many of the structural modifications is less extensive but with the increasing application of modern physiological and biochemical techniques many aspects of avian digestive function and metabolism are becoming clear. Some of the structural features pertinent to these studies are described in this chapter.

The general structure of the fowl digestive tract is illustrated in Fig. 1 and reference should be made to the detailed studies of Mangold (1929) and Nolf (1938) for information on the innervation of the different regions of the tract. Similarly, the extensive investigations of Nishida *et al.* (1969) on the vascular supply and drainage of the digestive tract and of Akester (1967) on the renal portal system should be consulted if vascular catheterization studies are contemplated.

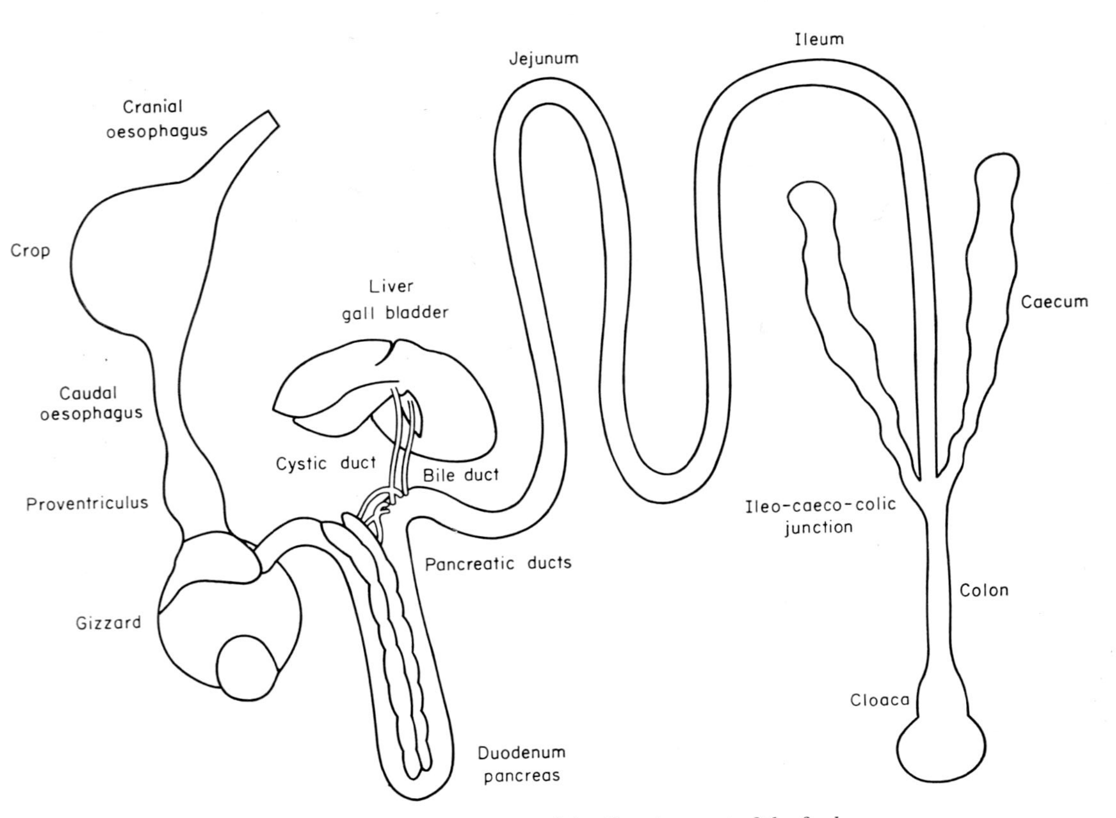

Fig. 1. The general structure of the digestive tract of the fowl.

II. Mouth

The general arrangement of the various parts of the mouth and pharynx is shown in Fig. 2 and the co-ordinated movement of some of these structures, which takes place during swallowing, is based on their innervation by the lingual and laryngo-lingual branches of the glossopharyngeal nerve. The

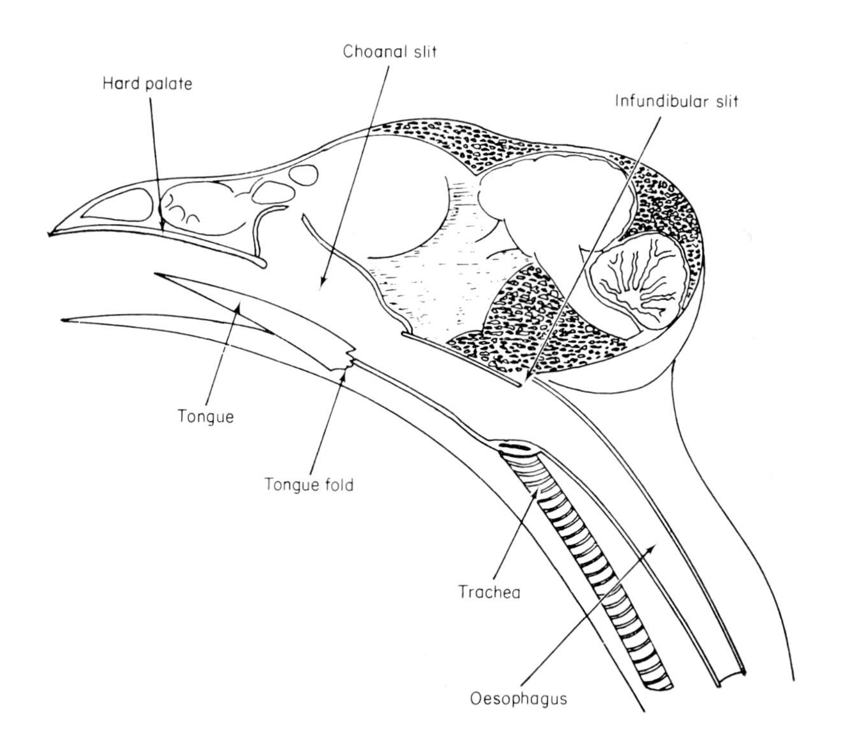

Fig. 2. Median sagittal section through the head of the fowl.

lingual nerve enters the tongue and supplies most of the anterior region, part of the sides and part of its posterior region whilst the laryngo-lingual nerve, which arises from the glossopharyngeal nerve distal to the lingual nerve, supplies branches to the internal and external surfaces of the larynx, posterior region of the tongue and the wings (Kitchell *et al.*, 1959).

The entire surface of the tongue is covered with thick stratified squamous epithelium which is heavily cornified on the dorsum and on the ventral surface. Cornification ceases at the tongue fold and it is posterior to this point, on the base of the tongue and on the floor of the pharynx, that the taste buds are distributed.

A. TASTE BUDS

These structures are intermediate in shape between those of fish and mammals and are few in number; about twelve in the young chick and double this number in the 3-month-old bird (Lindenmaier and Kare, 1959). The taste buds are flask-shaped structures which extend through the stratified squamous epithelium and consist of outer or sustentacular cells surrounding gustatory cells. A small pore allows communication between the oral cavity and the receptor cells but the fowl has none of the fine hair-like processes which, characteristic of mammalian taste buds, extend from these cells into the pore. The receptor cells respond to stimulation in a way similar to those of mammals as assessed by the occurrence of nerve impulses in fibres of the glossopharyngeal nerve when salt, bitter or acid solutions are applied to the taste buds (Kitchell *et al.*, 1959; Halpern, 1962). (See also Chapter 46.)

Other sensory organs and free nerve endings which may be concerned with pressure sensation are found on the hard palate and on the beak (Andersen and Nafstad, 1968).

B. SALIVARY GLANDS

The salivary glands are simple, branched or compound tubular glands which are widely scattered through the mouth and pharynx. Groupings of glands occur in certain regions and these have been described as:

1. Maxillary glands in the roof of the mouth.
2. Palatine glands present on either side of the common opening from the nasal chamber.
3. Spheno-pterygoid glands in the roof of the pharynx on either side of the common opening of the auditory tube.
4. Anterior submandibular glands in the angle formed by the union of the two halves of the mandible.
5. Posterior submandibular glands arranged in three groups.
6. Lingual glands in the tongue.
7. Crico-arytenoid glands about the opening of the larynx.
8. Small glands at the angle of the mouth.

In all these collections of glands the simple tubules open into a common cavity from which one or more ducts lead to the mouth. The ducts are lined with columnar epithelium which gradually merges into stratified squamous oral epithelium. Lymphoid tissue is often present in the connective tissue between the simple glands.

Each gland consists of columnar mucous cells with a small basal nucleus and a foamy cytoplasm (Fig. 3); Chodnik (1948) has described a physiological cycle of secretion which comprises alternate accumulation and discharge of mucus by the cells. During this cycle there is considerable variation in the size and shape of the cells, most frequently they appear as elongated cylinders with a faintly defined border completely filled with accumulated

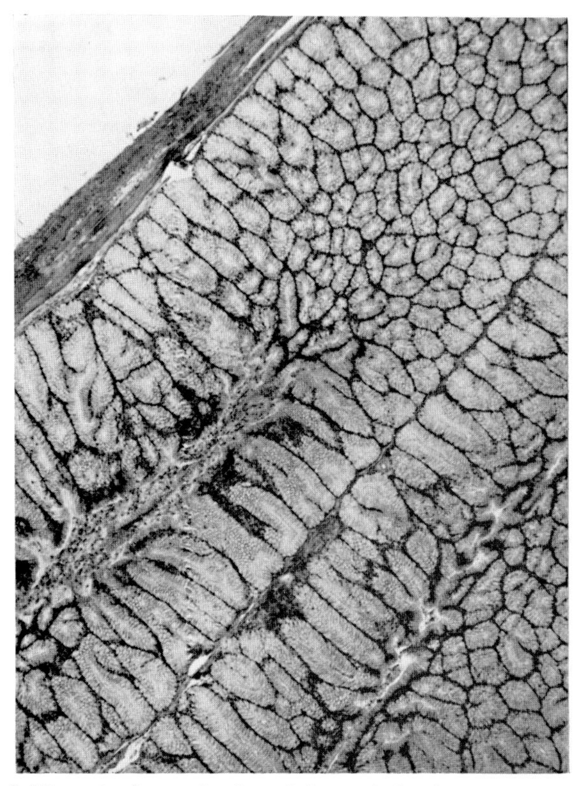

Fig. 3. Sublingual salivary gland consisting entirely of mucous cells. ($\times 50$)

secretion. This is present as vesicles of mucus separated by narrow cytoplasmic strips, giving a reticulated appearance. Mitochondria and Golgi material are distributed inside the cytoplasmic strips.

After feeding, most of the cells appear unchanged although a limited number are seen to have discharged their contents into the lumen of the gland. Secretion starts with rupture of the cell membrane next to the lumen and continues until the cell is empty when it presents a shrunken appearance. Gradual regeneration and distension with cytoplasm then occurs and is followed by a period of mucus production and accumulation within the cell, which is followed by further secretory activity. Phasic glandular secretory activity appears to be well suited to the intermittent eating habits of the fowl.

III. Oesophagus and Crop

The oesophagus is comparatively long and possesses a diverticulum or crop at its point of entry into the thoracic cavity which forms the dividing point between the upper and lower oesophagus. The stratified squamous epithelium which lines the whole oesophagus is thrown into longitudinal folds and these permit considerable distension without undue stretching of the mucosa. Mucous glands in the lamina propria discharge on to the oesophageal lining and have a lubricatory function (Fig. 4). The well-developed circular and

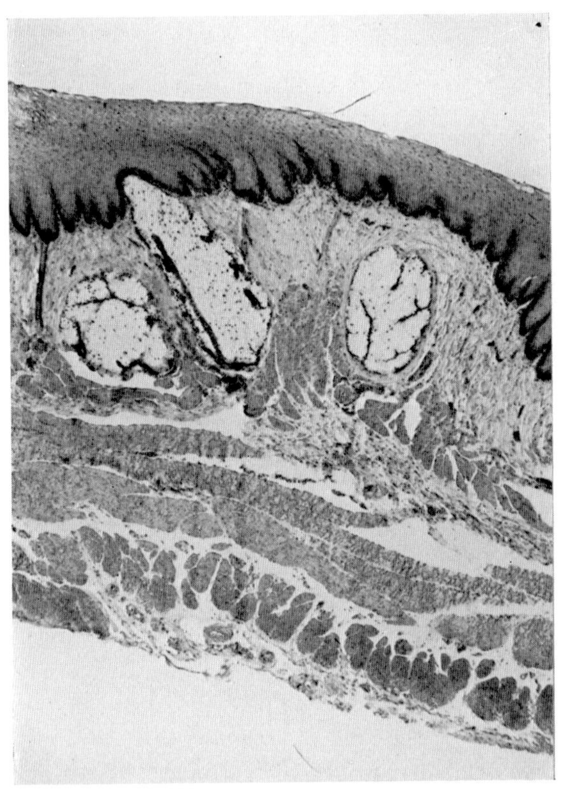

Fig. 4. Transverse section through upper oesophagus showing the stratified squamous epithelium and the sub-epithelial mucous glands. (×50)

longitudinal muscle coats are covered with a loose adventitia of elastic and fibrous tissue containing blood vessels and the vagus and sympathetic nerves. The latter connect with the ganglionated nerve plexuses lying in the submucosa and between the circular and longitudinal muscle coats.

The crop is a thin-walled storage pouch on the ventral surface of the oesophagus (Fig. 5) and possesses a similar type of lining which is also deeply folded to facilitate distension during food storage. Mucous glands are absent from the crop itself (Fig. 6) but, as in the oesophagus, extensive intramural plexuses are present in the submucosa and between the circular and longitudinal muscle coats.

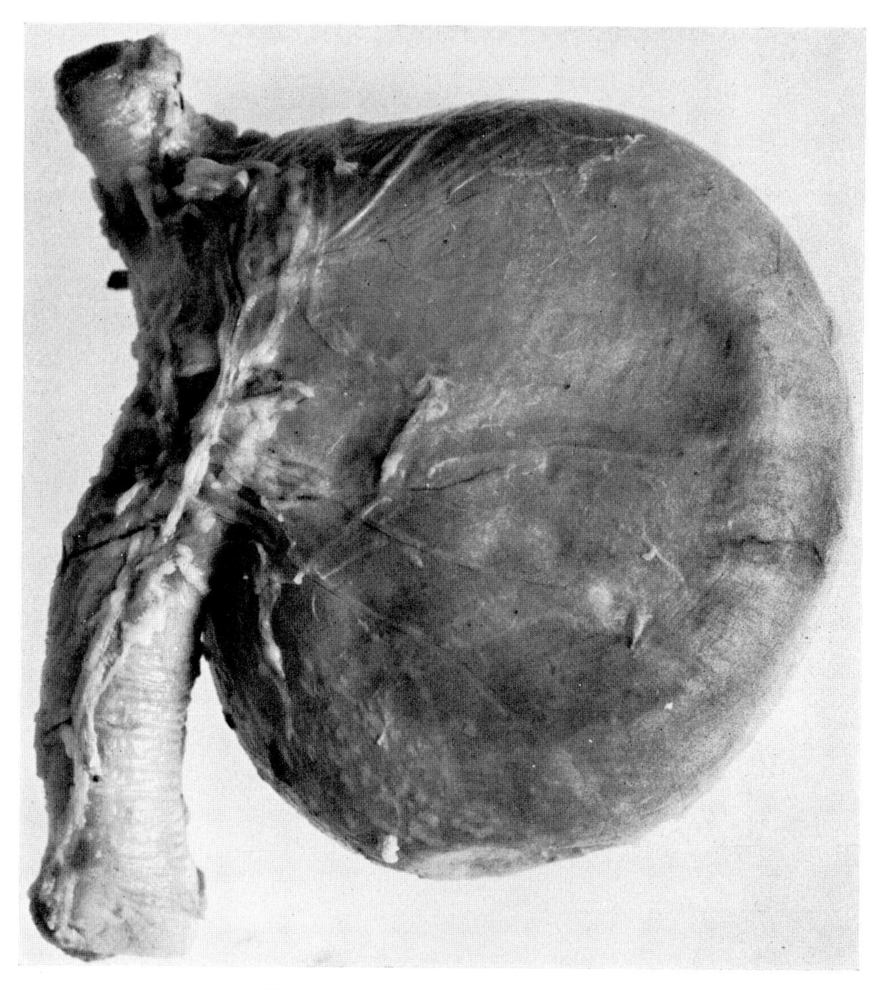

Fig. 5. The crop in its distended state.

IV. Proventriculus

The proventriculus is an ovoid structure placed between the lower oesophagus and the gizzard (Fig. 7) and it is lined with a glandular mucous membrane which contains the gastric secretory glands. It is distinguishable from the gizzard by the fifth day of incubation and its subsequent development has been fully documented (Dawson and Moyer, 1948). Fine structural differentiation of the gland cells begins as soon as the primitive glands are formed and the characteristic features of both acid—and enzyme—secretions are present in each cell from 9 days onwards (Toner, 1965b).

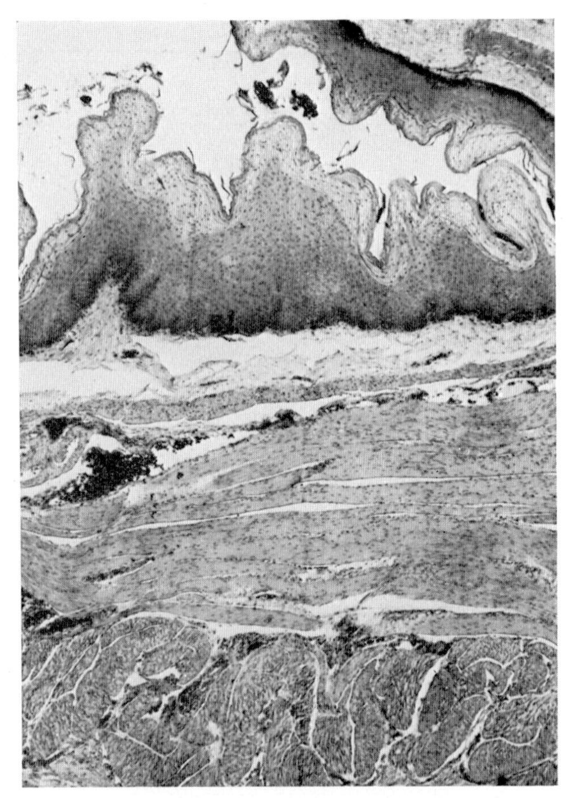

Fig. 6. Transverse section through the wall of the crop. Stratified squamous epithelium as in the oesophagus but mucous glands absent. (×50)

The general arrangement of the proventricular mucosa is shown in Fig. 8. The mucosal surface is covered with macroscopic papillae, each of which contains an opening leading from the proventricular glands; the surface of the mucosa consists of simple columnar epithelium, arranged concentrically around the gland opening. Much of the thickness of the proventricular wall is due to the proventricular glands which comprise a number of rounded lobules consisting of tubular alveoli. These drain by tertiary, secondary and primary ducts into the lumen of the proventriculus. Mucous neck-cells line the ducts and these are somewhat similar to the intestinal goblet cells in that a mass of mucus fills the supranuclear region of each cell. Following the intake of food most of this mucus is discharged and is subsequently replaced with secretion from the basal region of the cell (Chodnik, 1947).

The absence of well-characterized parietal or oxyntic (acid secreting) cells and peptic (enzyme secreting) cells in the proventriculus has been recognized for many years and it is now generally accepted that the alveolar cells are of a single type—oxyntico-peptic—which possess most of the features of the

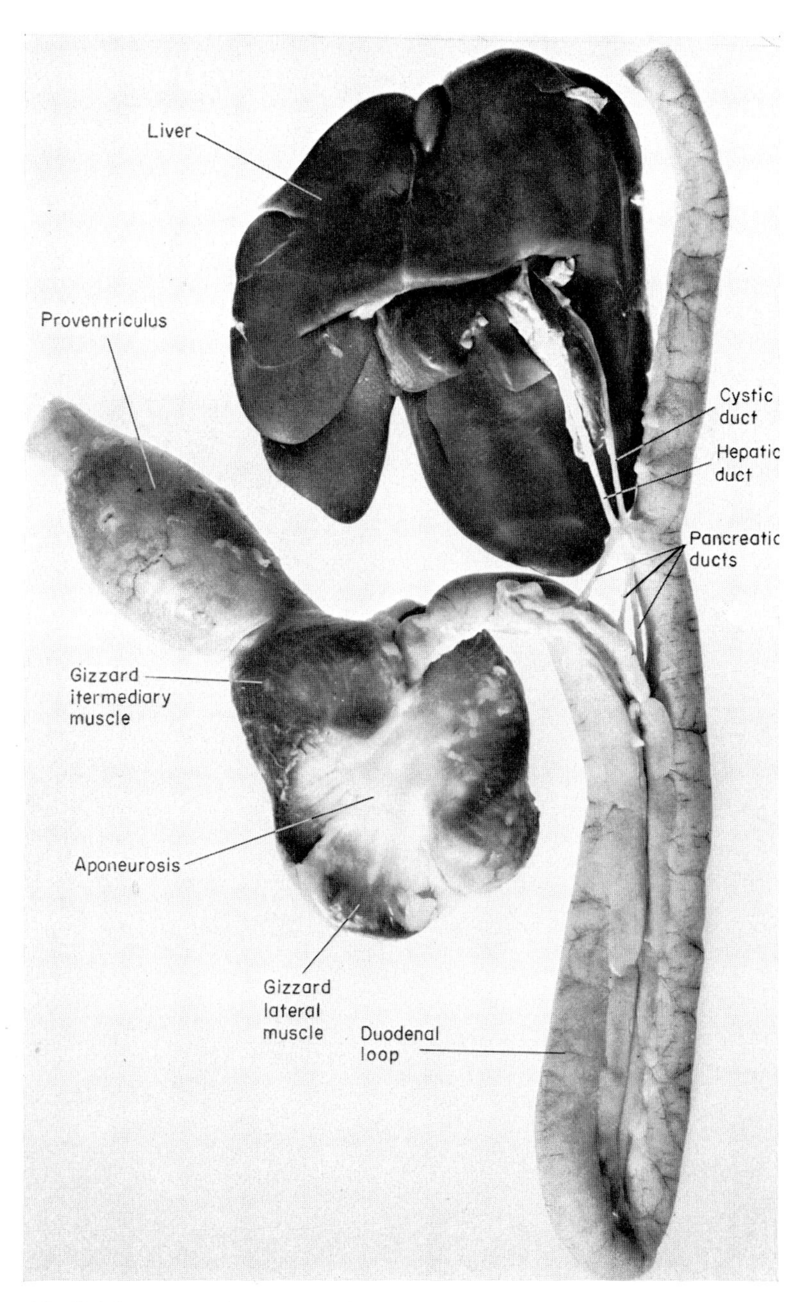

Fig. 7. The proventriculus–gizzard–duodenal region of the fowl's digestive tract.

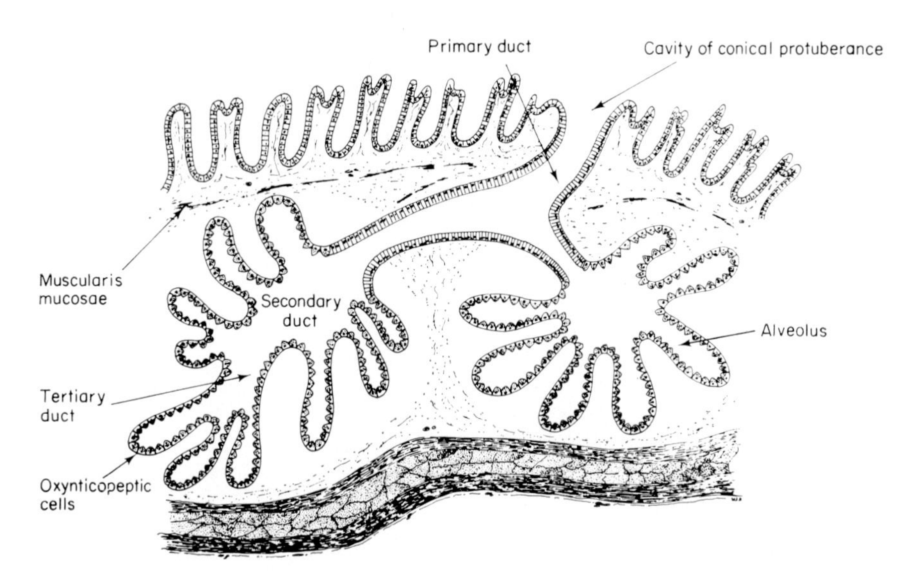

Fig. 8. Diagrammatic transverse section through the proventricular mucosa showing structure of the glandular alveoli.

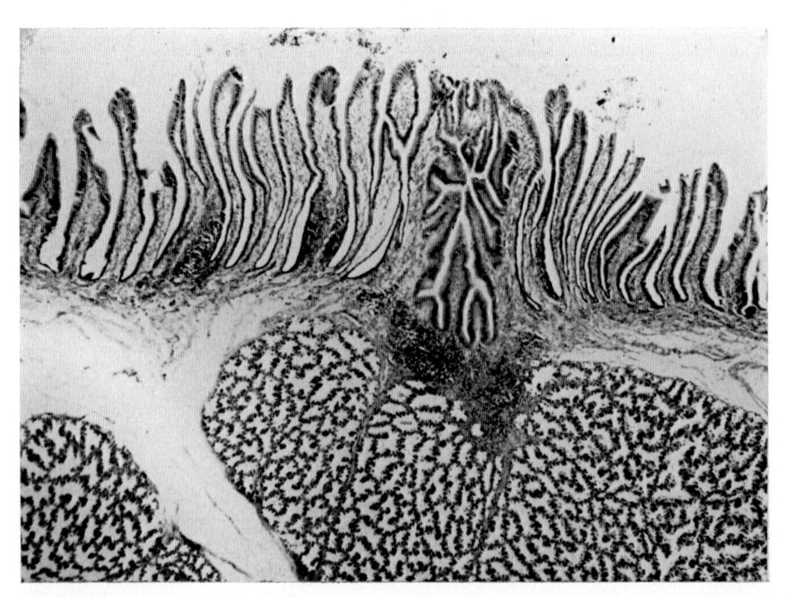

Fig. 9. Transverse section through proventriculus showing surface epithelium and glandular alveoli. (×50)

two separate types found in mammals (Table 1). Toner (1963) has, in fact, shown that the submucosal gland cells have a fine structure which is typical of both zymogenic and acid secretory activities and these observations are consistent with the view that the submucosal gland cells secrete both the acid and the proteolytic enzymes of the gastric juice.

Table 1

Comparison of the histological features of the avian oxyntico-peptic cell and the mammalian oxyntic cell and peptic cell. (From Menzies and Fisk, 1963)

Histological feature	Avian oxyntico-peptic cell	Mammalian oxyntic cell	Mammalian peptic cell
Presence of spherical granules	+	+	+
Presence of Bowie-positive granules	+	0	+
Discharge, after feeding, of Bowie-positive granules	+		+
Presence of Bowie-negative granules	+	+	0
Discharge after feeding of Bowie-negative granules	0	0	
Presence of PAS-positive precursor to granules	0	0	+
Presence of filiform mitochondria	0	0	+
Presence of Golgi material	+	0	+
Change of polarity of Golgi material during secretion	+		0

The oxyntico-peptic cells are arranged in a single layer which lines the alveoli. Their shape varies from low cuboidal to elongated columnar depending on their degree of functional activity and they are arranged in rows in such a way that their distal half is not in contact with the corresponding half of the neighbouring cells, giving the appearance of a serrated edge bulging into the lumen. The nucleus is round or ovoid and its position varies, depending on the functional state of the cell.

In all phases of food deprivation and digestion, most of the cells are densely packed with large, spherical secretory granules whose number increases during starvation. If the latter is prolonged, they extend from the luminal pole to the basal part of the cell where they form a mass of closely packed granules which gives the cell a cuboidal shape. Within 30 min of feeding a single meal, the number of zymogen granules decreases considerably and evacuation continues for a further 3 h. After this time the cells begin to refill with granules so that, 6 h after a meal, the number of granules reaches a level usually seen in birds with constant access to food. There is some evidence for phasic activity of the glandular cells as Chodnik (1947) has shown that only a certain number of the cells are involved in the secretory process at the one time, and that each cell appears to act as an independent unit. Vagal complicity in the secretory process is evidenced by the fact that pilocarpine injection induces partial discharge of the zymogen granules (Menzies and Fisk, 1963).

Argentophil cells are present in the proventriculus and gizzard, between the cells of the alveoli, and have a thread-like form. Unlike the argentophil cells of the intestine, they never acquire the characteristics of argentaffin cells and the argentophil cell in these locations, therefore, appears to be the definitive one (Dawson and Moyer, 1948; Aitken, 1958).

V. Gizzard

The gizzard presents two highly specialized morphological features; massive muscle development and a thick, hard covering over the mucous membrane which relates to its functions as a food grinding chamber and as a site for peptic proteolysis. Developmental aspects of these structures have been described by Hibbard (1942) and by Van Alten and Fennell (1957).

The main body of the gizzard comprises two thick, opposed, lateral muscles the ends of which are attached to a central aponeurosis, and two thin anterior and posterior intermediary muscles (Fig. 7). The gizzard lumen is larger than that of the proventriculus (Fig. 10) and, conventionally, contains food mixed with grit. A thick and relatively unyielding lining protects the mucosa from damage by the pressure of grit and food on its surface when the gizzard contracts and presumably provides some protection against the corrosive effect of the acid-enzyme mixture which flows into the gizzard from the proventriculus.

The unusual activity of the gizzard as a grinding organ has excited interest over many years and, as a result, extensive investigations have been made on the development and structure of the gizzard musculature (see Mangold, 1929; Farner, 1960; Calhoun, 1961). These have shown that the gizzard is composed entirely of smooth muscle fibres which are derived from the primordial circular muscle layer. The outer longitudinal muscle layer is lost during development and hence, Auerbach's plexus, which normally lies between the inner circular muscle layer and the outer longitudinal muscle layer, lies close to the outer surface of the gizzard, immediately under the serosa. It is possible that extensions of Auerbach's plexus into the musculature, may contain some sensory ganglion cells and that the large numbers of myelinated axons, seen throughout Auerbach's plexus, are afferent fibres (Bennett and Cobb, 1969c).

The vagal (cholinergic) innervation of the gizzard is established early in embryonic life (Bennett and Cobb, 1969b) and, as the density of innervation

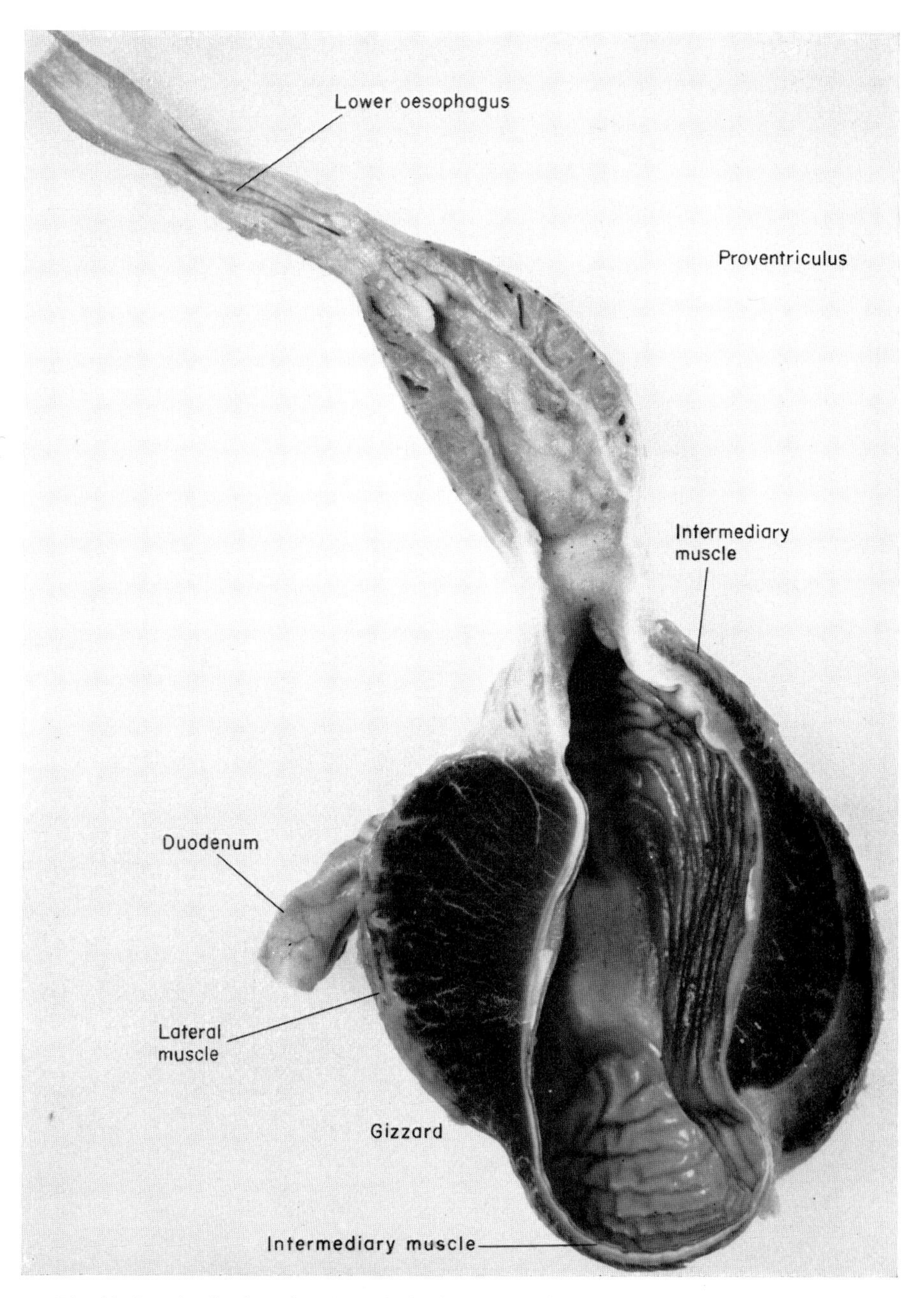

Fig. 10. Longitudinal section through the lower oesophagus, proventriculus and gizzard.

of the muscles appears to be the same throughout their mass, effective stimulation of the musculature may arise from ganglionic masses embedded in the lateral and intermediate muscles (Bennett and Cobb, 1969c).

There is extensive delineation of the muscle fibres into bundles by strands of connective tissue and there is also extensive communication between adjacent smooth muscle cells. Bennett and Cobb (1969a) suggest that this close connexion between the muscle cells provides an anatomical basis for the rapid propagation of impulses through the muscle mass and hence the massive, rapid contractibility of the whole organ.

The inner layer of the gizzard comprises a thin submucosa, a glandular mucous membrane and a thick, abrasion-resistant lining which is mainly composed of the hardened secretion of the gizzard glands and shows longitudinal ridges and grooves (Fig. 10). The nature of this secretion has intrigued avian histologists for many years and it has been generally concluded that it is a keratin-like substance which has been termed koilin. Attrition of this lining by the grinding action of the powerful muscle contractions, especially in the presence of grit, is countered by slow secretory activity of the gizzard glands which renews the koilin lining.

The glandular layer of the gizzard is made up of simple tubular glands, mainly arranged in groups of ten to thirty, each group opening into a crypt. Smaller groups of glands, or even single glands, tend to occur on the tops of the longitudinal ridges. Each gland consists of a neck, body and slightly expanded fundus which rests on the submucosal tissue (Fig. 11).

Fig. 11. Section showing gizzard glands and koilin lining. ($\times 50$)

The cells lining the glands and the crypts are mainly of one type, chief cells, and in the fundic region they tend to be short columnar ones with rounded nuclei. They contain cytoplasmic granules around the Golgi apparatus and in the apical pole of the cell. In the body and neck regions of the glands the granules are more numerous and are discharged at the cell-surface between the microvilli. The discharged material forms a dense layer over the microvilli (Chodnik, 1947; Toner, 1964a) and in the body and neck region forms filaments which become packed together to form bundles. The bundles of filaments from each gland are united in the crypt region to form a single large bundle or rod (Fig. 12).

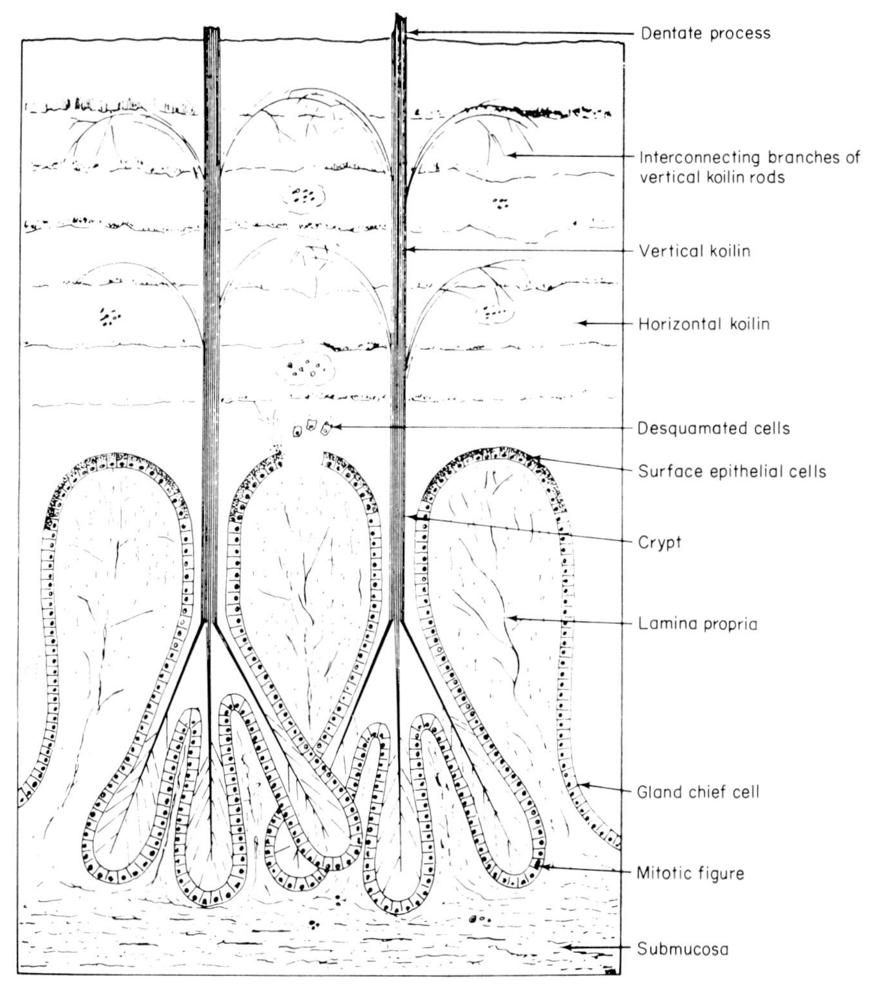

Fig. 12. Diagrammatic structure of gizzard glands and koilin layer.

The cells lining the mouth of the crypts and the surface epithelium are columnar and their apical granule-containing cytoplasm has been termed a zone of apocrine secretion by Eglitis and Knouff (1962). Toner (1964a) has shown, however, by electron microscopy, that the formation of this apical zone is due to the discharge of individual granules which remain and accumulate between the microvilli rather than to the shedding of a large cellular accumulation of secretory granules, as occurs in true apocrine secretion. There is a gradual transition from the chief cell to the crypt and surface epithelial cell and Chodnik (1947) has suggested that only those at the bottom of the crypts and in the gland are normal cells. Cytological and ultrastructural evidence (Chodnik, 1947; Toner, 1964a) support this view, in that the cells in the upper part of the crypt, and those forming the surface epithelium, show progressive degenerative changes which terminate in sloughing between adjacent crypts. Occasional mitotic figures occur in cells of the fundic region of the glands and it appears that the different appearance of the gland cells, crypt cell and surface epithelial cell merely represents a sequence in the progressive degeneration of the gland cell as it migrates towards the surface epithelium.

The filamentous secretion of the gland chief cells, the secretion of the cells at the mouth of the crypt and on the surface epithelium, and the sloughed, degenerated, surface epithelial cells constitute the gizzard lining (Fig. 12) and it is because of the physical arrangement of these components that the lining is able to withstand the considerable forces to which it is subjected. Thus, the filamentous bundles of chief cell secretion pass through and out of the crypts and form vertical columns or rods in the koilin layer with their tips projecting beyond the surface to form the so-called "gizzard teeth" or dentate processes. Lateral filaments from these rods join with filaments from neighbouring rods to form a three-dimensional supporting network for the somewhat different secretion derived from the crypt and surface cells (Fig. 13). Hardening of the secretion from the gland chief cells occurs in the lumen of the glands.

The horizontal matrix of the koilin layer, which stains less intensely than the vertical columns, is produced by cells of the crypt and surface epithelium as a secretion which spreads over the surface of the epithelium and around the vertical rods before it hardens. Differences in staining intensity of the horizontal matrix are frequently present and give the lining a banded appearance.

Historically, the gizzard lining has been considered to be keratin or keratin-like in composition, although the histological evidence that koilin is a true glandular secretion has cast doubt on this view. Moreover, recent chemical and histological studies have shown that, although the hardened secretion is a protein, it does not possess the characteristics of keratin and is in fact a polysaccharide-protein complex (Luppa, 1959; Eglitis and Knouff, 1962; Webb and Colvin, 1963).

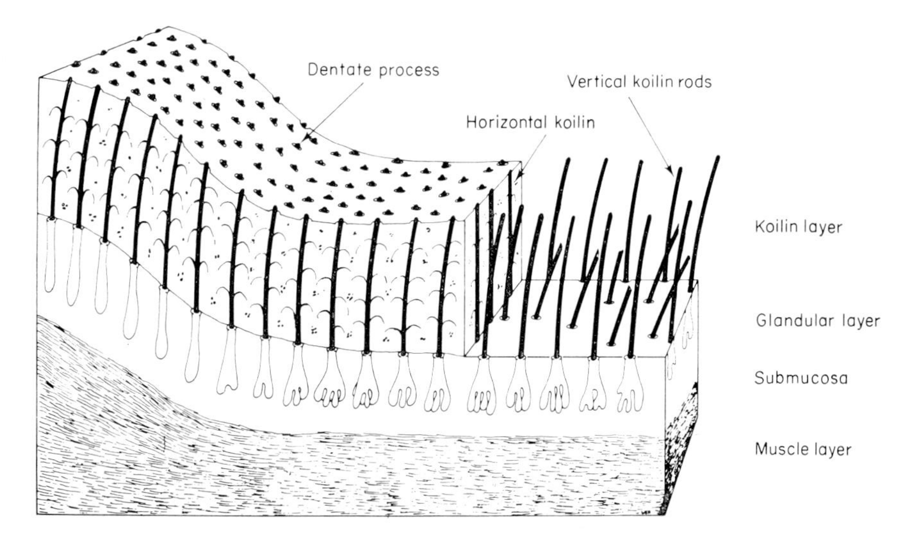

Fig. 13. Diagrammatic representation of the structure of the glandular and koilin layers of the gizzard.

On the basis of the chemical and physical properties of this complex, the latter authors have proposed that the horizontal koilin is formed in the following manner. Small droplets of secretion flow into the crypt mouth and on to the surface epithelium and spread out over the epithelial surface beneath layers previously secreted which have already hardened. Hardening is brought about by a reduction in the pH of the secretion due to the diffusion of hydrochloric acid from the gizzard lumen through the koilin layer. As the acid reaches the liquid secretion, protein is precipitated to form fresh layers or lamellae of solid material. Differences in the density of deposition of the precipitated protein may account for the differences in staining intensity observed in the horizontal koilin. As the upper layers of the gizzard lining are removed by abrasion, the more recently secreted material moves towards the surface and becomes progressively tougher and more enzyme-resistant.

As in the other regions of the alimentary tract, cell renewal and replacement occurs in the gizzard glands and a necessary part of this process is the shedding of dead cells from the epithelium. The main region of cell desquamation is from the epithelium between the glandular crypts; as cells are shed they are trapped in the horizontal matrix. Shedding is a cyclical process and hence the cell debris appears as a series of discontinuous vertical columns (Fig. 12).

In the adult the koilin lining presents a homogenous appearance with relatively little cellular debris trapped within its substance. The gizzard lining of the young bird exhibits far greater evidence of cell shedding, particularly in the young rapidly growing broiler.

VI. Small Intestine

The gizzard–duodenal junction is marked by the change from a tough koilin lining to a mucosa containing coiled glands and villi lined with tall mucous columnar cells. In the transitional zone the epithelium is covered by a thick layer of mucus which presumably has a protective function against the highly acid material which leaves the gizzard and enters the duodenum.

The small intestine is comparatively short compared with that of the mammal and has a uniform diameter throughout its length. Its general histological structure is comparable to that of the mammal although there are some minor differences. Thus, Brünner's glands are lacking from the duodenum and their mucus-secreting role is carried out by numerous goblet-type cells present between the columnar cells of the surface epithelium and in the superficial parts of the simple glands (Aitken, 1958). There is a high concentration of argentaffin cells in the upper duodenum although, unlike mammals where they are located deep in the glands, they occur in the surface epithelium and in the tops of the glands. Slight differences occur in their ultrastructure compared with the mammalian argentaffin cell (Toner, 1964b). Lymphatic tissue is fairly abundant in the lamina propria and the gland tubules tend to be widely separated from each other by this tissue.

The major unusual feature in the absorptive region of the small intestine is the absence of central lacteals in the villi, each villous core being occupied by a capillary bed. This correlates with the poorly developed lymphatic system of the fowl and with the biochemical evidence for lipid absorption into the portal blood (see also Chapters 3, 12).

Cell-turnover studies indicate that the avian intestinal epithelium has a regeneration time comparable to that of other classes, i.e. approximately 48 h (Imondi and Bird, 1966) and hence, that the desquamated epithelial cells contribute an appreciable amount of endogenous nitrogen to the digestive tract contents (Bird, 1968).

Electron microscope studies of the fowl intestine have shown that its cellular structure is similar to that of mammals (Toner, 1965a). Thus the absorptive cells are large and cylindrical with basal nuclei and with an apical membrane covered with long microvilli. Numerous elongated mitochondria surrounded by ergastoplasmic cisternae are present in the cytoplasm, as are numerous ribosomes. A conventional Golgi apparatus is present in each cell along with endoplasmic reticulum. Irregularly-shaped bodies with double membranes containing ferritin-like particles are present in the supranuclear zone (Hugon and Borgers, 1969).

A cell which is frequently present in the intestinal epithelium of the fowl is the globular leucocyte (Clara, 1926; Toner, 1965a). There is an extensive literature on the occurrence and possible role of these cells in other species (see Dobson, 1966) and the suggestion that they are concerned in local immunological defense mechanisms is perhaps relevant to the observation that their numbers are increased in association with the response of the chicken to coccidial infections (Pierce *et al.*, 1962).

VII. Large Intestine

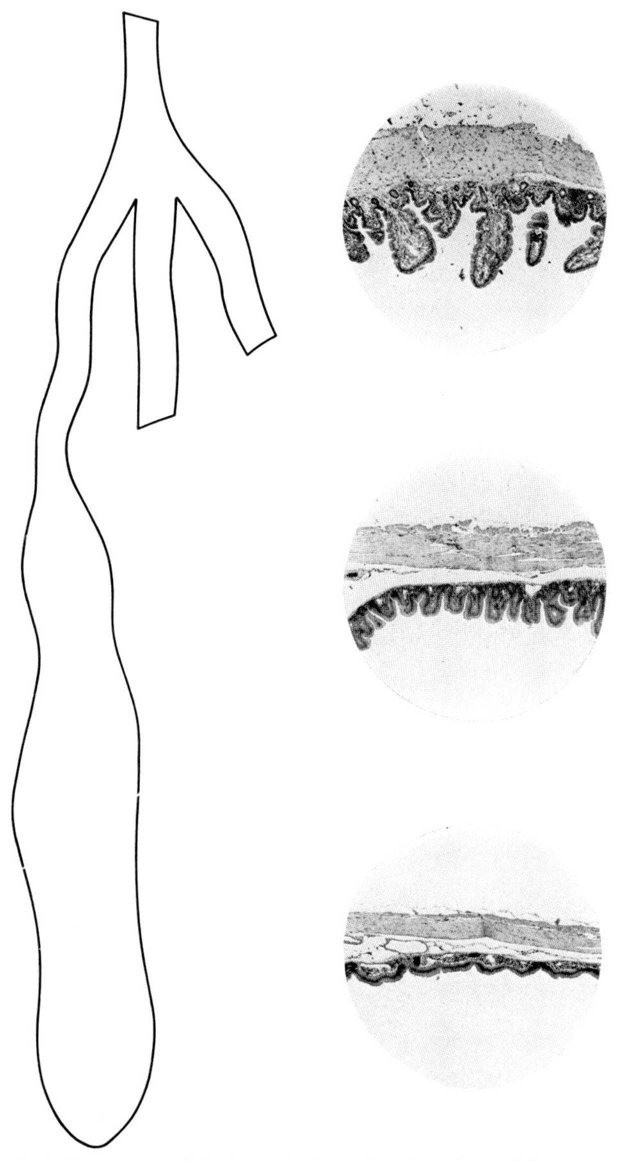

Fig. 14. Histological structure of the base, body and neck regions of the caeca showing the villus-like projections in the body and neck regions. (×50)

A. THE CAECA

These are paired, blind-ended tubes, which arise at the junction of the small and large intestine. They extend in a forward direction for about half their length and are then doubled back on themselves. Three regions are present in each caecum; a narrow neck region arising at the ileo-caeco-colic junction, a wider body region and a round-ended base. Well-defined muscle coats are present and each caecum is lined with columnar epithelium. In the base region this epithelium is relatively smooth but it is thrown into villus-like projections in the body region and the neck which are probably the major absorptive areas of the caeca (Fig. 14). Lymphoid tissue is scattered throughout the submucosa and occasional goblet cells are present in the lining epithelium.

B. THE COLON

The colon is a short narrow tube extending from the ileo-caeco-colic junction to the cloaca and, in keeping with its role as an absorptive organ, its mucosa is thrown into short broad, villus-like projections lined with columnar cells (Fig. 15). Numerous goblet cells also occur in the epithelium.

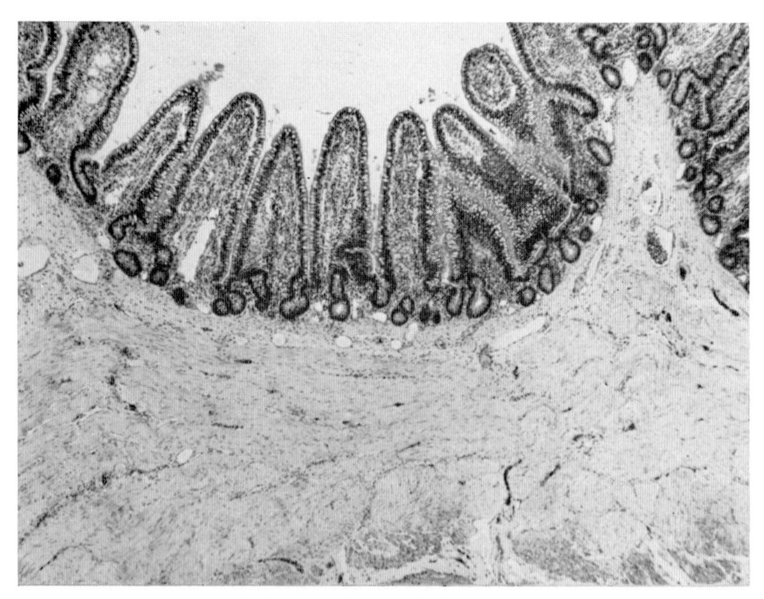

Fig. 15. Transverse section through the wall of the colon. (×50)

C. THE CLOACA

The digestive and uro-genital tracts converge to a common chamber, the cloaca, which is separated from the colon by a slight constriction formed by the thickened circular muscle coat and which opens to the exterior at the vent. Detailed studies on the development and histogenesis of this organ have been made by Boyden (1922).

For descriptive purposes the cloaca is conveniently divided into three areas, coprodeum, urodeum and proctodeum, each of which is defined by incomplete transverse folds, or flaps, of mucosa (Fig. 16). The coprodeum is

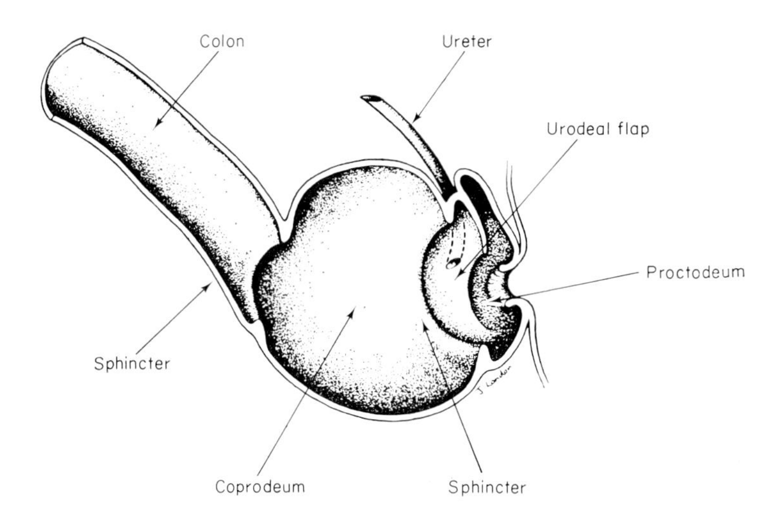

Fig. 16. The structure of the cloaca. (Reproduced with permission from Akester *et al.*, 1967.)

the largest chamber and serves as a reservoir for faeces and for the urine discharged from the ureteral openings which lie on the cranial side of the urodeal flap. The oviduct opens into the urodeum and connects, through the narrow proctodeum, with the vent. A mass of lymphoid tissue, the bursa of Fabricius, lies in the dorsal wall of the urodeum.

The whole of the cloaca is lined with columnar epithelium which forms short, cylindrical villus-like structures in the coprodeum and more flattened leaf-like structures in the urodeum. There is an abrupt transition to stratified squamous epithelium on the inner aspect of the upper and lower lips of the vent, and within the lips themselves there are structures which closely resemble Paccinian corpuscles.

VIII. Pancreas

The three lobes of this digestive gland occupy the space between the limbs of the duodenum. Three secretory ducts pass to the distal end of the duodenal loop and open into the duodenum on a common papilla with the bile duct.

Histologically, the pancreas is similar to that of the mammals apart from somewhat less well-defined islet tissue. The cells of the acini are irregularly pyramidal and in the starving state, zymogen granules fill the cell between the nucleus and the lumen. The individual zymogen granules vary in size and staining properties; shortly after feeding many of them disappear. Discharge of granules continues for about 2 h following a meal and restitution of the cell is complete some 5 h later. Chodnik (1948) observed that individual acini act as autonomous units, some containing few granules after feeding while others remain replete. As with the gastric glands, this appears to be a mechanism for the maintenance of the more or less continuous secretion which occurs during *ad libitum* feeding.

IX. Gall Bladder and Bile Ducts

The biliary system comprises two ducts, the cystic duct bearing the gall-bladder, which drains bile mainly from the right lobe of the liver, and the hepatic duct which drains the left lobe of the liver (Fig. 7). A connecting branch between the collecting ducts in the left lobe of the liver and those of the right lobe is present within the liver substance. Both ducts lie close together when they leave the liver and pass to the caudal end of the ascending limb of the duodenum where they open into the duodenum on a common papilla with the pancreatic ducts.

The gall-bladder possesses a thick, vascular serosa, outer and inner longitudinal muscle coats and a mucosa consisting of loose connective tissue thrown into folds covered with columnar epithelium. These folds are obliterated when the gall-bladder is distended with bile. Similar longitudinal folds thrown into villus-like projections covered with simple columnar epithelium are present in the bile-ducts and prevent damage to the mucosa during their vigorous contractions.

References

Aitken, R. N. C. (1958). *J. Anat.* **92**, 453–469.

Akester, A. R. (1967). *J. Anat.* **101**, 569–594.

Akester, A. R., Anderson, R. S., Hill, K. J. and Osbaldiston, G. W. (1967). *Br. Poult. Sci.* **8**, 209–212.

Andersen, A. E. and Nafstad, P. H. J. (1968). *Z. Zellforsch. mikrosk. Anat.* **91**, 391–401.

Bennett, T. and Cobb, J. L. S. (1969a). *Z. Zellforsch. mikrosk. Anat.* **96**, 173–185.

Bennett, T. and Cobb, J. L. S. (1969b). *Z. Zellforsch. mikrosk. Anat.* **98**, 599–621.

Bennett, T. and Cobb, J. L. S. (1969c). *Z. Zellforsch. mikrosk. Anat.* **99**, 109–120.

Bird, F. H. (1968). *Fedn Proc. Fedn Am. Socs exp. Biol.* **27**, 1194–1198.

Boyden, E. A. (1922). *Am. J. Anat.* **30**, 163–201.

Bradley, O. C. (1960). "The Structure of the Fowl". Revised by T. Grahame. 4th Edit. Oliver and Boyd, Edinburgh and London.

Calhoun, M. L. (1961). "Microscopic Anatomy of the Digestive System of the Chicken". State University Press, Ames, Iowa.

Chodnik, K. S. (1947). *Q. Jl microsc. Sci.* **88**, 419–443.

Chodnik, K. S. (1948). *Q. Jl microsc. Sci.* **89**, 75–87.

Clara, M. (1926). *Z. mikrosk. anat. Forsch.* **6**, 305–350.
Dawson, A. B. and Moyer, S. L. (1948). *Anat. Rec.* **100**, 493–514.
Dobson, C. (1966). *Aust. J. agric. Res.* **17**, 955–966.
Eglitis, I. and Knouff, R. A. (1962). *Am. J. Anat.* **111**, 49–66.
Farner, D. S. (1960). *In* "Biology and Comparative Physiology of Birds" (A. J. Marshall, ed.) Vol. I. pp. 411–467. Academic Press, New York and London.
Halpern, B. P. (1962). *Am. J. Physiol.* **203**, 541–544.
Hibbard, H. (1942). *J. Morph.* **70**, 121–149.
Hugon, J. S. and Borgers, M. (1969). *Acta histochem.* **34**, 349–359.
Imondi, A. R. and Bird, F. H. (1966). *Poult. Sci.* **45**, 142–145.
Kitchell, R. L., Ström, L. and Zotterman, Y. (1959). *Acta physiol. scand.* **46**, 133–151.
Lindenmaier, P. and Kare, M. R. (1959). *Poult. Sci.* **38**, 545–550.
Luppa, H. (1959). *Acta Anat.* **39**, 51–81.
Mangold, E. (1929). *In* "Handbuch der Ernährung und des Stoffwechsels der Landwirtschaftlichen Nutztiere." (E. Mangold, ed.). Springer, Berlin.
Menzies, G. and Fisk, A. (1963). *Q. Jl microsc. Sci.* **104**, 207–215.
Nishida, T., Paik, Y. K. and Yasuda, M. (1969). *Jap. J. vet. Sci.* **31**, 51–80.
Nolf, P. (1938). *Archs int. Physiol.* **46**, 1–85.
Pierce, A. E., Long, P. L. and Horton-Smith, C. (1962). *Immunology* **5**, 129–152.
Romanoff, A. L. (1960). "The Avian Embryo". The Macmillan Company, New York.
Toner, P. G. (1963). *J. Anat.* **97**, 575–583.
Toner, P. G. (1964a). *J. Anat.* **98**, 77–86.
Toner, P. G. (1964b). *Z. Zellforsch. mikrosk. Anat.* **63**, 830–839.
Toner, P. G. (1965a). *Acta anat.* **61**, 321–330.
Toner, P. G. (1965b). *J. Anat.* **99**, 389–398.
Van Alten, P. J. and Fennell, R. A. (1957). *Anat. Rec.* **127**, 677–695.
Webb, T. E. and Colvin, J. R. (1963). *Can. J. Biochem.* **42**, 59–72.

2

The Physiology of Digestion

K. J. HILL

Unilever Research Laboratory,
Colworth House,
Sharnbrook,
Bedford, England

I. Introduction

Of all our domestic species, the fowl has been subject to the most intensive change through genetic, dietary and environmental manipulation and as a result the productivity of the modern light hybrid (eggs) and broiler (meat) is much greater than that of the older, established breeds. It is probable that these changes have resulted in modification of digestive activity but there is little information on this point and the following account is therefore a generalized one based essentially on data derived from work on the older type of fowl.

II. The Mouth

A. DEGLUTITION

Information on the events which occur during deglutition has been obtained by radiographic techniques (Henry *et al.*, 1932; Halnan, 1949; Vonk and Postma, 1949) and by a combination of radiographical, pressure recording and fistulation procedures which have allowed study of transit time and pressure changes during swallowing (Pastea *et al.*, 1968b).

Deglutition is conveniently divided into active and passive phases which normally follow each other in rapid succession. In the active phase, after food has been conveyed to the mouth by the beak, there is initiation of its onward propulsion to the back of the mouth by movement of the tongue and by lateral pressure caused by contraction of the hyobranchio-lingual muscles. During this period the food is moistened by the secretions of the mouth and there is reflex closure of the choanal slit. Rapid extension of the neck, causing elevation and forward movement of the head, also occurs and the resultant change in the position of the larynx and trachea brings the glottis into apposition with the base of the tongue.

The food bolus is thus prevented from passing into the trachea and is conveyed, by the contraction of the hyobranchio-lingual muscles, to the cranial end of the oesophagus. In the passive phase of deglutition the food bolus is caught up by the oesophageal peristaltic movements and, aided by negative pressure, is carried along the oesophagus. During drinking the head is lowered and water allowed to flow into the mouth. The mouth is then closed and the head raised so that fluid flows into the oesophagus under the influence of gravity.

B. SALIVARY SECRETION

Each individual mouthful of food, consisting of single grains, pellets or small amounts of meal, is swallowed quickly. As mastication does not occur, the main requirement is for lubrication to assist the swallowing process which is met by the mucinous nature of the saliva. Estimates of the total volume of secretion produced have been obtained from fistulae placed high in the oesophagus and have ranged from 7 to 30 ml/24 h (Leasure and Link, 1940; Belman, 1962). Belman and Kare (1961) suggest that the volume may be greater, although it is difficult to separate the contribution of the oesophageal glands when collections are made from an oesophageal fistula. Quantitative collection of food boli and determination of their moisture content does not appear to have been carried out, nor indeed has the possible role of the crop in salivary stimulation been evaluated.

Because of the small size and diffuse nature of the saliva-secreting glands in the bird, it is difficult to carry out the classic experimental procedures of duct cannulation and nerve stimulation and the little information available on the stimulation of salivary secretion is based on histological changes in the secretory cells after feeding and pharmacological stimulation (Chodnik, 1948).

These studies have demonstrated that the secretion is mucinous and is markedly stimulated by parasympathetic stimulation. Evidence for amylase activity in the saliva is conflicting (Leasure and Link, 1940) although there appears to be no histological evidence for enzyme production.

Possibly because of the rapid transit of food through the mouth and pharynx, and the absence of mastication, the sense of taste appears to be poorly developed in the fowl. Taste buds, although few in number, are however present and neurophysiological studies have shown that they respond in a manner similar to those of mammals when adequately stimulated (Kitchell *et al.*, 1959; Halpern, 1962). Thus, impulses running in the peripheral branches of the glossopharyngeal nerve were detected after applying salt bitter or acid solutions to the taste buds and there was good correlation with the behavioural responses to solutions with these characteristics. Responses were also obtained to thermal stimuli and it is possible that receptors for textural assessment of food materials may also exist in the mouth. (For further discussion of taste see Chapter 46.)

III. Movements of the Digestive Tract

A. OESOPHAGUS AND CROP

Regulated passage of food boli from the mouth to the gizzard occurs as a result of an integrated sequence of movements of the upper and lower oesophagus, crop, proventriculus and gizzard. Overriding control appears to lie within the crop–proventricular–gizzard region since food boli pass directly into the gizzard after a period of deprivation, whereas they are diverted into, and stored in, the crop when the gizzard contains food. As the gizzard contents are discharged into the duodenum, intermittent release of food occurs from the crop, maintaining continuous digestive activity.

The passage of food along the oesophagus is facilitated by the distensibility of its wall and by the mucus secreted by the salivary and oesophageal glands. Inherent contractile ability is present in the oesophageal musculature and is present in the explanted oesophageal tissue of the very young embryo (Kuo, 1932), whilst spontaneous, regular contractions, are readily seen in *in vitro* preparations of oesophagus from older birds (Everett, 1966).

In the conscious intact bird, oesophageal contractions are greatly influenced by excitement, disturbance, and the presence or absence of food; carefully controlled conditions are therefore necessary to obtain reproducible data on the nature of the oesophageal movements. Under suitable conditions, the upper oesophagus exhibits peristaltic waves at about 15 s intervals whilst the lower oesophagus has a much slower contraction rate at 50 to 55 s intervals (Pintea *et al.*, 1957).

Co-ordination of peristalsis in the oesophagus is effected, as in mammals, by the influence of the vagal and sympathetic nerves on the myenteric plexuses, stimulation of the vagal trunks inhibiting the spontaneous contractions and producing strong contractions of short duration (Nolf, 1925). Parasympathetic drugs, such as carbachol, and acetylcholine produce a similar contractile response which, like the vagal effect, is abolished by atropine (Nolf, 1925; Hanzlik and Butt, 1928; Everett, 1966)–see Fig. 1. The

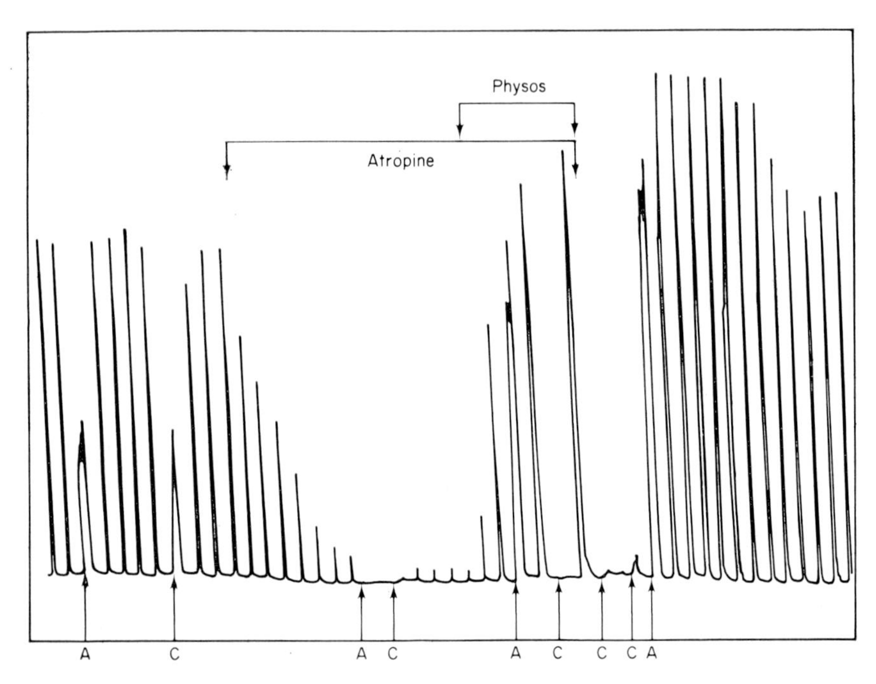

Fig. 1. Contractions of the isolated crop in response to vagal stimulation (unlabelled contractions), acetylcholine (A) and carbachol (c). The inhibitory effect of atropine and the restoration and potentiation of the response to acetylcholine, by physostigmine are clearly visible. (From Everett, 1966.)

essential similarity of the innervation of the oesophagus and of its diverticulum, the crop, to the mammalian oesophagus is indicated by these findings and by the fact that 5-hydroxytryptamine has been isolated from chicken intestine. The latter produces marked contractions of the chicken oesophagus and crop, although the sensitivity of the crop tissue to both 5-hydroxytryptamine and histamine appears to be greater in summer months than in winter (Everett, 1966). Peristaltic contractions carry food along the oesophagus and they either pass to the proventriculus directly or are diverted into the crop. Entry of boli into the crop is determined by the tonus of the oesophagus at its upper and lower levels, particularly in the region of the crop (oesophago-ingluvial fissura) and the functional state of the gizzard.

When the gizzard is empty, the oesophago-ingluvial fissura is largely obliterated by contraction of the longitudinal muscle layer of the oesophagus and crop, and freshly consumed food by-passes the crop, moving directly to the proventriculus. Relaxation of the fissura occurs when the gizzard contains food and the boli which pass down the oesophagus are then diverted into the crop, which acts as a temporary food storage organ (Ashcraft, 1930; Ihnen, 1928; Vonk and Postma, 1949).

The empty crop contracts at approximately 1 to $1\frac{1}{2}$ min intervals, each contraction being accompanied by an increase in tonus. After the entry of food into the crop, the contractions are inhibited, the duration of inhibition being related to the type of food material ingested and its rate of breakdown and removal from the gizzard. Onward movement of crop contents is brought about by a strong contraction which commences on the ventral surface of the crop and, as this moves towards its posterior extremity, contraction of the opposing oesophageal wall takes place so that a portion of the crop contents is separated off for passage down the oesophagus (Macowan and Magee, 1931). Factors which assist the passage of food from the gizzard, e.g. moist or wet food, finely ground food, also hasten the rate of crop evacuation. Wet mash, for example, is removed in 12 h compared with 20 h for dry mash (Heuser, 1945). Even when food has left the crop and passed along the oesophagus, it does not necessarily enter the gizzard. If the gizzard is partially contracted, the bolus remains in the oesophagus for several seconds and appears, under X-ray examination, as an oscillating mass. If the food mass is relatively large it may be divided into two portions by oesophageal contractions, of which the cranial portion passes back to the crop and the caudal portion to the gizzard (Vonk and Postma, 1949).

B. PROVENTRICULUS AND GIZZARD

The main function of the proventriculus is the production of gastric juice and the propulsion of juice and food into the gizzard. Peristaltic movements occur in the proventriculus of the 8 or 9-day-old embryo immediately after the differentiation of the muscle coats and, in the newly hatched chick, the contractions are regular and uniform (Kuo and Shen, 1936). Regular, rhythmical contractions, at about 1 min intervals, occur in the proventriculus of the adult fowl (Ashcraft, 1930). Passage of food to the proventriculus is dependent on the motor activity of the crop and of the lower oesophagus which are, in turn, regulated by the activity of the gizzard.

There is no appreciable residence time for food in the proventriculus, although there may be oscillation of contents between the gizzard and proventriculus and a pause of several seconds if food reaches the proventriculus before the gizzard is in the receptive stage.

Movements of the gizzard have been studied extensively by a variety of techniques (see Farner, 1960; Sturkie, 1965) and the type and rate of contraction have been found to vary, depending on whether the gizzard is empty or full, and on the nature of the food being fed. Hard grains tend to decrease the duration of individual contractions while softer foods are accompanied by contractions of longer duration; that is, the contractions are more frequent with material which requires grinding. The presence of grit in the gizzard appears to be associated with contractions of greater amplitude and, on a whole-grain diet, this is no doubt beneficial.

The typical gizzard contractions are grinding contractions and their genesis has been described by many workers (see Sturkie, 1965). A contraction sequence commences with the small intermediary muscles which propel the gizzard contents into the lumen between the main lateral muscles. The latter also contract towards the end of this process, the muscle fibres contracting towards the central aponeurosis of the gizzard. The lateral muscles are arranged asymmetrically and, as contraction occurs, the inner grinding surfaces of the gizzard are moved across each other with a sliding motion. As the gizzard is anchored firmly by its attachments to the proventriculus and duodenum, contraction of the massive lateral muscles imparts a rotary movement to the whole organ and ensures regular agitation of its contents.

The radiographic appearance of the proventriculus, gizzard and duodenum has been correlated with pressure recordings from within the gizzard lumen by Pastea *et al.* (1968b), who have described four stages of the contraction cycle. *Phase I*, in which food enters the gizzard and contraction begins. *Phase II*, when contraction is maximal and the asymmetric grinding surfaces of the gizzard lining are closely applied. Food is triturated during this period, aided by the presence of grit, and fluid and fine particles may pass into the proventriculus. *Phase III*, in which there is relaxation of the muscle walls. Fluid which has entered the proventriculus drains back into the gizzard and tends to pass to the duodenal orifice where the mucosal ridges, present at this point, act as a filter and prevent large particles, including particles of grit, from entering the duodenum. *Phase IV*, a receptive or resting stage, in which filling of the gizzard from the crop and proventriculus occurs. If, during a meal, the gizzard is in this receptive phase at the precise moment when food is swallowed, the bolus may pass straight to the proventriculus and gizzard (Fig. 2).

Pastea *et al.* (1968b), also differentiate two further types of contraction: a mixing contraction in which food oscillates between the gizzard and the proventriculus, and a contraction which assists the discharge of fluid and particles from the gizzard to the duodenum. Mangold (1906) has also described three different types of contraction to which those described by Pastea *et al.* (1968b) are comparable.

The gizzard possesses a degree of intrinsic contractile activity and modification of this activity is accomplished through the autonomic nervous system. Vagal section results in a marked slowing of gizzard movements and gradual reduction in the size of the lateral muscles (Nolf, 1925), whilst parasympathetic stimulants, such as acetylcholine and pilocarpine, stimulate

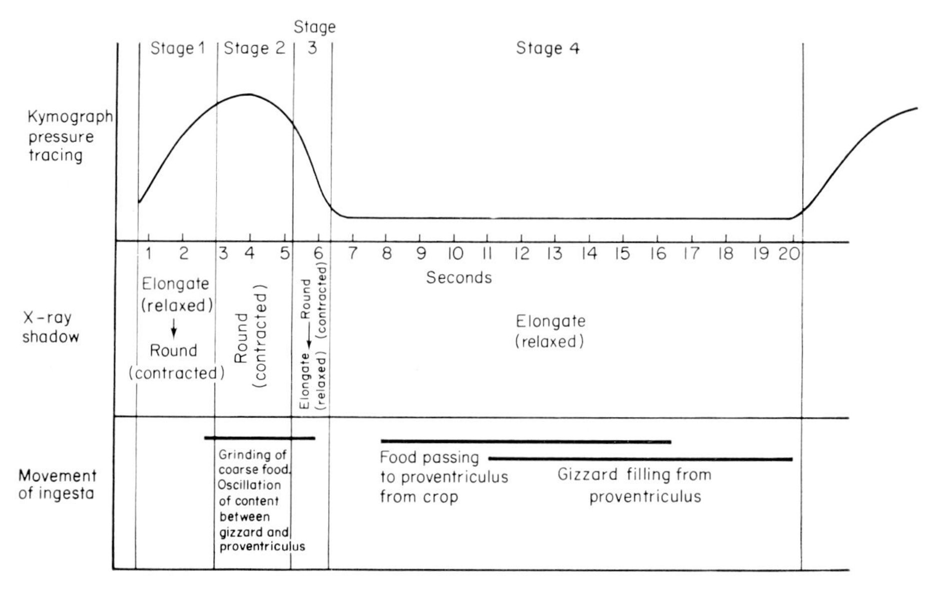

Fig. 2. Pressure changes, radiological appearance and movement of ingesta during a contraction-relaxation cycle of the gizzard.

gizzard contractions. Atropine inhibits gizzard movement and provides further evidence for the importance of the vagi in the control of gizzard movements.

C. INTESTINE

Contractile activity is possible as soon as mesenchymal differentiation of muscle fibre occurs and increases progressively, as does the response to drugs, with the development of the embryo.

Typical peristaltic and segmentation movements occur *in vitro* (Groebbels, 1930) and have been observed radiographically in the conscious fowl by Vonk and Postma (1949). The latter authors also noted that a barium-paste meal reached the duodenum within minutes of ingestion and was carried along the duodenum and jejunum by rapid peristaltic movements. Segmentation was more obvious in the intestine after a conventional meal of solid food.

The inherent rhythmicity of the small intestine is not affected by vagal section (Nolf, 1925) although, like the other regions of the digestive tract, modification of motility appears to be affected by the action of the autonomic nervous system on the intrinsic plexuses. This is exemplified by the inhibitory effect of excitement on intestinal movements (Henry *et al.*, 1932; Pastea *et al.*, 1968b). Trauma and climatic variations also produce alterations in tonus and amplitude of intestinal contractions (Pastea *et al.*, 1968a).

D. CAECA AND COLON

Despite the apparently minor role of the caeca in the digestive process, the mechanism by which they fill and discharge has intrigued many workers (see Browne, 1922; Olson and Mann, 1935) and recent confirmation of Browne's finding that material from the cloaca can pass in a retrograde manner into the caeca, where water, and possibly electrolyte absorption, can take place, has again underlined the need for an understanding of the physiological mechanisms involved in caecal function.

Retrograde flow of digesta from the colon into the caecal tubes has been observed radiographically in the conscious fowl (Akester *et al.*, 1967; Nechay *et al.*, 1968) and is the result of antiperistaltic movements of the colon (Fig. 3). Passage of colonic content into the caecal bodies is facilitated by contraction of the narrow caecal neck region, the contraction starting at the colonic-caecal tube junction and passing towards the caecal body. The caecal bodies exhibit mixing contractions and both caecal neck and body contractions are readily demonstrated *in vitro* (Pintea and Cotrut, 1958; W. S. Hardy and K. J. Hill, unpublished observations)–Fig. 4.

Colonic contents do not pass into the ileum, and it is apparent that the ileo-caeco–colic sphincter is closed during retro-peristalsis of the colon. Pintea and Cotrut (1958) have produced pharmacological evidence that the ileo-caeco-colic sphincter and the caecal tubes are reciprocally innervated and it is evident that there is an integrated sequence of reflexes which regulates the movement of the lower digestive tract.

The entrances to the caecal tubes are protected by valve-like structures which act as filters and permit only fluid and fine particles to pass into the caeca. As colonic retro-peristalsis commences at the cloaca, urine also enters the colon and the caeca, although the extent and frequency with which this occurs have not been established (Akester *et al.*, 1967).

The evacuation of the caeca appears to be the result of an occasional powerful contraction which commences at the base of each caecum, passes along the body and the neck and subsequently along the colon. Inhibition of the normal caecal contraction takes place during caecal evacuation.

The caeca are probably filled at regular intervals, as judged by the fairly uniform concentration of fermentation products in the caeca produced during *ad libitum* feeding. Frequency of emptying, as assessed by the production of caecal droppings, varies with diet, and may relate to the degree of caecal distension, or the hydrogen ion concentration or electrolyte concentration of the caecal content, as all these factors influence caecal contractions *in vitro*.

E. CLOACA

Discharge of cloacal content occurs at frequent intervals and appears to be caused by rapid contraction of the coprodeum, possibly provoked by distension of this region of the cloaca. Radiographic observations (Akester

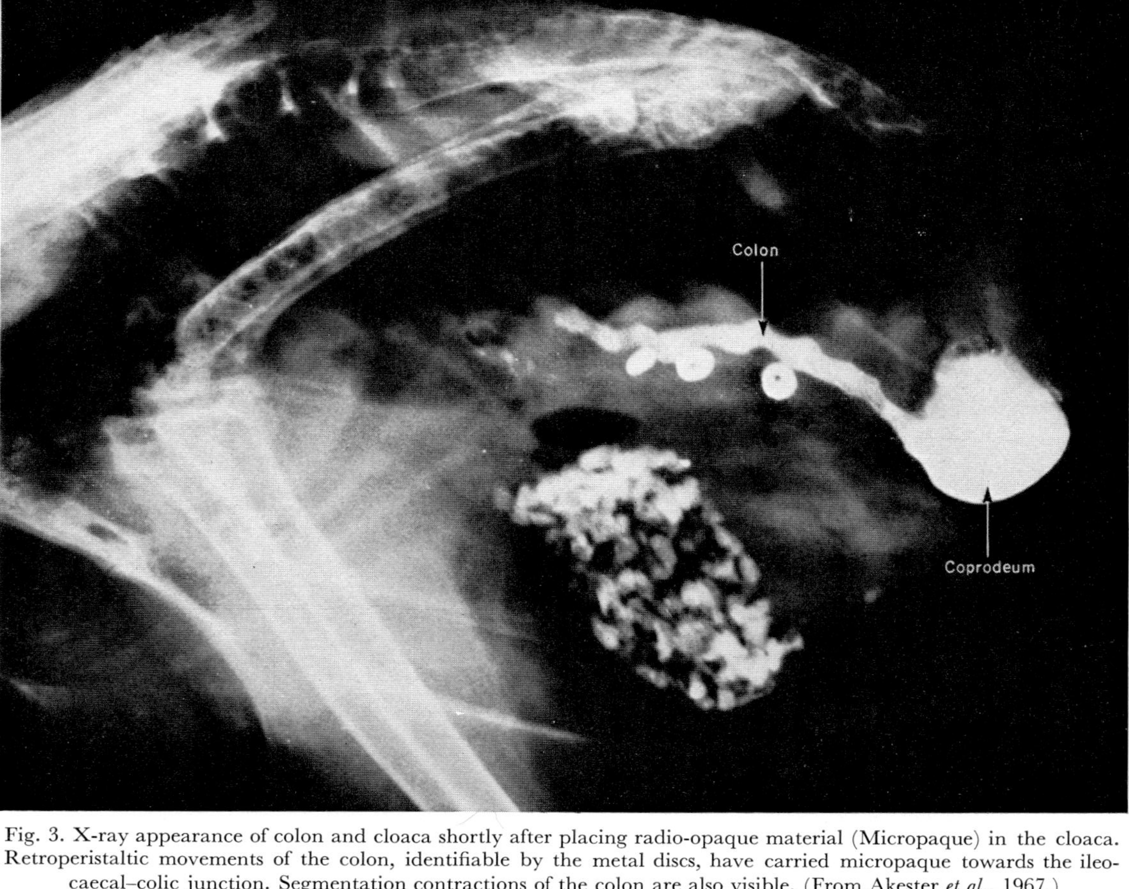

Fig. 3. X-ray appearance of colon and cloaca shortly after placing radio-opaque material (Micropaque) in the cloaca. Retroperistaltic movements of the colon, identifiable by the metal discs, have carried micropaque towards the ileo-caecal–colic junction. Segmentation contractions of the colon are also visible. (From Akester *et al.*, 1967.)

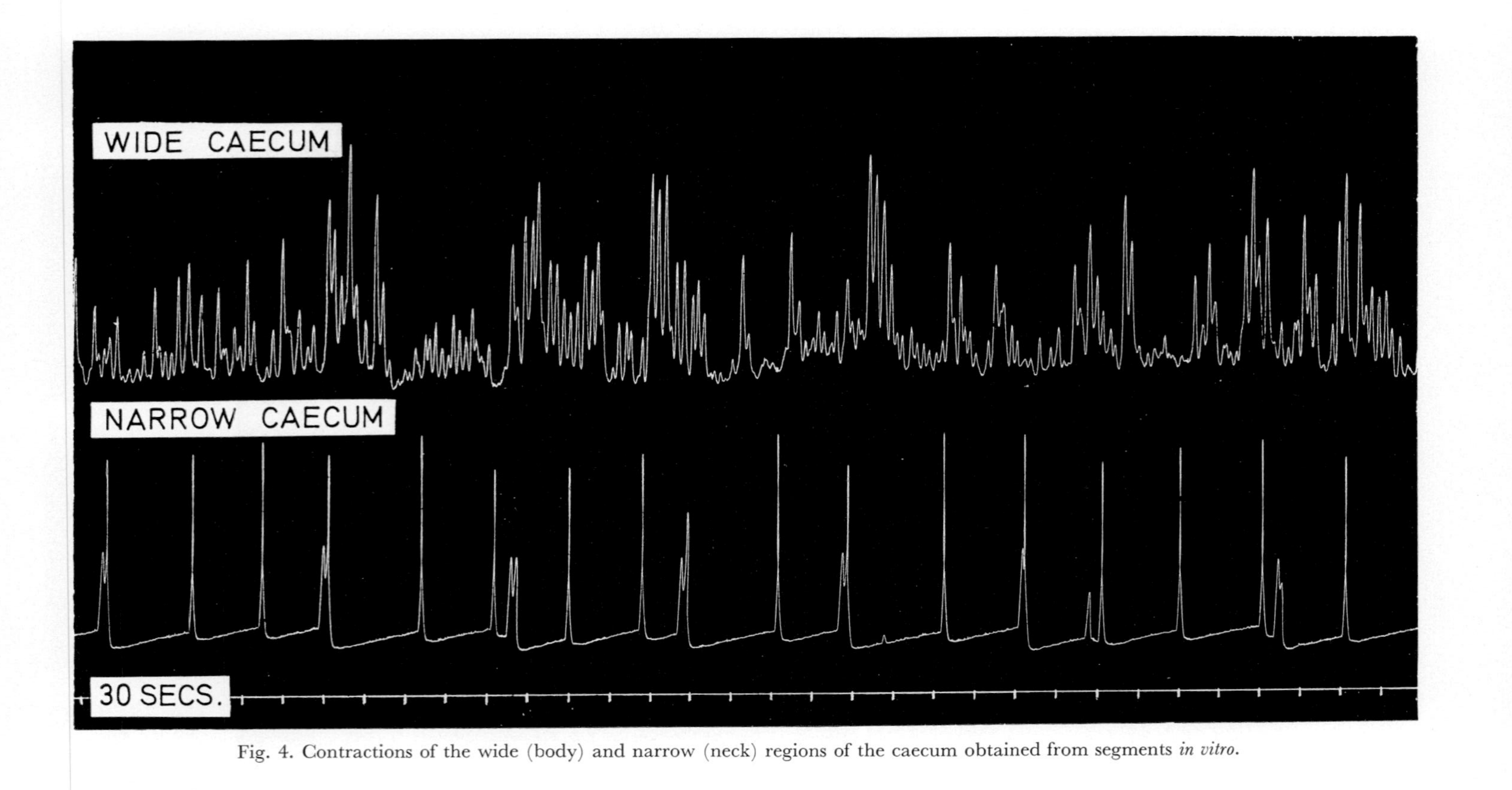

Fig. 4. Contractions of the wide (body) and narrow (neck) regions of the caecum obtained from segments *in vitro*.

et al., 1967) have shown that the coprodeum, although exhibiting periods of quiescence, is a highly active organ which not only undergoes powerful evacuatory contractions but also undergoes repeated peristaltic-type movements resulting in oscillation of the cloacal contents between the coprodeum and the distal end of the colon. At times the peristaltic waves spread cranially from the coprodeum to the colon and the ileo-caeco–colic junction and it is these contractions which are responsible for carrying colonic content into the caeca. Water and electrolytes are probably absorbed from the cloacal-colonic area (Skadhauge, 1967), a process which may be facilitated by the to-and-fro movement of the contents, and which results in the characteristic rounded shape and relatively dry appearance of the normal droppings.

IV. Secretory Activity of the Digestive Tract

A. PROVENTRICULAR SECRETION

The pH of the proventricular contents becomes markedly acid about the twentieth day of incubation and is indicative of a considerable secretion of hydrochloric acid by the proventricular glands. The actual onset of secretion occurs between the 11th and 13th day of incubation, in response to the ingestion of albumin by the embryo (Toner, 1965). Pepsin is present in the proventricular glands from the 12th to 16th day of incubation.

Studies on the physiology of gastric secretion by the adult bird have been limited by the relatively small size of the proventriculus, which has precluded the use of the gastric pouch. This classic preparation, used by the mammalian physiologist in investigations on the control of gastric secretion, has therefore been denied to the avian physiologist and knowledge has accrued solely from observations on the catheterized or cannulated proventriculus of the anaesthetized or conscious bird, and by post-mortem examination of tract contents (Cheney, 1938; Long, 1967).

During *ad libitum* feeding the reaction of the proventricular and gizzard contents is always acid and presumably reflects the continuous secretion of gastric juice. Both nervous and chemical factors are involved in the stimulation of the proventricular glands for production of this juice although the relative importance of each has not been established.

It has not been possible to demonstrate a secretory response to the sight of food (Farner, 1960) and it would seem that a conditioned, cephalic gastric secretory response is not easily established in the fowl. On the other hand, both Collip (1922) and Farner (1960) obtained a secretory response to sham-feeding, indicating the existence of a cephalic phase, while Long (1967) observed secretory activity in the starved fowl and concluded that it was due to vagal stimulation as it was abolished by atropine. Long (1967), in fact, found that there was a high rate of secretion, lasting several hours, even after 24-h starvation and related this to the "continuous" nature of feeding activity in the fowl. He further suggested that the neuronal activity directed towards feeding behaviour has a component directed towards the gastro-intestinal tract which is active, presumably, even in the absence of food.

The crop may influence proventricular secretion as distension of this organ has a slight stimulatory effect (Collip, 1922) and food in the crop, prevented from entering the proventriculus by a lower oesophageal fistula, markedly stimulates secretion (Farner, 1960).

Direct vagal (Friedman, 1939) and parasympathetic stimulation (Collip, 1922) is effective in provoking secretion as in mammals; however, insulin hypoglycaemia, which stimulates the vagal hypothalamic centres in mammals resulting in copious secretion of gastric juice, is not effective in the fowl. Possibly the vagal centres in the bird are not sensitive to hypoglycaemia.

Histamine is a potent stimulant of gastric secretion (Fig. 5) (Cheney, 1938; Friedman, 1939; Long, 1967) and in the histamine-induced secretion from

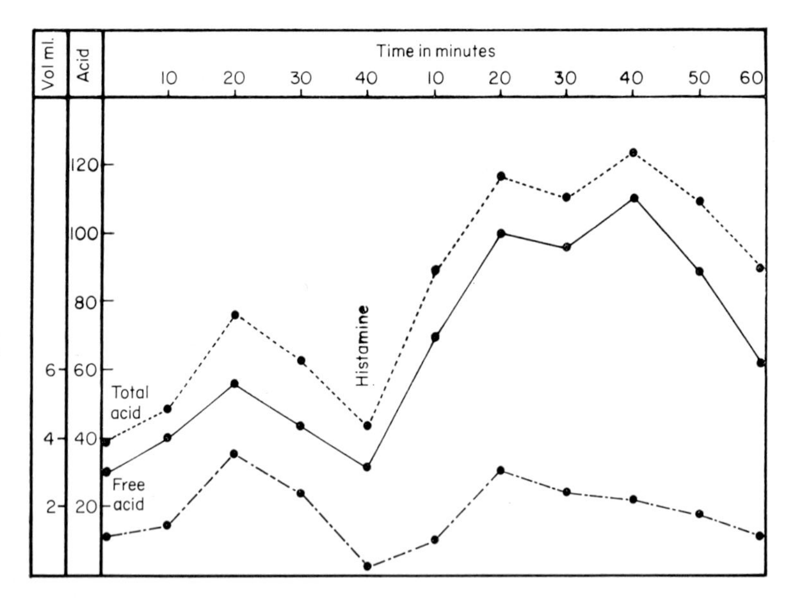

Fig. 5. Proventricular secretory response to histamine. Seven-month-old anaesthetized fowl, juice collected by aspiration. (From Cheney, 1938.)

the conscious bird there is usually a parallelism between acid and enzyme output which probably relates to the single-cell origin of acid and enzyme (Fig. 6). Injection of extracts of proventricular or duodenal tissue has been used to provoke secretion of gastric juice (Collip, 1922; Keeton *et al.*, 1920) but, as the extracts used by these authors probably contained histamine, it is

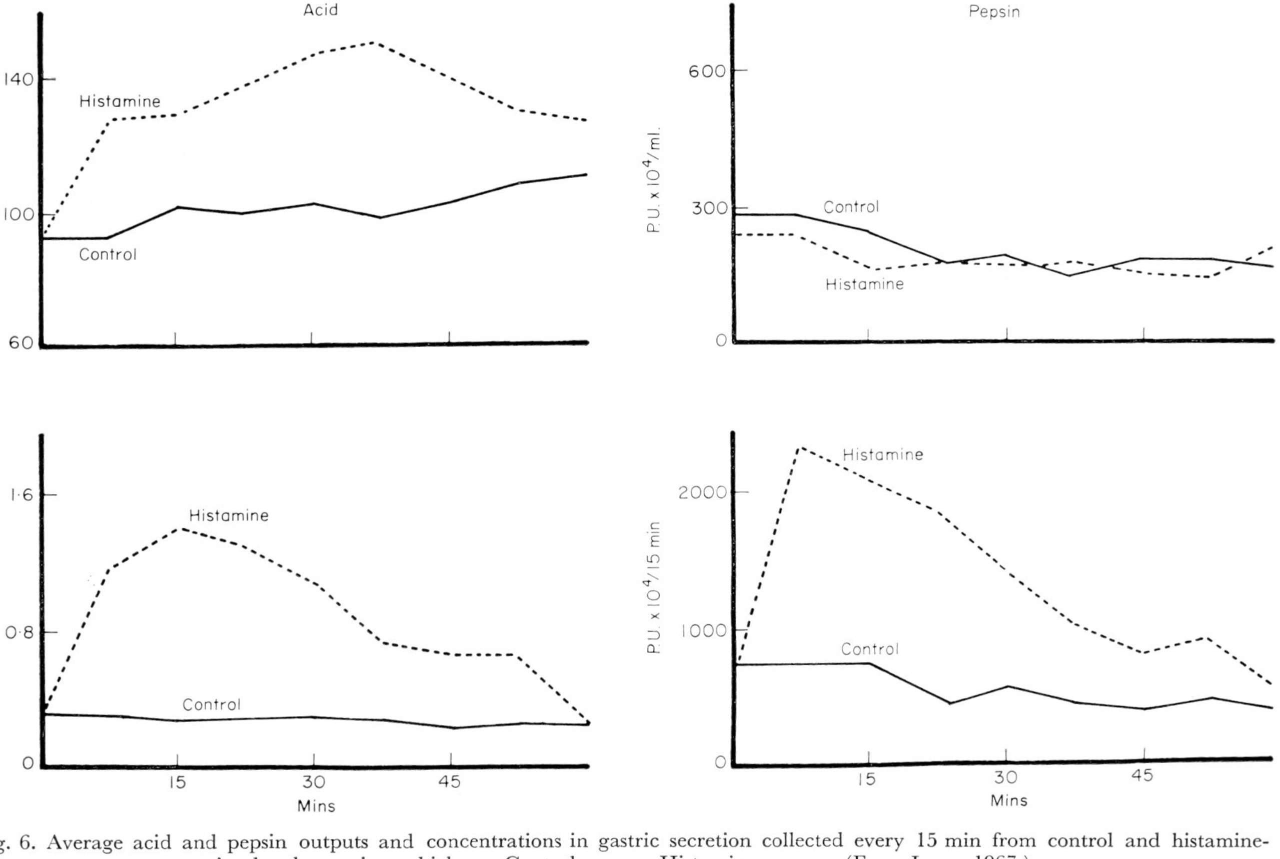

Fig. 6. Average acid and pepsin outputs and concentrations in gastric secretion collected every 15 min from control and histamine-stimulated conscious chickens. Control ——, Histamine – – – –. (From Long, 1967.)

not clear whether gastrin was responsible. There appears however, to be no reason to doubt that a gastrin mechanism is present in the fowl although it has not been specifically demonstrated and there is no evidence as to the site of gastrin production.

Similarly the presence of the gastric secretory inhibitory hormone, enterogastrone, has not been shown in the fowl and indeed, the absence of secretory inhibition after infusion of fat into the duodenum (Long, 1967), suggests that such a hormone may not be produced. Estimates of the volume of gastric juice produced by the proventriculus vary from 6·0 to 21·0 ml/h during starvation, and up to 38·8 ml/h after histamine stimulation. A total acidity of 145 mEq/l was obtained after histamine stimulation and such juice also possessed powerful proteolytic activity (Long, 1967).

There is no information on the gastric secretory response to a single meal, although undoubtedly differences in the residence time in the crop and the gizzard of different types of foodstuffs, will provide different secretory responses. In view of the storage function of the crop and the prevalence of *ad libitum* feeding, it may be more worthwhile to consider the secretion of gastric juice as continuous, with peak secretion rates of around 30·0 ml/h and much lower rates of around 5·0 ml/h during periods of darkness when the bird is eating little or no food.

The acidic and proteolytic properties of avian digestive juice have been recognized for many years and the fact that both acid and enzyme are secreted by the same gland cell is also well established. The proteolytic enzyme is a pepsin and its zymogen, pepsinogen, has been isolated from proventricular tissue (Herriott, 1938). Simultaneous secretion of pepsinogen and hydrochloric acid by the same cell is obviously highly effective for rapid conversion of the zymogen to the active enzyme.

B. DUODENAL SECRETION

There are no Brünner's glands in the duodenum of the fowl and the duodenal glands or crypts of Lieberkühn are the source of the duodenal secretion. They appear to secrete spontaneously although there is some indication that the enzyme content of the secretion is under vagal control as the cholinergic drug, mecholyl, provokes increased secretion of mucus and amylase. Hormonal control of secretion may also be important, as crude acid extracts of the duodenal and intestinal mucosa provoke increased secretion (Kokas *et al.*, 1967). The spontaneously secreted juice is poor in mucus and solids compared with that collected during stimulation and presumably reflects basal secretion by the duodenal glands.

The juice collected from a duodenal loop, from which the proventricular, biliary and pancreatic secretions have been excluded, is pale yellow and contains mucus, amylase, proteases and sucrose. Similar enzymes are present in extracts of the duodenal mucosa (Plimmer and Rosedale, 1922) although it seems probable that the only enzyme actually secreted in the duodenal juice is amylase. The other two enzymes are more likely to be derived from residual traces of gastric and pancreatic juice and from cellular breakdown.

C. PANCREATIC SECRETION

There is an extensive literature on the enzymes present in extracts of avian pancreatic tissue (see Farner, 1960; Sturkie, 1965) but it is only recently that attempts have been made to obtain pancreatic juice for analysis. The presence of three pancreatic ducts, however, almost precludes total quantitative collection and there are difficulties with cannulation of even the major duct, so that complete data on the volume of pancreatic juice secreted during feeding are not available (Heatley *et al.*, 1965; Dal Borgo *et al.*, 1968a).

Ivanov and Gotev (1962) successfully cannulated the major pancreatic duct in a one-year-old fowl and maintained it functional for a month. Secretion of pancreatic juice was continuous, although at a relatively low level of from 0·4 to 0·8 ml/h. Following a meal, the secretion rate increased to 3·0 ml/h, within an hour and gradually declined to the initial level over the ensuing 9 to 10 h. Secretion was low during the night when food was

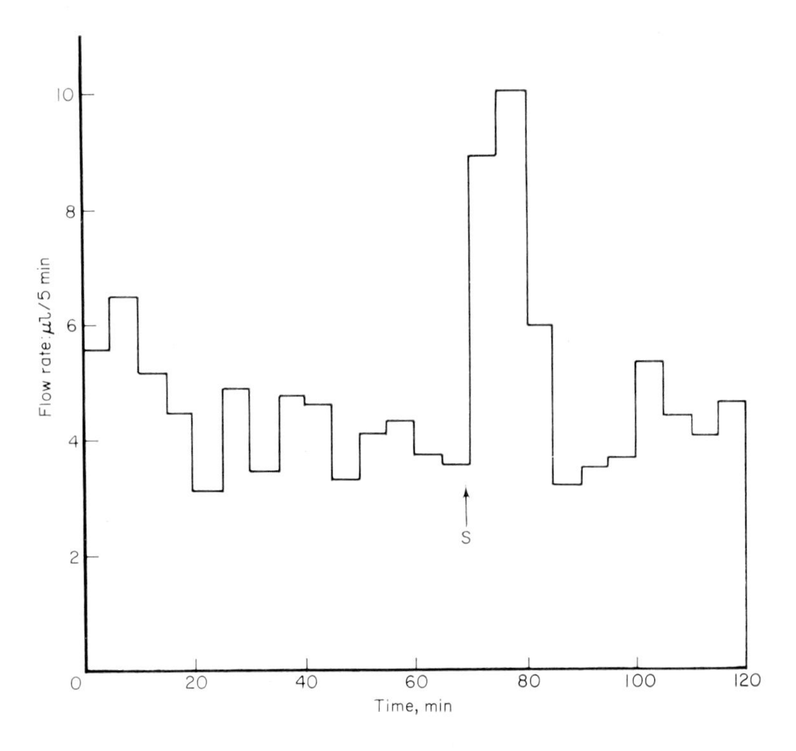

Fig. 7. Rate of pancreatic secretion from main duct. At S a neutralized crude acid extract of rat intestine was given intravenously. (From Heatley *et al.*, 1965.)

not available. Dal Borgo *et al.* (1968b) reported a flow of 9·7 ml/kg d based on the volume obtained from the cannulated main duct of the young chicken. There is no information on the total amount of juice produced by the pancreas under maximal stimulation.

The pancreatic ducts exhibit marked peristaltic activity and hence the juice is transmitted to the duodenum in surges at intervals of 10 to 100 s (Heatley *et al.*, 1965). Whether the juice normally passes into the duodenum or along the jejunum has not been established.

The stimulatory mechanisms for pancreatic secretion appear to resemble those of the mammal, in that humoral and nervous factors are involved. Intravenous injection of secretin provokes an increased flow of juice within a minute, maximal after 5 min with a gradual return to base-line flow within 20 min (Fig. 7). As secretin has been isolated from the duodenum and small intestine of the fowl there is no reason to doubt that, under acid conditions in the duodenum, there is liberation of secretin during normal digestion. Pancreozymin has not been identified in extracts of avian duodenal mucosa but the demonstration of increased and maintained enzyme output after feeding (Ivanov and Gotev, 1962) indicates that this hormone is likely to be present. Possible vagal effects on pancreatic secretion have not been investigated although the increase in enzyme output soon after feeding provides circumstantial evidence for vagal involvement.

Pancreatic juice collected from the conscious bird is a clear, slightly viscid, fluid with an alkaline pH. There is no information on its electrolyte composition although Farner (1942) states that, because of the relatively high concentration of buffering compounds, it has an important role in the neutralization of the duodenal chyme and therefore, by analogy with the mammal, the major factor is probably the bicarbonate ion.

The enzymic composition of freshly collected pancreatic juice is similar to that of mammalian juice. Ribonuclease, amylase, lipase and deoxyribonuclease are present in the non-activated juice and chymotrypsin, trypsin, carboxypeptidases A and B and elastase are present after activation with trypsin (Dal Borgo *et al.*, 1968b). Proteolysis occurs between pH 5·7 and 8·5 and is maximal at pH 7·0 (Hewitt and Schelkopf, 1955).

Some of these enzymes have been characterized and their appearance during development defined (Laws and Moore, 1963; Kulka and Duksin, 1964). Thus the pancreas of the newly-hatched chick has a high amylase content which declines gradually over the first three weeks of life and then remains constant. Heller and Kulka (1968) have shown that two amylase isoenzymes occur. Chicks may possess either one, or both of these, and it is suggested that this distribution may be genetically controlled.

The lipase activity of the pancreas is high in the newly-hatched chick and remains at a more or less constant level during growth while traces of esterase activity, which are present at hatching, show an increase during the first ten days after hatching.

D. BILE SECRETION

Bile secretion has been observed in chick embryos aged between 7 and 9 days and a secretory response to injected bile salts noted (Kuo, 1932). Comparable information on the effect of bile salts on bile secretion by the adult bird is not available although the nature of the bile acids has been defined by Anderson *et al.* (1957), and they are undoubtedly the major stimulus to bile production. Secretin, gastrin and pancreozymin are also likely to be concerned in stimulation of bile secretion and gall-bladder contraction.

Formation of bile by the liver is a continuous process and secretory rates of about 1·0 ml/h have been observed in the conscious, 14-week-old cockerel and, in the anaesthetized, 4 to 6-month-old bird (Clarkson *et al.*, 1957; Lind *et al.*, 1967). Starvation reduces the rate of secretion and is accompanied by concentration of the bile in the gall bladder (Schmidt and Ivy, 1937). Amylase activity, of unknown origin, has been found in gall bladder and hepatic duct bile (Farner, 1943).

The bile pigments, bilirubin and biliverdin, are present in avian bile although, following hepatic and bile duct ligation, only bilirubin accumulates in the plasma, suggesting that biliverdin is formed extrahepatically in the fowl (Lind *et al.*, 1967).

E. INTESTINAL SECRETION

The usual array of proteases, lipases and carbohydrases has been demonstrated in extracts of small-intestine tissue, in the intestinal contents (Siddons, 1969) and in the fluid collected from a Thiry fistula in the small intestine (Polyakov, 1958). The properties of these enzymes closely resemble those of mammalian enzymes. They are intracellular and are liberated into the intestinal lumen by disintegration of the shed cells. Pancreatic enzymes are also present in the intestinal contents.

Disaccharidases are present in the 12-day-old embryo in low concentration and show a marked increase during the last few days of incubation, at a time when there is elongation and increase in the number of villi. The newly hatched chick therefore possesses marked maltase activity, which is attributable to a complex of enzymes, and high sucrase activity due to a sucrase-isomaltase complex (Siddons, 1970).

The disaccharidases, maltase, sucrase and palatinase found in the caecum are entirely within the contents and originate from the small intestine. Lactase, also present in caecal contents, is derived from the micro-organisms present there.

V. The Large Intestine

A. THE CAECA

The caecal tubes constitute a region of the digestive tract where microbial breakdown, particularly of cellulose-containing materials, takes place. As such, their role in birds fed diets containing little fibre is a minor one, and even though fermentation of other dietary components, which escape enzymic digestion, also occurs in the caeca the contribution of the absorbed fermentation products as an energy source to the fowl is small (Annison *et al.*, 1968). Other possible roles for fermentation reactions within the caeca, *e.g.* vitamin synthesis, have not been adequately evaluated but, as caecectomy has negligible effect on performance, it would appear that caecal function is of little significance in the modern fowl.

B. THE COLON

The colon transports the small intestinal contents to the cloaca by peristaltic movements and participates in the filling of the caecal tubes. Segmentation contractions also occur in the colon and may relate to its role in the removal of water from the faeces. Absorption of electrolyte may accompany that of water (Hill and Lumijarvi, 1968).

VI. Digestion in the Fowl

The pattern of food intake and its passage through the digestive tract are the major factors influencing secretory, and hence digestive activity. Possibly because of the high metabolic rate of the fowl, a more or less continuous supply of food material to the digestive tract is required and this is largely met by the presence of the crop which acts as a food reservoir or storage area. Under natural or restricted feeding conditions food is diverted to the crop after the first few boli have passed to the gizzard and, because of its distensibility, it rapidly accumulates a considerable amount of food which is gradually supplied to the gizzard after feeding has ceased.

Crop function is less critical when food is always available and in the broiler the crop appears to be almost functionless. Feeding activity is affected by the availability of food and when provided *ad libitum* it is consumed throughout the 24 h, although there are some periods when intake is at a higher level than others (Table 1).

Food intake is affected by many factors, e.g. environment, temperature, physiological state of the bird, water intake and the physical and chemical nature of the diet. Pelleted food, for example, leads to hyperphagia in older birds, although chickens under two weeks of age prefer mash to pellets and eat less if the food is in pelleted form (Calet, 1965).

The rate at which food passes along the digestive tract is influenced by the same factors as food intake (see Hillerman *et al.*, 1953; Ferrando *et al.*, 1961). Compared to the mammalian alimentary tract that of the bird is relatively short and the rate of passage of a single meal of mash is rapid, e.g. 4 h in the pullet and growing hen; 8 h in the laying bird; 12 h in the broody hen (Kaupp and Ivey, 1923). Intact, hard grains tend to remain in the digestive tract for longer periods (Heuser, 1945).

Table 1

Food consumption[a] over a 16-h period by birds of different ages

Age of bird (weeks)	Period			
	04.00–08.00 h	08.00–12.00 h	12.00–16.00 h	16.00–20.00 h
4	13·4	11·3	12·2	17·8
5	18·4	10·8	13·7	20·6
6	23·2	17·7	17·0	20·6
7	27·0	17·4	20·8	25·0
8	24·8	21·9	20·1	20·8

[a] Food consumption (g), by birds of different ages, over four successive 4-h periods. Feeding occurred in each period with a diurnal rhythm in consumption. (From Siegel *et al.*, 1962.)

After ingestion, and admixture with saliva and oesophageal mucus, most of the food enters the crop and is thoroughly moistened with these secretions. In the modern layer, fed *ad libitum* on meal or pellets which disintegrate rapidly, emptying of the crop is accomplished rapidly but, in the farmyard fowl receiving whole grain and fresh vegetable material, food may remain in the crop for longer periods. Amylase, whether produced by the salivary or oesophageal glands, and the ingested enzymes in plant material may cause some breakdown under these conditions, as may the fermentative activities of the natural lactobacillary population in the crop, leading to production of lactic acid (Bolton, 1962; Ivorec-Szylit and Szylit, 1965). Selective absorption and active transport of sodium from the crop have been demonstrated (Soedarmo *et al.*, 1961) but their functional significance is not established.

The proventriculus and gizzard constitute the first important site of enzyme activity and, although the ingested food has a buffering effect, its gradual introduction from the crop, into the proventriculus, ensures that a low pH is maintained in this organ and in the gizzard which is the main site of gastric proteolysis (Fig. 8). The latter process is assisted by the rapid denaturation of the proteins in the highly acid environment. The powerful mixing and grinding contractions of the gizzard, assisted by the presence of small stones or grit, ensure that the peptide bonds are sufficiently exposed to proteolytic enzyme activity. Repeated oscillation of contents between the gizzard and proventriculus must also assist this process. The retention time of food in the gizzard is variable, finely divided food passing through to the duodenum within minutes whilst hard grains may remain in the gizzard for several hours. Removal of the superficial layers of the koilin lining of the gizzard, which is continuously produced, is also accomplished by the movements of the gizzard. Soluble grit, when present, slowly dissolves in the acid medium and presumably provides a continuous supply of calcium to the intestine. A

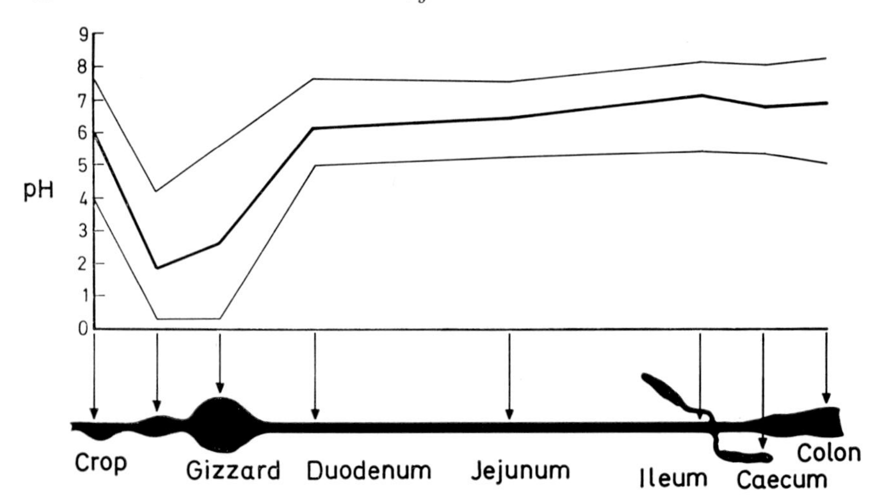

Fig. 8. Mean pH, with maximum and minimum deviation, of the contents of the different regions of the digestive tract of the fowl. (From Herpol and van Grembergen, 1967.)

clear picture of the rate of passage of gizzard contents into the duodenum is not available but it is obviously likely to vary with different feeding regimes and types of food. An adequate description of the contraction pattern of the digestive tract and the digesta flow characteristics can therefore only be meaningful if given in relation to these parameters although, in general, it is apparent that meals, crumbles and rapidly disintegrating food pellets are likely to pass through the proventriculus and gizzard comparatively quickly. The latter point has been demonstrated in an 11-week-old cockerel, weighing 2 kg with a duodenal re-entrant fistula, and receiving its diet in mash form. Duodenal flow rates varied between 40·0 and 60·0 ml/h and ingested food appeared quickly in the duodenal content as a suspension of relatively unchanged particles. Discharge of duodenal contents occurred at irregular intervals, sometimes there were several gushes in quick succession and, at other times, flow ceased for periods up to 5 min (P. J. Wilson personal communication). Evidence was also obtained that gastric proteolysis may continue in the duodenal loop as the duodenal contents were frequently between pH 3·0 and 4·0. Secretin and pancreozymin liberation were presumably optimal at this degree of activity, provoking maximal secretion of pancreatic juice.

Considerable amounts of nitrogen, as mucoprotein, are added to the duodenal content by the duodenal secretion (Imondi and Bird, 1955; Bolton, 1961) and it seems possible that it is the highly acid material from the gizzard which provokes this secretion (Table 2). Desquamated cells from the duodenal mucosa, bile and pancreatic juice must also add to the nitrogen content of the ingesta.

Table 2

Distribution of nitrogen in the digestive tract

Food or intestinal segment	%N	Cr_2O_3	N/Cr_2O_3
Food	3·55	0·458	7·75
Crop	3·55	0·352	10·03
Proventriculus	3·13	0·192	16·30
Gizzard	2·74	0·189	14·50
Upper duodenum	8·33	0·230	36·25
Lower duodenum	8·55	0·282	30·37
Upper jejunum	6·00	0·700	8·58
Lower jejunum	5·23	1·437	3·64
Upper ileum	4·54	1·247	3·65
Lower ileum	4·30	0·811	5·30

Percentage nitrogen, percentage chromic oxide and ratio of nitrogen to chromic oxide in the food and portions of the gastrointestinal tract of the 6-week-old chicken. (From Bird, 1968.)

Retrograde flow of bile, and probably of pancreatic juice, occurs along the duodenum of the fowl (Vonk and Postma, 1949) and these secretions, together with the duodenal secretion, buffer the contents. The extent or frequency of retrograde flow has not, however, been established and it seems possible, in view of the position of the bile and pancreatic ducts, that these secretions will tend to enter the first part of the jejunum rather than the duodenum. Certainly the pH conditions in the upper jejunum are more favourable for pancreatic enzyme activity and it is here that the polypeptides, formed as a result of peptic digestion, are further broken down.

Ingested carbohydrate and protein appear in the portal blood as glucose and amino nitrogen within 15 min of feeding and maximal digestive activity appears to last for about 2 h after feeding either a single meal or after *ad libitum* feeding (Aramaki and Weiss, 1962). Passage of protein and carbohydrate breakdown products and their absorption and further breakdown within the intestinal wall, is therefore a rapid process and correlates with the data on rate of passage which shows that food may enter the intestine within minutes of consumption (Henry *et al.*, 1932).

The high maltase activity present in the intestine of the young chick ensures that rapid breakdown of carbohydrate occurs in the young bird and Siddons (1969) has suggested that the intestinal disaccharidases of the fowl, as in mammals, develop in a way that allows utilization of the carbohydrates which the bird is likely to meet under natural feeding, i.e. starch containing grains. Elevated amylase output by the pancreas during the early weeks of life (Laws and Moore, 1963) is obviously in keeping with this suggestion.

Fats are utilized very efficiently by the fowl and up to 12% fat in the diet of growing chickens has little effect on the rate of passage of food or its digestibility (Tuckey *et al.*, 1958). Pancreatic and intestinal lipases are responsible for lipid digestion and their effectiveness is such that fat absorption is virtually complete half-way along the small intestine. Bile salts are essential for micelle formation and their importance is demonstrated by the fact that removal of bile from the alimentary tract by bile duct cannulation, in young chicks with low fat reserves, leads to a rapid drop in plasma lipoprotein level (Clarke *et al.*, 1964). The bile salts are also essential for the absorption of vitamins D, E and K and have a direct effect on calcium absorption (see also Chapter 3).

Food materials which escape enzymic breakdown along the tract and which are in a sufficiently divided state, pass into the caecal tubes and are subjected to bacterial breakdown. With fibre-rich diets and those containing types of carbohydrate which are resistant to, or for which the fowl does not possess the appropriate enzyme, e.g. lactose, the absorbed fermentation products may constitute a useful energy source to the bird. Conservation of water and electrolytes by the retrograde flow mechanism in the cloaca, colon and caeca may also be significant under certain conditions. (See also Chapter 9.)

Faeces from the digestive tract are mixed with the urine in the cloaca and constitute the droppings. Their appearance varies considerably but, typically, an individual dropping appears as a rounded, brown mass with a characteristic white cap of uric acid. The average daily weight of droppings produced by a laying hen is between 100 and 150 g although variations may occur largely because of fluctuations in the volume of urine produced.

The contents of the caecal tubes are also discharged from time to time and they appear as discrete masses of dark brown glutinous material. Röseler (1929) observed that the ratio of caecal to normal droppings varied, depending on the composition of the diet, thus it was 1 to 7 after feeding barley and 1 to 11 after feeding wheat. Although the factors responsible for provoking caecal discharge are not understood they presumably relate to the quantity of dead bacteria and undigestible material that accumulates in the caeca and has to be removed at intervals.

Colostomy techniques have been used extensively for the separate collection of urine and faeces (Ivy *et al.*, 1968), although their primary purpose has usually been to obtain uncontaminated urine and there is little information on the composition of the faecal material. Most of the data available have, therefore, been obtained by the analysis of droppings and hence they tend to reflect the variation in the amount of urine present in the droppings. This is particularly noticeable in relation to the droppings from the pre-lay and the laying bird where the moisture content has been shown to change from 69% to 82% (Anderson and Hill, 1968). A similar, though less obvious, difference is shown by the lower pH range (6·0 to 8·0) of droppings of the laying bird compared with those of the non-laying bird (7·0 to 8·0); again a reflection of differences in urine composition (Anderson, 1970).

The faecal components of the droppings are conveniently divided into those of dietary origin, i.e. the undigestible residues, and those of metabolic origin. The latter comprise the desquamated cells from the digestive tract mucosa, products of bacterial fermentation, residual digestive secretions and other endogenous proteins. Each of these components can be affected by dietary change. The output of bile acids, for example, is markedly affected by feeding different unsaturated and saturated lipids (Lindsay *et al.*, 1969) and similarly the output of endogenous faecal nitrogen is elevated when the protein level in the diet is increased.

Short chain volatile fatty acids formed by bacterial fermentation in the digestive tract are present in the droppings and occur in particularly high concentration in those of caecal origin (Annison *et al.*, 1968). Vitamin B_{12} is also present in the droppings and its concentration may be further increased by bacterial synthesis after the droppings have been voided (Coates, 1962).

The faecal components of dietary origin also vary in concentration. Thus variations in the digestibility of crude fibre result in different fibre concentrations in the droppings. Similarly starch grains of different origin are digested to a varying extent and, in relation to their dietary concentration, may be present in the faeces.

The presence of these and other valuable energy sources has stimulated considerable effort in recent years on methods of utilizing poultry droppings and their effective use as a source of food for ruminants and as substrate for gas production for heating and lighting purposes has been demonstrated (Hart, 1963).

References

Akester, A. R., Anderson, R. S., Hill, K. J. and Osbaldiston, G. W. (1967). *Br. Poult. Sci.* **8**, 209–212.
Anderson, I. G., Haslewood, G. A. D. and Wootton, I. D. P. (1957). *Biochem. J.* **67**, 323–328.
Anderson, R. S. (1970). *Annls Biol. anim. Biochim. Biophys.* **10**, 171–183.
Anderson, R. S. and Hill, K. J. (1968). *Proc. Nutr. Soc.* **27**, 3A.
Annison, E. F., Hill, K. J. and Kenworthy, R. (1968). *Br. J. Nutr.* **22**, 207–216.
Aramaki, T. and Weiss, H. S. (1962). *Archs int. Physiol. Biochim.* **70**, 1–15.
Ashcraft, D. W. (1930). *Am. J. Physiol.* **93**, 105–110.
Belman, A. L. (1962). *Thesis, Cornell University.* Abstracted in *Vet. Bull.* **32**, 2812.
Belman, A. L. and Kare, M. R. (1961). *Poult. Sci.* **40**, 1377.
Bird, F. H. (1968). *Fedn Proc. Fedn Am. Socs. exp. Biol.* **27**, 1194–1198.
Bolton, W. (1961). *Proc. Nutr. Soc.* **20**, xxvi.
Bolton, W. (1962). *Proc. Nutr. Soc.* **21**, xxiv.
Browne, T. G. (1922). *J. comp. Path. Ther.* **34–35**, 12–32.
Calet, C. (1965). *Wld's Poult. Sci. J.* **21**, 23–52.
Cheney, G. (1938). *Am. J. dig. Dis.* **5**, 104–107.

Chodnik, K. S. (1948). *Q. Jl microsc. Sci.* **89**, 75–87.
Clarke, G. B., Fukazana, K., Kummerow, F. A. and Nishida, T. (1964). *Proc. Soc. exp. Biol. Med.* **117**, 355–359.
Clarkson, T. B., King, J. S. and Warnock, N. H. (1957). *Am. J. vet. Res.* **18**, 187–190.
Coates, M. E. (1962). *In* "Nutrition of Pigs and Poultry" (J. T. Morgan and D. Lewis, eds) pp. 158–166. Butterworth, London.
Collip, J. B. (1922). *Am. J. Physiol.* **59**, 435–438.
Dal Borgo, G., Harrison, P. C. and McGinnis, J. (1968a). *Poult. Sci.* **47**, 1818–1820.
Dal Borgo, G., Salman, A. J., Pubols, M. H. and McGinnis, J. (1968b). *Proc. Soc. exp. Biol. Med.* **129**, 877–881.
Everett, S. D. (1966). *In* "Physiology of the Domestic Fowl" (C. Horton Smith and E. C. Amorosa, eds.) pp. 261–273. Oliver and Boyd, Edinburgh.
Farner, D. S. (1942). *Poult. Sci.* **21**, 445–450.
Farner, D. S. (1943). *Biol. Bull. mar. biol. Lab. Woods Hole* **84**, 240–243.
Farner, D. S. (1960). *In* "Biology and Comparative Physiology of Birds" (A. J. Marshall, ed.) Vol. I. pp. 411–467. Academic Press, New York and London.
Ferrando, R., Froget, J. and Heude, B. (1961). *Recl Méd. vét. Éc. Alfort* **137**, 357–365.
Friedman, M. H. F. (1939). *J. cell. comp. Physiol.* **13**, 219–234.
Groebbels, F. (1930). *Pflügers Arch. ges. Physiol.* **224**, 687–701.
Halnan, E. T. (1949). *Br. J. Nutr.* **3**, 245–252.
Halpern, B. P. (1962). *Am. J. Physiol.* **203**, 541–544.
Hanzlik, P. J., and Butt, E. M. (1928). *Am. J. Physiol.* **85**, 271–289.
Hart, S. A. (1963). *Wld's Poult. Sci. J.* **19**, 262–272.
Heatley, N. G., McElheny, F. and Lepkovsky, S. (1965). *Comp. Biochem. Physiol.* **16**, 29–36.
Heller, H. and Kulka, R. G. (1968). *Biochim. biophys. Acta* **165**, 393–397.
Henry, K. N., MacDonald, A. J. and Magee, H. E. (1932). *J. exp. Biol.* **12**, 153–171.
Herpol, C. and van Grembergen, G. (1967). *Annls Biol. anim. Biochim. Biophys.* **7**, 33–38.
Herriott, R. M. (1938). *J. gen. Physiol.* **21**, 575–582.
Heuser, G. F. (1945). *Poult. Sci.* **24**, 20–24.
Hewitt, E. H. and Schelkopf, L. (1955). *Am. J. vet. Res.* **16**, 576–579.
Hill, F. W. and Lumijarvi, D. H. (1968). *Fedn Proc. Fedn Am. Socs exp. Biol.* **27**, 421.
Hillerman, J. P., Kratzer, F. H. and Wilson, W. O. (1953). *Poult. Sci.* **32**, 332–335.
Ihnen, K. (1928). *Pflügers Arch. ges. Physiol.* **218**, 767–782.
Imondi, A. R. and Bird, F. H. (1955). *Poult. Sci.* **44**, 916–920.
Ivanov, N. and Gotev, R. (1962). *Arch. Tierernähr.* **12**, 65–73.
Ivorec-Szylit, O. and Szylit, M. (1965). *Annls Biol. anim. Biochim. Biophys.* **5**, 353–360.
Ivy, C. A., Bragg, B. D. and Stephenson, E. L. (1968). *Poult. Sci.* **47**, 1771–1774.
Kaupp, B. F. and Ivey, J. E. (1923). *J. agric. Res.* **23**, 721–725.
Keeton, R. W., Koch, F. C. and Luckhardt, A. B. (1920). *Am. J. Physiol.* **51**, 454–468.
Kitchell, R. L., Strom, L. and Zotterman, Y. (1959). *Acta physiol scand.* **46**, 133–151.
Kokas, E., Phillips, J. L. and Brunson, W. D. (1967). *Comp. Biochem. Physiol.* **22**, 81–90.
Kulka, R. G. and Duksin, D. (1964). *Biochim. biophys. Acta* **91**, 506–514.
Kuo, Z. Y. (1932). *J. exp. Zool.* **61**, 395–400.
Kuo, Z. Y. and Shen, T. C. (1936). *J. comp. Psychol.* **21,** 87–93.
Laws, B. M. and Moore, J. H. (1963). *Biochem. J.* **87**, 632–638.
Leasure, E. E. and Link, R. P. (1940). *Poult. Sci.* **19**, 131–134.
Lind, G. W., Gronwall, R. R. and Cornelius, C. E. (1967). *Res. vet. Sci.* **8**, 280–282.
Lindsay, O. B., Biely, J. and March, B. E. (1969). *Poult. Sci.* **48**, 1214–1221.

Long, J. F. (1967). *Am. J. Physiol.* **212**, 1303–1307.
Macowan, M. M. and Magee, H. E. (1931). *Q. Jl exp. Physiol.* **21**, 275–280.
Mangold, E. (1906). *Pflügers Arch. ges. Physiol.* **111**, 163–240.
Nechay, B. R., Boyarsky, S. and Catacutan-Labay, P. (1968). *Comp. Biochem. Physiol.* **26**, 369–370.
Nolf, P. (1925). *Archs int. Physiol.* **25**, 290–341.
Olson, C. and Mann, F. C. (1935). *J. Am. vet. med. Ass.* **87**, 151–159.
Pastea, E., Nicolau, A. and Popa, V. (1968a). *Lucr. stiint.* **11**, 189–198.
Pastea, E., Nicolau, A., Popa, V. and Rosca, I. (1968b). *Acta physiol. hung.* **33**, 305–310.
Pintea, V. and Cotrut, M. (1958). *Stiinte. Agricole* **5**, 69–83.
Pintea, V., Jurubescu, V. and Cotrut, M. (1957). *Lucr. stiint.* **1**, 297–310.
Plimmer, R. H. A. and Rosedale, J. L. (1922). *Biochem. J.* **16**, 23–26.
Polyakov, I. I. (1958). *Dokl. mosk. sel'.-khoz. Akad. K. A. Timiryazeva* **38**, 328–333. Cited by Kokas, E., Phillips, J. L. and Brunson, W. D. (1967). *Comp. Biochem. Physiol.* **22**, 81–90.
Röseler, M. (1929). *Z. Tierzücht. ZüchtBiol.* **13**, 281–310.
Schmidt, C. R. and Ivy, C. A. (1937). *J. cell. comp. Physiol.* **10**, 365–383.
Siddons, R. C. (1969). *Biochem. J.* **112**, 51–59.
Siddons, R. C. (1970). *Biochem. J.* **116**, 71–78.
Siegel, P. B., Beane, W. L. and Kramer, C. Y. (1962). *Poult. Sci.* **41**, 1419–1422.
Skadhauge, E. (1967). *Comp. Biochem. Physiol.* **23**, 483–501.
Soedarmo, D., Kare, M. R. and Wasserman, R. H. (1961). *Poult. Sci.* **40**, 123–128.
Sturkie, P. D. (1965). "Avian Physiology" 2nd Edit. Ballière, Tindall & Cassell, London.
Toner, P. G. (1965). *J. Anat.* **99**, 389–398.
Tuckey, R., March, B. E. and Biely, J. (1958). *Poult. Sci.* **37**, 786–792.
Vonk, H. J. and Postma, N. (1949). *Physiologia comp. Oecol.* **1**, 15–23.

3 Absorption from the Alimentary Tract

D. A. HUDSON,[1] R. J. LEVIN and D. H. SMYTH

Department of Physiology,
University of Sheffield,
Sheffield, England

[1] Present address: Dunn Nutritional Laboratory, Milton Road, Cambridge, England.

I. Introduction

While the past twenty years have seen an intense interest in mechanisms of absorption in the mammalian intestine, a parallel interest in the avian intestine has been lacking until recently, a somewhat surprising position in view of the increasing importance of poultry meat and eggs in human nutrition. In approaching the avian intestine it is natural to assume that many of the basic mechanisms of the mammalian intestine are likely to be present, and this has been shown by recent experimental evidence, particularly as regards amino acid and hexose absorption. In some cases, however, there are marked differences, e.g. the changes in the intestine passive ion permeability during ontogeny shown by Hudson and Levin (1969).

Because of limited space, this chapter deals mainly with the basic mechanism and features of absorption. Sugars, amino acids and calcium receive most attention simply because they have been most studied, and less space is given to fluid, other electrolytes, trace elements, vitamins, bile salts, fats and fatty acids.

II. Methods

A. TECHNIQUES

In general, the experimental study of absorption involves a combination of both *in vivo* and *in vitro* procedures. These techniques have been developed chiefly for mammalian studies, but nearly all are applicable to avian intestine. They have been discussed and criticized by Smyth (1961, 1963), Wilson (1962), Levin (1967) and Parsons (1968). The latter is particularly useful as a source for historical detail. Kimmich (1970) has developed a technique for obtaining from the small intestine of chickens isolated mucosal epithelial cells that are remarkably active in concentrating sugars and amino acids.

In absorption studies use is often made of non-absorbable markers. A large number of substances used for this purpose is listed by Smyth (1961). It is necessary to check that a marker used for one class of animal in a particular set of experimental conditions is equally suitable for another species in different conditions, and Miller and Schedl (1970) have drawn attention to some of the difficulties. The following have been reported satisfactory in the normal fowl: cellulose (Bolton, 1961), chromic oxide (Eyssen and Desomer, 1963; Imondi and Bird, 1965), polyethylene glycol (Webling, 1966), ^{144}cerium (Miller and Jensen, 1966), ^{91}yttrium (Hurwitz and Bar, 1965, 1966) and barium sulphate (Vohra and Kratzer, 1967).

B. PARAMETERS FOR EXPRESSING ABSORPTIVE CAPACITY

The parameters used for the mammalian intestine have been discussed in detail by Newey and Smyth (1967) and Levin (1967). These very often involve two kinds of measurement (1) total amounts of substance transferred by the intestine, and (2) the capacity of the intestine to concentrate a substance against a concentration gradient. In spite of the current preoccupation with the concentrating capacity of the cells, the total amount of substance absorbed is probably more important in terms of nutrition of the animal. It is quite certain that many discrepancies in the literature depend on the fact that different authors use different parameters for expressing their results. In general, it is much safer to use a number of parameters than to rely on one only. This is particularly important where intestinal absorption is studied in animals exposed to different environmental conditions or dietary and hormonal regimes which could affect both structure and function of the gut. The application of these concepts to the avian gut has been discussed by Fearon and Bird (1968), Hudson and Levin (1968a) and Hudson (1969). The second named authors have suggested the idea of structural and functional parameters in assessing absorption and transfer.

III. Amino Acids

A. SPECIFICITY

Kratzer (1944) studied amino acid absorption in the chicken by feeding, through a stomach tube, grossly hypertonic solutions of thirteen amino acids to conscious birds. He concluded that the rate of absorption was inversely proportional to the molecular volume of the amino acid, a finding in keeping with the then current view that no special mechanism existed for amino acid transfer, and that the L- and D-enantiomorphs were absorbed at the same rate. (For criticism of this technique see p. 56.) This erroneous conclusion was probably due to the very high concentration of amino acids used and the poor specificity of the chemical measurements. Gibson and Wiseman (1951) showed unequivocably in rats that L-amino acids were absorbed faster than

D-amino acids, and Wiseman (1953) showed that many L-amino acids could be transferred against a concentration gradient by sacs of everted rodent intestine. The large volume of subsequent work on amino acid absorption in the mammalian intestine has been reviewed by Wilson (1962) and Wiseman (1964, 1968).

Studies on the avian gut have shown the existence of a preference for L-amino acids similar to that in the mammalian intestine. Paine *et al.* (1959) found that L-methionine and L-histidine were absorbed faster than the D-enantiomorphs, while Tasaki and Takahashi (1966) showed that the rate of absorption of L-amino acids was not related to molecular weight. They also found that L-amino acids with large non-polar side chains (methionine, leucine, valine, isoleucine, trytophan and phenylalanine) were absorbed most rapidly, those with polar side chains (arginine, glutamic and aspartic acids and glycine) most slowly, while intermediate between these two groups were histidine, lysine, alanine, serine, threonine, tyrosine, cystine and proline.

B. ACTIVE TRANSFER OF AMINO ACIDS *IN VITRO*

Perry *et al.* (1956) were probably the first to demonstrate the energy-dependence of amino acid uptake in chick intestine; they showed that histidine accumulation by rings of chick jejuna could be severely reduced by dinitrophenol and enhanced by glucose. They were also the first to show that galactose depressed amino acid uptake and caused mucosal sloughing. Their data indicate that the histidine concentration in the gut tissue was above that of the medium, suggesting active accumulation. There was no evidence that the amino acid was in free solution in the cells, which, according to Levin (1967), is a prerequisite if active transfer is to be demonstrated. Holdsworth and Wilson (1967) have recently found glycine accumulation by chick intestine to a concentration greater than that in the medium. The first demonstration that chick intestine could concentrate an amino acid transcellularly was by Lin and Wilson (1960) using L-tyrosine. Fearon and Bird (1967a, b) and Lerner and Taylor (1967) have shown transcellular accumulation for L-lysine and L-methionine respectively, while Hudson (1969) found it with L-proline and sarcosine.

C. COMPETITION BETWEEN L-AMINO ACIDS

L-Amino acids compete with each other for transfer. Paine *et al.* (1959) found L-histidine absorption was depressed by L-methionine while Tasaki and Takahashi (1966) showed that L-methionine inhibited the absorption of glutamic acid, leucine and phenylalanine; strangely, glutamic acid enhanced that of L-methionine. Lerner and Taylor (1967) observed that L-methionine uptake was depressed by L-cystine, L-ethionine and *S*-methyl-L-cystine; glycine caused only a small inhibition. Recently Nelson and Lerner (1970) have shown that glycine transfer is mediated by a system not shared by a number of other amino acids. It is interesting to compare these findings with the sarcosine and methionine systems described by Daniels *et al.* (1969) in the rat.

D. COMPETITION BETWEEN L- AND D-ENANTIOMORPHS

Jervis and Smyth (1959) showed competition between D- and L-enantiomorphs in rat intestine; this is also observed in the chick. Paine *et al.* (1959) found that L-histidine absorption was slowed in the presence of both L- and D-methionine. Lerner and Taylor (1967) showed that L-methionine uptake was reduced by D-methionine and that D-methionine was reduced by L-methionine. The L-enantiomorphs had the greater affinity for the transfer process. The D-enantiomorphs of other amino acids (valine, phenylalanine, ethionine, histidine and aspartic acid) had little effect on L-methionine uptake.

E. KINETICS OF AMINO ACID TRANSFER

Amino acid transfer in mammalian intestine conforms superficially to a saturable process described by Michaelis–Menten kinetics. This has been shown to be the case in chick intestine for L-methionine (Lerner and Taylor, 1967), glycine (Hudson, 1969; Nelson and Lerner, 1970) and L-proline (Hudson, 1969).

F. SITES OF AMINO ACID AND NITROGEN TRANSFER

Fearon and Bird (1967b) investigated the transfer of lysine along the length of chick intestine *in vitro*. Movement against a concentration gradient was observed in the jejunum and ileum but not in the duodenum. Sites of maximal transfer depended on the parameter chosen; per unit length the capacity progressively decreased from duodenum to ileum but per unit dry weight, there was no difference between the upper and lower small intestine. Hudson (1969) found that L-proline, L-methionine and L-leucine were transferred best by the ileum and least by the upper jejunum, under *in vitro* conditions.

In other studies reviewed by Bird (1968) the disappearance of amino nitrogen from the small intestine was measured, and nitrogen was found to disappear most rapidly from the upper small intestine (Imondi and Bird, 1965). In addition to the uncertainties associated with the use of non-absorbable markers, these experiments are complicated by variations of the endogenous luminal protein content (pancreatic enzymes, sloughed cells, bacteria). Payne *et al.* (1968) have discussed the experimental difficulties of this technique.

G. ELECTROGENIC COUPLING OF AMINO ACID ACTIVE TRANSFER WITH ION TRANSFER

Actively transferred amino acids cause an increase in the transmural potential difference (p.d.) across the small intestine of many species (see reviews by Barry, 1967; Schultz and Curran, 1968). The only studies undertaken in chick small intestine (Hudson and Levin, 1968b) showed that in adults and embryos, glycine caused only a very small increase in the transmural potential despite its active accumulation by the gut.

IV. Sugars

A. *IN VIVO* STUDIES

All the published data on the absorption of sugars in the fowl *in vivo* have been obtained using the technique of Cori (1925) or some modification. Golden and Long (1942), although not the first to use the Cori technique in birds, described its use in these animals in some detail. The method has many disadvantages, not the least being the necessity to use starved animals. Solutions used are far outside physiological concentrations and greatly exceed the Km of the transfer process. Differences in the rate of emptying of the crop and proventriculus can occur according to the substance studied, in animals of different ages or in different experimental or nutritional states; all these could affect the rate at which substances enter the intestine. Finally the sugars are usually fed in Na^+-free solutions despite the importance of this ion for hexose transfer. Bearing in mind these criticisms, it is apparent that the information obtained by this technique is of limited value.

Elmslie and Henry (1933), feeding 50% solutions of sugars, found glucose absorbed faster than galactose or lactose thus indicating a certain selectivity in the absorptive process. Many years later Bogner (1961) re-investigated this selectivity in greater detail and showed that the rate of absorption of various sugars in young chick intestine was very similar to that previously observed in mammalian gut. Thus D-glucose and D-galactose were absorbed faster than D-xylose and D-fructose, while all of these sugars were absorbed faster than L- and D-arabinose, L-xylose, D-ribose, D-mannose and D-cellobiose. Phlorrhizin was just as effective in inhibiting the absorption of glucose and galactose as in mammals. Wagh and Waibel (1967) confirmed that, at low concentrations, D-xylose is absorbed at a slower rate than D-glucose but at high concentrations the reverse seems to be true. L-arabinose was absorbed more slowly than either of these sugars at all concentrations.

B. SUGAR TRANSFER *IN VITRO*—TISSUE ACCUMULATION

Hexose uptake by chick intestine *in vitro* has many of the characteristics of the mammalian hexose transfer systems (see Newey, 1967; Crane, 1968, for details of mammalian systems). Most studies, in the chick, however, were based on tissue accumulation techniques, and although the evidence from these experiments that the sugar in the intracellular compartment is osmotically free is not unequivocal, observations with mammalian intestine suggests that this is so (Schultz *et al.*, 1966). It seems clear that chick small intestinal wall can actively accumulate D-glucose, D-galactose, methyl-α-D-glucoside and 6-deoxyglucose in the intracellular fluid against a concentration gradient (Bogner and Haines, 1964; Bogner *et al.*, 1966; Holdsworth and Wilson, 1967; Alvarado and Monreal, 1967). Both oxygen (Bogner *et al.*, 1966) and sodium ions are necessary (Holdsworth and Wilson, 1967) and Michaelis–Menten kinetics are satisfied (Alvarado and Monreal, 1967; Holdsworth and Wilson, 1967; Hudson and Levin, 1968b). Competition has been demonstrated between glucose and galactose (Bogner and Haines, 1964), while phlorrhizin inhibits hexose accumulation *in vitro* as well as *in vivo*.

Although D-xylose is not transported against a concentration gradient by chick intestine, its entrance into the gut exhibits saturation kinetics suggesting a carrier mechanism (Alvarado, 1967; Alvarado and Monreal, 1967).

C. *IN VITRO* TRANSCELLULAR TRANSFER

While the above experiments indicate an active uptake of some sugars into mucosal cells they do not necessarily prove chick intestine to be capable of transcellular movement of sugars into the serosal fluid against a concentration gradient. Fearon and Bird (1968) reported that D-glucose accumulated in the serosal fluid only from an initial concentration of 1·8 mM or lower. Hudson (1969) has recorded active transcellular accumulation of both glucose and galactose from initial mucosal fluid concentration of 7·5 mM.

D. ELECTROGENIC COUPLING OF ACTIVE HEXOSE TRANSFER WITH ION TRANSFER

Evidence is now very strong that there is a coupling between active hexose movement and sodium ion movement across the mucosal cell of the small intestine and that this generates an electrical potential difference (p.d.) in many species (see Barry, 1967; Crane, 1968; Schultz and Curran, 1968). Hudson and Levin (1968b) showed that the addition of actively transferable hexoses (D-galactose, D-glucose, 3-*0*-methyl-D-glucose) to the mucosal fluid bathing chick intestine *in vitro* always caused a sharp increase in the transintestinal p.d. both in embryos (2–3 d before hatching) and in the hatched chick. Phlorrhizin inhibited this increase in electrical potential. Fructose, xylose and sorbose did not increase this potential but caused a small decrease.

E. GLYCOSIDES

Transfers of some glycosides (arbutin, salicin, helicin) are inhibited by phlorrhizin, D-glucose, D-galactose and D-xylose (Alvarado and Monreal, 1967; Alvarado, 1967) and are sodium sensitive (Alvarado, 1967). There may thus be a common pathway for sugar and glycoside absorption in chick intestine. It is of interest that salicin is concentrated by chick intestine but not by that of the hamster.

V. Fluid

Hudson (1969) has shown that fluid transfer by different regions of the chick small intestine like rat intestine (Barry *et al.*, 1961), has both a glucose dependent and a glucose independent (endogenous) fluid transfer; the former is more important in the upper parts of the intestine while the latter is prominent in the ileum. Fluid transfer can also be stimulated across chick intestine *in vitro* by supplying metabolizable hexoses (i.e. mannose and fructose) in high concentration to the serosal fluid just as in the rat (Duerdoth *et al.*, 1965.)

Little information exists about the absorption of fluid by chick intestine *in vivo*. Csaky (1968) has presented results from one animal showing net fluid secretion by the duodenum but absorption in the jejunum and ileum. This would agree with the duodenum's secretory role described by Kokas *et al.* (1967). A few studies have implicated the importance of parts of the alimentary tract other than the small intestine in relation to fluid absorption in the fowl (Lepkovsky *et al.*, 1967). Hart and Essex (1942) showed that the rectum is of importance in reabsorption of fluid and electrolytes while Hill and Lumijarvi (1968) suggested that the colon can function to conserve electrolytes. Skadhauge (1967) demonstrated water absorption *in vivo* from the cloacal lumen of fowls, apparently dependent on sodium transfer. Absorption of water by the caecum was found *in vivo* by Parhon and Bârză (1967) in anaesthetized fowls.

VI. Fat, Fatty Acids and Cholesterol

Because the intestinal lymphatics of the fowl are poorly developed (Chapter 1), it has been suggested that they do not represent an important pathway for fat or fatty acid absorption (Noyan *et al.*, 1964). These authors found that palmitic acid was absorbed from the gizzard and duodenum and that the portal blood was a significant pathway for fatty acid absorption. Renner (1965) using chromic oxide as a marker, showed that, *in vivo*, the middle section of the gut was most active in fat absorption, the terminal ileum having very little activity. Garrett and Young (1964) found that bile was essential for the maximum absorption of triglycerides and free fatty acids but that some absorption did occur in its absence. Bickerstaffe and Annison (1969) quote unpublished studies by C. P. Freeman that fat absorption in the chicken may be via micelles rich in monoglycerides and fatty acids, as in mammals.

Renner and Hill (1961) using lard triglycerides presented evidence that the absorbability of palmitic acid varied with its point of attachment to the triglyceride molecule, being highest at the 2 position. Cole and Boyd (1967) showed that fat absorption was greater in germ-free and monocontaminated chicks than in conventional ones, suggesting that the usual enteric flora exerts an inhibiting effect. Boyd and Edwards (1967) found that, while palmitic and stearic acids were absorbed better by germ-free chicks compared with conventional ones, no change was noted in the absorption of oleic or linoleic acids. (See also Chapter 4.) Parhon and Bârză (1967) reported significant *in vivo* absorption of acetic, propionic and butyric acids from the caecum.

Absorption of cholesterol by cockerels has been observed by Janacek *et al.* (1959), but the mechanism was not studied.

VII. Bile Salts

Glasser *et al.* (1965) showed that, *in vitro*, taurocholate was transferred against a concentration gradient only by everted ileal sacs from Leghorn fowl. *In vivo*, however, Webling (1966) found that taurocholate although absorbed most readily by the ileum was also absorbed by the mid-jejunum, and to a lesser extent by the duodenum. Lindsay and March (1967) in cockerels confirmed this generalized absorption of taurocholate and of glycholate which is not found in avian bile. The involvement of bile absorption with that of calcium and strontium has been discussed by Webling and Holdsworth (1965, 1966).

VIII. Calcium

A. *IN VIVO* STUDIES

A hen's body contains some 20 g calcium and, if it lays daily an egg containing 2 g Ca, approximately 10% of the body store will be turned over each day. This remarkable turnover underlines the importance of calcium absorption and its control during lay. The relation of intestinal calcium transfer to calcium homeostasis in the laying hen has been succinctly reviewed by Hurwitz and Bar (1969a).

Although the absorption of ^{45}Ca is greater in the duodenum than in the jejunum or ileum, the jejunum plays an important role physiologically since the element passes rapidly through the duodenum spending relatively more time in the middle and lower regions of the intestine (Hurwitz and Bar, 1965, 1966). Very little Ca is absorbed in the ileum or colon. Experiments to investigate the mechanism for calcium absorption in the laying hen were conducted by Hurwitz and Bar (1968). The activity, concentration and lumen-blood electrochemical potential difference (ECPD) of calcium were measured in different parts of the intestine. (The ECPD is defined as positive when the lumen electrochemical potential is greater than that of the blood.) It was found that the ECPD was unfavourable for absorption of calcium in the lower ileum and colon as compared to the duodenum and jejunum, a result in agreement with the fact that there is no net calcium absorption from these distal segments. The differences in ECPD, however, only partly explained the variation in different segments and it was suggested that complexed calcium may be a factor in this discrepancy. Further investigations were made by Hurwitz and Bar (1969b), in chickens and hens, into the relations between the calcium ECPD, calcium absorption and the calcium-binding protein (CBP) of the mucosal cells. With calcium contents in the diet varying from 0·6 to 3·9%, the ECPD in the hens was always positive in the duodenum and jejunum, and calcium was absorbed down this gradient. In the chick, the ECPD was positive with calcium concentrations of 0·71% and 1·11%, but negative with a concentration of 0·31%. Hence at the two higher concentrations absorption was down the gradient but at the lower concentration it was against the gradient. This suggests that in certain conditions active transport of calcium may be of importance (Hurwitz and Bar, 1969b). In the 3-week-old chick, Hurwitz and Bar (1970) found that most calcium is absorbed in the jejunum although some occurred in the upper ileum. A net secretion of calcium was observed in the duodenum.

B. *IN VITRO* STUDIES

Bar and Hurwitz (1969a) investigated the movement of ^{45}Ca into the epithelial cell using intestinal loops and slices and concluded that it was essentially by passive diffusion though they could not exclude some other form of transport. Lowering the concentration of sodium ions in the incubation medium resulted in increased entry of calcium into the cells. There was some accumulation of calcium when complexed with glucose and fructose, the rate being highest with fructose. Bar and Hurwitz (1969b) found little effect on calcium transfer *in vitro* was caused by changes in pH or temperature of the medium while nitrogen gas and metabolic inhibitors affected accumulation only slightly; these observations support the idea of passive calcium transfer. Accumulation was dependent upon the concentration of potassium, magnesium and phosphate, and the region of greatest accumulation depended upon the transfer parameter used.

The role of phosphate in calcium transport is not known with any certainty. Hurwitz and Bar (1969b) showed that increasing concentrations of calcium reduced phosphate absorption, the duodenum being the major site of this antagonism.

C. EFFECTS OF BILE AND DETERGENTS

Webling and Holdsworth (1965, 1966) have shown that calcium absorption is increased by bile, bile salts and detergents in both rachitic and vitamin D_3-treated chicks. They suggested that bile may have three separate actions—(a) by acting directly on the mechanism for absorption of soluble calcium, (b) by increasing vitamin D_3 absorption and hence indirectly affecting the Ca mechanism, and (c) by increasing the availability of Ca from sparingly soluble calcium phosphate.

D. EFFECTS OF VITAMIN D

The effect of vitamin D_3 on increasing calcium absorption has been reviewed by Wasserman (1968), Norman (1968), Taylor and Wasserman (1969b) and DeLuca (1969). The mechanism(s) of this enhancement is still not known. Wasserman and Taylor (1966) stimulated this field by their pioneering experiments suggesting that a metabolite of vitamin D_3 caused the formation of a new calcium-binding protein in the mucosal cells and Taylor and Wasserman (1969b) have recently summarized the evidence. The metabolite responsible for the synthesis of the protein is not known (Lawson *et al.*, 1969a, b) nor is the exact role of the Ca-binding protein (CBP). It has been suggested that, because CBP is found in the soluble fraction of the cell, it is unlikely to be concerned in membrane transport. Taylor and Wasserman (1969a, b) have reported, however, that the CBP can be located by immunofluorescent techniques in the brush border regions of intestines from vitamin D_3-treated rachitic chicks. There is also disagreement about both the time of onset of stimulation of calcium absorption and the maximal concentration of CBP. Harmeyer and DeLuca (1969) found a poor correlation between the two, the increase in absorption appearing many hours before the maximum increase in CBP. Hurwitz and Bar (1969) have shown that, in laying fowls, the CBP is independent of the level of calcium fed, but in chicks dietary levels below 0·7% increased CBP.

Preliminary studies by Martin *et al.* (1969) have indicated that the brush borders of mucosal cells possess a calcium-dependent ATPase that is markedly increased after administration of vitamin D_3 to deficient rats and chicks and suggest that this enzyme may be the factor responsible for the enhancement of calcium absorption. According to Holdsworth (1970), however, this enzyme is a non-specific phosphatase which he thinks is unlikely to play an important role in vitamin D control of calcium absorption.

Adams *et al.* (1969) have suggested that vitamin D_3 causes a basic change in the mucosal plasma membrane of the ileum, a region where *in vitro* they found active transfer of calcium. This activity, however, was present only in the vitamin D_3-treated deficient chick.

IX. Copper

Starcher (1969) injected ^{64}Cu into the lumens of the ligated proventriculus, gizzard and duodenum of anaesthetized chicks, and found absorption (measured by the Cu in the liver) to be about five times faster from the duodenum than from the proventriculus. The walls of the gizzard bound about 90% of an administered dose but EDTA reduced this binding thereby allowing more ^{64}Cu to pass through to the duodenum.

After feeding ^{64}Cu, the metal was firmly and specifically bound to a protein ($M \approx 10{,}000$) in the duodenal mucosa. Zinc and cadmium (inhibitors of copper absorption) appeared to act as antagonists to this binding process since they displaced copper from the same protein fraction. It is not known whether this protein binding is part of the physiological process of intestinal copper transport.

X. Iron

Sell (1965), examining the effects of calcium and phosphorus on iron absorption in male broiler-type chicks, found that an increase in the dietary load of calcium and phosphorus decreased the absorption of ^{59}Fe. Featherston *et al.* (1968) used a whole-body counting technique for ^{59}Fe to investigate the effects of diets containing either a deficiency of iron or excessive amount on its absorption and retention. They calculated that iron-deficient chicks absorbed some 78% of the ^{59}Fe given orally while those fed excess iron absorbed only 42%. Chicks fed marginal levels of iron in their diets were able to maintain normal body iron levels through "efficient" absorption.

XI. Zinc

Dietary EDTA increases zinc absorption in turkey poults (Kratzer and Starcher, 1963; Vohra and Gonzales, 1969) and chickens (Suso and Edwards, 1968). It is not known if the zinc-EDTA complex is absorbed into the blood intact (Koike *et al.*, 1964). Increased uptake of ^{65}Zn by many tissues (including the alimentary tract) was noted when ^{65}Zn was given after feeding a zinc-deficient diet (Zeigler *et al.*, 1964). A zinc-binding protein (ZBP) has been extracted from chick intestine; it is apparently different from the plasma ZBP (Suso and Edwards, 1969) and its importance is unknown.

Experimental infection with the coccidium *Eimeria tenella* has only a mild and variable effect on zinc absorption (Turk and Stephens, 1967a).

XII. Vitamins

A. VITAMIN A

Shellenberger *et al.* (1964), using ligatured intestinal loops, found that vitamin A (acetate) was absorbed more quickly from the duodenum than from the rest of the intestine. Chickens, selected for their egg laying ability, and young birds absorbed the vitamin better than broiler strains and older fowls. Ascarelli (1969), reviewing the effects of protein restriction on vitamin A absorption with special regard to the activity of vitamin A esterases, reported that protein deprivation in the chick, by reducing intestinal hydrolase activity, reduced absorption of vitamin A palmitate but not of the free alcohol.

It is interesting to note that vitamin A deficiency in chicks causes an increased thickness and weight of the colon, the villi being thicker and pitted with recesses (Bayer *et al.*, 1968). Whether this morphological change affects absorption processes is not known.

B. VITAMIN B_2 (RIBOFLAVIN)

Riboflavin is transported *in vitro* across everted sacs of chick intestine and the rate is increased in the presence of glucose. Intestines from oestradiol-treated chicks did not show any enhanced transfer (Cordona and Payne, 1967).

C. VITAMIN B_{12}

It is presumed that the chick, like the mammal, needs an intrinsic factor (IF) for vitamin B_{12} absorption (Wilson, 1962). A vitamin B_{12} binding factor extracted from chick stomach was found to promote growth of vitamin B_{12}-deficient chicks but hog IF and a rat stomach extract inhibited B_{12} absorption in these animals (Coates *et al.*, 1955). Edwards (1969) using a whole-body counting method and vitamin B_{12} labelled with ^{57}Co and ^{60}Co found that laying hens absorbed more of the vitamin than did roosters.

D. VITAMIN E (α-TOCOPHEROL)

The absorption and distribution of α-tocopherol-^{14}C has been studied by Krishnamurthy and Bieri (1963) in rat and chicken. Only 10% of a single oral dose of 500 μg of the vitamin was recovered in the tissues examined while some 15% was present in the excreta. The radioactivity in all the tissues examined up to 7 d after feeding was due to the unchanged vitamin.

E. VITAMIN K

By an *in vitro* tissue accumulation technique, Berdanier and Griminger (1968) showed that both the magnitude and pattern of uptake of K_3 (menadione) differed from vitamins K_1 (phylloquinone) and K_2 (menaquinone). The uptake of vitamin K_3 was less than that of vitamins K_1 or K_2 and was maximal at 60 min compared to 30 min for the other two analogues. The uptake of all three analogues by different regions of the intestine was large intestine $>$ ileum $>$ jejunum $>$ duodenum. *In vitro*, chick bile was found unnecessary for the optimal uptake of the three substances. With all three, the concentration in the tissue exceeded that in the incubation saline, particularly with vitamins K_1 and K_2. It was suggested that uptake of vitamins K_1 and K_2 may take place by means of a carrier-mediated active transport process. Although the large intestine gave the highest vitamin uptakes, this organ was not considered normally to play an important role in vitamin K absorption.

XIII. Development of Transfer Mechanisms during Ontogeny

A. MONOSACCHARIDES

Bogner and co-workers have studied the development of monosaccharide-transfer mechanisms during ontogeny. *In vivo* Bogner (1961) and Bogner and Haines (1964) used very concentrated solutions and their results are difficult to interpret in terms of changes in transfer mechanisms (see p. 56). According to Bogner (1966) and Bogner *et al.* (1966) *in vitro* evidence that embryonic intestine, 3 d before hatching, accumulated glucose was equivocal, but galactose was definitely not accumulated. It was concluded that active sugar transfer by embryonic intestine was "limited, if it occurred at all". At or after hatching both sugars were accumulated. On the other hand Holdsworth and Wilson (1967), *in vitro* g found methyl-α-D-glucoside to be actively accumulated 4 to 5 d before hatching; a maximum was reached 4 to 5 d after hatching. The affinity (as judged from the apparent Km) of the transfer mechanism did not change over the developmental period but its maximum transfer did, suggesting that intestinal hexose transfer is increased during development by production of more transferring units rather than changes in their affinity for hexoses. Similar conclusions were obtained by Hudson and Levin (1968b) using an electrical method of measuring active sugar transfer in intestine from embryos and hatched chicks.

B. AMINO ACIDS

Holdsworth and Wilson (1967) showed that embryonic intestine accumulated glycine some 4 d before hatching. The electrogenicity of this mechanism was poor (Hudson and Levin, 1968b). Hudson (1969) found active transcellular concentration of glycine, L-proline and L-methionine but not sarcosine by embryonic intestine 1 to 2 d before hatching. Active sarcosine transfer was recorded, however, in young chicks. Lotenkov and Podluzhnaya (1967) interpreted their data on the absorption of eight amino acids to indicate that absorption in 1-day-old chicks was proportional to the molecular weight of the amino acid but that in 1-month-old chicks, L-leucine and L-methionine were absorbed more quickly than the other amino acids. Hudson (1969) found evidence that L-methionine transfer in embryonic intestine was different from its transfer after hatching. The embryonic gut appeared to have two distinct mechanisms compared with a single one after hatching.

C. XANTHINE TRANSPORT SYSTEM

Taube and Berlin (1970) reported that the carrier for xanthine transport begins to appear 2 d prior to hatching. The experimental evidence suggested that it was located at the mucosal face in contradistinction to previous findings in mature hamster intestine.

D. CALCIUM BINDING PROTEIN (CBP)

Corradino *et al.* (1969) found CBP to be present on the day of hatching. Precocious formation of CBP (2 d before hatching) was induced by cortisone acetate, (but not by vitamin D_3) injected into the chorio-allantoic sac of embryonated eggs 5 d before hatching.

E. PERMEABILITY

Hudson and Levin (1969) studied changes in passive ionic permeability during ontogeny by inducing osmotic and diffusion potentials across the mid-jejunum *in vitro*. These potentials were taken to indicate a positively-charged membrane in embryonic intestine which changed to a negatively-charged one just before hatching and remained so. The electrical resistance of embryonic intestine was at least 2 to 3 times greater than that of intestine from hatched chicks. It is likely that this high resistance is the cause of the large potentials recorded across the intestine before hatching (Hudson and Levin, 1968b).

F. ANTIBODY AND MACROMOLECULAR ABSORPTION

Neither antibody (Brierley and Hemmings, 1956) nor polyvinyl pyrrolidone (Clarke and Hardy, 1970) are absorbed from the chick intestine during the first 48 h after hatching. Clarke and Hardy stated that the chick intestine showed none of the histological characteristics associated with macromolecular absorption that mammalian intestine exhibits.

XIV. Factors Reducing Absorption

Reduced absorption from the intestine *in vivo* or reduced transfer *in vitro* can be caused by (a) disease and (b) various chemical or physical agents. The term malabsorption is generally applied to pathological cases, while the term inhibitor can conveniently be used for the chemical or physical agents.

A. DISEASE

Pathology is outside the scope of this chapter but it is worth noting that the intensive analysis of human malabsorption syndromes has not been made in the fowl and would repay some attention. A few references to intestinal functions in diseased fowls do exist. Turk and Stephens (1967a, b, c, 1968, 1969) found that infections with the coccidia, *Eimeria necatrix* and *E. acervulina* had different actions on the absorptions of zinc, oleic acid, chlorella protein and amino acids. Coccidiosis slows the passage of food through the alimentary tract (Aylott *et al.*, 1968) and reduces the absorption of histidine and of glucose (Preston-Mafham and Sykes, 1967a, b); it increases the leakage of plasma protein into the gut lumen (Preston-Mafham and Sykes, 1967b). Eyssen and DeSomer (1963, 1967) reported impaired absorption of fats and carbohydrates when sucrose was fed to young chicks and postulated as a reason the combined action of bacterial and viral agents.

B. INHIBITORS

Agents which reduce absorption can do so in three ways (1) by acting on the columnar epithelial cell, (2) by acting on nutrient substances in the lumen of the intestine, (3) by acting in other ways to change the conditions in the lumen. A review of inhibitors by Sanford (1967) applies mainly to the mammalian intestine.

Substances acting on the cell can be divided into those which damage the cell and those which competitively block the transfer mechanisms. Damage might be to carrier sites on the cell surface, energy-producing mechanisms in the cell, changes in permeability of the cell membrane, or interference with the genesis of new cells in the crypts. Tannins in food, which cause a marked decrease in "feed efficiency" (Vohra *et al.*, 1966) might act by damaging the cell. Competitive blocking of transfer pathways can occur when two substances compete for one site. Polin *et al.* (1963) found that the coccidiostat amprolium caused thiamine deficiency in high doses because it competed for the thiamine absorption mechanism in the duodenum.

Substances given by mouth can bind or destroy nutrients in the lumen, making them unavailable for entry into the mucosal cells. Examples of the former effect are the binding of zinc, manganese and copper by diets containing soybean protein (Davies *et al.*, 1962), and the binding of calcium (and to some extent iron) by dietary phytates (McGillivray and Nelson, 1968). Nitrites in the food can destroy vitamin A (Roberts and Sell, 1963).

Some agents can alter the luminal milieu of the alimentary tract by effects on digestive gland secretion, gut motility or flora. Raw soybean meal protein is known to affect the secretion of enzymes from the pancreas (Saxena *et al.*, 1963) and this can reduce efficiency of absorption. Eyssen *et al.* (1966) have reported that neomycin lowers blood cholesterol levels in germ-free chicks by forming insoluble complexes with bile salts, thus hindering cholesterol absorption.

XV. Factors Enhancing Absorption

A. CHELATORS

The effect of chelating agents in increasing the absorption of some metallic ions has been previously discussed in the sections on zinc and other metals. Chelators have also been used to increase the absorption of tetracycline antibiotics as it has been suggested that such antibiotics have an affinity for calcium. Nakaue *et al.* (1967) found significant increases in plasma antibiotic levels when EDTA and terepthalic acid were added to the diet.

Sodium sulphate, when included in the diet at 1·5% caused an increased absorption of the antibiotic chlortetracycline in birds infected with *Myco-*

plasma gallisepticum (Gale and Baughn, 1968). The reason for this enhancement is unknown.

B. ANTIBIOTICS

Many antibiotics when fed increase the growth rate of conventional chicks and other animals, and reduce the weight and thickness of the gut (Coates *et al.*, 1955; Heth and Bird, 1962). It is argued that this growth response to antibiotics is due to inhibition of harmful intestinal bacteria, as antibiotics do not stimulate growth of the germ-free or clean-environment animal (Coates *et al.*, 1951a,b ; Forbes and Park, 1959). There is little doubt, however, that feeding antibiotics leads, in some cases, to an increased absorption of nutrients by the small gut. Draper (1958) found penicillin to decrease intestinal thickness and increase lysine absorption. More recently Lotenkov and Podluzhnaya (1967) reported that chlortetracycline (20 to 25 mg/kg food) promoted the absorption of eight amino acids tested; the antibiotic increased the metabolic activity of the gut wall. Whether this was due to enhanced amino acid absorption (see Bronk and Parsons, 1966) or to the antibiotic *per se* is not clear. Relations between the enteric flora and the actions of antibiotics on the mucosal cells are still uncertain. The literature has been surveyed by Madge (1969a, b) in relation to effects of antibiotics on mammalian absorption.

XVI. Conclusion

This attempt to review the information available on intestinal absorption in the fowl has revealed to us the great gaps which exist in this field, and it appears that there is much scope for investigations on the fowl along the lines of mammalian work. The fowl certainly seems to have been neglected as an experimental animal by gastroenterologists with the exception of its use in the study of calcium transfer-mechanisms and in developmental work. It is likely, however, that the next few years will produce much basic information on the fowl's intestine which could well be of considerable value in the economic problems of poultry production.

References

Adams, T. H., Wong, R. G. and Norman, A. W. (1969). *Fedn Proc. Fedn Am. Socs exp. Biol.* **28**, 759 (Abstr. 2799).

Alvarado, F. (1967). *Comp. Biochem. Physiol.* **20**, 461–470.

Alvarado, F. and Monreal, J. (1967). *Comp. Biochem. Physiol.* **20**, 471–488.

Ascarelli, I. (1969). *Am. J. clin. Nutr.* **22**, 913–922.

Aylott, M. V., Vestal, O. H., Stephens, J. F. and Turk, D. E. (1968). *Poult. Sci.* **47**, 900–904.

Bar, A. and Hurwitz, S. (1969a). *Biochim. biophys. Acta* **183**, 591–600.

Bar, A. and Hurwitz, S. (1969b). *Poult. Sci.* **48**, 1105–1113.

Barry, B. A., Matthews, J. and Smyth, D. H. (1961). *J. Physiol., Lond.* **157**, 279–288.

Barry, R. J. C. (1967). *Br. med. Bull.* **23**, 266–269.
Bayer, R. C., Foss, D. C. and Donovan, G. A. (1968). *Poult. Sci.* **47**, 1654.
Berdanier, C. D. and Griminger, P. (1968). *Int. Z. VitamForsch.* **38**, 376–382.
Bickerstaffe, R. and Annison, E. F. (1969). *Biochem. J.* **111**, 419–429.
Bird, F. H. (1968). *Fedn Proc. Fedn Am. Socs exp. Biol.* **27**, 1194–1198.
Bogner, P. H. (1961). *Proc. Soc. exp. Biol. Med.* **107**, 263–265.
Bogner, P. H. (1966). *Biologia Neonat.* **9**, 1–9.
Bogner, P. H. and Haines, I. A. (1964). *Am. J. Physiol.* **207**, 37–41.
Bogner, P. H., Braham, A. H. and McLain, P. L. (1966). *J. Physiol., Lond.* **187**, 307–321.
Bolton, W. (1961). *Proc. Nutr. Soc.* **20**, 26.
Boyd, F. M. and Edwards, H. M. (1967). *Poult. Sci.* **46**, 1481–1483.
Brierley, J. and Hemmings, W. A. (1956). *J. Embryol. exp. Morph.* **4**, 34–41.
Bronk, J. R. and Parsons, D. S. (1966). *J. Physiol., Lond.* **184**, 950–963.
Clarke, R. M. and Hardy, R. N. (1970). *J. Physiol., Lond.* **209**, 669–687.
Coates, M. E., Dickinson, C. D., Harrison, G. F., Kon, S. K., Cummins, S. H. and Cuthbertson, W. F. J. (1951a). *Nature, Lond.* **168**, 332.
Coates, M. E., Davies, M. K. and Kon, S. K. (1951b). *Br. J. Nutr.* **9**, 110–119.
Coates, M. E., Gregory, M. E., Harrison, G. F., Henry, K. M., Holdsworth, E. S. and Kon, S. K. (1955). *Proc. Nutr. Soc.* **14**, xiv–xv.
Cole, J. R. and Boyd, F. M. (1967). *Appl. Microbiol.* **15**, 1229–1234.
Cordona, N. A. and Payne, I. R. (1967). *Poult. Sci.* **46**, 1176–1179.
Cori, C. F. (1925). *J. biol. Chem.* **66**, 691–715.
Corradino, R. A., Taylor, A. N. and Wasserman, R. H. (1969). *Fedn Proc. Fedn Am. Socs exp. Biol.* **28**, 760 (Abstr. 2802).
Crane, R. K. (1968). *In* "Handbook of Physiology" (C. F. Code, ed.), Vol. III, Sect. 6, pp. 1323–1351. The Williams and Wilkins Co., Baltimore.
Csaky, T. Z. (1968). *In* "The Germ-free Animal in Research" (M. E. Coates, ed.), pp. 151–159. Academic Press, London.
Daniels, V. G., Newey, H. and Smyth, D. H. (1969). *Biochim. biophys. Acta* **183**, 637–639.
Davies, P. N., Norris, L. C. and Kratzer, F. H. (1962). *J. Nutr.* **77**, 217–223.
DeLuca, H. F. (1969). *Fedn Proc. Fedn Am. Socs exp. Biol.* **28**, 1678–1689.
Draper, H. H. (1958). *J. Nutr.* **64**, 33–42.
Duerdoth, J. K., Newey, H., Sanford, P. A. and Smyth, D. H. (1965). *J. Physiol., Lond.* **176**, 23*P*.
Edwards, H. M. (1969). *Poult. Sci.* **48**, 414–420.
Elmslie, A. R. G. and Henry, K. M. (1933). *Am. J. Physiol.* **27**, 705–710.
Eyssen, H. and DeSomer, P. (1963). *J. exp. Med.* **117**, 127–138.
Eyssen, H. and DeSomer, P. (1967). *Poult. Sci.* **46**, 323–333.
Eyssen, H., Evrard, E. and Bosch, J. van den (1966). *Life Sci., Oxford* **5**, 1729–1734.
Fearon, J. R. and Bird, F. H. (1967a). *Poult. Sci.* **46**, 1037–1041.
Fearon, J. R. and Bird, F. H. (1967b). *J. Nutr.* **93**, 198–202.
Fearon, J. R. and Bird, F. H. (1968). *Poult. Sci.* **47**, 1412–1416.
Featherston, W. R., Pockat, T. J. and Wallace, J. (1968). *Poult. Sci.* **47**, 946–950.
Forbes, M. and Park, J. T. (1959). *J. Nutr.* **67**, 69–84.
Gale, G. O. and Baughn, C. O. (1968). *Poult. Sci.* **47**, 342–344.
Garrett, R. L. and Young, R. J. (1964). *Fedn Proc. Fedn Am. Socs exp. Biol.* **23**, 340.
Glasser, J. E., Weiner, I. M. and Lack, L. (1965). *Am. J. Physiol.* **208**, 359–362.
Gibson, Q. H. and Wiseman, G. (1951). *Biochem. J.* **48**, 426–429.

Golden, W. R. C. and Long, C. N. H. (1942). *Am. J. Physiol.* **136**, 244–249.

Harmeyer, J. and DeLuca, H. F. (1969). *Archs Biochem. Biophys.* **133**, 247–254.

Hart, W. M. and Essex, H. E. (1942). *Am. J. Physiol.* **136**, 657–668.

Heth, D. A. and Bird, H. R. (1962). *Poult. Sci.* **41**, 755–760.

Hill, F. W. and Lumijarvi, D. H. (1968). *Fedn Proc. Fedn Am. Socs exp. Biol.* **27**, 421, (Abstr. 1165).

Holdsworth, E. S. (1970). *J. Membrane biol.* **3**, 43–53.

Holdsworth, C. D. and Wilson, T. H. (1967). *Am. J. Physiol.* **212**, 233–240.

Hudson, D. A. (1969). *Ph. D. thesis, University of Sheffield.*

Hudson, D. A. and Levin, R. J. (1968a). *J. Physiol., Lond.* **198**, 40*P*–41*P*.

Hudson, D. A. and Levin, R. J. (1968b). *J. Physiol., Lond.* **195**, 369–385.

Hudson, D. A. and Levin, R. J. (1969). *Life Sci., Oxford* **8**, 1271–1279.

Hurwitz, S. and Bar, A. (1965). *J. Nutr.* **86**, 433–438.

Hurwitz, S. and Bar, A. (1966). *J. Nutr.* **89**, 311–316.

Hurwitz, S. and Bar, A. (1968). *J. Nutr.* **95**, 647–654.

Hurwitz, S. and Bar, A. (1969a). *Am. J. clin. Nutr.* **22**, 391–395.

Hurwitz, S. and Bar, A. (1969b). *J. Nutr.* **99**, 217–224.

Hurwitz, S. and Bar, A. (1970). *Poult. Sci.* **49**, 324–325.

Imondi, A. R. and Bird, F. H. (1965). *Poult. Sci.* **44**, 916–920.

Janacek, H. M., Suzuki, R. and Ivy, A. C. (1959). *Am. J. Physiol.* **197**, 1341–1344.

Jervis, E. L. and Smyth, D. H. (1959). *J. Physiol., Lond.* **145**, 57–65.

Kimmich, G. A. (1970). *Biochemistry* **9**, 3659–3668.

Koike, T. I., Kratzer, F. H. and Vohra, P. (1964). *Proc. Soc. exp. Biol. Med.* **117**, 483–486.

Kokas, E., Phillips, J. J. and Brunson, W. D. (1967). *Comp. Biochem. Physiol.* **22**, 81–90.

Kratzer, F. H. (1944), *J. biol. Chem.* **153**, 237–247.

Kratzer, F. H. and Starcher, B. C. (1963). *Proc. Soc. exp. Biol. Med.* **113**, 424–426.

Krishnamurthy, S. and Bieri, J. G. (1963). *J. Lipid Res.* **4**, 330–336.

Lawson, D. E. M., Wilson, P. W., and Kodicek, E. (1969a). *Nature, Lond.* **222**, 171–172.

Lawson, D. E. M., Wilson, P. W. and Kodicek, E. (1969b). *Biochem. J.* **115**, 269–277.

Lepkovsky, S., Feldman, S. E. and Sharon, I. M. (1967). *In* "Handbook of Physiology" (C. F. Code, ed.), Vol. I. Sect. 6 pp. 117–128. The Williams and Wilkins Co., Baltimore.

Lerner, J. and Taylor, M. W. (1967). *Biochim. biophys. Acta* **135**, 991–999.

Levin, R. J. (1967). *Br. med. Bull.* **23**, 209–212.

Lin, E. C. C. and Wilson, T. H. (1960). *Am. J. Physiol.* **199**, 127–130.

Lindsay, O. B. and March, B. E. (1967). *Poult. Sci.* **46**, 164–168.

Lotenkov, M. I. and Podluzhnaya, L. I. (1967). *Khim. Sel. Khoz.* **48**, 775–779. (National Lending Library, R.T.S. No. 5728).

McGillivray, J. J. and Nelson, T. S. (1968). *Poult. Sci.* **47**, 1695.

Madge, D. S. (1969a). *Br. J. Nutr.* **23**, 637–646.

Madge, D. S. (1969b). *Comp. Biochem. Physiol.* **30**, 295–307.

Martin, D. L., Melancon, M. J. and DeLuca, H. F. (1969). *Biochim. biophys. Res. Commun.* **35**, 819–823.

Miller, J. K. and Jensen, L. S. (1966). *Poult. Sci.* **45**, 1051–1053.

Miller, D. L. and Schedl, H. P. (1970) *Gastroenterology* **58**, 40–46.

Nakaue, H. S., Thomas, J. M. and Reid, B. L. (1967). *Poult. Sci.* **46**, 417–421.

Nelson, K. M. and Lerner, J. (1970). *Biochim. biophys. Acta* **203**, 434–444.

Newey, H. (1967). *Br. med. Bull.* **23**, 236–240.

Newey, H. and Smyth, D. H. (1967). *Proc. Nutr. Soc.* **26**, 5–12.
Norman, A. W. (1968). *Biol. Rev.* **43**, 97–137.
Noyan, A., Lossow, W. J., Brot, N. and Chaikoff, I. L. (1964). *J. Lipid Res.* **5**, 538–541.
Paine, C. M., Newman, H. J. and Taylor, M. W. (1959). *Am. J. Physiol.* **197**, 9–12.
Parhon, C. C. and Bârză, E. (1967). *Rev. roum. Biol.* **12**, 109–115.
Parsons, D. S. (1968). *In* "Handbook of Physiology" (C. F. Code ed.), Vol. III. Sect. 6, pp. 1177–1216. The Williams and Wilkins Co., Baltimore.
Payne, W. L., Combs, G. F., Kifer, R. R. and Snyder, D. G. (1968). *Fedn Proc. Fedn Am. Socs exp. Biol.* **27**, 1199–1203.
Perry, J. W., Moore, A. E., Thomas, D. A. and Hird, F. J. R. (1956). *Acta paediat., Stockh.* **45**, 228–240.
Polin, D., Wynosky, E. R. and Porter, C. C. (1963). *Proc. Soc. exp. Biol. Med.* **114**, 273–277.
Preston-Mafham, R. A. and Sykes, A. (1967a). *Experientia*, **23**, 972–973.
Preston-Mafham, R. A. and Sykes, A. (1967b). *Proc. Nutr. Soc.* **26**, Abstract XXVII.
Renner, R. (1965). *Poult. Sci.* **44**, 861–864.
Renner, R. and Hill, F. W. (1961). *J. Nutr.* **74**, 254–258.
Roberts, W. K. and Sell, J. L. (1963). *J. Anim. Sci.* **22**, 1081–1085.
Sanford, P. (1967). *Br. med. Bull.* **23**, 270–274.
Saxena, H. C., Jensen, L. S., McGinnis, J. and Lauber, J. K. (1963). *Proc. Soc. exp. Biol. Med.* **112**, 390–393.
Schultz, S. G. and Curran, P. F. (1968). *In* "Handbook of Physiology". (C. F. Code, ed.), Vol. III sect. 6, pp. 1245–1275. The Williams and Wilkins Co., Baltimore.
Schultz, S. G., Fuisz, R. E. and Curran, P. F. (1966). *J. gen. Physiol.* **49**, 849–866.
Sell, J. L. (1965). *Poult. Sci.* **44**, 550–561.
Shellenberger, T. E., Parrish, D. E. and Sanford, P. E. (1964). *J. Nutr.* **82**, 99–105.
Skadhauge, E. (1967). *Comp. Biochem. Physiol.* **23**, 483–501.
Smyth, D. H. (1961). *In* "Methods in Medical Research Year Book" (J. H. Quastel, ed.), Vol. 9, pp. 260–272. Medical Publishers, Chicago.
Smyth, D. H. (1963). *In* "Recent Advances in Physiology", (R. Creese, ed.), pp. 36–68. J. & A. Churchill Ltd., London.
Starcher, B. C. (1969). *J. Nutr.* **97**, 321–326.
Suso, F. A. and Edwards, H. M. (1968). *Poult. Sci.* **47**, 1417–1423.
Suso, F. A. and Edwards, H. M. (1969). *Fedn Proc. Fedn Am. Socs exp. Biol.* **28**, 761 (Abstract 2808.)
Tasaki, I. and Takahashi, N. (1966). *J. Nutr.* **88**, 359–364.
Taube, R. A. and Berlin, R. D. (1970). *Am. J. Physiol.* **219**, 666–671.
Taylor, A. N. and Wasserman, R. H. (1969a). *Fedn Proc. Fedn Am. Socs exp. Biol.* **28**, 759 (Abstr. 2801).
Taylor, A. N. and Wasserman, R. H. (1969b). *Fedn Proc. Fedn Am. Socs exp. Biol.* **28**, 1834–1838.
Turk, D. E. and Stephens, J. F. (1967a). *Poult. Sci.* **46**, 939–943.
Turk, D. E. and Stephens, J. F. (1967b). *Poult. Sci.* **46**, 775–777.
Turk, D. E. and Stephens, J. F. (1967c). *J. Nutr.* **93**, 161–165.
Turk, D. E. and Stephens, J. F. (1968). *Poult. Sci.* **47**, 1728.
Turk, D. E. and Stephens, J. F. (1969). *Fedn Proc. Fedn Am. Socs exp. Biol.* **28**, 446.
Vohra, P. and Gonzales, N. (1969). *Poult. Sci.* **48**, 1509–1510.
Vohra, P., Kratzer, F. H. and Joslyn, M. A. (1966). *Poult. Sci.* **45**, 135–142.
Vohra, P. and Kratzer, F. H. (1967). *Poult. Sci.* **46**, 1603–1604.
Wagh, P. V. and Waibel, P. E. (1967). *Proc. Soc. exp. Biol. Med.* **124**, 421–424.

Wasserman, R. H. (1968). *Calc. Tiss. Res.* **2**, 301–313.
Wasserman, R. H. and Taylor, A. N. (1966). *Science, N.Y.* **152**, 791–793.
Webling, D. D'A. (1966). *Aust. J. exp. med. Sci.* **44**, 101–104.
Webling, D. D'A. and Holdsworth, E. S. (1965). *Biochem. J.* **97**, 408–421.
Webling, D. D'A. and Holdsworth, E. S. (1966). *Biochem. J.* **100**, 652–660.
Wilson, T. H. (1962). "Intestinal Absorption." W. B. Saunders Co., Philadelphia.
Wiseman, G. (1953). *J. Physiol., Lond.* **120**, 63–72.
Wiseman, G. (1964). "Absorption from the Intestine." Academic Press, London.
Wiseman, G. (1968). *In* "Handbook of Physiology" (C. F. Code ed.), Vol. III. sect. 6, pp. 1277–1307. The Williams and Wilkins Co., Baltimore.
Zeigler, T. R., Leach, R. M., Scott, M. L., Huegin, F., McEvoy, R. K. and Strain, W. H. (1964). *J. Nutr.* **82**, 489–494.

Addendum

The following papers were published after completion of the manuscript.

Calcium

Adams and Norman (1970) investigated calcium transport across isolated chick ileum in an apparatus which allowed the tissue to be short-circuited and flux measurements made. They found that calcium transport was an active process operating against an electropotential gradient in vitamin D-treated chicks; the ion, however, was not actively transferred in vitamin D-deficient ones. Vitamin D apparently altered the calcium carrier in the mucosal brush border making calcium uptake more efficient. Experiments on calcium transfer across isolated chick ileum, treated with the polyene antibiotic filipin, suggested to Adams *et al.* (1970) that the latter changed the mucosal microvillal membrane in vitamin D-deficient chicks and activated the latent calcium transport system. The effects of filipin were very similar to those indirect effects induced by vitamin D that are actinomycin sensitive. Norman *et al.* (1970) found that administration of vitamin D to deficient chicks greatly increased the activity of the alkaline phosphatase of the mucosal brush border. This increase could be prevented by cycloheximide. A brush border-calcium activated-ATPase was also found to be increased after vitamin D administration. The simultaneous time-course of the appearance of the increased activities of the enzymes with the increased rate of calcium absorption suggests that these enzymes are concerned in the vitamin D-induced increases in calcium transfer.

Amino acids

Wakita *et al.* (1970) studied the effects of sex, age, parameters of absorption and metabolic inhibitors on the transfer of lysine, methionine and glutamic acid by isolated chick intestine incubated in vitro.

Adams, T. H. and Norman, A. W. (1970). *J. biol. Chem.* **245**, 4421–4431.
Adams, T. H., Wong, R. G. and Norman, A. W. (1970). *J. biol. Chem.* **245**, 4432–4442.
Norman, A. W., Mircheff, T. H., Adams, T. H. and Speilvogel, A. (1970). *Biochim. biophys. Acta* **215**, 348–359.
Wakita, M., Hoshino, S. and Morimoto, K. (1970). *Poult. Sci.* **49**, 1046–1050.

4 The Influence of the Intestinal Microflora on Nutrition

D. J. JAYNE-WILLIAMS and R. FULLER

National Institute for Research in Dairying, University of Reading, Reading, England

I. Introduction

It has been long recognized that the intestinal flora may have either a harmful or beneficial influence on the nutrition of the host. Because it is difficult or impossible to determine the influence of a population of bacteria as complex as that in the animal gut without recourse to some means of simplifying the situation, little progress was made until antibiotics and gnotobiotic animals became available.

Germfree animals, particularly chickens, frequently thrive better than their conventional counterparts. This, coupled with beneficial effects observed when antibiotics are given *per os*, suggests that the activities of intestinal bacteria are, on balance, detrimental to the host.

The literature on the mode of action of dietary antibiotics on the growth of chickens is extensive and conflicting; references to this literature will be made when pertinent. Readers are referred to: Branion *et al.* (1952), Braude *et al.* (1953), Combs (1956), Taylor (1957), Luckey (1959), François (1962) and Bird (1969).

In certain areas there is relatively little information relating specifically to the fowl; here data obtained with mammals have been included in order to present as complete a picture as possible.

II. Quantitative and Qualitative Aspects of the Gut Flora

The alimentary tract of the newly-hatched, healthy chicken is sterile, but micro-organisms are rapidly picked up in the nest or within the incubator from egg-shell fragments and other debris, the complexity of this flora being dependent on the hygienic condition of the environment. This process continues with organisms which contaminate food, water and environment of farmyard or brooder until a so-called "normal" flora develops. "Normal" conveniently describes the intestinal population of animals reared in ordinary environments and showing no overt clinical signs: it should, however, not be interpreted as implying quantitative or qualitative identity between bacterial populations so described.

Some of the bacteria ingested by animals will not find favourable conditions in the gut (e.g. temperature, availability of oxygen, pH, etc.) and will either be destroyed by digestive secretions or voided in the faeces. Furthermore, other organisms ingested may be able to multiply in the gut only in certain circumstances. Some developing initially may later be cleared by the metabolic activities of other components of the flora; others not ingested until later may find that the flora already established has made the conditions unfavourable.

A. THE CROP

Although a proportion of the nutrients ingested may pass directly to the proventriculus (Halnan, 1949), most of the food enters the crop where some of it may remain for several hours and be altered by host and bacterial enzymes. The crop has been described (Blount, 1947) as a "... receptacle favouring the growth of acid-forming carbohydrate-splitting bacteria ..." and has been considered by Bolton (1965) to function "... in a manner which bears certain points of resemblance to the rumen". Compared with the amount of bacteriological and biochemical work done on other parts of the fowl's alimentary tract, particularly the caeca, the crop has been neglected. As this organ is anterior to the main digestive and absorptive sites of the intestine, any beneficial or harmful activities of its microflora might be expected subsequently to exert a major effect.

Many workers (e.g. Sieburth *et al.*, 1954; Wiseman *et al.*, 1956; Lev and Briggs, 1956; Eyssen *et al.*, 1962; Smith, 1965a) have shown that lactobacilli predominate in the crop (*c.* 10^9/g wet wt.) though they are sometimes absent from the crops of chicks less than 3 d of age. Eyssen *et al.* (1965) found a high proportion of crop lactobacilli to be variants of *Lactobacillus acidophilus* which had lost part of their nucleic acid-synthesizing capabilities and hence concluded that desquamated epithelial cells in the crop might play a major role in their nutrition.

Enterococci and coli-aerogenes bacteria (both up to about 10^5/g), and occasionally small numbers of micrococci, staphylococci and yeasts are usually present. Strict anaerobes of the enteric type (e.g. bacteroides, clostridia) do not normally occur in the crop where, presumably, the redox potential, the predominance of lactobacilli, the type of substrate and other factors are unfavourable.

Diet can influence the crop flora. With turkeys Wheeler *et al.* (1957) found that a diet rich in glucose encouraged the development of large numbers of yeasts; the copious amounts of gas produced by these organisms caused the crops to become characteristically enlarged and pendulous. Smith (1965b) fed chicks on diets composed solely of either ground wheat or meat-and-bone meal and found that those fed the latter diet had about 10^5 *Clostridium welchii* per gram of crop contents whereas the former had none. Counts of *Escherichia coli* were higher and those of lactobacilli lower in birds fed meat-and-bone meal.

According to Herpol and van Grembergen (1961) the crop pH generally lies between 6·0 and 7·0. Others (e.g. Bolton, 1965; Smith 1965b; D. J. Jayne-Williams, unpublished observations), including those whose work was reviewed by Herpol and van Grembergen (1961), have found somewhat lower values, ranging from 4·0 to 6·0. In some (unknown) circumstances it seems that the typical lactobacillus flora may be replaced by organisms producing large amounts of organic acids other than lactic, thereby lowering pH; Bolton (1965) has reported "sour crops" with a pH of $\approx$3·7.

Ivorec-Szylit and Szylit (1965) showed DL-lactic acid to be formed in the crop by bacterial activity; the concentration increased exponentially with time, reaching a maximum 5 h after feeding a starch-rich diet. Glucose (as little as 1%) added to a diet containing starch was shown by Ivorec-Szylit *et al.* (1965) to reduce starch breakdown by acting as a preferential energy source for the lactobacilli. The addition of penicillin to a mixture of crop contents and a carbohydrate substrate *in vitro* markedly reduced lactate production. Bolton (1962) analysed crop contents of adults which had had access to food for 1 h and then been killed at intervals. During the first $2\frac{1}{2}$ h after feeding the concentration of reducing sugar increased; after $4\frac{1}{4}$ h it decreased, the pH fell and the concentrations of alcohol, acetic and lactic acids rose. Bensadoun and Ichhponani (1968) found in chickens, 2 h after force feeding a diet containing 46·5% starch, that the crop contained high concentrations of both lactate and short-chain fatty acids (132 and 30·2 mg/100 g contents, respectively). The following data of D. J. Jayne-Williams (unpublished) illustrate the correlation between counts of crop lactobacilli and lactate concentration. A group of 8-week-old birds was trained to eat a commercial type diet during a 2 h period; 2 birds of each sex were then killed (by sodium pentobarbitone i.v.) at intervals of 3 h up to 15 h after feeding had ceased. The crop contents were separately collected aseptically, weighed, then bulked and well mixed before counting the lactobacilli (Rogosa *et al.*, 1951) and measuring lactic acid (Barker and Summerson, 1941; Pennington and Sutherland, 1956). Fig. 1 shows that as the weight of the contents steadily

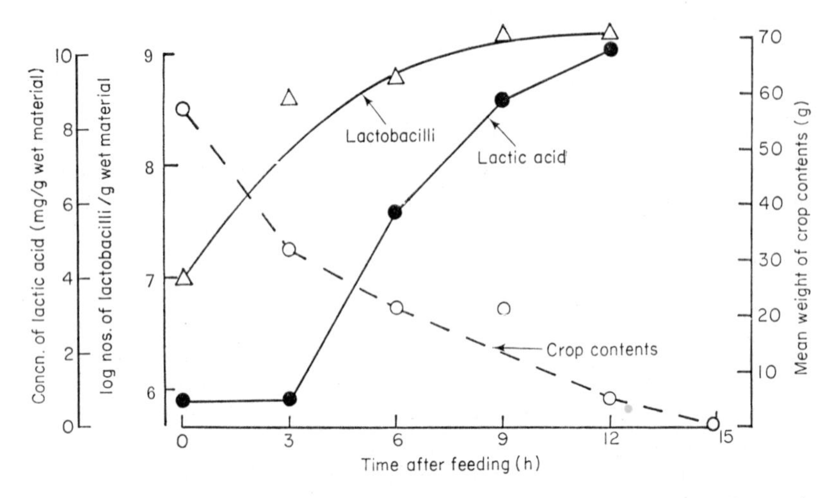

Fig. 1. The influence of time after feeding on the weight of crop contents, the numbers of lactobacilli and the concentration of lactic acid.

fell there was an increase in the number of lactobacilli and, after an initial lag, a steady increase in the amount of lactic acid.

THE GUT FROM PROVENTRICULUS TO ILEO-CAECO-COLIC JUNCTION

Bacteriological examinations of the contents of these sections of the gut are due to numerous workers (e.g. Barnes and Shrimpton, 1957; Ochi and Mitsuoka, 1958; Eissa, 1961; Ochi *et al.*, 1964; Timms, 1968) in addition to those who also examined the crop (see above). In general the qualitative findings were similar to those obtained for the crop.

Organisms discharged from the crop pass through the proventriculus and gizzard where they are subjected to the potentially damaging effects of acidity (pH 2·0 to 4·0) and are reduced in numbers by 10- to 100-fold. From the duodenum onwards, the pH of the contents rises to 5·8 to 7·5 (Herpol and van Grembergen, 1961; Smith, 1965a; Timms, 1968) and bacterial numbers increase (see, for example, Lev and Briggs, 1956). Whether this increase is due to recovery of acid-damaged organisms or to multiplication of undamaged bacteria is uncertain. However, because the chyme moves rapidly through the upper portions of the small intestine, there would be insufficient time for much bacterial multiplication unless organisms could become "anchored" to the intestinal mucosa. Eyssen *et al.* (1962) concluded that "The acid pH of the stomach prevents microbial growth, and the flora of the small intestine is merely a passage flora without active proliferation, and of the same composition as the crop flora". Furthermore, Smith (1965a) showed that when fowls were fed on a diet too acid (pH 2·0) to permit bacterial multiplication in the crop (though having little effect on the pH of the gizzard), the bacterial counts of the upper small intestine were drastically reduced. He obtained similar results by surgical ablation of the crop and concluded that ". . . the *majority* of the bacteria in the intestine of animals fed normally are the product of proliferation in the anterior compartment of the stomach [the crop] and not that of proliferation in the small intestine itself".

On approaching the ileo-caeco-colic junction the intestinal contents move more slowly and provide time for the proliferation of micro-organisms. The pH (6·5 to 7·8: Herpol and van Grembergen, 1961; Smith, 1965a; Timms, 1968) is also favourable. The increase in bacterial numbers at this point in the intestine is shown by the data of D. J. Jayne-Williams (unpublished). Five 3-week-old germfree chickens were inoculated *per os* with a pure culture of an unidentified *Clostridium* sp. isolated from chicken intestinal contents. An anaerobe was chosen to obviate confusion resulting from the rapid multiplication in the crop of a facultative anaerobe. After 1 week the birds were

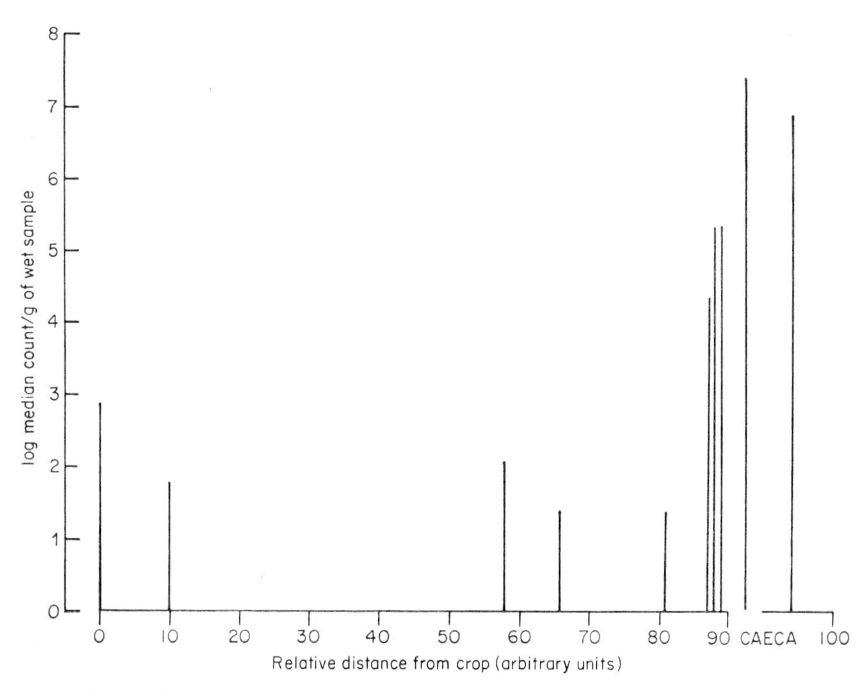

Fig. 2. The numbers of a *Clostridium* sp. occurring in different segments of the alimentary tract of monocontaminated chicks.

killed and samples taken from various sections of the alimentary tract for bacteriological analysis. Dilution counts (triplicate tubes/dilution) were performed in reinforced clostridial medium broth (Hirsch and Grinsted, 1954) and incubated anaerobically at 37° for 48 h before being examined for growth; the results were interpreted by the probability tables of Hoskins (1934). Median counts/g wet wt for the 5 birds are shown in Fig. 2. Organisms did occur in the crop, but whether these arose from multiplication or ingestion of faeces is not known. There was no increase in numbers/g throughout most of the small intestine but at a point just anterior to the ileo-caeco-colic junction a marked increase was found.

C. THE CAECA

These organs provide ideal conditions for bacterial proliferation; the substrate is liquid from colonic contents, possibly with some urine which presumably enters by retroperistalsis (see Chapters 2 and 9), the pH is favourable (6·5 to 7·5: Herpol and van Grembergen, 1961; Smith, 1965a; Timms, 1968), the conditions are anaerobic and, as the caeca evacuate themselves only every 6 to 8 h, there is intermittent stasis. Microscopically, the caecal contents are largely bacterial cells and numbers in excess of 10^9/g wet wt. can be detected on culture. Although many workers have reported the presence of bacteria only of types occurring in anterior parts of the alimentary tract (e.g. lactobacilli, streptococci, etc.), and occasionally anaerobic spore-formers, recent work (Ochi and Mitsuoka, 1958; Barnes and Goldberg, 1962; Ochi *et al.*, 1964; Mitsuoka *et al.*, 1965a, b; Smith 1965a) has shown that, by the use of special media and careful anaerobic techniques, large numbers of strictly anaerobic Gram-negative bacteria (bacteroides) occur in the caeca of normal birds. Barnes and Impey (1968) found that the total viable counts obtained with careful use of anaerobic jars were still lower than would be expected on the basis of microscopic counts. By applying the techniques of Hungate (1950), which have proved of great value in the study of rumen organisms, they have obtained much higher viable counts (Barnes and Impey, 1970). Hungate's techniques seek to prevent strict anaerobes coming into damaging contact with oxygen or oxidized substrates during bacteriological manipulation and incubation. Modifications of these techniques have been used to advantage in the examination of human intestinal contents (Drasar, 1967) and human gingival and mouse caecal samples (Aranki *et al.*, 1969). In addition to the bacteroides, the enumeration of other organisms known to occur in the chicken caecum (bifidobacteria, peptostreptococci: Mitsuoka *et al.*, 1965b; Barnes and Impey, 1970) would no doubt be improved by the use of Hungate's (1950) or similar techniques.

D. THE EXCRETA

Because the bird's droppings are a variable mixture of faeces proper, caecal contents, and urine bacteriological examination of such material is not to be recommended unless there is a requirement to keep the experimental birds alive. The large fluctuations in the counts of specific groups of bacteria in faeces reported by Smith and Crabb (1961) may have been due to this.

III. The Influence of Bacteria on Intestinal Function and Nutrition

A. INTESTINAL MORPHOLOGY

Weights of the small intestine of chicks fed ordinary diets are greater than those of corresponding germfree birds. Though the difference is less marked, they are greater than those of the gut of conventional birds given dietary antibiotic (Gordon, 1952; Pepper *et al.* 1953; Coates *et al.*, 1955; Jukes *et al.*, 1956; Draper, 1958; Coates and Jayne-Williams, 1966). As the length of the intestine seems to be little affected, it follows that the presence of microorganisms and/or their metabolic products causes increased thickness. Although it is tempting to infer from these findings that the improved growth rate observed with germfree and antibiotic-fed conventional birds may be associated with more efficient absorption of nutrients by the thinner gut, there is evidence to show that intestinal weights can be decreased in antibiotic-fed animals without there being a corresponding increase in body weight (in rats, Perdue *et al.*, 1957; in chicks, Hill *et al.*, 1957).

The increased gut weight of conventional birds not given an antibiotic can partly be explained by an increase in the amount of lamina propria, particularly the portion extending into the centre of the villi (Gordon and Bruckner-Kardoss, 1961), by an increase in the amount of lymphoid tissue (Gordon *et al.*, 1957–58), and by an increase in the numbers of free reticulo-endothelial cells in the ileal mucosa and submucosa (Gordon and Bruckner-Kardoss, 1958–59).

The presence of bacteria in the mammalian intestine causes a more rapid turnover of mucosal epithelium (Abrams *et al.*, 1963), the average transit time of germfree mouse duodenal crypt cells being about double that of conventional counterparts, due either to reduced attrition of the mucosa or to a lower proliferative capacity of the germfree cells (Lesher *et al.*, 1964). The period required for the epithelial cell renewal was reduced from 4 to 2 d when germfree mice were exposed to a normal flora (Khoury *et al.*, 1969) and corresponded to the time taken for a conventional-type bacterial flora to become established.

The turnover of intestinal epithelial cells in the chicken is about 2 d and is thus comparable with the rat (Imondi and Bird, 1966).

In considering the influence of bacteria on the regeneration times of intestinal mucosal cells, Visek (1969) has calculated that if, in a broiler chicken, the nutrients spared when epithelial cell-renewal is reduced by a dietary antibiotic, be diverted to other tissues, an increase in daily weight-gain could occur of the same order (*c.* 8%) as that found in practice. He concluded: "It would appear that cellular and tissue renewal resulting from the influence of micro-organisms on cell function, replacement, and life-span could embrace a number of mechanisms that have been proposed for the growth stimulatory actions of antibacterial substances". This hypothesis which embraces some of the many explanations of the mode of action of dietary antibiotics postulated in the past, would clarify the present very confused state of affairs.

B. CARBOHYDRATE

The organic acids (mainly lactic) and other metabolic products of the microflora may be partly absorbed through the crop wall together with certain dietary nutrients (e.g. glucose—Soedarmo *et al.*, 1961; lactic acid—Bolton and Dewar, 1962; Ivorec-Szylit and Sauveur, 1966; acetic, propionic and butyric acids—Bolton and Dewar, 1965); most, however, will move down the alimentary tract to the main absorptive sites of the small intestine. Whether such "predigestion" of carbohydrate by bacteria is of advantage to the host is not clear. Certain activities of bacteria in the crop are clearly detrimental (e.g. the "sour" crops and alcoholic fermentation reported by Bolton, 1965 and Wheeler *et al.*, 1957, respectively). Furthermore, it has been shown with rats that continual infusion of the small intestine *in vivo* with dilute solutions of lactic acid can give rise to steatorrhoea and alteration in the mucosal structure (Hamilton, 1967).

The presence of a microflora has little effect on amylase activities in different sites in the fowl's alimentary tract (Lepkovsky *et al.*, 1964) and no direct effect on disaccharidase production in the small intestine (Siddons and Coates, 1969).

For a variety of reasons dietary carbohydrate may escape digestion and absorption in the small intestine—the necessary enzymes may not be secreted, specific absorptive sites may be overloaded, or the intestinal motility increased when excessive amounts of carbohydrate (e.g. glucose or sucrose) are present. Carbohydrate capable of being utilized may be inside a protective covering as with raw tuber starches (Masson, 1954). In these circumstances the material provides a substrate for the vast numbers of bacteria that occur in the lower gut, particularly in the caeca. Though there is some evidence of bacterial digestion of fibre in the caecum (Radeff, 1928; Henning, 1929; Thornburn and Willcox, 1965a, b) it probably plays an insignificant part in the nutrition of the host compared with that occurring in the caeca of herbivorous mammals (e.g. the horse).

Little, if any, fibre from the colonic contents can enter the caecum through its narrow lumen, as comminuted fragments. The rest remains in the colon but for too short a time for much bacterial degradation to occur. The principal fermentation gases in the caecum are carbon dioxide and methane (Beattie and Shrimpton, 1958; Shrimpton, 1963), the concentration of the latter being influenced by the amount of dietary crude fibre (Shrimpton, 1958). Gas is produced at the rate of about 3 ml/h. Volatile fatty acids (VFAs) are characteristic microbial by-products in the caecum forming 0·2 to 1·0% of its contents (Shrimpton, 1963) or about 100 m-moles/kg (Annison *et al.*, 1968). In order of decreasing concentration, the VFAs are acetic, propionic, butyric (isobutyric) and higher acids (mainly valeric-isovaleric) (Shrimpton, 1963; Bensadoun and Ichhponani, 1968; Annison *et al.*, 1968). A similar pattern, except that the concentration of acetic acid was substantially lower, was reported by Moore *et al.* (1969). The marked reduction in output of VFAs in caecectomized birds and the almost complete absence of VFAs in the contents of the intestinal tracts of germfree birds (Annison *et al.*, 1968) amply demonstrate that these substances are of bacterial origin. Though produced at a distal site of the gut, VFAs are absorbed and carried to the liver where they are metabolized (Annison *et al.*, 1968).

C. PROTEIN

One obvious difference, with respect to protein utilization, between germfree and conventional birds and mammals is the form in which nitrogen is excreted. Much of the soluble nitrogen in germfree faeces would, were a microflora introduced, be rapidly metabolized. Some would be converted into bacterial cell protein, nucleic acids, etc. and some lost as gaseous ammonia. As a large proportion of the nitrogenous material of normal faeces is bacterial (as much as one-third of the total solids of human faeces—Bell *et al.*, 1959) misleading interpretations of digestibility trials can result (see Salter and Coates, 1969). Comparison of the activities of intestinal proteases in germfree and conventional chickens led to the conclusion that the gut microflora exerts little or no effect (Lepkovsky *et al.*, 1964). Dietary penicillin was reported by Anderson *et al.* (1952) to improve protein utilization in chickens. McGinnis (1951), Machlin *et al.* (1952) and Slinger *et al.* (1952a, b) have shown that dietary antibiotics can improve the growth of chickens or turkeys when the protein content of the diet is below that required for maximum growth. The uptake of amino acids (methionine in germfree mice, Hershovic *et al.*, 1967; lysine in penicillin-fed chicks, Draper, 1958) appears to be improved when the intestinal flora is absent or modified by dietary antibiotics. Miller (1967) found no difference between germfree and conventional birds in their efficiency of protein utilization when fed a diet containing ample (26%) readily digestible protein (casein and gelatin). However, the same may not hold if amino acids are less generously supplied, thus affording an opportunity for bacteria to compete with the host. Less easily digestible protein may provide a substrate for bacterial activity in the lower gut, but whether this leads to products utilizable by the host or to noxious substances is uncertain. Evidence available indicates the latter.

Bacterial activity may be responsible for the reduced growth of mammals and birds reported by numerous workers when raw soya meal or certain other leguminous seeds are fed; raw soya meal contains toxic principles including trypsin inhibitors which are inactivated by heat. Support for this view is provided by the observations (e.g. Braham *et al.*, 1959; Linerode *et al.*, 1961) that dietary antibiotics partially alleviate the growth reduction. Furthermore, both germfree chicks and Japanese quail grow only slightly less well when fed diets containing raw soya or navy bean meal than on similar diets containing heated meal, though a characteristic pancreatic hypertrophy occurs in both instances (Miller and Coates, 1966; Hewitt and Coates, 1969; Coates *et al.*, 1970).

Ammonia is one potentially harmful product of bacterial degradation of nitrogenous compounds in the gut. It could arise from the hydrolysis of urea by bacterial urease (in this connection Delluva *et al.* (1968) failed to demonstrate gastric urease in a number of different germfree animals including chickens), from the deamination of amino acids and from the autolysis of bacterial cells. The toxicity of ammonia has stimulated research on beneficial effects of antibiotics administered *per os* (e.g. Silen *et al.*, 1955; Phear and Ruebner, 1956). Ruebner and McLaren (1958) showed that Gram-negative strict anaerobes actively produced ammonia in the rabbit intestine. The control of bacterial urease in the intestine of chickens (and rodents) by means of dietary antibiotics and barbituric acid, and by immunization of animals with jackbean urease injected intraperitoneally, has been investigated by Visek and his co-workers (Dang and Visek, 1960; Visek, 1962; Harbers *et al.*, 1963). Such treatments reduced the concentrations of ammonia and urease activity in the gut and gave better growth. These findings have been discussed by Visek (1964) with particular reference to the mode of action of antibiotics and other growth stimulants.

Ammonia may also arise in the alimentary tract (crop, small intestine and caecum) through deamination of L-arginine by some enterococci (Fujita, 1968a). This is significantly reduced by antibiotics (Fujita, 1968b).

Another metabolic activity of some enterococci (e.g. *Streptococcus faecalis*) which, as has been suggested by Huhtanen and Pensack (1965), may have some influence on the growth rate of chickens, is the production of tyramine by the decarboxylation of tyrosine. Other amines are produced by many enteric micro-organisms (e.g. in the mammalian studies of Melnykowycz and Johansson, 1955; Phear and Ruebner, 1956; Larson and Hill, 1960; Cheeseman and Fuller, 1966; Michel, 1966). Many of the amines will be altered by tissue enzymes (amine oxidases); others are absorbed and subsequently excreted in their original or altered forms—at least in human urine (see Perry *et al.*, 1966).

Hydrogen sulphide and organic sulphur compounds are formed by bacterial degradation of protein in the chicken intestine (Shrimpton, 1963). These are toxic substances but, as is also the case with the amines, their effects on the growth and well-being of animals are unknown, at the concentrations at which they are normally found.

D. LIPID

There is ample evidence (see, for example, Gustafsson and Norman, 1962; Booth, 1965; Sacquet *et al.*, 1968; Rosenberg, 1969; Shimada *et al.*, 1969) that certain intestinal bacteria can split conjugated bile acids into taurine or choline and the constituent acid which can be further degraded. When this occurs in the upper part of the gut as a result of abnormal bacterial proliferation, impaired fat absorption and steatorrhoea result. The deconjugated acids *per se* may have undesirable effects: *in vitro* studies have shown that they can inhibit several mammalian mucosal cell functions including active transport of sugars and amino acids (see Rosenberg, 1969). Cholic acid in the intestine reduced the lifespan of murine mucosal cells (Ranken *et al.*, 1969) and lithocholic acid fed to germfree and conventional chicks caused elevation of serum cholesterol levels in both (Edwards and Boyd, 1963a).

Various species of *Bacteroides*, *Veillonella*, *Bifidobacterium*, *Sphaerophorus* and *Streptococcus*, most of which are strict anaerobes, are able to deconjugate glycocholic and taurocholic acids (Norman and Grubb, 1955; Drasar *et al.*, 1966; Hill and Drasar, 1968; Shimada *et al.*, 1969). The influence of bile acid deconjugation on the absorption of lipid appears to depend on the type of lipid. Comparison of the uptake by germfree and conventional rats of triglycerides (Wiech *et al.*, 1967) or ^{131}I-oleic acid and triolein (Tennant *et al.*, 1969) has shown absorption to be unaffected by the presence of a microflora. Dietary fat uptake by rats is improved in germfree conditions, however (Evrard *et al.*, 1964). Rearing chicks in a fumigated laboratory or feeding diets containing three antibiotics improved their absorption of lard fatty acids (Young *et al.*, 1963). Better retention of palmitic and stearic acids occurred in germfree compared with conventional chicks, but the uptake of oleic or linoleic acids remained unchanged (Boyd and Edwards, 1967). Fat absorption from the small intestine of conventional and gnotobiotic chicks was found by Cole and Boyd (1967) to be influenced in one of two ways, depending on the microflora. Compared with conventional chicks, germfree birds and those inoculated with pure cultures of either *E. coli*, or staphylococci or lactobacilli showed increased absorption of palmitic and stearic acids and total fat. In contrast conventional chicks and those inoculated with either *Strep. faecalis* or *Clostridium welchii* or both, and accidentally contaminated with a variety of bacteria, showed slightly increased uptake of oleic and linoleic acids.

Malsorption of fat was a sign, together with growth depression, observed in chicks fed a purified diet rich in sucrose (Eyssen and de Somer, 1963); the improvement in growth noted when an antibiotic was included indicated a bacterial aetiology. Subsequent gnotobiotic experiments have shown *Strep. faecalis* to be the causative organism (Eyssen and de Somer, 1965) the harmful effects of which can be increased by the presence of a filter-passing agent (possibly an enterovirus) from the gut of conventional birds (Eyssen and de Somer, 1967).

Bacterial degradation of cholesterol occurs in the mammalian gut and there are indications that the same may be true in the chicken. Diets containing added cholesterol (3%) and a high level of sucrose were found to have a hypercholesterolaemic effect on chicks, but not if the sucrose was replaced by glucose (Grant and Fahrenbach, 1959; Kritchevsky *et al.*, 1958). A rise in the blood cholesterol level was observed with glucose, however, when an antibiotic was also given, thus indicating involvement of bacteria in this effect. Confirmation of this was obtained (Kritchevsky *et al.*, 1959) when it was shown that germfree birds given the glucose-rich diet developed high blood cholesterol levels. With diets containing different carbohydrates but no added cholesterol, Coates *et al.* (1965) observed a small but consistent hypercholesterolaemia in germfree birds on a starch-rich diet; a concomitant decrease in the levels of hepatic cholesterol was noted, however, suggesting that the distribution of cholesterol between blood and liver was affected by the presence or absence of a microflora. Birds given a sucrose-rich diet and reared germfree had both higher blood and liver cholesterol levels. They concluded that there appeared to be some inter-relationship between dietary carbohydrate, the intestinal microflora and the metabolism of endogenous cholesterol in chickens.

Eyssen *et al.* (1969) found no significant differences between the cholesterol pools of germfree and conventional chicks fed diets without added cholesterol. Germfree birds given cholesterol supplements had higher pools than equivalent conventional birds, irrespective of the composition of the diet. Diet was shown to have a marked influence, however. When 0·25% of cholesterol was included in a semi-synthetic diet given to conventional birds, a three-fold increase in serum and liver cholesterol levels resulted. Inclusion of cholesterol in a practical type diet had no such effect. As similar results were obtained with germfree birds, it was concluded that a major part of the cholesterol-lowering effect noted with practical type diets was not due to the presence of bacteria: compounds having this hypocholesterolaemic effect were detected in the dietary fish meal. In addition, conventional chicks, irrespective of diet, excreted two to four times more faecal cholesterol and bile acids than their germfree counterparts; it was concluded that the intestinal microflora interferes with absorption of these steroids.

E. VITAMINS

Many common enteric bacteria have absolute requirements for vitamins and therefore compete with the host. Conversely many bacteria synthesize a wide variety of vitamins which might be utilized by the host.

It seems clear that, in general, bacterial competition with the host for available vitamins does not assume important proportions unless the diet contains limiting amounts of these vitamins. The elimination of bacteria competing for vitamins or encouragement of bacteria synthesizing vitamins are two of the many hypotheses put forward to explain the mode of action of dietary antibiotics (see March and Biely, 1967 for a useful reassessment of this subject). Many data on this topic are available, but, as they were largely obtained from animals fed limiting amounts of the vitamin, they will not be considered further.

Enteric syntheses of vitamins undoubtedly occur but whether they are available to the host is questionable. The vitamins synthesized in the ruminant fore-stomach (and possibly the chicken crop) can be utilized, as they are produced in sites anterior to the main absorptive sites of the intestine. Many of those formed in the lower gut, however, may not be absorbed to any extent. Any absorption which may occur in the lower gut seems to depend to a large extent on the characteristics of the material (possibly the molecular weight). Jackson *et al.* (1955) failed to detect uptake of isotopically-labelled vitamin B_{12} injected directly into occluded caeca of hens, though labelled methionine and sodium sulphate were recovered almost quantitatively from the blood.

Coprophagous animals (e.g. rats, mice, rabbits) can avail themselves of at least part of the vitamins synthesized by enteric bacteria (e.g. see Barnes *et al.*, 1960); some of the synthesized vitamins may be so strongly bound within the bacterial cell that they will not be liberated by digestive enzymes of the host (Wostmann *et al.*, 1962). Other animals including fowls ingest their faeces only accidentally, though under certain conditions of management (deep litter and free range) opportunity for so doing is increased with consequent benefit (Kennard *et al.*, 1948; Kennard and Chamberlain, 1948; Jacobs *et al.*, 1954).

High levels of biotin, pantothenic acid, riboflavin, nicotinic acid, and folic acid occur in the caecal contents of hens (Couch *et al.*, 1950). Caecal contents of conventional, but not germfree, chicks contain bacterially-synthesized vitamin B_{12} and other cobalamines (Coates *et al.*, 1963). Coates *et al.* (1968) have shown that vitamins synthesized in the lower gut of chickens are not generally absorbed from this site. Conventional and germfree chicks were reared on diets devoid of one of each of several B complex vitamins (riboflavine, pantothenic acid, pyridoxine, B_{12}, folic acid, nicotinic acid, thiamine and biotin). Livers and caecal contents of birds showing signs of vitamin deficiency were assayed for the vitamin omitted from the diet. Germfree caecal samples contained only negligible quantities of vitamin and though the conventional samples contained considerably more, little evidence was obtained that the deficiency signs were less marked in the conventional birds. Thus, with the possible exception of folic acid, these microbially-synthesized vitamins were of no benefit in the absence of coprophagy. The findings with folic acid accord with those of Miller and Luckey (1963) that its synthesis by *E. coli* in gnotobiotic chicks fed a folic acid-deficient diet led to higher haemoglobin values and tissue levels of the vitamin. Furthermore, Daft *et al.* (1963) found that rats prevented from coprophagy could utilize microbially-synthesized folic acid but not such pantothenic acid.

It has been known for some time that bacterial synthesis of vitamins occurs in voided chicken faeces (e.g. vitamin K, Almquist and Stokstad, 1936; riboflavine, Lamoreux and Schumacher, 1940); however, as pointed out by Jayne-Williams and Coates (1969) the vitamin-synthesizing activity in the excreta cannot be taken as a direct reflection of a similar procedure in the lumen of the gut.

Uptake of vitamins, as with other nutrients, can be adversely affected by bacterial interference with or damage to the absorptive mechanisms. Improved absorption of thiamine in germfree animals was reported by Gordon *et al.* (1960). Dietary penicillin increased the hepatic vitamin A in chicks (Burgess *et al.*, 1951; Coates *et al.*, 1952) and larger stores in germfree compared with conventional chicks were reported by Coates and Jayne-Williams (1966). By competing with the host or by destroying the intrinsic factor, abnormal bacterial populations in the small intestine resulting from artificially produced or naturally occurring anatomical defects (such as blind loops, diverticuli and stenoses) are known to give rise to vitamin B_{12} deficiency (see e.g. Booth, 1965): to what extent, if any, such activities occur in the intestine of normal animals is not known, however.

F. INORGANIC ELEMENTS

The enteric bacterial flora of animals can affect their metabolism of inorganic nutrients (see Coates, 1968). The following effects have been reported in chickens given dietary antibiotics: increased tibial deposition of ^{45}Ca (Migicovsky *et al.*, 1951); improved utilization of calcium and phosphorus (Lindblad *et al.*, 1952); higher concentrations of plasma calcium, but not phosphorus (Bogdonoff and Shaffner, 1954); enhanced bone calcification (Ross and Yacowitz, 1954); improved retention of calcium (Brown, 1957); increased blood calcium in laying hens (Gabuten and Shaffner, 1954). It is noteworthy that therapeutic doses of the same antibiotic administered parenterally was without effect on blood calcium levels (Sturkie and Polin, 1954).

Chickens reared in dirty, as opposed to fumigated premises, had a greater requirement for magnesium and zinc (Edwards *et al.*, 1960) and clean chickens reared in an isolator room required less magnesium and phosphorus than similar birds given chicken faeces orally or in the diet (Edwards and Boyd, 1963b). The absorption of calcium was found by Edwards and Boyd (1963c) to be increased in germfree chickens.

The uptake of water from the intestine of some animals is profoundly influenced by the presence of a microflora. Rodents reared in germfree conditions develop enlarged caeca the contents of which are very fluid. The extent of this enlargement is illustrated by the data of Gordon (1968); in conventional mice the mean weight of caecal content was 0·68 g/100 g of body weight whereas in germfree mice the figure was 12·77 g/100 g. Csaky (1968) showed the presence, in the contents of the upper alimentary tract, of a water absorption inhibitor. This could explain why no net water absorption occurs from the upper part of the gut. In the conventional animal this inhibitor is progressivly destroyed by bacteria so that in the large gut water absorption can take place normally. In the germfree animal, however, the inhibitor persists and prevents colonic water uptake.

Though the caecal contents of germfree chickens are more fluid than those of conventional counterparts (the absence of a dense suspension of bacteria could account for this), massive enlargement of the caeca does not occur.

G. OTHER CONSTITUENTS

Data relating to the influence of the intestinal microflora of mammals on other constituents of the diet and alimentary tract, e.g. orotic acid, choline, formate, bile pigments, are given by Coates (1968) who also deals with the growth-promoting effects of fish solubles when fed to chickens.

In addition to having a direct influence on the nutrition of the host, the micro-organisms that normally colonize the intestine may also have indirect effects: they may affect the transit time of chyme through the gut; they may in some circumstances encourage, in others discourage, the establishment of enteric pathogens or parasites (see e.g. Meynell, 1963; Abrams and Bishop, 1966; Balish and Phillips, 1966; Stefanski and Przyjalkowski, 1965); they are able to move (translocate) from the lumen of the intestine to other sites of the body, e.g. liver, peritoneum, yolk sac (Fuller and Jayne-Williams, 1968) where they may cause disease (Harry, 1957; Moore and Gross, 1968). Effects of this sort, however, are outside the scope of the present contribution.

References

Abrams, G. D., Bauer, H. and Sprinz, H. (1963). *Lab. Invest.* **12**, 355–364.
Abrams, G. D. and Bishop, J. E. (1966). *J. Bact.* **92**, 1604–1614.
Almquist, H. J. and Stokstad, E. L. R. (1936). *J. Nutr.* **12**, 329–335.
Anderson, G. W., Cunningham, J. D. and Slinger, S. J. (1952). *J. Nutr.* **47**, 175–189.
Annison, E. F., Hill, K. J. and Kenworthy, R. (1968). *Br. J. Nutr.* **22**, 207–216.
Aranki, A., Syed, S. A., Kenney, E. B. and Freter, R. (1969). *Appl. Microbiol.* **17**, 568–576.
Balish, E. and Phillips, A. W. (1966). *J. Bact.* **91**, 1736–1743.
Barker, S. B. and Summerson, W. H. (1941). *J. biol. Chem.* **138**, 535–554.
Barnes, E. M. and Goldberg, H. S. (1962). *J. appl. Bact.* **25**, 94–106.
Barnes, E. M. and Impey, C. S. (1968). *J. appl. Bact.* **31**, 530–541.
Barnes, E. M. and Impey, C. S. (1970). *Br. Poult. Sci.* **11**, 467–481.
Barnes, E. M. and Shrimpton, D. H. (1957). *J. appl. Bact.* **20**, 273–285.
Barnes, R. H., Kwong, E., Delany, K. and Fiala, G. (1960). *J. Nutr.* **71**, 149–155.
Beattie, J. and Shrimpton, D. H. (1958). *Q. Jl exp. Physiol.* **43**, 399–407.

Bell, G. H., Davidson, J. N. and Scarborough, H. H. (1959). "Textbook of Physiology and Biochemistry", p. 276. E. & S. Livingstone Ltd. Edinburgh and London.

Bensadoun, A. and Ichhponani, J. S. (1968). *Proc. Cornell Nutr. Conf. Feed Mfrs* pp. 115–118.

Bird, H. R. (1969). *Publ. natn. Acad. Sci., Washington* no. 1679, 33–41.

Blount, W. P. (1947). "Diseases of Poultry" p. 142. Baillière, Tindall and Cox, London.

Bogdonoff, P. D. and Shaffner, C. S. (1954). *Poult. Sci.* **33**, 1044.

Bolton, W. (1962). *Proc. Nutr. Soc.* **21**, xxiv.

Bolton, W. (1965). *Br. Poult. Sci.* **6**, 97–102.

Bolton, W. and Dewar W. A. (1962). *12th Wld's Poult. Congr., Sydney*, 117–119.

Bolton, W. and Dewar, W. A. (1965). *Br. Poult. Sci.* **6**, 103–105.

Booth, C. C. (1965). *In* "Recent Advances in Gastroenterology" (J. Badenoch and B. N. Brooke, eds.), pp. 162–201. Churchill, London.

Boyd, F. M. and Edwards, H. M. (1967). *Poult. Sci.* **46**, 1481–1483.

Braham, J. E., Bird, H. R. and Baumann, C. A. (1959). *J. Nutr.* **67**, 149–158.

Branion, H. D., Anderson, G. W. and Hill, D. C. (1952). *Poult. Sci.* **32**, 335–347.

Braude, R., Kon, S. K. and Porter, J. W. G. (1953). *Nutr. Abstr. Rev.* **23**, 473–495.

Brown, W. O. (1957). *J. Sci. Fd Agric.* **8**, 279–282.

Burgess, R. C., Gluck, M., Brisson, G. and Laughland, D. H. (1951). *Archs Biochem. Biophys.* **33**, 339–340.

Cheeseman, G. C. and Fuller, R. (1966). *J. appl. Bact.* **29**, 596–606.

Coates, M. E. (1968). *In* "The Germ-free Animal in Research" (M. E. Coates, ed.), pp. 161–179, Academic Press, London.

Coates, M. E., Harrison, G. F., Kon, S. K., Porter, J. W. G. and Thompson, S. Y. (1952). *Chemy Ind.* (7), 149.

Coates, M. E., Davies, M. K. and Kon, S. K. (1955). *Br. J. Nutr.* **9**, 110–119.

Coates, M. E., Gregory, M. E., Porter, J. W. G. and Williams, A. P. (1963). *Proc. Nutr. Soc.* **22**, xxvii.

Coates, M. E., Harrison, G. F. and Moore, J. H. (1965). *Ernährungsforschung* **10**, 251–256.

Coates, M. E., Ford, J. E. and Harrison, G. F. (1968). *Br. J. Nutr.* **22**, 493–500.

Coates, M. E., Hewitt, D. and Golob, P. (1970). *Br. J. Nutr.* **24**, 213–225.

Coates, M. E. and Jayne-Williams, D. J. (1966). *In* "Physiology of the Domestic Fowl" (C. Horton-Smith and E. C. Amoroso, eds.), pp. 181–188. Oliver and Boyd, Edinburgh and London.

Cole, J. R. and Boyd, F. M. (1967). *Appl. Microbiol.* **15**, 1229–1234.

Combs, G. F. (1956). *Publs natn. Res. Coun., Washington* no. **397**, pp. 107–125.

Couch, J. R., German, H. L., Knight, D. R., Sparks, P. and Pearson, P. B. (1950). *Poult. Sci.* **29**, 52–58.

Csaky, T. Z. (1968). *In* "The Germ-free Animal in Research" (M. E. Coates ed.), pp. 151–159, Academic Press, London.

Daft, F. S., McDaniel, E. G., Herman, L. G., Romine, M. K. and Hegner, J. R. (1963). *Fedn Proc. Fedn Am. Socs exp. Biol.* **22**, 129–133.

Dang, H. C. and Visek, W. J. (1960). *Proc. Soc. exp. Biol. Med.* **105**, 164–167.

Delluva, A. M., Markley, K. and Davies, R. E. (1968). *Biochim. biophys. Acta* **151**, 646–650.

Draper, H. H. (1958). *J. Nutr.* **64**, 33–42.

Drasar, B. S. (1967). *J. Path. Bact.* **94**, 417–427.

Drasar, B. S., Hill, M. J. and Shiner, M. (1966). *Lancet* i, 1237–1238.

Edwards, H. M. and Boyd, F. M. (1963a). *Proc. Soc. exp. Biol. Med.* **113**, 294–295.
Edwards, H. M. and Boyd, F. M. (1963b). *Poult. Sci.* **42**, 235–240.
Edwards, H. M. and Boyd, F. M. (1963c). *Poult. Sci.* **42**, 1030.
Edwards, H. M., Fuller, H. L. and Hess, C. W. (1960). *J. Nutr.* **70**, 302–306.
Eissa, Y. M. (1961). *J. Arab vet. med. Ass.* **21**, 433–458.
Evrard, E., Hoet, P. P., Eyssen, H., Charlier, H. and Sacquet, E. (1964). *Br. J. exp. Path.* **45**, 409–414.
Eyssen, H. and de Somer, P. (1963). *J. exp. Med.* **117**, 127–138.
Eyssen, H. and de Somer, P. (1965). *Ernährungsforschung* **10**, 264–273.
Eyssen, H. and de Somer, P. (1967). *Poult. Sci.* **46**, 323–333.
Eyssen, H., de Prins, V. and de Somer, P. (1962). *Poult. Sci.* **41**, 227–233.
Eyssen, H., Swaelen, E., Kowszyk-Gindifer, Z. and Parmentier, G. (1965). *Antonie van Leeuwenhoek* **31**, 241–248.
Eyssen, H., van Messom, G. and Van den Bosch, J. (1969). *In* "Germ-free Biology" (E. A. Mirand and N. Back, eds.), pp. 97–105, Plenum Press, New York.
François, A. C. (1962). *Wld Rev. Nutr. Diet.* **3**, 23–64.
Fujita, H. (1968a). *Jap. J. Poult. Sci.* **5**, 136–141.
Fujita, H. (1968b). *Jap. J. Poult. Sci.* **5**, 142–147.
Fuller, R. and Jayne-Williams, D. J. (1968). *Br. Poult. Sci.* **9**, 159–163.
Gabuten, A. R. and Shaffner, C. S. (1954). *Poult. Sci.* **33**, 47–53.
Gordon, H. A. (1952). *Colloqium Univers. Notre Dame, Lobund Institute.*
Gordon, H. A. (1968). *In* "The Germ-free Animal in Research" (M. E. Coates, ed.), pp. 127–150, Academic Press, London.
Gordon, H. A. and Bruckner-Kardoss, E. (1958–59). *Antibiotics A.* 1012–1019.
Gordon, H. A. and Bruckner-Kardoss, E. (1961). *Acta anat.* **44**, 210–225.
Gordon, H. A., Bruckner-Kardoss, E. and Kan, D. (1960). *5th Int. Congr. Nutr., Washington*, 13.
Gordon, H. A., Wagner, M. and Wostmann, B. S. (1957–58). *Antibiotics A.* 248–255.
Grant, W. C. and Fahrenbach, M. J. (1959). *Proc. Soc. exp. Biol. Med.* **100**, 250–252.
Gustafsson, B. E. and Norman, A. (1962). *Proc. Soc. exp. Biol. Med.* **110**, 387–389.
Halnan, E. T. (1949). *Br. J. Nutr.* **3**, 245–253.
Hamilton, J. R. (1967). *Pediat. Res.* **1**, 341–353.
Harbers, L. H., Alvares, A. P., Jacobson, A. I. and Visek, W. J. (1963). *J. Nutr.* **80**, 75–79.
Harry, E. G. (1957), *Vet. Rec.* **69**, 1433–1439.
Henning, H. T. (1929). *Landwn VersStnen* **108**, 253–286 quoted by Halnan (1949).
Herpol, C. and van Grembergen, G. (1961). *Annls Biol. anim. Biochim. Biophys.* **1**, 317–321.
Hershovic, T., Katz, J., Floch, M. H., Spencer, R. P. and Spiro, H. M. (1967). *Gastroenterology* **52**, 1136.
Hewitt, D. and Coates, M. E. (1969). *Proc. Nutr. Soc.* **28**, 47A.
Hill, C. H., Keeling, A. D. and Kelly, J. W. (1957). *J. Nutr.* **62**, 255–267.
Hill, M. J. and Drasar, B. S. (1968). *Gut* **9**, 22–27.
Hirsch, A. and Grinsted, E. (1954). *J. Dairy Res.* **21**, 101–110.
Hoskins, J. K. (1934). *Publ. Hlth Rep.* **49**, 393–405.
Huhtanen, C. N. and Pensack, J. M. (1965). *Poult. Sci.* **44**, 830–834.
Hungate, R. E. (1950). *Bact. Rev.* **14**, 1–49.
Imondi, A. R. and Bird, F. H. (1966). *Poult. Sci.* **45**, 142–147.
Ivorec-Szylit, O., Mercier, C., Raibaud, P. and Calet, C. (1965). *C. r. hebd. Séanc. Acad. Sci., Paris* **261**, 3201–3203.

Ivorec-Szylit, O. and Sauveur, M. (1966). *Annls Biol. anim. Biochim. Biophys.* **6**, 517–520.

Ivorec-Szylit, O. and Szylit, M. (1965). *Annls Biol. anim. Biochim. Biophys.* **5**, 353–360.

Jackson, J. T., Mangan, G. F., Machlin, L. J. and Denton, C. A. (1955). *Proc. Soc. exp. Biol. Med.* **89**, 225–227.

Jacobs, R. L., Elam, J. F., Fowler, J. and Couch, J. R. (1954). *J. Nutr.* **54**, 417–426.

Jayne-Williams, D. J. and Coates, M. E. (1969). *In* "International Encyclopaedia of Food and Nutrition" (D. Cuthbertson, ed.) vol. 17 Part 1. Pergamon Press, Oxford.

Jukes, H. G., Hill, D. C. and Branion, H. O. (1956). *Poult. Sci.* **35**, 716–723.

Kennard, D. C. and Chamberlin, V. D. (1948). *Poult. Sci.* **27**, 240–245.

Kennard, D. C., Bethke, R. M. and Chamberlin, V. D. (1948). *Poult. Sci.* **27**, 477–481.

Khoury, K. A., Floch, M. H. and Hersch, T. (1969). *J. exp. Med.* **130**, 659–670.

Kritchevsky, D., Grant, W. C., Fahrenbach, M. J., Riccardi, B. A. and McCandless, R. F. J. (1958). *Archs Biochem. Biophys.* **75**, 142–147.

Kritchevsky, D., Kolman, R. R., Guttmacher, R. M. and Forbes, M. (1959). *Archs Biochem. Biophys.* **85**, 444–451.

Lamoreux, W. F. and Schumacher, A. E. (1940). *Poult. Sci.* **19**, 418–423.

Larson, N. L. and Hill, E. G. (1960). *J. Bact.* **80**, 188–192.

Lepkovsky, S., Wagner, M., Furuta, F., Ozone, K. and Koike, T. (1964). *Poult. Sci.* **43**, 722–726.

Lesher, S., Walburg, M. E. and Sacher, G. A. (1964). *Nature, Lond.* **202**, 884–886.

Lev, M. and Briggs, C. A. E. (1956). *J. appl. Bact.* **19**, 224–230.

Lindblad, G. S., Slinger, S. J., Anderson, G. W. and Motzok, I. (1952). *Poult. Sci.* **31**, 923–924.

Linerode, P. A., Waibel, P. E. and Pomeroy, B. S. (1961). *J. Nutr.* **75**, 427–434.

Luckey, T. D. (1959). *In* "Antibiotics: Their Chemistry and Non-Medical Uses". (H. S. Goldberg ed.), pp. 174–321, D. van Nostrand Co., Inc., New Jersey.

Machlin, L. J., Denton, C. A., Kellogg, W. L. and Bird, H. R. (1952). *Poult. Sci.* **31**, 106–109.

March, B. E. and Biely, J. (1967). *Poult. Sci.* **46**, 831–838.

Masson, M. J. (1954). *10th Wld's Poult. Congr., Edinburgh*, 105–111.

McGinnis, J. (1951). *Poult. Sci.* **30**, 924.

Melnykowycz, J. and Johansson, K. R. (1955). *J. exp. Med.* **101**, 507–517.

Meynell, G. G. (1963). *Br. J. exp. Path.* **44**, 209–219.

Michel, M. C. (1966). *Annls Biol. anim. Biochim. Biophys.* **6**, 33–46.

Migicovsky, B. B., Nielson, A. M., Gluck, M. and Burgess, R. (1951). *Archs Biochem. Biophys.* **34**, 479–480.

Miller, H. T. and Luckey, T. D. (1963). *J. Nutr.* **80**, 236–242.

Miller, W. S. (1967). *Proc. Nutr. Soc.* **26**, x.

Miller, W. S. and Coates, M. E. (1966). *Proc. Nutr. Soc.* **25**, iv.

Mitsuoka, T., Sega, T. and Yamamoto, S. (1965a). *Zentbl. Bakt. ParasitKde* (I Orig.) **195**, 69–79.

Mitsuoka, T., Sega, T. and Yamamoto, S. (1965b). *Zentbl. Bakt. ParasitKde* (I Orig.) **195**, 455–469.

Moore, W. E. C. and Gross, W. B. (1968). *Avian Dis.* **12**, 417–422.

Moore, W. E. C., Cato, E. P. and Holdeman, L. V. (1969). *J. infect. Dis.* **119**, 641–649.

Norman, A. and Grubb, R. (1955). *Acta path. microbiol. scand.* **36**, 537–547.

Ochi, Y. and Mitsuoka, T. (1958). *Jap. J. vet. Sci.* **20**, 45–51.
Ochi, Y., Mitsuoka, T. and Sega, T. (1964). *Zentbl. Bakt. ParasitKde* (I Orig.) **193**, 80–95.
Pennington, R. J. and Sutherland, T. M. (1956). *Biochem. J.* **63**, 353–361.
Pepper, W. F., Slinger, S. J. and Motzok, I. (1953). *Poult. Sci.* **32**, 656–660.
Perdue, H. S., Spruth, H. C. and Frost, D. V. (1957). *Fedn Proc. Fedn Am. Socs exp. Biol.* **16**, 396.
Perry, T. L., Hestrin, M., MacDougall, L. and Hansen, S. (1966). *Clinica chim. Acta* **14**, 116–123.
Phear, E. A. and Ruebner, B. (1956). *Br. J. exp. Path.* **37**, 253–262.
Radeff, T. (1928). *Biochem. Z.* **193**, 192–196.
Ranken, R., Wilson, R. and Bealmear, P. (1969). *In* "The Germ-free Animal as a Tool in Research" pp. 39–40. Advance Study Institute, Leuven.
Rogosa, M., Mitchell, J. A. and Wiseman, R. F. (1951). *J. Bact.* **62**, 132–133.
Rosenberg, T. H. (1969). *Am. J. clin. Nutr.* **22**, 284–291.
Ross, E. and Yacowitz, H. (1954). *Poult. Sci.* **33**, 262–265.
Ruebner, B. and McLaren, J. R. (1958). *Br. J. exp. Path.* **39**, 85–89.
Sacquet, E., Garnier, H., Raibaud, P. and Eyssen, H. (1968). *C. r. Sci. nat.* **267**, 2238–2240.
Salter, D. N. and Coates, M. E. (1969). *8th Int. Congr. Nutr., Prague.*
Shimada, K., Bricknell, K. S. and Finegold, S. M. (1969). *J. infect. Dis.* **119**, 273–281.
Shrimpton, D. H. (1958). *11th Wld's Poult. Congr., Mexico City.*
Shrimpton, D. H. (1963). *J. appl. Bact.* **26**, i.
Siddons, R. C. and Coates, M. E. (1971). *Br. J. Nutr.* (in press).
Sieburth, J. M., Jezeski, J. J., Hill, E. G. and Carpenter, L. E. (1954). *Poult. Sci.* **33**, 753–762.
Silen, W., Harper, H. A., Mawdsley, D. L. and Weirich, W. L. (1955). *Proc. Soc. exp. Biol. Med.* **88**, 138–140.
Slinger, S. J., Bergey, J. E., Pepper, W. F., Snyder, E. S. and Arthur, D. (1952a). *Poult. Sci.* **31**, 757–764.
Slinger, S. J., Morphet, A. M., Gartley, K. M. and Arthur, D. (1952b). *Poult. Sci.* **31**, 881–887.
Smith, H. W. (1965a). *J. Path. Bact.* **89**, 95–122.
Smith, H. W. (1965b). *J. Path. Bact.* **90**, 495–513.
Smith, H. W. and Crabb, W. E. (1961). *J. Path. Bact.* **82**, 53–66.
Soedarmo, D., Kare, M. R. and Wasserman, R. H. (1961). *Poult. Sci.* **40**, 123–128.
Stefanski, W. and Przyjalkowski, Z. (1965). *Ernährungsforschung* **10**, 155–158.
Sturkie, P. D. and Polin, D. (1954). *Poult. Sci.* **33**, 209–210.
Taylor, J. H. (1957). *Vet. Rec.* **69**, 278–288.
Tennant, B., Reina-Guerra, M., Harrold, D. and Goldman, M. (1969). *J. Nutr.* **97**, 65–69.
Thornburn, C. C. and Willcox, J. S. (1965a). *Br. Poult. Sci.* **6**, 23–31.
Thornburn, C. C. and Willcox, J. S. (1965b). *Br. Poult. Sci.* **6**, 33–43.
Timms, L. (1968). *Br. vet. J.* **124**, 470–477.
Visek, W. J. (1962). *Am. J. vet. Res.* **23**, 569–574.
Visek, W. J. (1964). *Proc. Cornell Nutr. Conf. Feed Mfrs*, 121–133.
Visek, M. J. (1969). *Publs natn. Acad. Sci. Washington* no. 1679, 135–149.
Wheeler, H. O., Crawford, W. P. and Couch, J. R. (1957). *Poult. Sci.* **36**, 1167.
Wiech, N. L., Hamilton, J. G. and Miller, O. N. (1967). *J. Nutr.* **93**, 324–330.

Wiseman, R. W., Bushnell, O. A. and Rosenberg, M. M. (1956). *Poult. Sci.* **35**, 126–132.

Wostmann, B. S., Knight, P. L. and Kan, D. F. (1962). *Ann. N.Y. Acad. Sci.* **98**, 516–527.

Young, R. J., Garrett, R. L. and Griffith, M. (1963). *Poult. Sci.* **42**, 1146–1154.

5 The Anatomy of Respiration

A. S. KING and V. MOLONY

Department of Veterinary Anatomy,
University of Liverpool,
and Unit of Comparative Neurobiology,
University of Liverpool, Liverpool, England

I. Introduction

Many aspects of the physiology of avian respiration have been admirably presented, notably by Salt and Zeuthen (1960) and by Sturkie (1965). Recently there have also been reviews of the anatomy of the respiratory tract in birds generally (King, 1966) and in domestic birds in particular (King, 1972).

This chapter does not aim to duplicate these works but to discuss those aspects of respiratory anatomy which are of functional interest and to indicate the possible physiological significance of this information.

We have also attempted to emphasize a few particular growing points in functional respiratory anatomy. Areas recently attracting active attention include the blood-gas barrier, the bronchial smooth muscle and its innervation, and the afferent nerve supply of the lower respiratory tract.

II. Upper Respiratory Tract

A. NASAL CAVITY

This account of the nasal cavity is based on observations and a review of the literature by King (1972).

The external naris is guarded by three more or less parallel curved cartilages, the operculum, the vertical plate of the naris, and the ventral nasal concha (turbinate) (Fig. 1a). During nasal breathing the gases must pass through the narrow spaces between these plates. In all domestic birds (Bittner, 1925) the rostrally-pointing cone-shaped ventral concha is not a scroll but a simple C-shaped fold. The middle concha, much the largest of the three, is a scroll, forming about $1\frac{1}{2}$ turns in the fowl (Fig. 1b). The dorsal nasal concha in all the domesticated birds is a closed dome, the cavity of which connects with the infraorbital sinus but not with the nasal cavity (Fig. 1c, d).

In the fowl the ventral concha is lined by a stratified squamous epithelium, and the dorsal concha by an olfactory epithelium. The rest of the nasal cavity, including the nasal septum, is lined by ciliated columnar or pseudostratified columnar epithelium with many intraepithelial mucous glands. This nasal mucociliary system has been shown by Bang (1961) to consist of glands and ciliated cells which are arranged in alternating lines corresponding to the direction of flow of the mucous carpet. The flow is along two pathways. The wider ventral path sweeps the majority of the nasal cavity through the choanal opening. The narrow dorsal path sweeps towards the external naris, drawing mucus from the olfactory lining of the dorsal turbinate. Bang believed these dorsal and ventral pathways to be homologous in birds and mammals. The rate of flow of the mucous carpet is about 10 mm/min in both birds and mammals generally. The mucociliary lines are paralleled by the mucosal arteries but not by the veins.

The infraorbital sinus lies rostroventral to the eye beneath the skin and beak, being about 0·8 cm^3 in volume in adults (Hampl, 1957). In all the domesticated birds it has a wide connexion into the dorsal concha (Fig. 1d). The connexion to the nasal cavity (Fig. 1c) is much narrower, the movement of gases between the nasal cavity and the sinus presumably being limited largely to diffusion. In the region of its opening into the nasal cavity the sinus has a well-developed mucociliary system which appears to lead into the wide

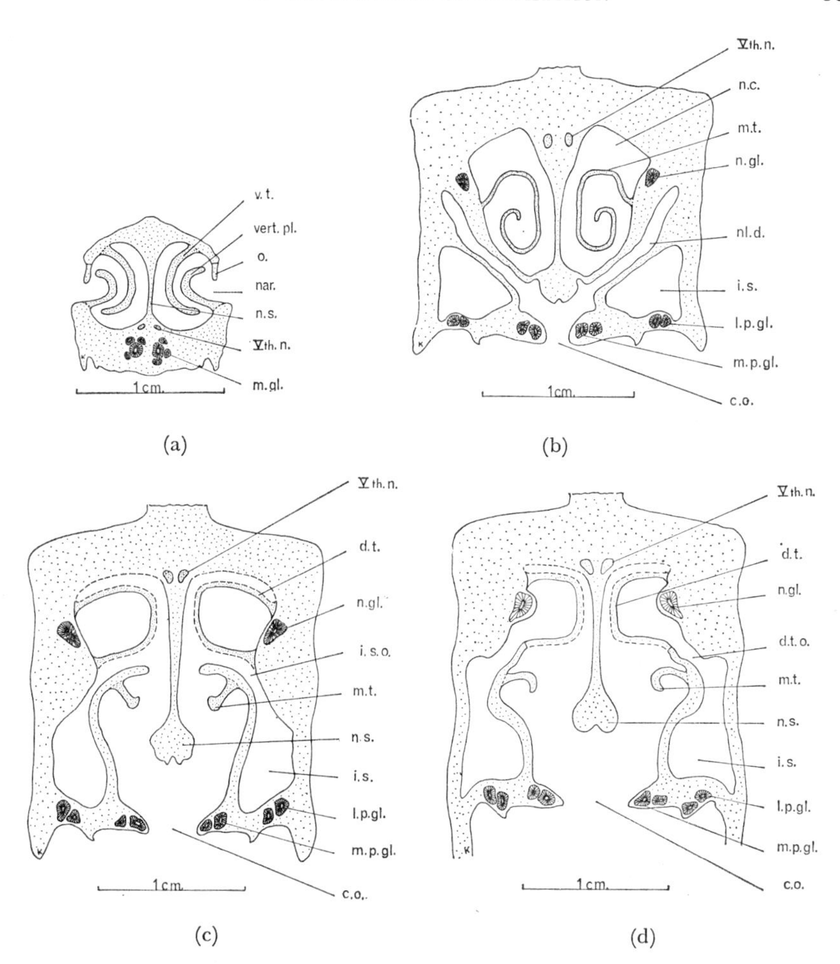

Figs. 1a, b, c, d. Four successive semischematic transverse sections through the nasal cavity of an adult. c.o., choanal opening; d.t., dorsal concha; d.t.o., opening of dorsal concha; i.s., infraorbital sinus; i.s.o., opening of infraorbital sinus; l.p.gl., lateral palatine salivary gland; m.gl., maxillary salivary gland; m.p.gl., medial palatine salivary gland; m.t., middle concha; nar., naris; n.c., nasal cavity; n. gl., nasal gland; nl. d., nasolacrimal duct; n.s., nasal septum; o., operculum; vert. pl., vertical plate of ventral concha; v.t., ventral concha; Vth. n., ophthalmic trunk of trigeminal nerve. The distribution of the olfactory mucosa, as claimed by Jungherr (1943), is indicated by a broken line in Figs. 1c and d. Fig. 1a is reproduced with permission from King (1972). Figs. 1b, c and d are reproduced with permission from McLelland *et al.* (1968).

ventral pathway of the nasal cavity; the more remote parts of the sinus have a simple squamous lining (Bang, 1961).

The olfactory region is said to be confined to the dorsal nasal concha in birds generally (Portmann, 1950). In the fowl it is distributed as in Figs. 1c, d, according to Jungherr (1943), but according to Juárez (1961) it extends rostrally in the dorsal part of the nasal cavity between the nasal septum and the middle concha.

The choanal opening (internal nares) is a median slit, widened caudally (Figs. 1b, c, d). By inserting a cine-endoscope into the oral cavity via the oesophagus in the semi-conscious fowl White (1968b) showed that the choanal opening closes when the roof of the pharynx is stimulated. This is essential during swallowing; otherwise the raking movements of the larynx against the palate would push the bolus into the nasal cavity.

B. OROPHARYNX

The extensive literature on the anatomy of the oral cavity and pharynx has been reviewed by McLelland (1972). The roof of the oral cavity is formed by the hard palate, which is pierced by the median choanal opening. The hard palate carries longitudinal ridges and transverse rows of caudally-pointing papillae. In the submucosa lie the paired maxillary and palatine salivary glands (Figs. 1a, b, c, d). The boundary between the oral cavity and pharynx lies immediately caudal to the choanal opening (see McLelland, 1972), being followed at once by the median infundibular slit which is the common opening of the Eustachian tubes. The many openings of the spheno-pterygoid salivary glands discharge on either side of the infundibular slit. There are various caudally-directed papillae on the roof of the pharynx. The most caudal part of the roof has a longitudinal layer of smooth muscle which is continuous with the inner muscle layer of the oesophagus. The floor of the oral cavity is covered by the free part of the tongue, while the floor of the rostral part of the pharynx carries the base of the tongue.

For a review of the complex anatomy of the tongue including its nerve endings, and of the many lingual, oral, and pharyngeal salivary glands, see McLelland (1972). Most authorities agree that the salivary glands have only mucous cells.

The main arteries and veins of the oral and pharyngeal cavities have been described by Richards (1967, 1968a). In view of the known importance of the oropharyngeal region in evaporative cooling in a number of birds (Bartholomew *et al.*, 1968) and the fact that its heavy vascularization is known to permit the cooling of large quantities of blood in at least one species (Schmidt-Nielsen *et al.*, 1969), the vascularity of the mucosa is of potential interest in the fowl. The innervation of the oropharynx by the Vth, VIIth and IXth nerves has been discussed by Cords (1904), Hsieh (1951) and Bubien-Waluszewska (1968).

C. LARYNX

Projecting dorsally from the caudal region of the pharyngeal floor is a conspicuous mound, the laryngeal prominence (Fig. 2), carrying the opening into the larynx. Here the stratified epithelium of the pharynx continues

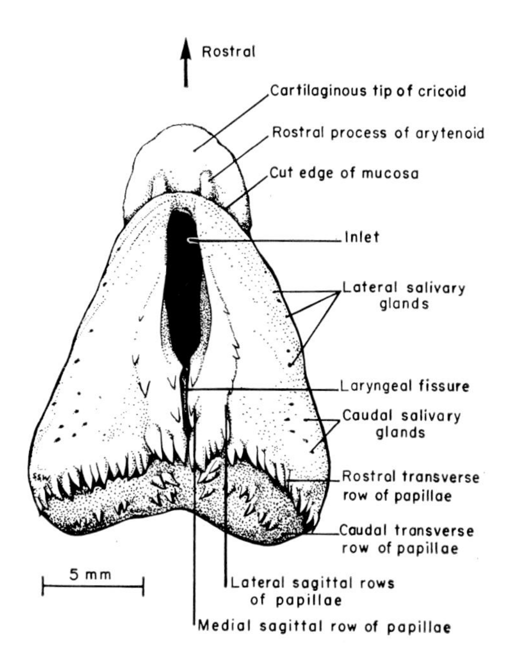

Fig. 2. Dorsal view of the laryngeal prominence of an adult fowl. Reproduced with permission from White (1972).

into the respiratory epithelium of the larynx. The anatomy of the larynx of the domestic birds has been surveyed by White (1972). In the hen the rostral part of the laryngeal prominence lies immediately ventral to the caudal part of the choanal opening, while in the male it lies well caudal to the latter, during eupnoea. The laryngeal inlet is a median slit supported on each side by the arytenoid cartilages. In a typical male the length of the inlet is about 11 mm, and its width during gasping reaches about 9 mm; in the female the corresponding measurements are about 8·5 mm and 7 mm. The inlet leads into the cavity of the larynx, which is lined by pseudostratified columnar ciliated cells and shallow mucous acini. The cavity is compressed dorsoventrally, measuring in its most caudal region in the resting position about 18 mm transversely and 3 mm dorsoventrally in the female. The capacity of the cavity is about 600 mm^3 in the male, and 260 mm^3 in the female. The above measurements relate to medium weight birds. It is possible that when the inlet widens, the dorsoventral measurement and hence the volume of the cavity are considerably increased. Vocal cords, epiglottis, and thyroid cartilage are absent. For an account of the arytenoid and cricoid cartilages and the attachments of their ligaments and muscles see White (1972). The superficial intrinsic muscles cause the roundness of the laryngeal prominence. The deep intrinsic muscles embrace the caudal part of the inlet like a horseshoe. Electrical stimulation shows that the superficial muscle dilates the inlet and the deep one closes it (White and Chubb, 1967). Being attached at points rostral and caudal to the larynx, the extrinsic muscles move the larynx rostrally and caudally (White and Chubb, 1967); for details of their attachments see McLelland (1965, 1968).

Because of anastomoses between the vagus, glossopharyngeal and hypoglossal nerves soon after they emerge from the cranial cavity (Cords, 1904; Hsieh, 1951; Watanabe, 1960, 1964; Bubien-Waluszewska, 1968) the source of motor fibres to the larynx is uncertain. The vagal-glossopharyngeal anastomosis sends fibres to the larynx. Brown (1970) demonstrated the presence of fibres of large diameter (between 10 and 12 μm) in this anastomosis, similar in size to those of the sciatic nerve. He suggested that they may be homologous with the group of large diameter vagal fibres in the cat which are known to enter the recurrent laryngeal nerve and supply the laryngeal muscles. In the bird vagal fibres may leave the vagus via the anastomosis with the IXth nerve, and go direct to the larynx (Watanabe, 1960; Bubien-Waluszewska, 1968). Hsieh (1951) called the fibres to the larynx the anterior laryngeal nerve of the glossopharyngeal nerve, and believed that "more than half" of them came from the vagus; this nerve supplies the oropharynx and oesophagus, as well as the mucosa and intrinsic muscles of the larynx. Section of the vagal-glossopharyngeal anastomosis causes retrograde degeneration in the dorsal vagal nucleus (Watanabe, 1968). The recurrent nerve fails to reach the larynx, being confined to the syrinx, oesophagus and crop (Hsieh, 1951; Fedde *et al.*, 1963). (When present, its pulmono-oesophageal branch goes to the lung as well as the oesophagus, see page 122.)

Laryngeal functions were investigated by White and Chubb (1967), and by White (1968a, b) with cineradiography and cine-endoscopy (see also White, 1972). In the conscious bird tactile stimulation of the laryngeal prominence produces little response, but stimulation of the interior of the inlet induces instant closure. The inlet can widen slightly during each eupnoeic inspiration. During inspiratory gasping wide dilation of the inlet with marked rostral movement of the whole prominence accompanies each inspiration. In the final stages of deglutition of solids, rapid rostrocaudal movements of the laryngeal prominence rake the bolus in a caudal direction between the prominence and the roof of the pharynx. The sticky mucous saliva and the caudally-directed laryngeal and oropharyngeal papillae aid this process. During crowing the larynx slides rapidly and caudally for several centimetres, as far as the base of the neck.

D. TRACHEA

The anatomy of the trachea has been described by McLelland (1965, 1966). Each of the 108 to 126 tracheal cartilages is a complete ring, overlapping its neighbours as in Fig. 3. This arrangement prevents compression but allows elongation and flexion of the trachea. The rings in the cranial third of the trachea are transversely oval; elsewhere they are circular except in the caudal few millimetres of the trachea, where they become vertically oval. Their diameter decreases progressively towards the thorax, from about 12 × 6 mm in the male and 9 × 4 mm in the female, down to about 3 × 5 mm

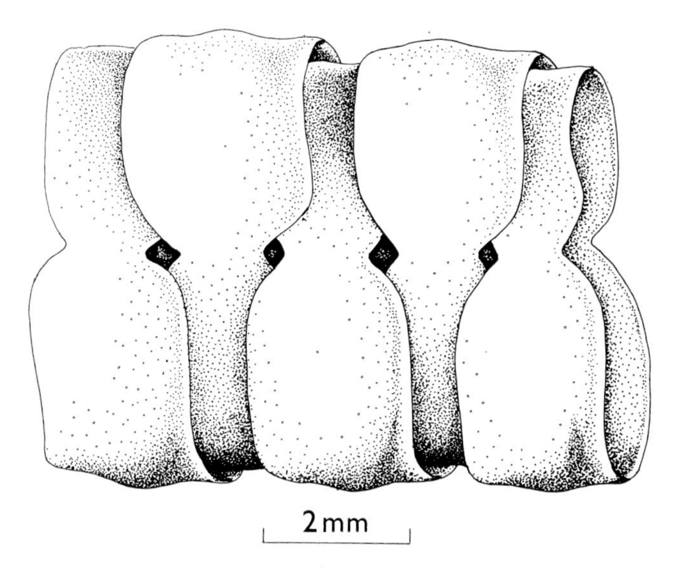

Fig. 3. Dorsal view of the middle of the trachea of an adult. to show the overlapping of the tracheal cartilages. Reproduced with permission from McLelland (1965).

in both sexes. A trachea which measures about 3 × 5 mm immediately cranial to the syrinx has a cross-sectional area at this point of about 12·5 mm^2. The total length of the excised trachea is about 17 or 18 cm in the adult male Rhode Island Red × White Leghorn and 15·5 or 16·5 cm in the adult female Golden Hampshire. The volume of the oral cavity, larynx, trachea and extra-pulmonary primary bronchus in dead Single Comb White Leghorn males averaging 2·1 kg in body weight was 8·4 ± 0·3 ml, as measured by filling with water (Fedde, 1970).

For a discussion of the muscles of the trachea see McLelland (1965, 1968) and King (1972). The sternolaryngeus (sternotrachealis) is of particular interest in that, since it arises from the sternum and inserts extensively on the rings of the trachea and the larynx, it appears capable of causing the trachea and syrinx to oscillate rostrocaudally in and out of the thoracic inlet (which may be important in song, see page 163). The arterial supply to the trachea (Westpfahl, 1961; Peterson, 1970) is via the ascending oesophageal, sterno-trachealis, and bronchial arteries. Since the trachea is a major site of evaporative cooling in some birds (Schmidt-Nielsen *et al.*, 1969), the vascular architecture of the tracheal mucosa should be studied in the fowl. The nerve supply is from the tracheal branches of the recurrent nerve (Watanabe, 1960; Peterson, 1970). Bennett and Malmfors (1970) found little smooth muscle in the trachea and very few varicose fluorescent nerves associated with this muscle. The internal lining of the trachea is mucociliary as in the larynx and nasal cavity, with a pseudostratified columnar ciliated epithelium having numerous shallow mucous crypts (Trautmann *et al.*, 1957).

E. SYRINX

The anatomy of the syrinx has been described by Myers (1917) and further details have been added by Appel (1929) and Morejohn (1966). A detailed study of the functional morphology of the avian syrinx has been made by Greenewalt (1968).

The syrinx, the vocal organ, lies at the end of the trachea just within the coelomic inlet and suspended within the clavicular air sac. At rest it is compressed laterally. The basis of its structure is the pessulus, a wedge-shaped

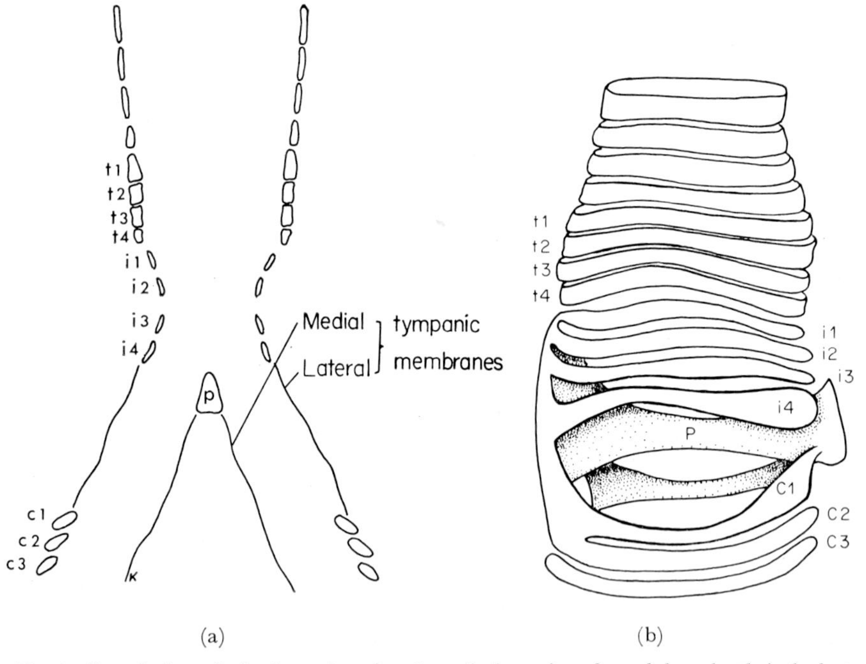

Fig. 4a. Dorsal view of a horizontal section through the syrinx of an adult cock; t 1–4, the last four (specialized) tracheal cartilages, the tympanum of Myers; i 1–4, four lateral syringeal cartilages; c 1–3, the first three (specialized) bronchial cartilages; p, pessulus. The remaining bronchial cartilages are ordinary C-shaped structures. Reproduced with permission from Myers (1917).

Fig. 4b. Left lateral view of the skeleton of the syrinx of an adult cock. Abbreviations as in Fig. 4a. Reproduced with permission from Myers (1917).

cartilage which lies dorso-ventrally in the midline and splits the lumen of the trachea into the two primary bronchi (Figs. 4a, b). Myers (1917) regarded the last three or four tracheal rings as the tympanum, a part of the syrinx. Caudal to this, the syrinx is supported on each lateral aspect by four to seven thin flexible lateral syringeal cartilages which are fused to the pessulus ventrally but not dorsally. These are the "intermediate cartilages" of Myers (1917), and the syringeal bars of Morejohn (1966). Caudal to the syrinx, the airway is founded on C-shaped bronchial cartilages which continue into the lung.

The walls of the syrinx are completed laterally and medially by two pairs of thin vibrating membranes, the lateral and medial tympanic membranes. The lateral membrane runs from the last lateral syringeal cartilage to the first bronchial cartilage. The medial tympanic membrane runs from the pessulus to the ends of the first three bronchial cartilages. Caudal to the third bronchial cartilage the medial membrane is continued by the relatively heavy fibrous tissue of the medial bronchial wall.

Since both medial and lateral surfaces of the syrinx are extensively membranous, and since the lateral syringeal cartilages which support the lateral aspect of the organ are very thin and flexible, its lumen is easily reduced or even obliterated by compression, as would occur if a pressure gradient arose between the clavicular sac and syringeal lumen (see page 153).

There are no intrinsic muscles in the syrinx but it can probably be moved craniocaudally by varying tensions in the tracheal muscles, particularly the sternolaryngeus (see Section IID). The innervation of the syrinx is from a branch of the recurrent nerve which also supplies the adjacent extrapulmonary bronchi, trachea, and tracheal muscles (the posterior laryngeal nerve of Hsieh, 1951).

III. Lung

The avian lung is a flattened quadrilateral structure occupying the roof of the cranial end of the coelom. It is usually described as being relatively small in relation to the rest of the body, but data compiled by Altman *et al.* (1958) on 90 species of mammal and 45 species of bird gave a mean of 1·38 g of fresh lung per 100 g body weight for the mammals and 1·62 g/100 g for the birds. On the other hand the volume of the avian lung is about one-tenth that of a mammal of equal body size (Burton and Smith, 1968).

On the ventromedial surface of the lung the stems of the four great craniomedial secondary bronchi are visible. In nearly all the avian species so far examined, but not in the fowl, the stems of the caudodorsal secondary bronchi also can be seen on the dorsolateral surface (Fischer, 1905; Juillet, 1912).

The dorsomedial border of the lung is deeply indented by furrows into which are embedded a corresponding number of ribs. Typically there are five furrows (Figs. 6a, 20), but one of these is usually only a shallow depression.

The relative depth of the furrows varies from specimen to specimen, the middle ones being the most consistently deep. The deepest point of each furrow is at the dorsomedial border of the lung, reaching a maximum of about 15 mm in casts of males of heavy breeds. The depth is halved midway along the furrow, and is then rapidly reduced to a shallow groove on the lateral surface of the lung. The region of the lung enclosed between these furrows is essentially the most dorsal part. Scharnke (1938) reckoned "a good third" of the lung to be embedded between the ribs. In casts of two lungs from different birds we found the volume of lung tissue enclosed by these furrows to be 14 and 22% of the total volume of the lung; these conditions may represent full inspiration. At full expiration an appreciably greater percentage of the lung may be enclosed between ribs. To obtain an estimate of this, we examined the lungs of two birds in which the viscera had been fixed with formalin *in situ*; the volume of lung tissue enclosed by ribs was between 28 and 32% of the total volume of the lung. The possibility that the embedding of ribs may support the lung and limit its compression during expiration is discussed on page 154.

Contemporary knowledge of the rather complex architecture of the bronchial system of the fowl's lung is founded mainly on the exhaustive researches of Campana (1875), Juillet (1912), Locy and Larsell (1916a, b) and Payne (1960). An impressive survey of the vast literature on the anatomy of the lungs and air sacs in a wide range of birds was compiled by Groebbels (1932). The field has been reviewed by King (1966, 1972).

A. BRONCHI

1. Primary Bronchus

The trachea divides at the syrinx into the left and right extrapulmonary primary bronchi, which continue through the lung as the left and right intrapulmonary primary bronchi (mesobronchi).

In resin casts from heavy breeds the cross-sectional area of the primary bronchus just caudal to the syrinx is typically about 15 mm^2 a value admittedly enlarged. Nevertheless, the combined cross-sectional area of the left and right primary bronchi together (about 30 mm^2) exceeds the cross-sectional area of the caudal end of the trachea (which is about 12·5 mm^2) by such a margin that the resistance to gas flow must be much less than in the trachea.

Many accounts of the avian lung have described a dilation of the first part of the intrapulmonary primary bronchus, at the origins of the four craniomedial secondary bronchi, and called it the vestibule. Such a dilation could have aerodynamic significance, and Hazelhoff (1951) incorporated it into his glass models of the lung. However, there is certainly no vestibule in the lung of the fowl (Juillet, 1912; Payne and King, 1959; Akester, 1960), and indeed it is doubtful whether it occurs in any species (see King, 1966). Each primary bronchus typically achieves its greatest cross-sectional area (about 23 mm^2) at a point outside the lung, about half-way between the syrinx and the first craniomedial secondary bronchus. Nor is Hazelhoff's guiding dam (a projection opposite the openings of the caudodorsal secondary bronchi) present in the primary bronchus. These and other anatomical objections to Hazelhoff's glass models were summarized by King and Payne (1960). These anatomical relationships have been established with enough certainty in only the fowl to enable the anatomical accuracy of aerodynamic models to be determined: in this species there is unfortunately virtually no anatomical basis for Hazelhoff's aerodynamic factors. The significance of efforts to substantiate their anatomical basis in other species (e.g. Duncker, 1968) cannot be assessed until further observations are available.

The typical anatomy of the primary bronchus is shown in Figs. 5a, 6a. In resin casts the cross-sectional area of each primary bronchus typically decreases more or less progressively from about 20 mm^2 at the first craniomedial

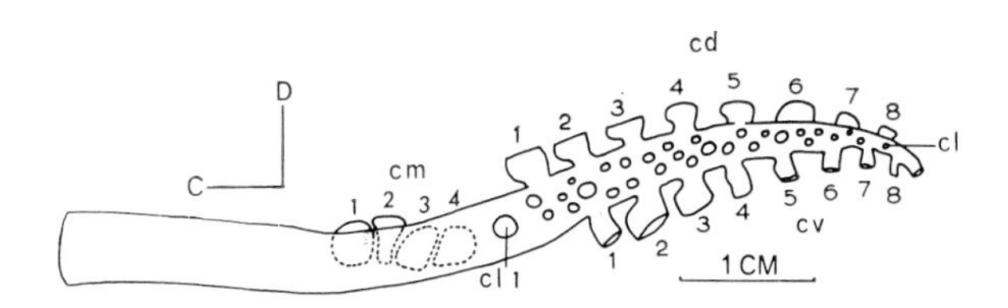

Fig. 5a. Lateral view of a typical primary bronchus of an adult fowl, left side, based on casts, dissections and radiographs. cm, craniomedial secondary bronchi; cd. caudodorsal secondary bronchi; cv, caudoventral secondary bronchi; cl, caudolateral secondary bronchi; D, dorsal; C, cranial. Reproduced with permission from King (1966).

secondary bronchus, to about 17 mm^2 between the fourth craniomedial and first caudodorsal secondary bronchi, to about 10 mm^2 at the third caudodorsal, and to about 2·5 mm^2 at the final opening into the abdominal air sac. The total volume of one primary bronchus from a typical resin cast from a heavy breed is about 1 ml.

The lumen is supported by C-shaped cartilages which occur in a continuous series from the syrinx to the last of the craniomedial secondary bronchi; a few more small pieces of cartilage are present as far as the beginning of the caudodorsal secondary bronchi. Otherwise the lung is devoid of cartilage.

The mucosa is thrown into longitudinal folds (Fig. 5b), which cannot be effaced even though there are numerous elastic fibres in the lamina propria. At the junction of the lamina propria with the muscle layer Cook (1970) has shown a close ultrastructural relationship between the elastic fibres and the muscle cells. Fibrils in the elastic fibres appear to be continuous with the surface membrane of the muscle cell.

In birds generally the mucosa is lined throughout by a ciliated epithelium

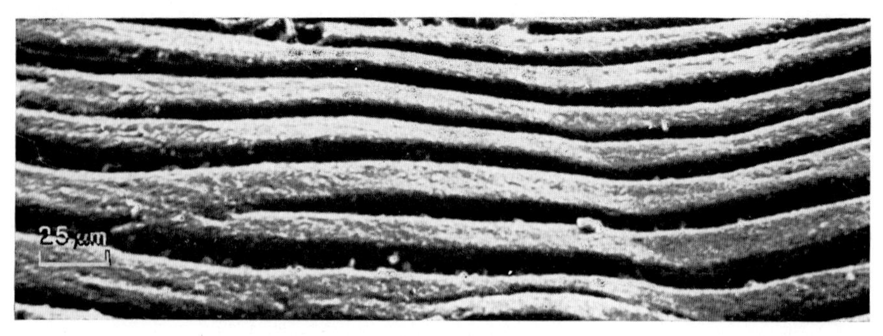

Fig. 5b. Scanning electromicrograph by R. D. Cook, Department of Veterinary Anatomy, University of Liverpool, of the interior of the primary bronchus showing the longitudinal folds. Cilia occur mainly on the folds. Mucous elements occur abundantly both on the folds and in the grooves. Adult White Leghorn male.

with mucous cells. In the fowl clusters of tall thin ciliated cells lie between simple mucous alveoli which do not enter the lamina propria. Isolated mucous cells are also present in the epithelium. In the roots of the larger secondary bronchi the epithelium remains the same for a short distance with a strong backing of elastic fibres, and then turns into simple squamous epithelium which extends throughout the tertiary bronchi. Thus only the primary bronchus and the roots of the larger secondary bronchi (and limited areas of the bronchi in the ostia of the air sacs) have a mucous lining. The absence of mucus from the vast majority of the airways greatly reduces the mucous carpet which, in mammals, acts as the first line of defence against infection by picking up and removing small particles. It might be expected that a second line of defence by a macrophage system, which in mammals functions more or less independently of the mucociliary apparatus (Kass *et al.*, 1966), might be proportionately better developed in the bird. However, Rigdon (1959) was unable to detect phagocytosis within the lumen of the airways of the duck and concluded that small particles which penetrate the epithelial lining of the airways either enter the lymphatics directly or are phagocytozed first in the lamina propria. The air sacs seem to be the main site of these methods of removal. On the other hand Nowell *et al.* (1970), using the scanning EM, observed roughened spherical particles about 5–10 μm in diameter which were widely distributed throughout the tertiary bronchi and atria of the quail, and considered it possible that these could be pulmonary macrophages (Fig. 7e).

2. *Secondary Bronchi*

Each primary bronchus gives rise to four series of secondary bronchi as in Figs. 5a and 6a, b. Two of these series, the craniomedial and caudodorsal secondary bronchi (ventrobronchi or entobronchi, and dorsobronchi or ectobronchi), are closely integrated through their anastomosing tertiary bronchi into one great functional unit comprising about two-thirds of the lung. There are two lesser series, the caudoventral and caudolateral secondary bronchi (also known as lateral and dorsal bronchi respectively). The orientations suggested by these terms (*not* by the older terms, which are in brackets) are correct for the fowl and probably for birds generally (King, 1966). The numbers and positions of these secondary bronchi are shown in Figs. 5a and 6a, b.

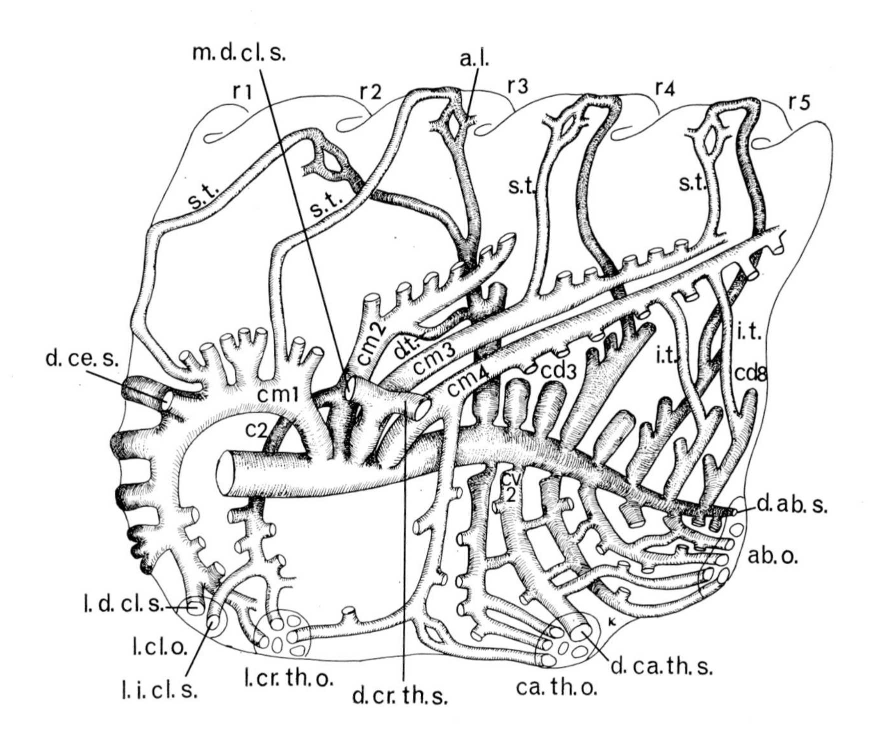

Fig. 6a. Ventromedial view of the right lung of an adult chicken drawn from casts and dissections, to show the primary and main secondary bronchi, examples of tertiary bronchi, and the connexions of the air sacs except that a complex group of indirect connexions of the clavicular sac to tertiary bronchi near the primary bronchus is not shown. The term "ostium" means the general area of connexion between sac and lung. a.l., anastomotic line caused by embryonic fusion of the primordial tertiary bronchi; ab.o., ostium of abdominal sac; ca.th.o., ostium of caudal thoracic sac; cd, caudodorsal secondary bronchi; cm, craniomedial secondary bronchi; c2, circumflex branch of cm2; cv, caudoventral secondary bronchi; d.ab.s., direct connexion of abdominal sac; d.ca.th.s., direct connexion of caudal thoracic sac; d.ce.s., direct connection of cervical sac; d.cr.th.s., direct connexion of cranial thoracic sac; d.t., a deep tertiary bronchus; i.t., tertiary bronchi of intermediate depth; l.cl.o., lateral ostium of clavicular sac; l.cr.th.o., lateral ostium of cranial thoracic sac, comprising indirect connexions only; l.d.cl.s., lateral direct connexion of clavicular sac; l.i.cl.s., lateral indirect connexion of clavicular sac; m.d.cl.s., medial direct connexion to clavicular sac; r 1–5, five impressions of vertebral ribs 2 to 6; s.t., superficial tertiary bronchi. Cranial is to the left and dorsal towards the top of the drawing. Reproduced with permission from King (1966).

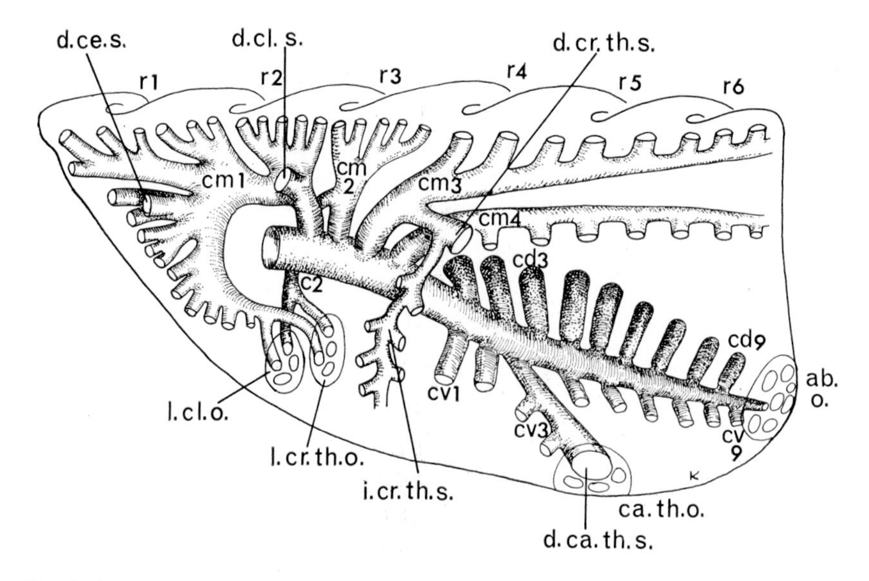

Fig. 6b. Semischematic ventromedial view of the right lung of an adult duck (*Anas platyrhynchos*), reconstructed from the illustrations and text of Juillet (1912) and Vos (1934). The bronchial connexions of the lateral ostium of the cranial thoracic sac (l.cr.th.o.) are uncertain; d.cl.s., direct connexion of clavicular sac; i.cr.th.s., indirect connexion of the cranial thoracic sac in the form of a special tube presumably arising from l.cr.th.o.; r 1–6, grooves for ribs 1 to 6. Other abbreviations as in Fig. 6a. Reproduced with permission from King (1966).

All the secondary bronchi, and specially the caudodorsals and caudoventrals, tend to be more or less constricted at their origins where some of them also curve a short distance cranially and then turn caudally as in Figs. 5a, 6a. Nevertheless the diameters of the openings are substantial. For instance in a typical resin cast from a heavy breed the cross-sectional area of the opening of the first craniomedial secondary bronchus is about 20 mm^2. Since the area of the primary bronchus itself at this point is also about 20 mm^2 it is apparent that the resistance to air flow should decrease at this point. Similar relationships (i.e. increasing total cross-sectional areas) exist along the primary bronchus at least until about half-way along the series of caudodorsal and caudoventral secondary bronchi. Thus at the level of the fourth caudodorsal and third caudoventral secondary bronchi (see Figs. 5a, 6a) the area of the primary bronchus is about 8 mm^2 and the openings of these two secondary bronchi are about 6 and 3 mm^2 respectively, giving the airway a total cross-sectional area of about 17 mm^2. It would appear that the resistance to flow should fall throughout most of the length of the primary bronchus. It is not easy to analyse the resistance of the various subsequent generations of bronchi, because their numerous interconnections make it difficult to measure diameter and length.

On the other hand the tendency for narrowing and bending of the origins of the secondary bronchi may have important aerodynamic effects on resistance. Hazelhoff (1951) placed the origins of the caudodorsal secondary bronchi along the most convex wall of the curved primary bronchus, and this relationship has aerodynamic consequences in his glass models. In fact it is the numerous and mainly small caudolateral series that arise from the most convex part of the primary bronchus.

3. *Tertiary Bronchi*

The secondary bronchi give rise to numerous anastomosing tertiary bronchi (parabronchi); none is blind-ending. Many join end to end to form long curved hoop-like bronchial circuits.

Payne (1960) showed the entire dorsal and cranial part of the lung to be composed of about 150 to 200 curved tertiary bronchi, running layer upon layer as a series of essentially parallel bronchial circuits up to 4 cm long between the craniomedial and caudodorsal secondary bronchi (Fig. 6a). These tubes comprise about two-thirds of the whole lung. The more superficial ones, the longest and greatest in diameter, pass through the region of the lung supported by the embedded ribs (see page 102). On the lateral surface of the lung there is a rather conspicuous area of relatively fine branching and anastomosing short tertiary bronchi, arising mainly from the caudolateral secondary bronchi. This is the "reseau anastomotique" of Campana (1875) and Locy and Larsell (1916a). The remaining caudoventral third of the lung is composed of anastomosing networks of tertiary bronchi joining the caudoventral and caudolateral secondary bronchi, both to each other and to the other two groups of secondary bronchi. In fact, each group of secondary bronchi is connected, by the tertiary bronchi at the periphery of its territory, to all of the other three groups of secondary bronchi.

A knowledge of the number, airway calibre, and average length of the tertiary bronchi has been required for calculations of gaseous diffusion between the tertiary bronchi and air capillaries (Zeuthen, 1942). Unfortunately, it is impossible to give exact figures for number and length because the tertiary bronchi in the networks of the caudoventral third of the lung are short, branching, and anastomosing, and therefore have no clear beginning or end. However, as a working basis it would be reasonable to assume a total number of about 400 to 500 tertiary bronchi in the whole lung, with an average length of about 20 mm. In resin casts from heavy breeds Payne (1960) found the length of most of the superficial and longest of the tertiaries of the craniomedial-caudodorsal system to be between 3 cm and 4 cm; the deepest and shortest were about 1 cm long. Measurements of the diameter of the airway depend on how the lung is prepared, and on the position in the lung, diameters being greatest in the superficial tertiary bronchi of the craniomedial-caudodorsal system and much smaller in the ventrolateral region of the lung. However, in females of heavy breeds which have been fixed intratracheally

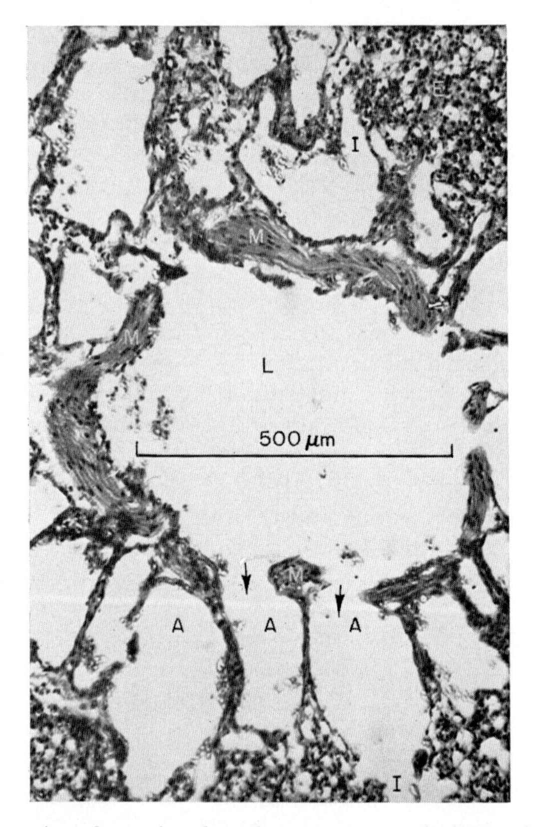

Fig. 7a. Paraffin section of a tertiary bronchus cut transversely from a 5-week-old White Leghorn hybrid. Fixation *in situ* by intratracheal perfusion with 5% formalin, staining by haematoxylin and chromatrope 2R. Since this is an immature bird the diameter of the airway has not reached its full size. L, lumen of tertiary bronchus; A, atria; I, infundibula; E, exchange area; M, bundles of bronchial muscle; arrows, atrial openings bounded by bronchial muscle. Reproduced with permission from King and Cowie (1969).

under low pressure and examined histologically (Fig. 7a) or with a dissecting miscroscope (Figs. 7b, c), the diameter of the airway of the tertiary bronchi (i.e. excluding the cavities of the atria) is typically about 1 mm. In Payne's resin casts from heavy breeds, the mean diameter of the superficial tertiary bronchi of the craniomedial-caudodorsal system was 1·86 mm ($\pm$0·23) in the male and 1·74 mm ($\pm$0·16) in the female; the diameter of the deepest and shortest tertiary bronchi of this system was 1·70 mm ($\pm$0·28) in the male and 1·39 mm ($\pm$0·16) in the female. These figures probably represent the maximum diameter of tertiary bronchi in a fully distended lung. Various other estimates of diameter are available. For instance Campana (1875) estimated the diameter in smaller breeds at about 1 mm. MacDonald (1970) put it at

0·48 mm (range 0·33 to 0·77 mm) in adult Brown Leghorn females which were fixed by vascular perfusion.

The airway lining of the tertiary bronchi is a thin squamous epithelium supported by a meshwork of elastic fibres. The elastic fibres appear to be closely associated with the surface membrane of smooth muscle cells in the bundles of bronchial muscle (Cook, 1970). Sometimes the lining epithelial cells appear rounded and have then been called cuboidal (e.g. Malewitz and Calhoun, 1958; MacDonald, 1970). However, this rounding is probably caused by muscular or elastic contraction within the lamina propria.

Each tertiary bronchus is surrounded by a roughly hexagonal zone of exchange area, enclosed by weak but clear connective tissue septa. These are distinct, but are incomplete in places allowing the air capillaries of adjacent tertiary bronchi to anastomose freely. The zone enclosed by the septa, i.e. a tertiary bronchus and its exchange area, is generally known as a lobule. The septa are therefore interlobular.

4. Bronchial Muscle

The smooth muscle of the primary bronchus in the fowl, and in birds generally, forms a layer immediately beneath the mucosa, the orientation of the muscle fibres being variously described as circular (Juillet, 1912; Groebbels, 1932), longitudinal (Trautmann *et al.*, 1957), or a combination of these (King, 1966). With the light microscope this sheet appears as a more or less continuous layer. With the electron microscope Cook (1970) has shown that the smooth muscle cells are enclosed by connective tissue into fascicles. Each fascicle appears to be an individual effector bundle. Close neuromuscular junctions with a minimum gap of 25 nm (Cook and King, 1970) can be found quite readily, without being as common as in the rat vas deferens where every cell is innervated directly by one or more close neuromuscular junctions. (Burnstock, 1970, defined "close" as between 20 and 50 nm). Cook (1970) has found several varieties of specialized myo-myal relationships ranging from areas of close contact including "bridges" and "intrusions", to areas of close apposition including nexi (for discussion of these terms see Burnstock, 1970). At least some of these, notably nexi, are generally accepted as sites for electrical coupling and the propagation of electrical activity through the muscle (Burnstock, 1970).

No quantitative data are available for the muscle of the avian primary bronchus, but a subjective impression of Cook's findings is that these specialized regions are about as common as in mammalian smooth muscle where they comprise between 3 and 6% of the total surface area of the cell (see Burnstock, 1970). Apart from the specialized regions, Cook has found the muscle cells to be separated over most of their surfaces by a gap of 20 to 250 nm occupied by the basement membrane of each cell and occasionally collagen fibres: in mammals the general intercellular gap is typically

50 to 80 nm, but reaches 300 nm in some tissues (see Burnstock, 1970). The arrangement and morphology of the muscle cells in the well-developed muscle layer at the roots of the secondary bronchi in the bird are essentially the same as in the muscle layer of the primary bronchus itself (Cook, 1970).

The smooth muscle of the tertiary bronchi is arranged in a network of large spiral bundles (bands) and short atrial bundles (Figs. 7a, b, c, 8), the

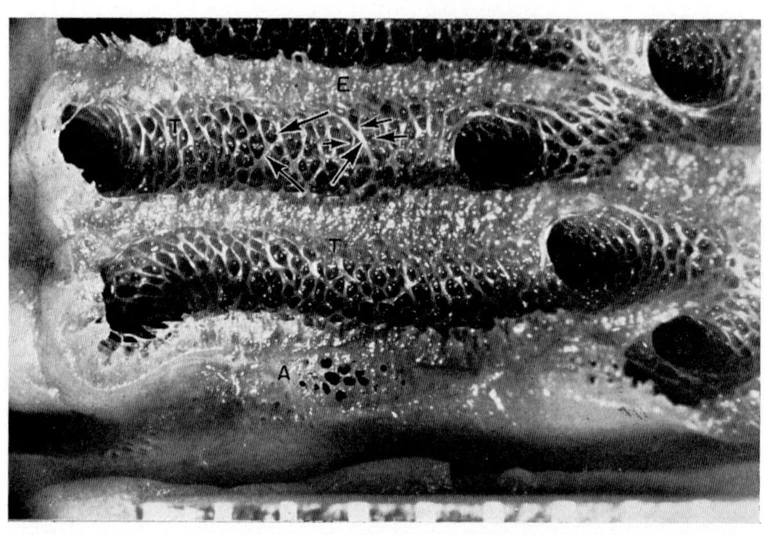

Fig. 7b. Dissected lung of an adult White Leghorn hybrid fixed *in situ* by intratracheal perfusion with 5% formalin. Two tertiary bronchi (T) have been opened longitudinally, but turn into the depth of the field on the left. Four more open on the right. The bundles of bronchial muscle form large spiral bands (three large arrows), and small irregular atrial bundles (three small arrows) which frame the atrial openings by joining one large bundle to another. The pockets between the muscle bundles are atria, a group of which have been cut tangentially at A. The atria lead into the exchange areas, E. Scale in mm. Reproduced with permission from King and Cowie (1969).

latter framing the atrial openings (King and Cowie, 1969). Cook (1970) observed that the smooth muscle cells in such a bundle are closely packed together in one compact mass. The whole bundle may be regarded as a single effector unit. Close neuromuscular junctions are scarce but they do occur, the minimum gap being 18 nm (Cook and King, 1970). The gap between the surfaces of adjacent muscle cells of the tertiary bronchi is consistently narrow; generally it is between 15 and 50 nm, and is often maintained at about 25 nm over quite extensive areas, the adjacent surface membranes then being parallel and separated only by fused basement membrane. Specialized areas of close contact or close apposition are fairly numerous. Many of these appear to have an intermembrane gap of 2 to 4 nm, but because of problems of preparation the detailed morphology of these areas is

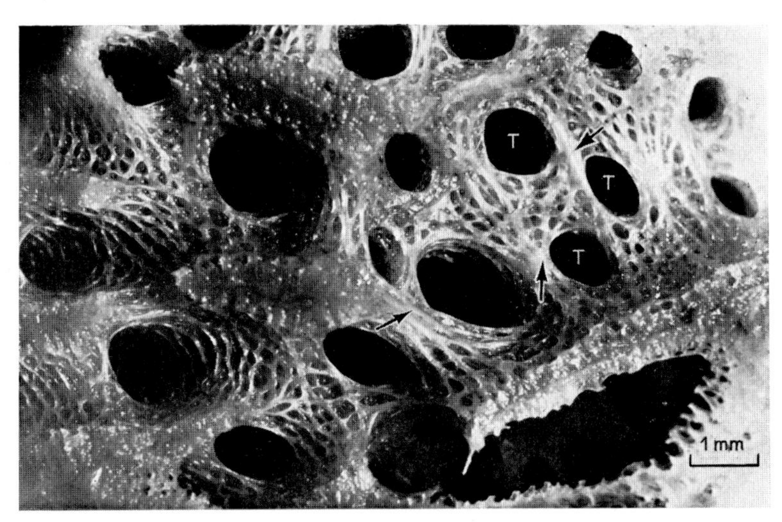

Fig. 7c. Dissected lung prepared as in Fig. 7b, showing the interior of a secondary bronchus. The openings of the many tertiary bronchi (T), are embraced by massive bundles of bronchial muscle (arrows). Reproduced with permission from King and Cowie (1969).

still uncertain, and Cook is unable as yet to determine whether or not they conform to the "gap junctions" of Uehara and Burnstock (1970).

Bundles of axons occur in the smooth muscle of both the primary and the tertiary bronchi, but are about twenty times more numerous in the former than in the latter (Cook and King, 1970). On the whole, the large gaps between the muscle cells of the primary bronchus, and the greater profusion of its innervation, suggest that the muscle of the primary bronchus might be under more precise neural control than that of the tertiary bronchi (Cook and King, 1970). The muscle of the primary bronchus might fit model B of Burnstock's (1970) scheme for the autonomic innervation of smooth muscle which envisages that only a moderate number (20 to 50%) of the muscle cells have a direct innervation by one or more close neuromuscular junctions. These innervated muscle cells are joined to the remainder by intercellular couplings (nexi, "gap junctions") which allow propagation of electrical activity between adjacent cells. Such smooth muscle should be capable of fairly fast contraction, but with some capacity for complex local regulation. The muscle of the tertiary bronchus might fit Burnstock's model C where only a few muscle cells have close neuromuscular junctions, but many have highly developed intercellular couplings. Such muscles are usually spontaneously active. Burnstock regarded model C as "geared for complex, slow, graded local tension changes". The muscle of the tertiary bronchus agrees with model C in having sparse neuromuscular junctions, spontaneous rhythmicity (see immediately below), and quite numerous specialized myo-myal relationships in which (subject to further investigation) the intermembrane gap may be as narrow as in "gap junctions". Since some of the muscle cells of both the primary and the tertiary bronchi are closely associated or even continuous with elastic fibres (see pages 103 and 109), a complex integration of smooth muscle action and elastic forces appears possible. (See page 114 for consideration of the functions of the elastic fibres, and page 161 for discussion of the possible role of the smooth muscle of the primary, secondary and tertiary bronchi in regulating airway resistance.)

The muscle of the secondary bronchial trunks forms a network of muscle bundles (Fig. 7c), the gross arrangement being essentially the same as in the tertiary bronchi except that the bundles are generally more massive; only the very thin medial (superficial) walls at the beginning of the four great craniomedial secondary bronchi are devoid of the network. Histologically the orifices of the secondary bronchi from the primary bronchus often show a marked thickening of the smooth muscle layer which Cook (1970) regarded as sphincter-like.

In the tertiary bronchi each muscle bundle is located on the internal surface, being separated from the airway only by simple squamous epithelium and a minimum of connective tissue. On the lateral surface of the lung, particularly between and within the grooves for the vertebral ribs, there are small

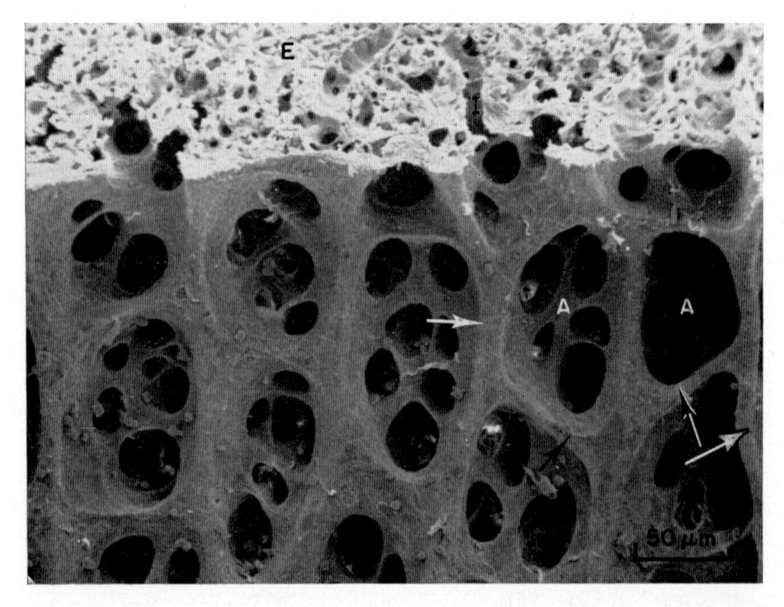

Fig. 7d. Scanning electronmicrograph by Janice A. Nowell, Department of Anatomy, University of California, Davis, California, of part of a tertiary bronchus of the quail (*Coturnix coturnix japonica*). A,A, two atria; I, an infundibulum in LS; E, exchange area. White arrows, two large spiral muscle bundles; black arrows, two short atrial muscle bundles. The depth of focus achieved by the scanning EM causes the atria to look shallow.

areas where the surface of the lung is transparent, allowing the bands of bronchial muscle to be observed *in vivo* (King and Cowie, 1969).

Anatomically these spiral bands appear able to constrict the airway calibre of secondary and tertiary bronchi, and perhaps more importantly the openings into the atria; their ability to act in these ways has been investigated by King and Cowie (1969). Strips and slices of lung *in vitro*, and surface areas of the lung *in vivo*, were contracted by cholinergic drugs, this response being blocked by atropine. Sometimes complete occlusion of tertiary bronchi was observed *in vitro*, and of atria on the surface of the lung *in vivo*. Adrenergic drugs usually induced relaxation; this may be by direct action on the smooth muscle or, possibly via adrenergic nerves, the presence of which is disputed (see page 128).

No evidence of vagal tone was found by Fedde *et al.* (1961) either by reversible blocking or section of the vagus during a unidirectional gas flow. An inability to induce local relaxation of the bronchial muscle by applying atropine to the transparent areas *in vivo* also indicates absence of vagal tone; on the other hand relaxation after adrenergic drugs, both *in vitro* and *in vivo* confirms the presence of an intrinsic smooth muscle tone (King and Cowie, 1969). Spontaneous rhythmic contractions were observed *in vitro* by Lewis (1924), and both *in vitro* and *in vivo* by King and Cowie (1969) following the application of excitatory drugs, the contractions being arrested by atropine.

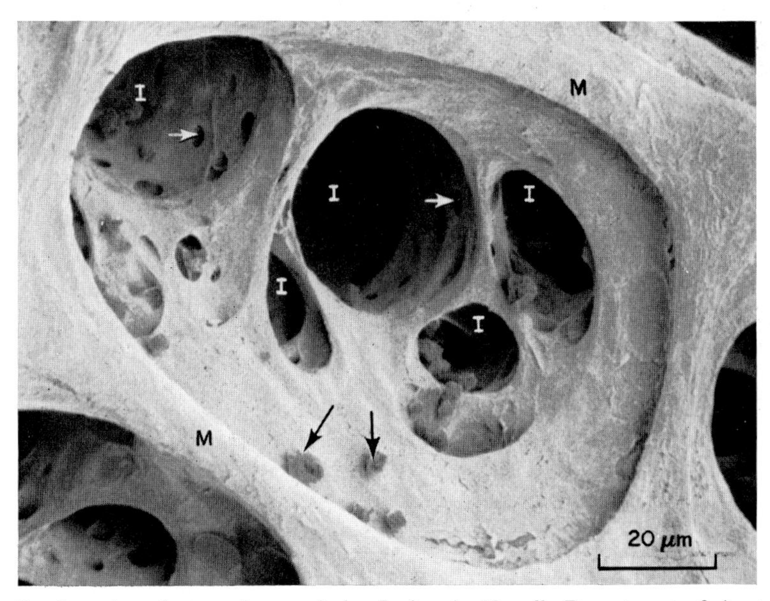

Fig. 7e. Scanning electronmicrograph by Janice A. Nowell, Department of Anatomy, University of California, Davis, California, of one atrium of the quail (*Coturnix coturnix japonica*). M, bundles of bronchial muscle; I, infundibula. White arrows, openings of air capillaries. Black arrows, roughened particles possibly pulmonary macrophages.

B. ATRIA AND EXCHANGE AREA

1. *Atria*

The wall of each tertiary bronchus is pierced by numerous openings (Figs. 7a, b, d) which lead into the atria. The atria are roughly spherical chambers between 100 and 200 μm in diameter (King and Cowie, 1969). Their walls are lined by squamous epithelium supported by an extensive network of elastic fibres (Fig. 8).

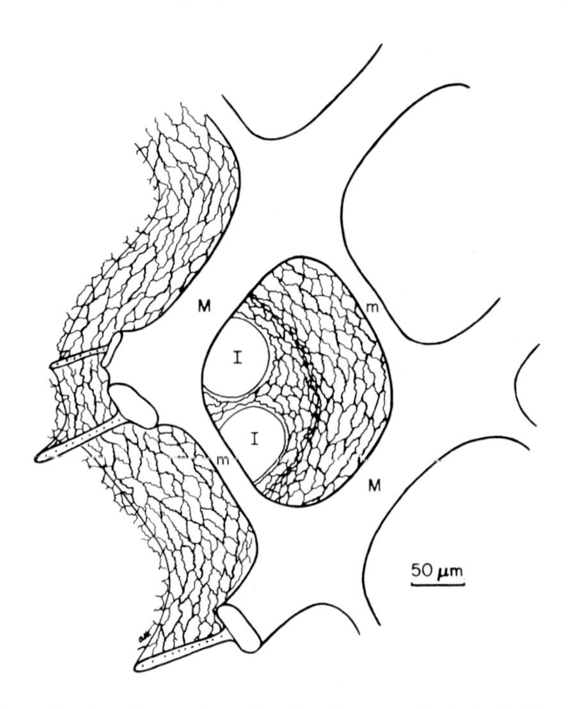

Fig. 8. Semischematic view of an atrium, showing the network of elastic fibres in its walls and floor. M, large spiral bundles of bronchial muscle; m, short atrial muscle bundles joining the two large bundles; I, infundibula.

The presence of elastic fibres suggests that the atria are an essentially distensible and mobile component of the lung. In particular the internal rim of a tertiary bronchus, i.e. the lining of the airway, should be mobile (King *et al.*, 1967), allowing the bronchial muscle to constrict the lumen of the bronchus in general and the openings of the atria in particular. The elastic fibres could act as an energy-storing system, tending to keep the airway of the tertiary bronchus open in opposition to the intrinsic tone of the bronchial muscle. Relaxation of the muscle would be followed by dilation of the airway of the tertiary bronchus (Fig. 9). The elastic atria may also act like a spring, balancing the tone of the bronchial muscle against the surface tension

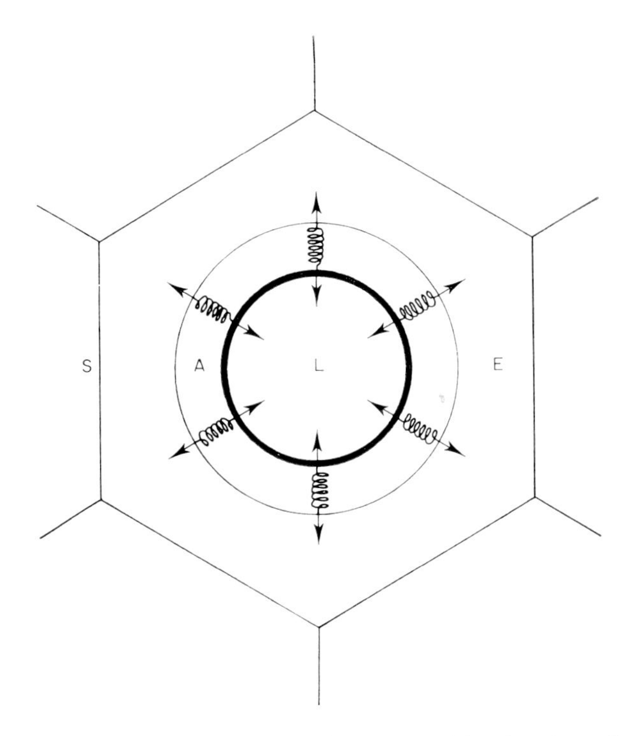

Fig. 9. Diagram of the forces which may be acting on the atrial region of a tertiary bronchus. L, lumen of the tertiary bronchus; A, region of the atria; E, exchange area; S, interlobular septa. The arrows pointing towards the lumen represent the tone of the bronchial muscle (the heavy black ring). The arrows pointing towards the exchange area represent the forces of surface tension in the air capillaries. The springs in the atrial region represent the elastic fibres in the walls of the atria balancing these forces.

of the air capillaries (see page 118). The complex free afferent nerve endings which occur in the atrial region (see page 131) may be signalling in response to atrial movement.

The floor of a typical atrium continues into two or three smaller and much less well-defined cavities, the infundibula (Figs. 7a, d, e).

2. Air Capillaries

Each infundibulum leads into a complex network of fine anastomosing tubules, the air capillaries, which are intimately interlocked with blood capillaries (Fig. 10a) to form the exchange area. There are reticular fibres in the exchange area (Ogawa, 1920), but virtually no elastic fibres (Ogawa, 1920; King *et al.*, 1967), suggesting that it is a relatively immobile part of the lung.

As Petrík (1967) and MacDonald (1970) pointed out, the architectural relationship between the air capillaries and blood capillaries has received relatively little attention. The blood capillaries are the more abundant (Baer, 1896; Petrík and Riedel, 1968b; R. D. Cook, personal communication) and MacDonald (1970) estimated their diameter to be 3·7 μm (range 2·8 to 4·8 μm) following perfusion of the pulmonary artery.

The air capillaries are usually said to be relatively wider (Bargmann and Knoop, 1961; MacDonald, 1970; R. D. Cook, personal communication). Their mean diameter is about 10 μm (Marcus, 1937; Nagaishi *et al.*, 1964; Akester, 1970) but is influenced by the technique of preparation like the mammalian alveolus (see Krahl, 1964); the reported range of diameter of the human alveolus is from 50 to 600 μm (Weibel, 1964).

3. *The Blood-Gas Barrier*

The avian blood-gas barrier is exceedingly thin (Fig. 10b). However, in spite of doubts about the absence of an epithelial lining (Bargmann, 1936; Schulz, 1959; Salt and Zeuthen, 1960; Nagaishi *et al.*, 1964) the continuity of these cells has now been firmly established in several birds including the domesticated ones (e.g. Groodt *et al.*, 1960; Bargmann and Knoop, 1961; Tyler *et al.*, 1961; Policard *et al.*, 1962; Schulz, 1962; Lambson and Cohn, 1968; Fujiwara *et al.*, 1970). The barrier consists of three components, the epithelial cell of the air capillary, a common basement membrane, and the endothelial cell of the blood capillary. On the grounds that it is the only anatomical component of the blood-gas barrier capable of bearing the brunt of the pressure differences between the pulmonary capillary vessels and the alveolar gas phase, Chinard (1966) regarded the basement membrane as the permeability barrier.

In the fowl R. D. Cook (personal communication) found the total thickness of the barrier to be frequently between 0·1 and 0·2 μm, the minimum being 0·089 μm, but measurements of 0·5 μm and upwards were not uncommon, and the arithmetic mean was 0·35 μm—thicker than that of the pigeon (0·10 to 0·14 μm) (Schulz, 1962) but considerably thinner than that of the rat (1·4 μm) (Weibel, 1966). The endothelial cell accounts for most of this thickness. There is no evidence for pores or fenestrations in the endothelial lining. The epithelial cell is exceedingly thin, being typically about 37 nm with a range of 19 nm to 55 nm. Nuclei of epithelial cells and junctions between adjacent epithelial cells are seldom seen (e.g. Fig. 10a); this indicates the very extensive area covered by each individual cell, a point noted by Okada *et al.* (1965).

4. *Surfactant*

Pulmonary surfactant is of interest in various aspects of the physiology and pathology of the vertebrate lung. It keeps the small cavities open and limits transudation (Pattle, 1965). It has also been cited as a possible agent in the clearance of foreign material from mammalian alveoli and as a mediator in

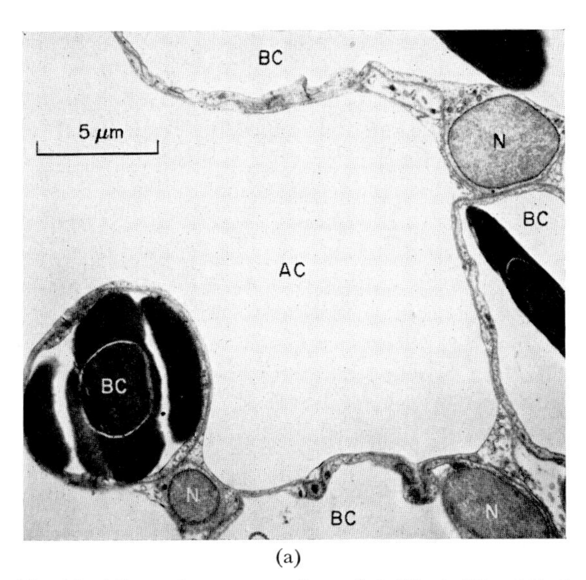

(a)

Fig. 10a. The exchange area of an adult Rhode Island Red male, fixed by perfusion of the pulmonary artery with glutaraldehyde. An air capillary (AC) is surrounded by blood capillaries (BC). The blood-gas barrier varies in thickness. Three nuclei (N) of endothelial cells are present, but no nuclei of epithelial cells are visible.

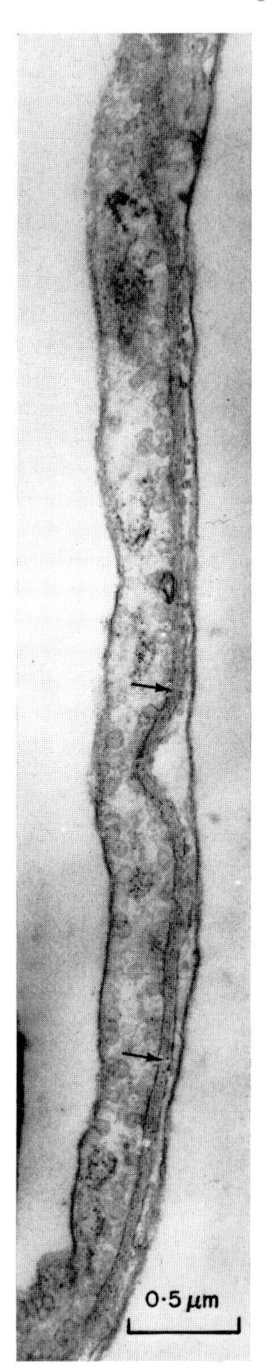

(b)

Fig. 10b. The blood-gas barrier in an adult Rhode Island Red male, fixed by perfusion of the pulmonary artery with glutaraldehyde. In this field the thickness of the barrier varies from about 0·3 μm (which is about average) at the bottom, to about 0·5 μm at the top, but the barrier is quite often reduced to 0·1–0·2 μm or less. The endothelial cell (on the left) typically has pinocytotic vesicles and is much thicker than the epithelial cell of the air capillary (on the right). Between the two cells is a common basement membrane (arrows). The unit membrane of the epithelial cell (the surface on the right of the tissue) shows signs of thickening, which may be a single lamina of surfactant. Electronmicrographs for Fig. 10 by R. D. Cook, Department of Veterinary Anatomy, University of Liverpool.

the exchange of respiratory gases (see Mendenhall and Sun, 1964; Chinard, 1966).

There has been considerable argument as to whether the avian lung possesses a surfactant. Pattle (1958) believed that he had demonstrated one in the pigeon's lung, but Miller and Bondurant (1961) and Klaus *et al.* (1962) concluded that distinctive surface activity was absent in the lungs of the fowl and pigeon. In 1963 Pattle and Hopkinson reaffirmed Pattle's original view. Subsequently a pulmonary phospholipid, chemically similar to the surfactant of mammals, was demonstrated in the lung of the fowl by Harlan *et al.* (1966).

The effects of surface tension are likely to be exerted at two points in the avian lung, the air capillaries and the atria. Pattle (1965) has suggested that the main function of surfactant in birds is to prevent excessive transudation. Certainly this would be of paramount importance in the air capillaries, where accumulation of fluid would depreciate the value of the exceptionally thin blood-gas barrier.

Pattle (1965) also suggested that surfactant is of relatively little importance in preventing collapse, since the air capillaries "are embedded in more or less rigid tissue". Possibly he was referring to the deep embedding of the ribs into the lung and the adhesion between the visceral and parietal pleurae, both of which may help to prevent the "external" collapse of the lung. However, the air capillaries within a lobule could collapse by pulling their atria outwards towards the interlobular septa. This outwards force on the atria must normally be opposed by the tone of the bronchial muscle aided by elastic fibres entwined among the muscle cells (see page 114). By limiting surface tension within the air capillaries, surfactant could help the muscle to balance these forces. On the other hand, a *thick* layer of surfactant would add significantly to the blood-gas barrier, and therefore only the minimal effective layer could be tolerated. The single surface lamina which is believed to line the air capillaries (see the next section) is consistent with these requirements.

The main function of the atria appears to be to provide elastic units which enable the bronchial muscle to regulate the access of air to the exchange tissue (see page 161). Being roughly spherical cavities with diameters like mammalian alveoli, the atria are likely to experience forces of surface tension comparable to those in the latter. These forces must be overcome by the bronchial muscle whenever the airway of the tertiary bronchus is being constricted. Abundant surfactant within the atria would therefore be an advantage, and the occurrence of many complex laminated structures within the atria (see the next section) is again consistent with these requirements.

5. Osmiophilic Laminae

Surfactant in the vertebrate lung appears to be closely related to osmiophilic laminated structures which are associated with cells lining certain parts of the lung (Nagaishi *et al.*, 1964). In birds they are much more extensive and numerous in the atria than anywhere else (Bargmann and Knoop, 1961; Tyler and Pangborn, 1964; Okada *et al.*, 1965; Lambson and Cohn, 1968; Petrík and Riedel, 1968a, b; Akester, 1970; Fujiwara *et al.*, 1970). It therefore seems likely that surfactant is more abundant in the atria than elsewhere. On the other hand the laminated structures and elastic fibres are minimal in the exchange area, suggesting that this region is relatively immobile. Laminated structures are absent in the upper respiratory tract (Petrík and Riedel, 1968b).

(*a*) *Varieties of laminated structures.* There seem to be at least four main varieties of laminated structures:

(i) Laminated intracellular inclusions. These occur within the epithelial cells lining the atria, and possibly those lining the tertiary bronchi, as in Fig. 11a (Tyler and Pangborn, 1964; Okada *et al.*, 1965; Petrík and Riedel, 1968a; Lambson and Cohn, 1968; Akester, 1970). Some reach the surface of the cell (Tyler and Pangborn, 1964; Akester, 1970), and these may be the pits visible with the scanning EM (Nowell *et al.*, 1970).

(ii) Multiple surface laminae. These lie on the surface of the atria as in Fig. 11b (Tyler and Pangborn, 1964; Lambson and Cohn, 1968), but have also been seen though rarely on the epithelial cells lining the air capillaries in young chicks (Petrík and Riedel, 1968a) and in the turkey (Fujiwara *et al.*, 1970). Nowell *et al.* (1970) suggested that the roughened surface of the atria, which they saw at higher magnifications with the scanning electron microscope, could be due to variations in the thickness of these laminae.

(iii) Intracellular invaginations of surface laminae. These comprise a few (typically 3 or 4) loosely-packed laminae (Fig. 11c) within the epithelial cells of the air capillaries (Bargmann and Knoop, 1961; Petrík and Riedel, 1968a, b).

(iv) Single surface lamina. This consists of a single lamina which lies very close, but externally to the unit membrane of the epithelial cell lining the air capillaries (Fig. 11d). It is, however, difficult to demonstrate (R. D. Cook, personal communication), but is believed to form a continuous film over the whole free surface of the air capillaries in the fowl (Petrík and Riedel, 1968b) and in the goose (Lambson and Cohn, 1968).

(*b*) *The source of the laminae.* Chemical, physicochemical and EM studies suggest that mammalian surfactant exists as a complex lamellar unit, each individual lamina being closely similar to the unit membrane of a cell (Mendenhall and Sun, 1964). The evidence of Petrík and Riedel (1968b) and Lambson and Cohn (1968) suggests that each individual lamina in the laminated structures of the avian lung consists of three layers (each being 3 nm to 5 nm thick) and resembles the unit membrane of a cell except for being relatively thicker and more electron dense. The lamina is probably lipoprotein. Therefore it is likely that surfactant and laminated structures are the same thing.

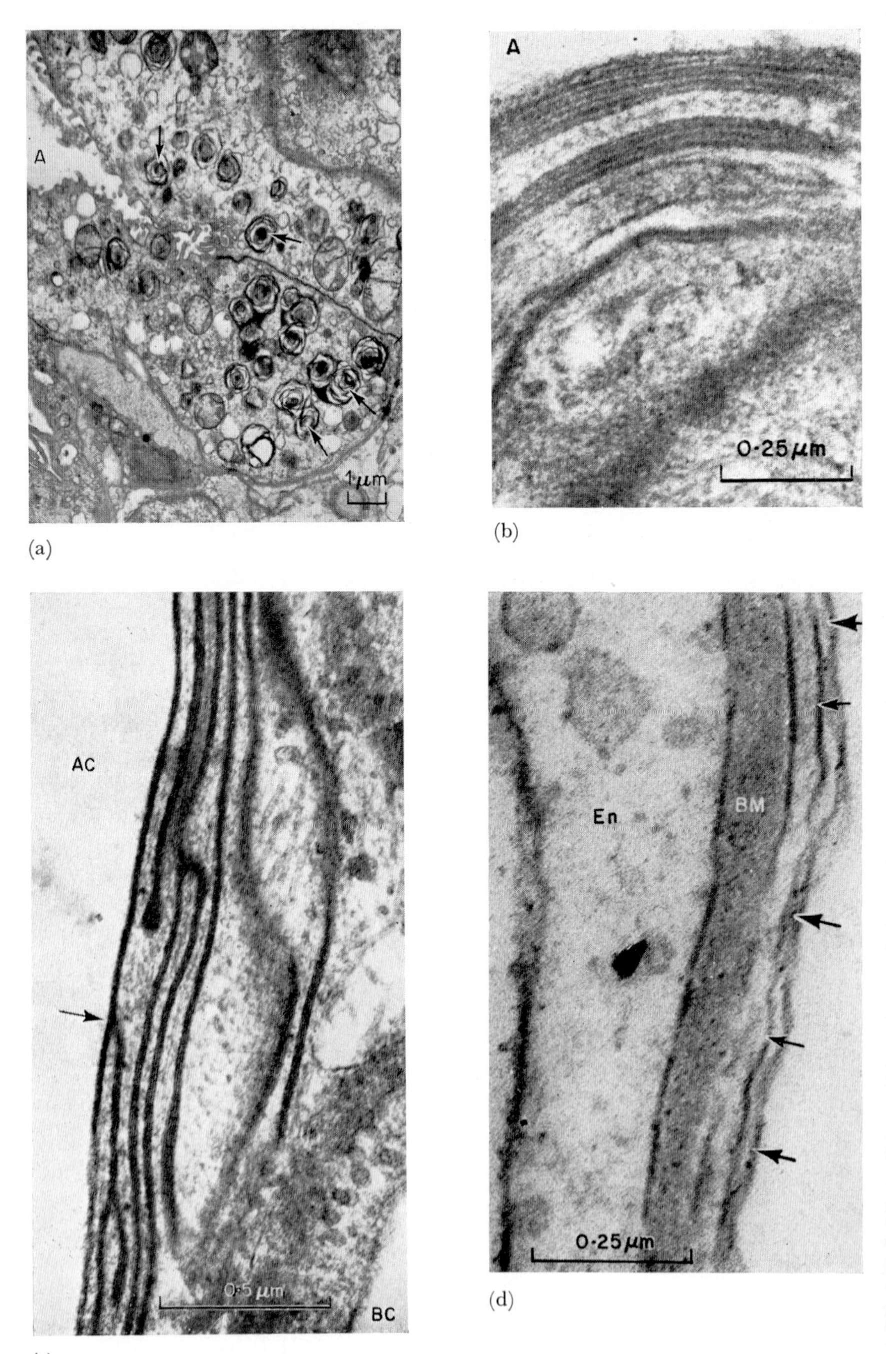

(a)

(b)

(c)

(d)

The source of the laminated structures is unknown. The earlier idea that they might arise from transformed mitochondria of alveolar epithelial cells (Klaus *et al.*, 1962) seems not to be favoured by those studying avian material. Tyler and Pangborn (1964) thought that the intracellular inclusions might arise from a penetration by surface laminae into the interior of the cell, while Petrík and Riedel (1968b) suggested that the single lamina is a merocrine secretion of the cells lining the air capillaries. Lambson and Cohn (1968) also believed that the lamina is formed inside the cell, and gave evidence that cisternae of the rough-surfaced endoplasmic reticulum and small vesicles of the Golgi apparatus may both be involved. Histochemical studies by Tyler and Pearse (1966) show that the atrial cells possess the requisite metabolic machinery to produce the quantities of phospholipids which are found there in the form of the laminated structures.

C. INNERVATION

1. Macroscopic Nerve Supply

It is generally agreed that the nerves of the avian lung arise mainly from the vagus, with a contribution from the thoracic sympathetic system. The anatomy of these nerves has been discussed by Cords (1904), Kaupp (1918), Hsieh (1951), Van Matre (1957), Watanabe (1960), Fedde *et al.* (1963), Peterson (1970) and McLelland (1970). The most topographically informative accounts are those of Hsieh (1951) and particularly McLelland (1970), and on these the following account is primarily based.

(*a*) *Vagal branches to the lung.* In their craniocaudal order of origin the thoracic branches of the vagus are: branches to the thyroid, parathyroid and ultimobranchial glands; the cranial cardiac nerves; the recurrent nerve; often the pulmono-oesophageal nerve; pulmonary nerves to the pulmonary plexus; the caudal cardiac nerves. The vagus supplies from three to seven branches to each lung. Figs. 12a, b show typical patterns. The first pulmonary nerve (P1) arises from the pulmono-oesophageal nerve, while the others (P2–6 in Fig. 12b) arise directly from the vagus.

Fig. 11a. Numerous laminated intracellular inclusions (arrows) in epithelial cells lining an atrium (A). Ten-week-old fowl; glutaraldehyde fixation.

Fig. 11b. Multiple surface laminae lying on the surface of an epithelial cell lining an atrium (A). Adult Rhode Island Red male, glutaraldehyde fixation.

Fig. 11c. Intracellular invaginations of surface laminae (arrow) within the epithelial cell of an air capillary (AC). BC, blood capillary. Adult White Leghorn male, glutaraldehyde fixation.

Fig. 11d. A single surface lamina (three large arrows) partly detached from the unit membrane (two small arrows) of an epithelial cell of the blood-gas barrier. BM, common basement membrane; En, endothelial cell. Seven-week-old chick fixed with phosphate-buffered formalin. (Electronmicrographs for Fig. 11 by R. D. Cook, Department of Veterinary Anatomy, University of Liverpool.)

(i) Pulmono-oesophageal nerve. This nerve was first described by Fedde *et al.* (1963). McLelland (1970) found it to arise either from the proximal part of the recurrent nerve (Fig. 12a), or directly from the dorsal aspect of the vagus a few millimetres caudal to the origin of the recurrent nerve (Fig. 12b). After a few millimetres it divides into medial and lateral branches. The medial branch goes to the oesophagus. The lateral, or pulmonary branch (P1 in Figs. 12a, b), supplies the cranial part of the pulmonary plexus.

(ii) Pulmonary nerves of vagal trunk. From two to six pulmonary nerves arise from the vagus (P2 to P6 in Fig. 12b). All are delicate (from 200 to 300 μm diameter) and close together (about 1 to 10 mm apart). The shortest are less than 0·5 mm long. All are eventually absorbed into the pulmonary plexus. Hsieh (1951) and Peterson (1970) classified the nerves to this plexus into cranial, middle, and caudal groups, of which the cranial "group" is the pulmonary branch of the pulmono-oesophageal nerve; however, McLelland (1970) concluded that the other pulmonary nerves (arising direct from the vagus) cannot be convincingly subdivided into two groups.

(iii) Pulmonary plexus. The pulmonary plexus is the variable network of very fine anastomosing nerves which is formed from the pulmonary nerves. It extends craniocaudally from the pulmonary artery to the pulmonary vein,

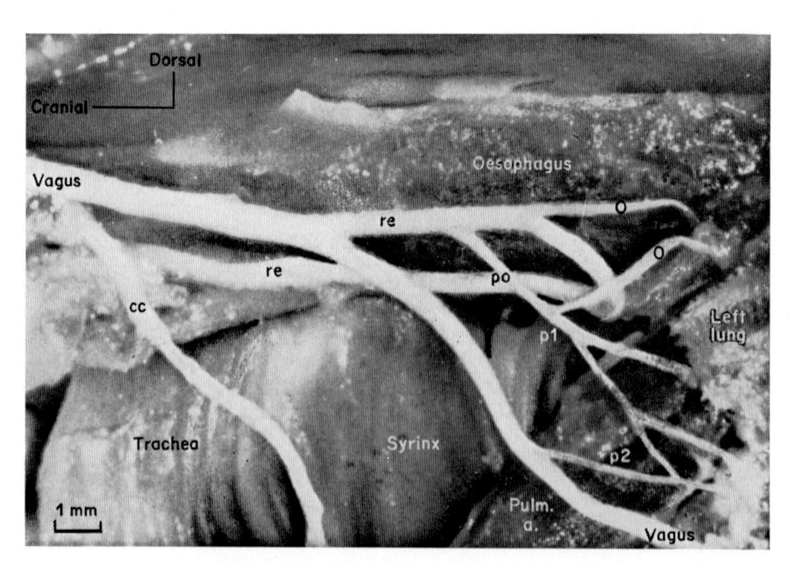

Fig. 12a. Lateral view of the left thoracic vagus of adult fowl fixed in formalin. In this specimen the pulmono-oesophageal nerve (po) arises from the recurrent nerve (re). Its pulmonary branch (p_1) breaks down into branches which join branches of the first pulmonary branch (p_2) of the vagus trunk, to form the rostral part of the pulmonary plexus. cc, cranial cardiac nerve; o, nerves to oesophagus; pulm. a., pulmonary artery. Reproduced with permission from McLelland (1970).

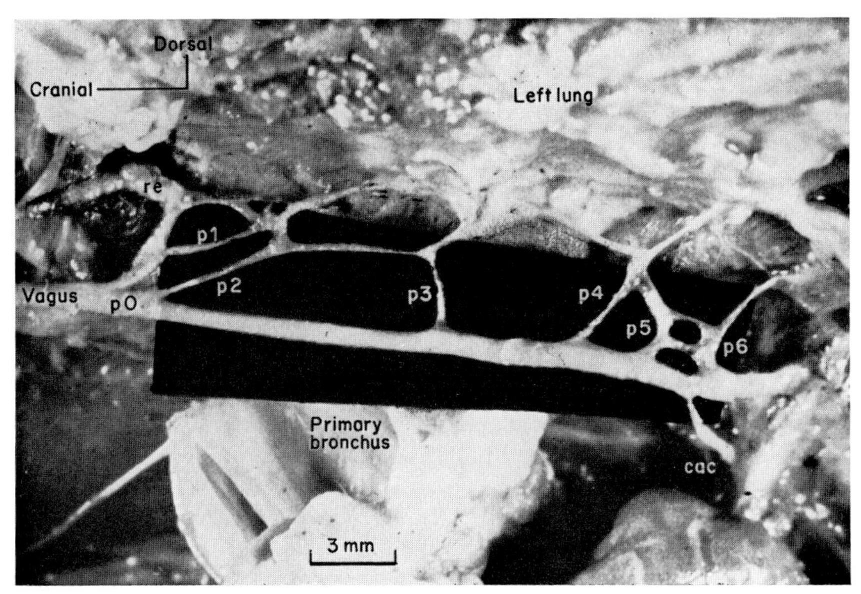

Fig. 12b. Lateral view of the left thoracic vagus of an adult fixed in formalin, with a black card beneath the nerve. In this specimen the pulmono-oesophageal nerve (po) arises directly from the vagus, lying on the root of the recurrent nerve (re). The pulmonary branch (p_1) of the pulmono-oesophageal nerve joins five pulmonary branches (p_2–p_6) from the vagal trunk to form the pulmonary plexus in the fat above the upper edge of the card. ca.c., caudal cardiac nerve. Reproduced with permission from McLelland (1970).

lying on the lateral surfaces of these vessels; ventrodorsally the plexus extends between the vagus ventrally and the hilus of the lung dorsally. It is often surrounded by adipose tissue. McLelland (1970) has found the plexus to be continuous rather than subdivided into cranial and caudal pulmonary plexuses as suggested by Fedde *et al.* (1963).

The lung itself finally receives fine nerves from the dorsal region of the plexus which enter the hilus by passing along the pulmonary artery and vein. Watanabe (1960) believed that those nerves which follow the artery are distributed to the cranial region of the lung, while those which follow the vein go to its middle and caudal regions.

(iv) Pleural plexus. This plexus lies immediately beneath the visceral pleura over the whole ventral surface of the lung (Hsieh, 1951). It arises mainly from the pulmonary branch of the pulmono-oesophageal nerve (Hsieh's anterior pulmonary branch of the vagus). After contributing to the pulmonary plexus, the pulmonary branch continues caudally beneath the visceral pleura across the whole surface of the lung, dividing the ventral surface into two nearly equal parts. It gives off numerous collaterals which form the pleural plexus. This is also reinforced by a substantial branch from the most caudal pulmonary nerves of the vagus (Hsieh's posterior pulmonary branch), and by the pulmonary branches of the cardiac (sympathetic) nerve. Nerve fibres from the pleural plexus pass into the lung (Hsieh, 1951).

Hsieh observed that a few very fine nerve fibres arise from the pleural plexus and enter the serous filaments which join the visceral to the parietal pleura over the ventral surface of the lung. These vagal nerve fibres pass through these serous filaments and enter the bronchopleural membrane (the pulmonary aponeurosis, or thoracic diaphragm of Hsieh). The numerous nerve fibres in the bronchopleural membrane (see page 145) could therefore include fibres of vagal origin. Hsieh found no nerve fibres in the serous filaments between the visceral and parietal pleurae on the medial (vertebral) surface of the lung.

(v) Branches to the extrapulmonary primary bronchus. Watanabe (1960) described a bronchial branch, which arose from the vagus at the level of the pulmonary artery and ascended the bronchus to the syrinx, but neither Fedde *et al.* (1963) nor McLelland (1970) could find it. Since direct stimulation of the wall of the extrapulmonary bronchi induced no changes in breathing, Fedde *et al.* (1963) concluded that if there are nerves on the extrapulmonary bronchus they contain no afferent respiratory fibres.

(vi) Respiratory afferent pathways in the vagus. Fedde *et al.* (1963) showed that, after unilateral cervical vagotomy, transection of the pulmonary branch of the contralateral pulmono-oesophageal nerve slowed breathing; surprisingly, so also did transection of its oesophageal branch. Stimulation of the central end of the pulmono-oesophageal nerve caused marked and prompt changes in breathing.

These authors also showed that breathing typical of bilateral vagotomy is obtained (after unilateral vagotomy) by transecting both the pulmonary branch of the pulmono-oesophageal (P1) and the "cranial pulmonary branches of the vagus" (possibly P2–3? in Figs. 12a, b), on the other side of the body. Evidently significant numbers of pulmonary respiratory afferent fibres leave the lung via both the pulmono-oesophageal nerve and the more cranial of the pulmonary nerves of the vagal trunk.

Pulmonary receptors which are strongly sensitive to CO_2 have been studied by Peterson and Fedde (1968), Peterson (1970) and Molony (1970) and appear to be located somewhere in the pulmonary airways (Peterson and Fedde, 1968; Peterson, 1970), the afferent pathway being in the vagus (Peterson and Fedde, 1968). Fedde (1970) reported that these CO_2-sensitive vagal afferents ceased firing when either the "anterior pulmonary" (pulmono-oesophageal) or the "posterior pulmonary" branch of the vagus was cut.

(*b*) *Sympathetic branches to the lung*. Hsieh (1951), Watanabe (1960), and Peterson (1970) mentioned a single sympathetic nerve to each lung, arising from the first thoracic ganglion on the sympathetic trunk. It blends with the pulmonary plexus. The functions of this sympathetic supply appear not to have been investigated.

2. Microscopic Nerve Supply.

Krahl (1964) has remarked that when the abundant literature is sifted for hard reliable facts "it becomes evident that no other area of (mammalian) pulmonary anatomy or physiology is so poorly known and understood". The innervation of the avian lung is probably even less well known or understood. The literature was sparse until the work of McLelland (1970) on which the following account is based.

Most of the nerve fibres in the lung are less than 1·5 μm in diameter, those of the largest being only 4 μm. According to Campenhaut (1956) the innervation is fully developed by 7 to 10 months of age, being distinctly less profuse at 3 or 4 months. A substantial increase in innervation between 1 and 12 weeks of age was also suspected by T. Bennett (personal communication 1970). These observations suggest that hazards may arise in electrophysiological experiments on the respiratory nerves of young birds.

(*a*) *Intrapulmonary primary bronchus.* The large nerves which enter the hilus of the lung alongside the pulmonary artery and vein leave these vessels and pass to the connective tissue on the outside of the intrapulmonary primary bronchus. There they form a coarse-meshed peribronchial plexus, in which

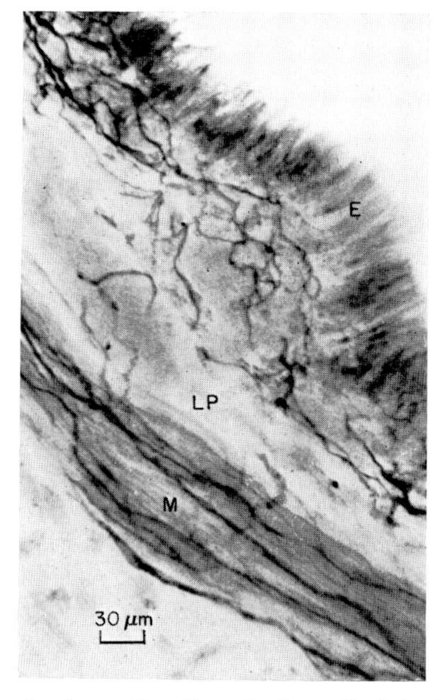

Fig. 13a. Photomicrograph of a section through the wall of the intrapulmonary primary bronchus. Nerve plexuses in the muscle layer (M) are continuous with plexuses in the lamina propria (LP), the latter being best-developed immediately under the epithelium (E). Uninhibited cholinesterase preparation (Gomori, 1952). Reproduced with permission from McLelland (1970).

the thickest bundles are about 300 μm in diameter. Numerous ganglia are present, containing up to 80 cells. A peribronchial plexus was mentioned also by Arimoto and Miyagawa (1930), Takino (1933a) and Muratori (1934).

Many nerves from the peribronchial plexus pass to the muscle layer of the primary bronchus (Fig. 13a) and form there a plexus of course bundles of axons within the network of which are narrower-meshed plexuses of finer nerve bundles. The course bundles contain small ganglia. Takino (1933a) and Muratori (1934) saw fine nerve fibres in the muscle layer. Fluorescence studies (Bennett and Malmfors, 1970) have shown that some nerve fibres in the muscle of the primary bronchus are adrenergic and have indicated that unlike the mammalian bronchial muscle, the adrenergic innervation is sparse; the main catecholamine transmitter in the fowl appears to be noradrenaline (Bennett and Malmfors, 1970; see also Chapter 28). On the basis of cholinesterase techniques with light microscopy, cholinergic nerve fibres appear to be common (McLelland, 1970). Observations on tissue treated with tritiated noradrenaline, 5-hydroxydopamine, α-methyl-meta-tyrosine, 6-hydroxydopamine and cholinesterase techniques (Cook, 1970) have confirmed the presence of both adrenergic and cholinergic axons in the muscle of the primary bronchus, but aminergic axons are in the majority.

The lamina propria of the primary bronchus possesses another nerve plexus, derived from the plexuses of the muscle layer. A few ganglia are present, each containing up to about four cells. The plexus is particularly dense near the epithelium (Fig. 13a). Nerve fibres apparently entering the epithelial layer have been reported by various light microscopists (Takino, 1933b; Muratori, 1934; Toussaint-Francx and Toussaint-Francx, 1959; McLelland, 1970). Using the electron microscope Cook (1970) quite often found single axons, completely bare of Schwann cell cytoplasm, between ordinary columnar ciliated epithelial cells (Fig. 14a). These enlarged axonal endings usually have many mitochondria, as well as agranular vesicles. The epithelial cell adjacent to these axons sometimes behaves like a Schwann cell, investing the axon via a "mesaxon" of surface membrane, as in the epithelium of the cornea and of the epidermis of the mammalian snout (Cauna, 1966) and ear pinna (Cauna, 1969). These intra-epithelial axonal endings appear to be similar to the afferent endings in the human nasal respiratory mucosa observed by Cauna *et al.* (1969), except that the latter ended just superficial to the basement membrane whereas in fowls they sometimes reached almost to the airway (Cook, 1970).

(*b*) *Secondary bronchi.* Campenhaut (1956) believed the secondary bronchi to be the most richly innervated of all the bronchi. He described a peribronchial plexus outside the muscle layer with relatively large ganglia, and a plexus in the lamina propria which gave branches towards the epithelium or into the muscle layer. McLelland (1970) found the plexuses at the roots of the secondary bronchi to be continuous with those of the primary bronchus.

Distal to their roots the walls of the secondary bronchi become thin, and the various plexuses cannot then be distinguished. Most of the ganglia are much smaller than those in the primary bronchus. Campenhaut (1956) suspected the presence of nerve fibres in the epithelial layer, but McLelland (1970) could find no evidence for this. On the other hand, with the electron microscope, Cook (1970) observed single enlarged axonal endings almost bare of Schwann cell cytoplasm, close to the thin squamous cells of the more distal part of the secondary bronchi, much as in Fig. 14b. Some of these axons are very near the airway. They have many agranular vesicles, but few mitochondria.

(*c*) *Tertiary bronchi and exchange area*. McLelland (1970) has shown that the nerves of the tertiary bronchi arise principally from a coarse plexus located in the interlobular connective tissue septa, which was also seen by Takino (1933a, b) and Muratori (1934). This interlobular plexus contains large ganglia. Large bundles arise from this plexus and directly penetrate the exchange area (Fig. 13c). Some of these penetrating bundles reach the rim of the tertiary bronchus itself and its associated atria, where they form a coarse plexus (Fig. 13b) at the rim of the tertiary bronchus but without ganglion cells. Many of the smallest nerve bundles of this plexus run along the bundles of bronchial smooth muscle and within the atrial walls (Fig. 13b). Ultrastructural

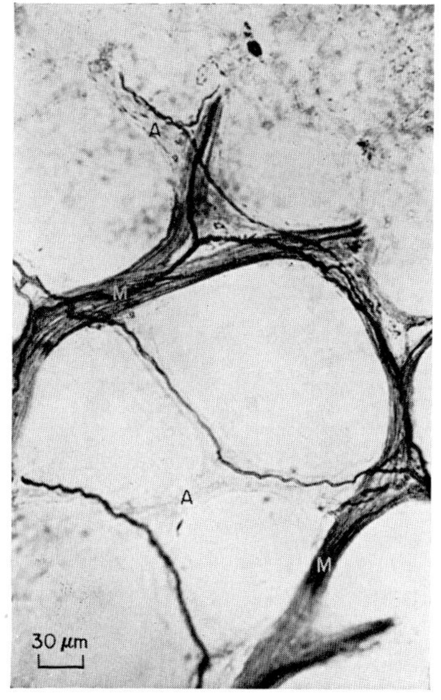

Fig. 13b. Photomicrograph of a section through the atria. Bundles of nerve fibres accompany the muscle bundles (M), and others travel in the walls of the atria (A). Uninhibited cholinesterase preparation (Gomori, 1952) intensified with silver. Reproduced with permission from McLelland (1970).

studies suggest that the muscle of the tertiary bronchi is less well-innervated than that of the primary and secondary bronchi (see page 111). On the basis of cholinesterase and fluorescence techniques Akester and Mann (1969) claimed the probable presence of cholinergic fibres and the absence of adrenergic ones in the muscle of the tertiary bronchi. The latter was confirmed by Bennett and Malmfors (1970), while McLelland (1970) confirmed the presence of many cholinergic fibres as did Cook (1970) who also demonstrated some adrenergic fibres.

Cook (1970) observed not uncommonly single axonal endings which were very closely applied to the squamous cells lining the tertiary bronchi and their atria (Fig. 14b). These endings resembled the axonal enlargements which were seen immediately beneath the squamous epithelium of the secondary bronchi.

The exchange area contains the large penetrating nerve bundles just described. Such bundles were also reported by Takino (1933b) and Hsieh (1951). In addition, McLelland (1969, 1970) showed that the coarse interlobular plexus in the interlobular septa gives rise to a plexus of fine nerve fibres which is distributed rather irregularly through the exchange area (Fig. 13c). Some, but not all, of these fine fibres in the exchange area are associated

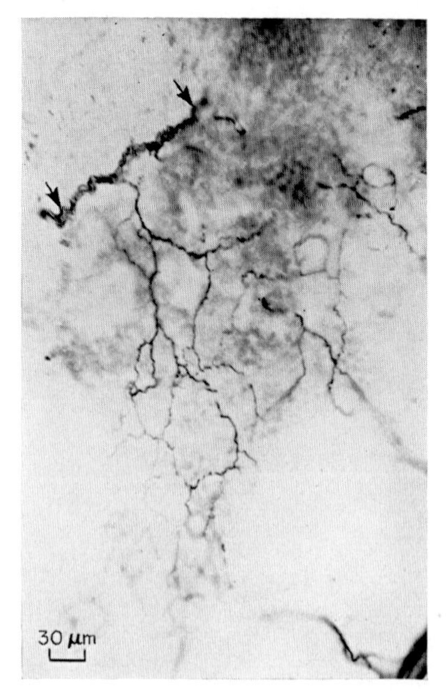

Fig. 13c. Photomicrograph of a section through the exchange area. A large nerve bundle (arrows) is travelling through the exchange area. It gives rise to a plexus of fine nerve fibres which ramify in the exchange area. Uninhibited cholinesterase preparation (Gomori, 1952) intensified with silver. Reproduced with permission from McLelland (1970).

with blood vessels. Unlike Takino (1933a), McLelland (1969, 1970) found no ganglion cells in the exchange area.

3. *Afferent Nerve Endings*

Discussions of the classification of afferent nerve endings following studies with the light microscope by Cauna (1959) and Miller and Kasahara (1964), and with the electron microscope by Cauna (1966) and Munger (1966), indicate three basic types of afferent nerve ending: (a) free nerve endings, (b) neurite-receptor cell complexes, and (c) encapsulated endings. Free nerve endings, the most elusive of all afferent endings especially at the ultrastructural level (Munger, 1966), can be subdivided into simple free endings which appear to be only slightly branched or unbranched, and complex free endings which result from repeated branching but are nevertheless distinctly discrete. Ultrastructurally, the features most indicative of afferent axonal endings are terminal enlargement, numerous mitochondria, and many agranular vesicles. These occur in free nerve endings in epithelial areas (Cauna, 1966, 1969; Cauna *et al.*, 1969) and in presumptive baroreceptors of the carotid sinus (Rees, 1967), in the axons of neurite-receptor cell complexes (see Cook and King,1969b), and in encapsulated endings such as Meissner's (Cauna, 1966) and Herbst corpuscles (Nafstad and Andersen, 1970). Nevertheless, there are no really conclusive criteria for identifying afferent nerve endings.

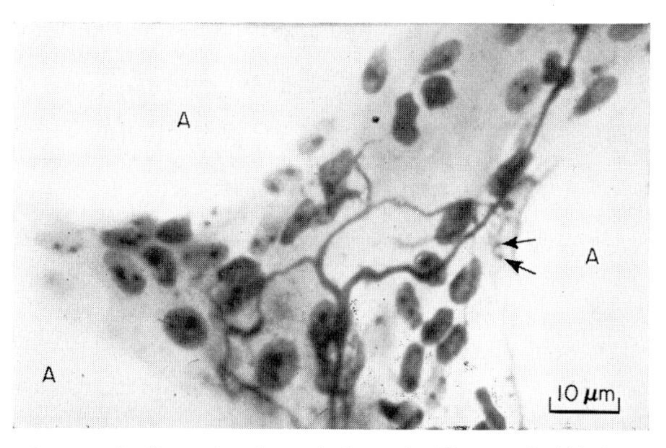

Fig. 13d. Photomicrograph of a section through the atria. Three atria (A) share a common wall at this point. In the wall a nerve fibre is branching into successively finer fibres. Some of the branches end in small enlargements (arrows); others extend well beyond this field. The appearance of this ending suggests that it could be a complex free afferent nerve ending. Modified Bielschowsky-Gros silver technique (Rintoul, 1960). Reproduced with permission from McLelland (1970).

(*a*) *Free nerve endings*. Both simple and complex free afferent nerve endings seem to have been found in the avian lung.

(i) Simple free endings. As described above (page 126), single bare axonal enlargements have been found within the epithelial layer of the primary bronchus (Fig. 14a). These are probably simple free afferent nerve endings. Single bare axons (see page 127) have also been observed close to the

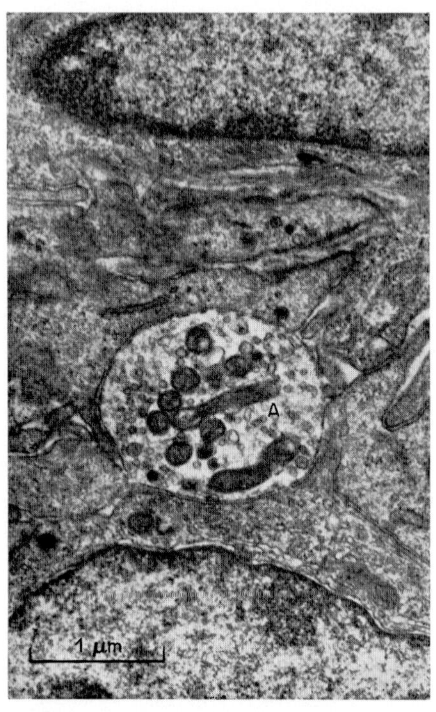

Fig. 14a. Electronmicrograph of a possible afferent axon (A) located between ordinary epithelial cells lining the airway of the intrapulmonary primary bronchus. The axon is enlarged and bare of Schwann cell cytoplasm, contains quite numerous mitochondria and agranular vesicles, and is closely applied to the processes of the epithelial cells which enclose it. Adult White Leghorn male, glutaraldehyde fixation. Reproduced with permission from Cook (1970).

simple squamous epithelial cells of the secondary and tertiary bronchi (Fig. 14b). These too are strongly suggestive of simple free afferent endings.

Several single axons of a specialized type, differing in general morphology from the numerous axons which are generally accepted as motor to the smooth muscle cells, were found by Cook (1970) between the smooth muscle cells of the primary bronchi and resembled those which Merrillees (1968) regarded as suggestive of a sensory ending in mammalian smooth muscle (see his Fig. 24). They are much enlarged, packed with mitochondria and agranular vesicles (Fig. 14c). A similar possible afferent axonal ending in mammalian bronchial smooth muscle appears in figure 4 of Fillenz and Woods (1970).

(ii) Complex free endings. Using silver nitrate and osmium tetroxide McLelland (1970) occasionally observed fibres of from 1·5 μm to 4 μm in

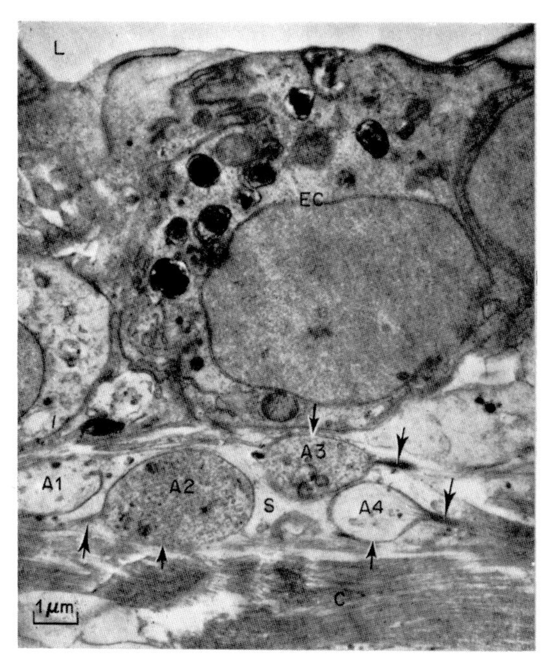

Fig. 14b. Electronmicrograph of a group of four axons (A1–A4) located close to the base of an epithelial cell (EC) lining the wall of an atrium. L, lumen of atrium. C, collagenous basis of the atrial wall, which was also bounded by squamous cells on the lower side of the field. The axons are partly enclosed by Schwann cell cytoplasm (S), but are exposed at several points (small arrows). All four axons appear enlarged. A2 and A3 contain numerous agranular vesicles. A3 also contains a few mitochondria. A3 and A4 appear to be veering off to the right of the field (two large arrows), and A2 to the left (large arrow). This may be the point of divergence of four branches of a free afferent ending. Rhode Island Red adult male, glutaraldehyde fixation. Reproduced with permission from Cook (1970).

diameter which divided rapidly several times into successively finer fibres (Fig. 13d) ending either freely or in minute knob-like swellings; the many branches spread over a large area. These structures were found in the lamina propria and muscle layer of the primary bronchus, and rather more often in the wall of the tertiary bronchus and its atria (a region likely to be relatively mobile, see page 114). The morphology of these nerve terminals is strongly suggestive of a complex free afferent ending.

(*b*) *Afferent neurite-receptor cell complex.* A possible neurite-receptor cell complex, consisting of specialized cells closely associated with unmyelinated axons, has been reported by Cook and King (1969a, b).

The specialized cells occur in small groups in the intrapulmonary primary bronchus, especially at the junction between the primary bronchus and the roots of the four craniomedial secondary bronchi (Cook, 1970). Each receptor cell is enclosed on all sides by the ciliated columnar epithelial cells which line the primary bronchus. So far none has been seen to reach the actual airway,

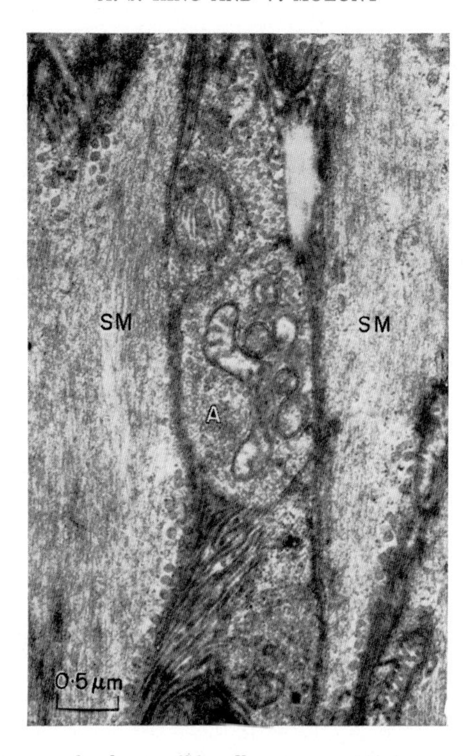

Fig. 14c. Electronmicrograph of a possible afferent axon (A) between two smooth muscle cells (SM) of the intrapulmonary primary bronchus. The axon is enlarged and contains quite numerous mitochondria and agranular vesicles. The axonal membrane is in close contact with the two muscle cells. The neuromuscular junction is close, the intermembrane gap being about 15 nm. Adult White Leghorn male, glutaraldehyde fixation. Reproduced with permission from Cook (1970).

but some approach very close to it as in Fig. 15a. The cytoplasm contains many dense-cored granular vesicles about 75 to 120 nm in diameter, often concentrated near axons associated with the cell. One or two bare axonal endings (Fig. 15b), packed with mitochondria and agranular vesicles, are often in very close contact with the receptor cell (Cook and King, 1969a; Cook, 1970).

These cells resemble the glomus cells of the mammalian carotid body (Lever *et al.*, 1959; Al-Lami and Murray, 1968; Biscoe and Stehbens, 1966), and Merkel cells of mammalian skin (Munger, 1965; Iggo and Muir, 1969) and in the avian palate (Andersen and Nafstad, 1968). Glomus cells are generally regarded as chemoreceptors, and Merkel cells as mechanoreceptors. There are therefore grounds for regarding these cells in the primary bronchus as another neurite-receptor cell complex, either chemoreceptor or mechanoreceptor in function. An apparently identical neurite-receptor cell complex

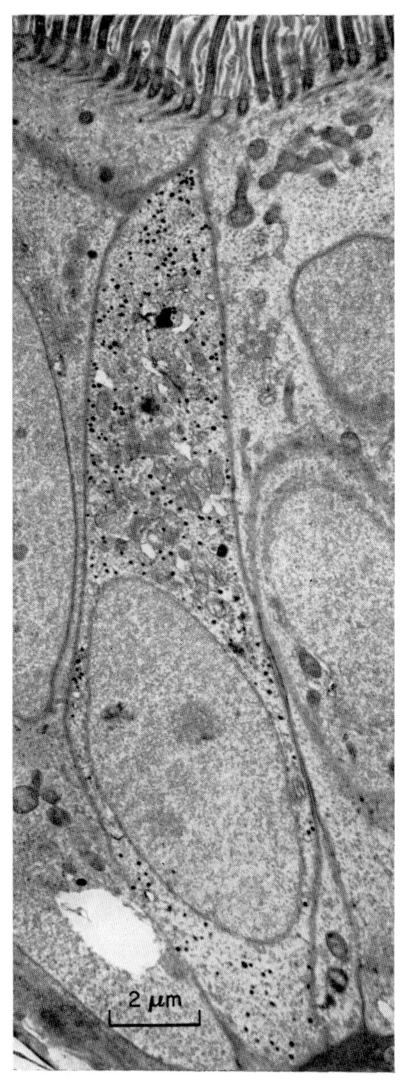

Fig. 15a. Electronmicrograph of a probable receptor cell lodged between two ordinary ciliated epithelial cells in the intrapulmonary primary bronchus. The receptor cell reaches almost to the airway. It contains many mitochondria (in its centre) and numbers of dense-cored granular vesicles. There are no axons in apposition to the receptor cell in this field. Identical cells occur in the bronchopleural membrane (pulmonary aponeurosis). Adult Rhode Island Red male, glutaraldehyde fixation. Reproduced with permission from Cook (1970).

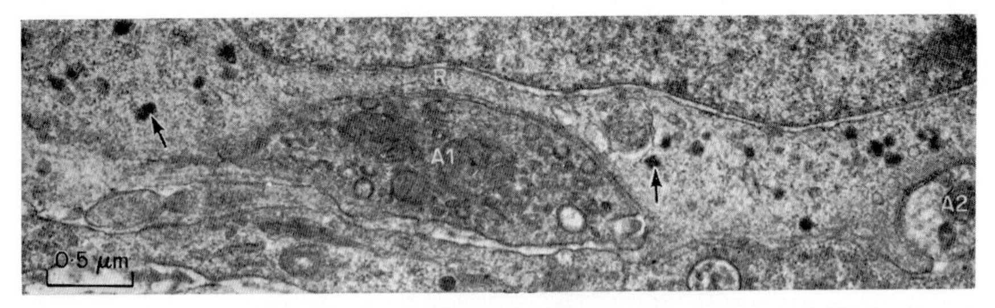

Fig. 15b. Two axons (A1, A2) closely applied to the base of a receptor cell (R) in the epithelium of the intrapulmonary primary bronchus. A1 is enlarged, bare of Schwann cell cytoplasm, has many agranular vesicles and three mitochondria, and is closely applied to the receptor cell. A2 is similar. About 20 dense-cored granular vesicles (arrows) are seen in the receptor cell. Reproduced with permission from Cook (1970).

occurs in the bronchopleural membrane (pulmonary aponeurosis, see page 145).

(*c*) *Encapsulated nerve endings.* There is no evidence for encapsulated nerve endings anywhere in the avian lung and the evidence for their presence in mammalian lung is equivocal (see Widdicombe, 1964 but compare Fisher, 1963).

4. Fibre Spectra and Conduction Velocities

Until the work of Brown (1970), the only papers on the fibre sizes and conduction velocities in the avian vagus were those by Dahl *et al.* (1964), Jones (1969), and Molony (1970). Jones showed that, in the cervical vagus of the duck, most of the fibres were below 5 μm, fibres of 6 to 10 μm being rare. Dahl *et al.* (1964) obtained compound action potentials from the cervical vagus of the fowl *in vitro* at temperatures of 20 to 26°C, the nerve being bathed in mammalian Ringer solution. Molony (1970) measured *in vivo* the conduction velocities of 46 single fibres in the cervical vagus of the fowl, all of which fired synchronously with breathing, about half of them being CO_2-sensitive. The fastest of Molony's units had a conduction velocity of 19 m/s and the slowest 2 m/s, with the main peak of distribution at 5 to 6 m/s (Fig. 16).

The fibre spectrum and compound action potential of the mid-cervical vagus, recurrent nerve, and the pulmono-oesophageal nerve were established as follows by Brown (1970), the number of unmyelinated fibres being estimated from electronmicrographs.

(*a*) *Mid-cervical vagus.* This contains about 10,000 myelinated fibres (Fig. 17). The largest are 6 to 7 μm, and nearly 90% are less than 3 μm in diameter, this being consistent with the general observations on the duck by Jones (1969). There are between 4,000 and 5,000 unmyelinated fibres.

The compound action potential of the mid-cervical vagus shows two elevations indicating two major groups of fibres. The faster group has a range of

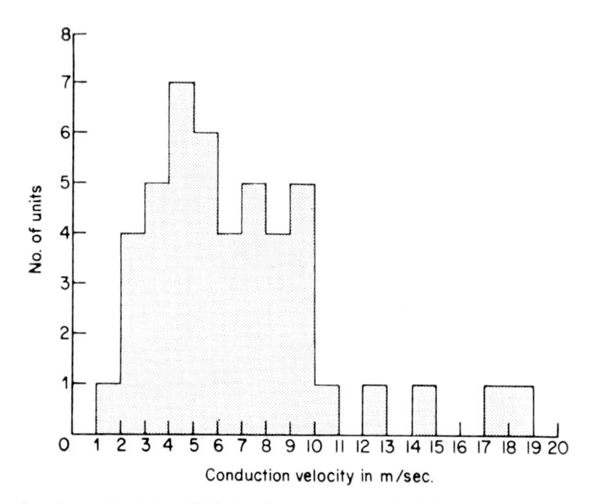

Fig. 16. Conduction velocities of 46 single units, recorded in the mid-cervical vagi of adult fowls, which fired synchronously with breathing, about half of them being CO_2-sensitive. Reproduced with permission from Molony (1970).

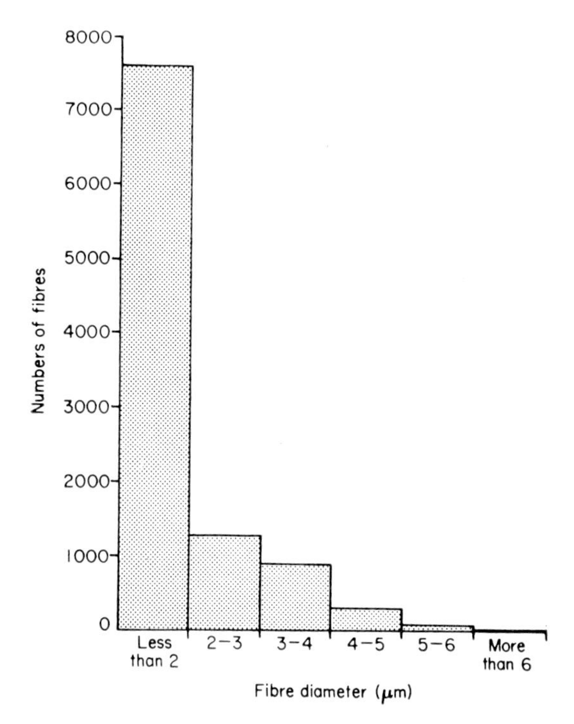

Fig. 17. Fibre spectrum of myelinated fibres in the mid-cervical vagi of adult female White Leghorn hybrids, mean of three birds. The total number of myelinated fibres is about 10,000. (There are also about 4,000 to 5,000 unmyelinated fibres.) Reproduced with permission from Brown (1970).

conduction velocities from between 1 and 17 m/s, and the slower from between 0·4 and 0·8 m/s. The latter fibres are presumably unmyelinated. These observations of Brown's differ from those of Dahl *et al.* (1964), who reported maximum conduction velocities of the faster group to be only 2·5 m/s, and the slower to be 0·4 m/s. These differences are probably due to the conditions of recording.

(*b*) *Recurrent nerve.* The recurrent nerve has about 3,500 myelinated fibres, of which about 95% are less than 3 μm in diameter and the largest about 6 μm. Unmyelinated fibres make up about the same proportion of the total number as in the mid-cervical vagus. The compound action potential of the recurrent nerve in the bird is essentially the same as for the mid-cervical vagus.

(*c*) *Pulmono-oesophageal nerve.* This fine nerve (Fig. 12a) has about 900 myelinated fibres, of which about 90% are less than 3 μm in diameter (Fig. 18),

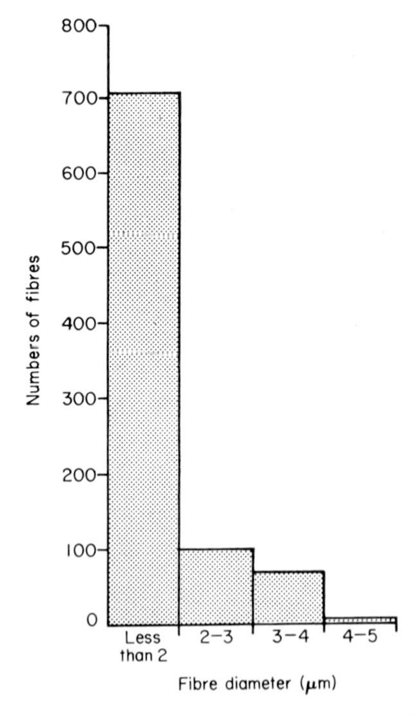

Fig. 18. Fibre spectrum of myelinated fibres in the pulmono-oesophageal nerve of one adult female White Leghorn hybrid. There are about 900 myelinated fibres altogether. (There are also between 300 and 450 unmyelinated fibres.) Reproduced with permission from Brown (1970).

while the largest are about 4 or 5 μm in diameter. The compound action potential shows an elevation with conduction velocities ranging from 1 to

18 m/s. The conduction velocities of Molony's 'respiratory' fibres are compatible with this group. Unmyelinated fibres are in about the same ratio to myelinated fibres as in the mid-cervical vagus, but Brown was unable to demonstrate their presence in the compound action potential.

This nerve is of particular interest since it evidently carries respiratory afferent fibres including CO_2-sensitive fibres (see page 124). These respiratory afferent fibres appear to be much thinner and slower in conduction than their mammalian counterparts (see Paintal, 1963).

IV. Air Sacs

The embryological studies of Locy and Larsell (1916b) established that the air sacs arise from six primordial pairs (Fig. 19). Two pairs fuse to form the median clavicular sac, and one other pair fuses to form the median cervical

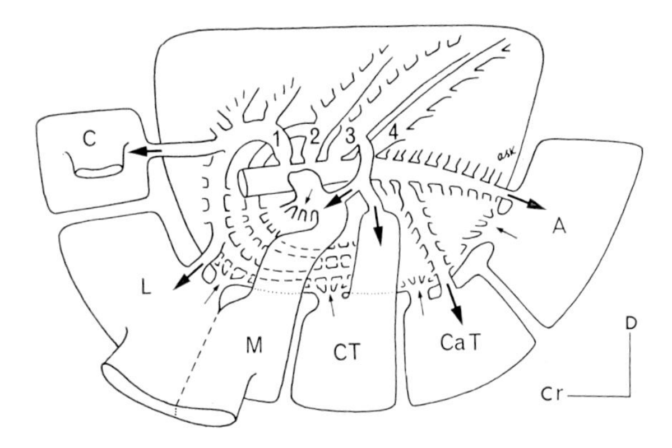

Fig. 19. Diagram of the basic air sacs on the right side of a fowl. Almost all birds have the six primordial pairs of air sacs which are shown, but in most species the lateral component of the clavicular sac (L) fuses with the medial component (M), on each side and across the mid-line, thus forming a single median sac. In most species the direct connexions of the sacs (large arrows) attach to secondary bronchi as shown. In a few species (including *Anas platyrhynchos*) the medial component of the clavicular sac arises from the root of the first craniomedial secondary bronchus (see Fig. 6b). The indirect connexions (small arrows) have not been established in detail except in fowls, but the available evidence is consistent with the connexions shown. The arrows do *not* indicate the direction of gas flow. C, cervical sac; CT, cranial thoracic sac; CaT, caudal thoracic sac; A, abdominal sac; 1 to 4, the four craniomedial secondary bronchi; Cr, cranial; D, dorsal. Reproduced with permission from King and Atherton (1970).

sac. Thus the definitive number of sacs in the fowl is eight, i.e. the median cervical sac, the median clavicular sac, and the paired cranial thoracic, caudal thoracic and abdominal sacs. This arrangement is very common among birds generally, though in many the cervical sacs remain paired giving a total of nine. For a summary of species variations, including minor modifications in the number and relative sizes of sacs, and of the many terminologies, see King (1966).

A. ANATOMY OF THE AIR SACS

The anatomy of the air sacs has been most fully described by Campana (1875) but see also King (1966, 1972). The arrangement of the sacs is shown in Fig. 20.

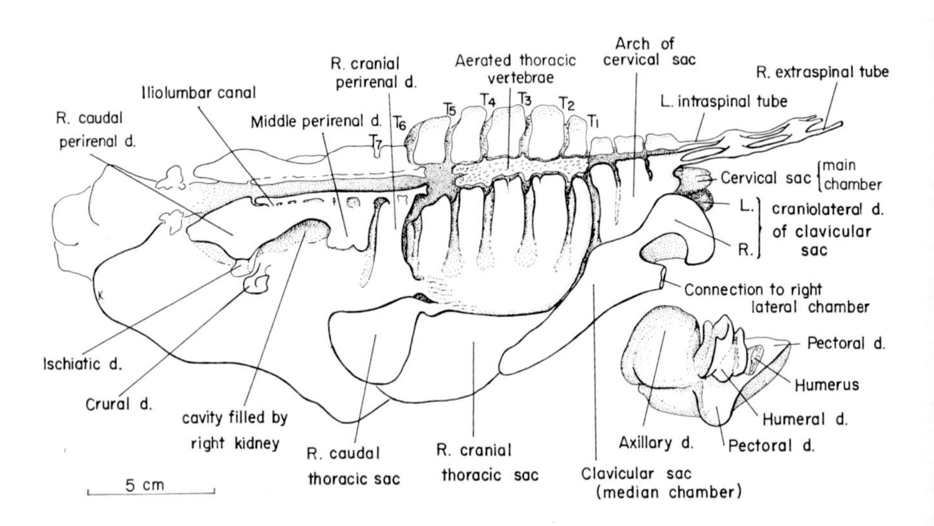

Fig. 20. Drawing of the right side of a cast of the lungs and air sacs of an adult. The right lateral chamber of the clavicular sac has been detached (lower right). T1–7, grooves for the seven vetebral ribs; d, diverticulum; L, left; R, right. Reproduced with permission from King (1972).

1. Cervical Sac

The small cervical sac has a median main chamber and tube-like diverticula associated with the vertebral column. The main chamber is in an extreme dorsal position in the rostral part of the coelom, between the oesophagus and vertebral column. The cervical sac aerates inclusively the cervical and thoracic vertebrae between C3 and T5, and the first two vertebral ribs.

2. Clavicular Sac

The clavicular sac plugs the thoracic inlet. It has a large median and paired lateral chambers. The median one occupies the base of the neck and the cranioventral region of the coelom. Dorsal to it are the oesophagus, median chamber of the cervical sac, and lungs; ventrally it lies on the sternum; caudally it is indented by the heart. The syrinx, and various vessels and nerves are suspended in it. It leads on each side, by a flattened canal compressed between muscles, into the paired lateral chambers. In vinylite casts from females of heavy breeds this canal is about 3 cm long, 1·5 cm wide, and 1 mm deep (Biggs and King, 1957). Each lateral chamber consists of potential spaces between the muscles of the shoulder joint. The median chamber aerates the sternum, coracoid and various ribs, and the lateral chamber aerates the humerus. These and the other small diverticula of the sir sacs which enter the bones or penetrate between muscles are unlikely to be ventilated, any gas movement being confined to diffusion only (Scheid and Piiper, 1969). The clavicular sac can be entered in the live bird immediately cranial to the ventral end of the first sternal rib, after passing through the pectoral muscles. The heart lies immediately medial to the sac at this point.

3. Cranial Thoracic Sac

This large paired sac occupies the dorsolateral region of the thoracic cage immediately ventral to the lung (Fig. 21), lying between the bronchopleural membrane (pulmonary aponeurosis) and bronchoperitoneal membrane (oblique septum). Its lateral surface is in contact with the five sternal ribs and the forked caudolateral process of the sternum. This sac possesses no diverticula. It can be entered in the live animal between the middle of the third and fourth sternal ribs. This point is just craniodorsal to the middle of the wide triangular-shaped pars lateralis of the caudolateral process of the sternum; the triangular-shaped pars lateralis can be palpated through the skin.

4. Caudal Thoracic Sac

This small paired sac is lateral in position, being in contact with the last two ribs and a small area of body wall caudal to the thoracic cage. It has no diverticula. It can be entered in the live animal between the ventral ends of the last two vertebral ribs.

5. Abdominal Sac

The paired abdominal sac consists largely of potential spaces only, though in making a cast its capacity is greatly distended. It reaches from the lung to the cloaca. Dorsolaterally it is attached to the body wall, but medially and ventrally it wanders between the viscera. It aerates the synsacrum and pelvis, but not the bones of the leg. In the live bird the sac can be entered, close to the ostium, in the angle between the caudal border of the dorsal end of the last vertebral rib and the lateral border of the ilium. This involves passing through a substantial mass of muscle.

B. CONNEXIONS TO THE LUNG

The general area where an air sac connects to the lung by one or several connexions is an ostium (Figs. 6a, b, 19). The cranial and caudal thoracic sacs have essentially a wide single direct connexion (about 2 or 3 mm in diameter in heavy breeds) to one of the secondary bronchi, and four to six narrower indirect connexions (about 1 to 2 mm in diameter in heavy breeds) which immediately become continuous with ordinary tertiary bronchi (Fig. 6a). The indirect connexions are sometimes called recurrent bronchi. Since it arises from two paired primordial sacs, the clavicular sac has two direct connexions (m.d.cl.s. and l.d.cl.s. in Fig. 6a) and two groups of indirect connexions (l.i.cl.s., and a cluster around the primary bronchus which is not shown in Fig. 6a). The direct connexion of the abdominal sac (d.ab.s.), via the primary bronchus itself, is atypical as it is of smaller diameter than each of the sac's six indirect connexions. The cervical sac possesses a direct connexion only (d.ce.s.).

Most of the ostia are located along the ventrolateral border of the lung (Figs. 6a,b, 19). The indirect connexions of these ventrolateral ostia connect all the sacs (except the cervical) to the tertiary bronchi of the ventral and lateral regions of the lung; these tertiary bronchi form an anastomosing network which links all the ostia on the ventral border of the lung. Potentially, therefore, the abdominal, caudal thoracic, cranial thoracic, and clavicular sacs are all interconnected across the ventrolateral region of the lung. For further details of the anatomy of the connexions of the sacs to the lung see Campana (1875), Juillet (1912), Locy and Larsell (1916b), Payne (1960) and King (1966).

Each direct and each indirect connexion, within each ostium, is surrounded partly or completely by smooth muscle. Most of this forms a thin, more or less continuous ring, but in places it is considerably thickened (J. L. Armstrong, personal communication). Whilst these muscles may or may not be acceptable as sphincters, they are undoubtedly capable of varying the calibres of their openings.

Being tubes of large calibre the direct connexions of all the sacs, except the abdominal, appear to offer pathways of low resistance between the primary bronchus and the air sacs. Some of the direct connexions, e.g. those of the clavicular and cranial thoracic sacs to the third craniomedial secondary bronchus, are not only wide but short, while the direct connexion of the caudal thoracic sac (d.ca.th.s.) is also quite straight. In contrast, the indirect connexions of all these sacs are essentially tertiary bronchi, and are therefore narrower, longer, more tortuous and suggest a relatively high resistance pathway. However, since a sac may have from three to six indirect connexions, the sum of the cross-sectional areas of the openings into all of its indirect connexions may exceed the cross-sectional area of its single direct connexion. For instance, in vinylite casts from females of heavy breeds (Biggs and King, 1957) the diameter of the direct connexion of the cranial thoracic sac is about 3 mm and its cross-sectional area about 7 mm^2. This sac may have as many as six indirect connexions (at l.cr.th.o. in Fig. 6a), each of which has the diameter of a tertiary bronchus (i.e. at least about 1·4 mm in casts from heavy breeds, see page 108). The sum of the cross-sectional areas of these indirect connexions would be about 9 mm^2. Broadly similar relationships exist in the clavicular and caudal thoracic sacs.

Evidently the resistance of the indirect connexions of these sacs is not necessarily prohibitively high (in comparison with the resistance of the direct connexions), and there is every possibility that gas will flow more or less freely within them. Possibly an ideal would be for the total resistance of the two alternative pathways, i.e. of the direct and indirect connexions, to be about equally balanced during eupnoea. A small adjustment of diameter by the bronchial muscle could then regulate the flow through the two pathways, according to the needs of metabolism or thermoregulation (see page 161).

In all birds in which the anatomy of these connexions is known (see King, 1966), the clavicular and cranial thoracic sacs have a much more direct access to the primary bronchus than do any other sacs, a point of functional interest which attracted the attention of Salt (1964). In most species they have a short wide common connexion to the third craniomedial secondary bronchus (to CM3 in Fig. 6a, and see Fig. 19). In vinylite casts from females of heavy breeds this common connexion is about 5 mm long and 3 mm wide (Biggs and King, 1957); it divides immediately into two equal tubes (m.d.cl.s. and d.cr.th.s. in Fig. 6a), each about 5 mm long and 3 mm in diameter, which open directly into the clavicular and cranial thoracic sacs respectively. Both of these sacs are therefore joined to the cranial part of the intrapulmonary primary bronchus by a short (1 cm long) wide tube. The possible functional significance of these connexions for channelling the dead space gas of the upper airway into the cranial sacs is discussed on page 160.

The relationship between the resistances of the direct and indirect connexions of the abdominal sac appear to be different from those of the clavicular and thoracic sacs. The direct connexion of this sac via the end of the primary bronchus is much tapered in birds generally (King, 1966), though the emperor penguin may be an exception (Duncker, 1968). In the fowl the primary bronchus is finally reduced at its caudal end to between 1·5 and 1·75 mm diameter in males and 1·0 mm diameter in females in casts from heavy breeds (Payne, 1960). The indirect connexions (Fig. 6a), on the other hand, are six or more in number and in Payne's casts all are over 2·0 mm in diameter. The resistance of the indirect connexions should almost always be lower than that of the direct connexion.

C. VOLUMES OF SACS AND LUNGS

Attempts to measure the volumes of the air sacs and lungs have depended mainly on filling the respiratory tract with an injection mass and measuring the volume of the component casts. Some of the details of these methods have been reviewed by King (1966), and their inherent inaccuracies have been discussed by Scheid and Piiper (1969). Estimates for the female range from 100 ml (Zeuthen, 1942) to 382 ml (King and Payne, 1958) and for the male the one published figure is 502 ml (King and Payne, 1962).

The total volume of the respiratory tract has been measured by gas dilution methods. Dehner (1946a, b) was the first to do this, and obtained values of between 162 and 277 ml for four species of duck. Scheid and Piiper (1969) found a mean volume of 170 ml in unanaesthetized White Leghorn hens of mean weight 1·6 kg. They regarded this figure as in rough agreement with the total volumes of the casts (275 ml) obtained by King and Payne (1962) if allowance is made for the larger body weight of 2·9 kg of the birds used by the latter. The volume of gas in the lungs of Single Comb White Leghorns *in situ* in the thoracic cage was estimated by Burton and Smith (1968) using a buoyancy method. This gave a value for both lungs together of 18·6 ml in males of about 2·4 kg and of 14·6 ml in females of about 1·7 kg. King and Payne (1962) estimated the maximum capacities of the air spaces inside the two lungs (by using casts) at about 70 ml in the male and 35 ml in the female of heavy breeds; the volume of air remaining in the two excised (collapsed) fresh lungs was about 8 ml in the male and 6 ml in the female. Gas dilution methods are usually regarded (e.g. see Sturkie, 1965 and Burton and Smith, 1968) as inapplicable to estimating the volumes of the lungs and air sacs individually in the live bird. However, Schmidt-Nielsen *et al.* (1969) have obtained promising results by applying this technique in the ostrich.

D. STRUCTURE OF THE AIR SAC WALL

The inner lining of the air sacs in birds generally is usually regarded as a simple squamous epithelium, except round the ostia where it is ciliated columnar (Schulze, 1872; Muller, 1908; Groebbels, 1932; Cover, 1953; Trautmann *et al.*, 1957); on the other hand according to Ross (1899) cilia occur everywhere except in the extracoelomic diverticula. McLeod and Wagers (1939) described the lining as "mucoserous", but Mennega and Calhoun (1968) could find neither mucous cells nor cilia in the duck. The general opinion is that intrinsic smooth muscle is absent (Baer, 1896; Muller, 1908; Groebbels, 1932), but Bennett and Malmfors (1970) found it occasionally in the fowl as did Mennega and Calhoun (1968) in the duck. The bronchial openings within the ostia possess rings of smooth muscle (see page 140). The wall of the sac is supported by elastic fibres (Baer, 1896; Muller, 1908; Kaupp, 1918; Groebbels, 1932; Cover, 1953) and its outer surface is attached to the body wall or is lined by a peritoneal epithelium if free in the coelom as in the case of the abdominal sac.

The vascularity of the avian air sac wall has usually been said to be slight (Baer, 1896; Muller, 1908; Schmidt-Nielsen *et al.*, 1969). It is supplied by systemic arteries (Hazelhoff, 1951, and see Baer, 1896, for details); substantial gaseous exchange through the wall is therefore unlikely (Hazelhoff, 1951). However, according to Mennega and Calhoun (1968) the walls are fairly vascular in the duck, with an abundance of capillaries. The air sacs, particularly the more cranial ones, may be an important site of evaporative cooling (see pages 160 to 161).

The walls of the air sacs are generally regarded as having a very small innervation (e.g. Fedde *et al.*, 1963), but the bronchopleural and bronchoperitoneal membranes (pulmonary aponeurosis and oblique septum respectively), both of which include the walls of air sacs, are profusely innervated (see page 144). Any smooth muscle in the walls is sparsely innervated by adrenergic fibres, but there are indications that the muscle of the ostia may receive a more extensive adrenergic innervation (Bennett and Malmfors, 1970).

V. Respiratory Mechanics

A. PLEURA

1. Pleural Cavity

The developmental and definitive anatomy of the avian pleural cavity has been reviewed (King, 1966; McLelland and King, 1972). The primordia of the cranial and caudal thoracic sacs penetrate the double-layered partition which first separates the pleural and peritoneal cavities (Fig. 21). The air

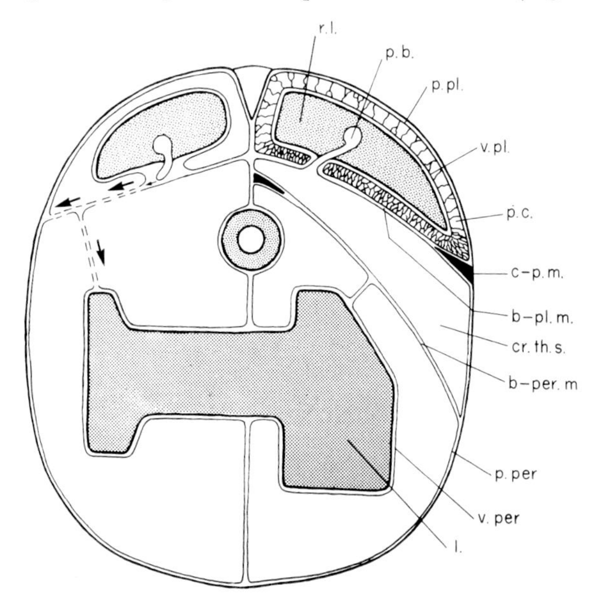

Fig. 21. Diagrammatic transverse section through the trunk to show the pleural cavities. The left side shows how the pleural cavity becomes separated from the peritoneal cavity in the embryo by the ventrolateral extension of the pulmonary fold to the lateral body wall and to the liver (dotted lines), the three large arrows indicating the direction in which these extensions grow. The single small arrow indicates the subsequent penetration by the air sacs. The right side shows the definitive condition. The black mass in the bronchoperitoneal membrane near the midline represents smooth muscle. r.l., right lung; p.b., primary bronchus; p.pl., parietal pleura; v.pl., visceral pleura; p.c., pleural cavity; c-p.m., costopulmonary muscle; b-pl.m., bronchopleural membrane; cr.th.s., cranial thoracic sac; b-per.m., bronchoperitoneal membrane; p.per., parietal peritoneum; v.per., visceral peritoneum; l, liver. Reproduced with permission from McLelland and King (1972).

sacs split this double-layered sheet into two membranes, the bronchopleural membrane (pulmonary aponeurosis, pulmonary diaphragm) dorsally, and the bronchoperitoneal membrane (oblique septum, thoraco-abdominal diaphragm) ventrally.

During development the pleural mesothelium is eroded and serous filaments unite the visceral and parietal pleurae. In the fowl, unlike many other birds including the duck, the pleural cavity is only partly obliterated in this way (see Fig. 21).

2. Bronchopleural and Bronchoperitoneal Membranes

Since there is no structure in any bird which is anatomically, embryologically, or physiologically comparable to the mammalian diaphragm, the term "diaphragm" should be avoided altogether in birds. For discussions of terminology see Fedde *et al.* (1964c), King (1966) and McLelland and King (1972).

The concave bronchopleural membrane (pulmonary aponeurosis) is situated ventral to the lung as in Fig. 21. Its wall is double, the dorsal layer being parietal pleura and the ventral being air sac wall. Juillet (1912) believed that the membrane proper begins immediately caudal to the hilus of the lung, the ventral covering of the lung rostral to the hilus being the wall of the cervical air sac. Caudally the membrane is pierced by the ostia of the caudal thoracic and abdominal air sacs.

The membrane attaches laterally by four fascicles of skeletal muscle, the costopulmonary muscle of Fedde *et al.* (1964c), to the junctions of vertebral and sternal ribs numbers 3, 4, 5 and 6. The fascicles have an intercostal nerve supply (de Wet *et al.*, 1967). From direct observations Soum (1896) discovered that the fascicles contract during expiration, and this has been confirmed electromyographically by Fedde *et al.* (1964c). Soum suggested that the tension on the membrane should remain more or less constant throughout the respiratory cycle, being caused passively by the lateral movements of the ribs during inspiration and actively by the four muscular fascicles during expiration. He proposed that the function of this tension is to keep open the ostia and their bronchi during expiration, when retraction of the lung might tend to obstruct the flow of gas from the sacs; Fedde *et al.* (1964c) concluded that this "may be correct". The possible functional significance of the bronchopleural membrane as a device for protecting the lung against compression during expiration is further discussed on page 154.

The bronchoperitoneal membrane (oblique septum) is arranged as in Fig. 21. Its dorsal layer is air sac wall and its ventral layer is parietal peritoneum. The medial border contains a sheet of smooth muscle about 1 cm wide.

3. Innervation of the Bronchopleural and Bronchoperitoneal Membranes

(*a*) *Adrenergic innervation.* Bennett and Malmfors (1970) found that the smooth muscle of the bronchoperitoneal membrane (oblique septum) is

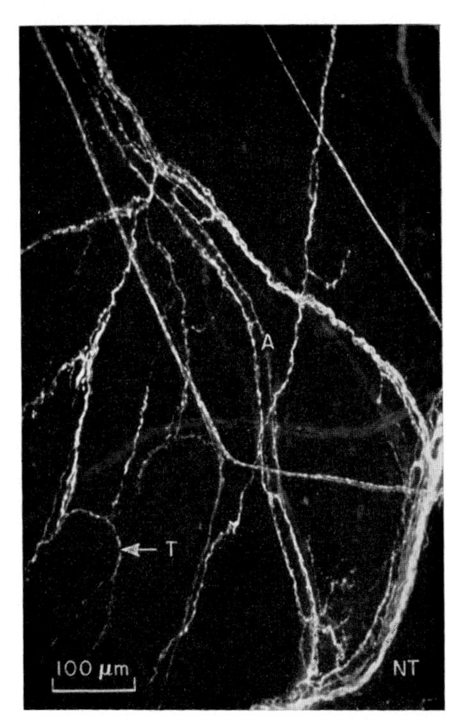

Fig. 22a. Fluorescent stretch preparation by T. Bennett, Department of Zoology, University of Melbourne, of the bronchopleural membrane (pulmonary aponeurosis). A non-terminal bundle (NT) and varicose terminal adrenergic axons (T) ramify over the membrane, some being associated with an arteriole (A).

densely innervated by terminal adrenergic varicose axons derived from large branching bundles of non-terminal axons which come from the sympathetic chain. In addition they also discovered a surprisingly dense network of adrenergic axons and their endings all over the membrane. Non-terminal and varicose terminal adrenergic axons were also found throughout the bronchopleural membrane (pulmonary aponeurosis). These were usually associated with blood vessels, but some varicose terminal axons ran freely over the membrane without association with any identifiable structures (Fig. 22a).

(*b*) *Cholinergic innervation.* A profuse network of axons positive to cholinesterase covers the bronchopleural membrane as in Fig. 22b (T. Bennett, personal communication). These fibres could be vagal in origin, from the pleural plexus (see page 124). As Bennett and Malmfors (1970) have pointed out, the bronchopleural and bronchoperitoneal membranes may be more important in respiration than previously supposed.

(*c*) *Neurite-receptor cell complex.* Following an initial observation by T. Bennett (personal communication), a receptor complex, identical to

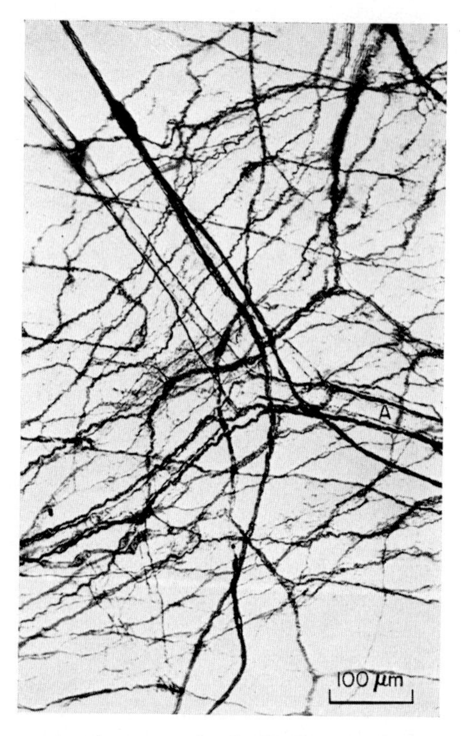

Fig. 22b. Cholinesterase stretch preparation by T. Bennett, Department of Zoology, University of Melbourne, of the bronchopleural membrane (pulmonary aponeurosis). The membrane possesses a profuse network of cholinesterase positive nerve bundles and axons, a few being associated with an arteriole (A) and its branches.

that in the primary bronchus (see Figs. 15a, b and page 121) has been found in an ultrastructural study by R. D. Cook (personal communication) in the bronchopleural membrane. The receptor cells, however, are much more numerous here than in the primary bronchus. They lie on the respiratory side of the membrane in groups, or possibly in bands (T. Bennett, personal communication) between the epithelial cells and are therefore close to the gas in the air sac. The axons associated with the receptor cells could again be vagal in origin via the pleural plexus (see page 124).

Morphologically these neurite-receptor cell complexes resemble both chemo- and mechano-receptors (see page 132). However, the presence of either type of receptor in this site is of potential physiological interest.

B. RESPIRATORY MUSCLES

From anatomical observation and experiments the earlier anatomists and physiologists (e.g. Magnus, 1869; Soum, 1896; Siefert, 1896; Groebbels, 1932) pieced together a list of the avian inspiratory and expiratory muscles.

Electromyography (Burkart and Bucher, 1961; Kadono and Okada, 1962; Kadono *et al.*, 1963; Fedde *et al.*, 1964a,b,c) has shown these lists to be remarkably accurate. Tables 1 and 2 summarize the actions of these muscles,

Table 1

Inspiratory muscles of the fowl as established by electromyography, and their nerve supply

Muscle	Nerve supply (de Wet *et al.*, 1967)	Electromyography
External intercostals except in 5th and 6th spaces	appropriate intercostal nerves	Kadono *et al.* (1963) Fedde *et al.* (1964a)
Intercartilaginous parts of internal intercostals except 6th		Kadono *et al.* (1963)
Scalenus	spinal nerves, C15–T2 inclusive, and intercostal nerve 1	Kadono *et al.* (1963) Fedde *et al.* (1964a, b)
Costisternalis pars major	intercostals 1–5 inclusive	Kadono *et al.* (1963) Fedde *et al.* (1964a,b)
Levatores costarum	intercostals 2–6 inclusive	Kadono *et al.* (1963)
Serratus dorsalis		Kadono *et al.* (1963)

The costisternalis pars major of Fedde *et al.* (1964a,b) and de Wet *et al.* (1967) is presumed to be the triangularis sterni of Kadono *et al.* (1963).

and include their nerve supply as established by de Wet *et al.* (1967) who have also clarified their structure and attachments. Fedde *et al.* (1964a) found no evidence of differences in the performance of the respiratory muscles in the erect and supine posture. Soum (1896) analysed the relative importance of the respiratory muscles by denervation and tenotomy. The triangularis sterni (presumably the costisternalis pars major of Fedde *et al.*, 1964a,b) is much the most important inspiratory muscle. When this is the only remaining inspiratory muscle, breathing continues but at a much reduced rate. Elimination of the external intercostals reduces ventilation considerably. Denervation of the external intercostals and triangularis sterni together arrests breathing at once, even when the innervation of all the more cranial muscles remains intact. Cutting the serratus group of muscles has no overt effect on breathing.

Table 2

Expiratory muscles of the fowl as established by electromyography, and their nerve supply

Muscle	Nerve supply (de Wet *et al.*, 1967)	Electromyography
External intercostal of 6th space and usually that of 5th space also	intercostal 6	Kadono *et al.* (1963) Fedde *et al.* (1964a)
Interosseous parts of internal intercostals of 3rd–6th spaces	intercostals 3–6 inclusive	Kadono *et al.* (1963)
Intercartilaginous part of internal intercostal of 6th space		Kadono *et al.* (1963)
Costisternalis pars minor[a]	intercostals 1 and 2	Fedde *et al.* (1964b)
External abdominal oblique	intercostals 2–6 incl., lumbar 1–3 incl.	Kadono *et al.* (1963) Fedde *et al.* (1964b)
Internal abdominal oblique	intercostal 6, lumbar 1 and 2	Kadono *et al.* (1963) Fedde *et al.* (1964a,b)
Transverse abdominal	intercostals 5 and 6, lumbar 1 and 2	Kadono *et al.* (1963) Fedde *et al.* (1964b)
Rectus abdominus	intercostals 5 and 6, lumbar 1 and 2	Kadono *et al.* (1963) Fedde *et al.* (1964b)
Serratus ventralis		Kadono *et al.* (1963)
Costopulmonary muscle	intercostals 3, 4 and 5	Fedde *et al.* (1964c)

[a]The term "costisternalis pars minor" of de Wet *et al.* (1967) is synonymous with "costisternalis pars anterior" of Fedde *et al.* (1964b)

C. MOVEMENTS OF THE BODY WALL

The cranial movement of the ribs during inspiration increases the diameter of the thoracic cage dorsoventrally, transversely, and craniocaudally as in Figs. 23a,b. The basis of these changes is that, as in mammals, the two vertebral articulations of each vertebral rib are so arranged that when the distal end of the rib is moved cranially by inspiratory muscles it must also move ventrally and laterally. The sternum is simultaneously pushed ventrally, pivoting on the coracoid. The ventral movement of the sternum draws the abdominal wall ventrally; similarly, the lateral movement of the last rib

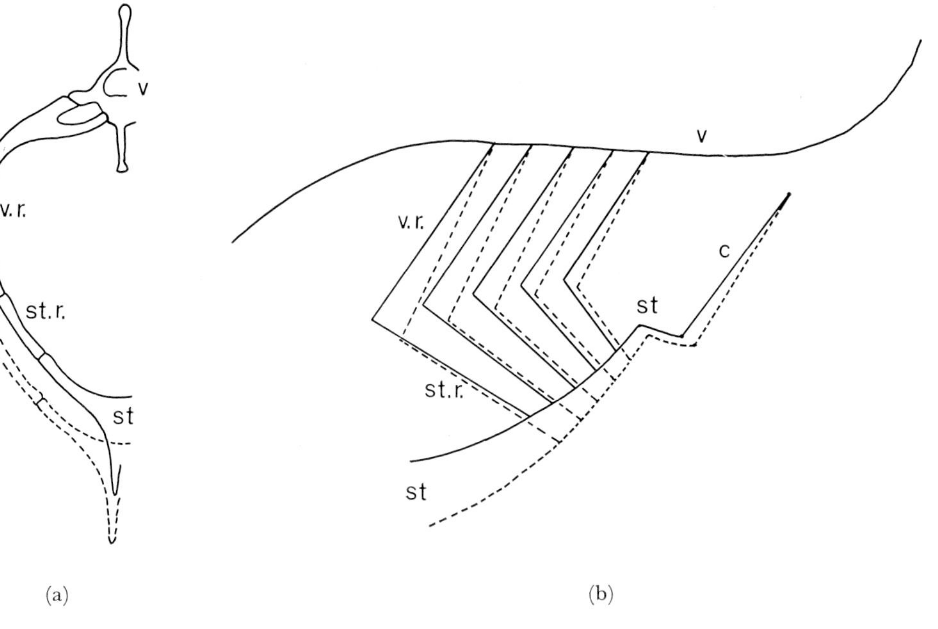

Fig. 23a. Diagrammatic cranial view of the right half of the thoracic cage to show the movements of the ribs and sternum during resting breathing in the erect, sitting, or supine postures. Solid lines, position at the end of expiration. Dotted lines, position at the end of inspiration. v, vertebra; v.r., vertebral rib; st.r., sternal rib; st, sternum. After Zimmer (1935), reproduced with permission from King (1966).

Fig. 23b. Diagrammatic lateral view of the thoracic cage to show the movements of the ribs and sternum during resting breathing in the erect posture. Solid lines, position at the end of expiration. Dotted lines, position at the end of inspiration. v, vertebral column; v.r., vertebral rib; st.r., sternal rib; st, sternum; c, coracoid. After Baer (1896), reproduced with permission from King (1966).

draws the abdominal wall laterally. These movements increase the dorsoventral and transverse diameters of the abdominal cavity synchronously with the equivalent changes in the diameters of the thoracic cage. These movements of the body wall were established, for a variety of birds by various workers including Bert (1870), Soum (1896), Baer (1896) and Zimmer (1935). Zimmer (1935) also found only minor differences when the posture was supine, which is consistent with the inability of Fedde *et al.* (1964a) to detect any differences electromyographically between these two postures. For general experimental purposes, however, it must be remembered that changing to the supine posture reduces the tidal air by 40 or 50% (King and Payne, 1964), probably because the viscera obstruct the ostia (Soum, 1896) or compress the thoracic and abdominal sacs (Vos, 1934). Both Soum (1896) and Salt and Zeuthen (1960) emphasized the need to take posture into account when making respiratory experiments on birds.

The over-all expansion of the body cavity causes a general fall in pressure within the coelom, and consequently within the air sacs, the pressure changes being usually regarded as synchronous throughout the whole system including the individual sacs and the airways generally (see Sturkie, 1965, for a review). As a result of this pressure gradient, air moves into the lungs and air sacs. On theoretical grounds (see Fig. 24), however, transient pressure gradients are to be expected because the resistance to flow into the sacs varies; there should be a craniocaudal gradient (high to low) during inspiration, and the reverse during expiration. Such pressure gradients have been recorded (Brackenbury, 1971), the pressure differentials between the caudal and cranial sacs reaching maxima of from 0·5 to 2 mm of water. Values for tidal volume, rate per minute, and minute volume in eupnoea are given in Table 1, Chapter 6. Piiper *et al.* (1970) believed their own values (Wt 1·6 kg, $\dot{V}_E$ 760 ml/min, f 23/min, $\dot{V}_T$ 33 ml) to be much higher on a body weight basis than the others cited, and ascribed this to probable differences in ambient temperature.

It has long been accepted from earlier work (e.g. Bert, 1870; Campana, 1875) that the air sacs are thus filled during inspiration. Soum (1896), however, produced experimental evidence that the lung itself undergoes some degree of distension during inspiration. This occurs firstly by means of the increased transverse diameter of the thoracic cage. Secondly it occurs because the areas of the intercostal spaces are enlarged as the ribs move cranially; since four or five (or six, in some birds) vertebral ribs are deeply embedded in the lung this enlargement must produce a corresponding enlargement of the dorsolateral surface of the lung up to one third of it being involved (see page 102). This factor was emphasized by Zimmer (1935) and Scharnke (1938). Thirdly the lung as a whole must expand, because of the pressure gradient between the bronchial airways and the air sacs. However, owing to the openings from the lungs into the air sacs a constant pressure gradient between the inside and the outside of the lung will not be maintained throughout the whole of the inspiratory phase. On the contrary, it will *decrease* as the peak of inspiration is approached; and if the position at the peak of inspiration is held the pressure gradient will disappear entirely as the air passes into the air sacs, thus bringing the whole system (coelom, air sacs, interior of lungs, trachea, atmosphere) into equilibrium. Similar considerations apply during expiration.

Stresses in the lower respiratory tract which directly result during the respiratory cycle from changes in the angulation of the ribs will be maintained for as long as the position of the ribs is maintained. In contrast, stresses which are dependent on the pressure gradient across the substance of the lung will decrease as the gradient decreases. These stresses, notably those arising from the pressure gradient, differ from those in the mammalian lung. Therefore it would not be expected that firing patterns in the bird of slowly adapting pulmonary stretch receptors (see Molony, 1970) would be the same as those found in mammals. The firing patterns of the main group of respiratory afferent units described by King *et al.* (1968) can be correlated reasonably well with the pressure gradients. These afferent units are therefore possibly to some extent mechanosensitive. However, other factors are also important, particularly the pCO_2 of the air with which they are in contact (Peterson, 1970; Fedde, 1970; Molony, 1970). The impossibility of sustaining an inflation of the avian lung (because of the flow of gas into the sacs) could be one of the reasons why the inflation reflex in birds is weak (see Richards, 1968b).

It was established by Siefert (1896) and Stübel (1910) that, in birds generally, after all muscular forces have been eliminated, the thoracic cage comes to rest roughly midway between the inspiratory and expiratory peaks. Soum (1896) concluded that the first phase of the respiratory cycle is due to the elastic recoil of the thoracic cage. However, electromyographic studies have established that although elastic recoil may be important it is not solely responsible for the initial phases of the respiratory cycle (Fedde *et al.*, 1964a). These authors showed that in lightly anaesthetized cocks electrical activity of the inspiratory muscles begins only a few milliseconds after the onset of the inspiratory movement of the sternum, while the electrical activity of the expiratory muscles actually begins a few milliseconds before the onset of the expiratory sternal movement. Furthermore the activity of both the inspiratory and the expiratory muscles overlaps into the next succeeding respiratory phase, indicating a direct antagonism between the two groups of muscles and producing a smooth transition between the two phases of the respiratory cycle.

VI. Functional Demands on the Avian Respiratory System

The following section is a brief discussion of some of the demands placed upon the avian respiratory system and some of the ways in which they may be met by the structure and function of the system. The complexity of the system reflects the complexity of the demands placed upon it, and as Salt (1964) said, "we have as yet no final answers on the functioning of this most complicated system".

The avian respiratory system performs three main functions, gaseous exchange, thermoregulation and vocalization.

A. EXTERNAL RESPIRATION

The avian respiratory system is more highly developed for obtaining large amounts of oxygen from its environment per unit of time than that of any other animal. The maximum oxygen requirements of birds are much greater than those of other animals due to the heavy expenditure of energy during flight (Zeuthen, 1942; Tucker, 1968a,b). Since most birds maintain an aerial existence, the demands placed upon the respiratory system during flight are the dominant factor in its design.

1. Oxygen Requirements of Flight

Tucker (1968a,b) has analysed the oxygen requirements of some flying birds and has shown how these are met. Flight can demand an increase of up to twenty-one times the resting oxygen consumption (Le Febvre, 1964; Hart and Roy, 1966; Tucker, 1968a,b). This is provided by a massive increase in the rate and depth of respiration, a concurrent increase in cardiac output, and probably an increase in the diffusing capacity of the lungs. A poor respiratory capacity to weight ratio prohibits sustained flight, as in the fowl which is unable to maintain flight even though its respiratory anatomy is essentially similar to that of flying birds. The resting minute volume in the cock of heavy breeds is 777 ml/min (King and Payne, 1964); a twenty-fold increase in this would give a minute volume of 15·5 l/min. The maximum minute volume quoted for the cock is 2·2 l/min (King and Payne, 1964) or 2·3 l/min (Frankel *et al.*, 1962).

The minute volume required to sustain flight could be achieved in the cock only by the adoption of both rates and depths approaching their maximum levels, i.e. tidal volumes of about 130 ml (King and Payne, 1964) and breathing rates of 140 to 150/min (Frankel *et al.*, 1962), which would give a minute volume of 18·2 l/min. It is not considered possible for the cock to sustain such high minute volumes, since the work rate of the respiratory apparatus would be far outside its limits of efficiency in terms of oxygen gained from work done. Such a bird creates an oxygen debt if it flies at all.

In birds which fly, although the task is simplified by a much higher respiratory capacity to weight ratio, the passage of large volumes of air at high flow rates across the exchange tissue must still occur. The attainment of a respiratory capacity capable of accommodating these large volumes of air is accomplished in the bird by the development of large air sacs, one notable advantage of this being the relatively small percentage change in volume of the coelom necessary to move the requisite tidal air. The bird has thus achieved the necessary vital capacity not by an increase in the size of the lungs, but by the development of the air sac system. Vital capacity is of particular importance at high altitudes where the high water vapour pressure decreases the effectiveness of the respiratory excursion. At an altitude of 19,000 m no air at all will enter the respiratory system no matter how deep inspiration may be (Fenn, 1964).

Calculations of gaseous diffusion between the tertiary bronchi and air capillaries (Zeuthen, 1942; Hazelhoff, 1951) suggest that this is adequate for the bird's needs both at rest and in flight. The oxygen made available at the exchange area must then be transported from the lungs to the tissues, carbon dioxide being transported simultaneously from the tissues to the lungs.

2. Resistance to Flow

A respiratory system which must transmit large volumes of air at high velocities demands a minimum resistance to the flow of air both in and out.

The resistance to flow is governed by the cross-sectional area, length and aerodynamic properties of the airways. Maximum efficiency is obtained if minimum turbulence is produced at maximum velocity. The aerodynamic features required to prevent turbulence are the avoidance of sharp bends and the development of airways wide enough to accommodate the flow rates encountered.

The distribution of resistance throughout the respiratory system of the bird is not likely to be the same as that seen in the mammal (see Widdicombe, 1966, for a review of this factor in the mammal). The resistance of the nasal passages will be ignored, since they are likely to be by-passed when demand for oxygen is high. In the bird the trachea is probably the most important source of resistance during inspiration (see pages 99 and 102 for the cross-sectional area of the trachea and other airways). Because of the presence of complete rings of cartilage, the diameter of the trachea cannot be varied. This diameter may be a compromise between the need for minimum resistance and minimum dead space. The need to minimize dead space is essential for "ideal" gaseous exchange, since the time and work involved in moving dead space air are wasted. The volume of this part of the upper respiratory dead space may also be influenced by the demands of heat dissipation (see page 159).

(*a*) *Resistance during inspiration.* The resistance of the secondary and tertiary bronchi during inspiration is probably decreased by expansion of the lungs (see page 150). This decrease is probably only a very small part of the total resistance and will not greatly affect the efficiency of the system. On the other hand the increase in cross-sectional area, due to the expansion of the lungs, will cause a decrease in the rate of airflow across the exchange tissue and will thus permit more exchange to occur.

(*b*) *Resistance during expiration.* During expiration the distribution of some of the components of resistance of the respiratory system is different from that in inspiration.

The resistance to airflow during expiration is probably concentrated at the syrinx and at the extra-pulmonary primary bronchus. This is due to collapse inwards of the thin parts of their walls, most particularly of the syrinx (see page 101), caused by the pressure difference between the clavicular air sac and the lumen of these airways. The importance of this valvular effect of the syrinx during expiration is that it reduces the pressure gradient across the lung. The "valve" will open when the pressure within the lumen of the syrinx and primary brochus approaches that in the air sacs. Something of this nature may have been envisaged by Scharnke (1938) when noting that the syrinx is involved not only in vocalization but also in a more general role as an accessory apparatus of respiration.

This appears to be one of several adaptations which minimize collapse or compression of exchange tissue and pulmonary airways during expiration, to maintain as low a resistance to airflow as possible. Other adaptations achieve the same effect by partly isolating the lung from the increased pressure in the coelom during expiration. For instance the location of the lung against the dorsolateral body wall leaves only one exposed surface, about two-thirds of which is covered by the broncho-pleural membrane (pulmonary aponeurosis) and its costopulmonary muscle (see page 144). The latter contracts during expiration (Soum, 1896; Fedde *et al.*, 1964c) and, because of its domed structure, opposes the pressure gradient across the lung; in Hazelhoff's (1951) words (translated) ". . . the contraction of the lungs on the contrary is prevented or at least counteracted by it". Contraction of this muscle may also maintain the patency of the ostia during expiration (see page 144). Further support is provided by the ribs which are deeply embedded in the lung—up to one-third of which being particularly well supported in this way (see page 102). Compression is unlikely to be eliminated completely, but increase in resistance in the lungs is not likely to be the limiting factor in expiration. The compression which does occur is not uniform, since the ventral parts of the lung, which are not supported by the ribs, are likely to be more readily compressed. This non-uniform compression may have important effects upon the pattern of airflow.

3. *Compliance and the Efficiency of the Respiratory Musculature*

The maximum compliance (9·5 ml/cm H_2O) of the air sacs and the surrounding body wall together, occurs between -10 cm and 0 cm of H_2O in paralysed hens weighing 1·6 kg (Scheid and Piiper, 1969). These authors remarked on the agreement in respect of relative compliance between the hen and man (4%/cm H_2O). This is static compliance. The efficiency of the bird's respiratory pump is determined by the dynamic compliance of the system, the diffusing capacity of the lungs, and the efficiency of the respiratory muscles especially at high work rates. The efficiency of the musculature depends upon the proportion of inspired oxygen required by the respiratory muscles. The latter factor probably becomes particularly important only when the bird is breathing air with a very low pO_2 as at high altitudes, or when it needs to increase its O_2 consumption to maximum during an all-out sustained effort. Avian respiratory mechanics are different, and may be more efficient than in the mammal, but no data are available to confirm or deny this.

4. *Flow Patterns*

To obtain maximum gaseous exchange, as much as possible of the air pumped into and out of the respiratory tract must be exposed to the exchange tissue. This could be achieved by a complex flow pattern through the lungs and air sacs. Such a pattern could be designed to maintain the highest possible diffusion gradients at the exchange tissue during the whole cycle.

The distribution of air between the lungs and the various air sacs, and the route which it takes within the lung itself, have been the subject of much experimentation and debate. For reviews of the extensive literature see Salt and Zeuthen (1960), Sturkie (1965) and King (1966).

(*a*) *Flow of air during inspiration.* The cranial air sacs are probably filled early in inspiration by air coming mainly from the upper respiratory dead space. Data for the ostrich indicate that no inspired tidal air enters the cranial sacs (Schmidt-Nielsen *et al.*, 1969). This may also occur in hens since these sacs receive only 8 to 16% of the tidal volume (Zeuthen, 1942), this being a volume which would rarely exceed the volume of the upper respiratory dead space (8·5 ml in the fowl; see page 99). Rapid influx of this dead space air into the cranial sacs is consistent with their wide and short direct connexions to the most cranial part of the intrapulmonary primary bronchus (see page 141). This could be tested by gradual elimination of the upper respiratory dead space while measuring the pO_2 of the air in the cranial sacs during inspiration.

The caudal sacs are filled by air from within the lungs and by part of the inspired tidal air. The rest of the inspired tidal air remains in the lungs and the upper respiratory dead space at the end of inspiration. On the basis of direct observations on pressure differentials between air sacs and on airflows within the caudodorsal secondary bronchi, Brackenbury (1971) concluded that the cranial sacs probably do fill more rapidly in early inspiration and also suggested that these sacs may draw end-inspiratory air from the lungs.

It was generally agreed by the earlier experimentalists (see King, 1966, for review) that the more caudal sacs in both the fowl and the duck have a higher pO_2 and lower pCO_2 than the more cranial sacs, and this finding has recently been confirmed by Cohn and Shannon (1968) and Piiper *et al.* (1970). [Shepard *et al.* (1959), reported that the gas composition was nearly the same in all the sacs, but, as Piiper *et al.* (1970) pointed out, this view is not quite consistent with their tracings.] Hence it was suggested that the caudal sacs are better ventilated. This was confirmed by experiments where the bird breathed hydrogen; Zeuthen (1942) concluded that in the fowl between 70 and 80% of the inspired air entered the abdominal sacs. Since the indirect connexions of these sacs seem to present a much lower resistance than their narrow direct connexions (see page 141), this air is readily available for gaseous exchange. Nevertheless, it does not achieve equilibrium with the blood in the time available because the diffusing capacity of the lungs during inspiration is insufficient to permit equilibrium to be attained. The diffusing capacity could be regulated by the action of the smooth muscle of the airways enabling the gas partly to bypass the exchange area (see page 161). According to Piiper *et al.* (1969) "the pulmonary diffusing capacity in birds able to fly should be much higher or should increase considerably during flight".

The increase in volume of each sac is dictated by the increase in volume of the particular part of the coelom in which the air sac lies. The pattern of filling the various air sacs is likely to be complex and governed by the resistance of the airways leading to each air sac and also by the elasticity of the septa between adjacent air sacs (see model, Fig. 24).

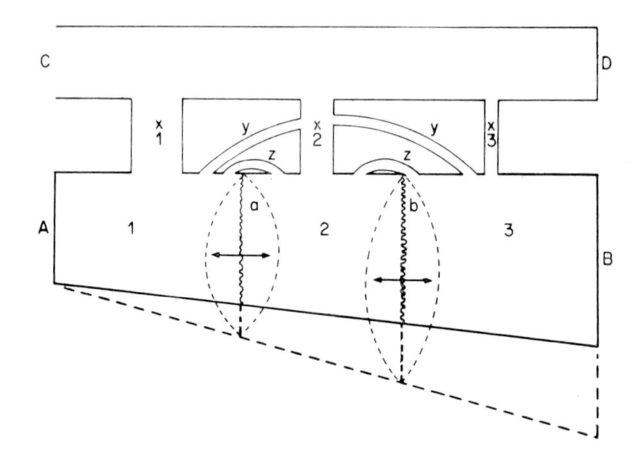

Fig. 24. Model of the coelom and air sacs. The chamber AB represents the coelom. Section 1 represents the clavicular and cranial thoracic sacs, section 2 the caudal thoracic sac, and section 3 the abdominal sac. CD is the primary bronchus. Each of the three sections has one large inlet (x1, x2, x3) representing the direct connexions of the sacs. These inlets are of different diameters, appropriate to their sacs. There are also smaller connexions (y, z), representing the indirect connexions of the sacs. The walls "a" and "b" are elastic membranes separating the sacs. If the whole chamber is increased in volume as shown by the broken lines, air will enter section 1 more rapidly than section 2, and section 2 more rapidly than section 3, since the resistance to flow through the inlet into section 1 is less than the resistance into section 2, and 2 is less than 3. Wall "a" will bulge into section 2, and wall "b" into section 3, until the stretching of the elastic walls counteracts the pressure gradient which has been set up. These pressure gradients are, however, only transient, and at the end of inspiration the elastic walls will return to their resting positions. When the resistances of all the inlets are low, the times involved are very short. The degree of bulging of the elastic walls will depend on the volume expansion of the sections, the different resistances of the inlets, and their elasticity. Some air will flow from one section to another through the small connexions y and z. The pressure gradients should be reversed during expiration, being now caudocranial in direction. Since the whole process is dynamic, pressure differences are not sustained and could be difficult to measure. Differentials between the caudal and cranial sacs of from 0·5 to 2 mm of water have been observed by Brackenbury (1971).

(*b*) *Flow of air during expiration.* Emptying the air sacs is governed, in general, by the same factors as described above for inspiration, the relationships being reversed as the coelom (Fig. 24) is compressed. The transient caudocranial pressure gradient proposed in Fig. 24 has been confirmed by Brackenbury (1971). The pattern of flow through the lungs during expiration is unlikely, however, to be the simple reversal of the inspiratory pattern of flow suggested by Zeuthen (1942) (one of his two proposals), since different aerodynamic conditions exist and a non-uniform compression of airways occurs. Increase in the resistance of airways due to compression and collapse during expiration, and the adaptations of the system to minimize this, have been described (see page 153).

Determination of the exact roles of individual airways is difficult, but the most widely accepted theories (Zeuthen, 1942; Hazelhoff, 1951) are compatible with the general outline just proposed, the essential point being that most of the tidal air can be available for exchange during both inspiration and expiration. The optimal length and calibre of a tertiary bronchus will be determined by the rates of flow of air with which it must cope under conditions of maximum demand. These rates of flow are in turn dependent on the number, lengths, and diameters of the tertiary bronchi. Under "ideal" conditions of gaseous exchange during maximum demand, the length of the tertiary bronchi should be such that the air *only* comes into equilibrium with pulmonary arterial blood as it *leaves* the exchange area for the second time.

5. *Gas Exchange and Transport*

Gas exchange in resting hens weighing 1·6 kg has been studied by Piiper *et al.* (1970). They obtained the following mean values: total ventilation, 760 ml BTPS/min; respiratory frequency, 23/min; oxygen uptake, 24 ml STPD/min; arterial pO_2, 87 mmHg; arterial pCO_2, 29 mmHg; cardiac output (Fick principle), 430 ml/min. The carbon monoxide diffusing capacity of the respiratory system ($0{\cdot}8 \pm 0{\cdot}07$) has been studied by Piiper *et al.* (1969). Gas exchange in other birds both at rest and in flight has been studied by Tucker (1968a,b).

The respiratory system must extract as much oxygen per unit time as possible from large volumes of rapidly moving air. The amount of exchange tissue required will depend on the maximum oxygen requirements to be met, the efficiency of the exchange process, and the maximum possible level of ventilation. Transport of such large amounts of oxygen and carbon dioxide demands very high cardiac outputs, which are also necessary to move enough oxygen to the tissues when the pO_2 in the inspired air is low and the oxygen carried per millilitre of blood is decreased. Under such conditions of maximum demand the distribution and flow pattern of the blood should allow as much gaseous exchange as possible.

The anatomical details of the pulmonary blood supply in the bird are largely unknown. Salt and Zeuthen (1960) concluded that there seems to be no possibility of a counterflow between the air and blood. Scheid and Piiper (1970), in an attempt to explain the paO_2 values which they found in resting hens, postulated the existence of pulmonary arterio-venous shunts. They also proposed that "the blood-flow units in the avian lungs are arranged in parallel but the ventilation for a given parabronchus is in series". This arrangement requires that blood of the same composition arrives at each part of the exchange tissue (in parallel), but that air which arrives at the exchange tissue passes successively from one region to the next (in series) along the tertiary bronchi, its composition being different in each region.

This results in "cross-over" exchange (Scheid and Piiper, 1970). These authors do not give any anatomical evidence for such a disposition of the blood supply, but observations by Salt and Zeuthen (1960) and by A. S. King (unpublished) on the distribution of these vessels support such an arrangement.

The isolation of the lung from expiratory pressure changes as discussed above will be as important in maintaining pulmonary blood flow as it is in maintaining air flow. The blood supply to the exchange tissue should therefore fluctuate much less during the breathing cycle in the bird than in the mammal.

Very high cardiac outputs occur during flight, exceeding 3·75 l/kg min in flying budgerigars (Tucker, 1966). This is more than seven times the maximum for man and dogs (Tucker, 1968a). The relatively large heart, 1·34% of body weight in the sparrow compared with 0·5% of body weight in the mouse (Tucker, 1968a), indicates how birds may achieve this relatively large cardiac output.

B. THERMOREGULATION

Since body temperature and thermoregulation are covered in Chapter 48, only aspects relating to the functional anatomy of thermoregulation are discussed here.

The respiratory system in the bird provides a means of heat dissipation. Zeuthen's (1942) proposal that in flying birds "the needs of heat regulation are so high that the ventilation of the lungs should be highly in excess of the needs of respiration" seems too extreme, and it is most likely that the basic design of the system is dictated by the requirements for gaseous exchange during flight rather than by the requirements of heat regulation. Heat dissipation by the respiratory system is, however, of major importance in many birds, being required under several different circumstances. The greatest demand is made when high ambient temperatures are combined with high metabolic activity. Nearly all the heat is then dissipated by evaporative cooling from the respiratory tract, loss through the skin being negligible (less than 2% of total water loss: Schmidt-Nielsen *et al.*, 1969). Where a high level of metabolic activity occurs at low ambient temperatures, for example in high altitude flight, only about 18% of the heat produced is dissipated via the respiratory tract, the major source of heat loss being by means of conduction and convection via the skin, especially that of the feet (Hart and Roy, 1966, 1967).

Maximum evaporative cooling from the respiratory tract can therefore be necessary under two different circumstances and the demands placed on the respiratory system are different in each case. The first of these is the combination of high ambient temperature and high metabolic activity, which demands both a high level of heat dissipation and a high level of gaseous exchange. The second is when high ambient temperatures are combined with low metabolic activity; this demands a high level of heat dissipation and a much lower level of gaseous exchange. The requirements of these and intermediate conditions must be met, under "ideal" circumstances, by means of a carefully controlled balance between heat dissipation and gaseous exchange. It is in the animal's best interest to avoid, if possible, the alkalosis which results from excessive elimination of carbon dioxide.

Evaporative cooling can occur from the whole respiratory tract, which will be considered in two parts, the upper airways and the lungs and air sacs.

1. Upper Airways

These include the nasal cavities, oropharynx, and trachea, constituting the upper respiratory dead space, and are of major importance in heat dissipation. Various special adaptations for this process, notably gular pouches (Bartholomew *et al.*, 1968), are found particularly in birds living at high ambient temperatures, such adaptation being designed to increase the surface area and its accompanying blood supply. Very little washing out of carbon dioxide should occur in these areas due to the thickness of the blood-air barrier. This barrier, however, must be relatively thin and the blood flow as large as possible to maintain the maximum rate of heat loss.

The volume of the upper respiratory dead space is important since control of the ratio of the tidal volume to the volume of the upper respiratory dead space is an important means of regulating the amount of gaseous exchange that occurs during heat dissipation. To permit such control, the arrival of "fresh" air in the lungs could be signalled to the central nervous control centres by means of carbon dioxide-sensitive receptors in the lungs (Peterson and Fedde, 1968; Peterson, 1970; Fedde, 1970; Molony, 1970).

2. Lungs and Air Sacs

The extremely large surface area and massive blood supply of the exchange tissue provides an excellent source of evaporative cooling, but hyperventilation of this tissue produces alkalosis unless the metabolic rate is correspondingly high. Therefore the bird can best employ this method of heat elimination when doing external work such as flying or running. If the metabolic rate is low and the heat dissipated from the upper respiratory dead space is insufficient, then the bird must either accept some alkalosis (Calder and Schmidt-Nielsen, 1968) or decrease the diffusing capacity of its lungs. The latter may be achieved by changes in the tone of the smooth muscle of the airways, particularly of the tertiary bronchi, or by changes in the smooth muscle tone of the blood vessels of the exchange tissue. Blood vessels which can act as arterio-venous shunts may also be important in heat dissipation, especially if these shunts could be shown to lie very close to the air stream. Birds seem to regulate their diffusing capacity with varying efficiency depending on the species. For example, ambient temperatures of 30°C (fowl) and 40°C (pigeon) induce marked increases in minute volume (three-fold in pigeons) without a fall in $paCO_2$ (Zeuthen, 1942; Calder and Schmidt-Nielsen, 1966), but alkalosis does develop in the pigeon if panting is induced by exposure to ambient temperatures of 45° to 50°C (Calder and Schmidt-Nielsen, 1966). The ostrich, however, shows no decrease in $paCO_2$ even after panting heavily for 8 h (Schmidt-Nielsen *et al.*, 1969).

Insensible heat dissipation from the air sacs will depend upon their ventilation, on the water vapour pressure of the air reaching them, and on the amount of sensible heat passing through their walls which is itself influenced by their thickness, surface area and the temperature gradient. The vascularity of the air sac walls and of the adjacent organs and tissues will dictate the temperature gradient. If the air has reached the air sacs after passing over a large amount of exchange tissue, its saturation with water vapour will be higher than if this tissue is bypassed in some way. A decrease in the diffusing capacity of the lung can serve two purposes in a bird with a low level of metabolic activity and a high requirement for heat dissipation, since the air arriving in the air sacs will be less saturated with water vapour and heat can thus be dissipated without CO_2 washout.

Although their relative importance in thermoregulation is uncertain, it is clear that the air sacs play a significant part in heat dissipation (Salt and Zeuthen, 1960; Salt, 1964; Schmidt-Nielsen *et al.*, 1969). Salt and Zeuthen considered the more cranial sacs to be of particular interest in respiratory evaporation. This conclusion they supported with various lines of experimental evidence, including Scharnke's (1938) analyses of the pCO_2 of air from the cranial thoracic and clavicular air sacs in the pigeon during hyperventilation in response to increased heat load, and to increased oxygen demand caused by fluttering or cooling; the pCO_2 fell during panting, but remained normal during fluttering and cooling. The decreased pCO_2 of the air in the cranial sacs during panting was interpreted by Salt and Zeuthen (1960) and Salt (1964) to indicate an increase in their ventilation. However, an alternative interpretation arises if, as proposed above (see page 155), these cranial sacs are filled mainly from the upper dead space. The pCO_2 of the air in the cranial sacs under these circumstances would then reflect the amount of gaseous exchange which occurred during the previous breath. The pCO_2 of air in these sacs should then be similar (it is higher in resting hens, see Piiper *et al.*, 1970) to that of end expired air. If gaseous exchange is decreased during panting, in the interests of delaying or avoiding a CO_2 deficit, the pCO_2 of air at the end of expiration will be decreased and that of air in the cranial sacs will also be correspondingly reduced.

The function of the cranial air sacs in heat regulation is probably related to their access to the primary bronchus (see page 141), which as Salt (1964) pointed out, is so direct that "it is theoretically possible that they could be actively ventilated without affecting the CO_2 level of the air in the lung capillaries". Use of the cranial sacs for the purpose of heat regulation by such a method would be even more efficient if the proportion of the inspired air which enters these sacs could be increased during the rapid shallow breathing which occurs particularly in panting. Such a phenomenon would demand that, during the hyperventilation of heat stress, the resistance to flow into these sacs be relatively lower than the resistance to alternative routes for a relatively greater proportion of the breathing cycle than is normal. This would be expected to follow automatically if they have the low resistance direct connexions described above, and if their filling capacity is restricted by their position at the rostral end of the coelom as in Fig. 24. The filling capacity of these sacs is an important consideration since, if the volume of air which they accept is similar to the volume of the tidal air during panting, a large part of this air could enter these sacs. Once these sacs are full, the rest of the inspired air must pass into the other parts of the respiratory tract.

Heat from the air sacs and lungs can be transferred to the surfaces of the upper respiratory tract by feedback heat transfer (Salt, 1964), the mechanism of which is passive and depends on condensation of water on the cooler surfaces of the upper airways.

3. Role of Smooth Muscle in Controlling the Diffusing Capacity of the Lung

Smooth muscle can play an important role in controlling the airflow through the avian lung (King and Cowie, 1969); little is known, however, about the role of smooth muscle in the control of avian pulmonary blood flow. King and Cowie (1969) have shown *in vivo* and *in vitro* how changes in smooth muscle tone can affect the diameter and thus the resistance of the airways in the avian lung (see page 113). Constriction of the tertiary bronchi, especially of the openings into the atria, reduces the ventilation of the exchange tissue and thus reduces the diffusing capacity of the lung. Effective decrease in the latter can be achieved in several ways by changes in smooth muscle tone, since these changes may be on an over-all, regional, or very local basis.

Uniform contraction of the bronchial muscle would diminish ventilation of the exchange areas, but would also increase to some extent the pulmonary resistance. This increase may be only a small proportion of the total resistance and therefore acceptable, but if it is at all substantial it will decrease the heat-dissipating efficiency of the system. Non-uniform changes in the smooth muscle tone could overcome this difficulty. Under "ideal" conditions constriction of the tertiary bronchi (to reduce ventilation of the exchange area) would occur simultaneously with a dilation of the secondary bronchi (to decrease the resistance to flow into the air sacs).

Very specific and localized changes in smooth muscle tone could also modify the "spatial orientation" (Schmidt-Nielsen *et al.*, 1969) of the openings into the airways, particularly the openings of the secondary bronchi from the primary bronchus, and thus manipulate the aerodynamic factors influencing the flow pattern. The dense innervation of the roots of the secondary bronchi (see page 126), the powerful muscle at their orifices (see page 112), and the ultrastructural morphology of the muscle cells here and their efferent nerve endings which are consistent with fast local contraction (see page 111) all point to the roots of the secondary bronchi as key points in the regulation of lung function. The concentration of neurite-receptor cell complexes at the orifices of the secondary bronchi could provide the necessary afferent information for integration of such complex smooth muscle activity.

King and Cowie (1969) have also obtained evidence for the existence of intrinsic tone of the bronchial muscle, which could act as a starting point for changes in calibre in either direction as in mammals. Non-uniform changes in the tone of the smooth muscle of the airways are regarded as a possibility in mammals (Widdicombe, 1966).

The effect of changes in concentration of oxygen and carbon dioxide on the tone of the smooth muscle of the airways or blood vessels, either directly or indirectly, could play an important part in the control of hypoxia and hypo- and hyper-capnia. The evidence cited by Widdicombe (1966) on the effects of these gases on smooth muscle in the mammalian lung indicates that they vary according to the particular experimental conditions and the state of the subject at the time. The most common effect of inhaled CO_2 in anaesthetized dogs and cats is broncho-constriction, a reflex effect which is eliminated by vagotomy. Shaw (1970) has shown that the effect of carbon dioxide on mammalian pulmonary vascular smooth muscle can be reversed by changing the experimental conditions, and has suggested a variable dilator effect of CO_2 or a variable balance between a vaso-constrictor and a vaso-dilator action. A demonstration of similar variations in the responses to CO_2 by mammalian bronchial smooth muscle might resolve some of the conflicting results obtained previously, and provide further evidence that the diffusing capacity of the lung is influenced by a combination of changes in both blood and air supply to the exchange surfaces.

Ray and Fedde (1969) found that addition of between 2 and 12% CO_2 to the gas flow in unidirectionally ventilated anaesthetized chickens, caused a decrease in the resistance of airways; with 12% CO_2 it is approximately 25%. Similar experiments carried out by V. Molony (unpublished) with the addition of 5% CO_2 to the gas stream (previously 100% O_2) in unanaesthetized chickens, produced no consistent change in resistance. The importance of the levels of CO_2 in the air and their effects on the resistance of the airways, and hence potentially on the diffusing capacity of the lung, should be studied further. Indirect control of these effects may be based on information received from the CO_2-sensitive receptors in the lung (Peterson, 1970; Fedde, 1970; King *et al.*, 1968; Molony, 1970). Some pilot observations (J. L. Armstrong, personal communication) carried out in our laboratory, indicate that in lung slices in an organ bath the bronchial muscle may relax fairly quickly when subjected to avian Ringer's solution in which varying amounts of carbon dioxide had been dissolved.

C. VOCALIZATION

Vocalization plays a major part in the behaviour of most birds. The extreme delicacy, duration, and flexibility of some bird song involves very rapid respiratory movements (Calder, 1970). The mechanics of bird song are beyond the scope of this discussion but it may be possible to discuss briefly the general part played in this process by the respiratory system.

Each note is made during an expiration, due possibly to the valvular effect produced at the syrinx as discussed above (see page 153). Movements of any elastic membranes in this region will be influenced by fluctuations in the pressures in the trachea and clavicular air sacs. The role of these pressure relationships in vocalization has been demonstrated by Gross (1964) by experiments on the syrinx *in vitro*.

The rapid "mini-breaths" described by Calder (1970) demand very rapid movements of air in and out of the coelom. Although the respiratory system of the bird is designed to work at high rates of muscle contraction it must be doubted whether the whole respiratory apparatus can be moved at the rate necessary for some "song" (e.g. 1,200 breaths/min). However, rapid fluctuations in coelomic volume could be produced by rapid craniocaudal movements of the trachea in and out of the thoracic inlet as observed in the chicken by Molony (1970) (Fig. 25). Tracheal muscles capable of producing these

Fig. 25. Recording from a whole body plethysmograph containing a decerebrate adult hen (head and most of the neck outside) weighing 1·3 kg. Inspiration is up. It was observed that the rapid fluctuations in volume (between the arrows) coincided with craniocaudal oscillations of the trachea. Tidal volume approximately 20 cc; mark, 2 s.

rapid movements have been mentioned on page 99. The pressure changes thus induced in the clavicular air sacs could produce oscillation of the syringeal membranes. As the trachea is pulled rostrally the syringeal membranes will bulge into the clavicular air sac, the coelomic pressure being reduced. When the trachea is released the syringeal membranes collapse inward, the pressure now being higher in the clavicular air sac. At the same time a small amount of air is forced through the syrinx to equalize the pressures. The work involved to achieve this is much smaller than that required to move the whole respiratory apparatus. Very high rates could theoretically be achieved, especially as active contraction could be required in one direction only, elastic recoil accounting for movement in the other direction.

The development of the trachea as a tube of constant diameter is important in vocalization, since sound projected into an elastic tube would suffer considerable damping.

Acknowledgements

We are specially grateful to J. McLelland, R. D. Cook, and C. M. Brown of this Department for permission to use much of their thesis material. All the respiratory studies carried out in this Department were made possible through grants from the British Egg Marketing Board from 1964 to 1971.

References

Akester, A. R. (1960). *J. Anat.* **94**, 487–505.
Akester, A. R. (1970). *J. Anat.* **107**, 189–190.
Akester, A. R. and Mann, S. P. (1969). *J. Anat.* **105**, 202–204.
Al-Lami, F. and Murray, R. G. (1968). *J. Ultrastruct. Res.* **24**, 465–478.
Altman, P. L., Gibson, J. F., Wang, C. C., Dittner, D. S. and Grebe, R. M. (1958). *In* "Handbook of Respiration", pp. 22–26. Saunders, Philadelphia.
Andersen, A. E. and Nafstad, P. H. J. (1968). *Z. Zellforsch. mikrosk. Anat.* **91**, 391–401.
Appel, F. W. (1929). *J. Morph.* **47**, 497–518.
Arimoto, K. and Miyagawa, R. (1930). *Mitt. med. Akad. Kiota* **4**, 100–103.
Baer, M. (1896). *Z. wiss. Zool.* **61**, 420–498.
Bang, B. G. (1961). *J. Morph.* **109**, 57–72.
Bargmann, W. (1936). *In* "Handbuch der mikroskopischen Anatomie des Menschen" (W. v. Möllendorff, ed.), Band 5, p. 835. Springer, Berlin.
Bargmann, W. and Knoop, A. (1961). *Z. Zellforsch. mikrosk. Anat.* **54**, 541–548.
Bartholomew, G. A., Lasiewski, R. C. and Crawford, E. C. (1968). *Condor* **70**, 31–34.
Bennett, T. and Malmfors, T. (1970). *Z. Zellforsch. mikrosk. Anat.* **106**, 22–50.
Bert, P. (1870), "Leçons sur la Physiologie Comparée de la Respiration", pp. 320–321. Baillière, Paris.
Biggs, P. M. and King, A. S. (1957). *J. Physiol., Lond.* **138**, 282–299.
Biscoe, T. J. and Stehbens, W. E. (1966). *J. Cell Biol.* **30**, 563–578.
Bittner, H. (1925). *Berl. tierärztl. Wschr.* **41**, 576–579.
Brackenbury, J. H. (1971). *J. Anat.* in press.
Brown, C. M. (1970). *B.Sc. (Hons.) thesis, Department of Veterinary Anatomy, University of Liverpool.*.
Bubien-Waluszewska, A. (1968). *Acta anat.* **69**, 445–457.
Burkart, F. von and Bucher, K. (1961). *Helv. physiol. pharmac. Acta* **19**, 263–268.
Burnstock, G. (1970). *In* "Smooth Muscle" (E. Bülbring, A. F. Brading, A. W. Jones and T. Tomita, eds.), pp. 1–69. Arnold, London.
Burton, R. R. and Smith, A. H. (1968). *Poult. Sci.* **47**, 85–91.
Calder, W. A. (1970). *Comp. Biochem. Physiol.* **32**, 251–258.
Calder, W. A. and Schmidt-Nielsen, K. (1966). *Proc. natn. Acad. Sci. U.S.A.* **55**, 750–756.
Calder, W. A. and Schmidt-Nielsen, K. (1968). *Am. J. Physiol.* **215**, 477–482.
Campana, A. (1875). "Anatomie de l'Appareil Pneumatique-Pulmonaire, etc., chez le Poulet". Masson, Paris.
Campenhaut, E. van (1956). *Archs Biol., Liège* **67**, 1–19.
Cauna, N. (1959). *J. comp. Neurol.* **113**, 169–209.

Cauna, N. (1966). *In* "Touch, Heat and Pain", Ciba Foundation Symposium (A. V. S. de Reuck and J. Knight, eds.), pp. 117–127. Churchill, London.
Cauna, N. (1969). *J, comp. Neurol.* **136,** 81–98.
Cauna, N., Hinderer, K. H. and Wentges, R. T. (1969). *Am. J. Anat.* **124**, 187–210.
Chinard, F. P. (1966). *In* "Advances in Respiratory Physiology" (C. G. Caro, ed.), pp. 107–147. Arnold, London.
Cohn, J. E. and Shannon, R. (1968). *Resp. Physiol.* **5**, 259–268.
Cook, R. D. (1970). *Ph.D. thesis, University of Liverpool.*
Cook, R. D. and King, A. S. (1969a). *J. Anat.* **105,** 202.
Cook, R. D. and King, A. S. (1969b). *Experientia* **25**, 1162–1164.
Cook, R. D. and King, A. S. (1970) *J. Anat.* **106**, 273–283.
Cords, E. (1904). *Arb. anat. Inst., Wiesbaden* **26**, 49–100.
Cover, M. S. (1953). *Am. J. vet. Res.* **14**, 239–245.
Dahl, N. A., Samson, F. E. and Balfour, W. M. (1964). *Am. J. Physiol.* **206,** 818–822.
Dehner, E. (1946a). *Science, N.Y.* **103**, 171.
Dehner, E. (1946b). *Ph.D. thesis, Cornell University.*
Duncker, H. R. (1968). *Anat. Anz.* **121**, 287–292.
Fedde, M. R. (1970). *Fedn. Proc. Fedn Am. Socs exp. Biol.* **29**, 1664–1673.
Fedde, M. R., Burger, R. E. and Kitchell, R. L. (1961). *Poult. Sci.* **40**, 1401.
Fedde, M. R., Burger, R. E. and Kitchell, R. L. (1963). *Poult. Sci.* **42**, 1224–1236.
Fedde, M. R., Burger, R. E. and Kitchell, R. L. (1964a). *Poult. Sci.* **43**, 839–846.
Fedde, M. R., Burger, R. E. and Kitchell, R. L. (1964b). *Poult. Sci.* **43**, 1119–1125.
Fedde, M. R., Burger, R. E. and Kitchell, R. L. (1964c). *Poult. Sci.* **43**, 1177–1184.
Fenn, W. O. (1964). *In* "Handbook of Physiology" (W. O. Fenn and H. Rahn, eds.), Sect. 3, Vol. 1, pp. 357–362. American Physiological Society, Washington D.C.
Fillenz, M. and Woods, R. T. (1970). *In* "Breathing: Hering-Breuer Centenary Symposium", Ciba Foundation Symposium (R. Porter, ed.), pp. 101–107. Churchill, London.
Fischer, G. (1905). *Zoologica, Stuttg.* **19**, 1–45.
Fisher, A. W. F. (1963). *J. Anat.* **97**, 300.
Frankel, H., Hollands, K. G. and Weiss, H. S. (1962). *Archs int. Physiol. Biochem.* **70**, 555–563.
Fujiwara, T., Adams, F. H., Nozaki, M. and Dermer, G. B. (1970). *Am. J. Physiol.* **218**, 218–225.
Gomori, G. (1952). "Microscopic Histochemistry". Chicago University Press, Chicago.
Greenewalt, C. H. (1968). "Bird Song: Acoustics and Physiology". Smithsonian Institution Press, Washington D.C. Cited by Calder (1970).
Groebbels, F. (1932). "Der Vögel" Vol. 1, pp. 40–81. Bornträger, Berlin.
Groodt, M. de, Sebruyns, M. and Lagasse, A. (1960). *Vlaams diergeneesk. Tijdschr.* **29**, 313–318.
Gross, W. B. (1964). *Poult. Sci.* **43,** 1004–1008.
Hampl, A. (1957). *Sb. vys. Sp. zeměd. les. Fac. Brne* **B1**, 23–29.
Harlan, W. R., Margraf, J. H. and Said, S. I. (1966). *Am. J. Physiol.* **211**, 855–861.
Hart, J. S. and Roy, O. Z. (1966). *Physiol. Zoöl.* **39**, 291–306.
Hart, J. S. and Roy, O. Z. (1967). *Am. J. Physiol.* **213**, 1311–1316.
Hazelhoff, E. H. (1951). *Poult. Sci.* **30**, 3–10.
Hsieh, T. M. (1951). *Ph.D. thesis, University of Edinburgh.*

Iggo, A. and Muir, A. R. (1969). *J. Physiol., Lond.* **200**, 763–796.
Jones, D. R. (1969). *Comp. Biochem. Physiol.* **28**, 961–965.
Juárez, P. J. S. (1961). *Col. vet. Espana* **8**, 785–789.
Juillet, A. (1912). *Archs Zool. exp. gén.* **9**, 207–371.
Jungherr, E. (1943). *Bull. Storrs agric. Exp. Stn* **250**, 1–36.
Kadono, H. and Okada, T. (1962). *Jap. J. vet. Res.* **24**, 215–223.
Kadono, H., Okada, T. and Ono, K. (1963). *Poult. Sci.* **42**, 121–128.
Kass, E. H., Green, G. M. and Goldstein, E. (1966). *Bact. Rev.* **30**, 488–497.
Kaupp, B. F. (1918). "The Anatomy of the Domestic Fowl" p. 274. Saunders, Philadelphia.
King, A. S. (1966). *Int. Rev. gen. exp. Zool.* **2**, 171–267.
King, A. S. (1972). *In* "The Anatomy of Domestic Animals" (R. Getty, S. Sisson and J. D. Grossman, eds.), 5th ed., Part VI, Chapter 6, in press. Saunders, Philadelphia.
King, A. S. and Atherton, J. D. (1970). *Acta anat.* **77**, 78–91.
King, A. S. and Cowie, A. F. (1969). *J. Anat.* **105**, 323–336.
King, A. S., Ellis, R. N. W. and Watts, S. M. (1967). *J. Anat.* **101**, 607.
King, A. S., McLelland, J., Molony, V. and Mortimer, M. F. (1968). *J. Physiol., Lond.* **201**, 35–36.
King, A. S. and Payne, D. C. (1958). *J. Anat.* **92**, 656.
King, A. S. and Payne, D. C. (1960). *Anat. Rec.* **136**, 223.
King, A. S. and Payne, D. C. (1962). *J. Anat.* **96**, 495–503.
King, A. S. and Payne, D. C. (1964). *J. Physiol., Lond.* **174**, 340–347.
Klaus, M., Reiss, O. K., Tooley, W. H., Piel, C. and Clements, J. A. (1962). *Science, N.Y.* **137**, 750.
Krahl, V. E. (1964). *In* "Handbook of Physiology" (W. O. Fenn and H. Rahn, eds.), Sect. 3, Vol. 1, pp. 213–284. American Physiological Society, Washington D.C.
Lambson, R. O. and Cohn, J. E. (1968). *Am. J. Anat.* **122**, 631–650.
Le Febvre, E. A. (1964). *Auk* **81**, 403–416.
Lever, J. D., Lewis, P. R. and Boyd, J. D. (1959). *J. Anat.* **93**, 478–490.
Lewis, M. R. (1924). *Am. J. Physiol.* **68**, 385–388
Locy, W. A. and Larsell, O. (1916a). *Am. J. Anat.* **19**, 447–504.
Locy. W. A. and Larsell, O. (1916b). *Am. J. Anat.* **20**, 1–44.
MacDonald, J. W. (1970). *Br. vet. J.* **126**, 89–93.
McLelland, J. (1965). *J. Anat.* **99**, 651–656.
McLelland, J. (1966). *M.V.Sc. thesis, University of Liverpool.*
McLelland, J. (1968). *Acta anat.* **69**, 81–86.
McLelland, J. (1969). *J. Anat.* **105**, 202.
McLelland, J. (1970). *Ph.D. thesis, University of Liverpool.*
McLelland, J. (1972). *In* "The Anatomy of the Domestic Animals" (R. Getty, S. Sisson and J. D. Grossman, eds.), 5th ed., Part VI, Chapter 5, in press. Saunders, Philadelphia.
McLelland, J. and King, A. S. (1972). *In* "The Anatomy of the Domestic Animals" (R. Getty, S. Sisson and J. D. Grossman, eds.), 5th ed., Part VI, Chapter 1, in press. Saunders, Philadelphia.
McLelland, J., Moorhouse, P.D.S., and Pickering, E.C. (1968). *Acta anat.* **71**, 122–133.
McLeod, W. M. and Wagers, R. P. (1939). *J. Am. vet. med. Ass.* **95**, 59–70.
Magnus, H. F. (1869). *Arch. Anat. Physiol.* **2**, 207–235.
Malewitz, T. D. and Calhoun, M. L. (1958). *Poult. Sci.* **37**, 388–398.

Marcus, H. (1937). *In* "Handbuch der vergleichende Anatomie der Wirbeltiere" (L. Bolk, E. Goppert, E. Kallius and W. Lubosch, eds.), Vol. III, p. 953. Urban and Schwarzenberg, Munich.
Matre, N. S. Van (1957). *Ph.D. thesis, University of California, Davis.*
Mendenhall, R. M. and Sun, C. N. (1964). *Nature, Lond.* **201**, 713–714.
Mennega, A. and Calhoun, M. L. (1968). *Poult. Sci.* **47**, 266–280.
Merrillees, N. C. R. (1968). *J. Cell Biol.* **37**, 794–817.
Miller, D. A. and Bondurant, S. (1961). *J. appl. Physiol.* **16**, 1075–1077.
Miller, M. R. and Kasahara, M. (1964). *Am. J. Anat.* **115**, 217–234.
Molony, V. (1970). *Ph.D. thesis, University of Liverpool.*
Morejohn, G. V. (1966). *Poult. Sci.* **45**, 33–39.
Muller, B. (1908). *Smithson. misc. Collns* **50**, 365–414.
Munger, B. L. (1965). *J. Cell Biol.* **26**, 79–97.
Munger, B. L. (1966). *In* "Touch, Heat and Pain", Ciba Foundation Symposium (A. V. S. de Reuck and J. Knight, eds.), pp. 129–130. Churchill, London.
Muratori, G. (1934). *Archo ital. Anat. Embriol.* **34**, 45–71.
Myers, J. A. (1917). *J. Morph.* **29**, 165–215.
Nafstad, P. H. J. and Andersen, A. E. (1970). *Z. Zellforsch. mikrosk. Anat.* **103**, 109–114.
Nagaishi, C., Okada, Y., Ishiko, S. and Daido, S. (1964). *Expl Med. Surg.* **22**, 81–117.
Nowell, J. A., Pangborn, J. and Tyler, W. S. (1970). Proc. 3rd. Annual Scanning Electron Microscope Symposium, Chicago.
Ogawa, C. (1920). *Am. J. Anat.* **27**, 333–393.
Okada, Y., Ishiko, S., Daido, S., Ikeda, S., Genka, K. and Kitano, M. (1965). *Acta tuberc. jap.* **14**, 89–95.
Paintal, A. S. (1963). *Ergebn. Physiol.* **52**, 74–156.
Pattle, R. E. (1958). *Proc. R. Soc.* **B148**, 217–240.
Pattle, R. E. (1965). *Physiol. Rev.* **45**, 1–79.
Pattle, R. E. and Hopkinson, D. A. W. (1963). *Nature, Lond.* **200**, 894.
Payne, D. C. (1960). *Ph.D. thesis, University of Bristol.*
Payne, D. C. and King, A. S. (1959). *J. Anat.* **93**, 577.
Peterson, D. F. (1970). *Ph.D. thesis, Kansas State University, Manhattan, Kansas.*
Peterson, D. F. and Fedde, M. R. (1968). *Science, N.Y.* **162**, 1499–1501.
Petrík, P. (1967). *Folia morph.* **15**, 176–186.
Petrík, P. and Riedel, B. (1968a). *Lab. Invest.* **18**, 54–62.
Petrík, P. and Riedel, B. (1968b). *Z. Zellforsch. mikrosk. Anat.* **88**, 204–219.
Piiper, J., Drees, F. and Scheid, P. (1970). *Resp. Physiol.* **9**, 234–245.
Piiper, J., Pfeiffer, K. and Scheid, P. (1969). *Resp. Physiol.* **6**, 309–317.
Policard, A., Collet, A. and Martin, J. C. (1962). *Z. Zellforsch. mikrosk. Anat.* **57**, 37–46.
Portmann, A. (1950). *In* "Traité de Zoologie" (P. P. Grassé, ed.), Vol. XV, pp. 257–269. Masson, Paris.
Ray, P. J. and Fedde, M. R. (1969). *Resp. Physiol.* **6**, 135–143.
Rees, P. M. (1967). *J. comp. Neurol.* **131**, 517–548.
Richards, S. A. (1967). *J. Zool., Lond.* **152**, 221–234.
Richards, S. A. (1968a). *J. Zool., Lond.* **154**, 223–234.
Richards, S. A. (1968b). *J. Physiol. Lond.* **199**, 89–101.
Rintoul, J. R. (1960). *M.D. thesis, University of St. Andrews.*
Rigdon, R. H. (1959). *Archs Path.* **67**, 215–227.
Ross. M. J. (1899). *Trans. Am. microsc. Soc.* **20**, 29–40.

Salt, G. W. (1964). *Biol. Rev.* **39**, 113–136.

Salt, G. W. and Zeuthen, E. (1960). *In* "Biology and Comparative Physiology of Birds" (A. J. Marshall, ed.), Vol. 1, pp. 363–409. Academic Press, New York.

Scharnke, H. (1938). *Z. vergl. Physiol.* **25**, 548–583.

Scheid, P. and Piiper, J. (1969). *Resp. Physiol.* **6**, 298–308.

Scheid, P. and Piiper, J. (1970). *Resp. Physiol.* **9**, 246–262.

Schmidt-Nielsen, K., Kanwisher, J., Lasiewski, R. C. and Cohn, J. E. (1969). *Condor* **71**, 341–352.

Schulz, H. (1959). "Die submikroskopische Anatomie und Pathologie der Lunge", pp. 22–23. Springer, Berlin.

Schulz, H. (1962). *In* "Pulmonary Structure and Function", Ciba Symposium, pp. 205–210. Churchill, London.

Schulze, F. E. (1872). *In* "Manual of Human and Comparative Histology" (S. Stricker, ed.), pp. 68–72. New Sydenham Society, London.

Shepard, R. H., Sladen, Brenda K., Peterson, N. and Enns, T. (1959). *J. appl. Physiol.* **14**, 733–735.

Siefert, E. (1896). *Pflügers Arch. ges. Physiol.* **64**, 428–506.

Shaw, J. W. (1970). *J. Physiol., Lond.* **207**, 75*P*–76*P*.

Soum, J. M. (1896). *Annls Univ. Lyon* **28**, 1–124.

Stübel, H. (1910). *Pflügers Arch. ges. Physiol.* **135**, 249–365.

Sturkie, P. D. (1965). "Avian Physiology", 2nd ed., pp. 152–172. Baillière, Tindall and Cassell, London.

Takino, M. (1933a). *Acta Sch. med. Univ. Kioto* **15**, 308–320.

Takino, M. (1933b). *Acta Sch. med. Univ. Kioto* **15**, 321–354.

Toussaint-Francx, J. P. and Toussaint-Francx, Y. (1959). *Acta tuberc. belg.* **50**, 179–197.

Trautmann, A., Fiebiger, J., Habel, R. E. and Biberstein, E. L. (1957). "The Fundamentals of the Histology of Domestic Animals". Comstock, New York.

Tucker, V.A. (1966). *Science, N.Y.* **154**, 150–151.

Tucker, V. A. (1968a). *J. exp. Biol.* **48**, 55–66.

Tucker, V. A. (1968b). *J. exp. Biol.* **48**, 67–87.

Tyler, W. S. and Pangborn, J. (1964). *J. Cell Biol.* **20**, 157–164.

Tyler, W. S., Pangborn, J. and Julian, L. M. (1961). *Am. Zool.* **1**, 268.

Tyler, W. S. and Pearse, A. G. E. (1966). *Poult. Sci.* **45**, 501–511.

Uehara, Y. and Burnstock, G. (1970). *J. Cell Biol.* **44**, 215–217.

Vos, H. J. (1934). *Z. vergl. Physiol.* **21**, 552–578.

Watanabe, T. (1960). *Jap. J. vet. Sci.* **22**, 145–154.

Watanabe, T. (1964). *Jap. J. vet. Sci.* **26**, 256–258.

Watanabe, T. (1968). *Jap. J. vet. Sci.* **30**, 331–340.

Weibel, E. R. (1964). *In* "Handbook of Physiology" (W. O. Fenn and H. Rahn, eds.), Sect. 3, Vol. 1, pp. 285–307. American Physiological Society, Washington D.C.

Weibel, E. R. (1966). *In* "Development of the Lung", Ciba Symposium (A. V. S. de Reuck and R. Porter, eds.), pp. 131–148. Churchill, London.

Westpfahl, U. (1961). *Z. Humboldt-Univ. Berlin Math. Nat. R.* **10**, 93–124.

Wet, P. D. de, Fedde, M. R. and Kitchell, R. L. (1967). *J. Morph.* **123**, 17–34.

White, S. S. (1968a). *J. Anat.* **103**, 390–392.

White, S. S. (1968b). *J. Anat.* **104**, 177.

White, S. S. (1972). *In* "The Anatomy of the Domestic Animals" (R. Getty, S. Sisson and J. D. Grossman, eds.), 5th ed., The Larynx, in Part VI, Chapter 6, in press. Saunders, Philadelphia.

White, S. S. and Chubb, J. C. (1967). *J. Anat.* **102**, 575.
Widdicombe, J. G. (1964). *In* "Handbook of Physiology" (W. O. Fenn and H. Rahn, eds.), Sect. 3, Vol. 1, pp. 585–630. American Physiological Society, Washington D.C.
Widdicombe, J. G. (1966). *In* "Advances in Respiratory Physiology" (C. G. Caro, ed.), pp. 48–82. Arnold, London.
Zeuthen, E. (1942). *Biol. Meddr.* **17**, 1–51.
Zimmer, K. (1935). *Zoologica, Stuttg.* **33**, 1–69.

6 Control of Respiration

M. G. M. JUKES

Department of Physiology,
Royal Veterinary College,
London, England

I. Resting Ventilation

The measurement of resting ventilation in the domestic fowl presents difficulties. It is well recognized (Kaupp, 1923; Sturkie, 1965) that anxiety or distress on the part of the bird will profoundly modify resting ventilation frequency, as will raised body temperature. Minor effects have been shown, or are to be expected, from differences in breed, age and sex, stage of reproductive activity, varying levels of sleep and wakefulness, and recent feeding. There appear to be no systematic studies reporting all the measurable factors, while one of the most important factors, anxiety or stress, cannot yet be accurately quantitated. More recent and complete observations are given in Table 1, but should be considered with the above reservations in mind.

A. LUNG AND AIR SAC VENTILATION

There have been many studies of the relation between the ventilation of the air sacs and that of the respiratory system as a whole (reviewed by King, 1966a, see summary of anatomical work).

One of the earliest and most important studies was that of Soum (1896) who demonstrated convincingly that no respiratory exchange of any magnitude took place across the walls of the air sacs. Thus, the air sacs are part

Table 1
Resting respiration in the unanaesthetized fowl

Source	Sex	n	Wt(Kg)	Breed	Age	$\dot{V}_E$	f	V_T	T_B	Methods
Zeuthen (1942)	F	2	1·85	Plymouth Rock "Crested"	?	337	19	18	—	Tracheal cannula, valves, Spirometer. (Given codeine to suppress cough reflex)
Frankel *et al.* (1962)	M	5	1·8–3·1	White Leghorn	Adult	655	20	33	41	Tracheal cannula, pneumotachograph
	F	8	1·5–2·5			568	38	15·3	41·5	
Weiss *et al.* (1963)	F	7	1·8	White Leghorn	16–18 months	554	37	15·4	41·5	Tracheal cannula, Pneumotachograph
King and Payne (1964)	M	10	4·2	Rhode Island Red and Light Sussex	Adult	777	17	46	—	Tracheal cannula, valves, Spirometer
	F	10	3·4			766	27	31	—	

n = number of birds.
Wt = body weight in kg. Means of groups or ranges.
$\dot{V}_E$ = pulmonary ventilation in cc/min.
f = respiratory frequency in breaths/min.
V_T = tidal volume in cc.
T_B = body temperature in degrees Centigrade.

of the respiratory dead space if they are ventilated by gas which does not participate in respiratory exchange in the lungs, during either inspiration or expiration.

Studies by Vos (1934), Zeuthen (1942), Shepard *et al.* (1959) and Cohn *et al.* (1963) have all shown that the air sac gas changes its composition fairly rapidly and in step with respiratory movements, either when the inspired gas composition is altered or when gases are introduced directly into the air sacs. It is therefore possible to visualize two extremes. On one hand, if all gas entering the air sacs has bypassed the lungs, the air sac gas would have a composition similar to that of the inspired air: it would in fact be inspired air mixed with some of that expired air which lies in the upper respiratory passages at the end of each expiration. On the other hand, if all the gas entering the air sacs has passed through the lungs and reached gaseous equilibrium with pulmonary capillary blood, then the tensions of CO_2 and O_2 in this gas should be similar to those of arterial blood.

Most earlier workers (see Salt and Zeuthen, 1960) were agreed that some air sacs, e.g. the interclavicular and anterior thoracic, had a relatively high CO_2 tension and low O_2 tension, while other air sacs, e.g. the abdominal, had a much lower CO_2 tension and higher O_2 tension, but in none of these reports were the gas tensions of arterial blood actually determined at the same time. Nevertheless, there is a fairly consistent agreement that the abdominal air sacs are ventilated by inspired gas which has bypassed the lungs, while the gas entering the anterior air sacs has already participated in pulmonary gaseous exchange.

Recent work has given different results. Rapid response apparatus has been used for analysis of expired air and air sac gases, together with blood gas analysis (Shepard *et al.*, 1959; Cohn and Shannon 1968). Both groups agree that the partial pressures of O_2 and CO_2 were similar in arterial blood, in end tidal expired air and in at least some air sacs; these air sacs therefore appear to be ventilated entirely by gas which has passed through the lungs where it has attained diffusion equilibrium with arterial blood. Shepard and his colleagues found, in the fowl, that all the air sacs had gaseous tensions similar to that of arterial blood, while Cohn and Shannon (1968) found in geese that this was true of the anterior air sacs, but that the posterior air sacs were somewhat diluted by pure inspired air. While the relatively minor inconsistencies between these two groups may be explained by different experimental conditions and species, there is a discrepancy between earlier and later results. Further work will have to be done before the issue is satisfactorily resolved.

It should be noted that this approach can only give information about the route by which gases enter the air sacs, and not about the pathway which is taken by gas leaving the air sacs. Moreover, with the exception of Scharnke's (1938) study on the pigeon, determinations of air sac gas composition have been made only in normothermic resting birds. There are no satisfactory data at present on what happens during hyperventilation due to various causes, e.g. exercise or thermal panting, and no evidence as to whether there is any selective control of the pathway of air within the avian respiratory system, a suggestion made by many workers in this field.

II. Chemical Control

Classical views on the chemical control of respiration stem initially from the demonstration by Miescher-Rüsch (1885) that a small rise in the CO_2 content of the inspired air caused a large increase in pulmonary ventilation, while a much larger diminution of the O_2 content caused only a small increase in respiration. The views of Haldane and Priestley (1935) on the dominant role of CO_2 developed from these findings. They were based on the relative constancy of the alveolar pCO_2 at rest, in the face of increased or decreased barometric pressure and in mild exercise and hypoxia, on the profound stimulation of respiration caused by a small increase in alveolar pCO_2 and on the apnoea caused by reducing alveolar pCO_2 by more than a small amount, e.g. after deliberate overbreathing. Haldane's views have been refined and elaborated since they were first put forward (Haldane and Priestley, 1905), but they still form a solid foundation for present theories on the chemical control of respiration in man. They have often been assumed, in the past, to apply to other mammals and to birds; yet for birds it is only in recent years that any adequate quantitative data have become available that might support or refute this commonly accepted assumption.

A. HYPERCAPNIA

Bert (1878) in his classic researches on a variety of different animals, showed that low concentrations of carbon dioxide in the respired air caused a stimulation of respiration, while higher concentrations led to coma and eventual death. Since that time there have been conflicting reports on the effects of hypercapnia upon the respiration of birds; the conflict seems partly due to the different types of experimental preparations used.

1. *Intact Unanaesthetized Birds.*

Andersen and Lövö (1964), using ducks, seem to have been the first to measure both respiratory frequency and tidal volume during hypercapnia in unanaesthetized birds. Increased CO_2 in the inspired air stimulated pulmonary ventilation, which reached a maximum in some ducks at between 6 and 11% CO_2 and then fell as the concentration of the inspired CO_2 rose further; respiratory frequency diminished during hypercapnia. Fowle and Weinstein (1966) obtained similar results in the fowl, pigeon and duck; the relation between pulmonary ventilation and inspired CO_2 was curvilinear over the lower range, as in mammals, and a peak response occurred at between 5 and 10% CO_2 with a decrease at greater CO_2 percentages. These authors did not detail all the effects of CO_2 on respiratory frequency, but when the CO_2 was high, the frequency decreased and this was attributed to local irritation by the high CO_2.

Earlier work gave only qualitative results for the effect of CO_2 on tidal volume and pulmonary ventilation. Dooley and Koppányi (1929) made ducks breathe various mixtures of CO_2 and O_2 and also administered pure CO_2 by intravenous infusion or through the humerus; in all cases there was increased frequency and depth of respiration, except when the concentration of the inspired CO_2 was very high when respiratory frequency slowed. Hiestand and Randall (1941) confirmed that CO_2 stimulated respiration in a variety of unanaesthetized birds; frequency and depth both increased; again, the frequency fell when the inspired CO_2 was high (6 to 10%).

2. Anaesthetized Birds.

Fowle and Weinstein (1966) showed clearly the difference between the respiratory responses of anaesthetized and unanaesthetized birds; the response is reduced by anaesthesia and the characteristic peak of pulmonary ventilation, seen at moderate CO_2 levels, is absent. Other experiments, on fowls anaesthetized with barbiturate, showed that CO_2 gives a moderate stimulation of tidal volume but, whereas Richards and Sykes (1967) found that respiratory frequency fell, Ray and Fedde (1969) noted little change in frequency in their more artificial preparations, ventilated by air introduced through the trachea and escaping through air sacs made patent. Hiestand and Randall (1941) also found that anaesthesia depressed the respiratory response to CO_2 in a variety of birds.

Orr and Watson (1913), in experiments on ducks and hens anaesthetized with ether, long ago suggested that the only effect of CO_2 upon respiration, even with inspired concentrations as low as 5%, was to cause apnoea. Although these results were disagreed with by later workers, Jukes *et al.* (1968), using unanaesthetized decerebrate ducks, found that even low concentrations of CO_2 may give an immediate, transient slowing of respiration or even apnoea, when first inhaled. Later however, respiration is stimulated. These findings stress the need to allow adequate time for equilibration when respiratory responses to changes in blood pCO_2 are investigated; care must be taken to distinguish between transient and steady state effects.

3. Unanaesthetized Decerebrate Birds.

To avoid the difficulties of using either anaesthetics which depress respiratory responses to CO_2 or unanaesthetized intact birds which have to be restrained, experiments have been carried out on unanaesthetized decerebrate birds. Decerebrate birds may survive for months without obvious impairment of the respiratory, cardiovascular or thermoregulatory mechanisms

(Rogers, 1919a, b) and without the decerebrate rigidity seen in mammals. Rogers and Wheat (1921) showed that the resting metabolism of such preparations is identical with that of intact resting birds, moreover they are not disturbed by the various sights and sounds which may be perceived by intact unanaesthetized birds and which may interfere in respiratory experiments in an uncontrolled way.

The ventilatory response of the decerebrate fowl to CO_2 (Fig. 1A, Johnston and Jukes, 1967) is similar to that of the intact unanaesthetized fowl (Fowle and Weinstein, 1966) and duck (Andersen and Lövö, 1964); both frequency and depth of respiration are increased. Plots of ventilation against tidal volume are linear, with an intercept on the tidal volume axis which is close to the volume of the dead space (Fig. 1B); similar results are found in man for a variety of respiratory stimuli (Hey *et al.*, 1966).

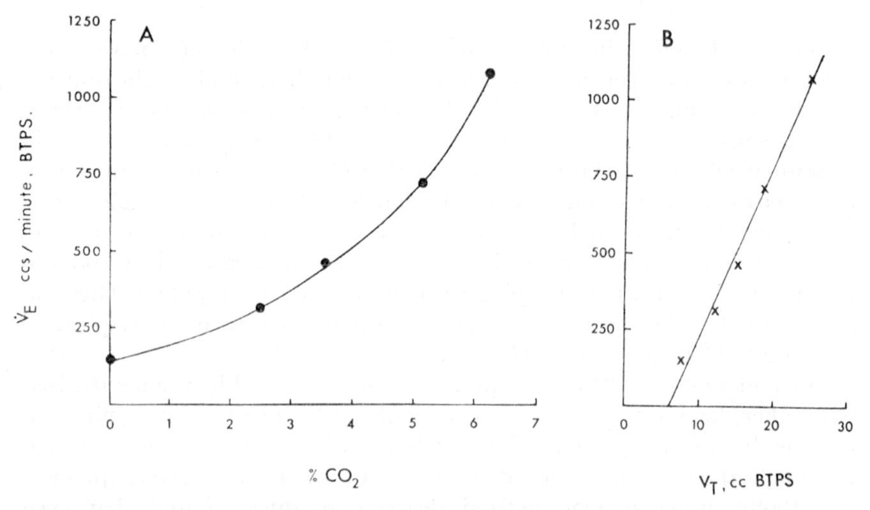

Fig. 1. Steady state respiratory responses of unanaesthetized decerebrate cockerel to hypercapnia. A. Relation between pulmonary ventilation ($\dot{V}_E$) and inspired CO_2: B. Relation between pulmonary ventilation ($\dot{V}_E$) and tidal volume (V_T). All volumes corrected to standard pressure and body temperature (BTPS). (Reproduced with permission from Johnston and Jukes (1967) and unpublished data.)

4. *Summary*

Most workers agree that low concentrations of CO_2 in the inspired air stimulate respiration, but that high concentrations reduce ventilation below peak values with a reduction in respiratory frequency. It is not yet clear whether this reduction is due to local irritant properties of CO_2 on non-specific receptors which have an inhibitory effect upon respiration, or to the general anaesthetic properties of CO_2 operating at lower levels in birds than in man and other mammals (Hill and Flack, 1908).

B. HYPOCAPNIA

If normal respiration in the fowl is adjusted to maintain a relative constancy of the arterial pCO_2, then lowering of the pCO_2 by deliberate overventilation should cause subsequent depression or cessation of respiration. It is doubtful whether it was possible to distinguish satisfactorily between this chemical apnoea and the vagal reflex apnoea of lung inflation before the works of Miescher-Rüsch (1885) and Haldane and Priestley (1905), although effects of imposed hyperventilation had been studied in birds before this time.

Foa (1911) opened the air sacs of the turkey and caused apnoea by continuously blowing in air through the trachea. He correctly saw the need to distinguish between the vagal inflation reflex and hypocapnic apnoea, and noted that apnoea persisted even when the blood became asphyxic and when the insufflating gas contained moderate concentrations of CO_2, thus supporting the idea that the apnoea which he studied was the vagal inflation reflex.

The only comparable modern studies appear to be those of Ray and Fedde (1969) using unidirectional flow in the anaesthetized fowl. They found that there were no respiratory movements unless the insufflating gas contained 3 or 4% CO_2, while a level of 5% CO_2 gave respiratory movements and blood pCO_2 which appeared normal. It thus seems, at least for the anaesthetized fowl, that hypocapnia causes depression or cessation of respiratory movements.

C. HYPOXIA

Moderate hypoxia stimulates respiration in all the different experimental preparations used. In ducks anaesthetized with ether, Orr and Watson (1913) showed both increased frequency and increased depth of respiration; this has been confirmed for the fowl anaesthetized with barbiturate (Richards and Sykes, 1967; Ray and Fedde, 1969). Butler (1967) found increased respiratory frequency in the unanaesthetized fowl but did not measure amplitude of respiration. In decerebrate unanaesthetized ducks, pulmonary ventilation is stimulated by hypoxia (Fig. 2A), both frequency and depth of respiration being increased. Plots of pulmonary ventilation against tidal volume are linear, similar to those obtained during hypercapnia (Fig. 2B).

Although the respiratory response to hypoxia is not usually very great until the inspired O_2 is reduced below about 15% (Butler, 1967; Ray and Fedde, 1969), there is some evidence for a small hypoxic drive to respiration in birds breathing air at sea level. Dooley and Koppányi (1929) infused pure O_2 into a vein of the unanaesthetized duck and showed a definite reduction in ventilation. Administration of one or two breaths of pure O_2 gives a transient depression of respiration in unanaesthetized decerebrate ducks (M. G. M. Jukes and M. M. R. Jukes, unpublished observations); similar findings in dogs and man have been taken as evidence for a hypoxic drive at normal blood oxygen tensions (Dejours, 1963).

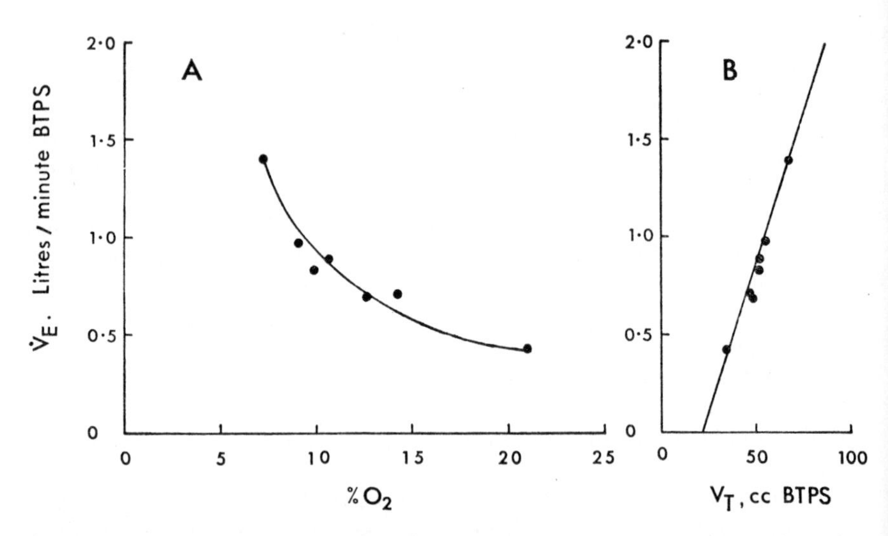

Fig. 2. Steady state respiratory responses of unanaesthetized decerebrate duck to hypoxia. A. Relation between pulmonary ventilation ($\dot{V}_E$) and inspired O_2: B. Relation between pulmonary ventilation ($\dot{V}_E$) and tidal volume (V_T). All volumes corrected to standard pressure and body temperature (BTPS). (From M. G. M. Jukes, and M. M. R. Jukes, unpublished data.)

D. PERIPHERAL AND CENTRAL CHEMORECEPTORS

Birds possess carotid and aortic chemoreceptors histologically similar to those of mammals (see, e.g. Nonidez, 1935). Little is known about their functional importance, perhaps because the intrathoracic position of avian carotid receptors makes them less accessible than those of mammals. A recent comprehensive review of mammalian chemoreceptors is given by Torrance (1968).

After bilateral cervical vagotomy, hypercapnia still causes increased frequency and depth of respiration in the duck (Artom, 1925), showing that vagally innervated chemoreceptors are not exclusively responsible for this effect. Probably birds, like mammals, have central chemoreceptors which monitor the pCO_2 and pH of both arterial and venous blood and cerebrospinal fluid (e.g. Mitchell, 1966).

In contrast, Butler (1967) has shown that bilateral cervical vagotomy in the fowl abolishes the increased frequency of respiration caused by hypoxia in the intact bird. This finding is consistent with the idea that vagally innervated chemoreceptors are responsible for the respiratory stimulation which occurs in hypoxia, but an increased amplitude of respiration was not excluded in these experiments. Fedde *et al.* (1961) report that removal of carotid receptors in the fowl does not alter the pattern of respiration; this is not unexpected since the contribution of the hypoxic drive to respiration is probably slight in the resting bird at sea level.

Petersen and Fedde (1968) have claimed the existence of receptors in the lungs, sensitive to the concentration of CO_2; such receptors were earlier suggested by Hiestand and Randall (1941) on what now appears to be very slender evidence. The case made out by Petersen and Fedde rests upon the rapidity with which their anaesthetized fowls altered their respiration in response to a decrease in the inspired CO_2. There is possibly some support from histological evidence of receptors close to pulmonary blood vessels (Cook and King, 1969), but these have yet to be shown to be chemoreceptors. Further supporting evidence for the existence of pulmonary chemoreceptors will be awaited with keen interest.

E. SUMMARY

The available experimental evidence provides general support for the idea that the chemical control of respiration in birds, and in the fowl in particular, is similar to that of mammals. Carbon dioxide is a more potent respiratory stimulus to respiration than lack of oxygen and, in the normothermic resting fowl respiration is largely governed by the arterial pCO_2 (or resultant pH) acting on both central and peripheral chemoreceptors. The central drive to respiration most likely depends upon the composition of arterial blood and of cerebro-spinal fluid, as in mammals.

III. Nervous Control

A. VAGAL REFLEXES

Reflexes from the lungs and air passages have been investigated either by inflation or deflation, or by clamping the trachea at the end of inspiration. Graham (1940) showed a definite inflation reflex in the anaesthetized fowl; inflation of the lungs gave expiration followed by inhibition of respiratory movements. This reflex was abolished by vagotomy. Foa (1911) had previously shown a similar reflex in the turkey. By inflating the respiratory system via the trachea and allowing the gas to escape via the air sacs, he made the important observation that apnoea persisted even when CO_2 was added to the inflating gas, thus showing that this effect was not due to lowering of the blood pCO_2. Richards (1968, 1969) has confirmed the existence of inflation reflexes in the anaesthetized fowl. Graham (1940) also showed a weaker deflation reflex.

In contrast, Fedde *et al.* (1961) and Sturkie (1965) state that sudden expansion of the lungs has no effect on respiration in anaesthetized fowls which are unidirectionally ventilated at an adequate rate; they discount any effect of pulmonary stretch receptors upon the regulation of normal respiration. These findings do not disprove the existence of vagal inflation or deflation reflexes, nor do they show whether these are active in the unanaesthetized fowl.

Reflexes have been elicited by air flow into the trachea (Graham, 1940). An outward flow of air stimulates expiration and reduces respiratory frequency; an inward stream of air increases respiratory frequency. Both reflexes are abolished by vagotomy, local anaesthesia of the tracheal mucosa and local denervation. Hiestand and Randall (1941) found rapid flows of air or other gases outwards to cause immediate apnoea, slow flows of irritant gases including CO_2 also caused inhibition. These two effects are presumably mediated through different sets of receptors.

1. Vagotomy

Bilateral cervical vagotomy in most avian genera decreases the frequency of breathing; this may also become irregular, and increase in depth (cf. King, 1966b). A transient initial stimulation of respiration may occur when the vagi are first cut (Graham, 1940; Sinha, 1958). Fedde *et al.* (1963a) made an important contribution in that the effects of vagotomy are particularly affected by anaesthesia, which may often mask the decrease in frequency of breathing seen in the unanaesthetized fowl. Age was also found to be significant; 32-week-old cocks were more affected by vagotomy than 16-week-old ones; older birds died within a few hours of vagotomy with pulmonary vascular congestion, pulmonary oedema and emphysema; younger birds might survive for several days (Fedde and Burger, 1963).

The usual explanation given for the effect of bilateral vagotomy is that vagal proprioceptive reflexes from the lungs and airways limit both inspiration and expiration, particularly the former; they thus permit a frequency of respiration faster than that set by the respiratory centre alone. Support for this concept comes from work on the reflex effects of lung inflation and deflation and the results of electrical stimulation of vagal afferents. However, a quite different explanation has been advanced by Fedde *et al.* (1963b) who consider that vagotomy removes a peripheral drive to respiration from pulmonary chemoreceptors sensitive to CO_2.

2. Vagal Afferents

King and his collaborators (1968a, b; 1969) have recorded activity of single afferent fibres in the cervical vagus of anaesthetized fowl during normal breathing. Different types of unit were described, usually discharging impulses in fixed relation to the respiratory cycle. Most were linked to inspiration, about a quarter to expiration and a few discharged during both inspiration and expiration. The authors believe afferents indicate the state of the respiratory cycle. It is not yet clear to what extent these afferents correspond to those of mammals which respond to mechanical stimulation of the lungs and airways (Widdicombe, 1964).

3. Vagal Afferent Stimulation

Electrical stimulation of the central end of the cut cervical vagus of birds has been carried out by many workers. The clearest results are those of Sinha (1958) on the pigeon; slow frequency stimuli (up to 20/s) of moderate intensity gave faster, deeper respiration, while high-frequency stimulation at just suprathreshold strengths gave slower, shallower respiration or even apnoea. The former effects may be the counterpart of the vagal deflation reflex, the latter corresponding to the vagal inflation reflex.

Apart from supporting the general idea that there are vagal respiratory reflexes in the bird, similar to those of mammals, at present vagal stimulation seems too crude a tool to elucidate the detailed mechanisms of vagal reflexes. The use of electrical stimulation requires much more detailed knowledge of the fowl than is presently available in relation to vagal nerve composition, nerve conduction velocities and of the electrical thresholds of different types of afferent nerve fibres. Further vagal influences on respiration in pigeons seem different from those in birds of other genera, e.g. thermal panting.

B. RESPIRATORY CENTRE

Decerebrate birds may survive for long periods with apparently normal frequency and depth of respiration (see Rogers, 1919b); the cerebral hemispheres and striatum are thus not essential for the regulation of respiration. Damage to the thalamus causes defective thermoregulation, but if the environmental temperature is appropriately regulated the birds may survive many days (Rogers, 1919a; Rogers and Wheat, 1921). The only studies known on the location of central respiratory control mechanisms appear to be those of von Saalfeld (1936) who found, in pigeons, that transection through the medulla, but not above this level, led to permanent abolition of respiratory movements. The results suggest the existence of medullary respiratory centres, anologous to those of mammals.

C. THERMAL PANTING

Rogers made a systematic study of those brainstem centres which regulated thermal polypnoea, and noted that decerebrate pigeons possess apparently normal physiological temperature regulation; destruction of the dorsal thalamus was seen to lead to the abolition of thermal polypnoea, with resulting hyperthermia in hot environments (Rogers, 1919a; Rogers and Wheat, 1921). Von Saalfeld (1936) sectioned the pigeon's brainstem at various levels and showed that some structure at the level of thalamus was necessary for thermal polypnoea. He also obtained panting on warming the anterodorsal aspect of the midbrain; thermal panting in response to raised body temperature could be abolished by local cooling of this region. Åkerman *et al.* (1960) produced polypnoea in pigeons by local electrical stimulation of the pre-optic area, and Richards (1970a) in fowls by stimulation of a more dorsal and caudal area of the midbrain. Discrete lesions of the hypothalamus, particularly of the antero-ventral part, reduce the ability of the fowl to lose heat in the face of hyperthermia by abolishing thermal panting (Feldman *et al.*, 1957; Kanematsu *et al.*, 1967).

Thus, there seems to be a mechanism at the level of the thalamus concerned with the polypnoeic response to raised body temperature; it is not yet clear whether this mechanism is confined to the antero-ventral part of the brainstem, or involves the antero-dorsal aspect as well. A more caudal mid-brain centre may also be involved in thermal polypnoea (Richards, 1970a).

1. *Central and Peripheral Thermoreceptors*

The experiments of Rogers (1928) on cooling the brainstem, and of von Saalfeld (1936) on its local heating suggest a central mechanism of thermal reception, as in mammals. The functional importance of cranial thermo-receptors was shown by Randall (1943), who subjected young chickens to body surface heating while cooling the neck, and thereby the blood supply to the head; in spite of a raised body temperature, thermal panting did not occur.

One technique used to investigate the possible influence of peripheral thermoreceptors has been to cool the skin while the deep body temperature is raised and thermal polypnoea is in progress. Randall and Hiestand (1939) and Randall (1943) found this procedure to inhibit panting in the unanaes-thetized fowl, but their evidence cannot be properly evaluated unless it is known whether this occurred before the brain temperature was lowered, and thus truly due to peripheral receptors.

When body temperature is quickly raised by external heating, thermal panting might be initiated by peripheral thermal receptors before the central nervous temperature has risen. Unfortunately, those authors who have tried to test this possibility (Randall and Hiestand, 1939) measured only cloacal temperatures and so their results remain inconclusive (see also Chapter 48).

2. *Vagal Afferents*

In the fowl (Hiestand and Randall, 1942; Richards, 1968), but not in the pigeon (von Saalfeld, 1936; Richards, 1968), bilateral vagotomy abolishes thermal panting. Stimulation of the central end of one vagus, in the fowl, enables restoration of thermal panting in response to raised body temperature (Richards, 1968). There is an important generic difference between the pigeon and other birds; the former has a greater degree of central control over thermal panting than the fowl, requiring information from the lungs to initiate and maintain thermal panting.

3. *Respiration and Blood Gases*

In the fowl there appears to be a definite critical body temperature at which the frequency of breathing starts to increase; above this body tem-perature breathing frequency rises proportionally until a maximum is reached; beyond this point the frequency falls and respiration becomes deeper (second-phase breathing) until body temperature reaches the upper lethal limit and the thermoregulatory mechanisms collapse (Randall and Hiestand, 1939; Frankel *et al.*, 1962). In the fowl there is an increase in total pulmonary ventilation during panting, until the onset of second-phase breathing, but the tidal volume is reduced (Frankel *et al.*, 1962). There are, however, difficulties in measuring tidal volume during panting, since any resistance imposed by respiratory valves, pneumotachographs, etc. reduces somewhat the frequency of respiration and may also alter tidal volumes.

The air sacs play some part in the increased evaporative water loss during panting. Soum (1896) found that destruction of the air sacs reduced the amount of water lost during hyperthermia and reduced the heat tolerance of the birds. It has been repeatedly suggested (e.g. Krogh, 1941) that the air sac system enables large volumes of air to be respired for evaporative cooling without increasing the ventilation of the lungs at the same time. However, thermal panting in the fowl is associated with alkalosis of the blood and lowering of the arterial pCO_2 (Linsley and Burger, 1964; Calder and Schmidt-Nielsen, 1967; Frankel and Frascella, 1968) and birds in general seem to be no better at preserving blood homeostasis during hyperthermia than panting mammals (Richards, 1970b for review). The CO_2 levels in the air sacs fall during panting (Scharnke, 1938) which has been taken to indicate their increased ventilation by atmospheric air. Other explanations are possible, e.g. that the air sac gases have passed through the lungs and reached equilibrium with arterial blood with a low pCO_2. In conclusion, there seems to be little evidence of a differential routing of atmospheric air to the air sacs during thermal panting.

References

Åkerman, B., Andersson, B., Fabricus, E. and Svensson, L. (1960). *Acta physiol. scand.* **50**, 328–336.

Andersen, H. T. and Lövö, A. (1964). *Comp. Biochem. Physiol.* **12**, 451–456.

Artom, C. (1925). *Archs neerl. Physiol.* **10**, 362–394.

Bert, P. (1878). "La Pression Barometrique." Balliere, Paris.

Butler, P. J. (1967). *J. Physiol., Lond.* **191**, 309–324.

Calder, W. A. and Schmidt-Nielsen, K. (1967). *Am. J. Physiol.* **213**, 883–889.

Cohn, J. E., Burke, L. and Markesbery, H. (1963). *Fedn Proc. Fedn Am. Socs exp. Biol.* **72**, 397.

Cohn, J. E. and Shannon, R. (1968). *Resp. Physiol.* **5**, 259–268.

Cook, R. D. and King, A. S. (1969). *J. Anat.* **105**, 202.

Dejours, P. (1963). *Ann. N.Y. Acad. Sci.* **109**, 682–695.

Dooley, Marian S. and Koppányi, T. (1929). *J. Pharmacl. exp. Ther.* **36**, 507–518.

Fedde, M. R. and Burger, R. E. (1963). *Poult. Sci.* **42**, 1236–1246.

Fedde, M. R., Burger, R. E. and Kitchell, R. L. (1961). *Poult. Sci.* **40**, 1401.

Fedde, M. R., Burger, R. E. and Kitchell, R. L. (1963a). *Poult. Sci.* **42**, 1212–1223.

Fedde, M. R., Burger, R. E. and Kitchell, R. L. (1963b). *Poult. Sci.* **42**, 1224–1236.

Feldman, S. E., Larsson, S., Dimick, M. K. and Lepkovsky, S. (1957). *Am. J. Physiol.* **191**, 259–261.

Foa, C. (1911). *Arch. ital. Biol.* **55**, 412–422.

Fowle, A. S. E. and Weinstein, S. (1966). *Am. J. Physiol.* **210**, 293–298.

Frankel, H. M. and Frascella, D. (1968). *Proc. Soc. exp. Biol. Med.* **127**, 997–999.

Frankel, H., Hollands, K. G. and Weiss, H. S. (1962). *Archs int. Physiol. Biochim.* **70**, 555–563.

Graham, J. D. P. (1940). *J. Physiol., Lond.* **97**, 525–532.

Haldane, J. S. and Priestley, J. G. (1905). *J. Physiol., Lond.* **32**, 225–266.

Haldane, J. S. and Priestley, J. G. (1935). "Respiration", 2nd ed. Oxford University Press, London.

Hey, E. N., Lloyd, B. B., Cunningham, D. J. C., Jukes, M. G. M. and Bolton, D. P. G. (1966). *Resp. Physiol.* **1**, 193–205.

Hiestand, W. A. and Randall, W. C. (1941). *J. cell. comp. Physiol.* **17**, 333–340.

Hiestand, W. A. and Randall, W. C. (1942). *Am. J. Physiol.* **138**, 12–15.

Hill, L. and Flack, M. (1908). *J. Physiol., Lond.* **37**, 77–111.

Johnston, A. M. and Jukes, M. G. M. (1967). *J. Physiol., Lond.* **184**, 38*P*–39*P*.

Jukes, M. G. M., Jukes, M. M. R. and Win, B. H. (1968). *Proc. 24th Int. Congr. Physiol.* **7**, 223.

Kanematsu, S., Kii, M., Sonada, T. and Kato, Y. (1967). *Jap. J. vet. Sci.* **29**, 95–104.

Kaupp, B. F. (1923). *Vet. Med.* **18**, 36–40.

King, A. S. (1966a). *Int. Rev. gen. exp. Zool.* **2**, 171–267.

King, A. S. (1966b). *In* "Physiology of the Domestic Fowl". (C. Horton Smith and E. C. Amoroso, eds.), pp. 302–310. Oliver and Boyd, Edinburgh.

King, A. S. and Payne, D. C. (1964). *J. Physiol., Lond.* **174**, 340–347.

King, A. S., McLelland, J., Molony, V., Bowsher, D., Mortimer, M. and White, S. S. (1968a). *J. Anat.* **104**, 182.

King, A. S., Molony, V., McLelland, J., Bowsher, D. and Mortimer, M. F. (1968b). *Experientia* **24**, 1017–1018.

King, A. S., McLelland, J., Molony, V. and Mortimer, M. F. (1969). *J. Physiol., Lond.* **201**, 35*P*–36*P*.

Krogh, A. (1941). "The Comparative Physiology of Respiratory Mechanisms". University of Pennsylvania Press, Philadelphia, Pennsylvania.

Linsley, J. G. and Burger, R. C. (1964). *Poult. Sci.* **43**, 291–305.

Miescher-Rüsch, F. (1885). *Arch. Anat. Physiol. Lpz.* 355–380.

Mitchell, R. A. (1966). *In* "Advances in Respiratory Physiology". (C. G. Caro, ed.), pp. 1–47. Arnold, London.

Nonidez, J. F. (1935). *Anat. Rec.* **62**, 47–73.

Orr, J. B. and Watson, A. (1913). *J. Physiol., Lond.* **46**, 337–348.

Peterson, D. F. and Fedde, M. R. (1968). *Science, N.Y.* **162**, 1499–1501.

Randall, W. C. (1943). *Am. J. Physiol.* **139**, 56–63.

Randall, W. C. and Hiestand, W. A. (1939). *Am. J. Physiol.* **127**, 761–767.

Ray, P. J. and Fedde, M. R. (1969). *Resp. Physiol.* **6**, 135–143.

Richards, S. A. (1968). *J. Physiol., Lond.* **199**, 89–101.

Richards, S. A. (1969). *Comp. Biochem. Physiol.* **29**, 955–964.

Richards, S. A. (1970a). *J. Physiol., Lond.* **207**, 57*P*–59*P*.

Richards, S. A. (1970b). *Biol. Rev.* **45**, 223–264.

Richards, S. A. and Sykes, A. H. (1967). *Comp. Biochem. Physiol.* **21**, 691–701.

Rogers, F. T. (1919a). *Am. J. Physiol.* **49**, 271–283.

Rogers, F. T. (1919b). *J. comp. Neurol.* **31**, 17–35.

Rogers, F. T. (1928). *Am. J. Physiol.* **86**, 639–650.

Rogers, F. T. and Wheat, S. D. (1921). *Am. J. Physiol.* **57**, 218–227.

Saalfeld, E. von (1936). *Z. vergl. Physiol.* **23**, 727–743.

Salt, G. W. and Zeuthen, E. (1960). *In* "Biology and Comparative Physiology of Birds" (A. J. Marshall, ed.), Vol. I, pp. 363–409. Academic Press, New York.

Scharnke, H. (1938). *Z. vergl. Physiol.* **25**, 548–583.

Shepard, R. H., Sladen, B. K., Peterson, N. and Enns, T. (1959). *J. appl. Physiol.* **14**, 733–735.

Sinha, M. P. (1958). *Helv. physiol. pharmac. Acta* **16**, 58–72.

Soum, J. M. (1896). *Ann. Univ. Lyon.* **28**, 1–126.

Sturkie, P. D. (1965). "Avian Physiology", 2nd ed. Comstock (Cornell University Press), Ithaca, N.Y.

Torrance, R. W. (Ed.) (1968). "Arterial Chemoreceptors". Blackwell, Oxford.

Vos, H. J. (1934). *Z. vergl. Physiol.* **21**, 552–578.

Weiss, H. S., Frankel, H. and Hollands, K. G. (1963). *Can. J. Biochem. Physiol.* **41**, 805–815.

Widdicombe, J. G. (1964). *In* "The Handbook of Physiology. Section 3: Respiration". (W. O. Fenn and H. Rahn, eds.), Vol. I, pp. 585–630. American Physiological Society, Washington, D.C.

Zeuthen, E. (1942). *Biol. Medd. Kbh.* **17**, 1–40.

7 Transport of Blood Gases

M. G. M. JUKES

Department of Physiology,
Royal Veterinary College,
London, England

I. Introduction

Despite the growing numbers of investigators interested in the transport of blood gases in the fowl and the changes which occur in blood chemistry during thermal panting and egg laying, surprisingly little reliable information is available on many basic aspects. For example, data are available only for the composition of blood in systemic arteries and veins and there are virtually no experimental studies on the pulmonary gas exchange, nor on exchange in peripheral tissues. There is an extensive theoretical treatment by Zeuthen (1942, see also Salt and Zeuthen, 1960) on exchange of gases in the lungs; this contains the interesting suggestion that air leaving the lungs has a pCO_2 corresponding to that of mixed venous blood rather than that of arterial blood. This suggestion was made on the basis of some individual measurements of expired air which were found to have a higher CO_2 content than that of air taken, in other experiments, from the humeri and which was thought to be in equilibrium with the CO_2 of arterial blood. Shepard *et al.* (1959) and Cohn and Shannon (1968) have shown, however, that the end tidal expired pCO_2 in the fowl and the goose is virtually the same as the pCO_2 of arterial blood and thus provide no support for Zeuthen's suggestion. Nevertheless, the avian lung might well provide an interesting contrast to the mammalian lung in the mechanisms of its respiratory exchange.

II. Blood Gas Tensions and pH

Routine determinations of pO_2 and pCO_2 are now facilitated through direct-reading electrodes. The values recorded depend upon the temperatures at which the measurements are made and should therefore be done at the appropriate body temperature; temperature correction factors have not been determined for the fowl and have to be assumed (e.g. Chiodi and Terman, 1965). Nightingale *et al.* (1968) report that values of pO_2 for fowl blood, using a direct-reading electrode, may be too low when the electrode is calibrated with humidified gases; no such error was found for dog blood. The error found was large (13%) and was related to the haematocrit value; it is not apparent whether the results of other workers are subject to this effect, which the authors think may be due in part to the fowls erythrocytes being nucleated.

Ideally, samples of blood for determination of normal gas values should be drawn anaerobically, without application of any appreciable negative pressure, from undisturbed, unanaesthetized and unrestrained animals. But many have found it either necessary or convenient to restrain unanaesthetized fowl, in some cases on their sides; moreover, it seems impossible to assess accurately the degree of stress or anxiety which the fowl may experience in an experimental situation, particularly when attached by leads and cannulae to recording and sampling apparatus. For this reason, there must still be some uncertainty about the acceptability of published results as those of the "ideal" normothermic resting fowl. Some recent results are given in Table 1.

The values of pO_2 for the arterial blood of the unanaesthetized fowl are similar to those of mammals but the pCO_2 is lower and the pH higher. It is not yet clear whether these differences occur in the completely undisturbed fowl; stress would be expected to lead to hyperventilation with a consequent lowering of arterial pCO_2 and increase of pH. Anaesthesia appears to cause some decrease in pO_2 and pH, with increased pCO_2, results consistent with a depression of the respiratory exchange. The results on venous blood are generally those on samples taken from a wing vein rather than mixed venous blood, but show reasonable agreement. Values previously quoted in reviews are usually those of Morgan and Chichester (1935) who calculated pO_2 and pCO_2 on the basis of their study of O_2 dissociation curves and the buffering capacity of fowl's blood, together with values of O_2 and CO_2 content of blood obtained by cardiac puncture. These authors expressed grave doubts about the accuracy of their pO_2 values; it is important to note that these doubts are not usually mentioned when their figures are quoted.

It should be noted that the pH of both arterial and venous blood in the laying hen varies with the stages of the egg-laying cycle, this is thought to be due to the hyperventilation and respiratory alkalosis which occurs as a compensation for the metabolic acidosis associated with egg shell formation (Mongin and Lacassagne, 1966a, b; Hodges, 1969).

Blood gas tensions and pH of blood of adult fowl

Reference	Sex	Temperature of measurement (°C)	pO_2 (mmHg)	pCO_2 (mmHg)	pH	Remarks
I. ARTERIAL BLOOD						
a. Unanaesthetized						
Morgan and Chichester (1935)	F	40	85 (calculated)	34 (calculated)	7·45	Cardiac puncture Properties of pooled venous blood
Chiodi and Terman (1935)	F	36·5 and 37 corrected to 42	99 (4)	32·8 (8)	7·49	Restrained on side
Frank and Burger (1965)	F	Not stated	—	28·3 (10)	7·48	Cardiac puncture
Frankel (1965)	M	Body temperature of fowl	88 (8)	23 (8)	7·50	
Butler (1967)	M	41·5	99 (5)	—	—	Restrained on side
Calder and Schmidt-Nielsen (1968)	F	Not stated		29·9 (2)	7·48 (2)	Unrestrained
	M	Not stated		29·2 (3)	7·53 (3)	Unrestrained
Frankel and Franscella (1968)	M	41	92 (6)	26 (6)	7·46 (6)	Restrained
b. Anaesthetized						
Shepard *et al.* (1959)	F	Not stated	100	40	—	Barbiturate Restrained on side
Richards and Sykes (1967)	F	41·5	92 (6)	—	7·46 (6)	Barbiturate
Richards (1969)	M and F	41·5	95 (4)	—	7·47	Barbiturate
II. VENOUS BLOOD, *Unanaesthetized*						
Morgan and Chichester (1935)	F	40	48 (calculated)	45 (calculated)	—	Cardiac puncture Properties of pooled venous blood
Mueller (1966)	F	Not stated	—	51	7·30	
Hunt and Simkiss (1967)	M	41·5	—	39·6	7·40	Brachial vein
	F	41·5	—	37·6	7·39	Brachial vein

Numbers in brackets refer to number of determinations, where given.

III. Oxygen Dissociation Curve

The relation between the percentage saturation of haemoglobin and the pO_2 of the blood of the adult fowl has the same characteristic S shape (Fig. 1) as that of mammals. The results of Christensen and Dill (1935),

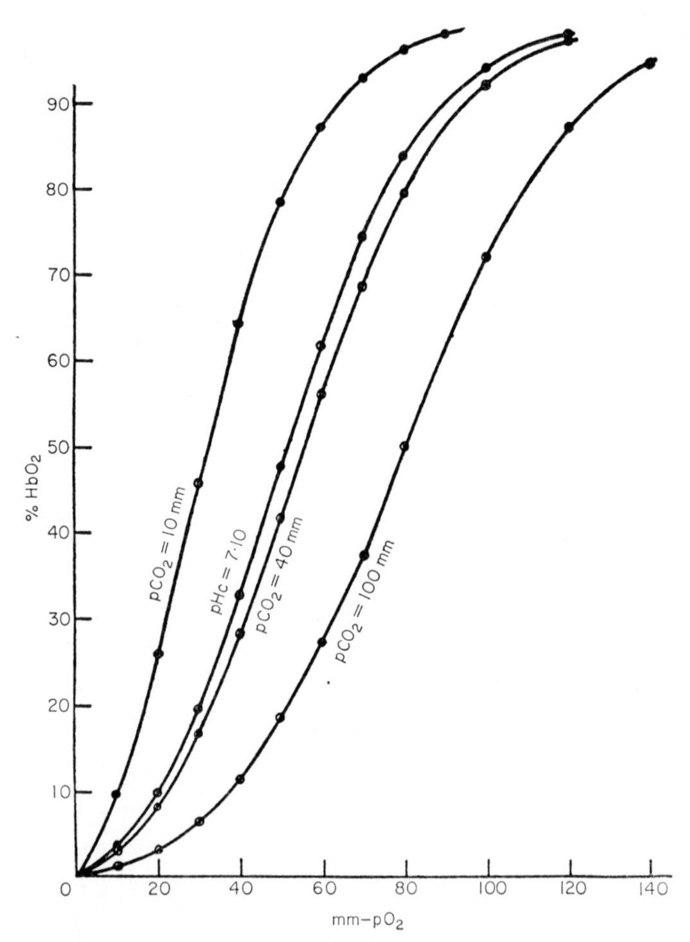

Fig. 1. Oxygen dissociation curves of adult fowls' blood at 40°C. Effects of variation of pCO_2.

Morgan and Chichester (1935), Rostorfer and Rigdon (1947) and Chiodi and Terman (1965) broadly agree on the overall shape of the curve and also on its displacement to the right compared to that of the human or dog blood (see Fig. 1). However, there are some differences in detail which are probably due to the differing temperatures and pH at which the various measurements were made.

Morgan and Chichester (1935) showed that the O_2 dissociation curve is influenced by the pCO_2 of the blood, increased acidity decreasing the affinity of haemoglobin for O_2 and displacing the curve to the right, i.e. the Bohr effect (Fig. 2). A convenient way of expressing effects upon the O_2 dissociation curve is to describe the change in the value of pO_2 which gives 50% saturation of haemoglobin (p_{50}). The results of Morgan and Chichester gave:

$$p_{50} = 5{\cdot}76 - 0{\cdot}57 \text{ pH}$$

The slope of this relationship, i.e. the shift of the O_2 dissociation curve for a given change in $[H^+]$, was the same for the fowl as for the dog and was close to that of human blood. Christensen and Dill (1935) showed a similar effect of acidity upon the O_2 dissociation curves of solutions of fowl's haemoglobin.

No direct measurements seem to have been made upon the effect of temperature on the O_2 dissociation curve of fowls' blood, but Wastl and Leiner (1931a) have shown that raised temperature shifts to the right the curves of blood from the duck and goose; this is similar to the effect upon mammalian blood and is generally assumed to apply also to fowls' blood.

The measurements of Chiodi and Terman (1965) at pH 7·50, corrected to 42°C, are those made at values closest to those of normal arterial blood and at present seem the most acceptable. Estimates of p_{50} by various workers are shown in Table 2. It should be noted that the comparison shown in Fig. 2 between the curves for man, dog and fowl were made at pH 7·1, but that fowls' blood is normally rather less acid than mammalian blood; if the comparison had been made between bloods of normal pH, pCO_2 and temperature for each species, the difference between the fowl and the mammals would be less than that shown.

Hall (1935) studied the oxygen dissociation curves of embryos and young chicks and compared them with those of the adult fowl, finding a shift to the right with increasing age; this was confirmed by Rostorfer and Rigdon (1947). Hall suggested that the shift might be due to progressive replacement of an embryonic haemoglobin by an adult form with a lesser affinity for oxygen. Christensen and Dill (1935) also postulated two types of haemoglobin in the adult fowl to explain the difference between the O_2 dissociation curves of whole blood and of solutions of haemoglobin. These predictions of different types of haemoglobin in the adult fowl have since been confirmed (Johnson and Dunlap, 1955; Rodnan and Ebaugh, 1956; Saha *et al.*, 1957) and several types have now been characterized (Washburn, 1968a; Denmark and Washburn, 1969). The replacement of embryonic by adult forms of haemoglobin during maturation has also been followed (Washburn, 1968b). Although the haemoglobin of the chick has a higher affinity for O_2 than that of the adult fowl, this contrasts with the duck where adult haemoglobin has a greater affinity than that of ducklings (Rostorfer and Rigdon, 1946; 1947); human haemoglobin variants may have either a lesser or a greater affinity for O_2

than normal haemoglobin (Parer, 1970). The significance of the differences between the duck and the fowl has yet to be explained.

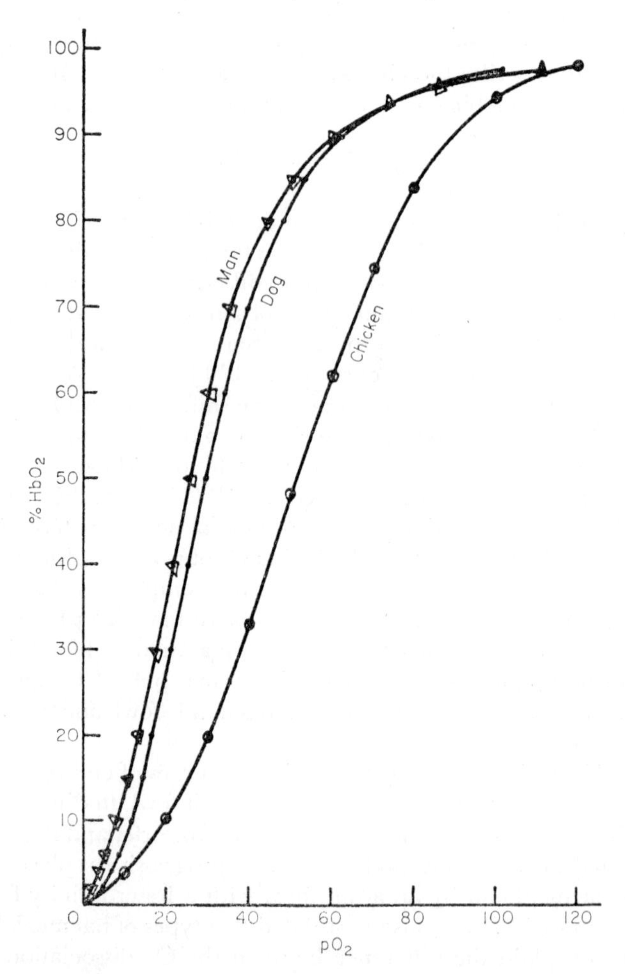

Fig. 2. Comparison of oxygen dissociation curves at pH 7·1. Equilibration temperatures for man and dog were 37·5°C and for chicken, 40·0°C.

IV. Carbon Dioxide Dissociation Curve

The full relationship between the CO_2 and the pCO_2 of fowls' blood has not been investigated. Wastl and Leiner (1931b) showed that the curves for duck and goose blood were similar to that of human blood and that oxygenated blood had a smaller CO_2 capacity than reduced blood at any particular pCO_2, the Haldane effect. Morgan and Chichester (1935) demonstrated that oxygenated fowls' blood at 40°C took up as much CO when the

Table 2

Oxygen pressure (p_{50}) at which haemoglobin of adult fowl is 50% saturated with O_2

	p_{50} (pO_2, mmHg)	Temperature (°C)	pH	pCO_2 (mmHg)
Morgan and Chichester (1935)	51	40	7·10	37
Hall (1935)	21	37	7·38	—
Christensen and Dill (1935)	52	37·5	7·10	—
Rostorfer and Rigdon (1947)	58	38	—	31
Chiodi and Terman (1965)	55	42	7·50	—

pCO_2 was raised from 30 to 60 mmHg as did human blood of comparable haemoglobin and bicarbonate content (at pCO_2 40 mmHg). They concluded that the buffering power of fowls' blood was similar to that of human blood under these circumstances but it should be remembered that human blood normally has a considerably higher haemoglobin content and therefore a higher buffering capacity.

The concentrations of H^+ and HCO_3^- and the pCO_2 of blood are related by the Henderson-Hasselbalch equation·

$$pH = pK' + \log \frac{(HCO_3^-)}{\alpha . pCO_2}$$

where α is the solubility coefficient of CO_2.

This equation can be used to calculate, for example, the pCO_2 of blood when the pH and total CO_2 content have been determined but an exact value of pK′ must be used. The value of the plasma pK′ for the blood of the fowl was investigated by Helbacka *et al.* (1964) and found to be 6·090 for true plasma at 41°C and pH 7·35; this value was altered by changes in either pH or temperature, effects which are similar to those found for mammalian blood (Severinghaus *et al.*, 1956).

V. Oxygen and Carbon Dioxide Content of Blood

Values for these are summarized in Table 3, but only some representative figures for CO_2 contents are given from the numerous determinations reported. The estimates of the O_2 capacity of blood are in reasonable agreement, but those for the O_2 content of arterial and venous blood show great variation. In view of the general concurrence that the arterial pO_2 is normally above 90 mmHg (Table 1) and that the O_2 dissociation curve shows nearly complete saturation at this value, it seems most likely that the arterial

Table 3
Oxygen and carbon dioxide content of blood of adult fowl

Reference	Sex	O_2 capacity (vols %)	O_2% saturation		CO_2 content in Eq/l[a]	Source
			Arterial	Venous		
Morgan and Chichester (1935)	F	13·5	88	40	20·2 (10)	Whole blood
Sykes (1960)	M	15·25 (2)	78 (2)	56·5 (1)	—	Whole blood
	F	13·1 (2)	56·5 (2)	22·4 (2)	—	Whole blood
Chiodi and Terman (1965)	F	14·0 (8)	91 (8)	—	21·2 (7)	Whole blood
Helbacka and Casterline (1963)	F (Laying)	—	—	—	19·7 (60)	True Plasma
					24·4 (8)	Fed
					18·6 (8)	Fasting

[a] Determination at 40°C and pCO_2 40 mmHg.
Numbers in brackets, where given, refer to number of determinations.

blood saturation is normally above 90%, despite the figures given by Sykes (1960). The values for saturation of venous blood seem very low and further measurements on mixed venous blood are clearly desirable. The work of Helbacka and Casterline (1963), showing the effects of feeding and starvation on plasma CO_2 content, illustrates the need for adequate control of the conditions of blood sampling. Variations can also occur with the stage of the egg-laying cycle.

VI. Blood Gases During Thermal Panting

During extreme hyperthermia, panting in the domestic fowl may drive the arterial pCO_2 below 15 mmHg, the CO_2 content below 10 mEq/litre and the pH above 7·70 (Linsley and Burger, 1964; Frankel and Frascella, 1968; Calder and Schmidt-Nielsen, 1968). The fall in the CO_2 content is chiefly due to the loss of CO_2 from the body due to the lung ventilation being increased by hyperthermia. In addition, there is an increase in the lactate level of the blood, which may rise some 2 mEq/litre or more (Frankel and Frascella, 1968); this effect seems to be due to the arterial hypocapnia and not to arterial hypoxia, for the arterial pO_2 is raised (Frankel, 1965).

The alkalosis which develops in arterial blood during thermal panting is evidence that no completely effective mechanism exists whereby respired air can ventilate air sacs, cooling the body by evaporation of water, but bypass the pulmonary gas-exchange system. Linsley and Burger (1964) demonstated that pCO_2 begins to fall, and pH to rise, as soon as body temperature and respiratory rate begin to increase, i.e. from the very start of the respiratory stimulation by hyperthermia; thus even under the most favourable circumstances there is not complete separation of lung and air sac ventilation (see also Chapter 6).

References

Butler, P. J. (1967). *J. Physiol., Lond.* **191**, 309–324.

Calder, W. A. and Schmidt-Nielsen, K. (1968). *Am. J. Physiol.* **213**, 883–889.

Chiodi, H. and Terman, J. W. (1965). *Am. J. Physiol.* **208**, 798–800.

Christensen, E. and Dill, D. B. (1935). *J. biol. Chem.* **109**, 443–448.

Cohn, J. E. and Shannon, R. (1968). *Resp. Physiol.* **5**, 259–268.

Denmark, C. R. and Washburn, K. W. (1969). *Poult. Sci.* **48**, 464–474.

Frank, F. R. and Burger, R. E. (1965). *Poult. Sci.* **44**, 1604–1606.

Frankel, H. M. (1965). *Proc. Soc. exp. Biol. Med.* **119**, 261–263.

Frankel, H. M. and Frascella, D. (1968). *Proc. Soc. exp. Biol. Med.* **127**, 997–999.

Hall, F. G. (1935). *J. Physiol., Lond.* **83**, 222–228.

Helbacka, N. V. and Casterline, J. L. (1963). *Poult. Sci.* **42**, 450–451.

Helbacka, N. V. L., Casterline, J. L., Smith, C. J. and Shaffner, C. S. (1964). *Poult. Sci.* **43**, 138–144.

Hodges, R. D. (1969). *Comp. Biochem. Physiol.* **28**, 1243–1257.

Hunt, J. R. and Simkiss, K. (1967). *Comp. Biochem. Physiol.* **21**, 223–230.

Johnson, V. L. and Dunlap, J. S. (1955). *Science, N. Y.* **122**, 1186.
Linsley, J. G. and Burger, R. E. (1964). *Poult. Sci.* **43**, 291–305.
Mongin, P. and Lacassagne, L. (1966a). *Annls. biol. anim. Biochim. Biophys.* **6**, 93–100.
Mongin, P. and Lacassagne, L. (1966b). *Annls. biol. anim. Biochim. Biophys.* **6**, 101–111.
Morgan, W. E. and Chichester, D. F. (1935). *J. biol. Chem.* **110**, 285–298.
Mueller, W. J. (1966). *Poult. Sci.* **45**, 1109.
Nightingale, T. E., Boster, R. A. and Fedde, M. R. (1968). *J. appl. Physiol.* **25**, 371–375.
Parer, J. T. (1970). *Resp. Physiol.* **9**, 43–49.
Richards, S. A. (1969). *Comp. Biochem. Physiol.* **29**, 955–964.
Richards, S. A. and Sykes, A. H. (1967). *Comp. Biochem. Physiol.* **21**, 691–701.
Rodnan, G. P. and Ebaugh, F. G. (1956). *Fedn Proc. Fedn Am. Socs exp. Biol.* **15**, 155.
Rostorfer, H. H. and Rigdon, R. H. (1946). *Am. J. Physiol.* **146**, 222–228.
Rostorfer, H. H. and Rigdon, R. H. (1947). *Biol. Bull.* **92**, 23–30.
Saha, A., Dutta, R. and Ghosh, J. (1957). *Science, N. Y.* **125**, 447–448.
Salt, G. W. and Zeuthen, E. (1960). *In* "Biology and Comparative Physiology of Birds" (A. J. Marshall, ed.), Vol. I, pp. 363–409. Academic Press, New York.
Severinghaus, J. W., Stupfel, M. and Bradley, A. F. (1956). *J. appl. Physiol.* **9**, 197–200.
Shepard, R. H., Sladen, B. K., Peterson, N. and Enns, T. (1959). *J. appl. Physiol.* **14**, 733–735.
Sykes, A. H. (1960). *Poult. Sci.* **39**, 16–17.
Washburn, K. W. (1968a). *Poult. Sci.* **47**, 561–564.
Washburn, K. W. (1968b). *Poult. Sci.* **47**, 1083–1089.
Wastl, H. and Leiner, G. (1931a). *Pflügers Arch. ges. Physiol.* **227**, 367–420.
Wastl, H. and Leiner, G. (1931b). *Pflügers Arch. ges. Physiol.* **227**, 460–474.
Zeuthen, E. (1942). *Biol. Medd. Kbh.* **17**, 1–40.

8 Structure of the Kidney

W. G. SILLER

Agricultural Research Council's Poultry Research Centre,
King's Buildings,
Edinburgh, Scotland

I. Gross Anatomy

In the fowl, as in birds generally, the two large kidneys, lying extraperitoneally and deep within the bony recesses of the pelvis, are drained by branched ureters which terminate in the urodeum of the cloaca. There is neither a renal pelvis nor a urinary bladder. Each kidney displays three ill-defined regions, frequently, but incorrectly, termed lobes. After Goodchild (1956), these are the anterior, middle and posterior divisions.

II. Organization of the Lobule

The avian kidney is lobulated, each division being made up of numerous lobules which on the surface produce delicate irregularities somewhat reminiscent of cerebral gyri (Fig. 1). Although each lobule consists of cortical and medullary tissue, the cortical component greatly exceeds in bulk that of the medulla. The disproportion is because only a small number of nephrons have medullary loops of unequal length.

It is not only this cortico-medullary relationship but also the dual afferent blood supply and the arrangement of the nephrons within the lobule, which makes the architecture of the avian renal lobule somewhat complex and strange to the histologist accustomed to the mammalian kidney.

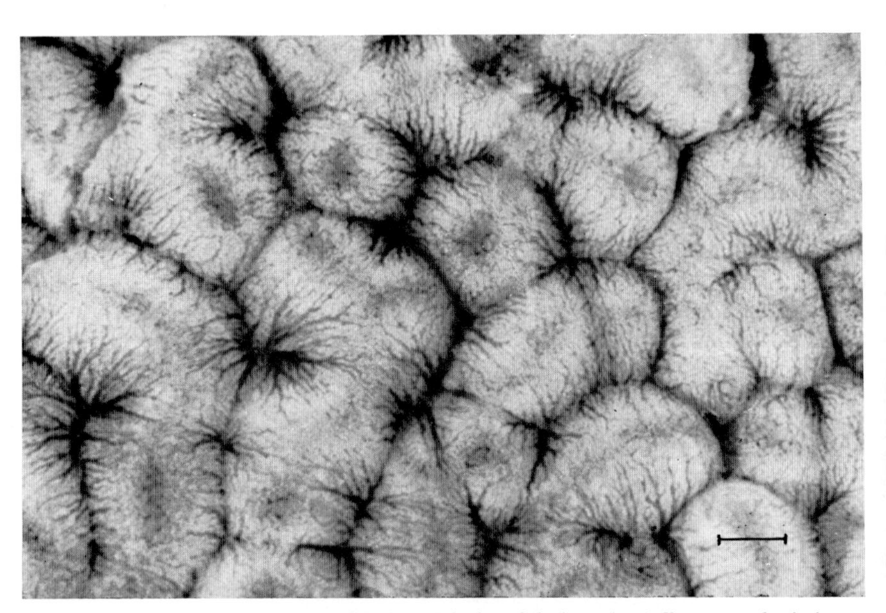

Fig. 1. Surface of the fowl's lobulated kidney. The interlobular veins (afferent renal veins) run between the lobules and on to their surface. In some lobules the dark central area denotes the position of the intralobular (efferent) vein. The scale marker indicates 1 mm.

A. CORTEX

The lobule is limited by the intricate ramifications of the interlobular veins, which are the terminal branches of the afferent renal portal vein and spread not only between the lobules but also onto the apical surface (Fig. 1). As the peritubular network of venous sinuses, they continue in a centripetal direction into the lobule from the periphery (Fig. 2) to join the very stout intralobular or efferent vein in the centre. The nephrons follow, or it would perhaps be better to say, dictate this orientation of the sinuses in that, doubling back upon themselves, they occupy a tortuous but generally radial course between central and intralobular veins. The freely branching perilobular collecting ducts, which receive the collecting tubules, the terminal segments of the nephrons, follow a peripheral course from the apex of the lobule to the medullary tract at the base (Fig. 3).

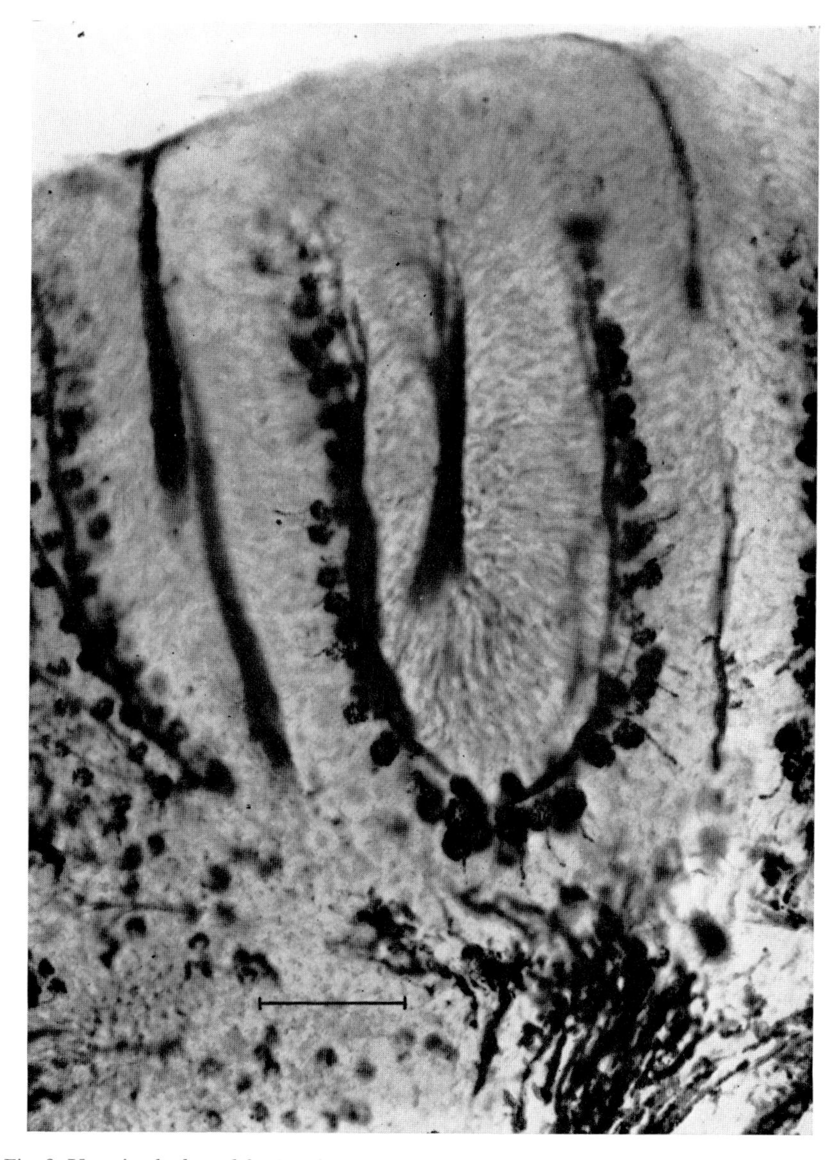

Fig. 2. Unstained, cleared freehand cross section of a kidney lobule in which the arteries had been injected to demonstrate the intralobular arteries, glomerular capillaries and efferent arterioles. Note the radiating course of the peritubular capillary sinus, connecting the interlobular veins with the centrally situated intralobular vein. The scale marker indicates 500 μm.

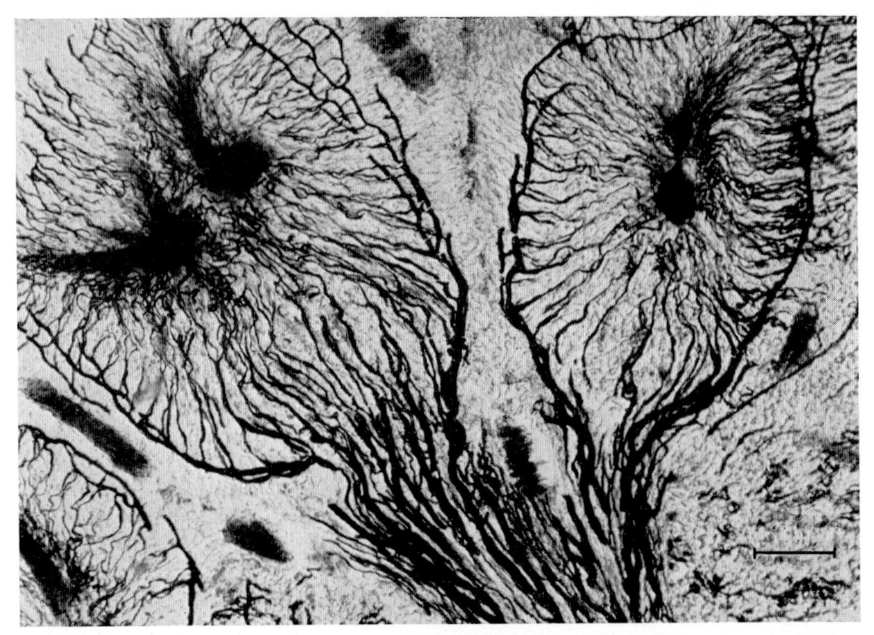

Fig. 3. Retrograde ureteral injection of Indian ink, illustrating the distribution of medullary and perilobular collecting ducts and intralobular collecting tubules. Being full of blood, both the afferent and efferent veins appear dark (this is not due, of course, to the Indian ink). The scale marker indicates 500 μm. Unstained, cleared frozen section about 200 μm thick.

In a vertical section of a lobule where the large cortex is at the top and the medullary tract at the base (Figs. 2 and 4) it may be seen that the glomeruli, situated approximately half-way between the inter- and intralobular veins, are arranged in roughly horse-shoe fashion around the central vein. The peripheral glomeruli are considerably smaller than the juxtamedullary ones. The bulk of the disproportionately large cortex is made up of intertwining loops of nephrons whose orientation is not random, for there is a concentration of distal convoluted tubules around the central vein while at the lobular periphery, where the afferent renal portal vein enters, the predominant segment is the proximal convoluted tubule.

B. MEDULLA

The branching perilobular collecting ducts aggregate in groups at the base of the lobule and form into a discrete, encapsulated cone-shaped tract into which project medullary loops of nephrons of mammalian type (Fig. 4). Although the number of lobules per medullary tract differs in various species of birds (Johnson and Mugaas, 1970), several neighbouring lobules share the same medullary tract (Fig. 3) and, as they progress distally, the tracts join to form large bundles which eventually terminate in the ureteral branches (Fig. 5).

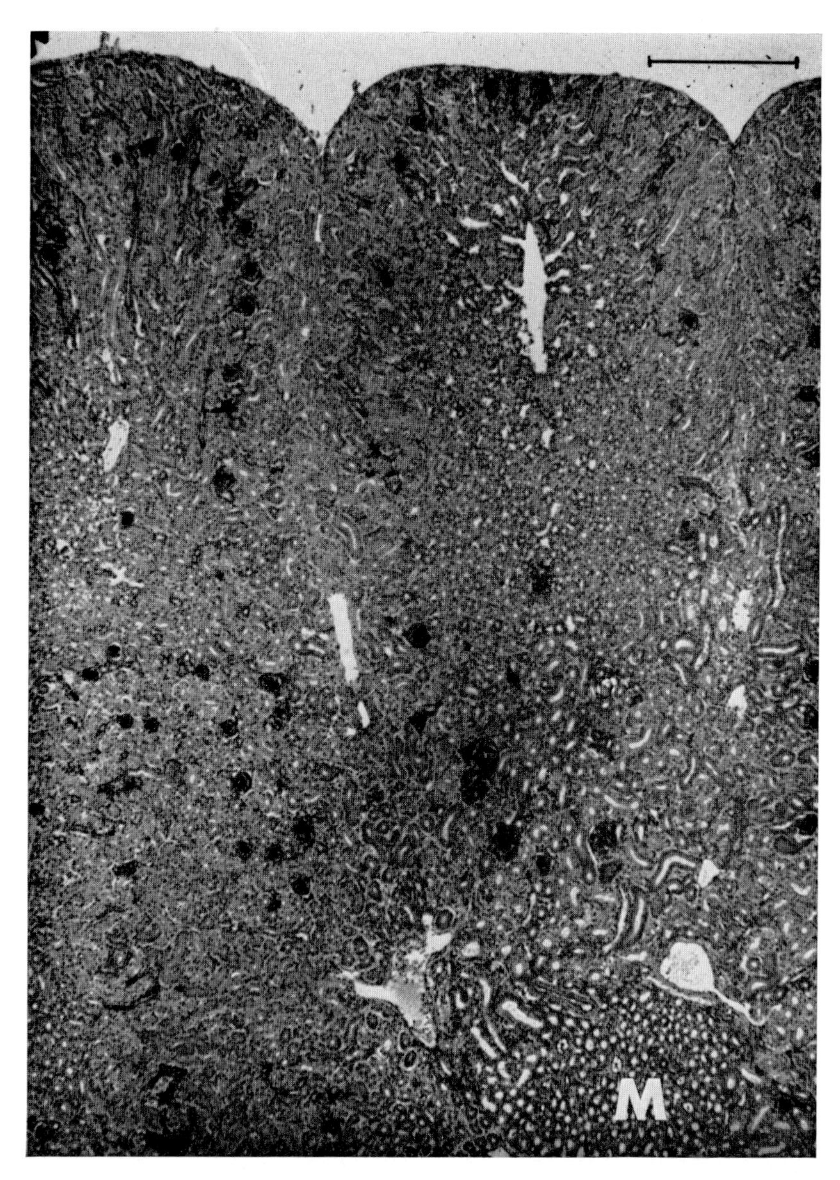

Fig . 4. Cross section of a kidney lobule, showing the large area of cortical tissue with a small medullary tract (M). Parts of the interlobular and intralobular veins and the characteristic arr angement of the glomeruli are illustrated. Stained with haematoxylin and eosin. The scale marker indicates 400 μm.

Fig. 5. Partial dissection of ureter branches and bundles of medullary tracts, following a retrograde ureteral injection of white neoprene. The scale marker indicates 1 mm.

Johnson and Mugaas (1970) considered a lobe to be the unit of lobules drained by a single medullary tract. Although the avian ureter-collecting duct complex is a continuous dendritic system (cf. Johnson and Mugaas, 1970), an avian ureter branch with its component bundles of medullary tracts has analogy to the minor calyx and its pyramid in mammalian lobed kidneys. Since, according to Bloom and Fawcett (1962) the mammalian pyramid defines the lobe, it is more correct and in agreement with Goodchild (1956), to consider the avian lobe as the total complement of lobules drained by a single ureteral branch.

In a cross-section of a medullary tract (Fig. 6) we see the large and prominent collecting ducts surrounded by the thick and thin segments of the medullary loop. Running between these and in the same general direction is the vasa recta, capillaries which form a closely knit network around the components of the medullary tract. Arrangement of these components in concentric rings as described in the house finch (Poulson, 1965) is not obvious in the fowl.

The whole tract is surrounded by a connective tissue capsule (Fig. 6) and although there are no septa, the medullary interstitium is rich in reticulum and collagen fibres.

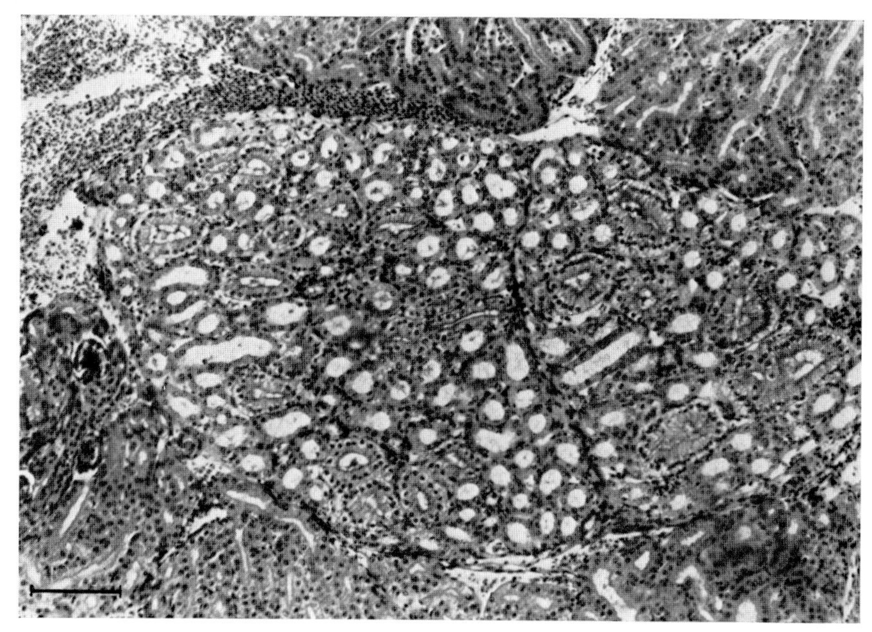

Fig. 6. Cross-section of two medullary cones within a common capsule. Note the large medullary collecting ducts, irregularly interspersed between segments of the medullary loops. Periodic Acid-Schiff reaction. The scale marker indicates 100 μm.

III. The Nephron

As in mammals, the basic renal unit is the nephron; its structure is fundamentally similar to its mammalian counterpart, although in birds two nephron-types are recognized, one cortical or reptilian, the other medullary or mammalian. The basic difference between the two is the absence or presence of a medullary loop (analogous to the loop of Henle). The vast majority of nephrons are of cortical type and confined entirely to the cortex; only a few are of mammalian type with a medullary component. Huber (1917) differentiated a third or intermediate type in which the medullary loop is very short.

The fowl's nephron consists of a glomerulus, proximal and distal convoluted tubules and collecting tubules. Medullary loops, as just mentioned, occur only in a minority of avian nephrons. In the following description the term *tubule* is strictly reserved for parts of the nephron while the term *duct* is applied to various parts of the collecting duct system.

A. GLOMERULUS

Apart from size, there is no fundamental structural difference between the small glomeruli of the cortical nephrons and the large juxtamedullary ones which belong to the mammalian type.

The glomerulus is surrounded by a layer of flattened epithelial cells—the parietal layer of Bowman's capsule (Fig. 7). At the urinary pole it is continuous with the first part of the proximal convoluted tubule (Fig. 7) while at the vascular pole it is reflected as the visceral layer onto the capillary loops of the tuft. Bowman's space lying between the visceral and parietal layer is thus continuous with the lumen of the nephron. The capillary loops are lined by endothelium separated from the visceral epithelium by a prominent, PAS-positive basement membrane (Fig. 7).

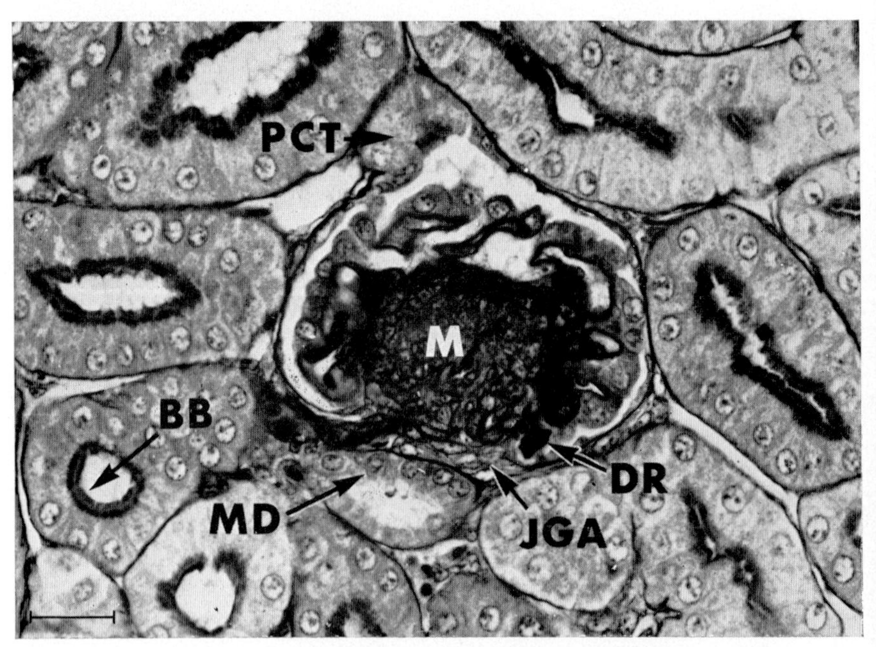

Fig. 7. Glomerulus showing a large central core of mesangial cells (M) and a juxtaglomerular apparatus (JGA) with macula densa (MD) and epithelial cell droplets (DR). The first part of the proximal convoluted tubule (PCT) has been cut somewhat tangentially. The PAS-positive brush borders (BB) of the proximal convoluted tubules are well shown. Periodic Acid-Schiff reaction. The scale marker indicates 20 μm.

In both small and large glomeruli the centre of the tuft is occupied by a compact mass of mesangial cells (Fig. 7) around which the capillary loops (Fig. 8) are arranged in very simple fashion (Siller and Hindle, 1969). These features are considered by Bulger and Trump (1968) to reflect a lower glomerular filtration rate. They constitute the main differences between glomeruli of birds and mammals. In the large glomeruli of mammals the tuft capillaries are more complex with only a few mesangial cells lying between them. The avian glomerulus is less vascular, and Smith (1951) considers it "degenerate", but Sperber (1960) rightly points out that, although the

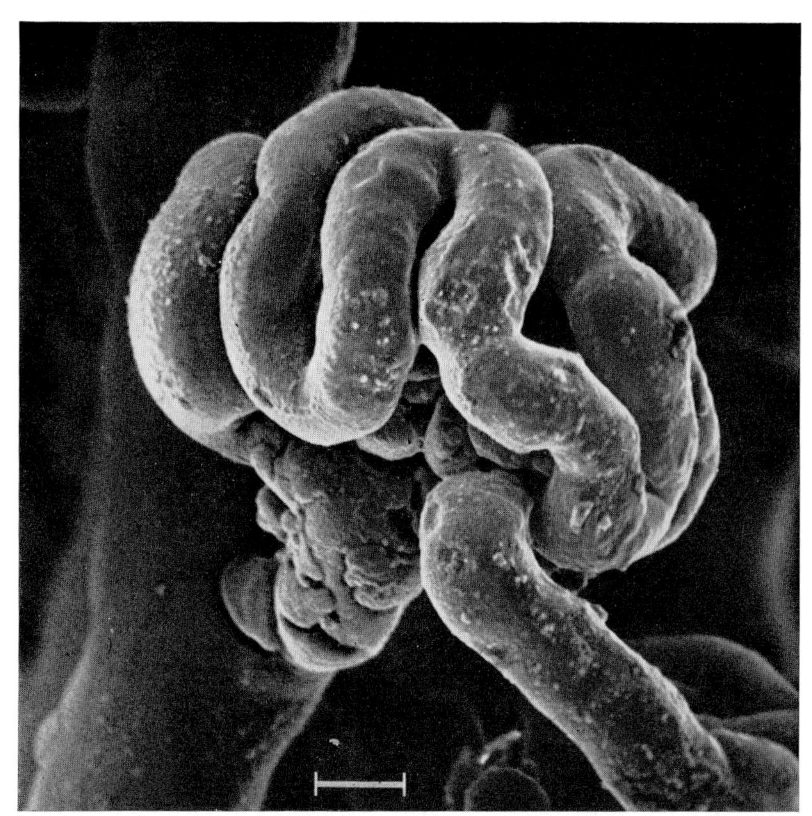

Fig. 8. Scanning electronmicrograph of a gold-paladium coated neoprene cast of a fowl's glomerulus. It shows the afferent arteriole arising from the intralobular artery in the background, the simple arrangement of the capillary loops and the efferent arteriole. The central core of mesangium has been removed in processing. The scale marker indicates 10 μm.

glomerular filtration rate in birds is lower than in mammals, the avian glomeruli function very efficiently. Because of their smaller size and simplicity, there is less filtration surface per glomerulus, but this is compensated by the larger number of glomeruli and because the excretion of waste nitrogen as urate depends to a great extent on tubular secretion, which is sustained by the massive renal portal blood supply.

The ultrastructure of the avian glomerulus is similar to that of the mammal except for the presence of the core of mesangial cells (Fig. 9) which Pak Poy and Robertson (1957) call the "central cell mass". The flattened epithelial cells of Bowman's capsule have oval or sometimes elongated nuclei with generally a single nucleolus. In the finely granular cytoplasm, the sparse organelles consist of a few mitochondria, endoplasmic reticulum and ribosomes. The visceral epithelium covering the glomerular tuft, although continuous with the parietal layer of Bowman's capsule, appears very

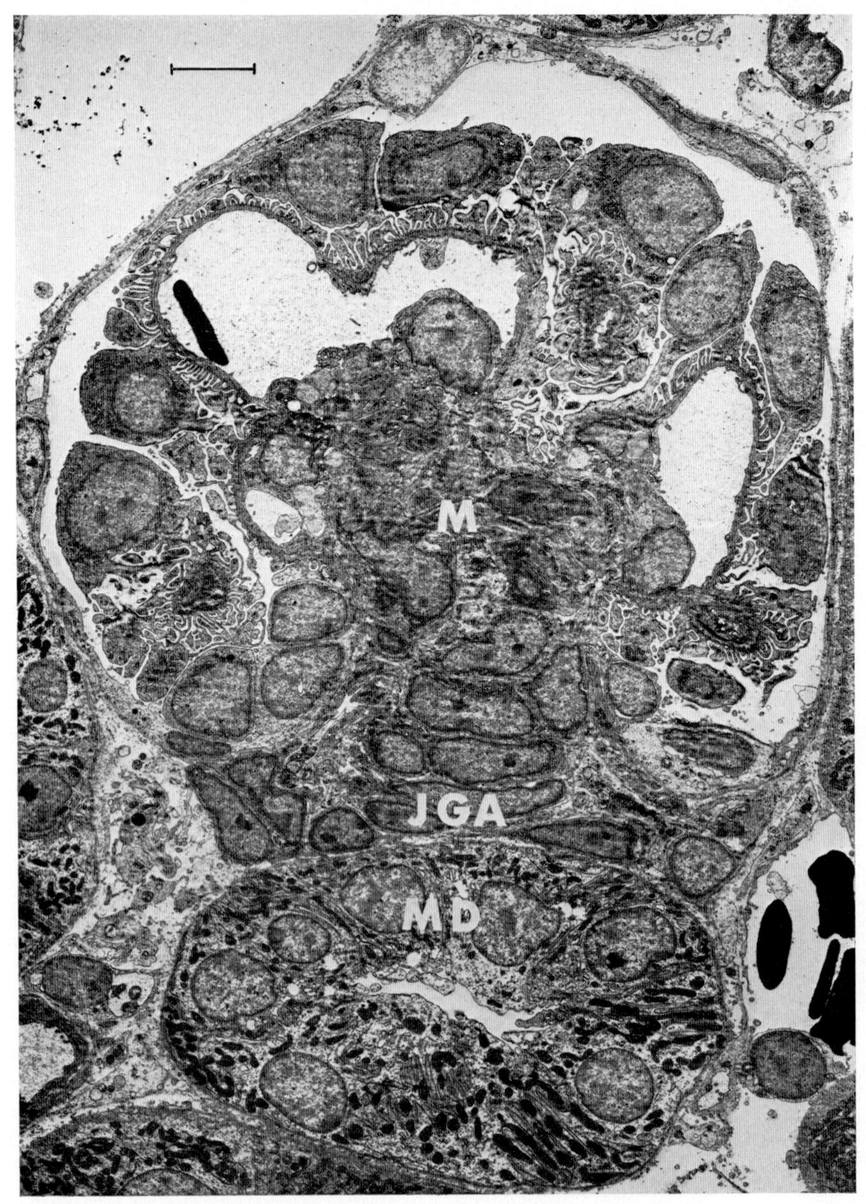

Fig. 9. Low power electronmicrograph of a fowl's glomerulus, with mesangium (M), juxtaglomerular apparatus (JGA) and macula densa (MD). 1% Dalton's osmium (pH 7·4). The scale marker indicates 5 μm.

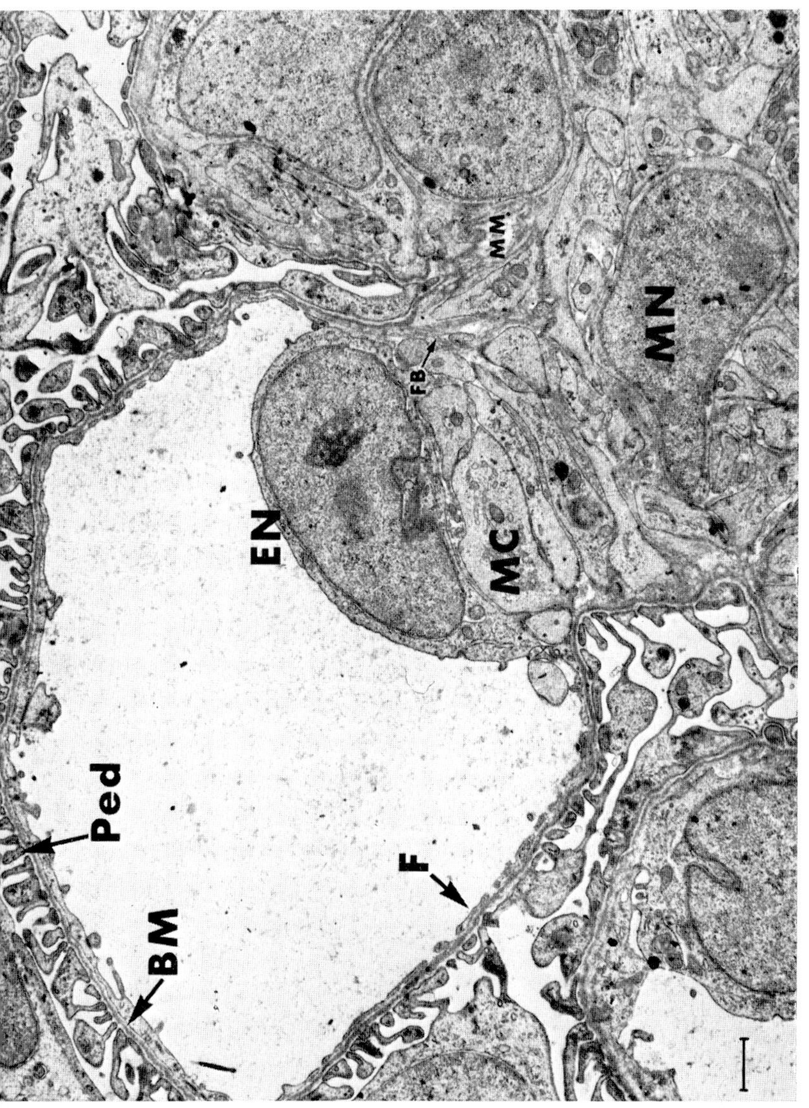

Fig. 10. Glomerular capillary, showing epithelial cell foot processes (Ped), capillary basement membrane (BM) and endothelial cell (EN) forming a delicate, fenestrated layer (F) around the lumen of the capillary. The endothelial cell makes close contact with the ramifying cytoplasmic processes of the mesangial cell (MC). A mesangial cell nucleus (MN) is also shown. The matrix of the mesangium (MM) is continuous with the capillary basement membrane and also has fibrillary components (FB). 1% Dalton's osmium (pH 7·4). The scale marker indicates 1 μm.

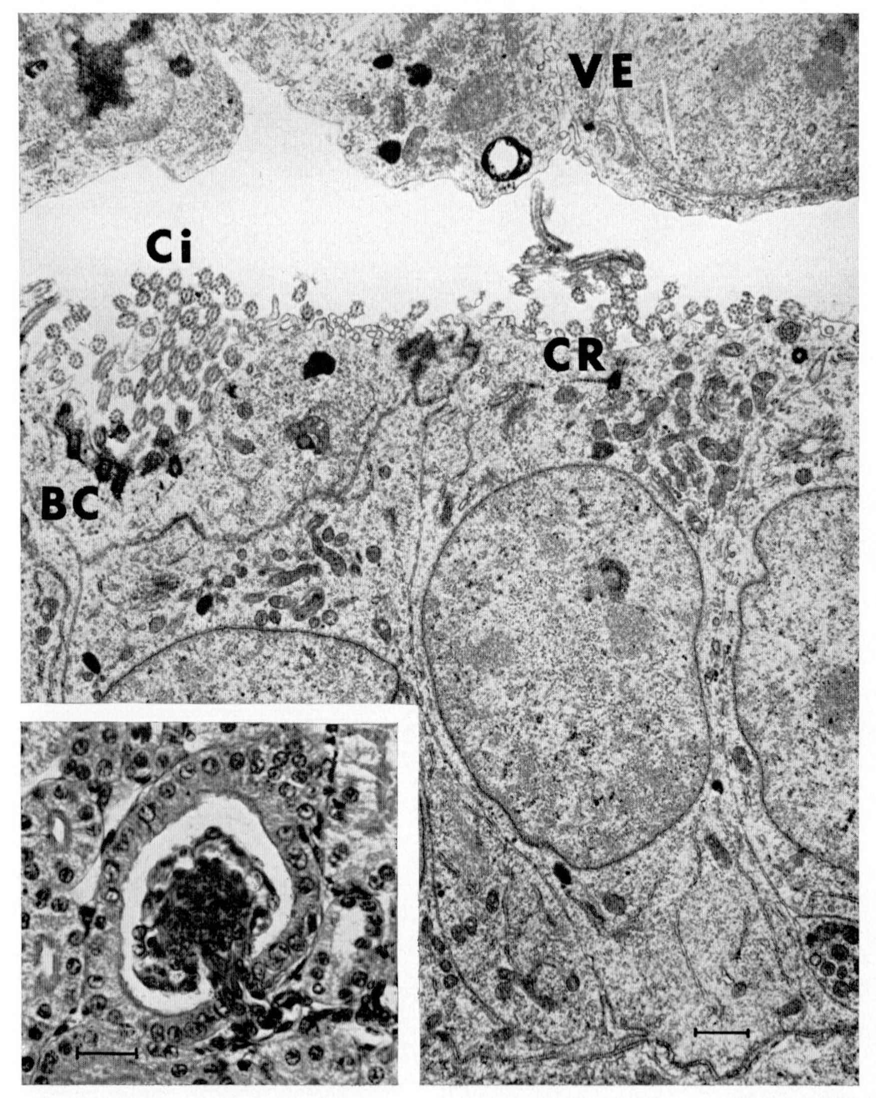

Fig. 11. Cuboidal metaplasia of Bowman's capsule. Under the light microscope (inset) the epithelial cells resemble those of the proximal convoluted tubule but their ultrastructure differs mainly in the paucity of mitochondria and other organelles, the simplicity of the plasma membrane with absence of microvilli and the presence on some cells of cilia (Ci) projecting into Bowman's space. Basal corpuscles (BC) and ciliary rootlets (CR) are shown, with the visceral epithelial cells (VE) of the glomerular tuft at the top of the picture. 1% Dalton's osmium (pH 7·4). The scale marker indicates 1 μm and 20 μm (inset).

different (Fig. 9); the cells are cuboidal with larger oval or round nuclei. Organelles, particularly mitochondria and rough endoplasmic reticulum are more prominent and there is much granular cytoplasm which contacts the capillary basement membrane through numerous characteristic foot processes (Fig. 9 and 10). By virtue of the spaces between them, the glomerular filtrate can enter the capsular space without having to traverse the layer of epithelial cells.

In otherwise normal glomeruli the epithelial cells of Bowman's capsule are sometimes cuboidal (Fig. 11). Such glomeruli are common in young males (Siller, 1962). A similar metaplasia of Bowman's capsule has been reported in mammals (Helmholz, 1935) and under various pathological conditions in man (*cf.* Siller, 1962). Crabtree (1941) has shown that cuboidal metaplasia is normal in male mice and constitutes a form of sex dimorphism.

A very unusual feature is the presence of typical cilia in such a metaplastic capsule (Fig. 11). Normally, both capsule and proximal convoluted tubule of birds are strictly non-ciliated but both cilia and mucin granules are seen in the neck region of the nephrons in lower species (Bulger and Trump, 1968; Miyoshi, 1970).

The prominent capillary basement membrane shows three layers (Fig. 10) the dense central one containing fibrous elements measuring some 120 Å in diameter. The oval or irregular nucleus of the endothelial cell (Fig. 10) is surrounded by sparse cytoplasm containing relatively few organelles. The cytoplasm spreads, in contact with the basement membrane, over the internal surface of the capillary loops and forms a continuous delicate sheet of endothelium, in some places no more than 400 Å thick and is fenestrated by pores 500 to 1000 Å across.

In mammals, intercapillary or so-called mesangial cells lie between the capillary loops; since their first description as supporting structures (Zimmerman, 1933) their function and origin have been variously interpreted (cf. Simon and Chatelanat, 1969). In the fowl the centre of the glomerular tuft is formed by a mass of such mesangial cells round which the capillary loops are arranged. These mesangial cells proliferate in glomerulonephritis in fowls (Siller, 1959a) as in mammals (Zuzuki *et al.*, 1963; Kazimierczak 1967). The matrix between the plasma membranes of adjacent mesangial cells, and their ramifying processes, resembles and is continuous with the basement membrane of the glomerular capillaries (Fig. 10). In the mesangial region it is much more irregular and can be 1 μm or more in section where it spreads out to fill a large intercellular space. A fibrillary component is present as in the capillary basement membrane. Dark lysosome-like bodies are sometimes seen within mesangial cells.

B. JUXTAGLOMERULAR APPARATUS

Edwards (1940) claims that the juxtaglomerular (JG) apparatus or "periarteriolar pad", as he terms it, is poorly developed in the bird. Although the granulated modified myocytes of the afferent arteriole are difficult to demonstrate in fowls, there can be no doubt that birds have a JG apparatus. There is a macula densa (Figs. 7 and 9) and, at the glomerular vascular pole, closely associated with the mesangium (Fig. 7 and 9), there is an accumulation of Goormaghtigh, Polkissen or Lacis cells. Although Smith (1966) saw typical JG granules in the fowl's kidney, these are sparse, very difficult to demonstrate and must not be confused with large PAS-positive droplets frequently found in the visceral epithelial cells of the glomerular tuft (Fig. 7). In birds on a salt-free diet there is a marked increase in JG granules (W. G. Siller, unpublished); it is interesting that these are then found not only in the wall of the afferent arteriole but also in substantial numbers in the mesangial cells (W. G. Siller, unpublished).

Dunihue and Boldosser (1963) have already suggested that, in the cat, the function of mesangial and JG cells may be similar. A renin-like substance is present in the fowl's kidney (Bean, 1942) and therefore the suggestion of Bing and Kazimierczak (1963) that renin is produced in the macula densa and stored as granules in the JG apparatus, may hold good in the fowl if its mesangium is also regarded as a store.

C. PROXIMAL CONVOLUTED TUBULE (PCT)

The proximal convoluted tubule arises at the urinary pole of the glomerulus as a continuation of Bowman's capsule. Huber (1917), pointing out that it is the longest segment forming about half the length of the nephron, found the PCT of fowls' cortical and medullary nephrons to be 3·5 to 4 mm and 7·5 to 8·5 mm, respectively. About half the cortical tissue thus consists of interwoven coils of PCT. The diameter of this segment measures 35–45 μm in histological sections. The distal part of the PCT (it would be incorrect in the fowl to talk of convoluted and straight parts) is appreciably thicker than the proximal part.

Surrounded externally by a distinctly PAS-positive basement membrane, the epithelium is a single layer of columnar cells with basally situated, spherical nuclei. The luminal surface is lined by a well developed PAS- and alkaline phosphatase-positive brush border below which a terminal web is clearly visible in appropriate preparations (Fig. 12). Numerous mitochondria are concentrated in the basal two-thirds of these cells (Fig. 12). Berger (1962) described a similar distribution of alkaline phosphatase in the kidney of the pigeon.

The proximal convoluted tubules consist of two morphologically distinct parts. The more proximally situated portion, which corresponds to the convoluted part in mammals, is of smaller diameter and, because of its more granular cytoplasm and greater mitochondrial content appears darker with such stains as Mallory's trichrome. The more distal portion, corresponding to the mammalian straight part, is considerably thicker, stains less intensely and has fewer mitochondria; the brush border and terminal web are readily

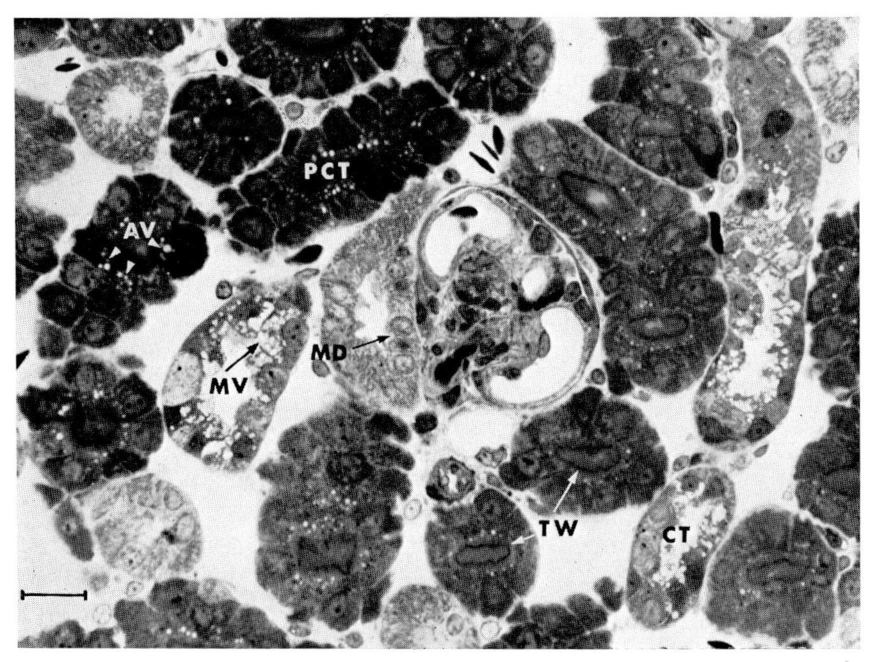

Fig. 12. Araldite-embedded section showing a glomerulus with macula densa (MD), proximal convoluted tubules (PCT) with brush border, terminal web (TW) and prominent apical vacuoles (AV). Numerous mucopolysaccharide-containing vacuoles (MV) are present in the collecting tubules (CT). Toluidine blue. The scale marker indicates 20 μm.

demonstrable. In nephrons with medullary loops the distal part of the PCT extends some way into the medulla, then gradually decreases in size to continue as the thin segment.

The ultrastructures of the avian and mammalian PCT are substantially similar, and it is interesting that the mesonephric tubule in 8-d embryos is little different (Gibley and Chang, 1967). The well-developed brush border lining the lumen is an array of regular and generally unbranched microvilli, about 1 to 1·5 μm long and some 800 to 1000 Å in diameter; there is a space of about 200 Å between them (Fig. 13).

The area immediately below the brush border is occupied by apical tubules and vacuoles which are closely associated with invaginations of the apical intervillous plasma membrane and are believed to be concerned with the uptake of larger molecules into the cell. The vacuoles vary greatly in size up to 1 μm or more in diameter, but the very long, tortuous and striated apical tubules are more uniform in diameter at 600 to 700 Å (Fig. 13). The cytoplasm below this apical region is rich in both smooth and rough endoplasmic reticulum and free ribosomes. It contains numerous mitochondria with prominent microbodies and a number of other profiles including lysosomes

(Fig. 14) and a variety of cytoplasmic bodies. The spherical nucleus is situated in the lower one-third of the cell; one or more nucleoli are usually present.

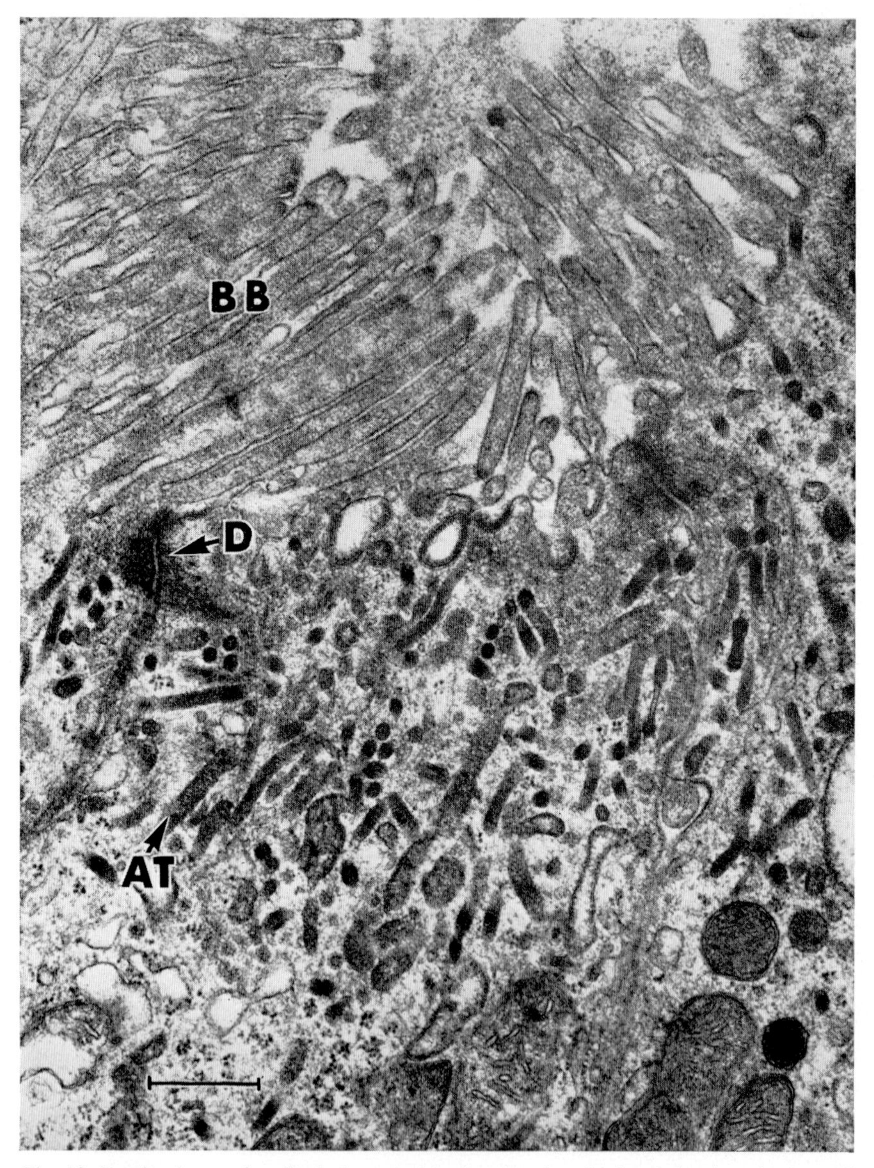

Fig. 13. Proximal convoluted tubule, apical region showing the brush border of microvilli (BB), two desmosomes (D) and striated apical tubules (AT). 1% Dalton's osmium (pH 7·4). The scale marker indicates 500 nm.

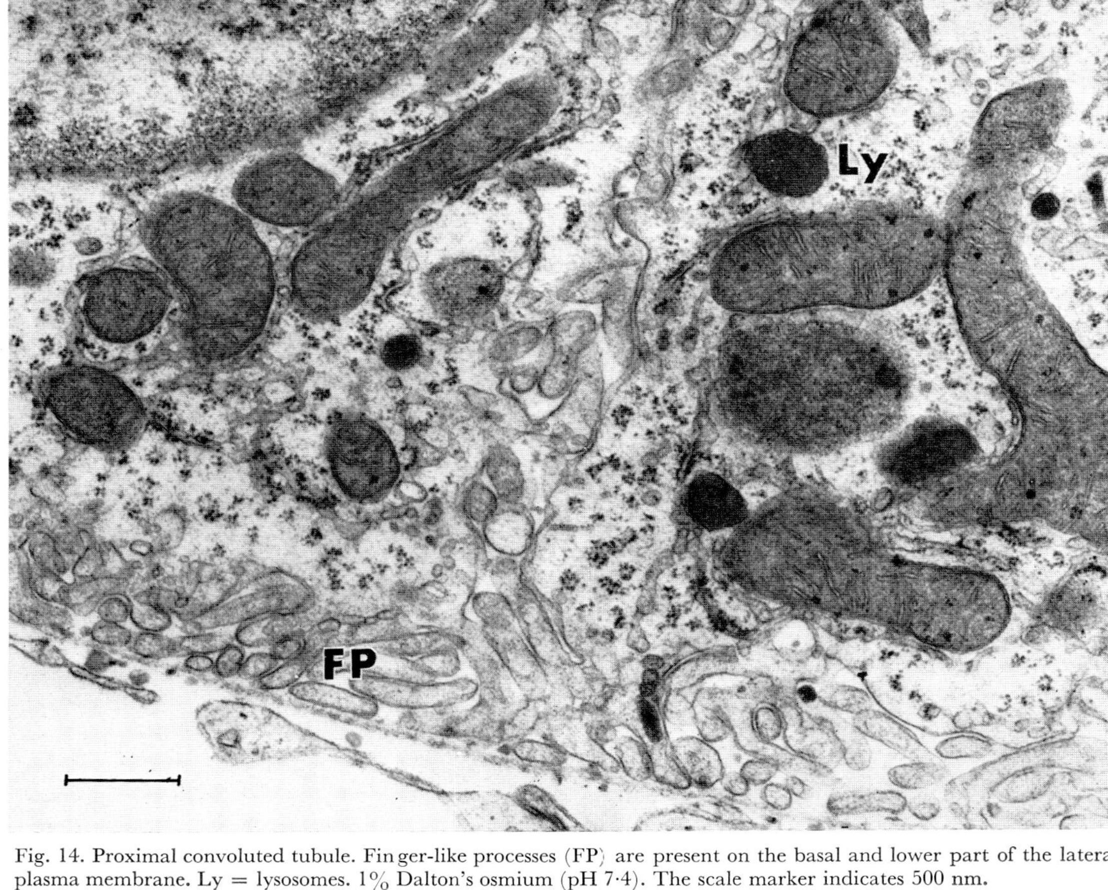

Fig. 14. Proximal convoluted tubule. Finger-like processes (FP) are present on the basal and lower part of the lateral plasma membrane. Ly = lysosomes. 1% Dalton's osmium (pH 7·4). The scale marker indicates 500 nm.

The basal plasma membrane shows numerous small, lightly curved, finger-like processes of about 1000 Å diameter (Fig. 14) but the deep infoldings of the mammalian basal membrane are not present in the fowl's PCT. However, as the complexity of the infoldings decreases in the distal parts of the mammalian tubule (Rhodin, 1958), so do the basal processes in the fowl become less numerous in more distal portions of the PCT.

In the proximal portion, interdigitating processes similar to the basal ones are prominent in the lower half of the lateral plasma membrane (Fig. 14), but they disappear in the distal portion of the PCT. Where processes are lacking, the lateral plasma membranes of adjacent cells are in close apposition, leaving a fairly uniform gap of about 250 Å which narrows towards the apical end, forming a well-defined desmosome.

D. MEDULLARY LOOP

This is analogous to the mammalian loop of Henle but in birds it occurs only in the so-called medullary or mammalian type of nephron. While the thick segment is comparable in structure with its mammalian counterpart, the thin segment differs considerably (Fig. 15).

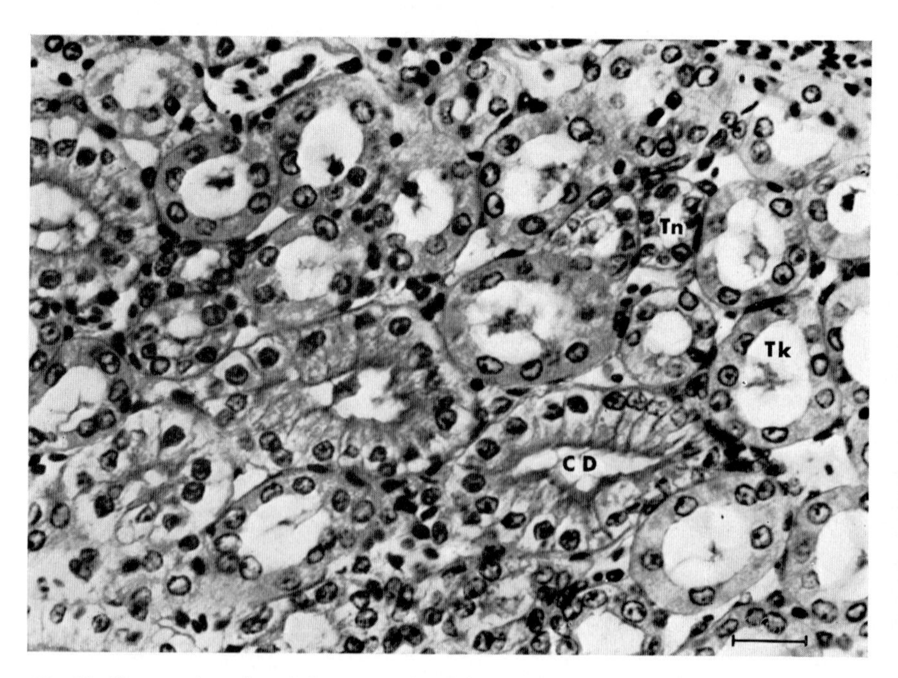

Fig. 15. Cross-section of medullary tract. Medullary collecting duct (CD) is surrounded by thick (Tk) and thin (Tn) segments of medullary loop. Periodic Acid-Schiff reaction. The scale marker indicates 20 μm.

1. Thin Segment

Serial cross-sections of the medullary tract show that there are very few thin segments compared with the numerous thick ones, especially in distal parts of the tract. One may conclude that the thin segment is also shorter and that the arch is probably formed solely by the thick component. This confirms Sperber's (1960) findings that in birds all medullary loops are of the "short" type (using mammalian terminology) and that the turn is in the thick segment. That the thin segment of the fowl has much taller cells (Fig. 15) might suggest that they are less water-permeable than those of the mammal. Furthermore, the gaps in the continuity of the squamous epithelium lining the descending thin limb in mammals, which are considered to play a part in the greater permeability of this segment (Darnton, 1969), are not seen in the fowl where a tight junction exists in the apical region between each cell (Fig. 17). The assumption of Ericsson and Trump (1969) "that the thin loops of Henle . . . appear to be similar to those in the mammalian kidneys" is therefore not borne out. It is all the more interesting that, despite the structural differences in the thin segment, the countercurrent mechanism is said by Poulson (1965) and Skadhauge and Schmidt-Nielsen (1967) to function in birds.

In mammals the junction between PCT and the thin limb of Henle is clear cut, due to the abrupt change not only in calibre but also in epithelial cell character. The same is not true in the fowl where, due to the very gradual but nevertheless substantial decrease in diameter of the terminal part of the PCT as it descends the medullary tract, it is difficult to decide exactly where the thin segment begins. The brush border, terminal web and apical tubules remain demonstrable, even in very small tubules (Fig. 16), in which the epithelium measures only about 6 μm in height.

The basal and lateral plasma membranes of these cells are simple without interdigitations or finger-like processes (Fig. 16). The apical membrane, however, shows not only microvilli but often also cilia (Fig. 16). Cytosomes, apical dense tubules and vacuoles are present, mitochondria, small and few in number, and free ribosomes, smooth and rough endoplasmic reticulum can be demonstrated.

Even the smallest tubules, in which the cells measure no more than 3 to 4 μm in height (Fig. 17), differ from the mammalian thin limb which, according to Möllendorff (1930), is lined by cells no thicker than the endothelium of the vasa recta.

2. Thick Segment

Size is the main differentiating feature between thin and thick segments in ordinary histological sections of the medullary tract (Figs. 6, 15), but with the electron microscope it is possible to follow the gradual transition in detail. In the first part of the thick limb (Fig. 18) the height of the cells increases to

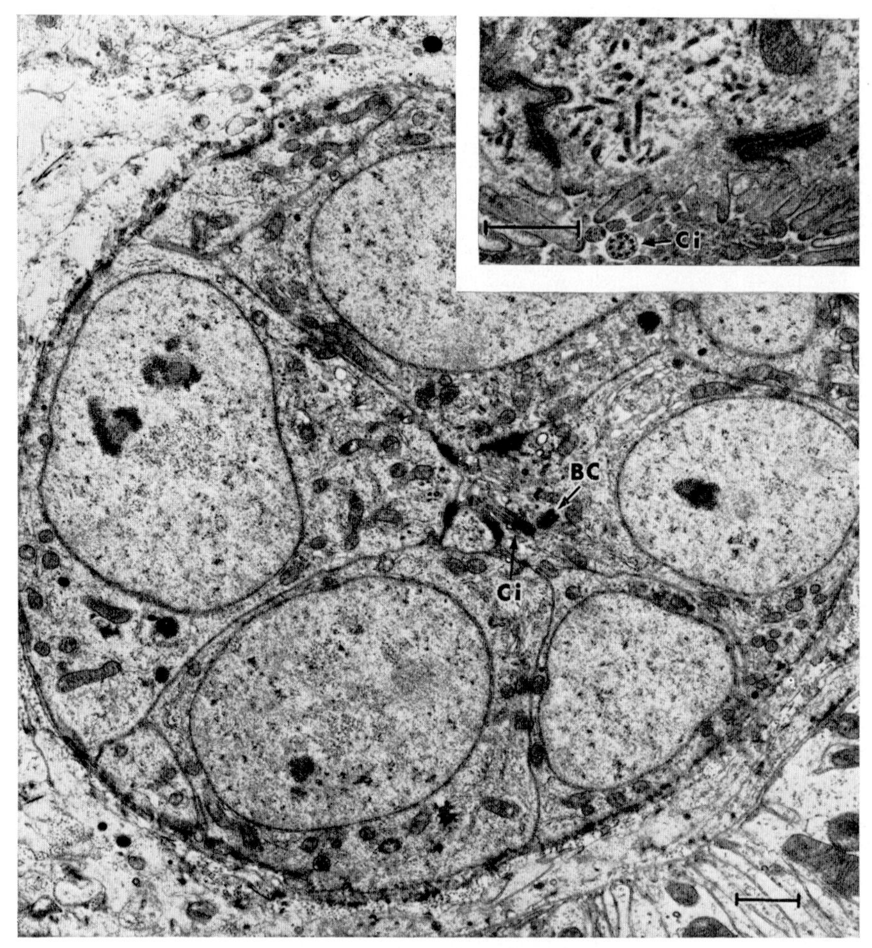

Fig. 16. Thin segment of medullary loop. Projecting into the closed lumen are a few microvilli and one cilium (Ci). Its basal corpuscle (BC) is also shown. Desmosomes, apical tubules and one cilium (Ci) in cross-section can be seen in the inset. 1% Dalton's osmium (pH 7·4). Both scale markers indicate 1 μm.

some 8 μm and the basal and lateral plasma membranes become somewhat more complex and show mitochondria enveloped within the invaginations.

The large round, centrally situated nucleus usually has one nucleolus and fairly evenly distributed chromatin. The apical membrane carries only a few insignificant microvilli. Apical dense tubules and vacuoles are completely lacking, endoplasmic reticulum and free ribosomes are sparse.

Eventually in the more distal parts of the thick segment the basal infoldings become more complex, and the mitochondria lying within them are large and elongated (Fig. 19). The apical part of the cells tends to be free from

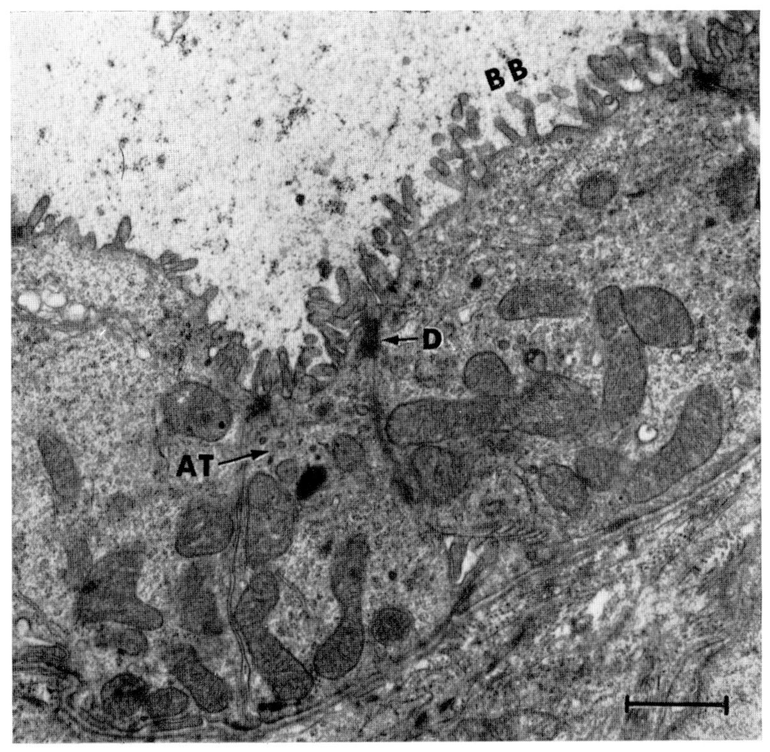

Fig. 17. Thin segment of medullary loop. Low epithelial cells with simple basal and lateral plasma membranes and relatively few microvilli (BB). Some cells still have apical tubules (AT). D = Desmosome. 1% Dalton's osmium (pH 7·4). The scale marker indicates 1 μm.

mitochondria and other organelles and forms tongue-like projections into the lumen (Fig. 24). Tight junctions are present where the lateral membranes meet in the apical region; generally there are no microvilli.

E. CORTICAL INTERMEDIATE TUBULE

Several authors, notably Huber (1917), Feldotto (1929) and Marshall (1934) describe the existence of an intermediate segment which forms the connexion between proximal and distal convoluted tubules in the cortical nephrons of birds. Huber believed it to correspond to the thick limb of the medullary loop and Sperber (1960) found the intermediate tubule to be short and sometimes lacking altogether. This segment, therefore, requires further study, but there can be no doubt that sometimes the PCT of cortical nephrons are seen to terminate, by an abrupt decrease in diameter (Figs. 20 and 21), in a short tubule, often very difficult to identify, which has neither brush border nor terminal web and is about half the thickness of the distal convoluted tubule.

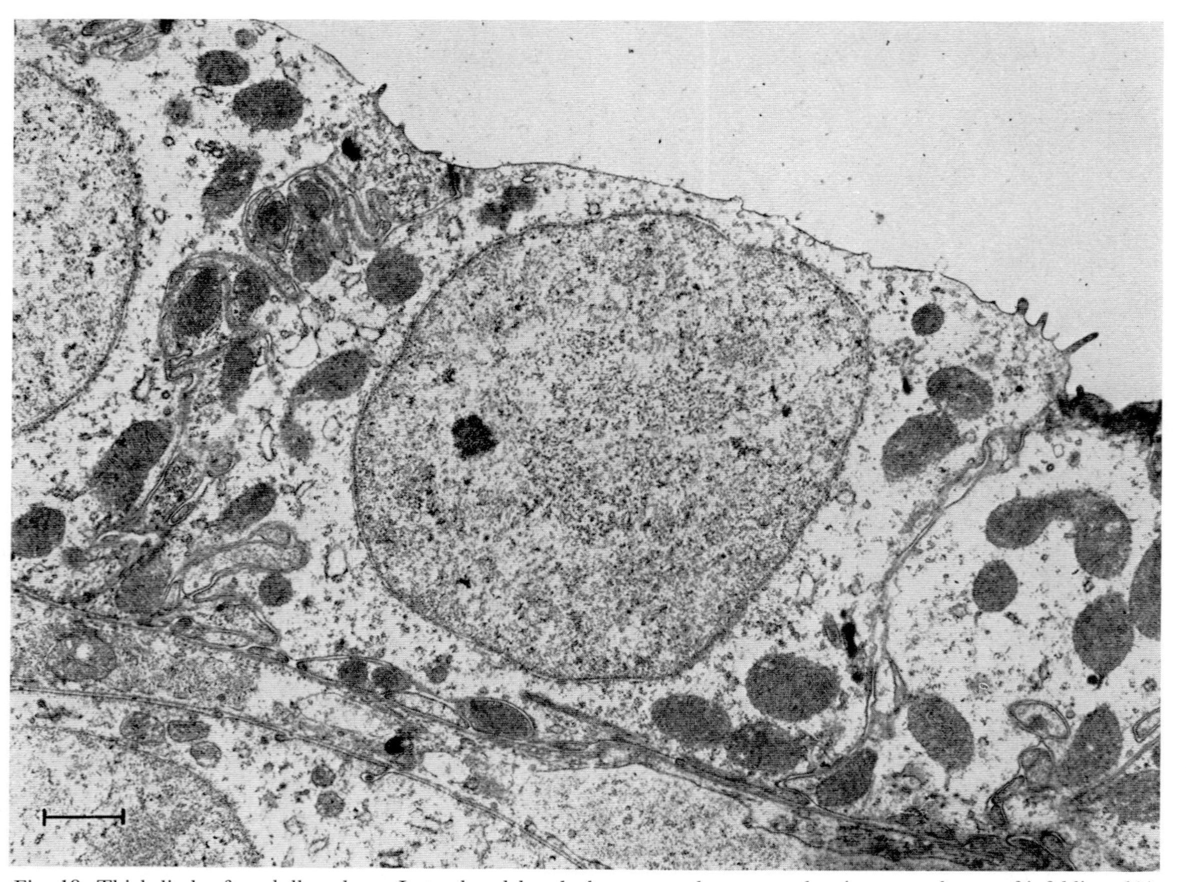

Fig. 18. Thick limb of medullary loop. Lateral and basal plasma membrane are showing some degree of infolding. 1% Dalton's osmium (pH 7·4). The scale marker indicates 1 μm.

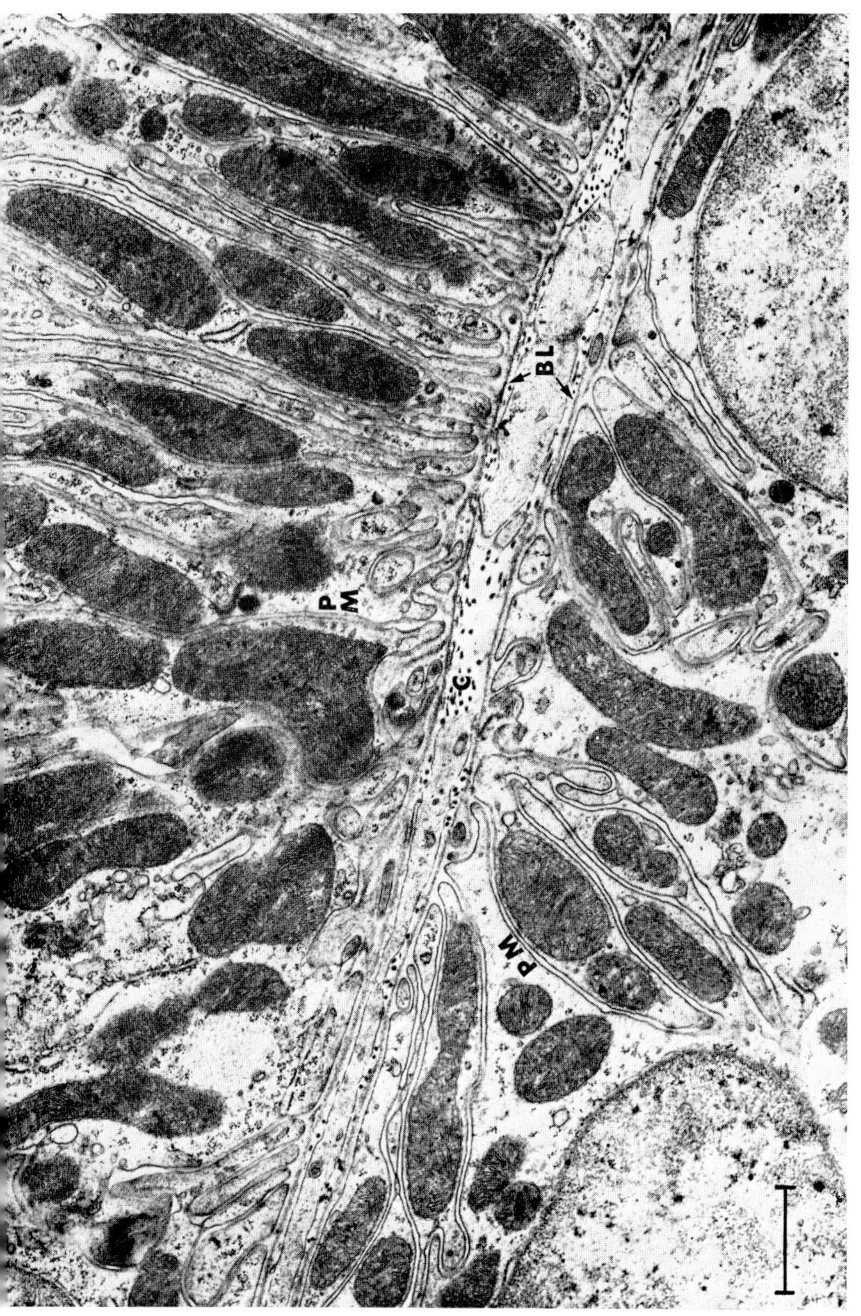

Fig. 19. Basal parts of two adjacent thick segments showing different degrees of infolding of basal plasma membranes (PM). BL = Basal lamina; C = collagen. 1% Dalton's osmium (pH 7·4). The scale marker indicates 1 μm.

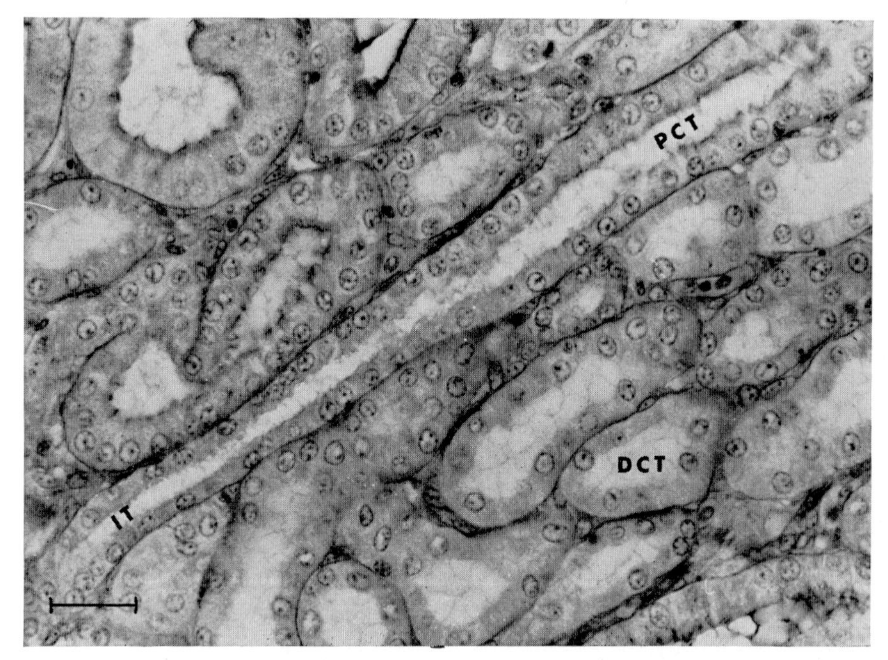

Fig. 20. Cortical intermediate tubule, running diagonally across the picture. The abrupt transition is obvious from PCT (top right) to the much thinner intermediate tubule (IT). Distal convoluted tubules (DCT) are of greater diameter. Haematoxylin and eosin. The scale marker indicates 20 μm.

Since its ultrastructural features have not yet been sufficiently investigated it is not possible to state whether this intermediate tubule resembles either the thick or the thin segment of the medullary loop.

F. DISTAL CONVOLUTED TUBULE (DCT)

This is confined to the cortex and although its convolutions appear to mingle freely with other parts of the nephron, they are concentrated around the central vein. Neither a brush border nor mucopolysaccharide granules are histologically demonstrable, so that the DCT is readily distinguishable from both PCT and collecting tubules (Figs. 12, 20).

The large, elongated mitochondria with numerous cristae are well shown in electronmicrographs to lie within the deep and complex involutions of the basal and lateral plasma membranes (Figs. 10 and 22). From the prominent tight junctions, the apical plasma membrane bulges some distance into the lumen and bears a few small microvilli, but micropinocytotic vesicles are frequent and form a prominent feature of the apical cytoplasm which is relatively free of organelles and of rather low electron density. The nucleus is situated centrally and shows one or two prominent nucleoli; the nucleo-

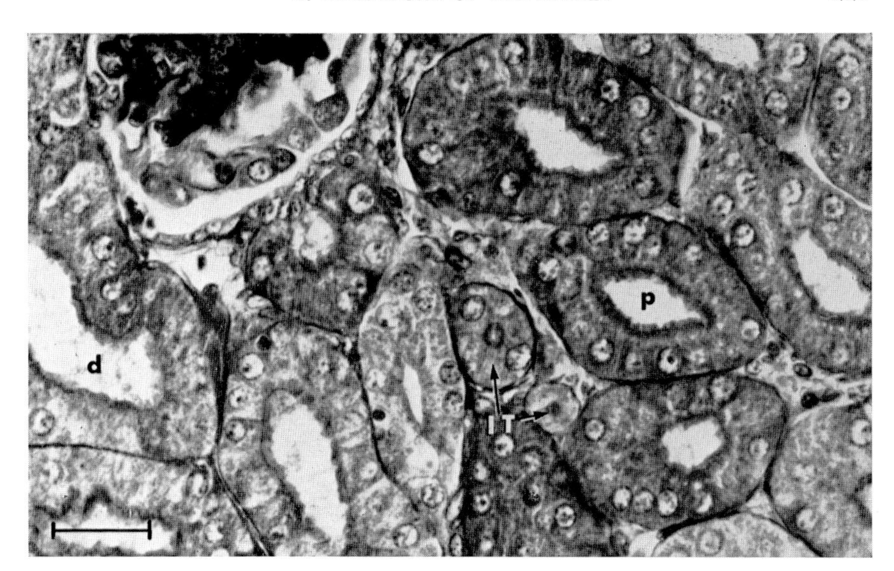

Fig. 21. Two small intermediate tubules (IT) are shown between proximal and distal convoluted tubules. Note the difference in size and staining intensity between the proximal (p) and distal (d) parts of the proximal convoluted tubule. Brush border and terminal web are present even in the intermediate tubule. Picro-Mallory's trichrome. The scale marker indicates 20 μm.

plasm is evenly distributed. Free ribosomes are scattered throughout the cytoplasm and both rough and smooth endoplasmic reticulum is present in the cells, some of which show prominent Golgi structures.

G. MACULA DENSA

In the region of the macula densa the nephric tubule is closely applied to the hilus of the glomerulus and plays an integral part in the JG complex. Structurally the macula densa is similar in both birds and mammals in that the cells nearest the glomerulus are tall and have crowded nuclei (Figs. 7 and 9). Although they show the characteristic basal involutions of the plasma membrane, as do the cells on the opposite side of the lumen and of the remainder of the DCT, they differ in that the mitochondria are smaller and irregularly arranged. It is generally accepted that the mammalian macula densa represents the initial part of the distal convoluted tubule (Ericsson and Trump, 1969); this may also be true in birds. It is interesting that dilation of this region and of the DCT generally is a characteristic feature of acute pyelonephritis in the fowl (Siller, 1964).

H. COLLECTING TUBULE

Huber (1917) termed this segment the "connecting" or "junctional" tubule. It is the terminal part of the nephron and the connexion between the

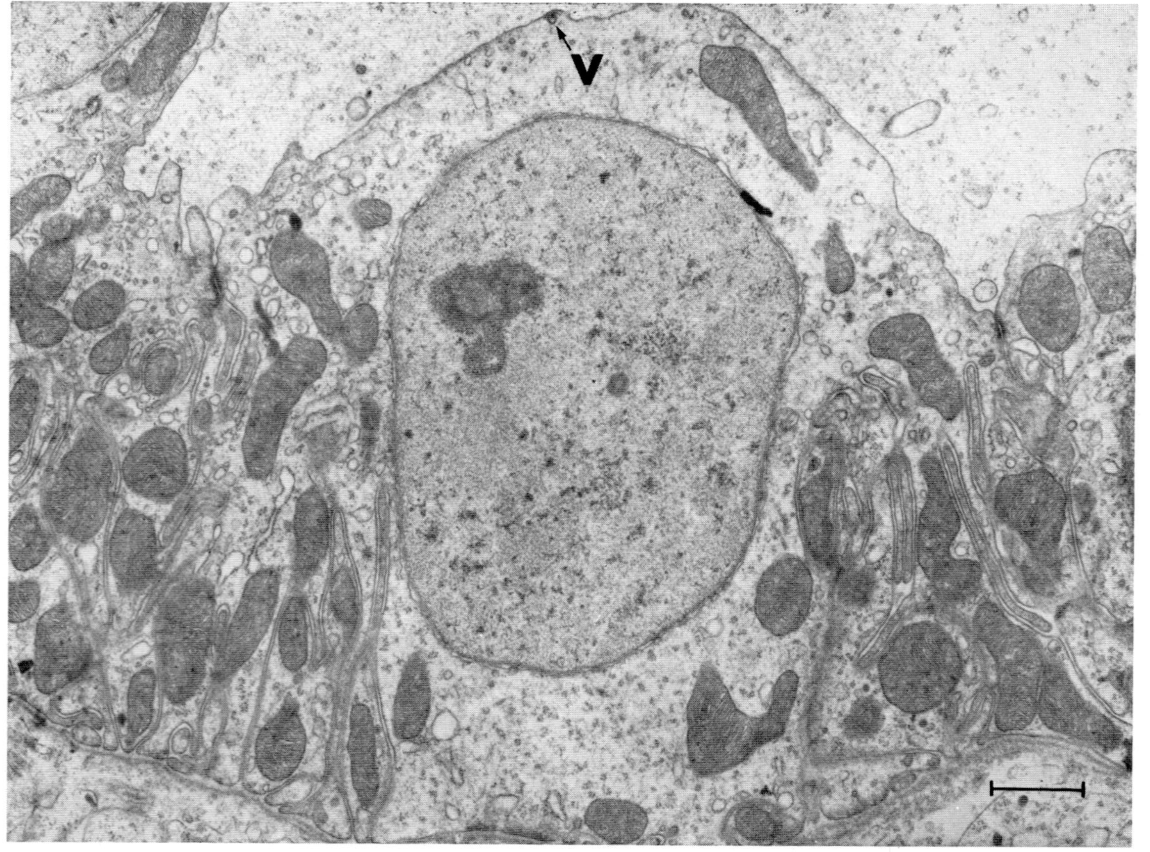

Fig. 22. Distal convoluted tubule showing complexity of lateral plasma membrane, large centrally placed nucleus and simple apical membrane with micropinocytotic vesicles (V). There are few organelles but the relatively large mitochondria are often enclosed within the folds of plasma membrane. 1% Dalton's osmium (pH 7·4). The scale marker

DCT and the collecting ducts. This segment is readily distinguished both in light and electronmicroscopical preparations by the presence of numerous granules of acidic mucopolysaccharide, concentrated in the apical part of the epithelial cells (Fig. 12). These granules are PAS positive and react with alcian blue, Hale's colloidal iron and show toluidine blue and neutral red metachromasia. The tubules are intermediate in size between PCT and DCT. They have basal nuclei and with Altmann's stain show few mitochondria. The lumen is irregular in outline and sometimes contains mucinous strands.

In electronmicrographs the mucin-containing vacuoles, which are sometimes devoid of limiting membranes and may coalesce into large groups or long chains (Fig. 23) often appear empty; if appropriately fixed, they show finely granular material. Rupture of the vacuoles with discharge of the contents into the tubular lumen may frequently be observed. The plasma membranes are comparatively simple with interdigitating basal finger-like processes, somewhat longer and thicker than those of the PCT and are more prominent in the proximal part of the segment. Interdigitations of the lateral membrane also decrease in complexity distally, and luminal microvilli appear to be present in numbers only in cells devoid of mucin vacuoles. These microvilli are responsible for the formation of numerous smaller pinocytotic vacuoles confined to the apical part of the cells.

The cytoplasm contains free ribosomes and the endoplasmic reticulum is of both rough and smooth types. Golgi structures and membrane-bound cytosomes are also present and the mitochondria of this segment differ little from those of the DCT. They are found in all parts of the cell and show well-developed cristae and microbodies.

IV. The Collecting Duct System

This system of ducts carries urine from nephron to ureter and if a countercurrent multiplier system operates in birds, it also plays a part in urine concentration.

The characteristic feature of the collecting ducts is that they branch freely; this differentiates them from all segments of the nephron. In common with the ureteral epithelium and the collecting tubules the ducts secrete mucin. The significance of this has been variously interpreted but it seems that it lubricates and ensures that insoluble urates, which may precipitate as water is withdrawn in the medullary tracts, do not obstruct the ducts. It is interesting that although urine stasis is often associated with some forms of renal disease (Siller, 1959b) urolithiasis is rare in the fowl.

On topographical grounds the collecting duct system can be divided into perilobular and medullary ducts.

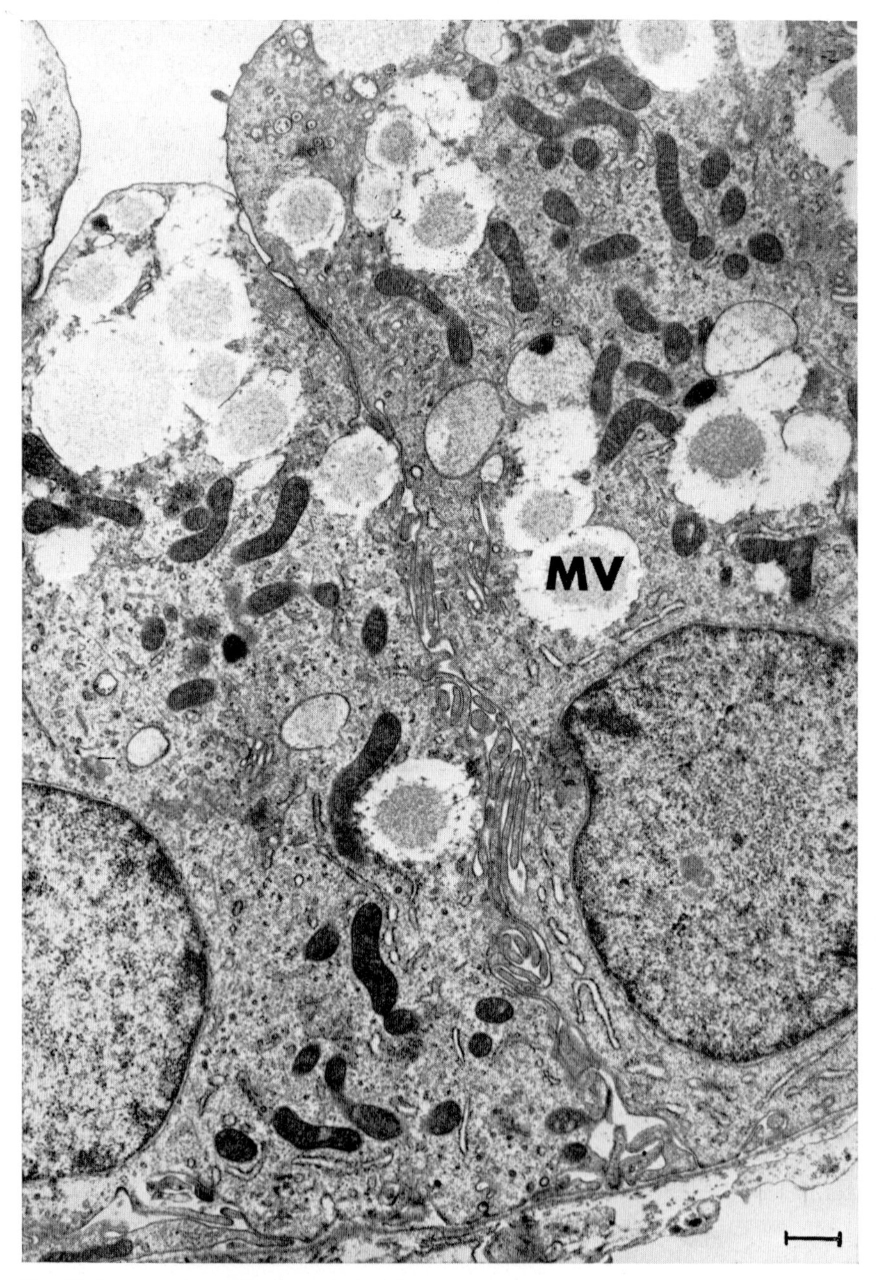

Fig. 23. Intralobular collecting tubule. The lateral, and to a lesser extent, the basal plasma membrane have interdigitating finger-like processes. The characteristic mucopolysaccharide-containing vesicles (MV) are often confluent and concentrated in the apical region. 5% GTA in 0·1 M phosphate buffer (pH 7·4). The scale marker indicates 1 μm.

A. PERILOBULAR COLLECTING DUCTS

The characteristic arrangement of these ducts around the periphery of the lobule has been recognized as long ago as 1863 by Hyrtl, who counted 12 to 20 per lobule and Voigt (1920) confirmed his findings. These branching ducts run from the apex of the lobule and near the periphery, receiving, all along their course, several collecting tubules and joining other similar ducts to congregate in the medullary tract where this fusion continues until the very large and distinctive medullary collecting ducts are reached.

The distribution within the lobule of the perilobular collecting ducts can readily be visualized by using acidic mucopolysaccharide stains such as toluidine or alcian blue or, as in Fig. 3, by retrograde intra-ureteral injections of indian ink.

Although somewhat greater in diameter (30 μm) and branching, they resemble the collecting tubules in that the apical part of the epithelial cell contains mucous granules. The nucleus is basally situated and the lumen frequently contains strands of mucous. The basal and lateral plasma membranes are simple and the few luminal microvilli are generally confined to the non-vacuolated cells.

B. MEDULLARY COLLECTING DUCTS

The large and very characteristic medullary collecting ducts are shown in Figs. 6 and 15. Their basally situated nuclei and distinct cell boundaries make them readily recognizable in sections whatever the stain. The subluminal region is deeply PAS- and alcian blue-positive. In cross-sections of the medullary tract these large ducts are scattered among the thick and thin segments of the medullary loops (Fig. 15).

The apical mucous granules are generally distinct in electronmicrographs (Fig. 24) which also illustrate the simple basal and comparatively simple lateral plasma membranes. Organelles are sparse; small mitochondria are often concentrated in the apical third of the cells with rough endoplasmic reticulum and few free ribosomes. Apical pinocytotic vesicles are common, microvilli are not. The basally situated nuclei with usually one nucleolus are striking.

V. Ureter

Each kidney is drained by a stout ureter which runs from about the anterior pole to the urodeum of the cloaca. Goodchild (1956) found each ureter to have about 8 primary branches of varying length and some secondary branches but Liu (1962) counted 13 primary branches. The histological features of the ureter and its branches are similar. In cross-section they are roughly circular in outline with generally a stellate lumen due to folding of the mucosa during fixation (Fig. 25).

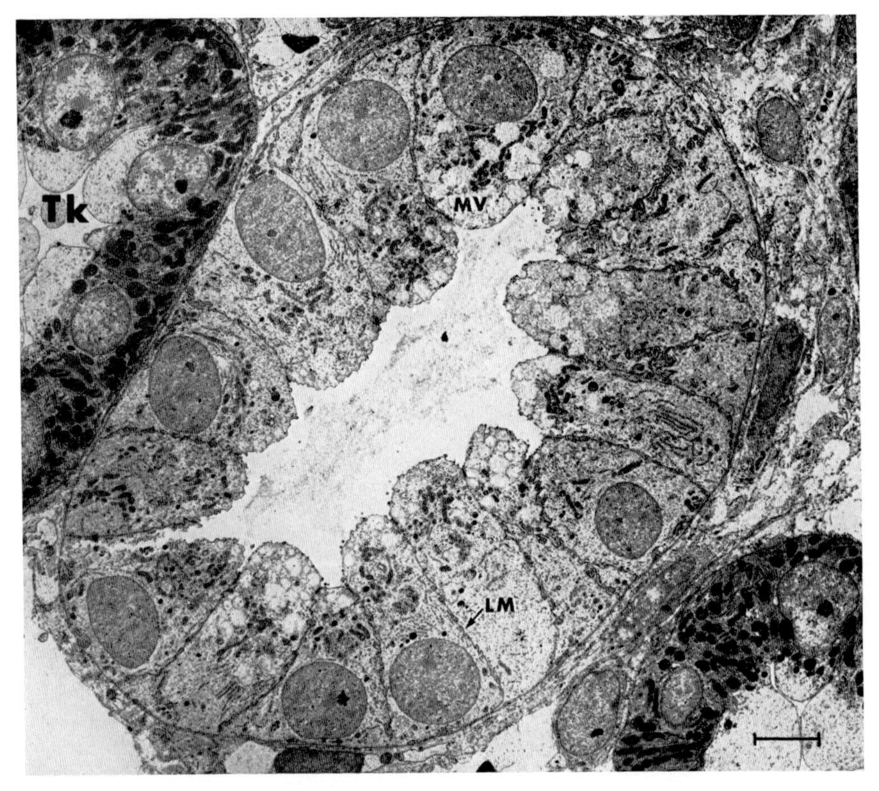

Fig. 24. Medullary collecting duct. The individual cells are clearly defined by the serpentine lateral plasma membrane (LM). The nuclei are basally situated. Few organelles are present but the apical mucous-containing vacuoles (MV) are prominent. Typical thick segments (Tk) of medullary loops are also shown. 1% Dalton's osmium (pH 7·4). The scale marker indicates 4 μm.

The mucosa is made up of a thick (about 40 μm) pseudostratified columnar epithelium which can be divided into two morphologically separate layers. The external one contains two or more rows of nuclei, one roughly above the other, and the internal layer is made up of tightly packed vacuoles containing mucopolysaccharide.

The submucosa varies considerably in thickness due to the presence of mucosal folds (Fig. 25). It consists of connective tissue, blood vessels and lymphatics and often large lymphoid aggregations. The latter are prominent as so-called lymphoid nodules in healthy birds but increase considerably in renal disease (Siller, 1959a).

The muscularis consists of an inner longitudinal and an outer circular layer of smooth muscle fibres although Liu (1962) has described, towards the cloacal end, a third outer coat of longitudinal muscle fibres.

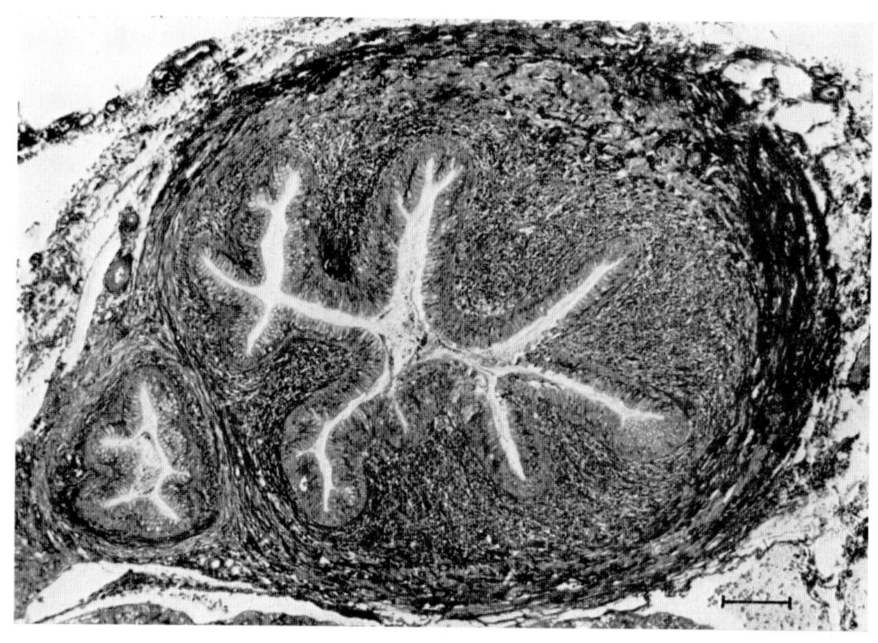

Fig. 25. Ureter and primary ureteral branch in cross-section. The stellate lumen is lined by bistratified epithelium surrounded by very cellular submucosa and muscle layers. Picro-Mallory's trichrome. The scale marker indicates 100 μm.

Peristaltic movement of the ureter was observed by Engelmann (1869) and shown in the fowl to be under sympathetic control (Gibbs, 1928).

VI. Blood Supply

The kidneys of the fowl have a dual blood supply in that they receive arterial blood through the renal arteries and afferent venous blood through the renal portal vein. The peritubular venous sinuses in which the various segments of the nephron are bathed contain post-glomerular arterial blood and venous blood. The sinuses are drained by the efferent renal vein which finally discharges into the posterior vena cava.

A. ARTERIAL SUPPLY

There are three pairs of renal arteries but only the most anterior pair originates directly from the aorta to supply the anterior division of the kidney. As they pass over the ventral surface of the organ, the sciatic arteries give off the middle and posterior renal arteries which supply the middle and posterior divisions.

The arterial branches within the kidney generally follow the efferent veins, closely. There is usually no spatial relationship between the arteries and branches of the ureter. On the contrary, an artery is not confined within the limits of an individual lobe. Within the lobule, however, the relationship between artery and collecting ducts, afferent and efferent veins is clear cut and constant. A number of intralobular arteries arise from the larger extralobular branches and run almost straight towards the apex of the lobule approximately half-way between the central and interlobular veins. Afferent arterioles leading to glomeruli are given off at intervals but always point towards the periphery. The afferent arterioles usually split into two capillaries which surround the core of mesangium as comparatively simple loops (Fig. 8) to emerge as the efferent arteriole which joins the peritubular network of sinuses some distance from the glomerulus (Fig. 2). In the juxtamedullary region some efferent arterioles as the vasa recta run towards the medullary tract.

Gadow (1891) was until recently the only author to describe "aglomerular" arterial branches in birds; he considered these to be part of a separate arterial system. The existence of such vessels was confirmed by Siller and Hindle (1969) who showed that such glomerular bypasses were seen most frequently in the juxtamedullary region where they ran towards the medulla to join the system of vasa recta. Sometimes they were also encountered in the cortical region.

B. RENAL PORTAL SYSTEM

With many eminent anatomists and physiologists hotly disputing for many years the existence of a functional renal portal system in birds, there was considerable confusion about the renal circulation until Spanner (1924/25) published his now famous paper. More recently it was shown by several authors and different methods (Sperber, 1948; Levinsky and Davidson, 1957; Siller and Carr, 1961) that this renal portal system was functional (see also Chapter 32). Although we know of the existence of this system and that it functions, we are still in some doubt as to its true significance. Spanner (1924/25) believed it to be associated with water conservation but it is more probable that, in uricotelic animals at any rate, it is essential to the secretion of uric acid. Mayrs (1924) had shown that glomerular filtration could not account for urate clearance and that tubular secretion must be involved. Sperber (1960) roughly calculated that about two-thirds of the blood to the kidney is supplied by the renal portal vein, and since the arterial plasma flow is insufficient to sustain normal urate clearance the importance of the renal portal system in urate excretion becomes obvious.

The flow of venous blood from the leg, carried in the external iliac vein, which traverses the ventral surface of the kidney, is impeded before entry into the posterior vena cava by a valve (Spanner, 1924/25; Akester, 1964). It has a rich nerve supply (Gilbert, 1961; Akester and Mann, 1969; Dolezel and Zlabek, 1969), is constricted by histamine or acetylcholine and dilated by adrenaline (Rennick and Gandia, 1954). Thus it can control the amount of blood flowing to the vena cava on the one hand and to the kidneys, via the afferent renal vein, on the other. The latter vessel doubles back along the ventral surface of the kidney and sends several branches into the organ. It meets the sciatic and vertebral veins and finally forms a posterior arch-like anastomosis with the afferent vein from the other side. In this region it is joined by the two hypogastric veins and the coccygeo-mesenteric vein, through which a link is established with the hepatic portal system.

Several large branches are given off the afferent vein to the main divisions of the kidney. These branch and rebranch and run between the lobules as the very conspicuous interlobular veins (Fig. 1). The afferent vein terminates in the peritubular network of capillary sinuses into which the efferent arteriole also discharges its post-glomerular arterial blood (Fig. 2). The mixed blood flows between the tubules across the lobule into the intralobular vein situated in the centre of the lobule, and thus into the branches of the efferent renal vein. Finally, the large efferent renal vein emerges onto the ventral surface of the kidney and runs in an anterior direction to anastomose with the external iliac vein of the same side, just distal to the renal portal valve and the two efferent veins unite to form the posterior vena cava.

The renal portal system plays a most important part in supplying the nephrons since it carries venous blood to the proximal convoluted tubules which are concentrated in the lobular periphery and are responsible for the tubular secretion of urate. However, it may have other functions of which we are as yet ignorant. Akester (1967) has shown that any or all parts of one or both kidneys can be excluded from the renal portal supply by the operation of what he calls the "renal portal shunt". Siller and Hindle (1969) have suggested that by excluding the vast venous bed of the kidney from the portal circulation, a substantial amount of "extra" blood would become available to the peripheral circulation in time of emergency. These authors have also pointed out that, since under conditions of renal portal shunt the arteries could maintain the nutrition of the entire lobule for a considerable period, it could prevent retrogressive changes in the renal parenchyma.

References

Akester, A. R. (1964). *J. Anat.* **98**, 365–376.

Akester, A. R. (1967). *J. Anat.* **101**, 569–594.

Akester, A. R. and Mann, S. P. (1969). *J. Anat.* **104**, 241–252.

Bean, J. W. (1942). *Am. J. Physiol.* **136**, 731–742.

Berger, C. (1962). *Inaugural Dissertation, Berlin.*

Bing, J. and Kazimierczak, J. (1963). *In* "Hormones and the Kidney." Memoirs of the Society for Endocrinology No. 13. (P. C. Williams, ed.), pp. 255–261.

Bloom, W. and Fawcett, D. W. (1962). "A Textbook of Histology" p. 525. W. B. Saunders, Philadelphia and London.

Bulger, R. E. and Trump, B. F. (1968). *Am. J. Anat.* **123**, 195–225.
Crabtree, C. (1941). *Anat. Rec.* **79**, 395–413.
Darnton, S. J. (1969). *Z. Zellforsch. mikrosk. Anat.* **93**, 516–524.
Dolezel, S. and Zlabek, K. (1969). *Z. Zellforsch. mikrosk. Anat.* **100**, 527–535.
Dunihue, F. W. and Boldosser, W. G. (1963). *Lab. Invest.* **12**, 1228–1240.
Edwards, J. F. (1940). *Anat. Rec.* **76**, 381–390.
Engelmann, T. W. (1869). *Pflüger's Arch. ges. Physiol.* **2**, 243–293.
Ericsson, J. L. E. and Trump, B. J. (1969). *In* "The Kidney: Morphology, Biochemistry, Physiology". (C. Rouiller and A. F. Muller, eds.), Vol. I. pp. 351–447. Academic Press, New York and London.
Feldotto, A. (1929). *Z. mikrosk. anat. Forsch.* **17**, 354–370.
Gadow, H. (1891). *In* "Klassen und Ordnungen des Thier-Reichs". Vol. 6. Sect. 4, pt. 1, p. 824. A. G. Bronn, Leipzig.
Gibbs, O. S. (1928). *Am. J. Physiol.* **87**, 594–601.
Gibley, C. W. and Chang, J. P. (1967). *J. Morph.* **123**, 441–461.
Gilbert, A. B. (1961). *J. Anat.* **95**, 594–598.
Goodchild, W. M. (1956). *M.Sc. thesis, University of Bristol, Bristol.*
Helmholz, H. F. (1935). *Proc. Staff. Meet. Mayo Clin.* **10**, 110–111.
Huber, G. C. (1917). *Anat. Rec.* **13**, 305–339.
Hyrtl, M. (1863). *Sber. Akad. Wiss. Wien.* **47**, Abt. 1, 146–204.
Johnson, O. W. and Mugaas, J. N. (1970). *Am. J. Anat.* **127**, 423–436.
Kazimierczak, J. (1967). *Acta path. microbiol. scand. Suppl.* **67**, 187.
Levinsky, N. G. and Davidson, D. F. (1957). *Am. J. Physiol.* **191**, 530–536.
Liu, H.-C. (1962). *Am. J. Anat.* **111**, 1–15.
Marshall, E. K. (1934). *Physiol. Rev.* **14**, 133–159.
Mayrs, E. B. (1924). *J. Physiol., Lond.* **58**, 276–287.
Miyoshi, M. (1970). *Z. Zellforsch. mikrosk. Anat.* **104**, 213–230.
Möllendorff, W. von (1930). *In* "Handbuch der mikroskopischen Anatomie des Menschen". (W. von Möllendorff, ed.), Vol. 7, pt. 1, pp. 225–234. Springer, Berlin.
Pak Poy, R. K. F. and Robertson, J. S. (1957). *J. Biophys. Biochem.* **3**, 183–192.
Poulson, T. L. (1965). *Science, N.Y.* **148**, 389–391.
Rennick, Barbara R. & Gandia, A. (1954). *Proc. Soc. exp. Biol. Med.* **85**, 234–236.
Rhodin, J. (1958). *Int. Rev. Cytol.* **7**, 485–534.
Siller, W. G. (1959a). *J. Path. Bact.* **78**, 57–65.
Siller, W. G. (1959b). *Lab. Invest.* **8**, 1319–1346.
Siller, W. G. (1962). *Ph.D. thesis, University of Edinburgh.*
Siller, W. G. (1964). *Res. vet. Sci.* **5**, 323–331.
Siller, W. G. and Carr, J. G. (1961). *Res. vet. Sci.* **2**, 96–99.
Siller, W. G. and Hindle, R. M. (1969). *J. Anat.* **104**, 117–135.
Simon, G. T. and Chatelanat, F. (1969). *In* "The Kidney, Morphology, Biochemistry, Physiology" (C. Rouiller and A. F. Muller, eds.), Vol. I, p. 261–349. Academic Press, New York and London.
Skadhauge, E. and Schmidt-Nielsen, B. (1967). *Am. J. Physiol.* **212**, 1313–1318.
Smith, C. L. (1966). *Stain Technol.* **41**, 291–294.
Smith, H. W. (1951). "The Kidney. Structure and Function in Health and Disease", p. 134. Oxford University Press, New York.
Spanner, R. (1924/25). *Morph. Jb.* **54**, 560–632.
Sperber, I. (1948). *Zool. Bidr. Upps.* **27**, 429–448.

Sperber, I. (1960). *In* "Biology and Comparative Physiology of Birds" (A. J. Marshall, ed.), Vol. I. p. 469–492. Academic Press, New York and London.

Voigt, F. (1920). *Veterinary Dissertation, Hannover.*

Zimmerman, K. W. (1933). *Z. mikrosk.-anat. Forsch.* **33**, 176–278.

Zuzuki, Y., Churg, J., Grishman, E., Mautner, W. and Dachs, S. (1963). *Am. J. Path.* **43**, 555–578.

9 Formation and Composition of Urine

A. H. SYKES

Wye College (University of London), near Ashford, Kent, England

I. Introduction

Since the avian kidney probably functions in principle like that of mammals it has not been considered in any depth by comparative physiologists who have found more interesting variations in renal function either amongst the poikilotherms or amongst the wide ecological range of the mammals. With one exception, the fowl's kidney possesses no feature of special interest to the wider problems of general physiology or medical science. The exception is the presence of a renal portal circulatory system which has been used by pharmacologists to investigate the excretory characteristics of a wide range of compounds but, unfortunately, without revealing very much about the function of this system in the fowl. (See also Chapter 32.)

Had man shared the same biochemical mutation as his Dalmation coach hound and consequently excreted considerably more uric acid, it is likely that greater interest would have been shown in the mechanisms of uric acid excretion. The fowl might then have been used as a convenient animal model in much the same way as the cockerel's ability to develop atheroma has been used and, incidentally, added much to our knowledge of the avian vascular system.

Much of this chapter, therefore, must necessarily appear superficial when compared with detailed textbooks and reviews devoted to mammalian renal physiology. At the same time, to include every possible reference of value would result in an indigestible list of observations often having little connexion with each other. Nevertheless, it is hoped that enough key references have been given to facilitate retrieval of other sources.

Two recent and valuable accounts of the avian kidney are available (Sperber, 1960; Sturkie, 1965) written with different emphases and containing some material which has not been included in this or Chapter 8). A suitable introduction to renal mechanisms would be an answer to the simple question "what is urine?". There is probably no better single reference than that of Szalagyi and Kriwuscha (1914). These authors undertook a comprehensive analysis of fowl's urine and identified many of the problems yet to be solved. The total daily nitrogen excretion was measured and partitioned into its major components; inorganic analyses were made and expressed in terms of both urinary concentration and daily output; pH and osmotic measurements were made and the chemical nature of urinary uric acid discussed.

The urine of the fowl may be defined as a semi-fluid product of the kidneys, in volume about 120 ml/d, which on standing separates into a white precipitate and a supernatant fluid. It contains about 0·75 g of nitrogen of which 85% is from uric acid and the rest is mainly from ammonia, urea and amino acids. The ash contains sodium, potassium, magnesium, calcium, phosphorus, chloride and sulphur and its reaction is slightly acid.

II. Methods of Urine Collection

Most investigations of the renal function involve collection of urine; this is particularly difficult in birds because of the intimate relationship between urine and faeces as expelled normally from the cloaca. Furthermore, there is some uncertainty about the physiological validity of the findings when steps are taken to separate urine from faeces. Nevertheless, this separation is necessary and although doubts may exist, particularly over the quantitative aspects of the excretion of individual compounds, the broad principles of renal function in the fowl have undoubtedly been established by the use of quite crude separation techniques.

Interest in urine composition first arose from nutritional considerations since estimates of the digestibility of protein feedingstuffs could not be made on the droppings which contained uric acid and other nitrogenous products of urinary origin. Early attempts to measure urinary nitrogen consisted in collecting the excreta after a period of starvation or attempting to separate urinary and faecal components by chemical means. These methods have serious limitations; hence surgical separation of the ureters and rectum was attempted. The artificial anus or colostomy first described by Weiner (1902) has been used fairly extensively. Operatively, the rectum is divided after laparotomy, the distal portion closed by ligature and the proximal portion brought out to the flank of the bird and anchored to the skin by sutures (Fig. 1). By everting the mucosa the growth of the surrounding skin is less

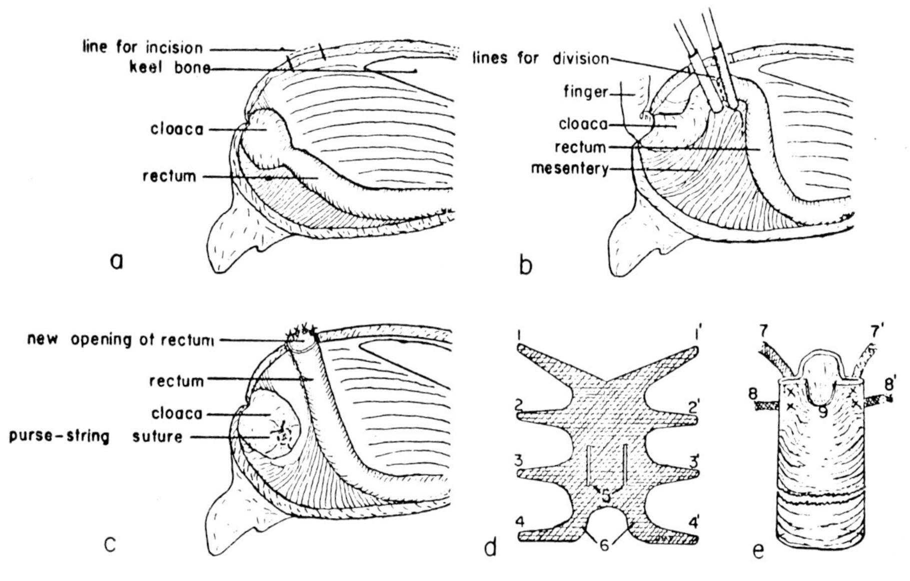

Fig. 1. Technique for preparation of a rectal fistula and a harness for attaching a vessel for urine collection. Reproduced with permission from Dixon (1958).

likely to obliterate the new anal opening (Fussell, 1969). The fistula thus created can remain functional indefinitely but the operation is not always so successful and closure or retraction can take place in a matter of days or weeks.

Faeces are usually collected into a container held over the fistula by means of a harness or sutured to the skin. The urine collects in the cloaca and is

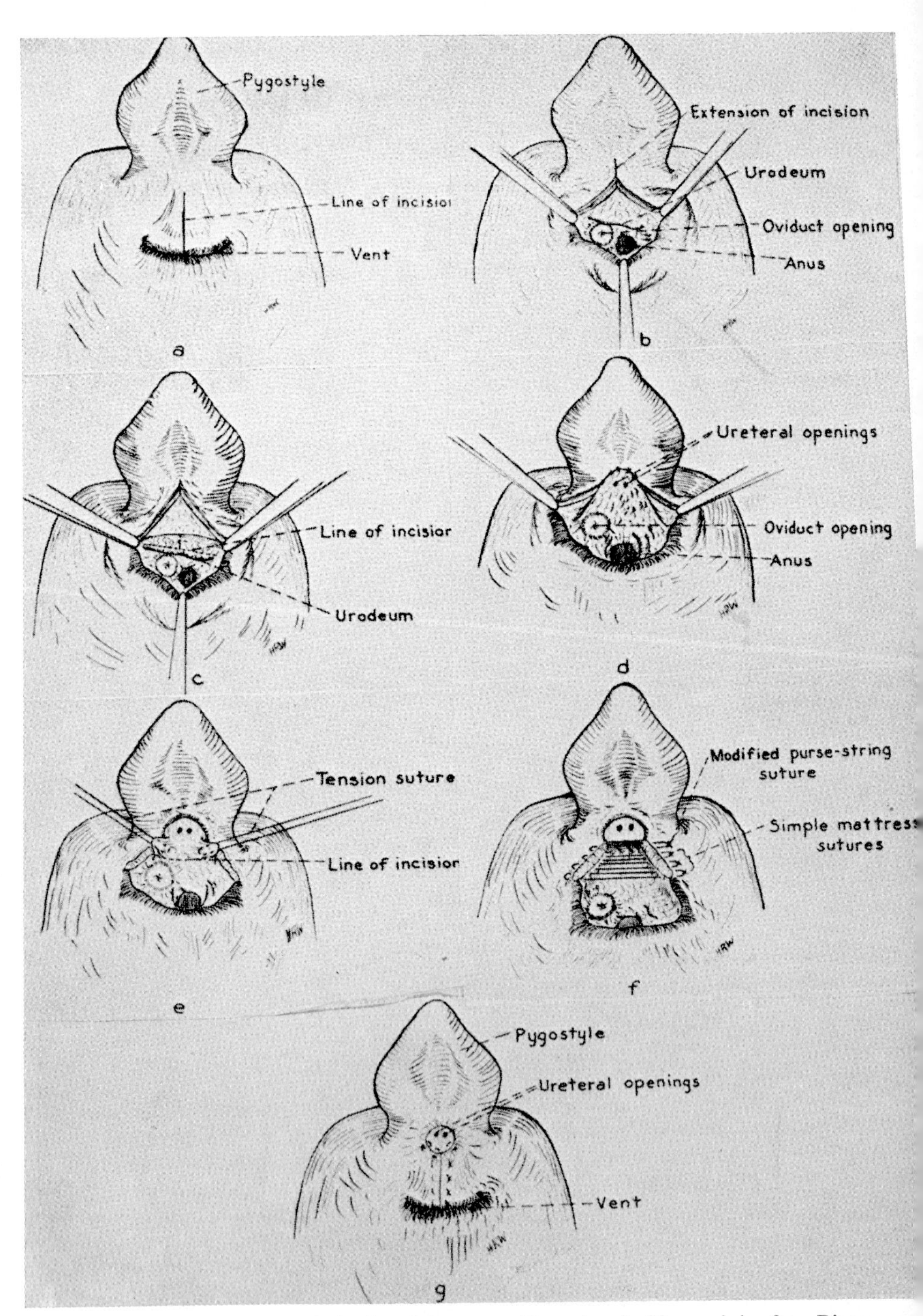

Fig. 2. Technique for exteriorization of the ureters. Reproduced with permission from Dixon and Wilkinson (1957).

expelled at intervals probably as a result of a stimulus arising from distension. Since urine ejection is spasmodic frequent sampling can give widely different values but over successive periods of 24 h reliable figures for urinary output can be obtained.

The objection to colostomy is that it is unsuitable for short-term collections, necessary, for instance, in the measurement of renal clearances it is not certain how far ligation of the rectum, by preventing possible reabsorption from the urine, alters its composition and so of the metabolic balance of the whole bird in respect of a particular metabolite. This question of cloacal and rectal reabsorption will be discussed later (Section VI).

An alternative surgical method is to separate the ureters from the cloaca and bring them to the body surface on the pygostyle (Fig. 2). Faeces are then collected on trays usually, but a special receptacle must be attached to collect the urine. This operation has not been so widely practised as colostomy and there are difficulties in ensuring healing of those incised areas of the cloaca in contact with urine. It has been used more for short-term collections, over a period of days, than as a permanent arrangement although some workers have maintained birds in a functional condition for much longer (Hester *et al.*, 1940; Hart and Essex, 1942; Dixon and Wilkinson, 1957).

Direct cannulation of the ureters, either through the cloaca or after laparotomy, is useful only for very short term collections; the cannulae appear to disturb the normal peristaltic movements of the ureters which quickly become blocked by precipitated uric acid and hydronephrosis may result.

Short-term collections may be made directly by means of funnels sutured over each ureteral orifice (Fig. 3, Sperber, 1948a) or by a specially designed cannula which fits into the cloaca so that urine drains away separately from the faeces (Fig. 4, Bokori, 1961). At very low rates of flow these cannulae can be flushed either continuously or at intervals with deionized water for greater accuracy. A more elaborate system has been devised (Campbell, 1960) which provides two constant streams of rinsing water combined with a

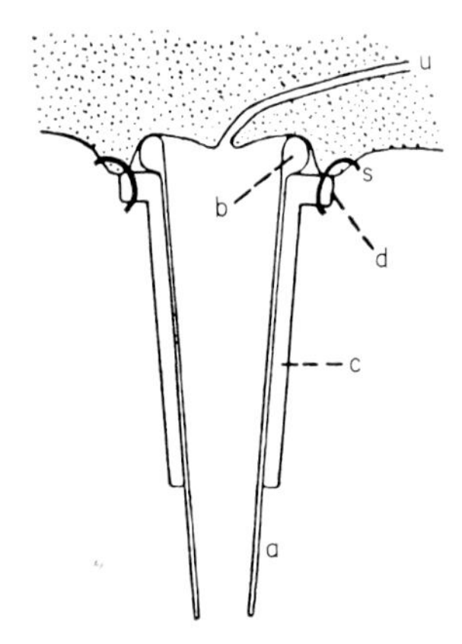

Fig. 3. Diagram of a funnel for the collection of urine from the ureter. a: inner funnel; b: annular thickening; c: outer funnel; d: its edge; s: stitch; u: ureter. Reproduced with permission from Sperber (1948a).

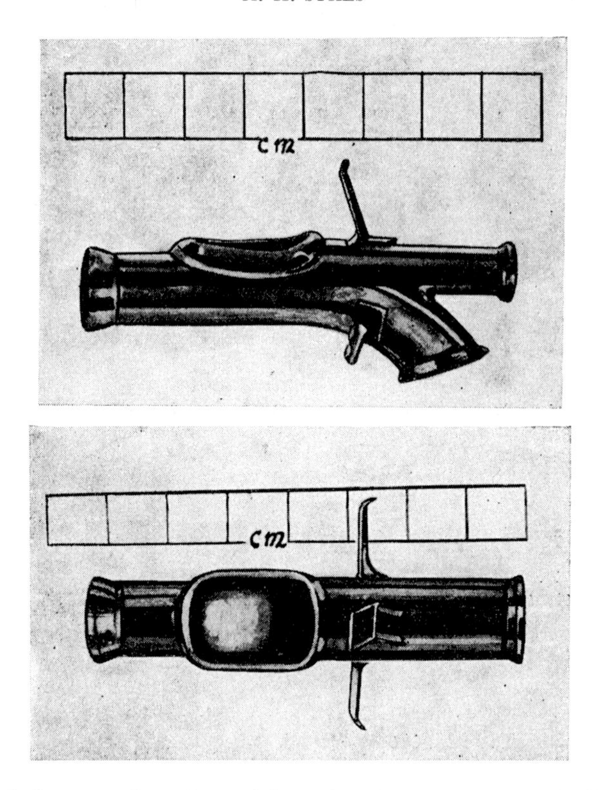

Fig. 4. Cannula for separating urine and faeces from intact, unanaesthetized fowls. Reproduced with permission from Bokori (1961).

fraction collector. Earlier workers merely plugged the rectum with cotton wool and collected the cloacal urine into a funnel.

Because the normal rate of flow is very low (0·02 ml/min kg) diuretics are frequently used. The most obvious diuretic is water *per os*; this gives very unpredictable results in the fowl since water is often retained in the crop for long periods. Hypotonic saline infusions are better as well as osmotic diuretics such as mannitol or urea. Other diuretics which have been used include chlorothiazide, hydrochlorothiazide and mersalyl.

Since these diuretics have different modes of action their effects on the composition of the urine will differ and the choice of diuretic will be influenced by the purpose of the experiment. For example, potassium excretion usually increases during diuresis but not after treatment with mersalyl which inhibits potassium secretion; sodium excretion usually rises but not after water loading. Caution is necessary when interpreting the results of experiments which involve urinary separation, diuretics, anaesthesia and sometimes abnormal postures or methods of restraint.

III. Renal clearances

The concept of renal clearances and their importance in relation to the quantitative basis of renal function is described in several textbooks. No extensive application of clearance techniques has been made in the fowl but it has been firmly established that the mechanisms of filtration, reabsorption and secretion are involved in urine formation.

A. GLOMERULAR FILTRATION

The search for a carbohydrate suitable for measuring filtration rate arose from the knowledge that glucose reabsorption can be abolished by phloridzin and its clearance then becomes a measure of the filtration rate. This was the first method used in the fowl (Marshall, 1932) when the clearance rate of 0·9 ml/min kg, although low, showed conclusively that filtration took place and that it alone could not account for the quantity of urate excreted. Subsequently, the clearances of xylose and sucrose were used to estimate filtration under more normal conditions (Burgess *et al.*, 1933) but it is now believed that both compounds are reabsorbed to some extent and therefore underestimate the true value.

Pitts (1938) showed that the mean inulin clearance of the fowl was 1·84 ml/min kg with a range of from 1·4 to 2·6. The inulin clearance was also compared directly with the clearance of glucose after treatment with phloridzin and the ratio of these two estimates of filtration rate was 1·01.

The clearance of inulin has been measured as part of many subsequent investigations and a range of values from 1·5 to 3·0 ml/min kg found. Whether the filtration rate in an individual fowl can alter under physiological conditions will be discussed later.

When the clearance of a compound is greater than the inulin clearance it is assumed that net secretion takes place; clearances less than the filtration rate imply net reabsorption. Thus any compound which might be used to measure filtration has to show a clearance ratio compared with inulin equal to one. By this criterion, endogenous creatinine (0·78), exogenous creatinine (1·46), ferrocyanide (1·59) and thiosulphate (3·0) are unsuitable in the fowl although they have been used in other animals (Sykes, 1960b). Polyethylene glycol has a ratio of 1·0 and could therefore replace inulin (Hyden and Knutsson, 1959); it also has the advantages of cheapness and ease of estimation.

B. REABSORPTION

The urinary constituent showing the greatest amount of tubular reabsorption is water; if the filtration rate is 2 ml/min and the urine flow 0·02 ml/min then, assuming that these volumes consist only of water, about 99 % of the filtered load is reabsorbed. Similar considerations apply to sodium, chloride, bicarbonate and other filterable constituents of the plasma.

Glucose is reabsorbed completely from the filtrate. Thus at plasma levels of 250 mg/100 ml and a filtration rate of 2 ml/min kg, glucose would be reabsorbed at the rate of 5 mg/min kg. The maximal reabsorption rate has been estimated (Sperber, 1960) at between 10 and 12·5 mg/min kg; in the presence of phloridzin, glucose reabsorption ceases.

C. SECRETION

Tubular secretion was first demonstrated with the indicator phenol red (Pitts, 1938). At plasma levels of 2 mg/100 ml, its clearance was 30 ml/min kg and declined as the plasma level increased reaching a minimum of 3 ml/min kg. This response, summarized in Fig. 5, is typical of a compound secreted by the tubules.

Uric acid was of obvious importance for investigation and there was already good evidence from the urine to plasma ratio that a secretory mechanism was involved (Mayrs, 1923). The clearance of administered uric acid ranged from 6 to 30 mg/kg min (Marshall, 1932) and, as with phenol red, the clearance was inversely proportional to plasma concentration. The highest clearance, at normal plasma levels, was about 15 times the filtration

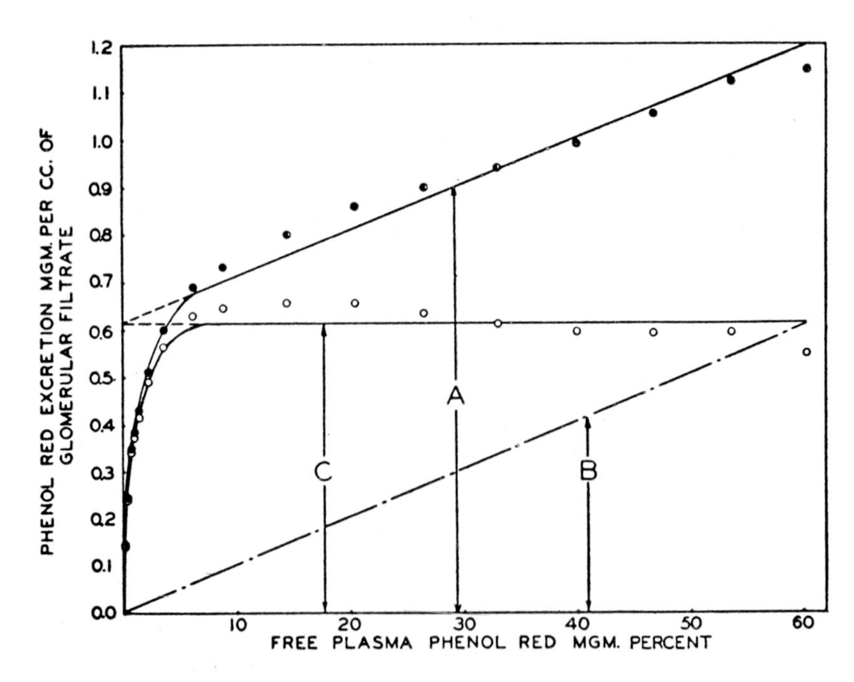

Fig. 5. The excretion of phenol red in relation to plasma concentration. A: total phenol red excreted; B: phenol red filtered; C: phenol red secreted by tubules. Reproduced with permission from Pitts (1938).

rate and thus it could be calculated that 93% of the excreted uric acid was secreted by the tubules (Shannon, 1938). The absolute amount of uric acid secreted rose as the plasma level increased and a tubular maximum of about 3·0 mg/min kg was found.

IV. The Renal Portal System

The existence of an incomplete portal system to the convoluted tubules has been long known but although its anatomy has been well described its functional significance remains obscure. (See also Chapter 32.)

The direction of the circulation in the major vessels, under physiological conditions has not been clearly determined. Observations at laparotomy (Clementi, 1945; Gordeuk and Grundy, 1950) suggest that blood flows down the coccygeo-mesenteric vein towards the kidney and there is no doubt that experimentally the entire hepatic portal blood can be so diverted. Because of this alternative route, intestinal stasis does not occur, as it would in mammals, upon the ligation of the hepatic portal vein, but it does not follow that even under these extreme conditions more blood necessarily perfuses the portal capillaries.

Observations have been made by Akester (1964, 1967) on the distribution of radio-opaque materials after injection into the renal circulation by different routes. At times the sphincter can close completely and some blood is diverted through the peritubular capillaries given off from both the cranial and caudal portal veins. The former vessel also offers what Akester terms a portal shunt in that blood can bypass the renal tissue and return to the heart via the vertebral venous sinuses as well as via the coccygeo-mesenteric vein. He has suggested that, in addition to the sphincter, blood flow to the tubules may be controlled by local vasomotor mechanisms at a point where the smaller branches of the renal portal veins enter the kidney tissue. This work demonstrates most elegantly the pathways open to the blood but it is not clear what significance they have in relation to urine formation.

The first demonstration of the functional nature of the renal portal capillaries was made by Sperber (1946, 1948a) who, in one simple but convincing experiment, advanced our knowledge of the fowl and provided a new approach to renal physiology generally. He collected urine from each kidney separately by means of small funnels sutured near the ureteral orifices and found that phenol red (which is secreted by the tubules), injected into the leg muscles on the left side appeared in greater amounts in the urine from the left kidney. Injection in the right limb caused an excess excretion from the right kidney whereas an injection into the breast muscle resulted in a similar rate of excretion from both kidneys (Fig. 6). The excretion of inulin from both kidneys remained the same regardless of which side was injected with phenol red and hence differences in filtration could not account for the differential excretion. The explanation is that the portal

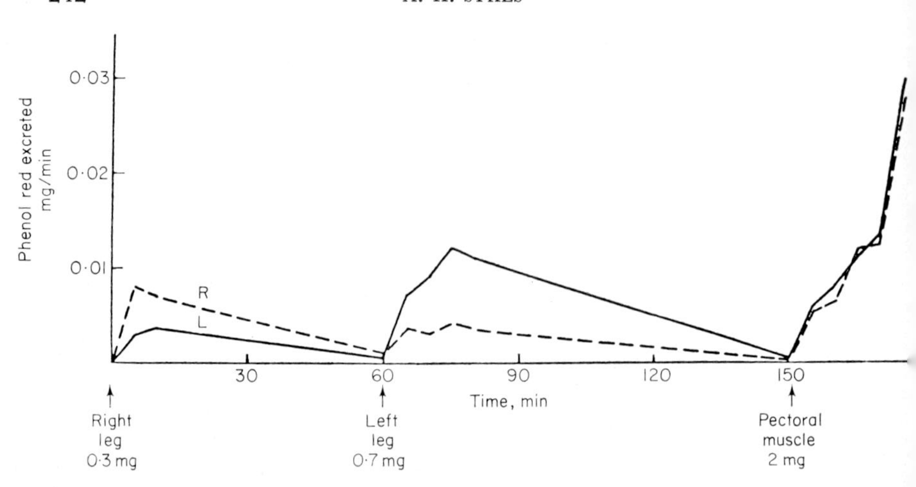

Fig. 6. The excretion of phenol red from each kidney following injections into one leg at a time and into the pectoral muscle. Continuous line indicates left kidney; dashed line indicates right kidney. Reproduced with permission from Sperber (1948a).

system on the injected side received phenol red directly from the depot and transferred some of it to the tubules. The remainder returned to the heart and was subsequently delivered equally to both kidneys. Similar results were found when paraamino hippuric acid (PAH) or uric acid was infused via one of the tibial veins. Unilateral infusions of inulin did not result in differential excretion and measurements of the glomerular filtration rate showed no differences. It would not be expected that venous blood, at low pressure, could make any contribution to glomerular filtration and anatomical observations on the distribution of indian ink after portal injection also showed that the portal capillaries do not penetrate the glomeruli (Spanner, 1925). This is not true of the frog which has a complete renal portal system.

Differential excretion of compounds following their unilateral infusion has now become one of the criteria for determining whether tubular secretion takes place. There are, however, certain limitations; if the substance exists in the circulation at concentrations above the tubular maximum, or if a competitive substance is present, then, with both kidneys fully saturated, no differential excretion can be established. The unilateral infusion of urea results in an increased excretion on the infused side although urea itself is reabsorbed (Sykes, 1962); the probable explanation is that at high concentrations urea produces a unilateral diuresis and its excretion is, to some extent, flow-dependent. The excretion of sodium can also be influenced by the osmotic conditions in the tubule and can be reduced by the ipsilateral infusion of albumin solutions (Vereerstraiten and Toussaint, 1968).

Some attempts have been made to determine the urinary function of the portal system; Cuypers (1959) found that urine flow ceased after ligation of the aorta above the suprarenal glands which at least confirmed that the portal system could not secrete *urine*, i.e. water, but only specific compounds into a filtrate.

Occluding the hepatic portal vein, so as to re-route the blood through the coccygeo-mesenteric and renal portal veins, did not increase the clearance of PAH which suggests that other factors may control how much blood actually enters the renal tissue. It is possible that compounds can find their way directly from the alimentary circulation into the urine via the renal portal system. This was shown by ligating the hypogastric vein on one side, which ensured that there would be access to only one kidney via the coccygeo-mesenteric vein. On infusing PAH into a mesenteric vein a slight excess was secreted from the kidney on the "open" side; this excess became marked when, on occluding the hepatic portal vein, the whole of the blood flowed in

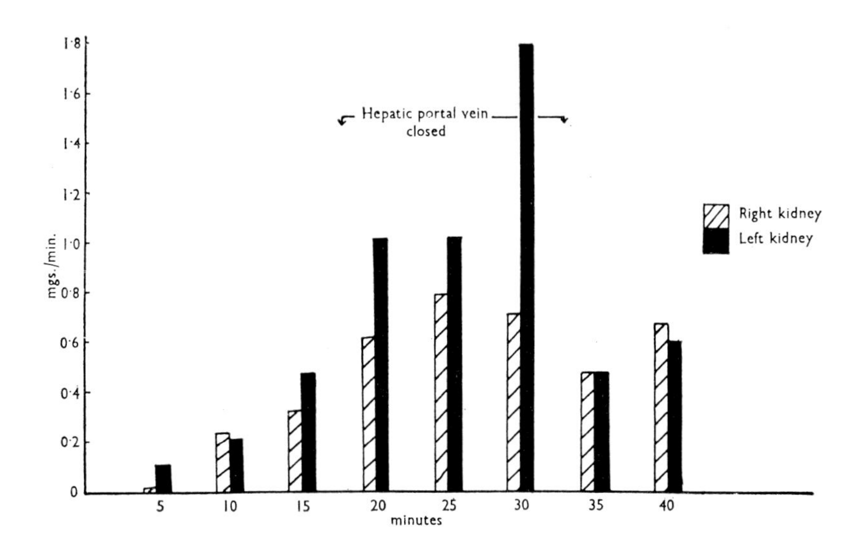

Fig. 7. The excretion of para-aminohippuric acid (PAH) from each kidney following its infusion into a mesenteric vein; the right hypogastric vein was ligated throughout; samples 4, 5 and 6 were collected after occlusion of the hepatic portal vein proximal to the infusion. The excess excretion of PAH from the left kidney indicates that hepatic portal blood can reach the renal portal capillaries directly. Reproduced with permission from Sykes (1966).

the reverse direction (Fig. 7). These are, of course, most unphysiological circumstances but it does raise the question of how far products from the digestive tract can reach the kidney directly without first passing through the liver (Sykes, 1966).

The sphincter at the confluence of the renal vein and the external iliac vein has aroused much interest. If it closed completely, presumably all the blood from the legs and tail region would then have to pass through the portal capillaries and into the renal vein or else return to the heart via the coccygeo-mesenteric vein and the hepatic portal system; both routes might be used simultaneously by a proportion of the total blood flow. With the sphincter completely open, less blood might perfuse the renal tubules.

The sphincter has been shown to contract *in vitro* in the presence of acetyl choline and histamine and to relax with adrenaline (Fig. 8). In the intact fowl, the differential excretion of PAH following injection into one leg was practically abolished by atropine (Fig. 9) and it was suggested that the inhibition of endogenous acetyl choline allowed the sphincter to open widely and shunt blood away from the peritubular capillaries (Rennick and

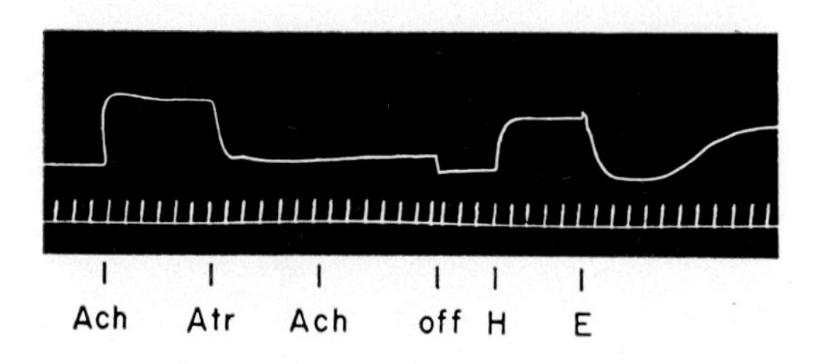

Fig. 8. The *in vitro* response of the renal portal sphincter of the turkey to drugs. The contraction in response to acetyl choline (Ach) is blocked by atropine (Atr); histamine (H) also causes contraction and adrenaline (E) relaxation. Reproduced with permission from Rennick and Gandia (1954).

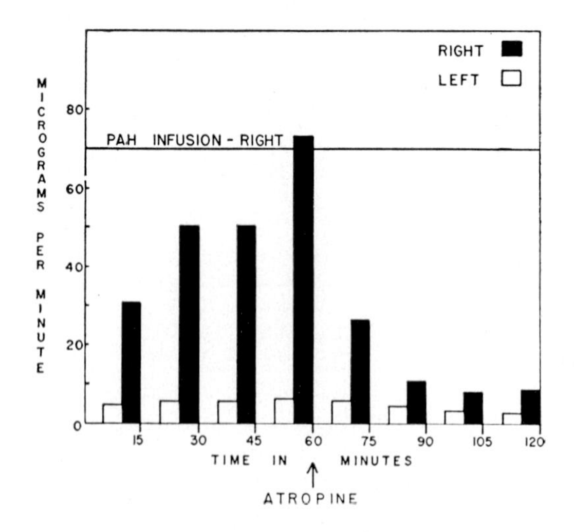

Fig. 9. The excretion of para-aminohippuric acid (PAH) from each kidney following infusion of 70 μg/min of PAH into the right tibial vein. The excess excretion on the infused side was considerably reduced after injecting atropine. Reproduced with permission from Rennick and Gandia (1954).

Gandia, 1954). It should be noted that large doses of atropine are required to demonstrate this effect and that the differential secretion is not always abolished even then. Subsequent histochemical examination has suggested the presence of both adrenergic and cholinergic nerves (Gilbert, 1961; Akester and Mann, 1969).

It is, then, far from clear exactly what role the renal portal system serves. It is easy to conclude that an additional blood supply to the tubules must be of value to an animal which depends upon tubular secretion for the elimination of its main nitrogenous end-product, uric acid; the functional significance may well be in relation to the vasomotor control of the vascular bed rather that to the processes of renal excretion.

Use has been made of the renal portal system to demonstrate the excretory characteristics of a wide range of compounds unrelated to normal physiological function in the fowl. This method was introduced by Sperber (1948b) in his study of the excretion of organic acids and bases and the results were expressed quantitatively in terms of the apparent tubular excretory fraction (ATEF).

This is an expression for the difference between the ipsilateral and contralateral excretion rates (UV) as a fraction of the infusion rate:

$$\text{ATEF} = \frac{\text{UV}_{\text{Ipsi}} - \text{UV}_{\text{contra}}}{\text{Infused}} \times 100$$

For example the ATEF of choline is 23% which is therefore considerably less able to cross the tubular epithelium than PAH with an ATEF of 65%. These values are of course affected by plasma concentration but they do have some value in making comparisons.

V. Renal Blood Flow

The kidney in mammals has a considerable blood supply, equivalent to about 20% of the cardiac output. The simplicity of the major blood vessels (a single renal artery and vein) and the encapsulated nature of the kidney facilitate estimation of blood flow by direct methods such as electromagnetic flowmeters placed around these vessels. It is also possible to use indirect methods, based upon renal clearances, since the arterio-venous difference or extraction ratio can be determined for a particular indicator and direct and indirect methods of estimation can be compared simultaneously in the same animal.

In the fowl, not only are the arterial and venous supplies more complex, which makes direct methods impossible to apply, but the presence of a renal portal system probably leads to considerable errors in estimating blood flow indirectly.

Gibbs (1928) attempted to measure blood flow in one kidney directly by tying off one branch of the vena cava and the coccygeo-mesenteric vein and collecting the venous outflow from a cannula in the ipsilateral femoral vein. He found a mean value of 27 ml/min for one kidney which is probably an underestimate since the normal venous drainage would have been impeded by the reverse flow through the renal portal sphincter. This figure would also include blood which returned from the caudal region through the hypogastric veins and which would not contribute to the effective renal blood flow, i.e. that which circulates to the nephrons.

The most commonly used indirect method depends upon the Fick principle as used for estimating cardiac output. If the excretion rate of an indicator and its extraction rate by the kidney, i.e. the arterio-venous difference, are known it is possible to estimate the minimum quantity of blood, or plasma, which would account for the observed quantity excreted. If extraction is complete, i.e. the renal venous concentration is zero, then the renal clearance (UV/P) of the indicator gives a measure of the effective renal plasma flow.

Diodone and PAH are most commonly used for this purpose and clearances of the order of 20 ml/min kg have been found in the fowl. In adult birds, weighing about 3 kg, values range from 30 to 130 ml/min with a mean of 60 ml/min. The clearance of PAH falls as the plasma concentration increases above about 2 mg/100 ml and hence underestimates of the blood flow can easily be obtained. There is competition with uric acid and thus any rise in plasma uric acid concentration would lower the clearance of PAH.

Endogenous uric acid, provided it is present at low plasma concentrations, can be used to estimate renal blood flow; its clearance then may be identical with that of PAH, about 68 ml/min (Sykes, 1960a).

An indirect method of estimating blood flow based upon the rate of deposition of radioactive rubidium in the kidney (Sapirstein and Hartman, 1959) has given values, for a 3 kg hen, of 69 ml/min. Assuming a packed cell volume of 33%, this is equivalent to a blood flow of 100 ml/min, nearly 50% of the cardiac output. Clearly the kidney has a very considerable blood flow but the only methods available for its measurement do not have the precision nor perhaps the sensitivity to determine variations in flow particularly those that might occur due to the activity of the portal valve or to the effects of anoxia.

VI. Cloacal Reabsorption

It has often been stated that the fowl produces a dilute urine, by means of which insoluble uric acid is safely passed through the ureters, and which is finally concentrated in the cloaca by the reabsorption of considerable amounts of water, leaving the semi-solid urine to be voided along with the faeces.

This view arose partly from observations of the paste-like urine in the droppings and from direct measurements of urine flow, either from ureteral cannulae (Sharp, 1912) or from birds with a colostomy (Weiner, 1902) which showed that when the urine did not have access to the alimentary canal more water was excreted than could be accounted for by normal drinking (as measured by the intake of surgically intact fowls). Furthermore, by calculating urine flow per 24 h on the basis of short sampling periods, values exceeding 1000 ml/d were obtained which were clearly too high.

Cloacal reabsorption was first examined critically by Korr (1938) who concluded from measurements of the osmotic pressure of cloacal fluid that some isosmotic reabsorption could take place. He appreciated, however, that if cloacal reabsorption were obligatory, it would be impossible for the bird to excrete a hypotonic urine if the conditions of hydration required it. Neither would there be any value in excreting hypertonic urine into the cloaca if subsequently both water and solute were reabsorbed.

Later work made it clear that the role of the cloaca had been over-emphasized because of technical difficulties; anaesthetics, adrenaline and the presence of cannulae in the ureter all provoke the copious flow of dilute, milky urine which had been commented on by earlier authors. Urine volume declined with successive collection periods and on successive days, e.g. from 25 ml/h to 5 ml/h when using a cloacal funnel for collection. Urine from birds with exteriorized ureters amounted to 87 ml/d and in colostomized birds 129 ml/d, values much more in line with data on water balance (Hester *et al.*, 1940). It was therefore concluded that there is no need to postulate an obligatory role for cloacal reabsorption.

Nevertheless it was noted (Hart and Essex, 1942) that birds with a colostomy or with exteriorized ureters, required additional salt to keep in good condition and although the amount given, 1% of the diet, was not more than might be included in a layers ration, this finding did suggest that the cloaca might still have a part to play. Collections of ureteral urine by means of Sperber funnels, which do not interfere with the ureters, show that volumes as low as 0·025 ml/min of hypertonic urine can be produced and, in terms of water loss alone, there is no need for further cloacal modification of the ureteral urine. A comparison of the water balance of hens with either a colostomy or exteriorized ureters showed no difference compared with intact controls; about 67% of the water intake could be recovered from the urine and faeces of all three types of bird (Dixon, 1958). Experiments which show that reabsorption of water can take place from solutions placed into the empty cloaca and rectum do not prove that this process can alter the composition of the small quantities of ureteral urine which are normally involved.

A new approach to the problem has been made with a perfusion technique which enables a solution to be circulated through the cloaca up to the ileo-caeco-colic junction and returned for subsequent analysis (Fig. 10, Skadhauge, 1968a, b). It was thus shown that sodium and chloride could be actively reabsorbed together with an equivalent amount of water. From calculations based upon known rates of ureteral urine flow it was concluded

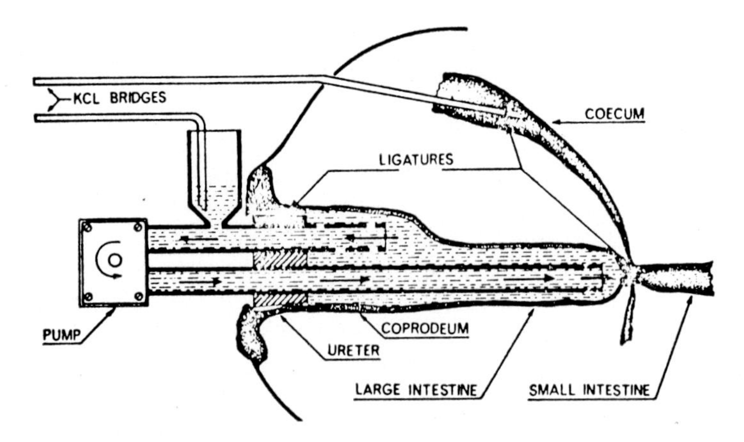

Fig. 10. Apparatus for the recirculation of fluids in the cloaca and colon. Samples may be withdrawn for analysis and the potential difference measured with the KCl bridges. Reproduced with permission from Skadhauge (1968a).

that normally 3% of the urine water and sodium could be reabsorbed in this way; during dehydration 15% of the water and up to 50% of the sodium could be reabsorbed.

These figures are valuable since they indicate the *capacity* for reabsorption but clearly perfusing an empty segment of intestine does not represent normal conditions. In further work more direct observations were made on the osmolal and ionic concentrations of ureteral and cloacal urine and it was concluded that under normal conditions of hydration there were no differences between them. During dehydration the ureteral urine was excreted at the rate of 18 μl/min kg compared with cloacal urine at 7·4 μl/min kg; the values for sodium excretion were 3·6 μ equiv./min kg and 1·3 μ equiv./min kg respectively. These figures suggest that the cloaca has been responsible for water and sodium reabsorption but it must be pointed out that this involves a comparison between a 24-h output and short-term ureteral collections extrapolated to 24 h, a calculation which can give rise to misleading results.

In a somewhat different approach, simultaneous urine samples were obtained (a) from a cannula placed high in one ureter, and (b) from the cloaca supplied directly by the other kidney (Nechay and Lutherer, 1968). As the results did not indicate any differences between the two samples during hypotonic or hypertonic saline infusions, it was concluded that the cloaca played no part in salt and water transport. It must be appreciated that this was a completely experimental situation involving laparotomy and ligation of the distal colon whereas the previous perfusion technique involved intact birds with the more normal stimulus of dehydration when the cloaca may assume greater significance.

It has been suggested that cloacal reabsorption of sodium and water has particular significance for marine and desert-living birds which are constantly faced with dehydration. Such are able to excrete the retained salt through their well-developed nasal glands and thus there is a net retention of water. In the absence of a functional salt gland, the fowl must excrete all its salt load through the kidney to the limit of its osmotic capacity and the possibility of conserving further water by cloacal reabsorption appears to be limited.

It is possible that the cloaca in the fowl may be more concerned with sodium reabsorption but no studies appear to have been made upon sodium-deficient birds. When a solution containing radioactive sodium was placed in the cloaca, a small quantity, less than 10%, appeared in the circulation (Weyrauch and Roland, 1958).

Thus, although the evidence is on the whole against an essential role for the cloaca in the maintenance of water balance, the intriguing fact remains that urine undoubtedly can travel from the cloaca into the rectum and even into the caeca. This has been shown by the appearance of inulin in the caeca in amounts too large to be explained by vascular leakage (Skadhauge, 1968b) and most convincingly by radiographs (Fig. 11) following intravenous injection of radiopaque contrast media which are rapidly excreted into the urine (Koike and McFarland, 1966; Akester *et al.*, 1967; Nechay *et al.*, 1968).

VII. The Excretion of Water

The kidney regulates the water balance of an animal by its ability to produce either a copious flow of dilute urine or a smaller amount of more concentrated urine according to the degree of hydration. Most animals can produce a hypotonic urine when faced with overhydration but the ability to produce hypertonic urine varies considerably from species to species. It is most highly developed in the desert-living mammals; birds are generally considered to be less well developed in this direction and the fowl is placed among those birds least able to concentrate their urine.

A. WATER LOADING

The rate of urine flow increases ten-fold, from 0·03 to 0·3 ml/min kg, following an oral dose of water equal to 3% of the body weight. The classical water diuresis curve, in which the whole of the extra water is recovered, is not always obtained in the fowl. The method for the separate collection of urine often imposes some restraint which may disturb the normal pattern of urine flow and water given orally may be retained in the crop and made available for absorption only gradually (Fig. 12).

As the urine volume increases its osmotic pressure falls from normal values of about 450 mOsm/litre to as little as 50 mOsm/litre, i.e. strongly hypotonic to the plasma which is usually taken to be 340 mOsm/litre.

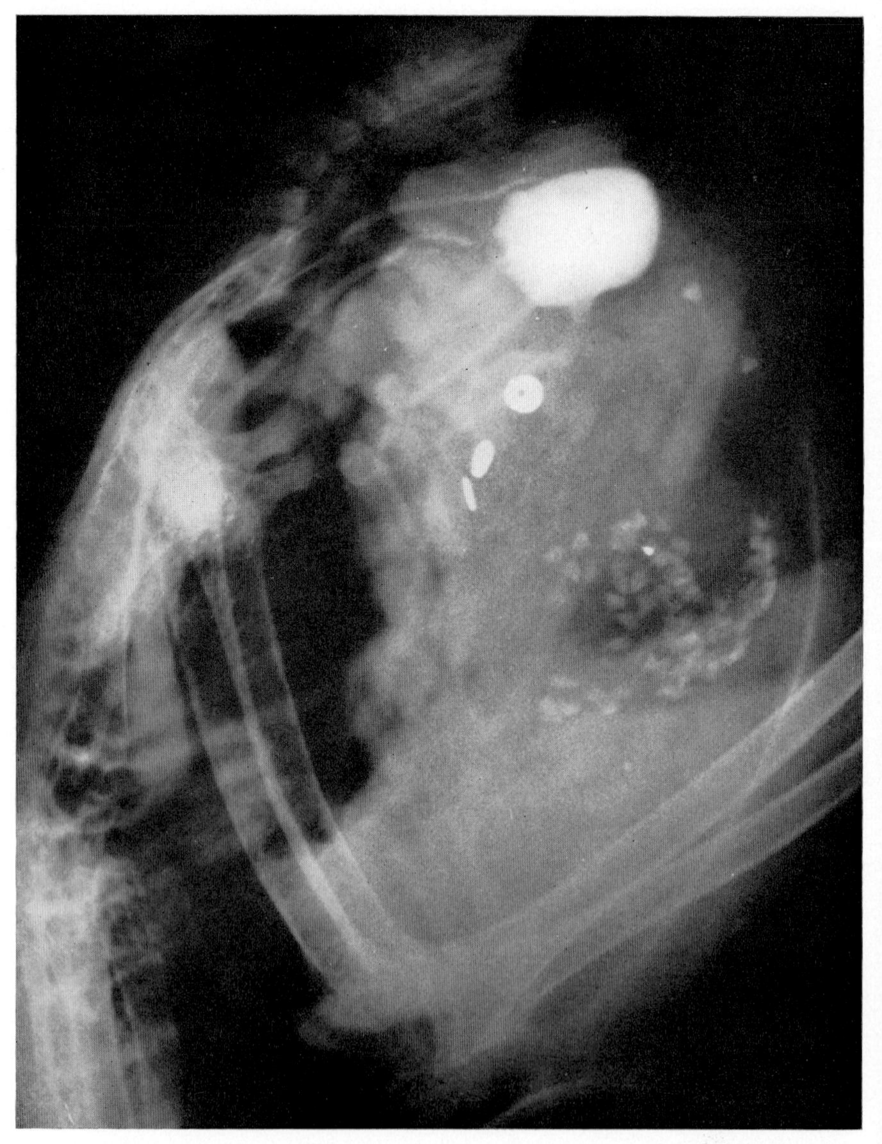

Fig. 11a. Radiograph showing the passage of a radio-opaque solution, injected intravenously, into the ureters, cloaca and, to a slight extent, the colon. The three discs are markers attached to the colon.

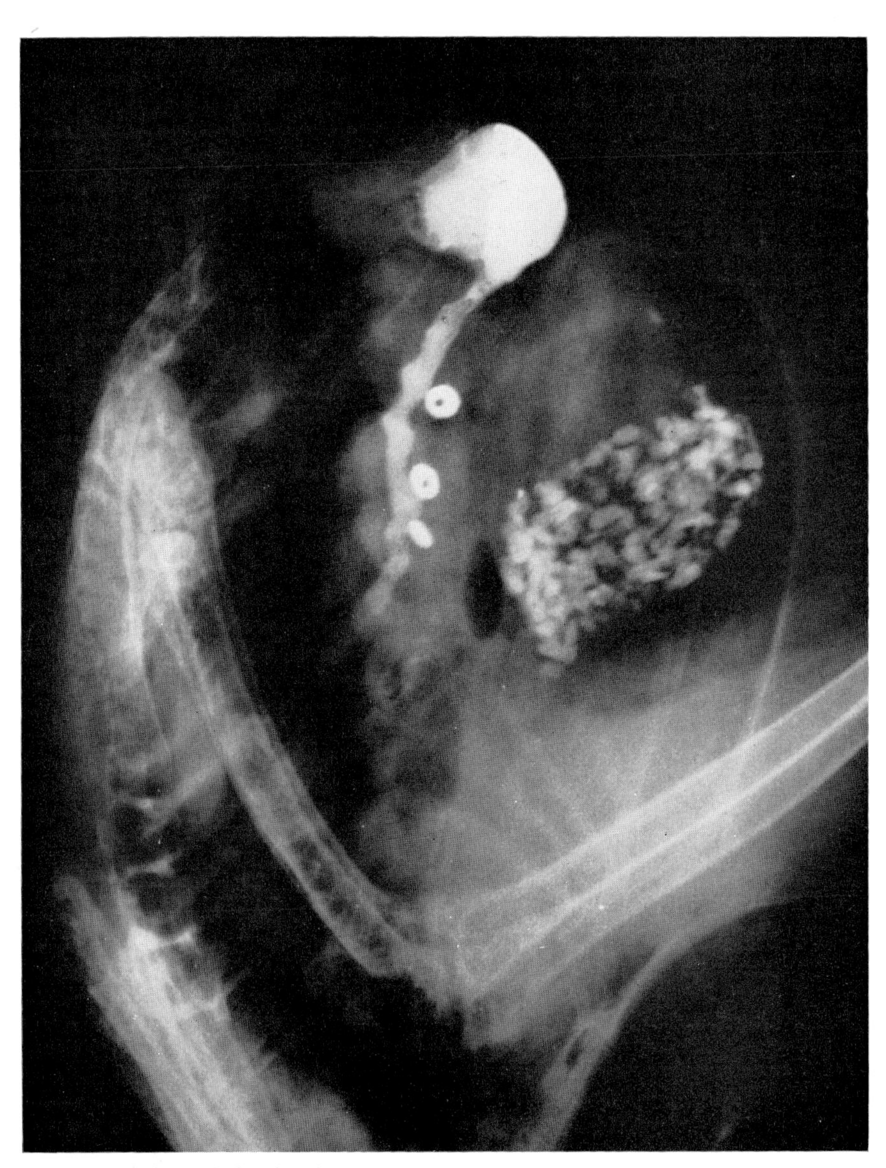

Fig. 11b. Radiograph showing the retrograde movement of a radio-opaque solution from the cloaca into the colon. Reproduced with permission from Akester *et al.* (1967).

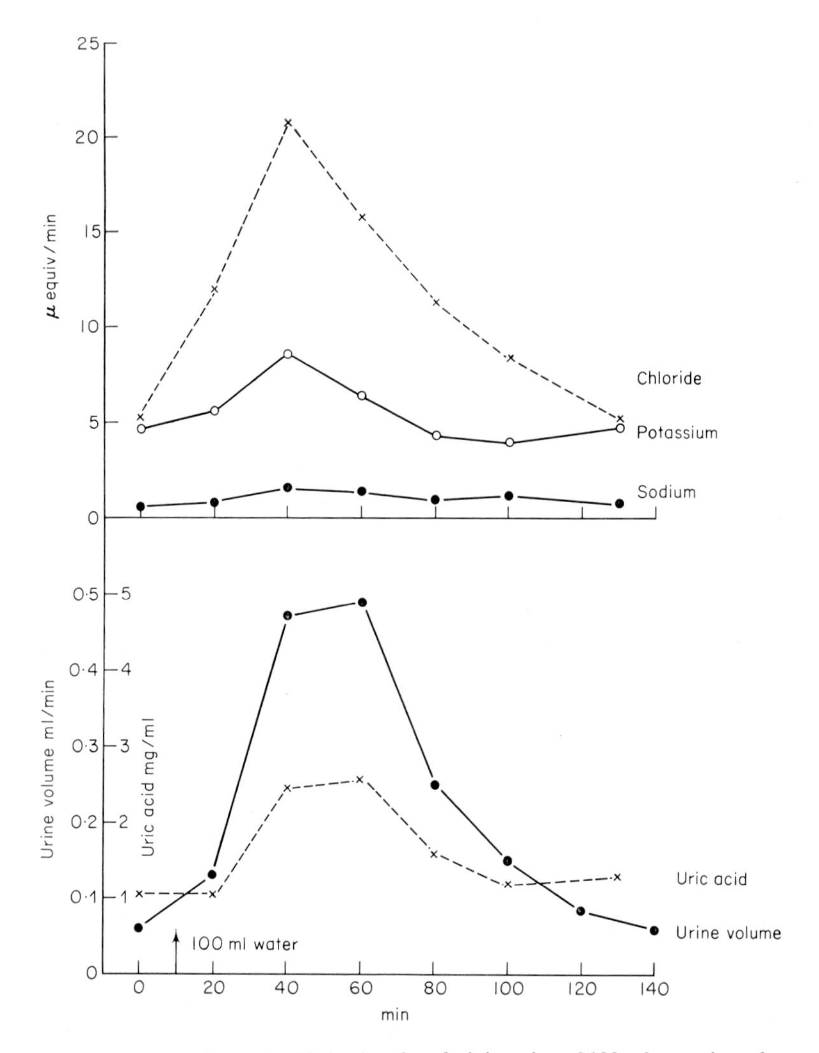

Fig. 12. Water diuresis in the fowl following the administration of 100 ml water into the crop. The response is slow owing to delayed emptying by the crop. From A. H. Sykes (unpublished).

Water diuresis is probably brought about by inhibition of the secretion of antidiuretic hormone from the posterior pituitary gland but there is some evidence which suggests that the filtration rate may increase during hydration. This in itself does not necessarily increase urine flow but it could do so if antidiuretic hormone activity remained constant or decreased. A filtration rate of about 2·1 ml/min kg was found in water-loaded fowls compared with an average of 1·8 ml/min kg (Korr, 1938; Skadhauge and Schmidt-Nielsen,

1967a) and, when individual values from many experiments are plotted against urine flow, a marked trend is observed. Changes in urine flow brought about by water-loading or dehydration are accompanied by a decrease or an increase in the osmotic concentration of the urine and, to a smaller extent, of the plasma (Table 1). The urine to plasma osmotic ratios range from 0·37

Table 1

Urinary excretion in relation to state of hydration

	Water loaded	Dehydrated
Osmolar concentration of plasma (mOsm/kg H_2O)	312	341
Osmolar concentration of urine (mOsm/kg H_2O)	115	538
Urine volume (ml/min kg)	0·300	0·018
Filtration rate (ml/min kg)	2·12	1·73
Percentage of filtrate reabsorbed (%)	86	99

From Skadhauge and Schmidt-Nielsen (1967a).

during water diuresis to 1·58 during dehydration. The osmotic concentrations varied from 50 mOsm/litre up to 600 mOsm/litre; the latter, equal to about twice the plasma concentration, is considerably lower than that found in many mammals and even in some birds. The fowl's kidney contains relatively few nephrons which possess a long loop of Henle (see Chapter 8). This loop dips down into the medullary tissue and forms the basis for the counter-current multiplier which allows the creation of a local area of very high osmotic concentration (Poulson, 1965). Hypertonic urine is formed as a result of fluid in the collecting ducts coming into equilibrium with this osmotic gradient, but, as a consequence of a relatively poor development of this section of the nephron, the fowl is not able to produce very hypertonic urine.

Experimental evidence of the existence of counter-current concentration in the fowl was provided by analyses of cortical and medullary tissue obtained after rapid freezing *in situ* (Skadhauge and Schmidt-Nielsen, 1967b). After dehydration, the osmotic pressure in the cortex was 413 mOsm/kg tissue water, in the medulla 467 and in the urine 542; the sodium concentrations were 80, 94 and 56—90 mEq/litre respectively. There were no changes in the tissue concentration of potassium or urea. The situation here is not so clearly defined as in mammals, partly due no doubt to the poor differentiation of cortical and medullary tissue in birds. The osmotic gradients are not very large and are accounted for mainly by sodium and chloride; urea makes little contribution to the total osmotic pressure whereas it accounts for about 50% in mammals (see Chapter 40).

It is interesting to note in this work that the hydrated fowl, excreting a hypotonic urine (167 mOsm/kg H_2O), still had hypertonic concentrations in the cortex and medulla (432 and 375 mOsm/kg H_2O respectively); it is not yet understood how this urinary dilution is achieved.

Table 2

Structure and activity of neurohypophyseal hormones

	S——————————————S | | Cys–Tyr–Phe–Glu(HN₂)–Asp(NH₂)–Cys–Pro–Arg–Glu(NH₂)
Vasopressin	Cys–Tyr–Phe–Glu(HN_2)–Asp(NH_2)–Cys–Pro–Arg–Glu(NH_2)
Oxytocin	,, ,, Ileu ,, ,, ,, ,, Leu ,,
Vasotocin	,, ,, Ileu ,, ,, ,, ,, Arg ,,

Activity in units per mg of hormone

	Vasopressin	Oxytocin	Vasotocin
Antidiuresis, fowl	300	—	880
Antidiuresis, dog	300	4	50
Vasodepressor, fowl	42	360	100
Vasopressor, rat	300	7	60
Oxytocic, fowl	240	29	640
Bladder, frog	20	360	19000

From Munsick *et al.* (1960).

B. ANTIDIURETIC HORMONE

Although changes in urine flow could be brought about by changes in filtration rate (either as an increased flow per glomerulus or as an increase in the number of functioning glomeruli) the extent to which this takes place under natural conditions is not known. It is clear, however, that the avian kidney is sensitive to the pituitary antidiuretic hormone (ADH) and that this hormone is released from the fowl's posterior pituitary gland in much the same way as in mammals.

The first work on ADH made use of crude extracts of the gland; this was followed by standardized preparations, then with purer fractions and finally with the synthesized hormones. The activity of one of the fractions from the avian neurohypophysis could not be matched for activity by either of the two mammalian hormones, vasopressin and oxytocin. However, a number of homologues had been synthesized and one of these, arginine vasotocin, was identical in activity with the avian antidiuretic hormone; chemical analysis showed subsequently that the two were in fact identical.

The amino-acid sequence of the avian hormones is shown in Table 2 together with the results of assays which show how avian ADH differs from that of mammals. Arginine vasotocin (AVT) has the ring structure of oxytocin but differs in having arginine, not leucine, in the side chain, which is thus identical with that of arginine vasopressin. The latter has not been identified in the fowl's neurohypophysis.

It will be seen that arginine vasotocin (AVT) has both antidiuretic and oxytocic properties in the fowl, the oxytocic response being shown by both the contraction of uterine muscle *in vitro* and the lowering of blood pressure. In frogs it is more potent than vasopressin in reducing urine flow and in altering bladder permeability.

Removal of the posterior pituitary gland brings about a polyuria which persists for several weeks and is accompanied by polydipsia (Shirley and Nalbandov, 1956). Injections of vasopressin (and presumably of AVT) bring about a prompt antidiuresis. The same effect may also be seen in short term experiments on decerebrate hens (Fig. 13). Chronic polydipsia of over a litre per day was observed as long ago as 1935 when destruction of the ADH mechanism was achieved by the insertion, through the orbit, of a

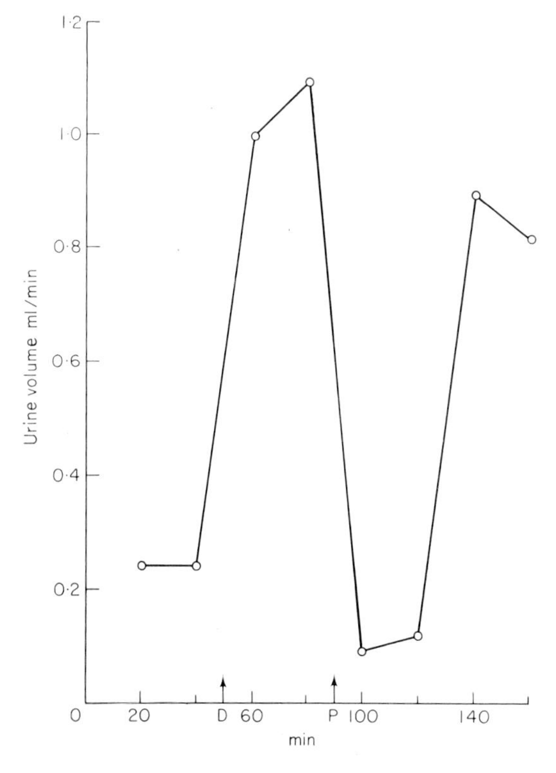

Fig. 13. Diuresis following decerebration (D) and the anti-diuretic action of pituitrin (P). From A. H. Sykes (unpublished).

tube of radium (Lacassagne and Nyka, 1935). The lesion caused by the radiation was not specific but it is now possible to place discrete, electrolytic lesions in the hypothalamus which prevent release of AVT whilst leaving the bird in a more normal condition. After such a lesion, the urine flow increased up to five times and there was persistent polydipsia, the classical symptoms of diabetes insipidus (Ralph, 1960; Koike and Lepkovsky, 1967).

The normal stimulus for the release of AVT is likely to be an increase in the plasma osmotic pressure; a number of other stimuli are known to be effective in mammals (posture, sleep, distension of the atria, haemorrhage) but it is not known whether these are also effective in the fowl. Water diuresis is inhibited during oviposition (Fig. 14) and it is possible that AVT is released

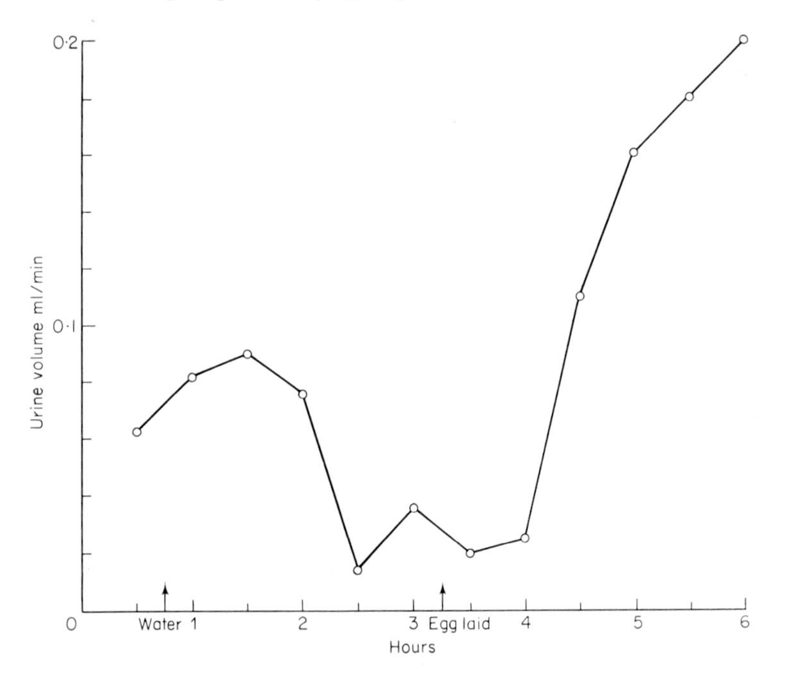

Fig. 14. Excretion of a water load administered shortly before oviposition. Maximum diuresis was delayed until after oviposition had taken place. From A. H. Sykes (unpublished).

at this time either as an ocytocic hormone itself or along with naturally occurring oxytocin.

It is generally believed that AVT increases reabsorption of water in the collective ducts and it has been shown, by means of unilateral infusion through the renal portal system, that AVT can bring about a unilateral antidiuresis (Skadhauge, 1964). Thus it can reach the tubule directly from the blood and need not be filtered first. No changes were found in the filtration rate or in the renal plasma flow. It had been observed earlier that the filtration rate could be lowered by vasopressin but very large doses were used and, with our present knowledge of the potency of AVT, there seems no reason to involve changes in filtration rate in the normal response to antidiuretic hormones.

VIII. The Excretion of Nitrogen

Since estimates of protein digestibility and of the protein requirements of poultry demand a knowledge of the composition and amount of nitrogen which is excreted there have been a number of investigations into the nitrogenous fractions of the fowl's urine. These show a wide range of variation but typical examples are given in Table 3.

Table 3

Partition of nitrogen in the urine of adult hen

	Fed	Starved
Total nitrogen (mg/100 ml urine)	440	237
Percentage of the above appearing as:		
Uric acid	84·1	57·8
Ammonia	6·8	23·0
Urea	5·2	2·9
Creatinine	0·5	4·3
Amino acid	1·7	2·8
Undetermined	1·7	9·2

From A. H. Sykes (unpublished).

Uric acid is clearly the major nitrogenous component and its excretion is considered in more detail below. Ammonia is second in importance and its proportion increases considerably during fasting. This may be a consequence of the acidaemia of starvation and since the physiological significance of ammonia lies in relation to hydrogen ion regulation its excretion will be considered in the section devoted to acid-base balance.

A. URIC ACID

Uric acid is synthesized in the liver of the fowl and the kidney excretes, by means of filtration and tubular secretion, the uric acid presented to it through the vascular system. The terminal step in uric acid synthesis is the oxidation of hypoxanthine first to xanthine and then to uric acid in the presence of xanthine oxidase (xanthine dehydrogenase). This enzyme is present in the kidney as well as the liver and it is possible that some uric acid could be synthesized there. In the classical experiments of Minkowski (1886) on the heptatectomized goose, uric acid did not completely disappear from the urine and some other extra-hepatic source was presumed to exist. If there is any renal synthesis of uric acid the significance of its renal clearance would be misleading but there is at present no means of assessing the renal contribution, if any, to the total excretion of uric acid. It should be emphasized that urea is *not* involved in uric acid synthesis; injected, labelled urea is not incorporated into uric acid *in vivo* and liver slices cannot utilize urea for uric acid synthesis *in vitro*.

In a bird weighing 2 kg, with plasma levels of uric acid of 2·5 mg/100 ml, about 0·15 mg/min of uric acid will reach the tubules from the glomerulus, assuming that the plasma content is completely filterable. The excretion rate is about 1·5 mg/min and therefore 90% of this is added to the urine by tubular secretion. It has been argued (Martindale, 1969) that if the filtration fraction is the same in the fowl and mammal, about 20% , then the renal arterial plasma flow would amount to 25 ml/min and could account for not more than half the excreted uric acid. The filtration fraction is the ratio of the glomerular filtration rate to the renal plasma flow, (C_{Inulin}/C_{PAH}) but in the fowl the arterial inflow is not the only source of blood since there is a variable, but probably a considerable, supply from the renal portal system. Estimates of the renal plasma flow in the fowl based on the clearance of PAH would include the portal contribution and there is no means of knowing what proportion comes directly from the arteries. The filtration fraction in the fowl is much lower than in mammals, between 8% and 10%; this is not the result of an abnormally low filtration rate (on a body weight basis) but of a greatly increased plasma flow. No doubt the portal supply contributes to the total uric acid excretion, indeed this is sometimes considered as its *raison d'être*, but there are no grounds for estimating this fraction to be as much as 50% or for assuming that the arterial blood flow must be five times the filtration rate as in mammals.

The clearance of uric acid in relation to plasma concentration was shown by Shannon (1938) to be similar to that of phenol red and is summarized in Fig. 15. This is the typical response of a compound which is secreted. At normal plasma levels, the clearance is about 25 ml/min kg and 1·0 mg/min of uric acid is secreted; as the plasma level rises the rate of secretion reaches a maximum of 3 mg/min and thereafter any increase in the total quantity excreted comes solely from the glomerular fraction. The tubular maximum is constant over a range of plasma concentrations from 31 to 66 mg/100 ml but at concentrations over 100 mg/100 ml the secretion rate falls; this may be a toxic effect. At normal plasma concentrations the clearance of uric acid is as much as 20 times that of inulin which indicates the importance of tubular secretion. As the plasma level increases the uric acid to inulin clearance ratio falls as the secretory mechanism becomes saturated (Fig. 16).

Uric acid secretion is depressed by the administration of PAH and, conversely, uric acid depresses the accumulation of PAH by slices of chicken

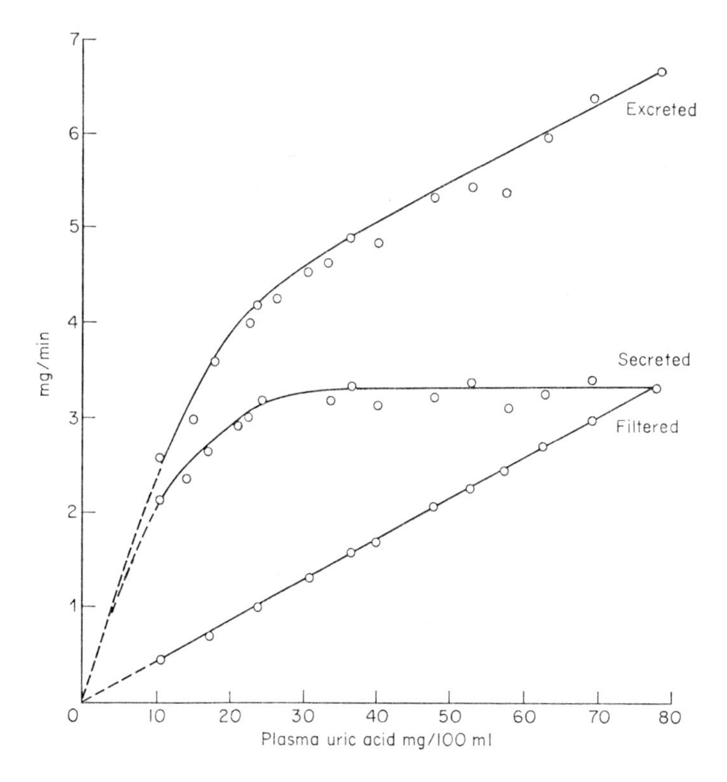

Fig. 15. The excretion of uric acid in relation to plasma concentration. Drawn from data of Shannon (1938).

or rabbit kidney. It is therefore likely that these two substances use the same secretory pathway. In man, and most other mammals, uric acid is actively reabsorbed and this can be inhibited by a number of drugs, particularly benemid. Other compounds which depress the secretion of uric acid are sulfinpyrazone, phenylbutazone, sodium salicylate and chlorothiazide. This inhibition results in uricosuria and is effective in lowering the plasma uric-acid level in cases of gout. The use of benemid in mammals has revealed the existence of a secretory as well as a reabsorptive mechanism for uric acid. In fowls benemid depresses uric-acid secretion with a consequent raising of the plasma level (Nechay and Nechay, 1959; Berger *et al.*, 1960).

The biochemical mechanisms responsible for the transport of uric acid across the renal tubular epithelium of the fowl have not yet been examined in any detail. Platts and Mudge (1961) showed that chicken kidney slices could concentrate uric acid from the medium to give tissue slice to medium ratios greater than 2 whereas in human kidney slices the ratio was always less than 1. This was an aerobic process and the ratio fell to unity in nitrogen or after treatment with dinitrophenol or benemid.

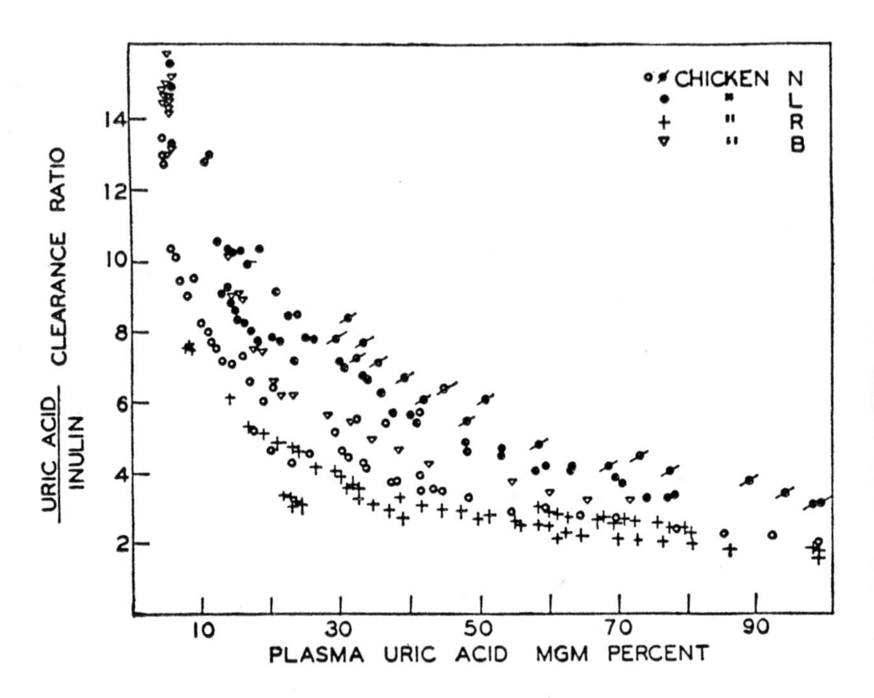

Fig. 16. Uric acid to inulin clearance ratio in relation to plasma concentration of uric acid. Reproduced with permission from Shannon (1938).

1. The Transport of Uric Acid

Uric acid ($C_5H_4O_3N_4$, MW 168·2) is potentially trivalent, but at physiological pH a monobasic salt is formed and the proportion of urate (assumed to be monosodium urate) to free uric acid will be given by the Henderson-Hasselbalch equation in which the pK is taken as 5·64. Thus in plasma at pH 7·4 98·9% of the total would be present as urate.

Uric acid is only slightly soluble in water but the salts are considerably more soluble (Table 4). Physiologically, however, the solvent is not water but a mildly alkaline, buffered solution, containing electrolytes and non-electrolytes and liable to relatively large changes in concentration during the passage of the glomerular filtrate through the tubules. Uric acid becomes less soluble with increasing acidity (Fig. 17) and this might be expected to make precipitation more likely as urinary acidification takes place.

Moreover, although the salts are more soluble than the free acid, excess of electrolyte, normally Na or K, depresses solubility owing to a common ion

Table 4
Solubility of uric acid in water at 37°C

	mM/litre	mg/100 ml	mg/100 ml as acid
Uric acid	0·39	—	6·5
Ammonium urate	2·92	54·0	49·1
Sodium urate	6·76	140·9	113·7
Potassium urate	12·06	284·0	203·0

From Peters and van Slyke (1946).

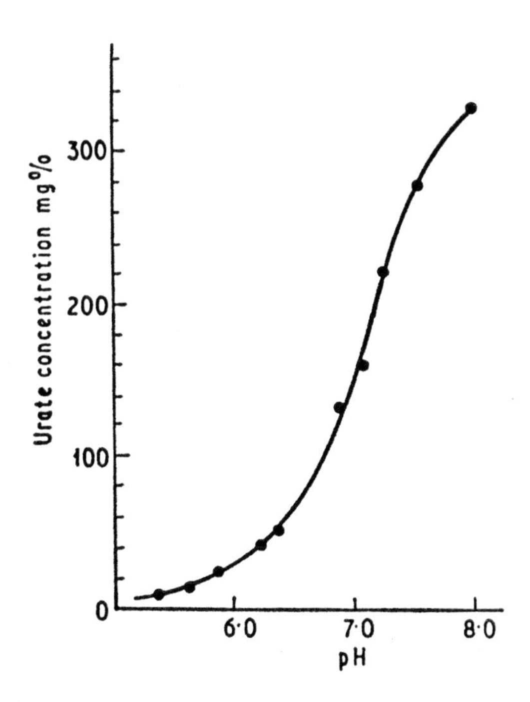

Fig. 17. The solubility of uric acid in 0·067 M-phosphate buffers over the physiological range. Reproduced with permission from Porter (1963).

effect. The solubility is also reduced as the temperature falls; dilute clear urine collected at body temperature becomes cloudy as it cools and eventually a precipitate forms.

However, uric acid and urates can form colloidal solutions with concentrations up to 2% and it is this property which allows transport through the tubules and collecting ducts to take place without the formation of a precipitate.

This ability to form sols may be seen in solutions of lithium urate (one of the more soluble salts) on the addition of electrolytes such as KCl, NaCl or NH_4Cl. The solution becomes more viscous as sol formation proceeds but the sol is not completely stable; if too much electrolyte is added, precipitation occurs.

Sol formation can be recognized from the fact that the concentration of urates is greater in urine or in phosphate buffer than can be accounted for by its solubility in water at the same temperature. Sol formation thus gives rise to a state of supersaturation but the concentrations concerned, about 300 mg/100 ml in phosphate buffer, are very much less than those found in avian urine. In mammalian urine (and also in fowl urine when it is so dilute as to be a clear fluid) this supersaturation exists without any cloudiness or viscosity to indicate the presence of urates; nevertheless, if precipitation occurred it could lead to obstruction of the collecting ducts. The composition of urine is not constant; in the defence of homeostasis the concentrations of hydrogen ions, ammonia and electrolytes vary considerably and yet these are the factors most likely to bring about urate precipitation. It appears that the stability of urate sols under conditions which could cause precipitation is due in some way to the protective action of other urinary colloids, probably proteins and glycoproteins.

Because of its instability in aqueous solutions colloidal urate is said to be lyophobic in contrast to the stable, or lyophilic, macromolecular colloids. (It was once thought that other urinary constituents, such as urea, hippuric acid or creatinine, might increase the stability of urate sols but there is no evidence that they play any such role.) Total urinary colloids have been determined by dialysis and in mammalian urine a value of 90 mg/litre was obtained of which 55 mg/litre were protein, the rest being glycoproteins.

The protective action of these lyophilic colloids has been examined by Porter (1966a, b) as part of a detailed investigation into the physicochemical properties of mammalian urinary urate. A urate sol was prepared in 0·067 M-sodium phosphate buffer at 37°C contained 274 to 286 mg/100 ml, about twice as much as would be expected from its solubility in water. The addition of 0·1 M-ammonium sulphate caused flocculation and the urate remaining in solution at any given time was determined in the supernatant after centrifuging off the precipitate. Within 30 min of adding NH_4^+ the urate concentration had fallen from 250 mg/100 ml to 50 mg/100 ml. When lyophilic colloid was added to the urate sol there was, at low protein concentrations, increased sensitivity to the flocculating action of ammonium sulphate but, as the concentration increased the urate sols became protected and did not flocculate as readily as the control samples. The degree of protection conferred by the protein increased over the range of approximately 20 to 200 mg/litre as shown by increases in Porter's protection coefficient (Fig. 18).

Flocculation involves the formation and growth of micellar particles,

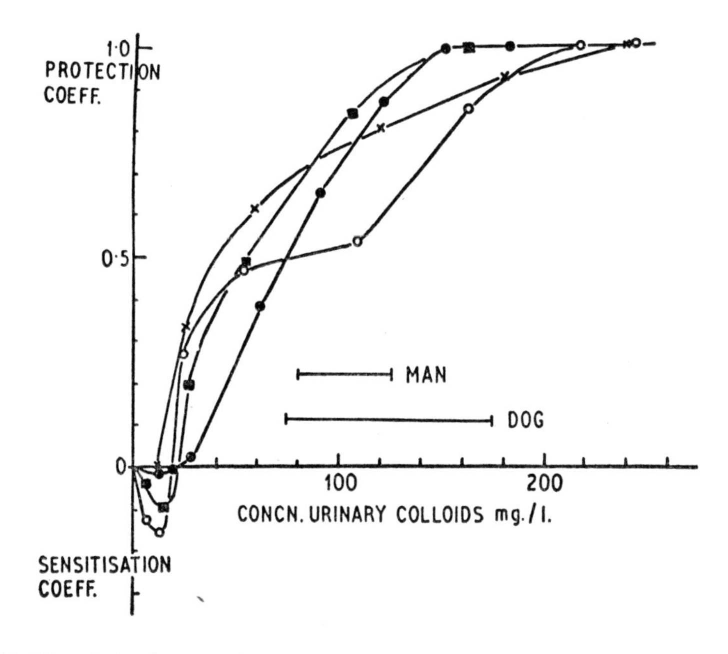

Fig. 18. The relation between the concentration of urinary colloids and the stability of urate sols. A protection coefficient of 1·0 indicates complete stability in response to the addition of ammonium sulphate as a precipitating agent. At very low concentrations of colloid precipitation occurs more readily than in the absence of colloids. The normal range of concentrations of urinary colloids in man and dog is shown. Reproduced with permission from Porter (1966a).

until the solution becomes milky and finally urate precipitates as coarse granules; this parallels exactly the appearance of dilute avian urine on standing in the cold. Unfortunately, Porter's work has not been applied to the fowl although earlier authors commented upon the viscous and stringy nature of ureteral urine and have implied that mucus (largely glycoprotein) may be a necessary lubricant for moving precipitated urate down the ureters. Young and Dreyer (1933) found that *clear* urine still contained up to 25% of its urate in a colloidal form; it was appreciated that the colloidal properties of urates could result in the formation of a gel which would be easier to transport than a frank precipitate. It does not necessarily follow that urate in avian urine is transported entirely as a protected colloidal solution; indeed, urinary colloids have not yet been identified. But ureteral urine, collected without the use of diuretics, may have a uric acid concentration as high as 20% which, although viscous does not contain granular material and will move freely down the ureters. In addition to the flocculating action of acidity and electrolytes, uric acid must be concentrated considerably by the reabsorption of water in the collecting ducts. Protection, if it occurs, may be necessary to prevent premature precipitation in the proximal tubule, the main site of secretion.

It is generally held that the excretion of nitrogen as urate is a particularly successful adaptation to terrestrial life which enables birds (and reptiles) to conserve water by excreting a semi-solid urine instead of the more obviously dilute watery solution excreted by amphibia and mammals. This view may be correct but it has not yet been expressed in quantitative terms with reference to the daily loss of water and nitrogen from a particular bird. Clearly if a given quantity of uric acid has to be removed entirely in solution, the water intake of the animal would be considerable. For example 2 g of uric acid, even as the most soluble salt, potassium urate, would require at least one litre of water for complete solution. This would amount to an intake of 500 ml/kg d for a hen compared with an actual intake of 50 to 70 ml/kg d. This shows the value of excreting a relatively insoluble form of nitrogen but it does not make it clear whether uric acid is necessarily superior, as a means of conserving water, to urea which, although very soluble, can be concentrated to a remarkable degree by the mammalian kidney. Thus both birds and mammals can survive in arid climates or at sea; it might even be argued that the mammalian kidney is better adapted for water conservation than that of birds since marine and desert-living birds also rely on (or at least undoubtedly possess) well-developed nasal glands for the excretion of salt. However, in the fowl the nasal glands are small and non-functional, even after salt feeding (McLelland and Pickering, 1969), and this aspect of water conservation need not be considered further.

The question still remains whether or not uric acid has any special advantages compared with urea. The volume of water excreted per gram of uric acid varies considerably; Hart and Essex (1942) give values of between 30 and 160 ml/g. If a conservative estimate of 66 ml/g is taken then an adult fowl could excrete 1·5 g of uric acid (0·5 g of nitrogen) per day in a total volume of 100 ml which amounts to 200 ml of water per gram of nitrogen. A normal man might excrete 20 g of urea per day in a volume of 1,500 ml which represents 10 g of nitrogen at the rate of 150 ml water per gram. On this basis, therefore, uricotelism shows no advantage over ureotelism. These values are typical of the normal, hydrated fowl rather than one which has been deprived of water and perhaps the selective advantage of uric acid would be more evident in desert-living birds but the evidence available indicates that they are no better adapted to water deprivation than desert-living mammals.

One might speculate upon the consequences of replacing urinary uric acid by urea in the fowl. Using the same values as given above, 0·5 g of urinary nitrogen could take the form of either 1·5 g of uric acid or 1·07 g of urea. The urea would form a 0·018 molar solution which is hypotonic to the plasma (0·3 molar) and there seems no reason why this small osmotic load could not be accommodated by the urine even with the fowl's limited capacity for producing hypertonic urine.

The conclusion that may be drawn from these calculations is that the fowl could maintain its water balance equally well if urea were substituted for uric acid. The biological significance of uric acid in the evolution of birds probably lies in its relation to the cleidoic egg. The adult bird, irrespective of cloacal reabsorption, loses urinary water from the body; the embryo on the other hand retains its urinary water in the allantois. Here the urate occurs, not as a watery, colloidal gel but as a crystalline, anhydrous deposit which leaves the transporting water and also sodium ions, free to be reabsorbed (Needham, 1931). It has been suggested that owing to the extreme solubility of urea it could not replace uric acid in the egg without at the same time holding a few millilitres of water in the allantois; such is the precarious water balance of the incubating egg that this extra loss would jeopardize the viability of the embryo.

B. UREA

The source of urea in the fowl is probably dietary arginine which is hydrolysed, in both the liver and the kidney, by arginase to yield urea and ornithine. (See also Chapter 40.) Ornithine is involved in detoxication reactions which will be considered in Section XI; urea is a metabolic end product which, in itself, as far as can be seen, serves no useful purpose. The amount of tissue arginase is very small in the chick, compared with the mammal. Determinations by Smith and Lewis (1963) gave activities of 380 and 4,000 units (μ moles urea/h g wet wt) for the chick's liver and kidney respectively; similar activities for the rat were 117,000 and 2,250 units.

The amount of urea in the plasma is only 1·4 mg/100 ml; this endogenous urea is filtered at the glomerulus and largely reabsorbed in the tubules giving a renal clearance of 1·5 ml/min kg. After infusions of urea which raise the plasma level to as high as 277 mg/100 ml the urea to inulin clearance ratio remains below unity i.e. net reabsorption takes place (Pitts and Korr, 1938).

When urea is infused through the renal portal system a unilateral diuresis can be produced which may, transiently, increase the apparent clearance above the filtration rate (Sykes, 1962); a similar clearance ratio was found following infusion of arginine into the portal system which apparently increased the renal synthesis of urea (Owen and Robinson, 1964). Under normal circumstances, however, there is no evidence of a secretory pathway for urea.

C. CREATININE

Endogenous creatinine is reabsorbed and cannot, therefore, be used to measure filtration rate; exogenous creatinine is secreted by the tubules (Shannon, 1938). Infusions into the renal portal system have given rise to an apparent tubular excretory fraction of about 6·5% compared with that of PAH of 65·0%. The tubular secretion of creatinine can be selectively inhibited by drugs which inhibit the transport of either organic bases, e.g.

priscoline, or organic acids, e.g. benemid. Since it is apparently transported by both pathways creatinine shows amphoteric properties (Rennick, 1967).

D. AMINO ACIDS

Small quantities of all the usual amino acids appear in the urine (Table 5) but since this represents nutritional waste it is not surprising, on teleological grounds, to find that a renal reabsorptive mechanism exists. There are at least two separate transport pathways involved, one for the diamino acids and one for the neutral amino acids (Boorman, 1969). The importance of the former may be its association with the lysine-arginine antagonism of nutritional origin; an excess of dietary lysine decreases plasma arginine probably as a result of their competing for the same reabsorptive pathway. An infusion of lysine will lower the percentage reabsorption of both lysine and arginine without affecting most of the other amino acids. The contribution of amino acids to the total daily nitrogen excretion is small, but it could be important in relation to the provision of minimal quantities of essential amino acids in the diet.

Table 5

The amino acid composition of chicken urine

Percentage of total amino acids			
Taurine	1·7	Isoleucine	1·4
Hydroxyproline	7·9	Leucine	1·6
Aspartic acid	7·4	Phenylalanine	1·3
Threonine	4·0	Ornithine	6·1
Serine	2·7	Lysine	6·5
Proline	11·1	Histidine	1·2
Glutamic acid	10·3	1-Methylhistidine	3·2
Glycine	20·9	Arginine	2·6
Alanine	3·6	Others	2·0
Valine	1·2	Amino acids as percentage of	
Cystine	1·2	total urinary nitrogen	2·2
Methionine	1·1		

From O'Dell *et al.* (1960).

IX. Acid-base Balance and the Excretion of Ammonia

A well-established renal function in mammals is the excretion of an acid or an alkaline urine in response to changes in the pH status of the blood. The effect of these compensatory changes in the excretion of acid-forming or base-forming ions is to restore to normal the carbonic acid-bicarbonate ratio of the blood and hence its pH. At a later stage the absolute quantities, as well as the ratio, of this buffer pair are also restored to normal.

The pH of fowl's urine can vary from pH 5·0 to 8·0; non-laying hens tend to have a more alkaline urine than layers. Since the fowl excretes between 1 and 2 g of uric acid per day it might be thought that the urine should always be strongly acid. However, uric acid and its salts are poorly dissociated and can act as efficient buffers. Moreover, since urate exists largely as a colloid in urine it can "extract" hydrogen ions completely therefrom. The proportion of the total uric acid which exists as an acid or as a salt is not clear since the Henderson-Hasselbalch equation applies only to dissolved substances but it seems likely that, in the non-laying bird, the bulk of the metabolically-produced protons can be buffered by uric acid leaving an excess of cations which are excreted with bicarbonate hence giving a slightly alkaline reaction to the urine.

The protons which pose a problem for the bird are those which, potentially at least, exist as -onium ions in solution. These are usually expressed as the number of milliequivalents of alkali needed to return the urinary pH to that of the blood, pH 7·4. If colloidal or semi-solid uric acid is present the acidity which it represents is not measured by this titration which may therefore underestimate the total hydrogen ion production of the kidney.

In mammals the urinary excretion of protons is largely in dihydrogen phosphate, among the chief urinary buffers. Phosphates occur in avian urine but do not account for so large a proportion of the titratable acidity.

In addition to the buffering action of phosphates there exists a second mechanism for the removal of hydrogen ions within the limits of the normal urinary pH. This is by the production of ammonia which diffuses into the acidic luminal fluid and there forms NH_4^+ which is excreted with anions, such as chloride, to preserve electrical neutrality. This process involves the formation of one bicarbonate ion for each ammonium ion and the reabsorption of one sodium ion. Both these mechanisms, direct buffering and ammonia synthesis, respond to experimentally induced acidaemia.

Milroy (1904) showed that urinary ammonia increases and uric acid decreases following the administration of HCl but these early observations were not reconsidered within the framework of acid-base balance until the work of Wolbach (1955) and Sauveur (1970). When dilute HCl is infused the urinary pH rapidly falls to as low as 5·0. As infusion proceeds the titratable acidity increases and the excretion of ammonia rises (Table 6); there is considerable variation between birds in the control values and hence in the response, but a clearer picture of the relation between urinary pH and the excretion of ammonia and acid (combined) may be seen in Fig. 19.

The titratable acidity of urine should be accounted for by the quantity of the appropriate buffer salts, urates and phosphates; infusion of any of these can increase the titratable acidity during acidaemia. From determinations of the titratable acidity and of phosphate, assuming a pK of 6·8, it is possible

Table 6
Urinary response to HCl infusion

		Control	Acidaemia
pH	Mean	6·9	6·1
	Range	5·68–8·34	5·13–7·44
Titratable acidity (μEq/min)	Mean	3·1	4·4
	Range	0–10·3	0–8·4
Ammonia (μEq/min)	Mean	2·2	4·9
	Range	0·5–4·6	2·0–8·4

From Wolbach (1955).

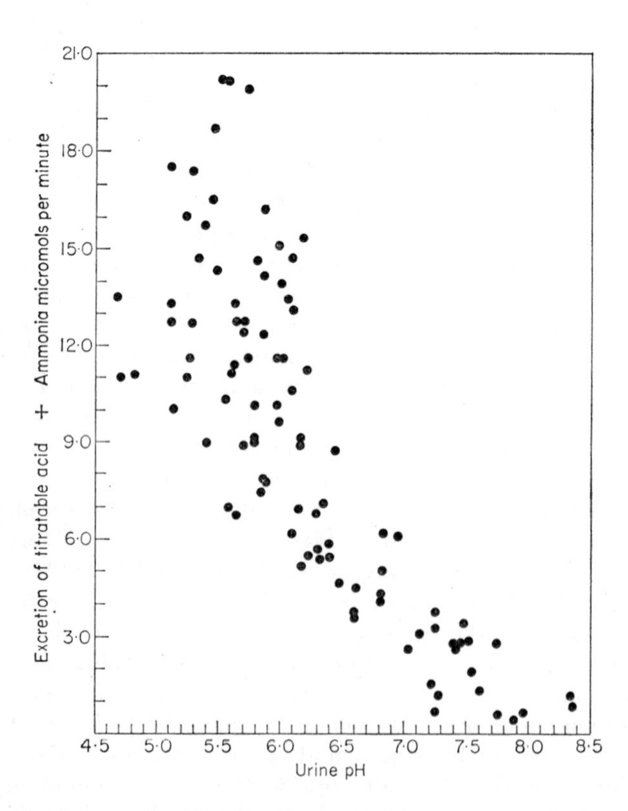

Fig. 19. The excretion of titratable acid and ammonia in relation to urinary pH. Reproduced with permission from Wolbach (1955).

to calculate the proportion of acidity accounted for by the dihydrogen phosphate and the difference can be attributed to urate (Table 7).

Table 7

The contribution of phosphate to titratable acid excretion during HCl acidaemia; results from two 10 min samples

	Acidaemia	Acidaemia + phosphate infusion
Urine flow (ml/min)	0·68	0·68
pH	5·46	5·60
Phosphate (μEq/min)	2·0	13·9
Titrable acid (μEq/min)		
Observed	5·0	13·2
Calculated from phosphate	1·5	10·3
Difference	3·5	2·9
Percentage from phosphate	29	78

From Wolbach (1955).

Under normal conditions, dihydrogen phosphate accounts for only 30% of the titratable acidity. Under suitable circumstances it can be responsible for nearly 80%.

The fowl undoubtedly possesses a renal mechanism which generates ammonia but the origin of the latter has not been investigated. In mammals the main source is plasma glutamine which can provide ammonia from either the amine or the amide radical, the latter reaction is catalysed by renal glutaminase. Other amino acids, e.g. alanine, can be deaminated to a slight extent but urea is not involved and neither is free blood ammonia. In fact the level of ammonia in the renal venous plasma is raised as a result of the synthesis.

Although glutamine is a very likely source of urinary ammonia it is possible that the fowl can also use an alternative pathway. It has been shown by Makarewicz and Zydowo (1962) that adenosine monophosphate can be a substrate for ammonia liberation in kidney homogenates and that, in addition to glutaminase, the enzymes adenylic aminohydrase and adenosine aminohydrase are present in the kidney.

This alternative pathway does not appear to be quantitatively significant in mammals but in fish and amphibia it is the principal source of their urinary ammonia. The fowl's kidney appears able to use both substrates in similar proportions *in vitro*. The importance of this phylogenetically older pathway in the intact fowl has yet to be determined.

A key mechanism in what amounts to the production of hydrogen ions by the kidney, or the regeneration of bicarbonate ions whichever view is preferred, is the formation and ionization of carbonic acid, a reaction catalysed by carbonic anhydrase. When this enzyme is inhibited, for example by acetazolamide, the urine becomes more alkaline, the titratable acidity falls and the excretion of ammonia and phosphate is reduced (Wolbach, 1955); the excretion of sodium increases considerably while that of potassium shows only a transient rise. Since chloride excretion does not rise proportionately some other anion, possibly bicarbonate, must become more important.

An important and interesting example of physiological acidaemia occurs during egg formation (Mongin, 1968). The production of carbonate for the shell results in an excess of hydrogen ions which are buffered by the blood but there is, nevertheless, a significant fall in blood pH. The reduction in the concentration of plasma bicarbonate suggests that the acidaemia is metabolic in origin; there is probably some degree of respiratory compensation since an increase in respiratory rate has been shown to occur during the period of shell formation. The urinary changes which occur are typically those of an acidaemia: a fall in pH, and bicarbonate, which is virtually completely reabsorbed (Fig. 20, Anderson, 1967) and an increase in ammonia, phosphate

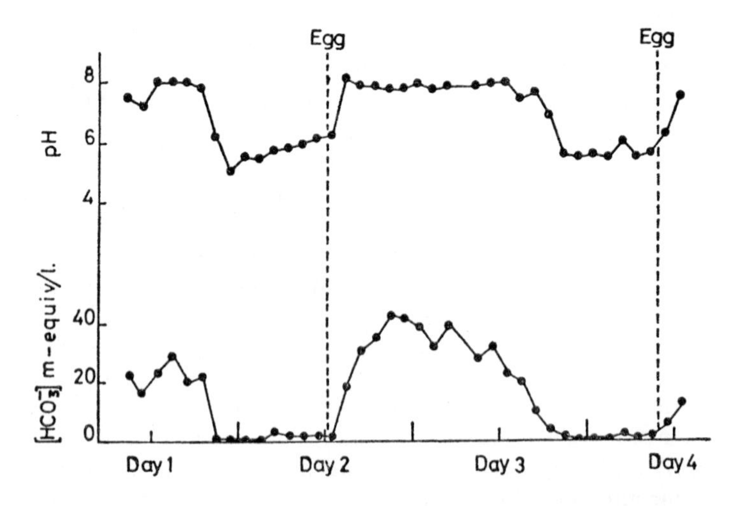

Fig. 20. Urinary pH and bicarbonate concentration in a laying hen. Egg—time of oviposition. Note the virtual disappearance of bicarbonate and the increasing acidity during the period of shell formation. Reproduced with permission from Anderson (1967).

and titratable acidity. The fact that the excretion of phosphate increases and that of urate decreases during shell formation is probably a result of the availability of phosphorus from bone when calcium, as calcium phosphate, is mobilized for the shell. When the egg is laid there is a dramatic change in the composition of the urine which becomes alkaline and now contains appreciable quantities of calcium.

Since shell formation takes about 20 h out of a total of 26 h for egg formation as a whole and an adult hen will lay on 250 d in a year, it can be appreciated that for half its productive life the fowl is faced with an incipient acidaemia and the kidney therefore must play a major role in the preservation of the neutrality of the blood.

Another situation in which acid-base homeostasis is endangered and where renal compensatory mechanisms are likely to be involved is the alkalaemia which frequently accompanies thermal panting (see Chapter 48). The increased ventilation, which is necessary for evaporative cooling, washes out CO_2 from the lungs and the pH of the blood becomes more alkaline rising at times to as high as pH 8. Under these circumstances it might be expected that the synthesis of ammonia would cease and that bicarbonate would no longer be reabsorbed leading to the excretion of a more alkaline urine.

X. The Excretion of Inorganic Ions

There have been few determinations of the total inorganic ions of avian urine partly, no doubt, because where osmolar concentrations are concerned it is simpler to measure a depression of freezing point rather than to attempt to sum the contributions made by the individual solutes. Urine composition is so variable that a single determination of its inorganic contents has little value unless accompanied by a knowledge of the dietary intake, state of growth or egg production, degree of hydration and other physiological variables. This may be seen from Table 8 which gives the concentrations of

Table 8

The ionic composition of hen's urine

		Bird No.	
	1	2	3
Urine flow (ml/min)	0·05	0·08	0·16
Cations			
Sodium (mEq/litre)	18	17	65
Potassium ,,	51	228	70
Ammonium ,,	186	53	30
TOTAL ,,	255	298	165
Anions			
Sulphate ,,	36	24	11
Phosphate ,,	116	131	75
Chloride ,,	20	111	38
TOTAL ,,	172	266	124

From A. H. Sykes (unpublished).

the principal anions and cations in three samples of urine obtained from normal birds without the use of anaesthetics or diuretics. Clearly the concentrations vary considerably and, since the determinations were not complete, the total anions do not balance the total cations as they must do in fact. The preponderance of potassium over sodium is probably a reflection of the high potassium intake from a diet composed mainly of plant materials. There is little specific information about the excretory mechanisms of individual ions and for details of the principles involved for the vertebrate kidney reference must be made to general textbooks and reviews.

A. SODIUM

The intravenous infusion of hypertonic saline results in a pronounced osmotic diuresis of up to 3 ml/min and an increased output of sodium; the excretion of potassium also rises. Because of the diuresis the urine osmolality does not increase much above that of the plasma (340 milliosmols/litre) whereas in dehydration, values up to nearly twice that of the plasma are found (Skadhauge and Schmidt-Nielsen, 1968a). With moderate rates of saline infusion the plasma sodium level does not rise; urinary concentration of sodium increases but does not reach the levels found in dehydrated birds. The higher sodium output is largely a reflection of the diuresis which results in 7 to 8% of the filtered load being excreted compared with only 1 to 2% during dehydration.

At higher rates of saline infusion the osmolality of the plasma increases to as much as 450 milliosmols/litre and there is a reduction in the glomerular filtration rate (Dantzler, 1966). The lability of filtration has already been considered and it is likely that this effect of very large saline infusions is a particular instance of a response to incipient dehydration. With only a limited ability to produce concentrated urine, body water is conserved by reducing filtration rate.

Although these findings give some indication of the capabilities of the fowl's kidney, such conditions are not likely to be found normally and it would be interesting to see how far the kidney can restrict sodium loss when faced with a chronic dietary lack of salt. This raises the question of the importance of adrenocortical hormones in the fowl in relation to sodium excretion; our knowledge here is very limited.

Adrenalectomy is rapidly fatal (Brown *et al.*, 1958b); birds survive not more than 60 h and determinations made on urine collected from exteriorized ureters during that time show a *reduced* excretion of sodium (between one third and a half that of intact controls). This is the reverse of what would be expected following loss of mineralocorticoids and one can only presume that the condition of the birds led to sodium retention from unknown causes. Adrenalectomized birds which receive injections of the unnatural mineralocorticoid deoxycorticosterone, show a similar low sodium output. Injections of cortisone, which in mammals has a greater glucogenic activity (but see

Chapter 39), restore sodium excretion to normal in the adrenalectomized bird. Similar effects are seen when these two steroids were injected into intact birds.

Hypophysectomy does not reduce sodium excretion in birds or mammals indicating that mineralocorticoid secretion is largely independent of the pituitary (Brown *et al.*, 1958a). In view of the large doses of steroids employed and the present uncertainty about the natural cortical hormones of birds, the physiological significance of these observations remains uncertain.

B. POTASSIUM

At normal plasma levels potassium shows a net reabsorption with an excreted to filtered ratio of the order of 0·2. Under a variety of circumstances, the amount of urinary potassium increases considerably and reveals the existence of a secretory mechanism; after infusion of K_2SO_4 the excreted to filtered ratio may exceed 3·0. When the infusion is made unilaterally into the renal portal circulation there is excess excretion on the infused side which also suggests the existence of a secretory mechanism (Rennick *et al.*, 1952; Orloff and Davidson, 1959).

C. PHOSPHATE

Variations in phosphate excretion are associated with changes brought about by egg-shell formation through the release of calcium phosphate from bone, when as much as 40 mg/h of phosphate may be excreted. At other times as little as 1 mg/h may be lost (Fussell, 1960). It is likely that parathyroid hormone is involved in the control of phosphate excretion; filtered phosphate is largely reabsorbed and the clearance ratio compared with inulin is about 0·1. After the infusion of parathyroid hormone (0·3 units/min) the ratio rises and can be as high as 3·6. This points to a secretory mechanism in addition to the inhibition of phosphate reabsorption (Levinsky and Davidson, 1957; Ferguson and Wolbach, 1967). (See also Chapter 38.)

D. SULPHATE

In an acute study of inorganic sulphate excretion Globus *et al.* (1961) showed that over a wide range of concentrations, the clearance increases but does not exceed the filtration rate. Sulphate shows the classical relationships of a compound which has a tubular maximum for reabsorption and which does not appear to be secreted. Radioactive sulphate when perfused into the portal circulation can show an excess excretion on the ipsilateral side but, as previously pointed out, this does not necessarily imply the existence of a secretory mechanism.

XI. Detoxication Reactions

Many drugs and compounds of dietary or intestinal origin are metabolized in the liver in such a way that their pharmacological or toxic activity is reduced. The compounds may be metabolized by oxidation, reduction, acetylation, methylation or conjugation reactions and the detoxication product is removed from the body by renal excretion.

In the mammal, benzoic acid can be conjugated with glycine to form hippuric acid but in the fowl it was shown in 1877 that ornithine, not glycine, was involved to yield ornithuric acid (dibenzoyl ornithine)

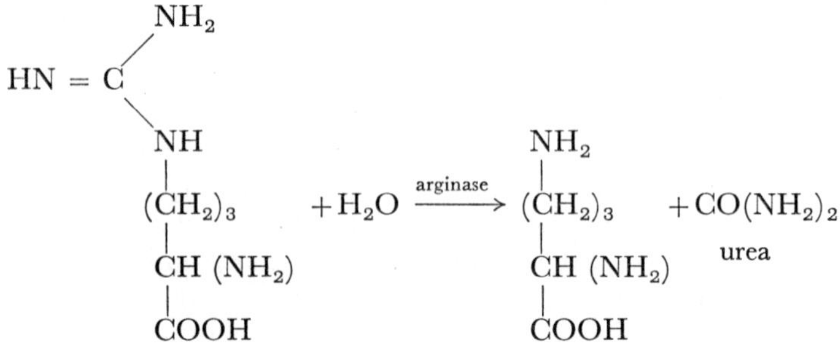

$$HN{=}C(NH_2){-}NH{-}(CH_2)_3{-}CH(NH_2){-}COOH + H_2O \xrightarrow{\text{arginase}} NH_2{-}(CH_2)_3{-}CH(NH_2){-}COOH + CO(NH_2)_2$$

arginine ornithine urea

$$C_6H_5{-}CO{-}NH \cdot CH_2 \cdot COOH$$

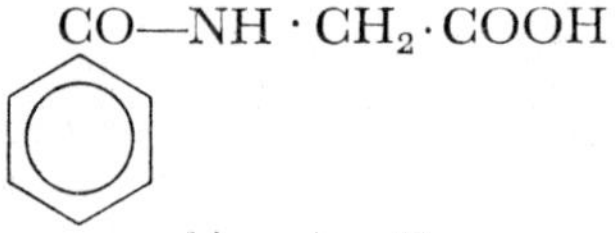

hippuric acid

$$C_6H_5{-}CO{-}NH \cdot (CH_2)_3{-}CH(COOH){-}NH{-}OC{-}C_6H_5$$

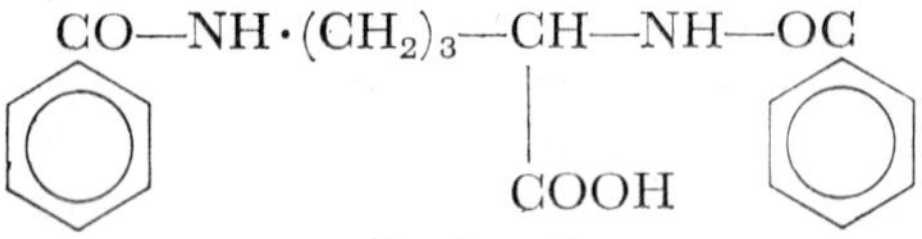

ornithuric acid

This conjugation takes place in the kidney and not in the liver, the more usual site of detoxication. The formation of ornithuric, instead of hippuric acid, was formerly thought to be a major qualitative difference between mammals and birds but more recent work has shown that this is not the case. The fowl undoubtedly synthesizes ornithuric acid but a considerable fraction, some 30%, of an ingested dose of benzoic acid, is excreted as the glucuronide and, moreover, some glycine conjugation also takes place and 5 to 20% of the benzoic acid appears as hippuric acid. There are in fact a diversity of conjugation reactions among birds; the pigeon does not synthesize ornithuric acid at all and the carrion crow and parrot do not conjugate benzoic acid with either ornithine or glycine; thus it is no longer possible to generalize about differences between birds and mammals (Bridges *et al.*, 1970).

The source of the ornithine is dietary arginine which is hydrolysed by arginase to form ornithine and urea. This is a well-known part of the ornithine cycle for urea synthesis in mammals but in uricotelic birds it appears to be solely concerned in detoxication and the urea produced has no significance as an end product of nitrogen metabolism. Although small quantities of arginase can be detected in the liver the main source is the kidney as previously mentioned.

Benzoic acid, administered daily, can be recovered quantitatively, mainly conjugated; during this period urea nitrogen increases from 1% to 9% of the total nitrogen excreted. Under these circumstances the detoxication mechanism is clearly operating and may adapt to meet the demands made upon it (Crowdle and Sherwin, 1923).

Ingested benzoic acid can depress the growth of chicks but the adverse effects can be minimized if additional arginine or ornithine are provided (Nesheim and Garlich, 1963). Moreover, since radioactive ornithine can be detected in the urine after feeding ^{14}C-labelled arginine its dietary origin is clear. Arginine is an essential amino acid for the fowl and ornithine excretion can account for at least 40% of the dietary arginine.

Ornithine, together with other diamino acids, is normally reabsorbed by the tubules and this is why little is found in the urine even when the presence of greater quantities of urea indicates that the arginase mechanism is active. When conjugated with benzoic acid, there is believed to be a net tubular secretion of ornithine in the form of ornithuric acid which is thus able to utilize an alternative excretory pathway.

It has already been mentioned that conjugation with glucuronic acid occurs in the fowl and Sperber (1947, 1948b) has shown that a wide range of compounds are also dealt with in this way, e.g. phenol and menthol. These compounds are also excreted as acid esters of sulphuric acid with the general formula $R—OSO_3H$, by another detoxication mechanism common to birds and mammals and which, quantitatively, may be more important than the formation of glucuronides. Both conjugates are excreted by tubular secretion as well as filtration and it is likely that there exists a common secretory mechanism which is also shared with innocuous compounds such as phenol red and PAH. The basis for this belief is the mutual inhibition of excretion which can be shown to follow simultaneous administration of two such compounds. Compounds with a different secretory mechanism, e.g. organic bases such as methyl nicotinamide, piperidine and guanidine are not affected by substances of the phenol red and hippuric acid group but show mutual interference between themselves (Sperber, 1948c).

It is noteworthy that different isomers of a compound do not necessarily use the same detoxication mechanism; thus ortho-aminobenzoic acid appears mainly as a glucuronide, most of the meta form is acetylated and most of the para compound is conjugated with ornithine.

References

Akester, A. R. (1964). *J. Anat.* **98**, 365–376.
Akester, A. R. (1967). *J. Anat.* **101**, 569–594.
Akester, A. R., Anderson, R. S., Hill, K. J. and Osbaldiston, G. W. (1967). *Br. Poult. Sci.* **8**, 209–212.
Akester, A. R. and Mann, S. P. (1969). *J. Anat.* **104**, 241–252.
Anderson, R. S. (1967). *Vet. Rec.* **80**, 314.
Berger, L., Yu, T. F. and Gutman, A. B. (1960). *Am. J. Physiol.* **198**, 575–580.
Bokori, J. (1961). *Acta vet. Med. Hung.* **11**, 415–422.
Boorman, K. N. (1969). *Ph.D. thesis, University of Nottingham.*
Bridges, F. W., French, M. R., Smith, R. L. and Williams, R. T. (1970). *Biochem. J.* **118**, 47–51.
Brown, K. I., Brown, D. J. and Meyer, R. K. (1958a). *Am. J. Physiol.* **192**, 43–50.
Brown, K. I., Meyer, R. K. and Brown, D. J. (1958b). *Poult. Sci.* **37**, 682–684.
Burgess, W. W., Harvey, A. M. and Marshall, E. K. (1933). *J. Pharmacol.* **49**, 237–249.
Campbell, D. E. S. (1960). *Acta pharmac. tox.* **17**, 205–212.
Clementi, A. (1945). *Boll. Soc. ital. Biol. sper.* **20**, 464–466.
Crowdle, J. H. and Sherwin, C. P. (1923). *J. biol. Chem.* **55**, 15–31.
Cuypers, Y. (1959). *Arch. int. Physiol.* **67**, 35–42.
Dantzler, W. H. (1966). *Am. J. Physiol.* **210**, 640–646.
Dixon, J. M. (1958). *Poult. Sci.* **37**, 410–414.
Dixon, J. M. and Wilkinson, W. S. (1957). *Am. J. vet. Res.* **18**, 665–667.
Ferguson, R. K. and Wolbach, R. A. (1967). *Am. J. Physiol.* **212**, 1123–1130.
Fussell, M. H. (1960). *Ph.D. thesis, University of Cambridge.*
Fussell, M. H. (1969). *Res. vet. Sci.* **10**, 332–337.
Gibbs, O. S. (1928). *J. Pharmac.* **34**, 277–291.
Gilbert, A. B. (1961). *J. Anat.* **95**, 594–598.
Globus, D. L., Becker, E. L. and Thompson, D. D. (1961). *Am. J. Physiol.* **200**, 1105–1110.
Gordeuk, S. and Grundy, M. L. (1950). *Am. J. vet. Res.* **11**, 256–259.
Hart, W. M. and Essex, H. E. (1942). *Am. J. Physiol.* **136**, 657–668.
Hester, H. R., Essex, H. E. and Mann, F. C. (1940). *Am. J. Physiol.* **128**, 592–602.
Hyden, S. and Knutsson, P. (1959). *K. LandtbrHögsk. Annlr* **25**, 253–259.
Koike, T. and Lepkovsky, S. (1967). *Gen. comp. Endocr.* **8**, 397–402.
Koike, T. I. and McFarland, L. Z. (1966). *Am. J. vet. Res.* **27**, 1130–1133.
Korr, I. M. (1938). *J. cell. comp. Physiol.* **13**, 175–194.
Lacassagne, A. and Nyka, W. (1935). *C. r. hebd. Séanc. Soc. Biol., Paris* **119**, 354–356.
Levinsky, N. G. and Davidson, D. G. (1957). *Am. J. Physiol.* **191**, 530–536.
Makarewicz, W. and Zydowo, M. (1962) *Comp. Biochem. Physiol.* **6**, 269–275.
Marshall, E. K. (1932). *Proc. Soc. exp. Biol. Med.* **29**, 971–973.
Martindale, L. (1969). *J. Physiol., Lond.* **205**, 24*P*.
Mayrs, E. B. (1923). *J. Physiol., Lond.* **58**, 276–287.
McLelland, J. and Pickering, E. C. (1969). *Res. vet. Sci.* **10**, 518–522.
Milroy, T. H. (1904). *J. Physiol., Lond.* **30**, 47–60.
Minkowski, O. (1886). *Arch. exp. Path. Pharmak.* **21**, 41–87.
Mongin, P. (1968). *Wld's Poult. Sci. J.* **24**, 200–230.
Munsick, R. A., Sawyer, W. H. and van Dyke, H. B. (1960). *Endocrinology* **66**, 860–871.
Nechay, B. R., Boyarsky, S. and Catacutan-Labay, P. (1968). *Comp. Biochem. Physiol.* **26**, 369–370.

Nechay, B. R. and Lutherer, B. D. C. (1968). *Comp. Biochem. Physiol.* **26**, 1099–1105.
Nechay, B. R. and Nechay, L. (1959). *J. Pharmac.* **126**, 291–295.
Needham, J. (1931). *Nature, Lond.* **128**, 152–153.
Nesheim, M. C. and Garlich, J. D. (1963). *J. Nutr.* **79**, 311–317.
O'Dell, B. L., Woods, W. D., Laerdal, O. A., Jeffay, A. M. and Savage, J. E. (1960). *Poult. Sci.* **39**, 426–432.
Orloff, J. and Davidson, D. G. (1959). *J. clin. Invest.* **38**, 21–30.
Owen, E. E. and Robinson, R. R. (1964). *Am. J. Physiol.* **206**, 1321–1326.
Peters, J. P. and van Slyke, D. D. (1946). *In* "Quantitative Clinical Chemistry Interpretations" Vol. I, pp. 949–955. Williams and Wilkins, Baltimore.
Pitts, R. F. (1938). *J. cell. comp. Physiol.* **11**, 99–115.
Pitts, R. F. and Korr, I. M. (1938). *J. cell. comp. Physiol.* **11**, 117–122.
Platts, M. M. and Mudge, G. H. (1961). *Am. J. Physiol.* **200**, 387–392.
Porter, P. (1963). *Res. vet. Sci.* **4**, 580–591.
Porter, P. (1966a). *Res. vet. Sci.* **7**, 128–137.
Porter, P. (1966b). *J. comp. Path.* **76**, 197–206.
Poulson, T. L. (1965). *Science, N.Y.* **148**, 389–391.
Ralph, C. L. (1960). *Am. J. Physiol.* **198**, 528–530.
Rennick, B. R. (1967). *Am. J. Physiol.* **212**, 1131–1134.
Rennick, B. R. and Gandia, H. (1954). *Proc. Soc. exp. Biol. Med.* **85**, 234–236.
Rennick, B. R., Latimer, C. and Moe, G. (1952). *Fedn Proc. Fedn Am. Socs exp. Biol.* **11**, 132.
Sauveur, B. (1970). *Annls Biol. anim. Biochem. Biophys.* **9**, 379–391.
Sapirstein, L. A. and Hartman, F. A. (1959). *Am. J. Physiol.* **196**, 751–752.
Sharp, N. C. (1912). *Am. J. Physiol.* **31**, 75–84.
Shannon, J. A. (1938). *J. cell. comp. Physiol.* **11**, 123–134.
Shirley, H. V. and Nalbandov, A. V. (1956). *Endocrinology* **58**, 477–483.
Skadhauge, E. (1964). *Acta endocr., Copenh.* **47**, 321–330.
Skadhauge, E. (1968a). *Comp. Biochem. Physiol.* **23**, 483–501.
Skadhauge, E. (1968b). *Comp. Biochem. Physiol.* **24**, 7–18.
Skadhauge, E. and Schmidt-Nielsen, K. (1967a). *Am. J. Physiol.* **212**, 793–798.
Skadhauge, E. and Schmidt-Nielsen, B. (1967b). *Am. J. Physiol.* **212**, 1313–1318.
Smith, G. H. and Lewis, D. (1963). *Br. J. Nutr.* **17**, 433–444.
Spanner, R. (1925). *Morph. Jb.* **54**, 560–596.
Sperber, I. (1946). *Nature, Lond.* **158**, 131.
Sperber, I. (1947). *K. LandtbrHögsk. Annlr* **15**, 108–112.
Sperber, I. (1948a). *Zool. Bidr. Upps.* **27**, 429–448.
Sperber, I. (1948b). *K, LandtbrHögsk. Annlr* **15**, 317–349.
Sperber, I. (1948c). *K. LandtbrHögsk. Annlr* **16**, 49–64.
Sperber, I. (1960). *In* "Biology and Comparative Physiology of Birds" (A. J. Marshall, ed.) Vol. 1, pp. 469–492. Academic Press, London.
Sturkie, P. D. (1965). "Avian Physiology" 2nd ed. pp. 372–405. Baillière, Tindall and Cassell, London.
Sykes, A. H. (1960a). *Res. vet. Sci.* **1**, 308–314.
Sykes, A. H. (1960b). *Res. vet. Sci.* **1**, 315–320.
Sykes, A. H. (1962). *Res. vet. Sci.* **3**, 183–185.
Sykes, A. H. (1966). *In* "Physiology of the Domestic Fowl" (C. Horton-Smith and E. C. Amoroso, eds.), pp. 286–293. Oliver and Boyd, Edinburgh.
Szalagyi, K. and Kriwuscha, A. (1914). *Biochem. Z.* **66**, 122–138.

Vereerstraiten, P. and Toussaint, C. (1968). *Pflügers Arch. ges. Physiol.* **302**, 13–23.
Weiner, H. (1902). *Beitr. chem. Physiol. Path.* **2**, 42–85.
Weyrauch, H. M. and Roland, S. I. (1958). *J. Urol.* **79**, 255–263.
Wolbach, R. A. (1955). *Am. J. Physiol.* **181**, 149–156.
Young, E. G. and Dreyer, N. B. (1933). *J. Pharmac.* **49**, 162–180.

10 Metabolic Energy and Gaseous Metabolism

B. M. FREEMAN

Houghton Poultry Research Station,
Houghton,
Huntingdon, England

I. Introduction

The energy requirements of the fowl are derived from the ingested food. Not all the potential energy of the food is available to the bird: the former is therefore usually termed the gross energy. In order to arrive at the net energy value—that is the energy available to the bird—the energy content of the faeces and urine and the heat increment (specific dynamic action) of the food have to be deducted from the gross energy value. Usually, the net energy content of the diet of the chicken is between 70 and 90% of the gross energy value.

Lavoisier first demonstrated a quantitative relationship between heat production and oxygen consumption. Heat results from the oxidative breakdown of carbohydrates, fats and protein. The complete oxidation of 1 g of carbohydrate in a bomb calorimeter yields an average of 4·2 kcal whilst 1 g of fat yields 9·5 kcal and 1 g of mixed protein (i.e. animal and vegetable protein)

5·3 kcal. Oxidation of carbohydrates and fats is usually complete *in vivo* and, therefore, similar amounts of heat are produced as *in vitro*. The nitrogen of protein, however, is not completely oxidized by birds, being synthesized into uric acid. The amount of heat yielded by pullets *in vivo* is about 4·2 kcal/g but this is increased by about 25 per cent on complete oxidation *in vitro*. The volumes of oxygen consumed and carbon dioxide produced, the resultant respiratory quotients (RQ), i.e. CO_2 output/O_2 intake, the heat produced *in vivo* and the thermal equivalents of oxygen are summarized in Table 1.

Table 1

Energy yields from oxidation of foodstuffs *in vivo*

Metabolic substrate (1 g)	O_2 consumed ml	CO_2 produced ml	Heat produced kcal	Thermal equivalent kcal/litre	RQ
Carbohydrate	829·3	829·3	4·2	5·047	1·000
Fat	2013·2	1431·1	9·5	4·686	0·707
Protein	894·4	657·6	4·2	4·750	0·735

II. Measurement of Energy Exchange

A. DIRECT CALORIMETRY

Heat production by an animal is best measured by direct means but the technique is complex and difficult. Aspects of direct calorimetry are discussed elsewhere (Chapter 48) but the essential features are that the sensible and insensible heat losses must be measured in order to find the total heat loss. The sensible heat loss is usually determined by either absorbing the heat into a medium and measuring the temperature change or by measuring the difference in temperature produced across a layer surrounding the animal. Both methods have their adherents. Insensible heat production is measured either by determining the change in the vapour pressure in the ventilating air, taking due note of the ventilation rate, or by maintaining the ingoing air at a constant humidity and determining the amount of water precipitated on lowering the temperature of the outgoing air to the dewpoint temperature of the ingoing air.

B. INDIRECT CALORIMETRY

The difficulties and complexities of designing and running direct calorimeters have meant that many workers still utilize the simpler indirect techniques. There are two basic methods: the first requires the calculation of heat production from the data of respiratory exchange and making the necessary adjustments for the catabolism of protein whilst the second relies on energy balance studies combined with carcass analysis.

1. Respiratory Calorimetry

a. The methods of respiratory calorimetry. Respiratory calorimetry requires, at least in theory, simultaneous measurements of oxygen consumption and carbon dioxide and urine productions. Both open- and closed-circuit respirometers are used. Open-circuit respirometers utilize either the classical Haldane technique when oxygen consumption is determined by gravimetric methods (for a modification for application to the fowl, see Zausch, 1965) or specific analytical techniques utilizing paramagnetic oxygen and infrared carbon dioxide analysers. The latter method is particularly valuable where measurement of isotopic carbon dioxide ($^{14}CO_2$) is envisaged. After determination of the total carbon dioxide production, the proportion of $^{14}CO_2$ is measured, either discontinuously by precipitation and scintillation counting (see, for instance, Annison *et al.*, 1966) or continuously (Simpson-Morgan, 1965). Closed-circuit respirometers are usually designed to allow the gravimetric determination of carbon dioxide production, the gas being absorbed as it is produced, and oxygen being injected at a rate sufficient to maintain the pressure of the system.

As already pointed out, in order accurately to determine the heat production by indirect means it is necessary to measure the RQ since the thermal equivalents of the respiratory gases vary with the metabolic mixture of foodstuffs. The RQ is also affected by concomitant protein catabolism. Since the non-protein RQ represents a specific mixture of carbohydrates and fats which, on oxidation, yields a specific thermal equivalent, it is usual to determine the protein being catabolized by urine analysis. However, because avian urine and faeces are voided together, the determination of urinary nitrogen output for the fowl is notoriously difficult (see Chapter 9). Separation of the excreta has been achieved by catheterization of the ureters (Coulson and Hughes, 1930), by exteriorization of the ureters (Dixon and Wilkinson, 1957) or by colostomy (e.g. by Fussell, 1960, 1969) but all have inherent disadvantages, the former because diuresis may result (Sturkie, 1958) whilst with the others the normal retroperistaltic movement of urine towards the caeca (Akester *et al.*, 1967) is prevented. Others have rejected surgical techniques and have developed physicochemical methods to separate urinary nitrogen from faecal nitrogen (Ekman *et al.*, 1949; Davidson and Thomas, 1969). The value for urinary nitrogen is multiplied by 6·25 to give the approximate amount of protein catabolized; then the volume of oxygen required for this catabolism and the volume of carbon dioxide produced are determined from the data included in Table 1 and deducted from total respiratory exchange in order to give the non-protein RQ. The thermal equivalent is then derived from the relevant tables such as those prepared by Lusk (1928) and the heat production determined.

This arduous procedure has been simplified, notably by Brouwer (1957) who produced a formula for the calculation of heat production by mammals:

$$T = 3{\cdot}869\ O_2 + 1{\cdot}195\ CO_2 - 0{\cdot}227P \qquad (1)$$

where T = heat production (kcal)
O_2 = oxygen consumption (litre)
CO_2 = carbon dioxide production (litre)
P = urinary nitrogen $\times$ 6·25 (g)

Romijn and Lokhorst (1961), from a study of the published data for the fowl, have derived from this equation one for the fowl:

$$T = 3{\cdot}871\ O_2 + 1{\cdot}194\ CO_2 - 0{\cdot}380P \qquad (2)$$

It is noteworthy that the coefficient for protein catabolism is the only significant difference between the "mammalian" and "avian" formulae. Romijn (1950a) and Romijn and Lokhorst (1961) have further shown that, if the protein component is ignored, an error of not more than 0·6 per cent is introduced into the calculated heat production. For practical purposes therefore the equation may be written as:

$$T = 3{\cdot}871\ O_2 + 1{\cdot}194\ CO_2 \qquad (3)$$

b. The validity of the method when the RQ value is abnormal. It has been variously argued that formulae such as those above are valid only when the RQ falls within the theoretical limits of 0·707 and 1·00. Respiratory quotients well below 0·707 and slightly above 1·00 have been observed in the fowl by many authors (Blobelt, 1926; Mitchell *et al.*, 1927; Bacq, 1929; Nichita and Mircea, 1933; Henry *et al.*, 1934; Dukes, 1937; Romijn, 1950b, c; Mellen and Hill, 1955; Romijn and Lokhorst, 1961, 1963, 1964; Romijn and Vreugdenhil, 1969) and consequently throw doubt on the validity of indirect calorimetry for birds.

It has been suggested that protein catabolism in uricotelic animals is the cause of the low RQ's but King (1957) has shown that the RQ for the production of urine in the fowl is 0·735. Gluconeogenesis seems the more likely cause.

Brouwer (1957) considering the special problems raised by these findings has shown the classical Zuntz formula and all derived from it (equations 1–3 above), to be valid under all conditions.

2. Energy Balance Studies

This method relies on the quantitative feeding and collection of excreta coupled with determinations, by bomb calorimetry, of their respective energy contents. Whole body analysis allows the energy gain to be determined. Comparison of energy gain with energy intake provides information on the heat loss during the period of observation.

The method is necessarily rather insensitive in that changes in heat production cannot be determined over short periods. It is often used, however, in conjunction with other calorimetric techniques (see Poczopko and Kowalczyk, 1965; Davidson *et al.*, 1968).

III. Thermoneutrality

The temperature, or range of temperatures, at which the metabolic rate is at a minimum is defined as the zone of thermal neutrality and is limited by the upper and lower critical temperatures. These are influenced by many factors including age, wind velocity, plane of nutrition, animal group size, etc.

Several publications describe ranges of thermal neutrality determined under standard or quasi-standard conditions. These data referring to the post-absorptive fowl have been collated in Table 2. It will be noted that, whilst the upper critical temperature declines little with age, the lower critical temperature shows a considerable fall.

Table 2

A comparison of the zones of thermal neutrality as determined by various authors

Author	Age (weeks)					
	1	2	4	8	12	Adult
Mitchell and Haines (1927a)[a]						23·9–26·6
Benedict *et al.* (1932)						15·0–28·0
Kleiber and Dougherty (1933)	38·0	38·0				
Barott *et al.* (1936)	35·5					
Barott and Pringle (1941)						22·8–27·7
Barott and Pringle (1946)	35·0	35·0	32·0–35·0	29·4–35·0	26·7–35·0	
Romijn (1950b)	35·0	35·0	35·0	35·0		27·0–32·0
Romijn and Lokhorst (1961)	32·0					28·0–32·0

[a] As recalculated by Barott and Pringle (1941).

The plane of nutrition influences the value of the lower critical temperature. As the environmental temperature falls, the heat resulting from the specific dynamic action of the food can be utilized to maintain body temperature. The zone of thermal neutrality is generally wider in fully-fed birds.

Little work has been carried out on this aspect of metabolism and the results of Freeman (1963a), the only ones available, cover only the first four weeks of life and are given in Table 3.

Table 3

Zones of thermal neutrality for the fully-fed individual fowl[a]

Age (weeks)	Zones of thermal neutrality (°C)
0	35
1	34–35
2	31–35
3	30–33
4	26–31

[a] Reproduced with permission from Freeman (1963a).

The thermal insulation afforded by the feathers is an important factor in determining the zone of thermal neutrality. In a group of birds social factors also affect the zone of thermal neutrality. By huddling, the bird is better able to withstand low environmental temperatures (Kleiber and Winchester, 1933).

IV. Metabolic Rate

A. METABOLIC RATE AND METABOLIC SIZE

When expressed in terms of heat produced per unit time, the basal metabolic rate (BMR) increases with increasing body weight but when expressed in terms of heat produced per unit weight then there is a decrease with increasing weight. Thus, a mouse produces 200 kcal/kg d whilst a cow produces only 13 kcal/kg d (Kleiber, 1965). Valid comparisons between birds of different body weight cannot be made simply on the basis of their metabolic rates per unit body weight.

The relationship between metabolic rate and body weight is expotential;

$$M \propto W^n$$

where M is the metabolic rate in kcal/d, W is the body weight in kg and n is a constant. The value of n has been shown to lie between 0·66 and 1·00 for all organisms (Zeuthen, 1953). Reviewing available data for the bird, King and Farner (1961) found that $n = 0{\cdot}744$ which is indistinguishable from Kleiber's (1947) value (0·756) for all homeotherms. Recently the value of 0·75 (or the three-quarters power as some prefer it to be expressed) has been adopted for describing metabolic size (Kleiber, 1965).

It must be noted that the exponent 0·75 does not always describe satisfactorily results obtained from growing animals.

B. METABOLIC RATE DURING THE NEONATAL PERIOD

The oxygen consumption of the mature (19 d old) embryo is lower than that of the chick immediately after hatching and is considerably lower than the chick aged 3 or 4 h. Typically, the mature embryo consumes about 25 ml O_2/h (Romijn and Lokhorst, 1951; Freeman, 1962, 1965a; Visschedijk, 1968), whilst immediately after hatching the rate has risen to approximately 40 ml/h (Freeman, 1962, 1964a, 1965b; Visschedijk, 1968). On escaping from the shell, the chick acquires the ability to regulate its metabolic rate and it becomes a homeotherm; there is a marked increase in its effective surface area as a result of hatching and as it is wet, it therefore is subjected to considerable cooling by surface evaporation. These factors provoke a large increase in the chick's metabolic rate so that about 5 or 6 h after hatching it rises to about 60–65 ml O_2/h (Freeman, 1964a, 1965b).

C. BASAL METABOLIC RATE DURING GROWTH

The concept of basal (or standard) metabolic rate (BMR), due to Krogh (1916) refers to the heat production per unit time determined when the animal is in a post-absorptive state, at rest and maintained in a thermally neutral environment. For comparative purposes, measurements should be made at similar times of day to minimize effects of diurnal rhythm.

About 48 h starvation brings the adult fowl to a post-absorptive state (Mitchell and Haines, 1927b; Mitchell and Kelley, 1933; Henry *et al.*, 1934; Dukes, 1937). With immature birds the period is more variable but is probably about 12 h (Hillerman *et al.*, 1953). When determining metabolic rates it is essential that the measurements are carried out within the zone of thermal neutrality. Some of the early work was done when details of the zones of thermal neutrality were still uncertain; others did not control the environmental temperature. As a result, although some experiments were begun at temperatures within the zone of thermal neutrality, they were completed with the temperature exceeding the upper critical level; the value of such studies is therefore doubtful.

Studies with the environment carefully controlled were initiated by Barott and his co-workers in the 1930's. These remain the classic studies on the energy metabolism of the fowl.

Barott *et al.* (1936) found that the metabolic rate of unfed chicks remained fairly constant during the first 5 d of post-embryonic life. If feeding were allowed, the BMR was found to increase progressively during the first 15 d (Barott *et al.*, 1938). This broadly confirmed the earlier findings of Mitchell *et al.* (1927) although in their study the maximum metabolic rate was achieved within 8 d. It is possible that the differences may have been due to breed.

The BMR of birds older than 15 d shows a progressive fall with age and reaches a relatively constant level after about 100 d (Barott *et al.*, 1938).

D. RESTING METABOLIC RATE DURING GROWTH

Measurement of the metabolic rate of the fully-fed but inactive bird in a thermally neutral environment provides information of its resting metabolic rate.

A marked increase in the metabolic rate of the fowl during the first weeks of life was first noted by Brody (1930) and was confirmed by Beattie and Freeman (1962) and Freeman (1964b), while Freeman (1965b) showed that part of this rise is due to the progressive replacement of space occupied by the yolk sac by metabolically active tissue (see also Chapter 48).

Changes in the metabolic pattern during growth have been examined by Kibler and Brody (1944) and by Freeman (1963b). They showed a positive correlation between metabolic rate and body weight and that there are perhaps three or four changes in the regression coefficients of oxygen consumption on body weight during growth. The pattern may also be influenced by diet (Freeman, 1963b)—see also Section IVE.

The differences between the BMR and the resting metabolic rate have been investigated by Barott *et al.* (1938) who found that, in male chicks, the latter was approximately 60% higher than the former during the first week of life and then progressively fell with age so that, at 20 weeks of age, *ad libitum* feeding caused only a 25% increase.

E. FACTORS AFFECTING METABOLIC RATE

1. Diurnal Rhythm

The fowl shows pronounced diurnal rhythm in its metabolic rate, accompanied by a corresponding rhythm in the deep body temperature (see Chapter 48). This rhythm was first described by Barott *et al.* (1938) who found a difference of approximately 24% between the maximum and minimum BMR's during the first week of life and that this variation declined with age. At 12 weeks of age the difference was 11% (Barott *et al.*, 1938) and in adults it was 9% (Deighton and Hutchinson, 1940). Barott *et al.* (1938) and Tasaki and Sakurai (1969) are agreed that the maximum metabolic rate of adults occurs at about 08.00 h with the minimum rate occurring 12 h later.

The variation in diurnal rhythm is somewhat larger in the fully-fed adult and declines as starvation proceeds (Tasaki and Sakurai, 1969).

2. Activity

The effects of activity on heat production and metabolic rate are well known. The metabolic rate is increased by about 40 to 45% immediately on standing (Deighton and Hutchinson, 1940) whilst the marked and continuous changes that occur during normal activity have been recorded by Hutchinson (1954).

3. Season

Seasonal changes in the metabolic rate have been reported (Winchester, 1940; Tasaki and Sakurai, 1969) and almost certainly result from alterations in the thyroid secretion rate. Stahl and Turner (1961), for instance, found at least a 50% increase in the secretion rate of the thyroid hormone during the autumn.

4. Sex and Reproductive Activity

Males probably have a higher metabolic rate than females (Mitchell *et al.*, 1927), although, during the growing period, it is difficult to separate the factors of metabolic size, age and endocrinological status.

It is generally agreed that the laying female has a higher metabolic rate than the non-laying female (Gerhartz, 1914; Mitchell and Haines, 1927b; Dukes, 1937) and recently it has been shown that the BMR of hens increases during the laying cycle (Leeson and Porter-Smith, 1970).

5. Breed

Marked differences in metabolic rate have been reported between different breeds (Nichita and Mircea, 1933; Ota and McNally, 1961; Huston *et al.*, 1962) but whether these differences are real is problematical since the factor of metabolic size has not always been considered. Mellen (1963) discussed the interpretation of such reports. Ota and McNally (1961) allowed for metabolic size by expressing their data in terms of surface area and so appear to have shown a valid difference between White Leghorns and Rhode Island Reds, the former having a higher rate. However, Bergman and Snapir (1965) found no differences in metabolic rate between White Leghorns, Plymouth Rocks and New Hampshire × White Leghorns when metabolic size was taken into account.

Tasaki and Sakurai (1969) found two distinct populations in a group of Plymouth Rock × White Leghorn hybrids; one had a mean metabolic rate of 67 $kcal/kg^{0.75}d$ and the other 86 $kcal/kg^{0.75}d$.

Breed differences noted in BMR's determined at elevated environmental temperatures (Bergman and Snapir, 1965) may be related to differences in heat tolerance.

6. Environmental Temperature

When the temperature falls below the lower critical temperature, the metabolic rate increases in order to maintain deep body temperature. The mechanisms of thermoregulation *per se* are discussed elsewhere (Chapter 48). The increase in metabolic rate for any given reduction of temperature below the lower critical temperature is dependent upon the age of the bird, the greatest change being seen in the young chick (Barott and Pringle, 1946) and probably reflects the larger surface-area-to-mass ratio of the younger animals.

The metabolic rate increases proportionally to the reduction in environmental temperature until it reaches a maximum. This is termed the summit metabolic rate and may be up to four times the minimal rate.

7. Diet

Claims that certain diets may promote a higher metabolic rate must be viewed in the light of metabolic size. At the same time, the adequacy of the diet in meeting the nutritional requirements of the bird must be confirmed (March and Biely, 1957).

Whilst the composition of the diet has little effect on the metabolic rate in the early post-hatching period (Freeman, 1964b) it seems likely that it may affect not only the absolute metabolic rate but also the metabolic pattern (Freeman, 1963b).

There is evidence that when the ratio of crude protein to metabolizable energy in a diet fall below about 62, growing birds are less able to utilize the food efficiently so that more energy is lost as heat (Davidson *et al.*, 1961, 1964, 1968).

8. Plane of Nutrition

As already noted, the resting metabolic rate is higher than the BMR. The effects of the plane of nutrition on the metabolic rate of two populations of hybrids ("high" and "low" metabolic rate) is illustrated in Fig. 1 taken from the work of Tasaki and Sakurai (1969). It is particularly interesting that the differences in metabolic rate between the two populations disappeared when the birds were receiving a maintenance ration.

V. Energy Sources

A. ENERGY SOURCES DURING THE NEONATAL PERIOD

The RQ remains at about 0·7 during the whole of the hatching period (Visschedijk, 1968). It may rise transiently, because the cooling effects of evaporation of surface water stimulate a rise in muscle tone immediately after hatching.

Although the 19-day-old embryo has moderate stores of carbohydrate (Muglia and Massuelli, 1934; Gill, 1938; Freeman, 1965a) these are extremely meagre when compared with the lipid stores (Romanoff and Romanoff, 1967). The carbohydrate is probably the main source of energy for the central nervous system (Freeman, 1969) as in mammals.

About 5 g of yolk remain after incubation and these are utilized by the chick within 5 d under normal conditions. During the perinatal period the

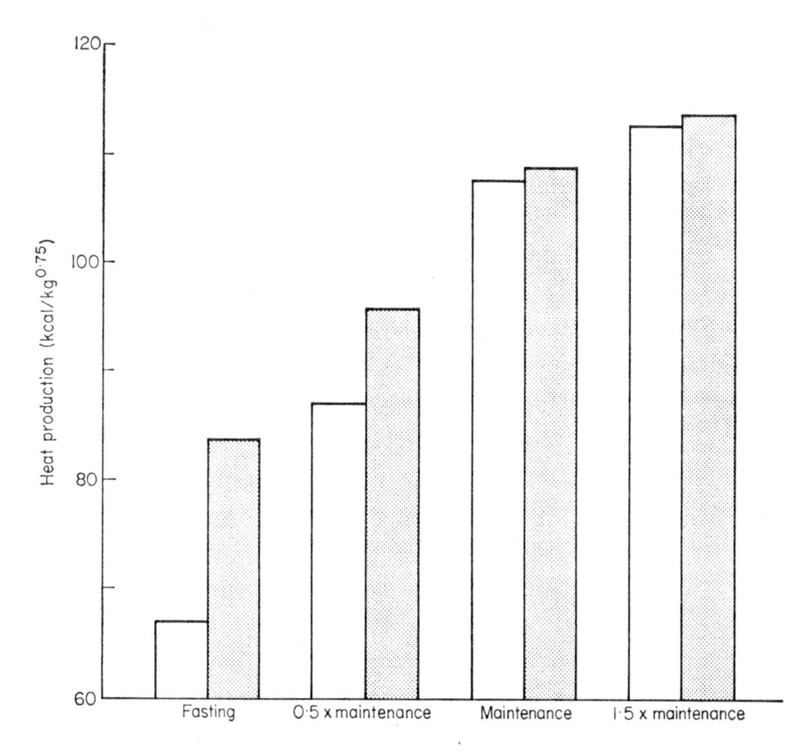

Fig. 1. The effect of the plane of nutrition on the heat production of the adult fowl. The unmarked histogram refers to the sub-group with low basal metabolic rate and the stippled histogram to the sub-group with high basal metabolic rate. (Redrawn from Tasaki and Sakurai, 1969 and reproduced with permission.)

RQ remains at about 0·7 indicating that yolk lipid is the major energy source.

That yolk is not essential for chicks to thrive and show an unimpaired growth rate has been shown by Barott *et al.* (1936) and Freeman (1964a) who removed the yolk sac on the first day after hatching.

Incomplete catabolism of lipid may occur during hatching when significant amounts of ketone bodies are found in the plasma (Best, 1966).

B. ENERGY SOURCES AFTER THE NEONATAL PERIOD

The nutrients of the fowl change dramatically at hatching from being predominantly lipid to predominantly carbohydrate. The RQ of the fully-fed mature fowl is between 0·85 and 0·95 (Henry *et al.*, 1934; Romijn, 1950b; Romijn and Lokhorst, 1963) indicating that the major energy source is carbohydrate. It has been shown that carbohydrate catabolism supplies about 87% of the energy required by the bird, and protein catabolism supplies virtually all the remainder (Tasaki and Sakurai, 1969).

After withdrawal of food, the RQ shows a rapid reduction and basal conditions (RQ = 0·7) are realized, at least in the adult, within 48 h (Mitchell and Haines, 1927b; Dukes, 1937). Within 24 h the proportion of heat provided by carbohydrate falls to below 60% (Henry *et al.*, 1934; Tasaki and Sakurai, 1969) and glycogen stores are severely depleted (Hazelwood and Lorenz, 1959). Fat catabolism increases to about 29% and, during this time, the protein component remains fairly constant (Tasaki and Sakurai, 1969). There is a marked rise in plasma free fatty acids (FFA) indicating that mobilization of lipids has begun (Langslow *et al.*, 1970).

By the second day of starvation, fat catabolism accounts for more than 80% of the total metabolism while protein catabolism remains at a normal level. Carbohydrate catabolism falls to about 4% of the total, when ketone bodies begin to appear in the plasma (Chubb and Freeman, 1963; Romijn and Lokhorst, 1963, 1964).

In starvation longer than 2 d the RQ may fall well below 0·70; values as low as 0·52 have been reported by Romijn and Lokhorst (1963, 1964) although Tasaki and Sakurai (1969) have been unable to confirm this. Romijn and Lokhorst suggest that such low RQ's indicate gluconeogenesis which can certainly occur (see Chapter 11) and the partial restoration of carbohydrate stores during prolonged starvation (Hazelwood and Lorenz, 1959) would support this conclusion.

Immediately following realimentation there is an extremely large increase in the RQ, and transient values in excess of 1·00 may be observed (Romijn and Lokhorst, 1963, 1964). After some hours the RQ returns to within the normal limits for the fully-fed fowl.

VI. Energy Intake and Storage

A. ENERGY INTAKE

The newly-hatched chick can survive for at least 5 d on the large stores of yolk which persist from its embryonic phase. Feeding normally begins immediately after hatching but does not prevent a slight fall in body weight in the neonatal period. The nutritional change at this time is marked, that of the embryo being predominantly lipid whilst that of the hatched bird will contain only 5 or 6% lipid and most of the energy requirements are supplied by carbohydrate.

Energy intake increases as the ratio of crude protein to the metabolizable energy content of the diet falls below about 62; Davidson *et al.* (1961, 1964, 1968) found that birds tended to consume more food but to use the extra energy less efficiently. Breed differences in energy intake have been noted (Davidson, 1965; Davidson *et al.*, 1968).

One of the major factors influencing energy intake is environmental temperature; as this falls so the metabolic rate rises to maintain the deep body temperature (see also Chapter 48). The energy requirements of the bird therefore increase as the environmental temperature falls and conversely its requirements fall with a rise in environmental temperature. Energy intake (appetite) is controlled by the hypothalamus (Feldman *et al.*, 1957).

B. ENERGY STORAGE

Ingested carbohydrate and lipid not immediately required for catabolic purposes are stored as glycogen or as triglycerides. Excess protein is deaminated and the amino groups are excreted as urate.

Glycogen stores are not large, the bulk being utilized within 24 h of the onset of starvation. The main energy reserve is fat derived from both carbohydrate and lipid and is stored in white adipose tissue. In the fowl there is little subcutaneous fat; unlike some birds the major adipose tissue sites are intra-abdominal.

References

Akester, A. R., Anderson, R. S., Hill, K. J. and Osbaldiston, G. W. (1967). *Br. Poult. Sci.* **8**, 209–212.

Annison, E. F., Hill, K. J., Shrimpton, D. H., Stringer, D. A. and West, D. E. (1966). *Br. Poult. Sci.* **7**, 319–320.

Bacq, A. M. (1929). *Annls Physiol. Physicochim. Biol.* **5**, 497–511.

Barott, H., Byerly, T. C. and Pringle, E. M. (1936). *J. Nutr.* **11**, 191–210.

Barott, H., Fritz, J. C., Pringle, E. M. and Titus, H. W. (1938). *J. Nutr.* **15**, 145–167.

Barott, H. G. and Pringle, E. M. (1941). *J. Nutr.* **22**, 273–286.

Barott, H. G. and Pringle, E. M. (1946). *J. Nutr.* **31**, 35–50.

Beattie, J. and Freeman, B. M. (1962). *Br. Poult. Sci.* **3**, 51–59.

Benedict, F. G., Landauer, W. and Fox, E. L. (1932). *Bull. Storrs agric. Exp. Stn* **177**.

Bergman, A. and Snapir, N. (1965). *Br. Poult. Sci.* **6**, 207–216.

Best, E. E. (1966). *Br. Poult. Sci.* **7**, 23–28.

Blobelt, P. (1926). *Biochem. Z.* **172**, 451–466.

Brody, S. (1930). *Res. Bull. Mo. agric Exp. Stn* **143**, 177–179.

Brouwer, E. (1957). *Acta physiol. pharmac. néerl.* **6**, 795–802.

Chubb, L. G. and Freeman, B. M. (1963). *Comp. Biochem. Physiol.* **10**, 95–102.

Coulson, E. J. and Hughes, J. S. (1930). *Poult. Sci.* **10**, 53–58.

Davidson, J. (1965). *In* "Energy Metabolism" (K. L. Blaxter, ed.), pp. 333–345. Academic Press, London.

Davidson, J., Hepburn, W. R., Mathieson, J. and Pullar, J. D. (1968). *Br. Poult. Sci.* **9**, 93–109.

Davidson, J., Mathieson, J., Williams, R. B. and Boyne, A. N. (1964). *J. Sci. Fd Agric.* **15**, 316–325.

Davidson, J., McDonald, I., Mathieson, J. and Williams, R. B. (1961). *J. Sci. Fd Agric.* **12**, 425–439.

Davidson, J. and Thomas, O. A. (1969). *Br. Poult. Sci.* **10**, 53–66.

Deighton, T. and Hutchinson, J. C. D. (1940). *J. agric. Sci.* **31**, 151–157.

Dixon, J. M. and Wilkinson, W. S. (1957). *Am. J. vet. Res.* **18**, 665–667.

Dukes, H. H. (1937). *J. Nutr.* **14**, 341–354.

Ekman, P., Emanuelson, H. and Fransson, A. (1949). *K. LantbrHögsk. Annlr* **16**, 749–777.

Feldman, S. E., Larsson, S., Dimick, M. K. and Lepkovsky, S. (1957). *Am. J. Physiol.* **191**, 259–261.
Freeman, B. M. (1962). *Br. Poult. Sci.* **3**, 63–72.
Freeman, B. M. (1963a). *Br. Poult. Sci.* **4**, 275–278.
Freeman, B. M. (1963b). *Br. Poult. Sci.* **4**, 169–178.
Freeman, B. M. (1964a). *Ph.D. thesis, University of Leicester.*
Freeman, B. M. (1964b). *Br. Poult. Sci.* **5**, 263–267.
Freeman, B. M. (1965a). *Comp. Biochem. Physiol.* **14**, 217–222.
Freeman, B. M. (1965b). *Br. Poult. Sci.* **6**, 67–72.
Freeman, B. M. (1969). *Comp. Biochem. Physiol.* **28**, 1169–1176.
Fussell, M. H. (1960). *Nature, Lond.* **185**, 332–333.
Fussell, M. H. (1969). *Res. vet. Sci.* **10**, 332–337.
Gerhartz, H. (1914). *Pflügers Arch. ges. Physiol.* **156**, 1–224.
Gill, P. M. (1938). *Biochem. J.* **32**, 1792–1799.
Hazelwood, R. L. and Lorenz, F. N. (1959). *Am. J. Physiol.* **197**, 47–51.
Henry, K. M., Magee, H. E. and Reid, E. (1934). *J. exp. Biol.* **11**, 58–72.
Hillerman, J. P., Kratzer, F. H. and Wilson, W. O. (1953). *Poult. Sci.* **32**, 332–335.
Huston, T. M., Cotton, T. E. and Carmon, J. L. (1962). *Poult. Sci.* **41**, 179–183.
Hutchinson, J. C. D. (1954). *In* "Progress in the Physiology of Farm Animals" (J. Hammond, ed.), Vol. 1, pp. 299–362. Butterworth, London.
Kibler, H. H. and Brody, S. (1944). *J. Nutr.* **28**, 27–34.
King, J. R. (1957). *NW Science* **31**, 155–169.
King. J. R. and Farner, D. S. (1961). *In* "The Biology and Comparative Physiology of Birds" (A. J. Marshall, ed.), Vol. 2, pp. 215–288. Academic Press, London.
Kleiber, M. (1947). *Physiol. Rev.* **27**, 511–541.
Kleiber, M. (1965). *In* "Energy Metabolism" (K. L. Blaxter, ed.), pp. 427–435. Academic Press, London.
Kleiber, M. and Dougherty, J. E. (1933). *J. gen. Physiol.* **17**, 701–726.
Kleiber, M. and Winchester, C. F. (1933). *Proc. Soc. exp. Biol. Med.* **31**, 158–159.
Krogh, A. (1916). "The Respiratory Exchange of Animals and Man". Longman, London.
Langslow, D. R., Butler, E. J., Hales, C. N. and Pearson, A. W. (1970). *J. Endocr.* **46**, 243–260.
Leeson, S. and Porter-Smith, A. J. (1970). *Br. Poult. Sci.* **11**, 275–279.
Lusk, G. (1928). "Elements of the Science of Nutrition". Saunders, Philadelphia.
March, B. E. and Biely, J. (1957). *Poult. Sci.* **36**, 1270–1277.
Mellen, W. J. (1963). *Agric Sci. Rev.* **1–8**.
Mellen, W. J. and Hill, F. W. (1955). *Poult. Sci.* **34**, 1085–1089.
Mitchell, H. H., Card, L. E. and Haines, W. T. (1927). *J. agric Res.* **34**, 945–960.
Mitchell, H. H. and Haines, W. T. (1927a). *J. agric Res.* **34**, 549–557.
Mitchell, H. H. and Haines, W. T. (1927b). *J. agric Res.* **34**, 927–943.
Mitchell, H. H. and Kelley, M. A. R. (1933). *J. agric. Sci.* **47**, 735–748.
Muglia, G. and Massuelli, L. (1934). *Bull. Soc. ital. Biol. sper.* **8**, 1772–1774.
Nichita, G. and Mircea, I. (1933). *Ann. Inst. zoot. Bucarest.* **2**, 45–66.
Ota, H. and McNally, E. H. (1961). *Agric. Res. Serv U.S.D.A.* **42–43**.
Poczopko, P. and Kowalczyk, J. (1965). *In* "Energy Metabolism" (K. L. Blaxter, ed.), pp. 347–352. Academic Press, London.
Romanoff, A. J. and Romanoff, A. L. (1967). "Biochemistry of the Avian Embryo". Interscience, New York.
Romijn, C. (1950a). *Tijdschr. Diergeneesk.* **75**, 465–480.

Romijn, C. (1950b). *Tijdschr. Diergeneesk.* **75**, 719–746.
Romijn, C. (1950c). *Acta physiol. pharmac. néerl* **1**, 517–519.
Romijn, C. and Lokhorst, W. (1951). *Physiologia comp. oecol.* **2**, 187–197.
Romijn, C. and Lokhorst, W. (1961). *Proc. 2nd Symp. Energy Metabolism, Wageningen*, 49–58.
Romijn, C. and Lokhorst, W. (1963). *Tijdschr. Diergeneesk.* **88**, 1521–1546.
Romijn, C. and Lokhorst, W. (1964). *Zbl. VetMed.* **A11**, 297–314.
Romijn, C. and Vreugdenhil, E. L. (1969). *Neth. J. vet. Sci.* **2**, 32–58.
Simpson-Morgan, M. W. (1965). *J. appl. Physiol.* **20**, 558–560.
Stahl, P. and Turner, C. W. (1961). *Poult. Sci.* **40**, 239–242.
Sturkie, P. D. (1958). *Poult. Sci.* **37**, 495–509.
Tasaki, I. and Sakurai, H. (1969). *Mem. Lab. Anim Nutr., Nagoya Univ.* **4**.
Visschedijk, A. H. J. (1968). *Br. Poult. Sci.* **9**, 173–184.
Winchester, C. F. (1940). *Poult. Sci.* **19**, 239.
Zausch, M. (1965). *In* "Energy Metabolism" (K. L. Blaxter, ed.), pp. 159–163. Academic Press, London.
Zeuthen, E. (1953). *Q. Rev. Biol.* **28**, 1–12.

11 Carbohydrate Metabolism

J. PEARCE and W. O. BROWN

Department of Agricultural Chemistry,
The Queen's University of Belfast
and Ministry of Agriculture,
Northern Ireland

I. Introduction

Until recently little information was available regarding carbohydrate metabolism in birds and this article attempts to review present knowledge of this aspect of metabolism in the fowl. Aspects of embryonic metabolism have been included where they are relevant and highlight particular facets of avian metabolism. Reviews on carbohydrate metabolism in the fowl have been written earlier by Hazelwood (1965) and Vohra (1967).

II. Development of Carbohydrate Metabolism in the Chick from Hatching

A. CARBOHYDRATE CONTENT OF THE CHICK EMBRYO

The egg contains about 500 mg carbohydrate (Romanoff and Romanoff, 1949; Romanoff, 1967); this is approximately 1% of its total weight.

Romanoff (1967) quotes figures which show that of 458 mg total carbohydrate, 145 mg is in the yolk and 313 mg in the albumen. Most of the albumen carbohydrate is "bound" (171 mg), with free glucose (133 mg) and glycogen (4 mg) making up the rest. In the yolk, 70% of the total carbohydrate is free glucose. The "bound" carbohydrates are combined with proteins, in both yolk and albumen, and also with lipid in the yolk (see also Romanoff and Romanoff, 1949).

Only one reference (Biasotti, 1932) has been found regarding the influence of the maternal diet on the carbohydrate content of the hen's egg. This states that the C_{22} fatty acid, behenic acid, decreased the glucose content of the egg.

B. PATHWAYS OF CARBOHYDRATE METABOLISM IN THE CHICK EMBRYO

1. *Glycolysis*

Anaerobic glycolysis in the embryo was first shown by Meyerhof and Perdigon (1940) who demonstrated the presence of its key enzymes and their observations were confirmed by Stumpf (1947) and Novikoff *et al.* (1948). The concentrations of the glycolytic intermediates in the musculature were determined by Arese *et al.* (1967) as well as the glycolytic flow rate which showed a progressive increase with time up to hatching. Increases in the activity of phosphofructokinase, fructose-1,6-diphosphate aldolase, glyceraldehyde-3-phosphate dehydrogenase, enolase and pyruvate kinase in the liver has been noted throughout the period of incubation (Rinaudo, 1962; Wallace and Newsholme, 1967).

2. *The Pentose Phosphate Pathway*

The aerobic pentose phosphate pathway is active, particularly in the early embryo. The activity of this pathway was investigated by measuring the rate of production of $^{14}CO_2$ from [1-^{14}C] glucose and [6-^{14}C] glucose. More $^{14}CO_2$ is obtained from [1-^{14}C] glucose than from [6-^{14}C] glucose if the pentose phosphate pathway is operative but approximately equal amounts of $^{14}CO_2$ are obtained if glycolysis is the sole route of glucose breakdown (Bloom *et al.*, 1953). Using this technique it was found that embryonic chick brain (Liuzzi and Angeletti, 1964), hearts and homogenates of whole embryos (Coffey *et al.*, 1964) utilize the pentose phosphate pathway to a greater extent than the glycolytic sequence. In general, the peak of activity was found between the 4th and 8th d of incubation. However, Liuzzi and Angeletti (1964) reported that the maximum pentose phosphate cycle activity in embryonic brain was between the 12th d and 15th d of incubation. After passing through this maximum level of activity, the pentose phosphate pathway declined in activity with increasing time of incubation until, at the point of hatching, it was virtually inactive as in the adult bird (see also Section II D).

Wenger *et al.* (1967) obtained the same pattern of activity using the chick embryo *in ovo*.

The activity of one of the main enzymes, glucose-6-phosphate dehydrogenase, and also the metabolism of ribose-5-phosphate by chick embryonic tissue have been found to mirror the overall activity of the pentose phosphate pathway (Cazorla and Guzmán Barrón, 1958; Burt and Wenger, 1961). This pathway in the chick embryo is important in providing both pentoses for nucleic acid biosynthesis and NADPH for reductive syntheses involved in differentiation.

3. *The Tricarboxylic Acid Cycle*

The activity of the tricarboxylic acid cycle in embryonic liver increases progressively throughout the period of development; malate dehydrogenase (Solomon, 1958), succinic dehydrogenase and cytochrome oxidase (Davidson, 1957), which can be taken as indicators of tricarboxylic acid cycle activity, show this effect.

4. *Gluconeogenesis*

In the development of the embryo, nutrients are supplied by the yolk independently of a maternal organism. The yolk contains large amounts of fat and only traces of carbohydrate so that the chick embryo grows on a high fat diet as virtually its sole source of energy. This is in contrast to the situation in the fetal mammal where development is supported by maternal glucose.

Glucose is necessary for the growth of the chick embryo and so it is not surprising that gluconeogenesis has been found to be active in its liver. The incorporation of [2-^{14}C]-glutamate and [U-^{14}C]-alanine into glucose, showing the overall activity of this pathway, has been observed in very young embryonic tissues (Yarnell *et al.*, 1966), and one of the key enzymes of gluconeogenesis has been detected in the liver at the 6th d (Kilsheimer *et al.*, 1960). On incubation, the two hepatic enzymes involved in the conversion of pyruvate to phosphoenol pyruvate, pyruvate carboxylase and phosphoenol pyruvate carboxykinase, increase in activity from the 13th d of incubation to maxima on the 17th and 16th d of incubation, respectively, and then show a decrease in activity towards the time of hatching (Nelson *et al.*, 1966; Felicioli *et al.*, 1967). Two other key enzymes of hepatic gluconeogenesis, fructose-1,6-diphosphatase and glucose-6-phosphatase, also increase in activity during incubation (Nelson *et al.*, 1966) and have peaks of activity on the 16th d (Ballard and Oliver, 1963; Kilsheimer *et al.*, 1960).

5. *Glycogen Metabolism*

Glycogen first appears in the chick embryo on the 6th d of incubation (Dalton, 1937); Ballard and Oliver (1963) observed that the liver glycogen content increased from 7 mg/g of tissue at 10 d to 26 mg/g at 19 d. The development of the hepatic enzymes concerned in glycogen synthesis was also investigated by Ballard and Oliver (1963) who found that all the glycogenic enzymes investigated increased in specific activity as incubation progressed. The activity of uridine diphosphate glucose α-glucan glucosyl transferase mirrors the increase in glycogen content and increases three-fold between the 10th and 19th d of incubation. α-Glucan phosphorylase, glucose-1-phosphate uridyl-transferase and phosphoglucomutase showed similar increases in activity over this period of time. The *in vitro* incorporation of [2-^{14}C]-pyruvate and [U-^{14}C]-glucose into glycogen by embryonic chick liver slices was variable and Ballard and Oliver (1963) concluded that this was due to the extremely fragile nature of the liver slices. However, pyruvate proved as good a substrate as glucose for glycogen formation; this would be expected since the gluconeogenic pathway for the formation of glucose from such precursors as pyruvate is active in chick embryonic liver (see Section II B4). It is noteworthy that fetal rat liver does not have the ability to synthesize phosphoenolpyruvate from malate; this pathway develops only after birth (Ballard and Oliver, 1963). The absence of the gluconeogenic pathway in fetal rat liver is also reflected in the lack of incorporation of pyruvate into glycogen by fetal tissue. In the rat, gluconeogenesis and glycogenesis develop rapidly after birth.

C. DEVELOPMENT OF PATHWAYS OF CARBOHYDRATE METABOLISM AFTER HATCHING

1. *Glycolysis*

Both phosphofructokinase and pyruvate kinase activities increase in embryonic liver during development so that the capacity to utilize carbohydrate is available after hatching when the animal is presented with a carbohydrate-rich diet (Wallace and Newsholme, 1967). The activities of these enzymes increased up to at least 8 d after hatching; much lower activities were found in adult birds (Table 1).

Rinaudo (1962) found that the activities of fructose-1,6-diphosphate aldolase, glyceraldehyde-3-phosphate dehydrogenase, phosphohexose isomerase and enolase were lower in the livers from 3-month-old birds than in newly-hatched chicks. Goodridge (1968a) also noted that hepatic glyceraldehyde-3-phosphate dehydrogenase and α-glycerophosphate dehydrogenase decreased in activity between 16 and 28 to 30 d of age.

However, hexokinase was very inactive in embryonic liver and there was no detectable glucokinase activity (Wallace and Newsholme, 1967). Hexokinase activity increased 4 d after hatching and increased progressively throughout the period of the experiment. It is noteworthy that the activity of hexokinase was higher in the adult than in the newly-hatched bird whereas, with all the other enzymes investigated, the converse applied. Wallace and Newsholme (1967) found the glucokinase activity in the adult bird to be only

Table 1

Changes in hepatic glycolytic enzymes with age

Age of chick (days)	Hexokinase	Phosphofructokinase	Pyruvate kinase	Glyceraldehyde-3-phosphate dehydrogenase (Change in extinction at 340 nm/2 min/0·1 ml of extract)	Phosphohexiosomerase (μg fructose-6-phosphate formed/4 min/0·1 ml extract)	Fructose-1,6-diphosphate aldolase (μg P formed/5 min/0·3 ml extract)
	(μmoles/h mg liver)					
−5	—	35 ± 8	222 ± 41	0·0200	—	47·0
−1	5·2 ± 1·1	48 ± 19	215 ± 24	0·0410	—	—
0 (day of hatching)	—	—	—	—	110·5	—
+1	—	—	—	0·0162	—	—
+1½	—	—	—	0·0110	—	—
+2	2·8 ± 0·5	68 ± 13	243 ± 33	—	—	—
+4	12·7 ± 2·4	136 ± 24	365 ± 52	—	—	—
+8	16·4 ± 1·9	300 ± 48	720 ± 110	—	—	—
3 months	—	—	—	0·0156	87·9	11·1
Adult	25·5 ± 3·5	81 ± 21	390 ± 32	—	—	—

Results from Rinaudo (1962) and Wallace and Newsholme (1967).

25% that of hexokinase; in the rat, however, glucokinase exceeds hexokinase about three-fold (Viñuela *et al.*, 1963). Pearce (1970) has also noted that glucokinase activity was only 30% that of hexokinase in 8-week-old pullets.

2. *Pentose Phosphate Pathway and Tricarboxylic Acid Cycle*

Very little information is available on the variation of activity of these pathways with age. The activities of liver and adipose tissue pentose phosphate pathway dehydrogenases change little between the 10th d of incubation and 26 d after hatching (Goodridge, 1968a). Duodenal succinic dehydrogenase activity increased at 20 d of incubation and continued to rise to a peak 5 d after hatching (Nunnally, 1962). Its activity then declined but remained much higher than in the embryo. The activity of this enzyme did not vary greatly between 6 d and 1 month after hatching and was equivalent to 4 to 5 times the embryonic level of activity.

However, as stated above, information on the activities of these two pathways may be obtained by investigation of the rate of release of $^{14}CO_2$ from [1-^{14}C]-glucose and [6-^{14}C]-glucose. Such studies have shown that glycolysis and the tricarboxylic acid cycle are the predominant pathways of glucose catabolism in the hatched bird with the pentose phosphate pathway playing only a minor role. Examination of the absolute amounts of $^{14}CO_2$ evolved from these labelled substrates (Table 2) indicate that the specific activities of these pathways do not vary appreciably with age (see also Section II E).

3. *Gluconeogenesis*

In the mammalian fetus a constant concentration of plasma glucose is maintained by the maternal blood supply and the fetus uses this monosaccharide as an energy source for growth.

After birth, the mammalian organism is presented with a high fat, low carbohydrate diet (milk) and for some months forms much glucose by

Table 2

The oxidation of [1-^{14}C]-glucose and [6-^{14}C]-glucose by liver slices from immature pullets, laying pullets and cockerels

Physiological state	Oxidation[a] of [1-^{14}C]-glucose to $^{14}CO_2$	Oxidation[a] of [6-^{14}C]-glucose to $^{14}CO_2$	$\frac{\text{C-1}}{\text{C-6}}$ ratio
Immature pullet (11–12 weeks)	743 ± 194	726 ± 177	1·02 ± 0·06
Laying pullet	772 ± 324	824 ± 301	0·94 ± 0·09
Cockerel	1144 ± 203	1053 ± 195	1·09 ± 0·01

[a] Counts per min per 10 mg dry tissue per h.
Results from Duncan and Common (1967).

gluconeogenesis. It is, therefore, not unexpected that the hepatic enzymes associated with gluconeogenesis (pyruvate carboxylase, phosphoenol pyruvate carboxykinase, glucose-6-phosphatase and fructose-1,6-diphosphatase) increase in activity during this stage of development and decrease again after the suckling period (Ballard and Oliver, 1965; Yeung *et al.*, 1967). Glycogenesis is also high during this period in the mammal (Ballard and Oliver, 1963).

In the chick the reverse situation holds since the diet of the embryo has a high fat content with only traces of carbohydrate; consequently the gluconeogenic and glycogenic pathways are highly active and decrease after hatching. Thind *et al.* (1966) found that the activity of liver fructose-1,6-diphosphatase decreased from the 18th day of incubation to about 20 d after hatching and then remained more or less constant up to 6 months of age. Hepatic glucose-6-phosphatase showed a similar pattern of activity. In both the liver and the kidney these two enzymes had much lower activities at 2 years of age than at 2 weeks. Nelson *et al.* (1966) and Wallace and Newsholme (1967) also found glucose-6-phosphatase and fructose-1,6-diphosphatase to be less active in adult birds than in embryos (Table 3).

Table 3

Changes in activity of gluconeogenetic enzymes after hatching

Age of chick	Glucose-6-phosphatase	Fructose-1,6-diphosphatase
	(μmoles/g h)	
20th d of incubation	296 ± 31	2177 ± 199
2 d after hatching	349 ± 6	2540 ± 145
8 d after hatching	229 ± 9	1446 ± 150
Adult	108 ± 9	1149 ± 121

Results from Wallace and Newsholme (1967).

4. *Glycogen Metabolism*

At the end of incubation the stored glycogen is rapidly mobilized (Muglia and Massuelli, 1934; Gill, 1938; Freeman, 1965, 1969) falling from 19·0 mg/g liver in the embryo after 18½ d incubation to 1·6 mg/g liver 1 d after hatching (Freeman, 1965); there is a concomitant rise in blood sugar over this period (Table 4). The cardiac glycogen also decreased between 18½ d incubation and 1 d after hatching. Glycogen probably provides energy for hatching.

Shulyak (1967) showed that hepatic glycogen in chicks was low immediately after hatching and remained low throughout the first month; the maximum

Table 4

Changes in tissue glycogen and blood sugar content over the period of hatching

Age of birds	Blood sugar[a] (mg/100 ml)	Liver glycogen (mg/g)	Cardiac glycogen (mg/g)
$18\frac{1}{2}$ d incubation	107 ± 2·0	19·0 ± 0·7	4·9 ± 0·4
Just hatched	129 ± 3·4	3·0 ± 0·5	3·3 ± 0·4
1 d after hatching	133 ± 2·3	1·6 ± 0·5	2·0 ± 0·3

[a] These results must assume no marked change in the volume occupied by cells in the blood (see Chapter 39).

Results from Freeman (1965).

level was observed with the attainment of sexual maturity. During ovulation, hepatic glycogen decreased; Shulyak (1967) suggested that this was due to the increased energy demands of egg laying. He found that the utilization of liver glycogen is also reflected in the activity of hepatic phosphorylase which breaks down glycogen to glucose-1-phosphate. The activity of phosphorylase was high in chicks aged 1 to 3 d but in sexually mature birds was lower. During ovulation, an increase in phosphorylase activity occurred; during moulting it is reduced by over 50% (Table 5).

Table 5

Variation of hepatic glycogen and phosphorylase activity in birds of different ages

Age of birds	Glycogen (mg/g liver)	Phosphorylase activity (in μg P utilized/100 mg liver)
1 month	0·98 ± 0·14	114·45 ± 8·48
4 to 5 months (attainment of maturity)	5·33 ± 0·88	72·89 ± 4·96
14 months (in lay)	0·50 ± 0·04	105·85 ± 7·66
18 months (moulting)	1·71 ± 0·03	43·46 ± 3·35

Results from Shulyak (1967).

The metabolism of oviducal glycogen was also investigated by Shulyak (1967) who found that both the utilization of glycogen and phosphorylase activity were higher during laying than during moulting.

D. CARBOHYDRATE METABOLISM AFTER HATCHING IN RELATION TO LIPOGENESIS

Lipogenesis in mammals is active in fetal liver, but after birth it decreases markedly, presumably because of the high fat content of the mother's milk

(Villee and Hagerman, 1958; Roux, 1966; Ballard and Hanson, 1967). In the chicken, the reverse situation occurs and the rate of embryonic lipogenesis is very low (Schoenheimer and Rittenberg, 1930; Kilsheimer *et al.*, 1960).

The incorporation of the carbon of [U-^{14}C]-glucose into fatty acids and cholesterol is very low in embryonic chicken liver but increases rapidly on hatching and feeding (Goodridge, 1968b); in the embryonic liver this incorporation was from 0·5 to 0·8 ng atoms of C/mg N/h. After 24 h of feeding, glucose-carbon incorporation into fatty acids increased 58-fold and continued to increase until a plateau was reached 7 to 8 d after hatching when it was about 1,000 times that in embryonic tissue. Cholesterol synthesis from tracer glucose is also very low in embryonic liver and remains low until 5 d after hatching when the rate increases 40-fold from 6 to 30 d of age.

The oxidation of glucose during this period follows the same pattern; the rate in livers of embryonic and newly-hatched chicks is low and increases five-fold 24 h after hatching when the birds were fed. In 5-day-old chicks the rate of glucose oxidation reached a plateau at about 10 times that of the embryo.

Fatty acid synthesis, cholesterol synthesis, glucose oxidation and glycogen synthesis were all found to be highly active in chick liver (Goodridge, 1968b) and it is notable that these were respectively, 100, 20, 4 and 11 times as high as in rat liver (Cahill *et al.*, 1958; Ballard and Hanson, 1967).

In rats and some other mammals fatty acid synthesis can take place in both liver and adipose tissue with the latter as the predominant site (Feller, 1954; Hausberger *et al.*, 1954; Favarger, 1965; Jansen *et al.*, 1966; Leveille, 1967). In the mouse and rat, adipose tissue accounts for at least 50% and in some cases as much as 95% of the fatty acids synthesized. Gibson and Nalbandov (1966) and O'Hea and Leveille (1968) observed that chicken adipose tissue could incorporate [U-^{14}C]-glucose and [1-^{14}C]-acetate into fatty acids. However, a comparison of rat and chicken adipose tissues (Table 6) showed

Table 6

Relative rates of fatty acid biosynthesis from different substrates by chick and rat adipose tissue

Substrate	Animal	Fatty acid (μmoles/100 mg tissue/2 h)
[U-^{14}C]-glucose	Chick	0·5 ± 0·1
	Rat	34 ± 6
[1-^{14}C]-acetate	Chick	116 ± 24
	Rat	888 ± 87
[2-^{14}C]-pyruvate	Chick	20 ± 2
	Rat	914 ± 127

Results from O'Hea and Leveille (1968).

that radioactive glucose, pyruvate and acetate were utilized for fatty acid biosynthesis much more rapidly in rat adipose tissue (O'Hea and Leveille, 1968). The incorporation of [U-^{14}C]-glucose into fatty acids was much higher in chick liver than in its adipose tissue (Goodridge, 1968b; Leveille *et al.*, 1968) the rate being ten-fold greater than in adipose tissue in 7-day-old chicks and 160 times greater in 28- to 30-day-old chicks (Goodridge, 1968b). Goodridge (1968b) also noted that the incorporation of glucose carbon into fatty acids in developing chickens is 1,000 times greater in 7-day-old birds than in embryos whereas in adipose tissue the increase in lipogenesis over this period was only fourteen-fold. Goodridge (1968c) and O'Hea and Leveille (1969) calculated that the lipogenic contribution of the liver was between 90 and 95% in the chick. Similar results were obtained in the pigeon by Goodridge and Ball (1966, 1967a) who concluded that adipose tissue was not an important site of lipogenesis and might account for only 4% of the fatty acids synthesized.

These observations were extended by Goodridge (1968a) who examined the activities of several liver and adipose tissue enzymes involved in glucose metabolism and the conversion of glucose to lipid. In non-ruminant animals, citrate is the main source of carbon for fatty acid synthesis and is converted by ATP-citrate lyase to yield oxaloacetate and acetyl-CoA which is then used for lipogenesis (Srere, 1959; Spencer and Lowenstein, 1962). The activity of ATP-citrate lyase varies with the nutritional state of the animal (Lowenstein, 1966); in general it is high in situations where fatty acid synthesis is taking place and is low when a diet containing a high proportion of fat is fed. Its activity has thus been taken as an indication of the extent of lipogenesis in a tissue. Fatty acid biosynthesis requires NADPH and Wise and Ball (1964) suggested that this could arise through the activity of the "malic" enzyme.

Goodridge (1968c) found the activities of ATP-citrate lyase and the "malic" enzymes to be very low in adipose tissue and remained low from the late-embryonic stage until at least 28 d after hatching. Although the hepatic activities of these enzymes were low in embryonic tissue there was a rapid increase on feeding the hatched bird reaching maximum activities at 1 week of age (6 d of feeding). After this time Goodridge (1968c) found no significant variation in activity throughout the rest of the experimental period (up to 4 weeks of age). Similar patterns of development of both ATP-citrate lyase and the "malic" enzyme (Fig. 1) in both pullets and cockerels have been observed by W. O. Brown (unpublished) and of ATP-citrate lyase by Felicioli and Gabrielli (1967). Ryder (1970) has also demonstrated that the pattern of development of the key lipogenic enzyme, acetyl-CoA carboxylase, is parallel to those of ATP-citrate lyase and the "malic" enzyme. O'Hea and Leveille (1968) confirmed that "malic" enzyme activity is greater in the liver than in the adipose tissue.

The increases in hepatic lipogenesis which occur when the newly-hatched chick is fed on a normal high carbohydrate diet can be duplicated by force-feeding 1-day-old birds with glucose (Goodridge, 1970). In an attempt to elucidate the mechanism Goodridge (1970) found that a single glucose meal stimulated the incorporation of acetate into fatty acids by liver slices but if

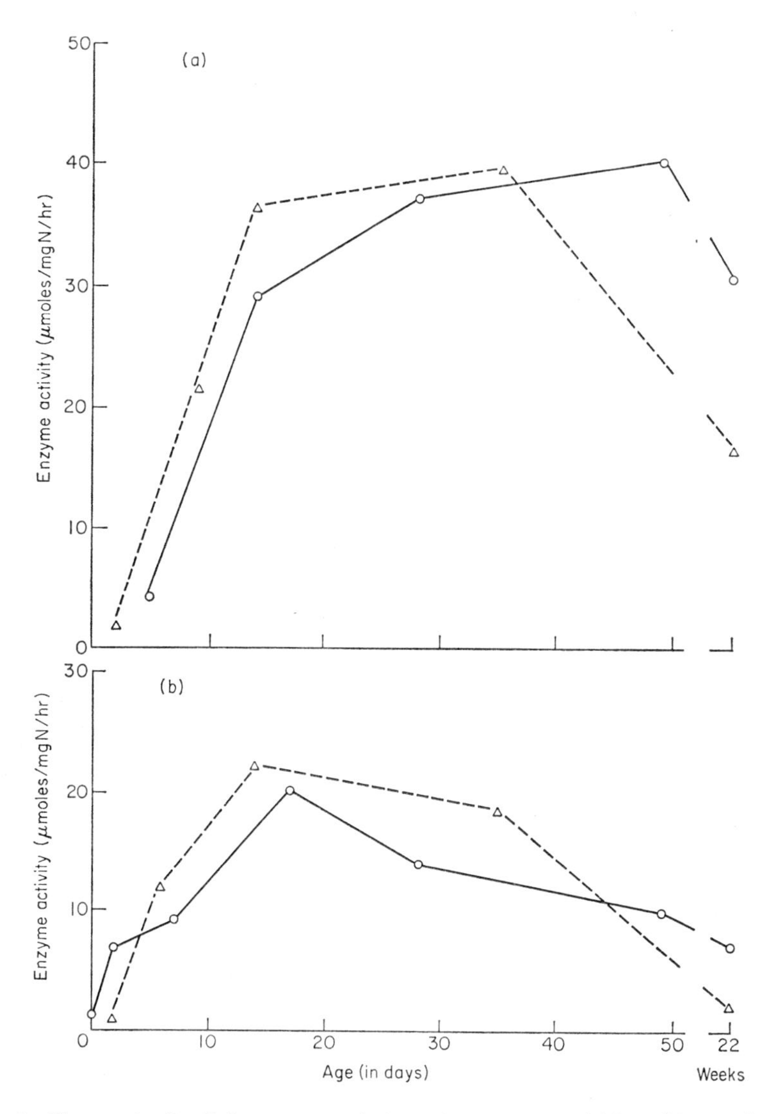

Fig. 1. Changes in "malic" enzyme and ATP-citrate lyase activities with age in the fowl. (W. O. Brown, unpublished.) (a) "Malic" enzyme activity in pullets (–○–○–) and cockerels (–△–△–); (b) ATP-citrate lyase activity in pullets (–○–○–) and cockerels (–△–△–).

the glucose were added to liver slices from unfed birds *in vitro* it had no effect. However both fructose and glycerol *in vitro* stimulated lipogenesis from acetate by liver slices suggesting that these substrates bypassed a rate-limiting step between glucose and triose phosphate. The presence of such a bypass in the metabolism of fructose by chick liver is shown by the results of Pearce (1970) (see Section III D). The administration of glucose probably stimulated or overcame this rate-limiting step *in vivo*. The stimulation of fatty acid synthesis caused by fructose did not require the synthesis of new enzyme protein. This observation was unexpected and Goodridge (1970) suggested that fatty acid synthesizing enzymes were present in a latent form in the liver from unfed 1-day-old chicks.

NADPH is essential for fatty acid biosynthesis and Duncan and Common (1967) investigated the pentose phosphate pathway by determining the $^{14}CO_2$ production from [1-^{14}C]-glucose and [6-^{14}C]-glucose by chicken liver slices. These workers concluded that the pentose phosphate pathway is not operative to any great extent in immature and laying fowl and that this pathway is not an important source of NADPH for fatty acid formation. Enzyme studies also suggest that the pentose phosphate pathway plays a minor role in hepatic lipogenesis in the chicken (Goodridge, 1968a, d) and in the pigeon (Goodridge and Ball, 1966, 1967b).

E. CARBOHYDRATE METABOLISM ASSOCIATED WITH DIFFERENT PHYSIOLOGICAL STATES

Fructose stimulates oxygen uptake by chicken liver slices from birds aged between 5 and 48 weeks and in both pullets and cockerels (Heald, 1962, 1963). Aerobically, the "composite curve" of the summation of CO_2 output and O_2 uptake was consistently positive when fructose was the substrate and usually negative in the absence of added substrate or in the presence of glucose. The same general pattern was found with young pullets of 4 to 10 weeks of age, mature laying birds, birds in moult and mature and immature cockerels. Similarly the respiratory quotient (RQ) of liver slices when fructose was the substrate was greater than unity in immature and laying pullets and in cockerels (Table 7).

Hawkins and Heald (1966) found that more triglyceride was synthesized by liver slices in the presence of fructose than in the presence of glucose and that a greater amount of palmitate was incorporated into neutral lipids by mature laying birds (aged 22 to 27 weeks) than by immature pullets (aged 8 to 17 weeks). Immature pullets (aged 11 weeks) injected intramuscularly with oestradiol (2 mg on alternate days over 8 d) gave similar results in that more palmitate was incorporated into triglyceride when compared with untreated controls.

Duncan and Common (1967) investigated the production of $^{14}CO_2$ from [1-^{14}C]-glucose and [6-^{14}C]-glucose by liver slices in relation to the production of NADPH for lipogenesis; the ratio of $^{14}CO_2$ from the two substrates

Table 7

Respiratory quotient (RQ) of liver slices from the domestic fowl

Age of birds (weeks)	Physiological state	Substrate	RQ
4·5–13·5	Non-laying	Fructose	1·13 ± 0·04
	Fed	Glucose	0·94 ± 0·01
		Nil	0·86 ± 0·03
20–55	In lay	Fructose	1·10 ± 0·02
	Fed	Glucose	0·85 ± 0·12
		Nil	0·77 ± 0·10
16–30	Cockerel	Fructose	1·05
	Fed	Glucose	0·85
		Nil	0·73

Results from Heald (1963).

$\left(\frac{\text{C-1}}{\text{C-6}}\text{ ratio}\right)$ was close to unity for both immature and mature pullets and cockerels (Table 2). These workers also determined the $\frac{\text{C-1}}{\text{C-6}}$ ratio in pullets over the period from 15 to 21 weeks of age but found no evidence of any increase in this ratio with the onset of maturity (Table 2). Similarly, pretreatment of sexually immature pullets with oestradiol (1 mg/d for periods of 1 to 10 d) by intramuscular injection, or the *in vitro* addition of oestradiol (8 or 80 μg) to liver slices, had no stimulatory effect on the oxidation of [1-^{14}C]-glucose and the $\frac{\text{C-1}}{\text{C-6}}$ ratio remained close to unity (Table 8). Duncan (1968) found no differences in the incorporation of ^{14}C from [1-^{14}C] glucose, [6-^{14}C]-glucose and [U-^{14}C]-glucose into CO_2, total lipid, fatty acids and glyceride-glycerol by liver slices from immature and laying birds. These results suggest that little or no modification of carbohydrate metabolism occurs when sexual maturity is reached.

The activities of ATP-citrate lyase enzyme and the "malic" enzyme in the livers of both pullets and cockerels were essentially similar during development. However, with the onset of sexual maturity, quantitative differences in the activities of these enzymes were observed between the sexes (W.O. Brown, unpublished). In the laying hen the specific activities of the ATP-citrate lyase and "malic" enzyme were (mean ± SE) 5·31 ± 0·97 and 77·95 ± 3·87 μmoles/mg N/h whereas in the mature cockerel the corresponding activities were 1·85 ± 0·29 and 45·25 ± 4·23 μmoles/mg N/h (each result was the mean of 8 observations). In the case of one laying bird which became broody and stopped laying, ATP-citrate lyase activity was

Table 8

The effect of oestradiol on the oxidation of [1-^{14}C]-glucose and [6-^{14}C]-glucose by liver slices from immature pullets (12–13 weeks)

	$\frac{C\text{-}1}{C\text{-}6}$ ratio of $^{14}CO_2$
Control (no oestradiol)	1·16
	0·99
Oestradiol-treated	1·04
	1·18

Results from Duncan and Common (1967).

1·97 μmoles/mg N/h and the "malic" enzyme activity was 44·74 μmoles/mg N/h, figures very similar to those for the mature cockerel. These results suggest that the activities of these enzymes reflect demands for hepatic lipogenesis. The laying hen has a massive lipid requirement for egg laying whereas there is no such demand in the cockerel or the non-laying pullet.

III. Metabolism of Specific Sugars

A. EFFECTS OF DIFFERENT CARBOHYDRATES ON CHICK GROWTH

In the chicken, the protein content of the diet has been shown to be involved in the comparative utilization of different carbohydrates. Monson *et al.* (1950) found that the growth-rate followed a decreasing order with carbohydrate provided by dextrin, "cerelose" (glucose monohydrate), sucrose and lactose when the protein supplied was low; but this difference was not shown when the dietary protein content was increased (Monson *et al.*, 1954). Similar results were obtained by Rutter *et al.* (1953) and Griminger and Fisher (1963) who found no differences in growth rate when the dietary carbohydrate was supplied by corn, dextrin, sucrose, maltose, glucose or fructose.

Lactose at 20% or above depressed the growth rate and caused diarrhoea and crooked toes (Rutter *et al.*, 1953). Examination of the blood showed a normal glucose content and also the presence of small amounts of lactose. This sugar at a dietary concentration of 15% depressed egg production; at a concentration of 63% egg production was completely inhibited. Shannon *et al.* (1969) found that lactose stimulated the fasting respiratory rate indicating that this sugar can be utilized by the laying hen although the effect was very slight. These authors also noted that, during starvation, the "metabolizable energy" (ME) of lactose was inefficiently utilized to replace body tissue as a source of energy and suggested that the energy which was obtained may be attributable to intestinal fermentation. Siddons (1969) detected lactase activity (equivalent to 426 μmole metabolized/h/total intestine) in the chick intestine. However, caecal lactase activity was much lower in germ-free chicks than in normals which confirms the view that lactose is mainly metabolized in the gut by bacterial lactase. Lactose is presumably unable to support growth because of the limited rate of breakdown to its component monosaccharides, D-glucose and D-galactose.

A number of polysaccharides including guar gum, locust gum, gum tragacanth, gum karaya, carrageenin and pectin depress growth in the chick (Sathe and Bose, 1962; Vogt and Penner, 1963; Vohra and Kratzer, 1964a; Couch *et al.*, 1965). The addition of suitable enzymes which can hydrolyse guar gum, locust gum and pectin were found to alleviate the depression of growth (Vohra and Kratzer, 1964a, b, 1965) by improving the "metabolizable energy" of the diets (Kratzer *et al.*, 1967).

B. GALACTOSE TOXICITY

In contrast to other dietary carbohydrates, D-galactose can be tolerated by the hen only up to a level of 10%. Above this, the sugar causes epileptiform seizures and sometimes death with a high level of galactose in the blood (Rutter *et al.*, 1953). Very high dietary galactose (55%) causes, also, a low hepatic glycogen content, serious kidney damage, convulsions and death but no noticeable change in blood glucose or muscle glycogen (Dam, 1944; Sondergaard *et al.*, 1957; Rigdon *et al.*, 1963). Considering the importance of UDP-monosaccharide derivatives in carbohydrate metabolism, Hansen *et al.* (1956) isolated the UDP-hexose from the livers of chicks fed galactose or glucose and found that the ratio of UDP-glucose to UDP-galactose was 3 to 1 for glucose-fed birds but in those fed galactose, UDP-glucose was found only in trace amounts.

It has been observed that female birds are far more sensitive to galactose than males (Nordin *et al.*, 1960). Diethylstilboestrol reduced the sensitivity of female chicks to galactose while testosterone increased it. In male chicks diethylstilboestrol and testosterone had no effect on either growth or survival. The rates of intestinal absorption of D-galactose by starved male and female chickens are very similar but the male is able to utilize the sugar far more readily than the female. Nordin *et al.* (1960) reported that red blood cells from male chickens metabolize galactose more readily than do the red blood cells from female birds, whereas glucose was metabolized at approximately the same rate by both sexes. These observations may have been due to contamination of the red blood cells with white blood cells since Shields *et al.* (1964) and Bell and Culbert (1968) have shown that erythrocytes from the domestic fowl do not metabolize measurable amounts of glucose.

The phenomenon of greater mortality of females on feeding galactose has been investigated by Mayes *et al.* (1970). A diet containing 18% galactose was fed to newly-hatched chicks and it was observed that galactose-1-

phosphate accumulated in the brains of birds of both sexes but a much greater concentration was found in the brains of the females. Brain was the tissue studied because the symptoms of galactose toxicity are mainly neurological. In order to try to explain this sex difference these workers determined the activity of the enzymes galactokinase, UDP-galactose-4-epimerase and galactose-1-phosphate uridyl transferase in the brain of male and female chickens. Galactokinase and UDP-galactose-4-epimerase had similar activities in both male and female chicks but galactose-1-phosphate uridyl transferase activity was significantly lower in tissue extracts from the female chick. Mayes *et al.* (1970) suggested that this difference would account for the increased accumulation of galactose-1-phosphate in females and thereby explain the higher mortality in females.

The intestinal absorption of D-galactose has been shown to be an active process and Bogner (1957) found that the absorption coefficient for this sugar in the 1-d-old chick was greater than for glucose. The toxic effect of galactose may thus be due to a depressant effect on glucose utilization. Administration of uridine and UDP-glucose (the coenzyme involved in the interconversion of galactose and glucose) reduces the blood galactose concentration but it returns to its high values afterwards and there is no alteration in the toxic state (Rutter *et al.*, 1953). This would indicate that galactose poisoning is not due simply to a lack of co-factors or the absence of an enzyme necessary for the conversion of galactose to glucose.

Kozak and Wells (1969) investigated the biochemistry of galactose toxicity in the chick brain. Male chicks were fed a diet containing 40% galactose and it was found that the ATP and phosphocreatine contents of the brains of galactose-fed birds decreased compared with control birds; at the same time increases in AMP, ADP and inorganic phosphate were found suggesting that ATP and phosphocreatine are metabolized more rapidly in the galactose-poisoned chick. The concentrations of all the glycolytic intermediates examined (glucose, glucose-6-phosphate, fructose diphosphate, 3-phosphoglycerate, α-glycerophosphate and lactate) were decreased. The formation of brain phospholipids was lower in galactose-fed chicks than in corresponding controls which Kozak and Wells (1969) suggest may be linked to the seizure and quivering syndrome of galactose toxicity.

These results suggest that one of the main reasons for galactose toxicity is that this sugar interferes with the normal glycolytic and other energy-yielding processes of the cell.

C. METABOLISM OF XYLOSE AND ARABINOSE

D-Xylose can be tolerated in the diet of the chicken up to a level of 10% without significantly affecting growth and food efficiency (Wagh and Waibel, 1966) but at the 20% level xylose depresses growth. On the other hand, 20% L-arabinose has no effect on the growth and food efficiency whereas a 40% dietary level does depress growth. Severe glycogen depletion occurs in the liver and muscles on feeding these pentoses and a probable explanation is that xylose and arabinose cannot be utilized or converted into glucose by the chick, at least not at rates sufficient to satisfy its requirement for glucose, and the deficit is made up by breaking down the glycogen reserves. Feeding diets containing 40% xylose or arabinose to chickens leads to an increased plasma uric acid level; this may point to increased gluconeogenetic protein catabolism in a situation of energy deficiency.

The ingestion of xylose and arabinose by chickens results in their appearance in the blood in direct relation to the dietary pentose levels (Wagh and Waibel, 1967a). Greater amounts of xylose than arabinose are found in the blood and this may be due to the faster absorption of xylose from the intestine (Wagh and Waibel, 1967b). Alvarado (1967) has also shown that xylose is taken up by the chicken's small intestinal mucosa by an active mechanism. Moreover, D-xylose can clearly be metabolized by the bird since this pentose increased the RQ of starved birds although their increase in heat production indicated that the sugar is poorly utilized (Shannon *et al.*, 1969).

The occurrence of both xylose and arabinose in hen's urine has been demonstrated by Bell (1967). These sugars may arise from either dietary or endogenous sources in the hen; in the human these pentoses are found in urine from fasting subjects (Hartmann *et al.*, 1953; Woolf and Norman, 1957; D. J. Bell and M. Q. K. Talukder unpublished), suggesting that they arise endogenously.

Very little is known about the metabolism of specific pentoses in the hen but D-xylose is incorporated into a glycoprotein by a particulate system from the hen oviduct (Grebner *et al.*, 1966). The synthesis of UDP-xylose from UDP-glucuronic acid has been shown in avian oviduct (Bdolah and Feingold, 1965) and the UDP-xylose formed can be a xylosyl donor for glycoprotein synthesis. UDP-xylose functions as an allosteric, feedback inhibitor of its own synthesis by influencing the activity of UDP-glucose dehydrogenase (Bdolah and Feingold, 1968); this enzyme catalyses the formation of UDP-glucuronate, the immediate precursor of UDP-xylose.

D. FRUCTOSE METABOLISM

The metabolism of fructose has been of great interest because of its relative absence in avian tissues and, on the other hand, its appreciable rate of absorption in the intestine. Heald (1962, 1963) and Hawkins and Heald (1966), during investigations on the carbohydrate and lipid metabolism of mature and immature domestic fowls, noted that the metabolisms of fructose and glucose by avian liver slices were quantitatively different. Fructose stimulated the rate of respiration of liver slices more than did glucose and also maintained a respiratory quotient (RQ) of greater than unity. Liver slices incubated in the presence of fructose accumulated more lipid than in the presence of glucose indicating a greater rate of lipogenesis from fructose than from glucose (Hawkins and Heald, 1966). Kritchevsky *et al.* (1959) and

Grant and Fahrenbach (1959) have shown that the hepatic cholesterol concentration in chickens was greater after feeding sucrose than after feeding glucose or starch, suggesting increased cholesterol synthesis due to the fructose moiety of the sucrose.

The relative rates of lipogenesis from fructose and glucose were investigated at the enzymic level by Pearce (1970). Although feeding fructose resulted in a greater total lipid content of the liver, there was no significant variation in the specific activities of the lipogenic enzymes, acetyl CoA-carboxylase, ATP-citrate lyase and the "malic" enzyme, or in the associated glycolytic enzymes, pyruvate kinase and lactic dehydrogenase. This indicated that none of these enzymes were rate-limiting steps in lipogenesis from fructose and glucose. However, ketohexokinase activity was significantly increased on feeding fructose as compared with glucose, and this enzyme had a much higher activity than hexokinase and glucokinase combined. It may be presumed that fructose was predominantly phosphorylated to fructose-1-phosphate far more rapidly than glucose was phosphorylated to glucose-6-phosphate. The presence of an aldolase capable of splitting fructose-1-phosphate was detected in chicken liver in which Heinz and Weiner (1969) have also reported the presence of ketohexokinase. The chicken's liver appears not to differ qualitatively from mammalian liver where fructose is similarly phosphorylated to fructose-1-phosphate and subsequently split by an aldolase to yield dihydroxyacetone phosphate and glyceraldehyde.

The increased rates both of respiration and of lipogenesis with fructose as compared with glucose, may thus reflect the relative ease with which the chicken can convert fructose to acetyl-CoA. The metabolism of fructose via fructose-1-phosphate would account for these differences.

IV. Effect of Dietary Modification on Carbohydrate Metabolism

A. PHYSIOLOGY AND BIOCHEMISTRY OF FEEDING "CARBOHYDRATE-FREE" DIETS TO THE CHICK

Carbohydrate can be completely replaced in the diet of the domestic fowl by triglyceride (soybean oil) without affecting the growth rate (Renner, 1964); however, if the sole non-protein energy source in the diet is soybean oil *fatty acids*, there is a marked depression in the growth rate suggesting that the absence of the glycerol moiety of the triglycerides was responsible (Renner and Elcombe, 1964; Brambila and Hill, 1966). The addition of glycerol to the soybean fatty acid diet increased the rate of growth to that of birds receiving a glucose control diet or the soybean oil diet. The same response was obtained by addition of glucose to the soybean fatty acid diet (Renner and Elcombe, 1964).

Subsequently, Renner (1966) found that dietary fructose, galactose, sorbitol, sucrose, xylose, dextrin and starch were as efficient as glucose or glycerol in alleviating the depressed growth rate which arose on feeding soybean fatty acids. Complete absence of dietary carbohydrate in the soybean fatty acid diet also leads to dermatitis and beak deformity in addition to the decreased growth rate (Brambila and Hill, 1966). These conditions are not relieved by the addition of glucose and indicate the essential nature of carbohydrates in the diet of the chick.

Chicks fed carbohydrate-free diets containing either soybean oil or its fatty acids maintained normal levels of blood glucose and muscle glycogen but the liver glycogen was markedly decreased. Also, the level of ketone bodies increased and the amount of liver fat decreased (Table 9) when the soybean *fatty acid* diet was fed whereas these parameters remained constant when soybean oil was the sole source of non-protein energy (Renner and Elcombe, 1967a). That the liver glycogen is reduced indicates that gluconeogenesis is not sufficient to maintain the former on a carbohydrate-free diet. The muscle glycogen remained unaltered indicating that in the chick, as in other species (Bollman *et al.*, 1925) muscle cannot supply glucose when there is a demand for carbohydrate in other parts of the body.

Renner and Elcombe (1967b) investigated the effectiveness of extra dietary protein in promoting the growth of chicks receiving a soybean fatty acid diet; the birds' requirement for carbohydrate could be met by protein but this was much less efficient in meeting the demand than was glucose. The removal of excess glucogenic amino acids from the diet by substitution with a suitable mixture of amino acids resulted in both a marked reduction in growth and nitrogen retention; this was overcome by the addition of glucose indicating that the excess glucogenic amino acids in soybean protein contribute to the carbohydrate requirement of the chick. It also shows that the ability of the chick to synthesize carbohydrate from fatty acids must be very limited, if it is indeed present.

Non-essential amino acids are synthesized from carbohydrate precursors and Renner (1969) investigated whether or not chicks fed carbohydrate-free diets could synthesize these amino acids when provided with a non-essential source of nitrogen. Glutamic acid and aspartic acid proved to be the main sources of non-essential nitrogen when non-protein energy was supplied by soybean oil or glucose, but the requirement was greater when the soybean oil diet was fed. L-Aspartic acid was as effective a nitrogen source as L-glutamic acid whereas diammonium citrate was less effective; DL-serine and DL-alanine, like ammonium acetate, could not act as the sole source of non-essential nitrogen in chicks fed the carbohydrate-free diet. Since non-essential amino acids arise from carbohydrate precursors these observations further support the view that the chick is very limited in its ability to convert lipid to carbohydrate.

As pointed out by Vohra (1967), the acetyl-CoA formed from fatty acids as a result of β-oxidation can be metabolized through the Krebs cycle to yield energy without any net synthesis of carbohydrate from acetate.

Table 9

The effect of feeding "carbohydrate-free" diets on various tissue constituents and enzyme activities

Source of non-protein energy	Muscle glycogen (% wet weight)	Blood glucose (mg/100 ml)	Liver glycogen (% wet weight)	Blood ketone bodies (mg/100 ml)	Liver fat (% wet weight)	Acetoacetate synthesis (nmoles/m mg protein)	Citrate synthase (nmoles/m mg protein)
Soybean fatty acids	0·29	197	0·10	32	6·23	5·95	49·0
Soybean oil	0·30	210	0·93	12	7·56	6·72	41·7
Glucose	0·22	230	2·28	5	7·04	6·80	44·6

Results from Renner and Elcombe (1967a), Allred (1969).

Experiments with ^{14}C-fatty acids have shown that this carbon is transferred to carbohydrates in the chick tissues due to randomization of the ^{14}C in the Krebs cycle reactions. In the Krebs cycle two carbon atoms enter as acetyl-CoA and two other carbons are lost as CO_2 so that there is no net synthesis of carbohydrate from fatty acids.

Allred (1969) studied the effects of feeding carbohydrate-free diets on the concentration of liver metabolites and the activities of some enzymes involved in acetyl-CoA production in relation to ketogenesis. There was a general increase in those compounds associated with ketogenesis and a decrease in those connected with gluconeogenesis. There was also no significant effect on the specific activities of citrate synthase and ATP-citrate lyase or the acetoacetate-synthesizing system (Table 9). Allred (1969) suggested that the ketogenesis which arose on feeding carbohydrate-free diets was the result of acetyl-CoA being diverted from oxidation and being transported from the mitochondria into the cytoplasm where acetoacetate formation takes place. He also found that the carbohydrate-free diet decreased the NAD/NADH ratio in the mitochondria; a similar decrease in the NAD/NADH ratio has been shown in relation to gluconeogenesis and ketogenesis (Krebs, 1966). To investigate the increased NADH content of liver mitochondria from birds fed the carbohydrate-free diet the RQ and P/O ratios of liver mitochondria were measured using either succinate, malate, α-ketoglutarate or citrate as the electron donor (Allred and Roehrig, 1969). There was no difference in these parameters between mitochondria from chicks on the carbohydrate-free diet and those fed a glucose reference diet. Also the rate of NADH oxidation in uncoupled mitochondria was greater in birds on the carbohydrate-free diet. These results indicate that the decreased NAD/NADH ratio observed on the carbohydrate-free diet is not due to a simple biochemical lesion in the mitochondria.

The effects of feeding diets in which all non-protein calories were supplied as either soybean oil or soybean oil fatty acids on neonatal gluconeogenic and glycolytic enzyme activity were examined by Allred and Roehrig (1970). In general the activities of gluconeogenic enzymes increased in response to "carbohydrate-free" diets whereas those concerned in glycolysis decreased. This situation is similar to that described in the Section IV C. These alterations would lead to the net hepatic production of glucose. The gluconeogenic enzyme glucose-6-phophatase was shown to play a major role in controlling blood glucose concentration and its activity was related to glycerol availability to the liver. Allred and Roehrig (1970) suggest that under these gluconeogenic conditions there is competition for the available carbon sources for glucose and glyceride-glycerol synthesis especially when all the non-protein calories in the diet are provided by soybean fatty acids. This could limit the amino acids available for the protein synthesis and so explain the restricted growth resulting from "carbohydrate-free" diets.

B. EFFECT OF FEEDING FAT ON LIPOGENESIS FROM CARBOHYDRATE SOURCES

It is well established that lipogenesis in the liver is dependent on the dietary regime of the animal. Hepatic lipogenesis is increased on a high carbohydrate diet and is decreased when the animal is starved or on fat-supplemented diets (see review, Masoro, 1962). There is considerable information on the effect of dietary manipulation on lipogenesis in mammalian liver but, in contrast, little work has been done with avian liver systems, particularly in the domestic fowl, where the situation is more complex due to the demands of egg production.

The effect of feeding fat on lipogenesis in laying hens was investigated by Weiss *et al.* (1967); the rate of lipogenesis from acetate in mature laying hens was decreased. At the enzyme level in laying hens, Pearce (1968) and Balnave and Pearce (1969) observed that this decrease in lipogenesis was reflected by the activities of both ATP-citrate lyase, which channels citrate into fatty acid, and acetyl-CoA carboxylase. However, Goodridge (1969) reported that feeding fat affected neither the incorporation of radioactive glucose into lipid nor the activity of the "malic" enzyme in 7-d and 4-week-old chicks. Yeh and Leveille (1969), Yeh *et al.* (1970) and J. Pearce (unpublished) have shown that dietary fat depresses hepatic lipogenesis and also the specific activities of ATP-citrate lyase and "malic" enzyme in the young chick. These results are contrary to the observations of Goodridge (1969) but agree with those of Weiss *et al.* (1967), Pearce (1968) and Balnave and Pearce (1969) and indicate that the situation in the young chick is the same as that in the mature laying hen.

C. EFFECT OF FEEDING FAT ON GLUCONEOGENESIS

A situation similar to that in the chick embryo can be attained by feeding fat-rich diets to the hatched bird. Kidney slices from 15-week-old chickens fed on boiled eggs showed an increased rate of gluconeogenesis (Krebs and Yoshida, 1963). Studies with 3-week-old chickens fed boiled egg yolks have shown an increase in the activities of some of the enzymes which participate in the synthesis of glucose from pyruvate. These include fructose-1,6-diphosphatase, glucose-6-phosphatase, triose phosphate isomerase and aldolase (Rinaudo and Galletti, 1966a). These increases in enzyme activity were inhibited by administration of ethionine showing that the increases were due to synthesis of enzymic protein (Rinaudo and Galletti, 1966b).

References

Allred, J. B. (1969). *J. Nutr.* **99**, 101–108.
Allred, J. B. and Roehrig, K. L. (1969). *J. Nutr.* **99**, 109–112.
Allred, J. B. and Roehrig, K. L. (1970). *J. Nutr.* **100**, 615–622.
Alvarado, F. (1967). *Comp. Biochem. Physiol.* **20**, 461–470.

Arese, P., Rinaudo, M. T. and Bosia, A. (1967). *Eur. J. Biochem.* **1**, 207–215.
Ballard, F. J. and Hanson, R. W. (1967). *Biochem. J.* **102**, 952–958.
Ballard, F. J. and Oliver, I. T. (1963). *Biochim. biophys. Acta* **71**, 578–588.
Ballard, F. J. and Oliver, I. T. (1965). *Biochem. J.* **95**, 191–200.
Balnave, D. and Pearce, J. (1969). *Comp. Biochem. Physiol.* **29**, 539–550.
Bdolah, A. and Feingold, D. S. (1965). *Biochem. biophys. Res. Commun.* **21**, 543–546.
Bdolah, A. and Feingold, D. S. (1968). *Biochim. biophys. Acta* **159**, 176–178.
Bell, D. J. (1967). *Comp. Biochem. Physiol.* **20**, 523–534.
Bell, D. J. and Culbert, J. (1968). *Comp. Biochem. Physiol.* **25**, 627–637.
Biasotti, M. (1932). *Archs int. Pharmacodyn. Ther.* **42**, 305–301.
Bloom, B., Stetten, M. R. and Stetten, D. (1953). *J. biol. Chem.* **204**, 681–694.
Bogner, A. P. H. (1957). *Diss. Abstr.* **17**, 2654.
Bollman, J. L., Mann, F. C. and Magath, T. B. (1925). *Am. J. Physiol.* **74**, 238–248.
Brambila, S. and Hill, F. W. (1966). *J. Nutr.* **88**, 84–92.
Burt, A. M. and Wenger, B. S. (1961). *Devl Biol.* **3**, 84–95.
Cahill, G. F., Hastings, A. B., Ashmore, J. and Zottu, S. (1958). *J. biol. Chem.* **230**, 125–135.
Cazorla, A. and Guzman Barron, E. S. (1958). *Expl Cell Res.* **14**, 68–79.
Coffey, R. G., Cheldelin, V. H. and Newburg, R. W. (1964). *J. gen. Physiol.* **48**, 105–112.
Couch, J. R., Bakshi, Y. N., Prescott, J. M. and Cregor, C. R. (1965). *Fedn Proc. Fedn Am. Socs exp. Biol.* **24**, 687.
Dalton, A. J. (1937). *Anat. Rec.* **68**, 393–409.
Dam, H. (1944). *Proc. Soc. exp. Biol. Med.* **55**, 57–59.
Davidson, J. (1957). *Growth* **21**, 287–295.
Duncan, H. J. (1968). *Can. J. Biochem.* **46**, 1321–1326.
Duncan, H. J. and Common, R. H. (1967). *Can. J. Biochem.* **45**, 979–989.
Favarger, P. (1965). *In* "Handbook of Physiology, Adipose Tissue" Sect. 5, chapter 4, pp. 19–23. American Physiological Society, Washington, D.C.
Felicioli, R. A. and Gabrielli, F. (1967). *Experientia* **23**, 1000–1001.
Felicioli, R. A., Gabrielli, F. and Rossi, C. A. (1967). *Eur. J. Biochem.* **3**, 19–24.
Feller, D. D. (1954). *J. biol. Chem.* **206**, 171–180.
Freeman, B. M. (1965). *Comp. Biochem. Physiol.* **14**, 217–222.
Freeman, B. M. (1969). *Comp. Biochem. Physiol.* **28**, 1169–1176.
Gibson, W. R. and Nalbandov, A. V. (1966). *Am. J. Physiol.* **211**, 1352–1356.
Gill, P. M. (1938). *Biochem. J.* **32**, 1792–1799.
Goodridge, A. G. (1968a). *Biochem. J.* **108**, 663–666.
Goodridge, A. G. (1968b). *Biochem. J.* **108**, 655–661.
Goodridge, A. G. (1968c). *Am. J. Physiol.* **214**, 897–901.
Goodridge, A. G. (1968d). *Biochem. J.* **108**, 667–673.
Goodridge, A. G. (1969). *Can. J. Biochem.* **47**, 743–746.
Goodridge, A. G. (1970). *Biochem. J.* **118**, 259–263.
Goodridge, A. G. and Ball, E. G. (1966). *Am. J. Physiol.* **211**, 803–808.
Goodridge, A. G. and Ball, E. G. (1967a). *Am. J. Physiol.* **213**, 245–249.
Goodridge, A. G. and Ball, E. G. (1967b). *Biochemistry, N.Y.* **6**, 2335–2343.
Grant, W. C. and Fahrenbach, M. J. (1959). *Proc. Soc. exp. Biol. Med.* **100**, 250–252.
Grebner, E. E., Hall, C. W. and Neufeld, E. F. (1966). *Biochem. biophys. Res. Commun.* **22**, 672–677.
Griminger, P. and Fisher, H. (1963). *Poult. Sci.* **42**, 1471–1473.
Hansen, R. G., Freedland, R. A. and Scott, H. M. (1956). *J. biol. Chem.* **219**, 391–397.

Hartmann, A. F., Grunwaldt, E. and James, D. H. (1953). *J. Pediatrics* **43**, 1–8.
Hausberger, F. X., Milstein, S. W. and Rutman, R. J. (1954). *J. biol. Chem.* **208**, 431–438.
Hawkins, R. A. and Heald, P. J. (1966). *Biochim. biophys. Acta* **116**, 41–55.
Hazelwood, R. L. (1965). *In* "Avian Physiology" (P. D. Sturkie, ed.), 2nd edn. pp. 313–371. Cornell University Press, New York.
Heald, P. J. (1962). *Nature, Lond.* **195**, 603.
Heald, P. J. (1963). *Biochem. J.* **86**, 103–110.
Heinz, F. and Weiner, F. (1969). *Comp. Biochem. Physiol.* **31**, 283–296.
Jansen, G. R., Hutchinson, C. F. and Zanetti, M. E. (1966). *Biochem. J.* **99**, 323–332.
Kilsheimer, G. S., Weber, D. R. and Ashmore, J. (1960). *Proc. Soc. exp. Biol. Med.* **104**, 515–518.
Kozak, L. P. and Wells, W. W. (1969). *Archs Biochem. Biophys.* **135**, 371–377.
Kratzer, F. H., Rajaguru, R. W. A. S. B. and Vohra, P. (1967). *Poult. Sci.* **46**, 1489–1493.
Krebs, H. A. (1966). *In* "Advances in Enzyme Regulation" (G. Weber, ed.), Vol. 5, pp. 409–434. Pergamon Press, London and New York.
Krebs, H. A. and Yoshida, T. (1963). *Biochem. J.* **89**, 398–400.
Kritchevsky, D., Kolman, R. R., Guttmacher, R. M. and Forbes, M. (1959). *Archs Biochem. Biophys.* **85**, 444–451.
Leveille, G. A. (1967). *Proc. Soc. exp. Biol. Med.* **125**, 85–88.
Leveille, G. A., O'Hea, E. K. and Chakrabarty, K. (1968). *Proc. Soc. exp. Biol. Med.* **128**, 398–401.
Liuzzi, A. and Angeletti, P. V. (1964). *Experientia* **20**, 512–513.
Lowenstein, J. M. (1966). *In* "Control of Energy Metabolism" (B. Chance, R. W. Estabrook and J. R. Williamson, eds.), pp. 261–265. Academic Press, New York and London.
Masoro, E. J. (1962). *J. Lipid Res.* **3**, 149–164.
Mayes, J. S., Miller, L. R. and Myers, F. K. (1970). *Biochem. biophys. Res. Commun.* **39**, 661–665.
Meyerhof, O. and Perdigon, E. (1940). *Enzymologia* **8**, 353–362.
Monson, W. J., Dietrich, L. S. and Elvehjem, C. A. (1950). *Proc. Soc. exp. Biol. Med.* **75**, 256–259.
Monson, W. J., Harper, E. A., Denton, D. A. and Elvehjem, C. A. (1954). *J. Nutr.* **53**, 563–573.
Muglia, G. and Massuelli, L. (1934). *Boll. Soc. ital. Biol. sper.* **8**, 1772–1774.
Nelson, P., Yarnell, G. and Wagle, S. R. (1966). *Archs Biochem. Biophys.* **114**, 543–546.
Nordin, J. H., Wilken, D. R., Bretthauer, R. K., Hansen, R. G. and Scott, H. M. (1960). *Poult. Sci.* **39**, 802–812.
Novikoff, A. B., Potter, V. R. and LePage, G. A. (1948). *J. biol. Chem.* **173**, 239–252.
Nunnally, D. A. (1962). *J. exp. Zool.* **149**, 103–112.
O'Hea, E. K. and Leveille, G. A. (1968). *Comp. Biochem. Physiol.* **26**, 111–120.
O'Hea, E. K. and Leveille, G. A. (1969). *Comp. Biochem. Physiol.* **30**, 149–159.
Pearce, J. (1968). *Biochem. J.* **109**, 702–704.
Pearce, J. (1970). *Int. J. Biochem.* **1**,306–312.
Renner, R. (1964). *J. Nutr.* **84**, 322–326.
Renner, R. (1966). *Fedn Proc. Fedn Am. Socs exp. Biol.* **25**, 303.
Renner, R. (1969). *J. Nutr.* **98**, 297–302.
Renner, R. and Elcombe, A. M. (1964). *J. Nutr.* **84**, 327–330.
Renner, R. and Elcombe, A. M. (1967a). *J. Nutr.* **93**, 31–36.

Renner, R. and Elcombe, A. M. (1967b). *J. Nutr.* **93**, 25–30.
Rigdon, R. H., Couch, J. R., Creger, C. R. and Ferguson, T. M. (1963). *Experientia* **19**, 349–352.
Rinaudo, M. T. (1962). *Enzymologia* **24**, 230–236.
Rinaudo, M. T. and Galletti, L. (1966a). *Ital. J. Biochem.* **15**, 395–401.
Rinaudo, M. T. and Galletti, L. (1966b). *Ital. J. Biochem.* **15**, 381–386.
Romanoff, A. L. (1967). "Biochemistry of the Avian Embryo." pp. 177–232, Wiley, New York, London and Sydney.
Romanoff, A. L. and Romanoff, A. J. (1949). "The Avian Egg", pp. 311–364. Wiley, New York.
Roux, J. F. (1966). *Metabolism* **15**, 856.
Rutter, W. J., Kritchevsky, P., Scott, H. M. and Hansen, R. G. (1953). *Poult. Sci.* **32**, 706–715.
Ryder, E. (1970). *Biochem. J.* **119**, 929–930.
Sathe, B. S. and Bose, S. (1962). *Indian J. vet. Sci.* **32**, 74–84.
Schoenheimer, R. and Rittenberg, D. (1930). *J. biol. Chem.* **114**, 381–396.
Shannon, D. W. F., Waring, J. J. and Brown, W. O. (1969). *In* "Energy Metabolism of Farm Animals" (K. L. Blaxter, J. Kielanowski and G. Thorbek, eds.), pp. 349–358. Oriel Press Limited, Newcastle upon Tyne.
Shulyak, V. D. (1967). *Fiziol. i Biokhim. Sel'skokhoz. Zhivotnykh.* **5**, 72–75.
Shields, C. E., Herman, Y. F. and Herman, R. H. (1964). *Nature, Lond.* **203**, 935–936.
Siddons, R. C. (1969). *Biochem. J.* **112**, 51–59.
Solomon, J. B. (1958). *Biochem. J.* **70**, 529–535.
Sondergaard, E., Prange, I., Dam, H. and Christensen, E. (1957). *Acta path. microbiol. scand.* **40**, 303–308.
Spencer, A. F. and Lowenstein, J. M. (1962). *J. biol. Chem.* **237**, 3640–3648.
Srere, P. A. (1959). *J. biol. Chem.* **234**, 2544–2547.
Stumpf, P. K. (1947). *Fedn Proc. Fedn Am. Socs exp. Biol.* **6**, 296–297.
Thind, S. K., Singh, A. and Sarkar, N. K. (1966), *J. Geront.* **21**, 93–96.
Villee, C. A. and Hagerman, D. D. (1958). *Am. J. Physiol.* **194**, 457–464.
Viñuela, E., Salas, M. and Sols, A. (1963). *J. biol. Chem.* **238**, 1175–1177.
Vogt, H. and Penner, W. (1963). *Arch. Geflügelk.* **27**, 43–47.
Vohra, P. (1967). *Wld's Poult. Sci. J.* **23**, 20–31.
Vohra, P. and Kratzer, F. H. (1964a). *Poult. Sci.* **43**, 502–503.
Vohra, P. and Kratzer, F. H. (1964b). *Poult. Sci.* **43**, 1164–1170.
Vohra, P. and Kratzer, F. H. (1965). *Poult. Sci.* **44**, 1201–1205.
Wagh, P. V. and Waibel, P. E. (1966). *J. Nutr.* **90**, 207–211.
Wagh, P. V. and Waibel, P. E. (1967a). *Proc. Soc. exp. Biol. Med.* **124**, 421–424.
Wagh, P. V. and Waibel, P. E. (1967b). *J. Nutr.* **92**, 491–496.
Wallace, J. C. and Newsholme, E. A. (1967). *Biochem. J.* **104**, 378–384.
Weiss, J. F., Naber, E. C. and Johnson, R. M. (1967). *J. Nutr.* **93**, 142–152.
Wenger, E., Wenger, B. S. and Kitos, P. A. (1967). *J. exp. Zool.* **166**, 263–270.
Wise, E. M. and Ball, E. G. (1964). *Proc. natn. Acad. Sci. U.S.A.* **52**, 1255–1263.
Woolf, L. I. and Norman, A. P. (1957). *J. Pediatrics* **50**, 271–295.
Yarnell, G., Nelson, P. and Wagle, S. R. (1966). *Archs Biochem. Biophys.* **114**, 539–542.
Yeh, Y. Y. and Leveille, G. A. (1969). *J. Nutr.* **98**. 356–366.
Yeh, Y. Y., Leveille, G. A. and Wiley, J. H. (1970). *J. Nutr.* **100**, 917–924.
Yeung, D., Stanley, R. S. and Oliver, I. T. (1967). *Biochem. J.* **105**, 1219–1227.

12 Lipid and Acetate Metabolism

E. F. ANNISON

Unilever Research Laboratory,
Colworth House,
Sharnbrook, Bedford, England

I. Introduction

The lipids, a complex and heterogeneous class of substances comprising triglycerides (the major components of fats and oils), phospholipids, sphingolipids and steroids are largely common to mammals and birds. In spite of these similarities in the nature and composition of tissue lipids there are major differences in several aspects of lipid metabolism in the bird. These include the mode of transport of absorbed dietary fat, the relative significance

Abbreviations used in this chapter: ATP = adenosine triphosphate; CoA = Coenzyme A; EFA = essential fatty acids; FFA = free fatty acids; NAD = nicotinamide adenine dinucleotide; NADP = nicotinamide adenine dinucleotide phosphate.

of hepatic and extra-hepatic lipogenesis, lipoprotein metabolism and the hormonal control of lipid metabolism.

II. Fat Digestion, Absorption and Transport

The chick is able to utilize relatively large amounts of dietary fat. Diets containing 34% animal fat (Donaldson *et al.*, 1957) or 30% soybean oil (Isaaks *et al.*, 1963) will support normal growth in chicks; the poor results experienced by some investigators with high fat diets must be attributed to imbalances in ratios between essential amino acids and energy, to inadequate intakes of vitamins or trace elements, or to the occurrence of toxic factors in the dietary fat. Carpenter (1968) has convincingly shown that oxidized fats are not toxic, but their inclusion in diets might possibly increase the requirement for vitamin E or selenium (see Scott, 1962).

Renner and Elcombe (1963) showed that soybean oil could completely replace all the carbohydrate in chick diets, without affecting growth rate or nitrogen retention. When soybean oil fatty acids were included in a carbohydrate-free diet for chicks, growth rate and food intake were severely depressed, showing that the glycerol moiety of the fat is essential for fatty acid uptake and utilization in the absence of carbohydrate (Hill and Brambila, 1965; Allred, 1968). Glycerides, by providing monoglycerides, aid the absorption of fat (see below), and in the absence of dietary carbohydrate may make an essential contribution to the overall glucose requirement of the bird.

There is substantial indirect evidence that the processes of fat digestion and absorption in chickens are closely similar to those known to occur in the dog, pig and man (see Chapters 2 and 3). The key feature is the formation of micellar fat (see Senior, 1964), which has been found in gut contents in the chicken (Garrett, 1967). The major components of the fat micelles are monoglycerides and free fatty acids which are absorbed into the epithelial cells of the small intestine. Resynthesis of triglycerides, in the chicken, occurs by both monoglyceride and glycerol-3-phosphate pathways (Bickerstaffe and Annison, 1969a), and, as in other animals, the triglyceride synthetase activity is located in the microsomal fraction of the intestinal epithelium. Glycerokinase activity has also been detected in this tissue (Bickerstaffe and Annison, 1969b), implying that the glycerol liberated in the small intestine during fat digestion may contribute to glycerol-3-phosphate synthesis.

In mammals, fat is secreted from the mucosal cells into lymph as chylomicrons which enter the systemic circulation via the thoracic duct (see Senior, 1964). The intestinal lymphatic system in the chicken, however, is less well developed, and differs in its structure. Kiyasu (cited by Noyan *et al.*, 1964) examined microscopic sections of the small intestinal villi of chickens after perfusion of the small intestine with Indian ink. The blood capillary network was found to occupy almost all of the central core, and there was no evidence of a central lacteal. The suggestion that all fat might be absorbed via the portal system was later confirmed by Noyan *et al.* (1964), who showed that at least 90% of absorbed fat entered the portal system as very low density lipoproteins. A close examination of the blood of chickens fed relatively high fat diets has failed to detect chylomicrons (D. R. Husbands, personal communication).

III. The Efficiency of Utilization of Dietary Fat

Several groups of workers have shown that when dietary glucose is replaced by certain fats the efficiency of utilization of metabolizable energy (ME) is somewhat improved (Rand *et al.*, 1958; Carew and Hill, 1964). This effect was not observed with hydrogenated coconut oil (Carew *et al.*, 1964), almost certainly because this fat is rich in fatty acids of medium chain-length (mainly lauric acid) which are not deposited directly in adipose tissue. Gosling *et al.* (1971a) have measured the efficiency of utilization of ME of diets containing 0, 5, 10 and 15% of added soybean oil when fed to broiler chicks (age 1 to 4 weeks). Maximum efficiency of utilization was reached at the 5% level of fat inclusion.

The explanation for the high efficiency of utilization of certain fats lies in the low energy cost of their hydrolysis in the intestine and absorption from it of triglyceride resynthesis in the intestinal epithelium, and of transport and deposition relative to the efficiency with which fatty acids and carbohydrates supply acetylCoA for energy requirements or fat synthesis (see Blaxter, 1962). When the level of dietary fat exceeds the capacity of the animal to deposit adipose tissue, the excess fat is utilized for energy production (Gosling *et al.*, 1971a).

IV. Lipogenesis

A. LIPOGENESIS IN GENERAL

Adipose tissue has a central role in overall energy metabolism. Most animals eat intermittently and are inevitably faced with the problem of storing the chemical energy of food ingested in excess of immediate requirements. Fat, with its high energy content per unit mass or volume, is the preferred form of stored chemical energy. The homeostatic control of lipogenesis and of fat mobilization in response to changing patterns of energy availability is equally important in both mammalian and avian systems.

The remarkable capacity of birds to synthesize and deposit fat just prior to migratory flights led Goodridge and Ball (1966, 1967) to investigate the sites of fat synthesis in the pigeon. Pigeon adipose tissue was found to be much less active in *de novo* lipogenesis than was that of the rat. The liver was shown to be the main site of fatty acid synthesis, in contrast to the dominant role of adipose tissue in mammalian lipogenesis (see Favarger, 1965). The low levels of certain enzymes active in lipogenesis, namely acetylCoA carboxylase, ATP-citrate lyase, malate enzyme and the hexose monophosphate dehydrogenases, accounted for the limited capacity of pigeon adipose tissue to synthesize fatty acids (Goodridge and Ball, 1966). Similar findings on the relative importance of lipogenesis in the liver and adipose tissue of the chicken were reported by Leveille and his colleagues (O'Hea and Leveille, 1968; Leveille, 1969). Studies with ^{14}C-labelled glucose and acetate showed that, although adipose tissue incorporated glucose into triglyceride glycerol, neither substrate was an important source of fatty acids which were largely derived from circulating triglycerides (O'Hea and Leveille, 1968).

Goodridge (1968) has shown that the liver of the embryonic chick has only slight lipogenic activity; but hepatic lipogenesis develops rapidly after newly-hatched chicks are fed. The dominant role of the liver in lipogenesis is maintained throughout the life of the bird, although influenced by the fat content of the diet, and by certain hormonal factors.

B. ENZYMIC ASPECTS OF LIPOGENESIS

A major breakthrough in avian lipid biochemistry was achieved when Wakil (1958) showed that an enzyme system located in the particle-free fraction of pigeon liver converted acetylCoA to long chain fatty acids in the presence of ATP, Mn^{2+}, bicarbonate and reduced NADP. MalonylCoA was identified as an intermediate, and in later studies two separate enzymes were recognized, acetylCoA carboxylase and fatty acid synthetase. Factors which stimulate or inhibit these enzymes have been intensively studied in efforts to provide an enzymic basis for the regulation of lipogenesis.

The stimulatory effects of tricarboxylic acid cycle intermediates, especially citrate, on fatty acid synthesis *in vitro* (Brady and Gurin, 1952) have been shown to be due to a direct effect on acetylCoA carboxylase (Waite and Wakil, 1962; Martin and Vagelos, 1962). Subsequent studies suggested that citrate exerts an allosteric effect on the enzyme (Lynen *et al.*, 1963). In contrast, there is considerable evidence that long chain fatty acids, and their acylCoA derivatives, inhibit acetylCoA carboxylase (Numa *et al.*, 1965a, b). Bortz *et al.* (1963) have reported that the control of lipogenesis in the fed rat was mediated through acetylCoA carboxylase, and there is ample evidence that levels of this enzyme in chicken liver are well correlated with lipogenic activity (Goodridge, 1968; Balnave and Pearce, 1969).

Fatty acid synthetase, which converts malonylCoA to long chain fatty acids in the presence of reduced NADP (Majerus and Vagelos, 1967), is a multi-enzyme complex bound to "acylcarrier protein" (Lynen, 1961). Active sub-units have been obtained from the bacterium *Escherichia coli*, and the individual enzyme steps are well documented (Majerus and Vagelos, 1967), but attempts to fractionate the avian synthetase complex were not successful (Yang *et al.*, 1965). Factors governing the chain length of the resultant fatty acids are largely unknown, but the major product of pigeon liver synthetase is palmitic acid (Bressler and Wakil, 1962).

Fatty acid synthetase probably has a minor role in the regulation of lipogenesis, since it is not involved in the early stages of fatty acid synthesis, and in most animals, the activity of the enzyme complex is high relative to that of acetylCoA carboxylase (Ganguly, 1960).

Fatty acid synthesis occurs in the cytoplasm of the cell, but the pyruvate formed from glucose, a major fat precursor in the bird, gives rise to acetylCoA within the mitochondria. Since the diffusion of acetylCoA out of the mitochondria is not rapid enough to support lipid synthesis, alternative mechanisms have been proposed (see Lowenstein, 1963). It is now generally established that acetylCoA generated within the mitochondria is converted to citrate, which leaves the mitochondria and is cleaved to acetylCoA and oxaloacetate in the cytoplasm by ATP-citrate lyase (Spencer and Lowenstein, 1966). The oxalacetate formed in the cleavage of citrate is probably reduced to malate via NAD-malate dehydrogenase and NADP-malate dehydrogenase (Young *et al.*, 1964). Pyruvate formed from the malate can then enter the cell to give rise to additional acetylCoA. Ballard *et al.* (1969) point out that these reactions result in the "transhydrogenation" of NADP to NADPH which augments the available NADPH for the support of lipogenesis. These series of reactions require the continuous replenishment of intramitochondrial oxalacetate, which arises from pyruvate via pyruvate carboxylase, and this enzyme is sufficiently active to account for the oxalacetate required for lipogenesis in rat tissue (Ballard and Hanson, 1967).

The low activity of ATP-citrate lyase in chicken adipose tissue (Goodridge and Ball, 1966) is consistent with the minor role of this tissue in lipogenesis.

C. THE EFFECT OF FEEDING FAT ON LIPOGENESIS

The incorporation of acetate into fatty acids was decreased by raising the level of dietary fat in mammals (Hausberger and Milstein, 1955; Hill *et al.*, 1958) and birds (Marion and Edwards, 1962; Husbands and Brown, 1965; Weiss *et al.*, 1967), but the nature of the dietary fat has been shown to influence the results. Donaldson (1965) showed that dietary unsaturated fatty acids were more effective than saturated fatty acids in reducing lipogenesis.

The influence of fat feeding on the activities of several enzymes involved directly or indirectly in lipogenesis has been studied in the laying hen by Balnave and Pearce (1969). The enzymes acetylCoA carboxylase and ATP-citrate lyase fell rapidly when corn oil was fed to laying hens maintained on a low-fat diet, but NADP-linked isocitrate dehydrogenase was less affected. A second enzyme system involved in the production of reduced NADP, the hexose monophosphate pathway of glucose metabolism was also poorly correlated with lipogenesis (Goodridge, 1968), and O'Hea and Leveille (1968) considered that the NADP-linked malic dehydrogenase was the major source of reduced NADP in the chick liver.

D. DESATURATION OF FATTY ACIDS

The interrelationships of saturated and unsaturated fatty acids of endogenous and dietary origin are shown in Fig. 1. Oleic and palmitoleic acids

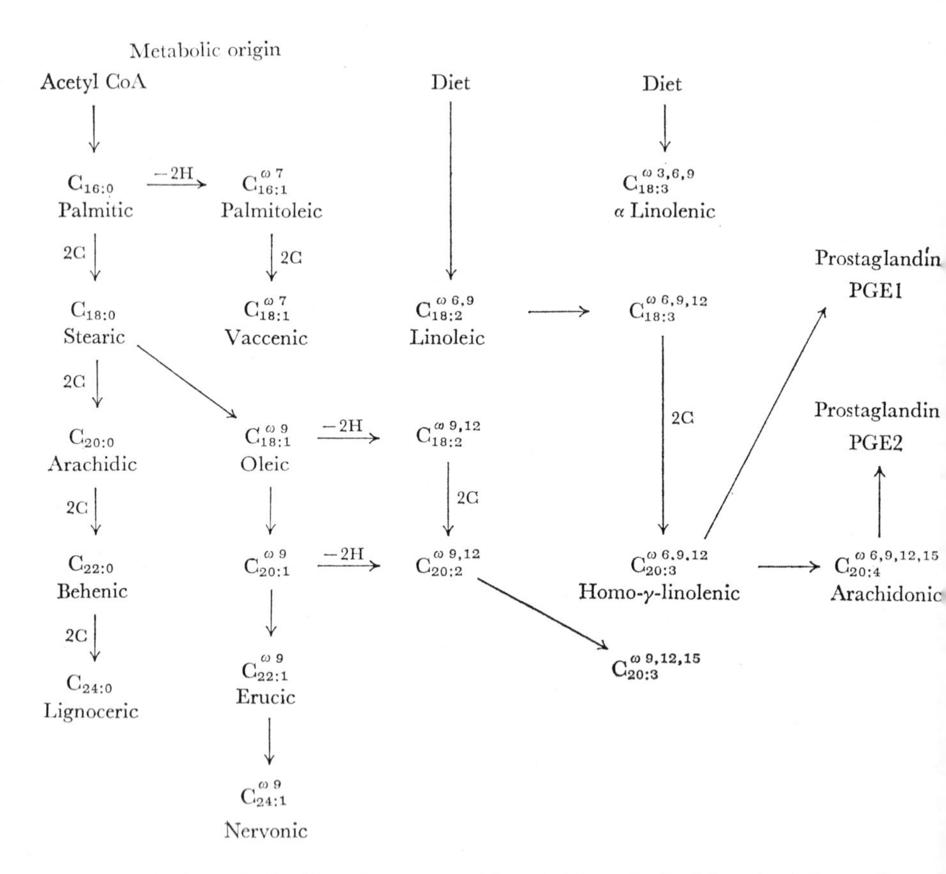

Fig. 1. The interrelationships of unsaturated fatty acids synthesized in animal tissues, the essential fatty acids (of dietary origin) and the prostaglandins.

(Carbon chain length and number of double bonds are shown as subscripts: the double bond position (ω) as a superscript, e.g. $C^{\omega 6,9}_{18:2}$ is linoleic acid with 18 carbon atoms and 2 double bonds located between carbon atoms 6,7 and 9,10 (numbering from —CH_3 group).)

largely arise by the desaturation of stearic and palmitic acids respectively. Chicken liver preparations have been found to desaturate fatty acids of chain-length C_{12}-C_{20}, and in each case the corresponding 9, 10 mono-enoic fatty acid was the major product formed (Johnson *et al.*, 1969). Maximum desaturation occurred both with the C_{14} fatty acid, and with the fatty acids C_{17} and C_{18}, suggesting the presence of at least two desaturating systems.

The desaturase system is inhibited by cyclopropene fatty acids and alcohols (Johnson *et al.*, 1969), and the sterculic and malvalic acids which occur in cotton are a potential hazard to egg quality. When cotton seed products are fed to hens, the pink discoloration of the albumen is invariably associated with abnormally high levels of stearic acid in the yolk (Allen *et al.*, 1967).

Raju and Reiser (1967) obtained evidence suggesting that the synthesis of oleic acid from acetate was less inhibited by sterculic acid than was the desaturation of stearic acid, and postulated the existence of an alternative pathway of oleic acid synthesis from acetate.

The desaturase activity of chicken liver preparations is inversely related to the unsaturated fatty acid content of the diet (Gosling *et al.*, 1971b) and is analogous to the inhibition of lipogenesis induced by feeding high levels of fat (see Section IVC).

E. LIPOGENESIS IN THE LAYING HEN

The onset of lay in the hen is accompanied by marked increases in the concentrations of plasma lipids and phosphoproteins. The changes are greatest in the 14 d preceding egg laying, when the values for plasma FFA may increase from between 0·2 and 0·5 meq./l to 4·0 meq./l (Heald and Badman, 1963) and total plasma lipids from 0·2 to 0·5g/100 ml to 10 to 14 g/100 ml (McIndoe, 1959). When laying starts, the plasma FFA has been found to fall sharply to between 0·75 and 1·5 meq./l (Heald and Badman, 1963), and the total lipids to between 1·5 and 3·0 g/100 ml.

The egg yolk lipids are derived from circulating plasma lipoproteins (see Section VI), which are synthesized in the liver in response to the hormonal changes which accompany the onset of lay. Hawkins and Heald (1966) found that triglyceride synthesis by liver slices from the laying hen was considerably greater than that shown by similar tissue from immature birds.

V. Free Fatty Acid Metabolism

Plasma FFA, which usually account for about 5% of total plasma lipids, are of major metabolic importance in both mammals and birds. Dole (1956) and Gordon and Cherkes (1956) independently recognized the physiological significance of fluctuations in plasma FFA which occurred in response to changes in nutritional status. Subsequent studies in many animals have confirmed that plasma FFA concentrations are low in fed animals (0·2 to 0·3 mmole/litre), and raised (0·5 to 1·5 mmole/litre) during starvation. The major source of plasma FFA is adipose tissue. Although adipose tissue triglycerides are derived from many sources, e.g. blood lipoproteins, and by synthesis from glucose and acetate, they are mobilized primarily, and probably exclusively, in the form of plasma FFA (Steinberg, 1963). Isotope dilution studies in many animals have shown that turnover rates of plasma FFA are high during starvation,

The effects of nutritional status and of the level of dietary fat on the concentration and composition of plasma FFA have been examined in broiler chicks (aged 8 weeks) fed low-fat (1·2% w/w) and high-fat (10·0 w/w) diets containing 8% (w/w) soybean oil (Gosling *et al.*, 1971a). The plasma FFA concentration in birds fed the low-fat diet (0·22 ± 0·002 (S.E.) mmole/litre) was significantly lower ($P < 0{\cdot}05$) than in the birds fed the high-fat diet (0·36 ± 0·02 mmole/litre). On starvation (48 h), the increased levels of plasma FFA observed in both groups of birds (0·59 and 0·62 mmole/litre) were not significantly different. The wide differences in the composition of plasma FFA observed in the two groups were largely accounted for by the much greater levels of linoleic acid (a major component of soybean oil) in the birds fed the high-fat diet (Gosling *et al.*, 1971a). Entry rates of palmitic, stearic, oleic and linoleic acids, which together accounted for about 80% of plasma FFA, were measured separately in these birds by isotope dilution using the continuous infusion procedure (see Annison *et al.*, 1969). When ^{14}C-labelled fatty acids were used, the patterns of excretion of $^{14}CO_2$ were followed during the measurement of total CO_2 production and, from the data, rough estimates were obtained of the contribution of each fatty acid to total CO_2 production and of the proportion of the fatty acid entering plasma that was promptly oxidized (Gosling *et al.*, 1971a). The rates of entry and oxidation of each plasma fatty acid were concentration-dependent. The lower values observed in fed birds than in starved ones were in line with the many observations made in man, dogs and ruminants but the amounts of the fatty acids oxidized in fed birds were much higher than in fed mammals (see Annison, 1964).

Plasma FFA have a special role in lipid metabolism in the laying hen. Heald and Badman (1963) and Heald *et al.* (1964) showed that the onset of lay is preceded by large increases in plasma FFA, total lipids and phosphoproteins, and that the quantities of these components decrease when laying commences. These changes in plasma constituents were not seen in male birds reaching maturity. The increased levels of plasma lipids and phosphoproteins were attributed to the secretion of oestrogens by the growing ovarian follicles under the influence of pituitary gonadotrophins. The marked decreases observed when laying commenced could not be accounted for by the uptake of plasma lipids and phosphoprotein by the egg, but were probably related indirectly to the decreased gonadotrophic stimulation of the ovary following ovulation (Heald and Badman, 1963).

Isotope dilution experiments based on a single injection of [1-^{14}C]-palmitate into groups of laying birds showed that the rate of entry of plasma FFA was relatively high (30 to 40 g/d) and was concentration dependent (Heald and Badman, 1963).

The effects of hormones on plasma FFA in the fowl are fully discussed elsewhere (Chapter 21), the most striking observation being the absence of an adrenaline response (Rudman, 1963; Carlson *et al.*, 1964).

VI. Lipoprotein Metabolism

Lipids, which are too apolar to circulate as free molecules in blood, form a range of soluble complexes containing varying proportions of protein; these constitute the transport media for lipids in mammals and birds (Frederickson *et al.*, 1967). Lipoproteins have a central role in the special features of avian lipid metabolism referred to earlier; dietary fat is absorbed as low-density lipoproteins, and not in the form of chylomicrons. Fat synthesized in the liver, the major site of lipogenesis, is transported to other tissues as lipoproteins, and plasma lipoproteins are the precursors of egg yolk in the laying hen.

Lipoproteins are classified largely on the basis of their density determined by ultracentrifugation. Comprehensive data on the lipoproteins of chicken plasma obtained by Husbands (1970) are shown in Table 1. The high lipid content of the blood of the laying hen has long been recognized (see McIndoe, 1959) and Vanstone *et al.* (1955) demonstrated the appearance of new lipoproteins in the hen at the onset of lay using electrophoretic techniques.

Schjeide and Urist (1956), Schjeide *et al.* (1963) and Schjeide and Wilkens (1964) have used analytical ultracentrifugation to characterize the lipoproteins of the hen at the onset of lay, and have shown that similar changes in plasma proteins and lipoproteins occur in the immature bird and cockerel given oestrogens. These changes include a large increase in the level of very low density lipoprotein and a decrease in the high density component. There is evidence that these changes are not entirely due to the effect of oestrogens, but that the synergistic effects of oestrogens, androgens and progesterone are also involved (Common *et al.*, 1946). Florsheim *et al.* (1963) found a marked increase in the incorporation of [^{35}S]-methionine into low density lipoprotein when cockerels were treated with oestrogen, suggesting increased lipoprotein synthesis, but adrenaline and cortisone were without effect.

The yolk of the egg contains large amounts of a lipoprotein which has similar density characteristics to plasma low density lipoproteins (Cook, 1968), but the precise identity of the two materials has not been established.

VII. Adipose Tissue

Chicken adipose tissue has little capacity for fatty acid synthesis (Section IV), but is able to take up fatty acids transported from the liver as lipoproteins. The fatty acids become available during the hydrolysis of lipoproteins by lipoprotein lipase, an enzyme closely associated with the walls of the blood capillaries (Robinson, 1964). Resynthesis of triglycerides occurs in the adipose tissue cells, but the structure of chicken adipose tissue triglycerides shows less evidence of the fatty acid positional specificity of the triglyceride synthetase enzyme system than is seen in mammalian tissue

Table 1

The composition of plasma lipoproteins (very low density, < 1·006, VLDL; low density, 1·006–1·063, LDL; high density, 1·063–1·21, HDL) of the chicken (from Husbands, 1970)

State	Lipoprotein class	Concentration (mg/100 ml serum)	Lipid content (relative to protein content = 100)			
			Cholesterol		Phospholipid	Triglyceride
			Free	Esterified		
Male, aged 9 weeks	VLDL	33·1	9·9	8·9	76·8	203
	LDL	69·9	11·8	19·6	78·1	30·9
	HDL	590	3·4	26·2	86·2	5·1
Laying hen	VLDL	5480	11·1	1·8	175	481
	LDL	198	9·7	6·8	115	194
	HDL	144	4·3	12·4	100	25·6

(Brockerhof *et al.*, 1966). In chicken liver, the fatty acid specificity of triglyceride synthetase closely paralleled the mammalian systems; the unsaturated fatty acid mainly occupied the 2-position while the saturated fatty acids and unsaturated fatty acids of *trans* configuration occurred at the 1, 3 positions of the triglycerides (Bickerstaffe and Annison, 1970). Egg yolk triglyceride showed a similar distribution of fatty acids (Leclerc *et al.*, 1966).

VIII. Essential Fatty Acids

Essential fatty acids (EFA) may be defined as those fatty acids which will correct certain pathological conditions which arise when a fat-free diet is fed. In the chick, EFA deficiency results in decreased growth rate (Dam *et al.* 1959) and an increased incidence of disease (Hopkins *et al.*, 1963). High mortality rates, and the production of fewer and smaller eggs with an increased incubation time and lower hatchability were characteristic of deficiency in hens (Menge, 1967).

The most important EFA are linoleic (Δ9,12-octadecadienoic), γ-linolenic (Δ6,9,12-octadecatrienoic) and arachidonic (Δ5,8,11,14-eicosatetraenoic) acids. Animal tissues recognize the EFA by their double bond configuration in relation to the terminal —CH_3 group, and EFA all have double bonds at the ω6,7 and ω9,10 positions (ω system, numbering from the terminal —CH_3 group). These acids cannot be synthesized by animals and must be supplied in the diet. Vegetable oils, particularly soybean, corn and safflower, are good sources of linoleic acid, which is elongated and desaturated to arachidonic acid by mammalian and avian tissues (Mead and Howton, 1957). The interrelationships between the essential fatty acids are shown in Fig. 1.

On EFA-deficient diets, the levels of linoleic and arachidonic acids in body tissues fall, but the rate of depletion of individual tissues is not constant (Aaes-Jorgensen and Holman, 1958). Rates of synthesis of palmitic, oleic and palmitoleic acids are increased in EFA deficiency (Allman and Gibson, 1965); the triglyceride content of the liver rises sharply in rats and mice (Allman, 1964) and in chicks (Hopkins and Nesheim, 1967). In addition, appreciable levels of 5,8,11 eicosatrienoic acid appear in the tissues; this acid is synthesized from oleic acid by the enzyme system which normally converts linoleic acid to arachidonic acid. The affinity of the enzyme for linoleic acid is much higher than that for oleic acid, and in the non-deficient animal there is little synthesis of eicosatrienoic acid (Brenner and Peluffo, 1966). Since the production of the trienoic acid is a sensitive indicator of EFA deficiency, the ratio of eicosatrienoic to arachidonic acids in tissues has been used as a measure of the adequacy of EFA intake when neither acid is present in the diet (Holman, 1960; Hill, 1966). Holman (1960) concluded that if the ratio exceeded 0·4, the diet was EFA deficient. The linoleic acid requirements of chicks and laying hens has been shown to be between 1·3 and 1·5% of the diet (see Hathaway, 1968), but the level and nature of the dietary fat, the sex of the bird and the previous EFA status may all influence the EFA requirement (Aftergood and Alfin-Slater, 1965).

Polyunsaturated fatty acids occur in the tissues largely as constituents of cholesterol esters and phospholipids; EFA appear to play a key role in the lipoproteins of cell membranes. Many of the signs of EFA deficiency, which include epidermal cell proliferation (Nasr and Shostak, 1965) and abnormal mitochondria in certain tissues (Levin *et al.*, 1957; Wilson and Leduc, 1963) may be attributed to defects in membrane structure.

A most intriguing development has been the recognition that essential fatty acids are the precursors of the prostaglandins (Van Dorp *et al.*, 1964; Bergstrom *et al.*, 1964). (See also Chapter 24.) The relationships of EFA to the more widely occurring prostaglandins, PGE_1 and PGE_2, are shown in Fig. 1. Prostaglandins show a wide variety of biological activities, and there is evidence that they are synthesized at or close to their target sites, and rapidly destroyed when they enter the circulation (van Dorp *et al.*, 1967). In general, prostaglandins appear to either increase or decrease the spontaneous activity of neuro-muscular tissue such as uterine smooth muscle, but effects on fat and carbohydrate metabolism in the whole animal have been recognized. Prostaglandin PGE_1, for example, has insulin-like effects in the whole animal, increasing lipogenesis from glucose and stimulating glucose oxidation. In the intact animal, prostaglandins may be essential agents in a number of local control mechanisms, where they may supplement the overall control by endocrine and nervous systems (Pickles, 1967). The interrelationships of EFA and prostaglandins in the bird are probably similar to those in mammals, but little work has so far been reported.

IX. Acetate Metabolism

Blood acetate concentrations in fed and starved chickens, laying hens and germfree chickens are in the range 0·3 to 1·0 mmole/litre (Annison *et al.*, 1968; E. F. Annison, unpublished). Blood acetate is derived from the alimentary tract (exogenous acetate) and from tissue metabolism (endogenous acetate). In herbivores, acetate produced by the microbial fermentation of dietary carbohydrate in the rumen or caecum constitutes an important source of energy. Acetate is produced in the chicken caecum (Shrimpton, 1963; Annison *et al.*, 1968) and enters the portal blood but indirect estimates of the amounts which do this suggested that exogenous acetate accounted for not more than 25% of the total acetate entry rate (Annison *et al.*, 1968). The relatively minor role of alimentary acetate in the overall acetate metabolism in chickens was demonstrated when the blood levels in germfree and normal chickens were shown to be similar in spite of the absence of acetate in the gut contents of the germfree birds (Annison *et al.*, 1968).

Isotope dilution techniques have been used to measure the rates of entry and oxidation of acetate in chickens in relation to overall energy expenditure (Annison *et al.*, 1969). In two dissimilar groups of chickens (fed males and starved non-laying hens) the oxidation of acetate accounted for between 9 and 15% of total CO_2 output, and between 6 and 10% of total energy expenditure.

The sources of endogenous acetate have not been clearly defined. In chickens considerable transfer of radioactivity from [U-^{14}C]-labelled glucose to acetate was observed (Annison *et al.*, 1969) and it is reasonable to assume that all metabolic processes which give rise to acetylCoA are potential sources of blood acetate. The high turnover rate of acetate in mammals and birds implies the ready activation of acetate to acetylCoA, as shown by the rapid appearance of $^{14}CO_2$ in blood when [^{14}C]-acetate is injected, and by the existence in certain tissues of acetyl CoA deacylase, but information on the latter is not available for avian tissues.

X. Plasma Lipids

The concentrations of the major lipid classes in chicken plasma are generally within the range encountered when comparisons between genera are made; except in the case of laying hens, where relatively high levels of the very low density lipoproteins and plasma free fatty acids occur in both fed and fasted birds (Heald and Badman, 1963). Circulating lipids unique to the bird have not been reported.

A. ANALYSIS OF PLASMA LIPIDS

Procedures devised for mammalian plasma have been successfully applied to chicken plasma (see Gosling *et al.*, 1971b). Lipids may be completely extracted with a mixture of chloroform and methanol (2:1 v/v), and separated into the main lipid classes (triglycerides, diglycerides, monoglycerides, phospholipids, free and esterified sterols and FFA) by thin layer chromatography (TLC). Some separation of the components of each class may be effected by further TLC, e.g. the glycerides may be separated into groups containing fatty acids of varying degrees of unsaturation by TLC on silica gel containing silver nitrate (Morris, 1966). Similarly, the phospholipids may be split into phosphatidyl choline, phosphatidyl ethanolamine and sphingolipids. The fatty acid composition of each fraction may be determined relatively easily by gas liquid chromatography (GLC). Details of the relevant TLC and GLC techniques are fully described by Morris and Nichols, (1967). The use of internal markers greatly facilitates the quantitative assay of plasma lipids (West and Rowbotham, 1967).

B. FREE FATTY ACIDS

The factors which influence the concentration and composition of plasma FFA are discussed elsewhere (Section V and Chapter 21).

C. TRIGLYCERIDES

Plasma triglycerides are associated with a range of lipoproteins, and the factors which influence plasma levels are discussed elsewhere.

Plasma triglyceride levels in broilers (aged 8 weeks) fed low-fat (1·2% w/w) and high-fat (10% w/w) diets containing 8% soybean oil was the same in both groups of birds, and fell significantly ($P < 0{\cdot}01$) on fasting for 24 h. (Gosling *et al.*, 1971a). The fatty acid composition of the plasma triglyceride in these birds was greatly influenced by the proportions of palmitic, stearic, oleic and linoleic acids in the high-fat and low-fat diets, which were 35, 13, 33 and 10% and 20, 9, 27 and 35% respectively.

Triglyceride synthesis occurs almost entirely in the liver (see Section II).

D. PHOSPHOLIPIDS

Plasma phospholipids occur as components of lipoproteins (see Section VI).

Much early work on plasma phospholipids was hampered by the lack of methods for the separation and analysis of lipoproteins, although the presence of high levels of phospholipid-rich lipoproteins in the plasma of laying hens was clearly recognized (see McIndoe, 1959). Data on the levels plasma phospholipids of the broiler and laying hen (Husbands, 1970) are shown in Table 1.

Husbands (1970) has confirmed that the plasma phospholipids of the chicken closely resemble those of mammals. In the broiler and laying hen phosphatidyl choline, phosphatidyl ethanolamine and phosphatidyl serine accounted for about 65, 30 and 5% of the total phospholipid; only traces of sphingolipids were observed.

There is evidence that phospholipid synthesis in the bird occurs mainly, if not entirely, in the liver (Vanstone *et al.*, 1955).

E. ACETATE

See Section X.

F. CHOLESTEROL

The data of Husbands (1970) shown in Table 1 clearly demonstrates the low levels of free and esterified cholesterol in chicken blood relative to that of rat and man.

References

Aaes-Jorgensen, E. and Holman, R. T. (1958). *J. Nutr.* **65**, 633–641.

Aftergood, L. and Alfin-Slater, R. B. (1965). *J. Lipid Res.* **6**, 287–294.

Allen, E. A., Johnson, A. R., Fogerty, A. C., Pearson, J. A. and Shenstone, F. S. (1967). *Lipids* **2**, 419–423.

Allman, D. W. (1964). *Diss. Abstr.* **25**, 2737.

Allman, D. W. and Gibson, D. M. (1965). *J. Lipid Res.* **6**, 51–62.

Allred, J. B. (1968). *Fedn Proc. Fedn Am. Socs exp. Biol.* **27**, 257.

Annison, E. F. (1964). *In* "Metabolism and Physiological Significance of Lipids" (R. M. C. Dawson and D. N. Rhodes, eds.), pp. 289–324. Wiley, London, New York, Sydney.
Annison, E. F., Hill, K. J. and Kenworthy, R. (1968). *Br. J. Nutr.* **22**, 207–216.
Annison, E. F., Shrimpton, D. H. and West, C. E. (1969). *In* "Energy Metabolism of Farm Animals" (K. L. Blaxter, J. Kielanowski and G. Thorbek, eds.), pp. 339–347. Oriel Press, Newcastle.
Ballard, F. J. and Hanson, R. W. (1967). *Biochem. J.* **102**, 952–958.
Ballard, F. J., Hanson, R. W. and Kronfeld, D. S. (1969). *Fedn Proc. Fedn Am. Socs exp. Biol.* **28**, 218–231.
Balnave, D. and Pearce, J. (1969). *Comp. Biochem. Physiol.* **29**, 539–550.
Bergstrom, S., Daniellson, H. and Samuelsson, B. (1964). *Biochim. biophys. Acta* **90**, 207–210.
Bickerstaffe, R. and Annison, E. F. (1969a). *Biochem. J.* **111**, 419–429.
Bickerstaffe, R. and Annison, E. F. (1969b). *Comp. Biochem. Physiol.* **31**, 47–54.
Bickerstaffe, R. and Annison, E. F. (1970). *Biochem. J.* **118**, 433–442.
Blaxter, K. L. (1962). "The Energy Metabolism of Ruminants". Hutchinson, London.
Bortz, W., Abraham, S. and Chaikoff, I. L. (1963). *J. biol. Chem.* **238**, 1266–1272.
Brady, R. O. and Gurin, S. (1952). *J. biol. Chem.* **199**, 421–431.
Brenner, R. R. and Peluffo, R. O. (1966). *J. biol. Chem.* **241**, 5213–5219.
Bressler, R. and Wakil, S. J. (1962). *J. biol. Chem.* **237**, 1441–1448.
Brockerhof, H., Hoyle, R. J. and Wolmark, N. (1966). *Biochim. biophys. Acta* **116**, 67–72.
Carpenter, K. J. (1968). *In* "Proceedings of University of Nottingham 2nd Conference for Food Manufacturers" (H. Swan and D. Lewis, eds.), pp. 54–67. J. and A. Churchill, London.
Carew, L. B. and Hill, F. W. (1964). *J. Nutr.* **83**, 293–299.
Carew, L. B., Hopkins, D. T. and Nesheim, M. C. (1964). *J. Nutr.* **83**, 300–306.
Carlson, A., Liljedahl, S., Verdy, M. and Wirsen, C. (1964). *Metabolism* **13**, 227–231.
Common, R. H., Bolton, W. and Rutledge, W. A. (1946). *J. Endocr.* **5**, 623–627.
Cook, W. H. (1968). *In* "Egg Quality" (T. C. Carter, ed.), pp. 109–132. Oliver and Boyd, Edinburgh.
Dam, R., Leach, R. M., Nelson, T. S., Norris, L. C. and Hill, F. W. (1959). *J. Nutr.* **68**, 615–632.
Dole, V. P. (1956). *J. clin. Invest.* **35**, 150–154.
Donaldson, W. E. (1965). *Poult. Sci.* **44**, 1365.
Donaldson, W. E., Combs, G. F., Romoser, G. L. and Supple, W. C. (1957). *Poult. Sci.* **36**, 807–815.
Favarger, P. (1965). *In* "Handbook of Physiology–Adipose Tissue", pp. 19–23. American Physiological Society, Washington.
Florsheim, W. H., Faircloth, M. A., Graff, D., Austin, N. S. and Velicoff, S. M. (1963). *Metabolism* **12**, 598–607.
Frederickson, D. S., Levy, R. I. and Lees, R. S. (1967). *New Eng. J. Med.* **12**, 598–602.
Ganguly, J. (1960). *Biochim. biophys. Acta* **40**, 110–118.
Garrett, G. L. (1967). *Diss. Abstr.* **28**, 2010.
Goodridge, A. G. (1968). *Biochem. J.* **108**, 663–666.
Goodridge, A. G. and Ball, E. G. (1966). *Am. J. Physiol.* **211**, 803–808.
Goodridge, A. G. and Ball, E. G. (1967). *Am. J. Physiol.* **213**, 245–249.
Gosling, J., Lewis, D. and Annison, E. F. (1971a). *Br. J. Nutr.* (in press).

Gosling, J., Lewis, D., Annison, E. F. and Bickerstaffe, R. (1971b). *Biochem. J.* (in press).
Gordon, R. S. and Cherkes, A. (1956). *J. clin. Invest.* **35**, 206–208.
Hathaway, H. D. (1968). *In* "Proceedings of University of Nottingham Second Nutrition Conference for Feed Manufacturers" (H. Swan and D. Lewis, eds.), p. 22. J. and A. Churchill, London.
Hausberger, F. X. and Milstein, S. W. (1955). *J. biol. Chem.* **214**, 483–488.
Hawkins, R. A. and Heald, P. J. (1966). *Biochim. biophys. Acta* **116**, 41–55.
Heald, P. J. and Badman, H. G. (1963). *Biochim. bipohys. Acta* **70**, 381–388.
Heald, P. J., Badman, H. G., Wharton, J., Walwik, C. M. and Hooper, P. I. (1964). *Biochim. biophys. Acta* **84**, 1–7.
Hill, E. G. (1966). *J. Nutr.* **89**, 465–470.
Hill, F. W. and Brambila, S. (1965). *Fedn Proc. Fedn Am. Socs exp. Biol.* **24**, 501.
Hill, R., Lingzasoro, J. M., Chevallier, F. and Chaikoff, I. L. (1958). *J. biol. Chem.* **233**, 305–310.
Holman, R. T. (1960). *J. Nutr.* **70**, 405–410.
Hopkins, D. T. and Nesheim, M. C. (1967). *Poult. Sci.* **46**, 872–881.
Hopkins, D. T., Witter, R. L. and Nesheim, M. C. (1963). *Proc. Soc. exp. Biol. Med.* **114**, 82–86.
Husbands, D. H. (1970). *Biochem. J.* (in press).
Husbands, D. H. and Brown, W. O. (1965). *Comp. Biochem. Physiol.* **14**, 445–451.
Isaaks, R. E., Davies, R. E., Reisier, R. and Couch, C. R. (1963). *J. Am. Oil Chem. Soc.* **40**, 747–749.
Johnson, A. R., Fogerty, A. C., Pearson, J. A., Shenstone, F. C. and Bersten, A. M. (1969). *Lipids* **4**, 265–269.
Leclerc, B., Argot, A. and Blum, J.-C. (1966). *C.r. hebd. Séanc. Acad. Sci. Paris* **262**, 2540–2542.
Leveille, G. A. (1969). *Comp. Biochem. Physiol.* **28**, 431–435.
Levin, E., Johnson, R. M. and Albert, S. (1957). *J. biol. Chem.* **228**, 15–21.
Lowenstein, J. M. (1963). *Biochem. Soc. Symp.* **24**, 57–61.
Lynen, F. (1961). *Fedn Proc. Fedn Am. Socs exp. Biol.* **20**, 941–951.
Lynen, F., Matsuhashi, M., Numa, S. and Schweizer, E. (1963). *Biochem. Soc. Symp.* **24**, 43–56.
Majerus, P. W. and Vagelos, P. R. (1967). *Adv. Lipid Res.* **5**, 1–33.
Marion, J. E. and Edwards, H. M. (1963). *J. Nutr.* **79**, 53–61.
Martin, D. B. and Vagelos, P. R. (1962). *J. biol. Chem.* **237**, 1787–1792.
McIndoe, W. M. (1959). *Biochem. J.* **72**, 153–159.
Mead, J. F. and Howton, D. R. (1957). *J. biol. Chem.* **229**, 575–582.
Menge, H. (1967). *J. Nutr.* **92**, 148–152.
Morris, L. J. (1966). *J. Lipid Res.* **7**, 717–732.
Morris, L. J. and Nichols, B. W. (1967). *In* "Chromatography" (E. Heftmann, ed.), pp. 466–509. Reinhold, New York.
Nasr, A. N. and Shostak, S. (1965). *Nature, Lond.* **207**, 1395.
Noyan, A., Lossow, W. J., Brot, N. and Chaikoff, I. L. (1964). *J. Lipid Res.* **5**, 538–541.
Numa, S., Bortz, W. M. and Lynen, F. (1965a). *Adv. Enzyme Reg.* **3**, 407–429.
Numa, S., Ringleman, E. and Lynen, F. (1965b). *Biochem. Z.* **343**, 243–247.
O'Hea, E. K. and Leveille, G. A. (1968). *Comp. Biochem. Physiol.* **26**, 111–120.
Pickles, V. R. (1967). *Biol. Rev.* **42**, 614–652.
Rand, N. T., Scott, H. M. and Kummerow, F. A. (1958). *Poult. Sci.* **37**, 1075–1085.
Raju, P. K. and Reiser, R. (1967). *J. biol. Chem.* **242**, 379–384.

Renner, R. and Elcombe, A. M. (1963). *Fedn Proc. Fedn Am. Socs exp. Biol.* **22,** 490.
Robinson, D. S. (1964). *In* "Metabolism and Physiological Significance of Lipids" (R. M. C. Dawson and D. N. Rhodes, eds.), pp. 275–285. D. W. Wiley, London, New York and Sydney.
Rudman, D. (1963). *J. Lipid Res.* **4**, 119–129.
Schjeide, O. A. and Urist, M. R. (1956). *Science, N.Y.* **124**, 1242–1244.
Schjeide, O. A. and Wilkens, M. (1964). *Nature, Lond.* **201**, 42–44.
Schjeide, O. A., Wilkens, M., McCandless, R. G., Munn, R., Peterson, M. and Carlsen, E. (1963). *Am. Zool.* **3**, 167–184.
Scott, M. L. (1962). *Nutr. Abstr. Rev.* **32**, 1–8.
Senior, J. R. (1964). *J. Lipid Res.* **5**, 495–521.
Shrimpton, D. H. (1963). *J. appl. Bact.* **26**, i–ii.
Spencer, A. F. and Lowenstein, J. M. (1966). *J. biol. Chem.* **237**, 3640–3648.
Steinberg, D. (1963). *In* "The Control of Lipid Metabolism" (J. K. Grant ed.), pp. 111–123. Academic Press, London.
Van Dorp, D. A., Beerthuis, R. K., Nagteren, D. H. and Vonkeman, H. (1964). *Biochim. biophys. Acta* **90**, 204–207.
Van Dorp, D. A., Jouvenaz, G. H. and Struijk, C. B. (1967). *Biochim. biophys. Acta* **137**, 396–399.
Vanstone, W. E., Maw, W. A. and Common, R. H. (1955). *Can. J. Biochem. Physiol.* **33**, 891–903.
Waite, M. and Wakil, S. J. (1962). *J. biol. Chem.* **237**, 2750–2757.
Wakil, S. J. (1958). *J. Am. chem. Soc.* **80**, 6465.
Weiss, J. F., Naber, E. C. and Johnson, R. H. (1967). *J. Nutr.* **93**, 142–152.
West, C. E. and Rowbotham, T. R. (1967). *J. Chromat.* **30**, 62–76.
Wilson, J. W. and Leduc, E. H. (1963). *J. Cell Biol.* **16**, 281–296.
Yang, P. C., Bock, R. M., Hsu, R. Y. and Porter, J. W. (1965). *Biochim. biophys. Acta* **110**, 608–615.
Young, J. W., Shrago, E. and Lardy, H. A. (1964). *Biochemistry, N.Y.* **3**, 1687–1695.

13 Protein Metabolism

K. N. BOORMAN and D. LEWIS

Department of Applied Biochemistry and Nutrition, School of Agriculture, University of Nottingham, Sutton Bonington, Loughborough, Leicestershire, England

I. Introduction

Quantitatively, the most important fate of amino acids in the fowl's body is their incorporation into protein. Protein synthesis is an area of active research and it is felt a description of the process would be out of place since specifically avian characteristics do not seem to have emerged. This account is therefore limited to a discussion of the metabolism of the amino group and of individual amino acids, particularly where these processes are thought to differ from those in mammals. In many cases the reader's knowledge of established metabolic pathways and current developments in mammalian biochemistry has been assumed, to avoid unnecessary duplication. For general information, reference can be made to the very exhaustive reviews by

Abbreviations used in this account are as follows: AMP, ADP and ATP = adenosine monophosphate, diphosphate and triphosphate respectively; FAD = flavin-adenine dinucleotide; FH_4 = tetrahydrofolic acid; FMN = flavin mononucleotide; GTP = guanosine triphosphate; NAD = nicotinamide-adenine dinucleotide; NADP = nicotinamide-adenine dinucleotide phosphate; UDP = uridine diphosphate.

Meister (1965), Sallach and Fahien (1969), Greenberg (1969), Rodwell (1969) and Kun (1969).

To establish that a metabolic pathway exists it is necessary to demonstrate that an end-product appears in response to the administration of a particular substrate and to substantiate this by identifying intermediates in the pathway and by establishing that enzymes exist which can account for the overall metabolic conversion. There are few cases in which such complete information is available for the bird. In general therefore, the approach taken here has been to establish whether all or some of the enzymes of a pathway which is known to occur in mammals exist also in the bird. The existence of appropriate enzymes does not establish that a particular pathway necessarily occurs *in vivo* and it certainly provides little information on the overall quantitative significance of such a pathway. The general lack of information, however, necessitates this approach and this lack is keenly felt by those familiar with current developments in mammalian biochemistry, where greater understanding of the quantitative significance of pathways and endocellular control mechanisms allows metabolic pathways to be seen in their true context.

Concerning the knowledge of biochemical pathways in the fowl it should be recognized that considerable information has been gathered by experimental embryologists; where this is considered relevant it is quoted. (For extensive reviews see Weber, 1965, 1967; Romanoff and Romanoff, 1967.) But the existence of a pathway in the embryo does not necessarily imply that the same pathway exists in the post-natal animal. As will emerge, marked changes in the activities of some enzymes are associated with hatching.

During the final preparation of this chapter Brown (1970) published a review of the metabolism of nitrogen-containing compounds in birds in which metabolism in the embryo, detoxification mechanisms, tissue metabolite levels and nitrogen excretion are included are reviewed at greater length.

II. Transport of Amino Acids

Amino acid transport across cell membranes can occur by passive diffusion, but of greater significance is the active process which allows the accumulation of amino acids by cells in the absence of a favourable concentration gradient. The mechanism of active transport is not understood. The kinetics of the process are consistent with the idea that an amino acid associates with a "carrier" molecule at the extracellular side of the membrane, traverses the membrane as a complex and dissociates from the carrier at the intracellular surface of the membrane. Active transport requires the provision of energy; the mechanism by which energy is made available is not fully understood. There also appears to be a requirement for sodium ions. H. N. Christensen (1962) has presented a comprehensive review of the characteristics of amino acid transport.

With regard to the specificity of transport, it is generally thought that anionic, neutral and cationic amino acids are transported by discrete systems although the specificity of each system is not absolute. There is evidence to suggest that several different systems exist for the transport of neutral amino acids, each system having a degree of specificity for a group of chemically similar neutral amino acids (H. N. Christensen, 1962; Christensen *et al.*, 1967; Eavenson and Christensen, 1967). All evidence is, however, not in accord with this hypothesis (Matthews and Laster, 1965; Jacquez *et al.*, 1970).

Investigations on avian tissues do not suggest marked differences, in respect of transport, from other animal tissues. Active transport of amino acids has been observed in several studies and the existence of more than one transport system for neutral amino acids in fowl intestine (Paine *et al.*, 1959; Lin and Wilson, 1960; Tasaki and Takahashi, 1966), pigeon erythrocytes (Vidaver *et al.*, 1964) and embryonic chick bone (Adamson and Ingbar, 1967) has been indicated. An active transport system for lysine in fowl intestine (Bird, 1968) and a discrete system for the cationic amino acids in the kidney tubule (Boorman, 1971) have been demonstrated. (See also Chapter 3.)

III. Metabolism of the Amino-Group

A. GENERAL PATHWAYS

The amino-group of most amino acids passes readily into metabolism. It can either be transferred from an amino acid to a suitable acceptor (an α-ketoacid) to produce another amino acid without the appearance of free ammonia (transamination), or the amino-group can be removed as free ammonia (deamination). The cell can thus exert a measure of control over the relative amounts of each amino acid present and hence maintain a suitable pattern of amino acids for protein synthesis and other metabolic processes. Animal cells contain a wide variety of enzymes which catalyse the transfer of the amino-group (transaminases or aminotransferases). The process is important not only in changing the relative proportions of the amino acids but also in the ultimate removal of nitrogen from the body (transdeamination—see below). Deamination can occur oxidatively or non-oxidatively. Enzymes having a relatively low substrate specificity exist which can catalyse oxidative deamination producing ammonia from a wide range of amino acids. This ammonia can then be rendered into a form suitable for temporary storage, translocation or elimination. In animal cells there is one special case of oxidative deamination in which the highly specific enzyme L-glutamate dehydrogenase catalyses the removal of the amino-group of L-glutamic acid. Thus, amino-groups from a variety of amino acids can be transferred to α-ketoglutaric acid by transamination, producing glutamic acid which is then deaminated. This linked process constitutes an important pathway for the elimination of nitrogen and has been termed transdeamination. In the case of the hydroxyamino acids (serine and threonine) non-oxidative deamination is possible by means of enzymes termed dehydratases, these will be described when the individual amino acids are discussed.

B. NON-SPECIFIC OXIDATIVE DEAMINATION

1. L-*Amino Acid Oxidases [EC 1.4.3.2]*[1]

Boulanger and Osteux (1955a, b, 1956) demonstrated and purified an L-amino acid oxidase from turkey liver which deaminated several amino acids. The enzyme showed highest activity with the cationic amino acids (arginine, ornithine and lysine); glutamic acid, aspartic acid, serine, valine and isoleucine were not attacked. The cationic amino acids produced their respective α-keto analogues; the reaction was typical of an oxidase. Boulanger *et al.* (1957), using ^{15}N, confirmed that the α-amino-groups of the cationic amino acids were liberated and it has since been shown (Mizon *et al.*, 1970) that activity is associated with the mitochondrial fraction of the cell, that FMN is a co-factor and manganese is a powerful activator of the enzyme. An L-amino acid oxidase occurs in chicken liver but appears to differ from the turkey liver enzyme. Struck and Sizer (1960) purified an oxidase from the microsomal fraction of chicken liver which deaminated several amino acids, showing its greatest activity with leucine. Further studies with an essentially similar preparation (Shinwari and Falconer, 1967) demonstrated that although the enzyme exhibited low activity, inhibition and activation could occur *in vivo*. The chicken liver enzyme is apparently also a typical oxidase but its substrate specificity appears to resemble mammalian amino acid oxidases more than the turkey liver enzyme.

2. D-*Amino Acid Oxidases [EC 1.4.3.3]*

It is generally accepted that animals do not utilize D-amino acids for protein synthesis or other essential metabolic pathways and, therefore, the existence of relatively active D-amino acid oxidases in tissues is not readily explicable. Whatever the reason for the existence of D-amino acid oxidases their presence in fowl tissues is of importance because they are believed to participate in the conversion of D-amino acids to their corresponding L-isomers in the following manner;

$$\text{D-amino acid} \xrightarrow[\text{oxidase}]{\text{D-amino acid}} \alpha\text{-keto analogue} \xrightleftharpoons{\text{transaminase}} \text{L-amino acid}$$

$$\qquad\qquad\downarrow \text{ammonia} \qquad\qquad\qquad\qquad \text{amino-group} \nearrow$$

[1] For convenience the numerical system and advised trivial names of enzymes of the International Union of Biochemistry (see Florkin and Stotz, 1965) have been used. The use of these conventions is not meant to imply that an enzyme in avian tissue has exactly the same properties as a similarly named enzyme in mammalian tissue.

Natural D-amino acids are known only as components of materials of microbiological origin.

Oxidative deamination of D-amino acids has been demonstrated in chicken liver (Williams *et al.*, 1949; Trufanov and Pavlova, 1951) and the need for a flavin co-factor has been demonstrated (Walaas and Walaas, 1956). The activity of the enzyme is higher in kidney homogenates than in liver (Bauriedel, 1962) and dietary supplements of D-methionine do not enhance the activity of the enzyme to a greater extent than supplements of L-methionine (Bauriedel, 1963). In mammalian tissue there is some doubt as to whether D-amino acid oxidase is a different enzyme from glycine oxidase, current evidence suggests that the two are identical (see Meister, 1965 p. 300; Sallach and Fahien, 1969 pp. 52–53). This aspect does not appear to have been examined in avian tissues.

The presence of the D-isomers in numerous synthetic DL-amino acids has stimulated many nutritional and a few metabolic studies of the conversion of D- to L-amino acids in the fowl. Different D-amino acids are converted with different efficiencies. Some of the findings of the most comprehensive nutritional study to date are shown in Table 1 and generally accord with earlier

Table 1

Utilization of dietary D-amino acids by the growing chick[a]

Amino acid	Utilization
Methionine Phenylalanine Leucine Proline	Equivalent to or almost equivalent to corresponding L-isomers
Valine	Approximately half as potent as L-isomer
Tryptophan Histidine Alloisoleucine	Little nutritional value
Lysine Threonine Arginine	No nutritional value but no evidence of toxicity[b]

[a] Modified from Sugahara *et al.* (1967).

[b] Studies of the utilization of the D-isomers of the non-essential amino acids showed that D-aspartic acid and to a lesser extent D-alanine caused growth retardations.

studies. It should be noted that the nutritional potency of a particular amino acid is not simply a measure of the ease with which D-amino acid oxidase catalyses its deamination. Rapid removal of the α-keto analogue by an alternative metabolic pathway, or failure to participate in a transamination reaction would also interfere with conversion. In this connexion it is interesting that both D-lysine and D-threonine are not converted to the L-isomers (Table 1) and that neither appear to participate readily in transaminations in animal tissues.

There has been interest in the use of α-hydroxy-γ-methylmercaptobutyric acid (a methionine α-hydroxy analogue) as a substitute for L-methionine. Several studies have shown complete or near nutritional equivalence between these two compounds (Bird, 1952; Machlin and Gordon, 1959; Calet and Melot, 1961; Bauriedel, 1963; Gordon and Sizer, 1965). This indicates that the α-hydroxy analogue of methionine can be oxidized and aminated. Enzyme systems capable of catalysing the conversion of α-hydroxyacids to α-ketoacids have been demonstrated in chicken liver (Gordon, 1965).

C. TRANSAMINATION AND TRANSDEAMINATION

1. Aspartate Transaminase [*EC 2.6.1.1*]

Of the transaminases of animal tissues the most frequently studied has been aspartate transaminase which catalyses the reaction:

$$\text{Glutamic acid} + \text{oxaloacetic acid} \rightleftharpoons \alpha\text{-ketoglutaric acid} + \text{aspartic acid}$$

and, like all known transaminases, requires pyridoxal phosphate as a co-factor. The enzyme has been obtained in a pure form from pigeon breast muscle and the amino acid sequence close to the point of attachment of the pyridoxal phosphate has been elucidated (Polyanovskii and Vorotnitskaya, 1965; Vorotnitskaya *et al.*, 1968). The enzyme has also been isolated from chicken liver (Sanchez de Jimenez *et al.*, 1967). Activity was restricted almost entirely to the soluble fraction of the cell, a feature which is in contrast to the situation in rat liver where as least two isozymes exist, one in the soluble fraction and the other in the mitochondria. The enzyme from chicken liver showed some similarities with the rat mitochondrial isozyme (pH optimum, chromatographic properties on DEAE-cellulose and serological properties) but was similar to the supernatant isozyme in other respects (heat stability and apparent affinity for α-ketoglutaric acid). The significance of the existence of two different forms of the same enzyme in different localities of the cell in the control of metabolism in mammals is currently emerging. The failure to demonstrate isozymes of the enzyme in chicken liver may indicate a lower level of sophistication in the control of metabolism in the fowl, but where cellular fractionation techniques are involved a weight of evidence is needed before conclusions can be drawn. Bertland and Kaplan (1970) have clearly demonstrated the existence of two forms of aspartate transaminase in chicken heart. It is of interest that Sheid and Hirschberg (1967) found that aspartate transaminase activity of chick embryo liver was associated with the mitochondrial fraction. It may be that the distribution of enzymes within the cell is different in embryonic tissue (see also glutamate dehydrogenase—Section III C.4).

2. Other Transaminases

Other transaminases of avian tissue have not been isolated and studied to the same extent as aspartate transaminase. Some general studies on the extent of transamination are referred to elsewhere (Section III D) and others have been reported in which the activities of specific transaminases have been measured; in that the latter provide evidence of the existence of other transaminases in avian tissue they are described below. Alanine transaminase [EC 2.6.1.2] activity has been found in the tissues of chickens (Gessler, 1965; Zimmerman *et al.*, 1968) and in embryonic tissues (Koivusalo *et al.*, 1963; Ponomareva and Drel, 1964; Sheid and Hirschberg, 1967). This enzyme catalyses the reversible transfer of the amino group of glutamic acid to pyruvic acid, producing alanine.

Tyrosine transaminase [EC 2.6.1.5] activity has been shown in chick embryos in studies concerned with enzyme changes during development (Chan and Cohen, 1964). The enzyme in embryonic tissue is specific for α-ketoglutaric acid as amino group acceptor (Litwack and Nemeth, 1965; Constantsas and Knox, 1967). Leucine transaminase [EC 2.6.1.6] activity has been measured in various tissues of the chicken (Shiflett and Haskell, 1969). In mammals it is uncertain whether separate transaminases exist for leucine, isoleucine and valine and the activity of leucine transaminase is higher in the kidney than in the liver (Ichihara and Koyama, 1966). The substrate specificity of the chicken enzyme was not studied but it was shown that activity was higher in the kidney than in the liver. It appears that the branched-chain amino acids can participate in another transamination reaction in the chick where these amino acids serve as amino donors for the amination of the α-keto analogue of methionine, a function which the α-amino-group of glutamine serves in the rat (Gordon, 1965). Phosphoserine transaminase has also been demonstrated in chicken liver (Walsh and Sallach, 1966). This enzyme is important in serine biosynthesis (see Section IV B.2) and utilizes α-ketoglutarate as amino-group acceptor.

Transaminases do not invariably catalyse the removal of an α-amino-group. Quastel and Witty (1951) demonstrated the participation of ornithine in the transamination reaction with pyruvate in pigeon liver and Vecchio and Kalman (1968) partially purified a transaminase [EC 2.6.1.13] from chicken liver which catalyses the reversible removal of the δ-amino group of ornithine, utilizing α-ketoglutarate as amino acceptor, thus:

$$\text{Ornithine} + \alpha\text{-ketoglutarate} \rightleftharpoons \text{glutamic } \gamma\text{-semialdehyde} + \text{glutamic acid}$$

The activity of this enzyme was lower in chick kidney than in liver, in contrast to the situation in the rat. A similar reaction in which the γ-amino group of γ-aminobutyric acid (see Section IV A) is removed with the formation of succinic semialdehyde has been demonstrated in developing chick brain (Van Den Berg *et al.*, 1965).

3. *Tissue Distribution and Activities of Transaminases*

Zimmerman *et al.* (1968) showed that the activity of aspartate transaminase (ASPT) was generally higher in all tissues in chickens and pigeons than that of alanine transaminase (ALAT). ASPT activity was highest in the plasma and heart; kidney and liver showed approximately equal activity. ALAT activity was highest in plasma and liver. The distributions and activities of the enzymes did not differ markedly or in any readily interpretable way from their distributions and activities in mammalian tissues. In the chick embryo ASPT activity is also higher than ALAT activity, the latter being very low (Ponomareva and Drel, 1964). The activities of both transaminases increase during the second half of incubation and rise sharply at hatching (Sheid and Hirschberg, 1967). Tyrosine transaminase shows a similar pattern of development (Litwack and Nemeth, 1965).

The activity of transaminases can be influenced by dietary factors. A deficiency of pyridoxine in the diet causes a decrease in the activities of ASPT (Goswami and Robblee, 1958) and leucine transaminase (Shiflett and Haskell, 1969). Gessler (1965) found that increasing the dietary protein level did not influence the activities of either ASPT or ALAT in plasma, but food restriction caused a sharp increase in ASPT activity and a decrease in ALAT activity.

4. *Transdeamination and Glutamate Dehydrogenase [EC 1.4.1.3]*

The importance of glutamate dehydrogenase in transdeamination and the elimination of nitrogen from the body has been described previously (Section III A). Glutamate dehydrogenase catalyses the coupled dehydrogenation and deamination of glutamic acid thus:

$$\text{Glutamic acid} + H_2O + NAD(P)^+ \rightleftharpoons \alpha\text{-ketoglutaric acid} + NAD(P)H + NH_4{}^+$$

In animal cells, in contrast to the situation in micro-organisms where specificity is shown for either NAD or NADP, either pyridine nucleotide can be utilized as hydrogen acceptor although conditions within the cell at any one time exert a measure of control over which nucleotide is utilized. The reaction is reversible and, if equilibrium is attained, strongly favours glutamate formation. Glutamate dehydrogenase can therefore provide a pathway for the incorporation of ammonia (non-protein nitrogen) into amino acids and for net glutamate catabolism to occur, the products of catabolism must be rapidly removed.

Because glutamic acid can be produced by several transamination reactions and because product-removal is required for net glutamate catabolism, the relationship between the activities of transaminases, glutamate dehydrogenase and the enzymes initiating uric acid production is critical in controlling the direction of nitrogen metabolism. Efficient nitrogen elimination by the transdeamination pathway requires high activity of all these enzymes. High transamination activity in the absence of high dehydrogenase activity and/or uric acid synthesizing activity would lead to amino-group conservation and protein anabolism. A change in the direction of nitrogen metabolism of this type appears to occur at hatching in the chick as evidenced by the fact that transamination activity increases (Section II C.3) and glutamate dehydrogenase activity decreases (Sheid and Hirschberg, 1967).

The control of glutamate dehydrogenase activity in ureotelic vertebrates is complex. Apart from simple mass action considerations such as those referred to above, activity can be modified allosterically by a number of metabolites including purines, purine nucleotides and pyridine nucleotides (see Sallach and Fahien, 1969 pp. 30–32). In ureotelic vertebrates sufficient is known about these modifiers and their effects on the relationship between glutamate dehydrogenase activity and carbamoyl phosphate synthase (the enzyme initiating urea synthesis) activity to account for significant ammonia production by glutamate dehydrogenase and its utilization by carbamoyl phosphate synthase. Thus a complex pattern of control of the transdeamination pathway is emerging (see Sallach and Fahien, 1969 p. 67). In view of the different form of nitrogen elimination and the relative lack of information it remains to be seen whether a similar pattern of control can be postulated for birds. As will emerge, the activity of glutamate dehydrogenase in birds is affected by similar modifiers, but the enzyme is not exactly similar to the glutamate dehydrogenase of ureoteles.

Glutamate dehydrogenase occurs in chicken liver and has been crystallized from this source (Snoke, 1956). It is generally accepted that this is a mitochondrial enzyme and therefore the finding that in embryo liver activity is higher in the cytoplasm than in the mitochondria during the 7th to 14th day of incubation and that thereafter the situation is reversed (Solomon, 1959) was taken to indicate that the intracellular location of an enzyme could change during development. Mason and Hooper (1969), however, have re-examined this finding and report that the activity is higher in the mitochondria than in other cellular fractions throughout embryogenesis. Ecobichon (1966) examined several tissues from mature cockerels and found activity restricted to the liver and kidney, although Freedland *et al.* (1966) found activity in the heart and brain of quail. The molecular weight of the enzyme has been variously quoted as $5{\cdot}0 \times 10^5$ (Rogers *et al.*, 1963), $4{\cdot}3 \times 10^5$ (Frieden, 1962) and $3{\cdot}26 \times 10^5$ (Anderson and Johnson, 1969). Frieden (1962) noted that in concentrated solutions the molecular weight was higher, indicating some association of sub-units. Anderson and Johnson (1969) noted that their value was similar to that found for the active sub-unit of glutamate dehydrogenase from other species (ox and dogfish). In both these investigations it was found that the tendency for the sub-units of the chicken liver enzyme to associate was markedly less than that found for the enzyme from other species. This was regarded as indicating a significant distinction between the avian enzyme and that from non-uricotelic vertebrates but this is not evident in marked differences between the amino-acid compositions of the enzymes from the chicken, ox and dogfish (Corman *et al.*, 1967); an antiserum prepared from rabbit did not distinguish between the enzymes from frog, tadpole, ox and chicken (Wiggert and Cohen, 1966). Similarity was also shown between the chicken and bovine enzymes in respect of the manner in which inactivation by silver ions occurs (Rogers, 1969).

In their activation or inhibition by purine nucleotides there is similarity between the glutamate dehydrogenases from chicken and pigeon and both differ from mammalian enzymes (see Frieden, 1965). The complexity of comparative studies has been demonstrated by Freedland *et al.* (1966) who have shown the existence of isozymes of glutamate dehydrogenase in the liver, kidney, brain and heart of quail. Both isozymes showed typical glutamate dehydrogenase characteristics (activation by ADP and inhibition by GTP) in all tissues except heart where the activity of one isozyme was not modified by purine nucleotides. Freedland *et al.* (1967) have further shown that an isozyme similar to that found in quail heart occurs in rat and ox, but in these species it occurs in the liver. The isozymes in chicken liver were similar to those found in quail liver.

D. OXIDATIVE DEAMINATION VERSUS TRANSDEAMINATION

As to the relative importance of direct oxidative deamination (*via* L-amino acid oxidase(s)) and transdeamination in amino acid degradation in avian tissues observations conflict. Bassler and Hammar (1958) studied the deamination of leucine, isoleucine, methionine, α-aminobuytric acid, valine and norvaline in cell fractions from livers and kidneys of chickens and found that α-ketoglutaric acid stimulated deamination; they could find no evidence of direct oxidative deamination. Chen and Li (1963) found that transamination activity was similar in rat, pigeon, tortoise and toad livers for 22 amino acids. In contrast, Efimochkina (1958, 1959) found evidence of oxidative deamination for several amino acids in the livers and kidneys of chickens, pigeons and turkeys. In these studies transamination was inhibited by hydroxylamine or fluoroacetate and deamination of aspartate and alanine was found to be inhibited. Deamination of leucine and phenylalanine was not consistently affected, the extent of inhibition varying markedly between individual birds; deamination of other amino acids (glycine, isoleucine, valine, tryptopan, histidine and the cationic amino acids) was not affected. As might be expected, inhibition of transamination did not reduce the rate of deamination of glutamate.

The findings of Efimochkina (1958, 1959) are consistent with the idea that, for many amino acids, direct oxidative deamination can occur in avian tissues. The existence of avian L-amino acid oxidase(s) has been referred to previously (Section III B.1). These observations do not necessarily mean, however, that oxidative deamination is an important pathway *in vivo*: this pathway may only become active when transdeamination is inhibited. It would be of interest to determine whether significant oxidative deamination occurs in mammalian tissues when transamination is inhibited. On the basis of existing evidence it would be unwise to conclude that oxidative deamination is of greater significance in birds than in mammals; it appears, however, that when transdeamination cannot occur, deamination of many amino acids can proceed *via* the direct oxidative route in birds.

E. ELIMINATION OF NITROGEN

1. Pathway of Nitrogen Elimination

In the bird the major nitrogen excretory product is uric acid. The pathway of biosynthesis of uric acid has been elucidated and the metabolic precursors of the nitrogen in the purine ring have been identified. Two of the nitrogen atoms arise from the amide-nitrogen of glutamine and the other two nitrogen atoms arise from the amino-nitrogens of glycine and aspartic acid. Aspartate can be formed from oxaloacetic acid by transamination; glycine, if not supplied in the diet in sufficient quantity may be synthesized, although whether this synthesis proceeds at a rate sufficient to meet demands is not entirely clear (see Section IV B.3). Glutamine is synthesized from glutamic acid and ammonia arising from the deamination of amino acids. It should be emphasized that it is the nitrogen from ammonia that appears in uric acid. Glutamine acts as a "carrier form"of ammonia and in addition to providing ammonia for purine synthesis, glutamine synthesis can prevent accumulation of ammonia. Glutamine also appears to be important in the transport of potential ammonia within the body. The functions of glutamine in the chick were illustrated by Olsen *et al.* (1963) who showed that incorporation of diammonium citrate and glutamic acid into the diet caused marked increases in blood glutamine concentration and uric acid excretion.

2. Glutamine Synthesis

Glutamine synthesis occurs as follows:

$$\text{Glutamic acid} + NH_3 + \text{ATP} \rightleftharpoons \text{glutamine} + \text{ADP} + \text{phosphate} + H_2O$$

The reaction is catalysed by glutamine synthetase [EC 6.3.1.2] and is reversible, glutamine or one of the other products must therefore be removed continually for net glutamine synthesis to proceed. This enzyme is widely distributed in living tissues and has been isolated from pigeon (Speck, 1949) and chicken liver (Fazekas and Denes, 1966). The avian system is similar to that from other sources in that metal ions (Mg^{2+}, Mn^{2+}, Co^{2+}) are required for glutamine synthesis. Glutamine synthetase activity can be detected in the brain, liver, heart and pancreas of birds but not in kidney, spleen, skeletal muscle, intestine or whole blood. The distribution in mammals differs in that the activity in liver is lower, is absent in the heart and pancreas but is present in the kidney. It is of interest that the distribution in snakes which are also uricoteles is similar to that in birds (Wu, 1963). The wider distribution in uricoteles is consistent with the fact that this enzyme serves in ammonia detoxification and purine synthesis for both nucleic acid production and nitrogen elimination in these animals; the lack of activity in the kidney seems worthy of further investigation.

It has been observed that the amide-nitrogen of asparagine can act as a precursor for the amide-nitrogen of glutamine; in mammalian tissues and pigeon liver the amide group of asparagine is first hydrolysed, releasing ammonia which is then incorporated into glutamic acid (T'ing-Sen, 1959; Lestrovaya, 1961). Asparaginase, which catalyses the hydrolysis of asparagine, has been isolated from chicken liver (Ohnuma *et al.*, 1967). A similar relationship between asparagine and glutamine has been demonstrated in the chick embryo liver. In the embryo asparagine may be important in ammonia metabolism since, in the first half of the incubation period, the activity of asparaginase is very high and the concentration of asparagine exceeds that of glutamine (Drel, 1964). Synthesis of asparagine has been demonstrated in embryonic chick liver (Arfin, 1967). The role of asparagine in the post-natal bird has not been studied and its role generally in animals is not well understood.

Glutamine synthetase has been of great interest to embryologists since it appears abruptly in the retina at about the 17th day of incubation and its activity increases rapidly, reaching a peak after hatching (Rudnick and Waelsch, 1955). The mechanism of this increase and its premature induction by certain steroid hormones has been the subject of several studies (see Reif-Lehrer, 1968).

Although the reaction catalysed by glutamine synthetase is reversible, it is probable that the hydrolysis of glutamine, releasing glutamate and ammonia is catalysed by glutaminase [EC 3.5.1.2]. This reaction is of importance in the conservation of "fixed base" cations by substitution by ammonium ions in the kidney.

3. *Uric Acid Synthesis*

The sequence of chemical reactions involved in the synthesis of purines has been elucidated. The biosynthetic pathway usually described refers to the pigeon liver and most of these reactions have been observed in the liver of the fowl in the several studies of Buchanan, Hartman and co-workers. Since most biochemistry texts include descriptions of this pathway a complete description will not be given here; the reader is referred to Meister (1965, pp. 629–636) for a concise account and Schulman (1961) for a detailed review. A summary of the chemical changes is shown in Fig. 1. It should be noted that an alternative reaction has been demonstrated for the initiation of purine biosynthesis (reactions 1a and 1b, Fig. 1) in avian tissue in which 5-phosphoribosylamine is formed from ribose-5-phosphate, ATP and ammonia in a reaction catalysed by an enzyme that is distinct from phosphoribosyl pyrophosphate amidotransferase (Reem, 1968). Inhibition of reaction 1b by purine nucleotides is thought to provide a control mechanism for purine synthesis. Hartman (1963) purified phosphoribosyl pyrophosphate amidotransferase from chicken liver and found that the enzyme was not inhibited by purine nucleotides. While acknowledging that this lack of inhibition might be the result of alterations to the enzyme caused by the method of purification, Hartman did question the need for a control system for purine synthesis in view of the fact that quantitatively the major function of purine biosynthesis in avian liver is the formation of uric acid for nitrogen elimination. It does, however, appear that this enzyme is subject to allosteric inhibition by the end-products of the pathway in the pigeon liver and that the availability of substrates exerts effects on the conformation of the enzyme and thus its activity (Rowe *et al.*, 1970). This evidence of a complex control system in the pigeon liver suggests that further investigation of the chicken liver enzyme might reveal a similar system.

The purine ring first appears in inosinic acid (hypoxanthine ribonucleotide) and the overall reaction from primary precursors can be summarized as follows:

$$\begin{aligned}&2NH_4^+ + 2HCOO^- + HCO_3^- + \text{glycine} + \text{aspartate}\\&\quad + \text{ribose-5-phosphate} + 9ATP \longrightarrow \text{inosinic acid} + \text{fumarate}\\&\quad + 8ADP + 8\ \text{phosphate} + AMP + PP + 9H^+\end{aligned}$$

This equation includes 2 moles of ATP for glutamine formation and 2 moles for rendering the one-carbon units into the appropriate reduced states for incorporation into the purine ring.

There are two pathways which might account for the conversion of inosinic acid to hypoxanthine (reaction 11, Fig. 1). Inosinic acid pyrophosphorylase [EC 2.4.2.8] transfers the phosphoribosyl group to pyrophosphate producing hypoxanthine and phosphoribosyl pyrophosphate. 5′-Nucleotidase [EC 3.1.3.5] catalyses the hydrolysis of ribonucleotides to ribonucleosides and purine nucleoside phosphorylase [EC 2.4.2.1] catalyses the transfer of the ribose from a nucleoside to phosphate, releasing the free purine. It is not known which of these pathways is of greater physiological significance in the fowl.

The oxidation of hypoxanthine to uric acid is catalysed by xanthine dehydrogenase and proceeds *via* xanthine. Either hypoxanthine or xanthine can act as substrate for this enzyme which utilizes NAD as hydrogen acceptor. It has been established that the avian enzyme is a dehydrogenase (Richert and Westerfeld, 1951); it was previously thought that this reaction was catalysed by an oxidase in mammalian species, but recently evidence has been presented that the enzyme exists both in oxidase and dehydrogenase forms in rat liver (Della Corte and Stirpe, 1970). Xanthine dehydrogenase has been purified from chicken liver (Remy *et al.*, 1955; Rajagopalan

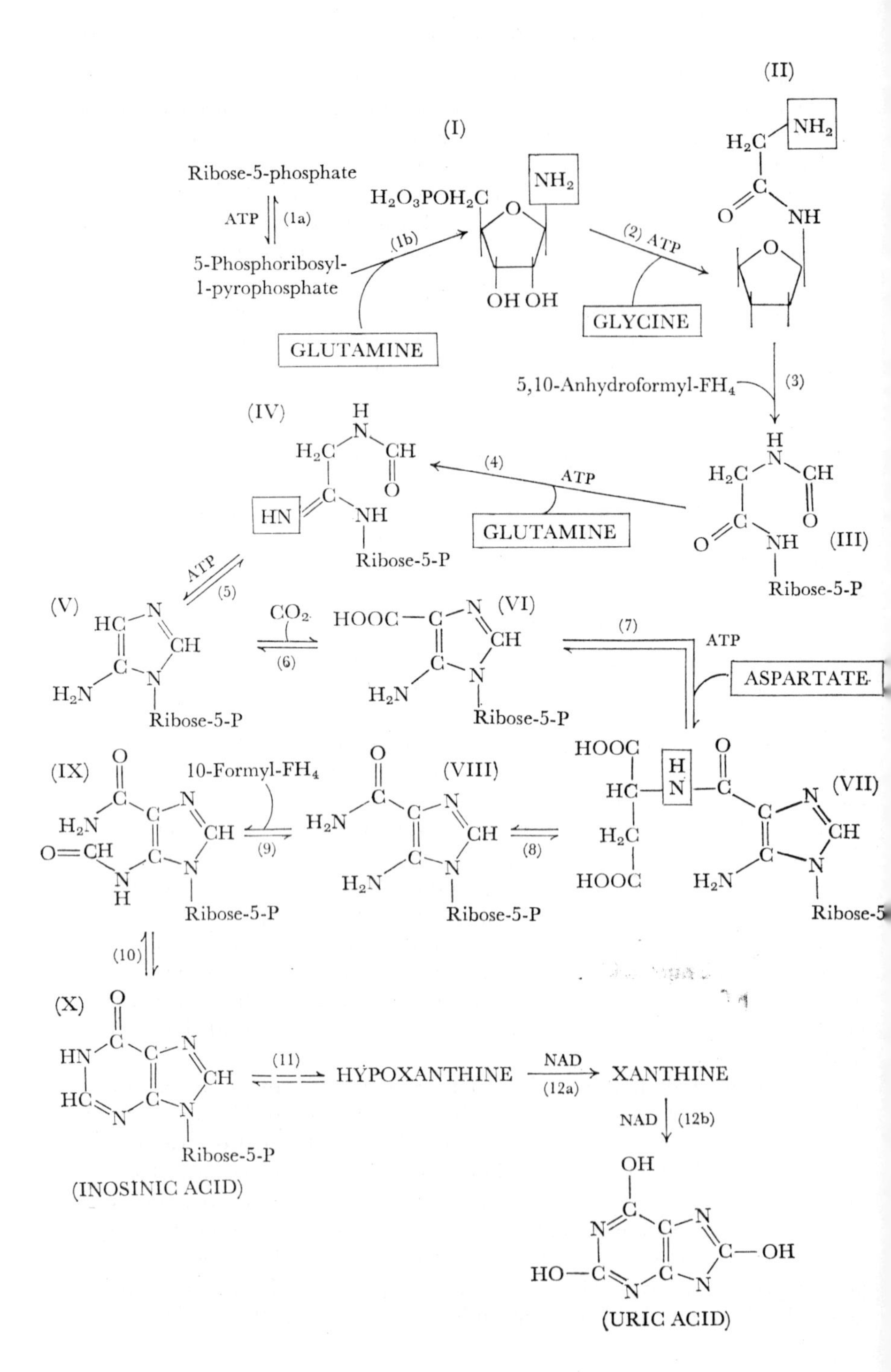
Ribose-5-phosphate
ATP
(1a)
5-Phosphoribosyl-1-pyrophosphate
(1b)
GLUTAMINE
(I)
H2O3POH2C
NH2
OH OH
(2) ATP
GLYCINE
(II)
5,10-Anhydroformyl-FH4
(3)
(III)
Ribose-5-P
(4)
ATP
GLUTAMINE
(IV)
HN
Ribose-5-P
ATP
(5)
(V)
Ribose-5-P
CO2
(6)
(VI)
HOOC
Ribose-5-P
(7)
ATP
ASPARTATE
(VII)
(8)
(VIII)
Ribose-5-P
10-Formyl-FH4
(9)
(IX)
Ribose-5-P
(10)
(X)
Ribose-5-P
(INOSINIC ACID)
(11)
HYPOXANTHINE
NAD
(12a)
XANTHINE
NAD
(12b)
(URIC ACID)

and Handler, 1967) and kidney (Landon and Carter, 1960) and has also been demonstrated in turkey liver; it is absent from pigeon liver (Richert and Westerfeld, 1951). Uric acid production from hypoxanthine is therefore extra-hepatic in the pigeon. Specific inhibition of the chicken liver enzyme by reduced NAD has been shown; the dehydrogenase form of the rat liver enzyme also exhibits this characteristic (Della Corte and Stirpe, 1970).

Xanthine dehydrogenase is present with relatively low activity in the embryo's liver but the activity increases considerably after hatching (Murison, 1969). Inosine but not xanthine nor hypoxanthine causes an increase in the activity of xanthine dehydrogenase in the liver (Della Corte and Stirpe, 1967). Dietary changes can also influence activity, 24-h starvation causing an apparent increase in activity which is the result of the maintenance of total liver xanthine dehydrogenase activity while total liver nitrogen decreases (Della Corte and Stirpe, 1967; Scholz and Featherstone, 1969). Longer periods of starvation cause depletion of liver xanthine dehydrogenase in some but not all breeds of domestic fowl (Scholz, 1970). Diets rich in protein cause increased activity in the liver which is not maintained on subsequent starvation (Scholz and Featherstone, 1969). The activity of xanthine dehydrogenase in the kidney is lower than in the liver but responds to alterations in dietary protein and short periods of starvation in a similar manner, activity in the kidney however remains relatively unchanged, irrespective of breed, during longer periods of starvation (Scholz, 1970). Starvation also causes a decrease in xanthine dehydrogenase activity in the pancreas (Curtis and Fisher, 1970), although whether this occurs in all breeds is not known.

IV. Metabolism of the Individual Amino Acids

A. ALANINE, ASPARTIC ACID AND GLUTAMIC ACID

These three amino acids are similar in that they can be formed by transamination from the ketoacids, pyruvate, oxaloacetate and α-ketoglutarate respectively, all of which can arise from carbohydrate metabolism. It is probable that transamination initiates the dissimilation of alanine and

---

Fig. 1. Biosynthesis of uric acid. Arabic numerals refer to the sequence in which the reactions occur. Roman numerals refer to the chemical intermediates in the pathway, the names of the intermediates are as follows:

I: 5′-phosphoribosylamine; II: 5′-phosphoribosyl-glycineamide;
III: 5′-phosphoribosyl-*N*-formylglycineamide;
IV: 5′-phosphoribosyl-*N*-formylglycineamidine;
V: 5′-phosphoribosyl-5-aminoimidazole;
VI: 5′-phosphoribosyl-4-carboxy-5-aminoimidazole;
VII: 5′-phosphoribosyl-4-(*N*-succinocarboxamide)-5-aminoimidazole;
VIII: 5′-phosphoribosyl-4-carboxamide-5-aminoimidazole;
IX: 5′-phosphoribosyl-4-carboxamide-5-formamidoimidazole.

aspartate and that glutamate catabolism is initiated *via* glutamate dehydrogenase (see Section III D). Avian glutamate dehydrogenase, like the enzyme from other species, shows some alanine dehydrogenase activity (Frieden, 1965).

Aspartate serves as a source of nitrogen in the synthesis of purines, pyrimidines and argininosuccinic acid (Section IV D.1). Studies of the avian enzymes initiating pyrimidine synthesis are due to Bowers and Grisolia (1962); full discussion of this pathway is beyond the scope of this treatment.

Glutamate, in addition to being a precursor of glutamine, also gives rise to γ-aminobutyric acid by decarboxylation. γ-Aminobutyrate has marked effects on the metabolism of nervous tissue and its production and effects have been studied in chick embryonic tissue (Sisken *et al.*, 1961; Van Den Berg *et al.*, 1965; Kuriyama *et al.*, 1968). The transamination of γ-aminobutyrate has also been observed in embryonic tissue (see also Section III C.2). The relationship between glutamate, arginine, ornithine and proline in the bird is discussed elsewhere (Section IV D.2).

B. GLYCINE AND SERINE

1. *Glycine*

Glycine can be formed from serine by serine hydroxymethyltransferase [EC 2.1.2.1] when the hydroxymethyl group of serine is transferred to tetrahydrofolic acid leaving glycine. The reaction is reversible and interconversion of glycine and serine has been observed in avian liver (Kisliuk and Sakami, 1954; Blakley, 1954; Greenberg, 1954). The hydroxymethyl of serine in the hydroxymethyltransferase reaction appears as 5,10-methylenetetrahydrofolate and from this compound glycine can be formed by a pyridoxal phosphate-containing enzyme system which utilizes NAD as a cofactor. This reaction is also reversible. The enzyme(s) catalysing the latter change and hydroxymethyltransferase appear to occur in a multi-enzyme system which accomplishes the interconversion of one mole of serine and two moles of glycine, thus:

$$\text{(a)}\quad \text{Serine} + \text{FH}_4 \rightleftharpoons \text{glycine} + \text{methyleneFH}_4$$

$$\text{(b)}\quad \underline{\text{methyleneFH}_4 + \text{NH}_3 + \text{CO}_2 + (2\text{H}) \rightleftharpoons \text{glycine} + \text{FH}_4 + \text{H}_2\text{O}}$$

$$\text{serine} + \text{NH}_3 + \text{CO}_2 + (2\text{H}) \rightleftharpoons 2\ \text{glycine} + \text{H}_2\text{O}\ \text{(net conversion)}$$

This net conversion has been observed in the livers of ducks, chicks and pigeons (Richert *et al.*, 1962).

In some living tissues glycine can also be formed by transamination from glyoxylic acid; it is uncertain, however, whether this reaction occurs in avian tissues. Weissbach and Sprinson (1953) administered isotopically labelled glyoxylate to pigeons and found that radioactivity appeared in the atoms of the uric acid molecule arising from glycine. Sanadi and Bennett (1960), however, found that unlabelled glyoxylate did not cause a significant decrease in the incorporation of ^{14}C from glycine into serine in chicken liver mitochondria and Richert *et al.* (1962) using chicken, duck and pigeon livers found that the small amount of radioactivity appearing in the glyoxylate pool from glycine-1-^{14}C could be accounted for by non-enzymatic transamination. Both these observations argue against the existence of an active glycine transaminase in avian liver and although the possibility that an active system exists in another tissue cannot be ignored, this is in agreement with the nutritional observation that glyoxylate does not stimulate the growth of chicks fed a glycine-deficient diet (Baker and Sugahara, 1970). The tissue distribution of this reaction should be investigated.

Glycine is involved in numerous anabolic pathways; uric acid synthesis, porphyrin synthesis, creatine synthesis (Section IV D.1), glutathione synthesis and, by interconversion with serine, in one-carbon group metabolism. These pathways provide a variety of routes for the dissimilation of glycine.

2. Serine

In addition to its synthesis from glycine serine is elaborated from an intermediate in glycolysis, which can give rise to serine by two similar pathways as shown below: reactions 1, 2 and 3 constitute the "phosphorylated"

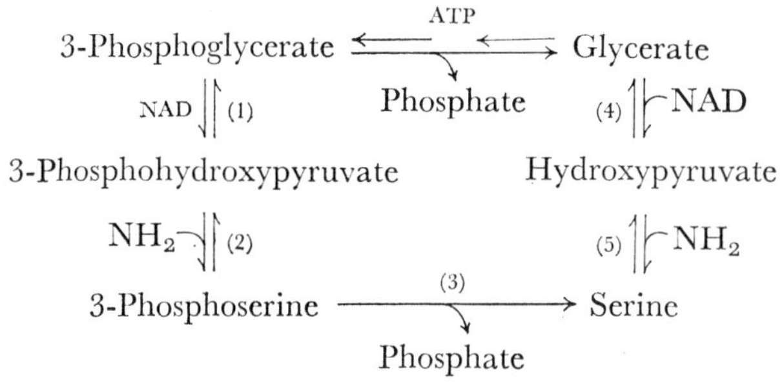

$$
\begin{array}{ccc}
\text{3-Phosphoglycerate} & \xrightleftharpoons[\text{Phosphate}]{\text{ATP}} & \text{Glycerate} \\
\text{NAD} \; \updownarrow (1) & & (4) \updownarrow \; \text{NAD} \\
\text{3-Phosphohydroxypyruvate} & & \text{Hydroxypyruvate} \\
\text{NH}_2 \; \updownarrow (2) & & (5) \updownarrow \; \text{NH}_2 \\
\text{3-Phosphoserine} & \xrightarrow[\text{Phosphate}]{(3)} & \text{Serine}
\end{array}
$$

pathway and reactions 4 and 5 the "non-phosphorylated" pathway. In chicken liver both pathways exist but the enzymes of the phosphorylated pathway show greater activity (Willis and Sallach, 1964; Walsh and Sallach, 1966). 3-Phosphoglycerate dehydrogenase (reaction 1 above) has been purified from chicken liver (Walsh and Sallach, 1965; 1967). Phosphoserine transaminase (reaction 2) utilizes glutamate as amino-donor in contrast to serine transaminase (reaction 5) which utilizes alanine (Walsh and Sallach, 1966). A specific enzyme for the catalysis of reaction 3, phosphoserine phosphatase [EC 3.1.3.3], occurs in chicken liver and has been partly purified and studied (Neuhaus and Byrne, 1959a, b). The enzyme is inhibited by serine (Neuhaus and Byrne, 1960). In goose and pigeon livers this reaction is catalysed by a less specific phosphatase (Grillo and Coghe, 1966); in the latter species the non-phosphorylated pathway is absent (Grillo *et al.*, 1966). In other vertebrates Walsh and Sallach (1966) found that the relative activities of the two pathways of serine biosynthesis differed in different tissues and thus the predominance of the phosphorylated pathway in avian liver does not preclude the possibility that the non-phosphorylated pathway predominates in other avian tissues.

By virtue of the hydroxymethyltransferase reaction serine provides one-carbon groups and *via* the cystathionine synthetase reaction participates in cysteine metabolism (Section IV C.2). Serine is a precursor of the bases of phospholipids (aminoethanol and lecithin) and in vertebrates other than mammals is a precursor of serine ethanolamine phosphate (Rosenberg and Ennor, 1966), a phosphodiester of uncertain function.

Serine can be catabolized by serine dehydratase [EC 4.2.1.13] a pyridoxal phosphate-containing enzyme, which produces pyruvate and ammonia. This enzyme is present in the liver of chickens and shows moderate activity. A partly purified preparation also showed threonine dehydratase and cystathionine synthetase activities (Nagabhushanam and Greenberg, 1965). These three activities are closely related in mammalian systems but specific enzymes have been isolated for each function in various tissues (see Greenberg, 1969 p. 121); further investigations of the avian enzyme may reveal a similar situation. Ascarelli and Bruckental (1969) have shown greater activity of serine dehydratase in the kidney of the chicken than in the liver which is of interest in view of the relatively low activity in the liver noted above. Grillo *et al.* (1965) demonstrated the existence of a highly active D-serine dehydratase in chicken kidney.

3. *Glycine-Serine Interconversion and the Dietary Need for Glycine*

It has been customary to assume that glycine should be provided in the diet of the chick. This is thought to reflect the fact that the rate of utilization of glycine is greater than the rate of its biosynthesis. The biochemical evidence for the formation of glycine from glyoxylate is conflicting (see Section IV B.1) although nutritional evidence (Baker and Sugahara, 1970) suggests that this pathway does not contribute significantly to glycine synthesis.

Under conditions of dietary insufficiency the major known source of glycine is serine. Recent nutritional investigations have tended to support this view. A study of work which claimed a dietary need for glycine revealed that sufficient attention had not been paid to the serine, or non-essential nitrogen, content of the diets (Anonymous, 1968); several reports indicating the non-essential nature of glycine, when the serine requirement is satisfied, have appeared (Klain *et al.*, 1960; Sugahara and Ariyoshi, 1967a; Baker *et al.*, 1968). Current thinking suggests that the primary need is to satisfy a serine requirement (Akrabawi and Kratzer, 1968) in that it is the rate of serine biosynthesis from phosphoglycerate that is limiting when diets inadequate both in serine and glycine are fed. These conclusions depend upon the assumption that serine can be readily converted to glycine and, as has emerged, this is generally accepted. If this is so, the observation of Sugahara and Ariyoshi (1967b) that feeding a diet containing a high level of serine caused an increase in plasma serine concentration but did not elevate plasma glycine needs further investigation.

C. METHIONINE AND CYSTEINE (CYSTINE)

1. Methionine

In addition to participating in protein synthesis, methionine provides cysteine and methyl-groups for methylation reactions. In order for a methylation reaction to occur, methionine is converted to *S*-adenosylmethionine in a reaction which utilizes ATP and is catalysed by methionine adenosyltransferase [EC 2.5.1.6]. The "active" methyl group is then transferred to an acceptor (e.g. guanidinoacetate, nor-adrenaline) by a methyltransferase, generating a methylated product (e.g. creatine, adrenaline) and *S*-adenosylhomocysteine. *S*-adenosylhomocysteine can then be hydrolysed enzymatically releasing adenosine and homocysteine (adenosylhomocysteinase [EC 3.3.1.1]).

Homocysteine arising from the above reaction can be utilized to re-form methionine. Thus, methionine is an essential amino acid in so far as homocysteine cannot be synthesized in animal tissues. The re-synthesis of methionine requires the *de novo* synthesis of a methyl-group through a complex enzyme system which contains vitamin B_{12}. The methyl-group arises from 5-methyltetrahydrofolic acid, formed by reduction of 5,10-methylenetetrahydrofolic acid, and is transferred *via* the B_{12}-containing enzyme system to homocysteine. Reduced FAD, NAD and *S*-adenosylmethionine are also required to effect this transfer; the methyl-group of *S*-adenosylmethionine is not transferred to homocysteine but is apparently incorporated into, and retained in, the B_{12} molecule (see Weissbach and Dickerman, 1965 and Greenberg, 1969 pp. 264–267 for detailed accounts). Methyl group transfer and methionine resynthesis are summarized below.

The re-synthesis of methionine has been shown in chicken liver and there is sufficient evidence to assume that it occurs in the manner described above (Weissbach *et al.*, 1963; Dickerman *et al.*, 1964). The requirement for vitamin B_{12} in *de novo* methyl-group synthesis explains the nutritional observation that provision of labile methyl-groups in the diet, particularly from methionine, partly alleviates the growth impairment caused by vitamin B_{12} deficiency (Gillis and Norris, 1949a, b; Spivey Fox *et al.*, 1957; Langer and Kratzer, 1964). A full discussion of the biochemical and nutritional interaction between methionine, vitamin B_{12} and folic acid has been presented by Weissbach and Dickerman (1965).

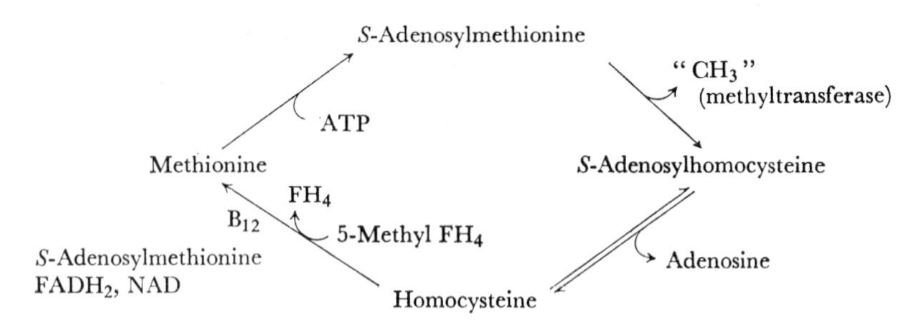

2. Cysteine

Homocysteine arising from methionine may react with serine producing cystathionine, thus:

$$\text{Homocysteine} + \text{serine} \longrightarrow \text{cystathionine} + H_2O$$

The enzyme catalysing this reaction is either identical to or closely associated with serine dehydratase. Cystathionine is hydrolysed producing cysteine and α-ketobutyric acid;

$$\text{Cystathionine} + H_2O \longrightarrow \text{cysteine} + \alpha\text{-ketobutyrate} + NH_3$$

This reaction is catalysed by an enzyme closely associated with homoserine dehydratase [EC 4.2.1.15]. Both the enzymes concerned in cysteine production contain pyridoxal phosphate. Serine dehydratase with cystathionine synthesizing activity has been isolated from chicken liver (Section IV B.2). In view of the higher activity of serine dehydratase in chicken kidney, the cystathionine synthesizing activity of this organ might repay investigation.

Although cysteine can be reversibly oxidized to cystine in the cell, it is generally thought that cysteine represents the intracellular form. Conversion to cystine and subsequent cleavage of this compound by homoserine dehydratase can account for the desulph-hydrative catabolism of cysteine, although there are many other reactions that can also account for apparent "desulph-hydrase" activity (see Meister, 1965 pp. 793–797).

Cysteine is important in the regulation of intracellular conditions and undergoes numerous reactions, few of which have been studied in the fowl. A full description of these reactions is beyond the scope of this treatment, detailed accounts have been presented recently (Meister, 1965 pp. 789–818; Kun, 1969). One aspect that has been studied in the bird is that of sulphur incorporation into organosulphur compounds. Incorporation of sulphate into organosulphur compounds has been observed in the chick embryo (Machlin *et al.*, 1955; Lowe and Roberts, 1955; Johnston *et al.*, 1966) and in the young chick (Machlin *et al.*, 1954; Machlin and Pearson, 1956; Miraglia *et al.*, 1966). The compound into which the greatest proportion of sulphur is incorporated is taurine and reports differ as to whether sulphate-sulphur appears in cysteine.

Sentenac *et al.* (1963), Fromageot and Sentenac (1964), Sentenac and Fromageot (1964) and Chapeville and Fromageot (1967) have described a pathway for the incorporation of sulphate into taurine in the chick embryo. This pathway involves the reduction of sulphate to sulphite and the formation of cysteic acid from the sulphite and cysteine by a substitution reaction: the cysteic acid is then decarboxylated forming taurine and the sulphide produced in the substitution reaction is utilized for the regeneration of cysteine. It appears that the enzymes catalysing the reduction of sulphate and the substitution reaction are unique to the yolk sac of the embryo and disappear with the resorption of this organ. The enzyme catalysing the substitution reaction between sulphite and cysteine has been purified (Tolosa *et al.*, 1969). Enzymes catalysing the formation of adenosine-3′-phosphate-5′-phosphosulphate (Miraglia and Martin, 1969) and the generation of cysteine from serine and sulphide (Braunstein *et al.*, 1969; Nguyen Dinh Lac, 1969) have been demonstrated in the hatched fowl: both these enzyme systems are elements of the pathway described by Chapeville and Fromageot (1967) for the chick embryo.

Cysteine is also incorporated into the peptide glutathione, another compound important in the regulation of intracellular conditions. The enzyme catalysing the reversible reduction of glutathione, glutathione reductase [EC 1.6.4.2], has been purified 600-fold from chicken liver (Biswas and Johnson, 1967).

It is thought that oxidative deamination is the main pathway of cysteine degradation in mammals. Two systems appear to exist, one in the cytoplasm in which cysteine sulphinate is an intermediate, and one in the mitochondria, of which little is known (see Kun, 1969 pp. 378–381). The ultimate products of catabolism are pyruvate and ammonia. These pathways have not been investigated in the fowl.

D. ARGININE, ORNITHINE AND PROLINE

1. *Arginine*

In ureotelic organisms arginine is formed from ornithine, ammonia and the amino-nitrogen of aspartate by the enzymes of the urea cycle. In the fowl arginine synthesizing enzymes are absent from the liver but some of the enzymes of the pathway occur in the kidney. Ornithine carbamoyltransferase [EC 2.1.3.3], argininosuccinate synthetase [EC 6.3.4.5] and argininosuccinate lyase [EC 4.3.2.1] are present in the kidney but carbamoylphosphate synthase [EC 2.7.2.5] is absent (Tamir and Ratner, 1963a) and has not been found in any avian tissue (see also below) (Brown and Cohen, 1960; Bowers and Grisolia, 1962). Since carbamoylphosphate synthase is necessary for ammonia fixation in the initiation of arginine synthesis its absence accounts for the absence of arginine synthesis in avian tissues. This is in accord with the finding that there is a dietary requirement for arginine and that ornithine does not substitute for arginine in the diet of chicks (Klose *et al.*, 1938). The presence of both argininosuccinate synthetase and lyase accounts for the fact that citrulline can replace arginine in the diet (Klose and Almquist, 1940; Tamir and Ratner, 1963b).

The absence of carbamoylphosphate synthase from avian tissues raises the question of how carbamoyl phosphate is produced for pyrimidine synthesis (Rochovansky, 1970). It has recently been shown that two forms of carbamoylphosphate synthase exist in mammalian tissue, one is present in mitochondria and requires acetylglutamate as a co-factor, the other shows no dependence on acetylglutamate and is located in the soluble fraction of the cell. It seems probable that the acetylglutamate-dependent enzyme is associated with the urea cycle, while the other enzyme is concerned with pyrimidine synthesis (Hager and Jones, 1967; Tatibana and Ito, 1967). A carbamoylphosphate synthase which is located in the soluble fraction and does not require acetylglutamate has been demonstrated in chicken (Maresh *et al.*, 1969) and pigeon (Peng and Jones, 1969) livers. This enzyme appears to be associated with pyrimidine synthesis and presumably carbamoyl phosphate from this source is incorporated exclusively into the pyrimidine ring.

In view of the general acceptance of the fact that arginine is an essential amino acid for the bird the finding that radioactivity from glucose rapidly appears in arginine in the pigeon (Reinking and Steyn-Parvé, 1964) needs further investigation, particularly since a similar investigation in the fowl (Nesheim and Garlich, 1963) yielded no evidence of arginine formation from labelled glucose as indicated by the lack of radioactivity in ornithine (see also Section IV D.2).

Creatine is formed by methylation of guanidinoacetic acid (guanidinoacetate methyltransferase EC 2.1.1.2]); guanidinoacetate arises by a transamidination reaction between arginine and glycine catalysed by glycine amidinotransferase [EC 2.1.4.1]) thus:

$$\text{Arginine} + \text{glycine} \rightleftharpoons \text{guanidinoacetate} + \text{ornithine}$$

Creatine formation has been demonstrated in the kidney and liver of birds (Borsook and Dubnoff, 1941; Walker, 1960) although whether the pathway in the bird is the same as that described above is uncertain (Alekseeva and Arkhangel' skaya, 1964). Dietary administration of creatine inhibits creatine synthesis by causing a decrease in the activity of amidinotransferase in ducks and chicks and it has been proposed that this represents a control system for creatine synthesis by end-product repression (Walker, 1960; 1961). A similar system of control has been postulated to exist in the chick embryo (Walker, 1963). Recently Ramirez *et al.* (1970) have pointed out that the concentrations of creatine necessary to exert this control are considerably in excess of normal physiological concentrations and have presented evidence to suggest that the system of control of creatine synthesis in the embryo and in the hatched bird is more complex than end-product repression of amidinotransferase.

Arginase exists with moderate activity in the liver and higher activity in the kidney of chickens (Smith and Lewis, 1963; Tamir and Ratner, 1963a) and provides another pathway for arginine dissimilation, producing ornithine and urea. The relatively low liver arginase activity found in chicks is not a characteristic of all birds (Brown, 1966). The liver arginase of uricotelic species differs from that of ureotelic species, in that the enzyme from uricoteles is not inhibited by high concentrations of its substrate. There are also differences in the Km values and molecular weights, although both require Mn^{2+} and are specific for L-arginine. The kidney arginase of fowls resembles the liver arginase of ureoteles (Mora *et al.*, 1965). The significance of this finding is uncertain (Mora *et al.*, 1966; Rossi and Grazi, 1969) but the presence of a relatively high arginase activity in the kidney of the fowl has been interpreted in terms of the need to provide ornithine for detoxification (see below). The possibility that kidney arginase is required to produce urea for the maintenance of a concentration gradient in the renal medulla for urinary concentration seems unlikely. Skadhauge and Schmidt-Nielsen (1967) found that urea contributed little, if at all, to concentrative function in the kidney of fowls.

Kidney arginase activity in the chick can be markedly influenced by dietary changes and it has been observed that excesses of certain amino acids in the diet cause increases in the activity of the enzyme (O'Dell *et al.*, 1965; Austic and Nesheim, 1969). It may be that such increases in activity cause enhanced catabolism of arginine and that this results in the apparent increase in the arginine requirement of chicks receiving dietary excesses of amino acids. Although dietary excesses of several amino acids cause increases in the arginine requirement (Snetsinger and Scott, 1961; Boorman and Fisher, 1966) particular attention has been given to the effect of an excess of lysine. Jones *et al.* (1967) found that excess lysine caused an increase in kidney arginase activity but that this increase was not evident until a few days after a decrease in the plasma concentration of arginine had been observed. To explain the initial depletion of arginine Jones *et al.* (1967) suggested that the increased plasma concentration of lysine might inhibit the renal tubular reabsorption of arginine, causing increased urinary excretion of arginine. A common transport system for the cationic amino acids does exist in the renal tubule of the fowl (Section II) and enhanced urinary excretion of arginine has been observed both in response to the intravenous infusion of lysine (Boorman *et al.*, 1968) and the feeding of excess lysine (Nesheim, 1968). It may be that dietary excesses of several amino acids, including lysine, cause increases in the arginine requirement by stimulating arginase activity and that lysine causes a further depletion of arginine by virtue of inhibition of renal reabsorption.

2. Ornithine and Proline

In addition to being formed from arginine by the action of amidinotransferase and arginase, ornithine can be formed from glutamic acid in mammals. This pathway involves reduction of the γ-carboxyl-group of glutamate to produce glutamate γ-semialdehyde and transamination of the latter to produce ornithine. Proline can be formed by reduction of Δ^1-pyrroline-5-carboxylic acid, a cyclic compound that may arise spontaneously from glutamate γ-semialdehyde. These conversions are summarized below:

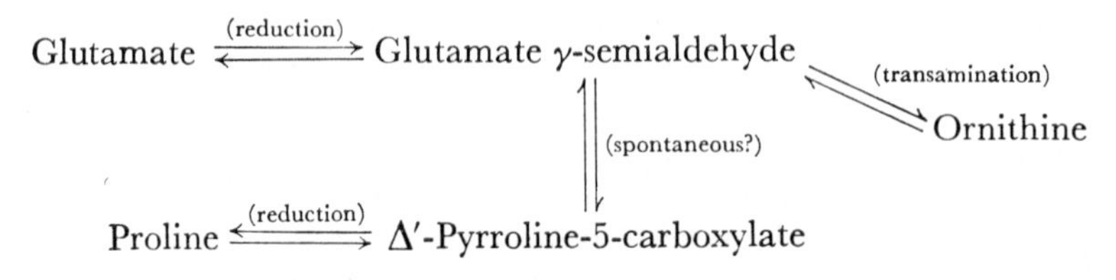

It may be that acetylation of the α-amino-group of glutamate occurs before conversion to ornithine to prevent spontaneous formation of Δ^1-pyrroline-5-carboxylate. The reaction sequence for the formation of proline would then be: glutamate $\rightarrow$ *N*-acetylglutamate $\rightarrow$ *N*-acetylglutamate γ-semialdehyde $\rightarrow$ *N*-acetylornithine $\rightarrow$ ornithine $\rightarrow$ glutamate γ- semialdehyde $\rightarrow$ Δ^1-pyrroline-5-carboxylate $\rightarrow$ proline. Such a pathway occurs in micro-organisms but has not been conclusively shown in vertebrates (see Meister, 1965 pp. 715 and Rodwell, 1969 pp. 317–331). Attempts to demonstrate the conversion of glutamate to ornithine in the fowl, using tracers *in vivo*, have been unsuccessful (Nesheim and Garlich, 1963; Tamir and Ratner, 1963b). A transaminase catalysing the interconversion of glutamate γ-semialdehyde and ornithine has been demonstrated in chick liver (Vecchio and Kalman, 1968) suggesting that the blockage to ornithine synthesis is caused either by the lack of enzymes involved in the acetylation pathway or the lack of an enzyme to reduce glutamate.

Proline formation has been little studied in the bird. Fitzsimmons and Waibel (1965) with chick embryos, reported in abstract, found that ornithine was converted to proline but that glutamate was not. Klain and Johnson (1962) using young chicks, found that ^{14}C from universally labelled arginine appeared in glutamate and proline and concluded that the pathway: arginine $\rightarrow$ ornithine $\rightarrow$ glutamate γ-semialdehyde $\rightarrow$ glutamate accounted for this conversion, proline arising from glutamate γ-semialdehyde. Radioactivity in ornithine was apparently not measured. The results of both these studies are in accord with those for ornithine formation in that transamination of ornithine occurs but formation of ornithine from glutamate does not. If proline and ornithine can be formed directly from glutamate (i.e. without *N*-acetylation) in most vertebrates then it would seem that the apparent blockage in these pathways in birds is at the formation of glutamate γ-semialdehyde from glutamate.

If proline is not formed from glutamate in the fowl it appears that the formation of proline from arginine (Klain and Johnson, 1962—see above) is its sole source in conditions of dietary insufficiency. Several nutritional studies have indicated that proline is required for optimal growth (Benton *et al.*, 1955; Roy and Bird, 1959; Greene *et al.*, 1962; Sugahara and Ariyoshi, 1967a) and in accord with biochemical findings Graber *et al.* (1969; 1970) found that glutamate did not substitute for proline in a proline-deficient diet. At variance with biochemical findings, however, Graber *et al.* (1970) found that arginine did not substitute for proline in the diet. On the basis of these findings therefore it appears that although proline can arise from arginine in the fowl, this pathway is not of nutritional significance in the young chick.

In mammals both ornithine and proline are converted to glutamate, a pathway that can account for their catabolism. Fitzsimmons and Waibel (1965) demonstrated that these conversions occur in embryonic tissue but could not find evidence of conversion of proline to ornithine, which might be expected if the conversion of proline to glutamate can occur. The findings of Klain and Johnson (1962), reported previously, also indicate that ornithine can be converted to glutamate. It is clear that further studies of the interconversions of these amino acids are needed.

Proline is also converted to hydroxyproline and is therefore important in collagen synthesis (see Meister, 1965 pp. 715–729; Rodwell, 1969 pp. 322–324). Hydroxyproline is oxidatively catabolized in a manner analogous to proline, producing γ-hydroxyglutamic acid which is catabolized further possibly via α-keto-γ-hydroxyglutaric acid to glyoxylate and pyruvate. In the mammal hydroxyproline is oxidized by both the liver and the kidney; in birds however, it appears that the kidney shows high activity while little activity is exhibited by the liver (Bannister and Burns, 1970). Ornithine does not occur in proteins and, in the fowl, apart from the role discussed above, appears to be concerned exclusively in detoxification of aromatic acids (e.g. benzoate). Conjugation with ornithine is an important mechanism for detoxification in the fowl (Crowdle and Sherwin, 1923). Enzymes for the synthesis of these conjugates occur in the kidney (McGilvery and Cohen, 1950) and ATP and co-enzyme A are required (Marshall and Koeppe, 1964). Detoxification by conjugation with glucuronic acid also occurs in the fowl (Sperber, 1947; Baldwin *et al.*, 1959). The enzymes for glucuronide conjugation occur in liver, kidney and the alimentary tract. Conjugation is by glucuronyl transfer from UDP-glucuronic acid (Dutton and Ko, 1966). The above remarks refer to the fowl, the situation is similar in the turkey, goose and duck but differs in other species of birds; in the pigeon glucuronate and glycine are utilized for detoxification of benzoate while ornithine is not (Baldwin *et al.*, 1960).

E. OTHER AMINO ACIDS

1. General Remarks

The state of knowledge of the metabolism of other amino acids in the fowl is limited. In general the pathways appear to be similar to those in the mammal and detailed accounts will not be given here; the reader is referred to the relevant sections of the reviews listed in Section I. Some specific observations on the metabolic pathways of other amino acids in the fowl are reported in the following section.

2. Threonine

Threonine is an essential amino acid for birds and mammals and is thought not to participate readily in transamination. Comments on the observation of threonine dehydratase activity in chicken liver have been presented elsewhere (Section IV B.2) and in view of the higher activity of serine dehydratase in the kidney of the fowl it is possible that this organ shows a higher threonine dehydratase activity. Threonine catabolism may also proceed by aldol cleavage in a reaction similar to that catalysed by serine hydroxymethyltransferase and oxidative decarboxylation to aminoacetone is now recognized as an important pathway in vertebrates.

3. Histidine

That the catabolism of histidine, an essential amino acid, probably proceeds by a route similar to that in mammals is indicated by the observation of Spivey Fox *et al.* (1961) that vitamin B_{12} deficiency in chicks causes enhanced excretion of formiminoglutamic acid, an intermediate in histidine catabolism. Formiminoglutamate is further catabolized by transfer of the formimino-group to tetrahydrofolic acid. It is probable that vitamin B_{12} deficiency causes tetrahydrofolate to become limiting for this reaction by preventing the formation of tetrahydrofolate from 5-methyltetrahydrofolate in the conversion of homocysteine to methionine (Section IV C.1). Formiminoglutamate therefore tends to accumulate and is excreted (see Weissbach and Dickerman, 1965). The synthesis of the histidine-containing dipeptides, carnosine and anserine has been studied in embryonic and young chicks (Kalyankar and Meister, 1959; McManus and Benson, 1967; Dobrynina and Gorbunova, 1968).

4. Phenylalanine and Tyrosine

Phenylalanine is an essential amino acid and gives rise to tyrosine by the action of phenylalanine 4-hydroxylase [EC 1.14.3.1]. Since this is the only pathway of tyrosine formation, phenylalanine is required to compensate for any dietary inadequacy in tyrosine. Phenylalanine 4-hydroxylase activity has been demonstrated in chicken liver (Udenfriend and Cooper, 1952; P. J. Christensen, 1962). A survey of the activity of this enzyme in the livers of vertebrates including the fowl, goose and pigeon has been performed and the enzyme was found to be universally present (Voss and Waisman, 1966).

Adrenaline, melanin and the iodotyrosines arise from tyrosine; ultimately, tyrosine is catabolized to fumarate and acetoacetate. The overall catabolic route has been observed in pigeon liver (Michalek-Morrica, 1965). Tyrosine transaminase, which initiates the catabolic pathway has been demonstrated in embryonic and young chicks (Section III C.2). Ascorbic acid functions in the conversion of p-hydroxyphenylpyruvic acid to homogentisic acid and it is of interest that in the young chicken, a species that does not exhibit a dietary requirement for vitamin C, high dietary levels of tyrosine induce symptoms that can be alleviated by dietary vitamin C (Sanford *et al.*, 1954).

5. *Tryptophan*

In the fowl interest in tryptophan, an essential amino acid, has been stimulated by its relationship with nicotinamide. Several studies have shown that tryptophan replaces nicotinamide in the diet of the fowl (Briggs, 1945; Anderson *et al.*, 1950). Jackson *et al.* (1959) and Wilson and Henderson (1960) have demonstrated the conversion of tryptophan to nicotinamide in chick embryos using tracers. Decker and Henderson (1959) showed that 3-hydroxyanthranilate would replace nicotinamide in the diet and as might be expected from the pathway Fisher *et al.* (1955) showed that nicotinamide would not replace tryptophan.

The enzymes initiating the nicotinamide pathway have also been studied in the embryo. The activity of tryptophan oxygenase [EC 1.13.1.12] remains constant at hatching but can be increased by administration of tryptophan before or after hatching. The activity of formamidase [EC 3.5.1.9] rises markedly at hatching (Knox and Eppenberger, 1966; Wagner *et al.*, 1969).

Other pathways of tryptophan metabolism in the fowl have not been studied to the same extent. Kido *et al.* (1965) found 6-hydroxykynurenic acid in addition to the more commonly occurring metabolites of tryptophan, kynurenic and xanthurenic acids in the urine of hens given oral doses of tryptophan. The 6-hydroxykynurenic acid was found to be a metabolic product of 5-hydroxykynurenine a compound not previously known to arise from tryptophan metabolism (Kido *et al.*, 1967). Kido *et al.* (1968) have since found the enzyme system in hen's liver which catalyses the conversion of kynurenine to 5-hydroxykynurenine.

6. *Leucine, Isoleucine and Valine*

Leucine transaminase activity has been demonstrated in chick tissues and this has been discussed elsewhere (Section III C.2). The branched-chain amino acids serve as amino donors in the amination of the α-keto analogue of methionine in the chick (Section III C.2).

7. Lysine

The catabolic pathway of this amino acid in mammals is the subject of controversy at present and there is some doubt as to whether the pathway originally proposed (*via* pipecolic acid) is of physiological significance. Alternative pathways have been suggested, one of which involves saccharopine (Higashino and Lieberman, 1965). It has also been suggested that acylation of the ε-amino-group occurs during catabolism (Paik, 1962); in this connexion an acylase which hydrolyses the acylated ω-amino-groups of ornithine, lysine and homolysine occurs in chicken kidney (Seely and Benoiton, 1968).

V. Concluding Remarks

On the basis of existing evidence it appears that many similarities exist between mammalian and avian species in respect of amino acid and protein metabolism. Many of the differences which do emerge appear to be associated directly or indirectly with the uricotelic mode of nitrogen excretion. In a direct sense, uricotelism results in the existence of a dual role for purine synthesis in birds and it is possible that differences will emerge between the control mechanisms for purine synthesis in birds and in mammals. Associated also with uricotelism is the absence of a hepatic urea-synthesizing pathway and thus, indirectly, the lack of a biosynthetic pathway for arginine.

In considering avian protein metabolism the extent to which known physiological differences between birds and mammals may influence a particular aspect under study should always be considered. In this context, the recent discoveries that in the fowl blood may flow from the hepatic portal system to the renal portal system (Akester, 1967) and that in certain instances blood from the absorptive area of the small intestine may flow directly to the kidney (Purton, 1970), might indicate that the kidney is an organ of greater metabolic significance in the bird than in the mammal. Such considerations suggest that in addition to studying the activities of metabolic pathways in the liver, the metabolic activity of the kidney should also be examined in the fowl.

References

Adamson, L. F. and Ingbar, S. H. (1967). *J. biol. Chem.* **242**, 2646–2652.
Akester, A. R. (1967). *J. Anat.* **101**, 569–594.
Akrabawi, S. S. and Kratzer, F. H. (1968). *J. Nutr.* **95**, 41–48.
Alekseeva, A. M. and Arkhangel'skaya, G. G. (1964). *Biokhimiya* **29**, 179–184.
Anderson, J. O., Combs, G. F. and Briggs, G. M. (1950). *J. Nutr.* **42**, 463–471
Anderson, P. J. and Johnson, P. (1969). *Biochim. biophys. Acta* **181**, 45–51.
Anonymous, (1968). *Nutr. Rev.* **26**, 279–280.
Arfin, S. M. (1967). *Biochim. biophys. Acta* **136**, 233–244.

Ascarelli, I. and Bruckental, I. (1969). *Comp. Biochem. Physiol.* **29**, 1277–1279.
Austic, R. E. and Nesheim, M. C. (1969). *Poult. Sci.* **48**, 1780.
Baker, D. H. and Sugahara, M. (1970). *Poult. Sci.* **49**, 756–760.
Baker, D. H., Sugahara, M. and Scott, H. M. (1968). *Poult. Sci.* **47**, 1376–1377.
Baldwin, B. C., Robinson, D. and Williams, R. T. (1959). *Biochem. J.* **71**, 638–642.
Baldwin, B. C., Robinson, D. and Williams, R. T. (1960). *Biochem. J.* **76**, 595–600.
Bannister, D. W. and Burns, A. B. (1970). *Br. Poult. Sci.* **11**, 505–508.
Bassler, K. H. and Hammar, C. H. (1958). *Biochem. Z.* **330**, 555–564.
Bauriedel, W. R. (1962). *Poult. Sci.* **41**, 1616–1617.
Bauriedel, W. R. (1963). *Poult. Sci.* **42**, 214–217.
Benton, D. A., Spivey, H. E., Harper, A. E. and Elvehjem, C. A. (1955). *Archs Biochem. Biophys.* **57**, 262–263.
Bertland, L. M. and Kaplan, N. O. (1970). *Biochemistry* **9**, 2653–2665.
Bird, F. H. (1952). *Poult. Sci.* **31**, 1095–1096.
Bird, F. H. (1968). *Fedn Proc. Fedn Am. Socs. exp. Biol.* **27**, 1194–1198.
Biswas, D. K. and Johnson, B. C. (1967). *Indian J. Biochem.* **4**, 159–163.
Blakley, R. L. (1954). *Biochem. J.* **58**, 448–462.
Boorman, K. N. (1971). *Comp. Biochem. Physiol.* **39A**, 29–38.
Boorman, K. N. and Fisher, H. (1966). *Br. Poult. Sci.* **7**, 39–44.
Boorman, K. N., Falconer, I. R. and Lewis, D. (1968). *Proc. Nutr. Soc.* **27**, 61–62A.
Borsook, H. and Dubnoff, J. W. (1941). *J. biol. Chem.* **138**, 389–403.
Boulanger, P. and Osteux, R. (1955a). *C.r. hebd. Séanc. Acad. Sci., Paris* **241**, 125–127.
Boulanger, P. and Osteux, R. (1955b). *C.r. hebd. Séanc. Acad. Sci., Paris* **241**, 613–615.
Boulanger, P. and Osteux, R. (1956). *Biochim. biophys. Acta* **21**, 552–561.
Boulanger, P., Bertrand, J. and Osteux, R. (1957). *Biochim. biophys. Acta* **26**, 143–145.
Bowers, M. D. and Grisolia, S. (1962). *Comp. Biochem. Physiol.* **5**, 1–16.
Braunstein, A. E., Goryachenkova, E. V. and Nguyen Dinh Lac (1969). *Biochim. biophys. Acta* **171**, 366–368.
Briggs, G. M. (1945). *J. biol. Chem.* **161**, 749–750.
Brown, G. W. (1966). *Archs Biochem. Biophys.* **114**, 184–194.
Brown, G. W. (1970). *In* "Comparative Biochemistry of Nitrogen Metabolism" (J. W. Campbell, ed.), Vol. 2, pp. 711–793. Academic Press, London.
Brown, G. W. and Cohen, P. P. (1960). *Biochem. J.* **75**, 82–91.
Calet, C. and Melot, M. (1961). *Annls Zootech.* **10**, 205–213.
Chan, S.-K. and Cohen, P. P. (1964). *Archs Biochem. Biophys.* **104**, 335–337.
Chapeville, F. and Fromageot, P. (1967). *Adv. Enzyme Reg.* **5**, 155–163.
Chen, H.-L. and Li, L. (1963). *Acta biochim. biophys. sin.* **3**, 1–12.
Christensen, H. N. (1962). "Biological Transport". pp. 45–70. W. A. Benjamin, Inc., New York.
Christensen, H. N., Liang, M. and Archer, E. G. (1967). *J. biol. Chem.* **242**, 5237–5246.
Christensen, P. J. (1962). *Scand. J. clin. Lab. Invest.* **14**, 623–628.
Constantsas, N. S. and Knox, W. E. (1967). *Biochim. biophys. Acta* **132**, 178–181.
Corman, L., Prescott, L. M. and Kaplan, N. O. (1967). *J. biol. Chem.* **242**, 1383–1390.
Crowdle, J. H. and Sherwin, C. P. (1923). *J. biol. Chem.* **55**, 365–370.
Curtis, J. L. and Fisher, J. R. (1970). *Biochim. biophys. Acta* **201**, 26–32.
Decker, R. H. and Henderson, L. M. (1959). *J. Nutr.* **68**, 17–24.
Della Corte, E. and Stirpe, F. (1967). *Biochem. J.* **102**, 520–524.
Della Corte, E. and Stirpe, F. (1970). *Biochem. J.* **117**, 97–100.

Dickerman, H., Redfield, B. G., Bieri, J. G. and Weissbach, H. (1964). *J. biol. Chem.* **239**, 2545–2552.

Dobrynina, O. V. and Gorbunova, A. V. (1968). *Biokhimiya* **33**, 570–575.

Drel, K. A. (1964). *Biokhimiya* **29**, 685–689.

Dutton, G. J. and Ko, V. (1966). *Biochem. J.* **99**, 550–556.

Eavenson, E. and Christensen, H. N. (1967). *J. biol. Chem.* **242**, 5386–5396.

Ecobichon, D. J. (1966). *Can. J. Biochem.* **44**, 1277–1283.

Efimochkina, E. F. (1958). *Biokhimiya* **23**, 683–689.

Efimochkina, E. F. (1959). *Biokhimiya* **24**, 53–62.

Fazekas, S. and Denes, G. (1966). *Acta biochim. biophys. Acad. Sci. Hung.* **1**, 45–54.

Fisher, H., Scott, H. M. and Johnson, B. C. (1955). *Br. J. Nutr.* **9**, 340–349.

Fitzsimmons, R. C. and Waibel, P. E. (1965). *Fedn Proc. Fedn Am. Socs exp. Biol.* **24**, Abst. 2060.

Florkin, M. and Stotz, E. H. (1965). "Comprehensive Biochemistry". Vol. 13. Elsevier, London.

Freedland, R. A., Martin, K. D. and McFarland, L. Z. (1966). *Poult. Sci.* **45**, 985–991.

Freedland, R. A., Martin, K. D. and McFarland, L. Z. (1967). *Biochem. J.* **103**, 6*P*.

Frieden, C. (1962). *Biochim. biophys. Acta* **62**, 421–423.

Frieden, C. (1965). *J. biol. Chem.* **240**, 2028–2035.

Fromageot, P. and Sentenac, A. (1964). *J. Biochem.* **55**, 659–668.

Gessler, M. (1965). *Zentbl. VetMed.* **12A**, 471–478.

Gillis, M. B. and Norris, L. C. (1949a). *J. biol. Chem.* **179**, 487–488.

Gillis, M. B. and Norris, L. C. (1949b). *Poult. Sci.* **28**, 749–750.

Gordon, R. S. (1965). *Ann. N.Y. Acad. Sci.* **119**, 927–941.

Gordon, R. S. and Sizer, I. W. (1965). *Poult. Sci.* **44**, 673–678.

Goswami, M. N. D. and Robblee, A. R. (1958). *Poult. Sci.* **37**, 96–99.

Graber, G., Allen, N. K. and Scott, H. M. (1969). *Poult. Sci.* **48**, 1813.

Graber, G., Allen, N. K. and Scott, H. M. (1970). *Poult. Sci.* **49**, 692–697.

Greenberg, D. M. (1969). *In* "Metabolic Pathways" (D. M. Greenberg, ed.), Vol. 3, pp. 95–190 and 237–315. Academic Press, London.

Greenberg, G. R. (1954). *Fedn Proc. Fedn Am. Socs exp. Biol.* **13**, 745–759.

Greene, D. E., Scott, H. M. and Johnson, B. C. (1962). *Poult. Sci.* **41**, 116–120.

Grillo, M. A. and Coghe, M. (1966). *Comp. Biochem. Physiol.* **18**, 169–173.

Grillo, M. A., Fossa, T. and Coghe, M. (1965). *Enzymologia* **28**, 377–388.

Grillo, M. A., Fossa, T. and Coghe, M. (1966). *Comp. Biochem. Physiol.* **19**, 589–596.

Hager, S. E. and Jones, M. E. (1967). *J. biol. Chem.* **242**, 5674–5680.

Hartman, S. C. (1963). *J. biol. Chem.* **238**, 3024–3035.

Higashino, K. and Lieberman, I. (1965). *Biochim. biophys. Acta* **111**, 346–348.

Ichihara, A. and Koyama, E. (1966). *J. Biochem.*, Tokyo **59**, 160–169.

Jackson, J. T., Pearson, P. B. and Denton, C. A. (1959). *Archs Biochem. Biophys.* **79**, 131–135.

Jacquez, J. A., Sherman, J. H. and Terris, J. (1970). *Biochim. biophys. Acta* **203**, 150–166.

Johnston, P. M., Harvey, J. A. N. and Bowers, V. F. (1966). *Am. J. Physiol.* **210**, 1122–1126.

Jones, J. D., Petersburg, S. J. and Burnett, P. C. (1967). *J. Nutr.* **93**, 103–116.

Kalyankar, G. D. and Meister, A. (1959). *J. biol. Chem.* **234**, 3210–3218.

Kido, R., Noguchi, T., Sakurane, T. and Matsumura, Y. (1965). *Wakayama Med. Rep.* **10**, 5–9.

Kido, R., Noguchi, T. and Matsumura, Y. (1967). *Biochim. biophys. Acta* **141**, 270–275.
Kido, R., Noguchi, T., Kaseda, H., Kawamoto M. and Matsumura, Y. (1968). *Archs Biochem. Biophys.* **125**, 1030–1031.
Kisliuk, R. L. and Sakami, W. (1954). *J. Am. chem. Soc.* **76**, 1456–1457.
Klain, G. J. and Johnson, B. C. (1962). *J. biol. Chem.* **237**, 123–126.
Klain, G. J., Scott, H. M. and Johnson, B. C. (1960). *Poult. Sci.* **39**, 39–44.
Klose, A. A. and Almquist, H. J. (1940). *J. biol. Chem.* **135**, 153–155.
Klose, A. A., Stokstad, E. L. R. and Almquist, H. J. (1938). *J. biol. Chem.* **123**, 691–698.
Knox, W. E. and Eppenberger, H. M. (1966). *Devl Biol.* **13**, 182–198.
Koivusalo, M., Penttila, I., Raina, A. and Tenhunen, R. (1963). *Acta physiol. scand.* **57**, 454–461.
Kun, E. (1969). *In* "Metabolic Pathways" (D. M. Greenberg, ed.), Vol. 3, pp. 375–401. Academic Press, London.
Kuriyama, K., Sisken, B., Ito, J., Simonsen, D. G., Haber, B. and Roberts, E. (1968). *Brain Res.* **11**, 412–430.
Landon, E. J. and Carter, C. E. (1960). *J. biol. Chem.* **235**, 819–824.
Langer, B. W. and Kratzer, F. H. (1964). *Poult. Sci.* **43**, 127–135.
Lestrovaya, N. N. (1961). *Biokhimiya* **26**, 505–510.
Lin, E. C. C. and Wilson, T. H. (1960). *Am. J. Physiol.* **199**, 127–130.
Litwack, G. and Nemeth, A. M. (1965). *Archs Biochem. Biophys.* **109**, 316–320.
Lowe, I. P. and Roberts, E. (1955). *J. biol. Chem.* **212**, 477–483.
Machlin, L. J. and Gordon, R. S. (1959). *Poult. Sci.* **38**, 650–652.
Machlin, L. J. and Pearson, P. B. (1956). *Proc. Soc. exp. Biol. Med.* **93**, 204–206.
Machlin, L. J., Jackson, J. T., Lankenau, A. H. and Pearson, P. B. (1954). *Poult. Sci.* **33**, 234–238.
Machlin, L. J., Pearson, P. B. and Denton, C. A. (1955). *J. biol. Chem.* **212**, 469–475.
Maresh, C. G., Kwan, T. H. and Kalman, S. M. (1969). *Can. J. Biochem.* **47**, 61–63.
Marshall, F. D. and Koeppe, O. J. (1964). *Biochemistry* **3**, 1692–1695.
Mason, T. L. and Hooper, A. B. (1969). *Devl Biol.* **20**, 472–478.
Matthews, D. M. and Laster, L. (1965). *Am. J. Physiol.* **208**, 601–606.
McGilvery, R. W. and Cohen, P. P. (1950). *J. biol. Chem.* **183**, 179–185.
McManus, I. R. and Benson, M. S. (1967). *Archs Biochem. Biophys.* **119**, 444–453.
Meister, A. (1965). "Biochemistry of the Amino Acids". Academic Press, London.
Michalek-Moricca, H. (1965). *Acta biochim. pol.* **12**, 167–177.
Miraglia, R. J. and Martin, W. G. (1969). *Proc. Soc. exp. Biol. Med.* **132**, 640–644.
Miraglia, R. J., Martin, W. G., Spaeth, D. G. and Patrick, H. (1966). *Proc. Soc. exp. Biol. Med.* **123**, 725–730.
Mizon, J., Biserte, G. and Boulanger, P. (1970). *Biochim. biophys. Acta* **212**, 33–42.
Mora, J., Tarrab, R. and Bojalil, L. F. (1966). *Biochim. biophys. Acta* **118**, 206–209.
Mora, J., Tarrab, R., Martuscelli, J. and Soberon, G. (1965). *Biochem. J.* **96**, 588–594.
Murison, G. (1969). *Devl Biol.* **20**, 518–543.
Nagabhushanam, A. and Greenberg, D. M. (1965). *J. biol. Chem.* **240**, 3002–3008.
Nesheim, M. C. (1968). *Fedn Proc. Fedn Am. Socs exp. Biol.* **27**, 1210–1214.
Nesheim, M. C. and Garlich, J. D. (1963). *J. Nutr.* **79**, 311–317.
Neuhaus, F. C. and Byrne, W. L. (1959a). *J. biol. Chem.* **234**, 109–112.
Neuhaus, F. C. and Byrne, W. L. (1959b). *J. biol. Chem.* **234**, 113–121.
Neuhaus, F. C. and Byrne, W. L. (1960). *J. biol. Chem.* **235**, 2019–2024.
Nguyen Dinh Lac (1969). *Biokhimiva*, **34**, 861–867.

O'Dell, B. L., Amos, W. H. and Savage, J. E. (1965). *Proc. Soc. exp. Biol. Med.* **118**, 102–105.
Ohnuma, T., Bergel, F. and Bray, R. C. (1967). *Biochem. J.* **103**, 238–245.
Olsen, E. M., Hill, D. C. and Branion, H. D. (1963). *Poult. Sci.* **42**, 1177–1181.
Paik, W. K. (1962). *Biochim. biophys. Acta* **65**, 518–520.
Paine, C. M., Newman, H. J. and Taylor, M. W. (1959). *Am. J. Physiol.* **197**, 9–12.
Peng, L. and Jones, M. E. (1969). *Biochem. biophys. Res. Commun.* **34**, 335–339.
Polyanovskii, O. L. and Vorotnitskaya, N. E. (1965). *Biokhimiya* **30**, 619–627.
Ponomareva, T. F. and Drel, K. A. (1964). *Biokhimiya* **29**, 185–190.
Purton, M. D. (1970). *J. Anat.* **106**, 189.
Quastel, J. and Witty, R. (1951). *Nature, Lond.* **167**, 556.
Rajagopalan, K. V. and Handler, P. (1967). *J. biol. Chem.* **242**, 4097–4107.
Ramirez, O., Calva, E. and Trejo, A. (1970). *Biochem. J.* **119**, 757–763.
Reem, G. H. (1968). *J. biol. Chem.* **243**, 5695–5701.
Reif-Lehrer, L. (1968). *Biochim. biophys. Acta* **170**, 263–270.
Reinking, A. and Steyn-Parvé, E. P. (1964). *Biochim. biophys. Acta* **93**, 54–63.
Remy, C. N., Richert, D. A., Doisy, R. J., Wells, I. C. and Westerfield, W. W. (1955). *J. biol. Chem.* **217**, 293–305.
Richert, D. A. and Westerfeld, W. W. (1951). *Proc. Soc. exp. Biol. Med.* **76**, 252–254.
Richert, D. A., Amberg, R. and Wilson, M. (1962). *J. biol. Chem.* **237**, 99–103.
Rochovansky, O. (1970). *Archs Biochem. Biophys.* **138**, 574–581.
Rodwell, V. W. (1969). *In* "Metabolic Pathways" (D. M. Greenberg, ed.), Vol. 3, pp. 191–235 and 317–373. Academic Press, London.
Rogers, K. S. (1969). *Enzymologia* **36**, 153–160.
Rogers, K. S., Geiger, P. J., Thompson, T. E. and Hellerman, L. (1963). *J. biol. Chem.* **238**, PC481–482.
Romanoff, A. L. and Romanoff, A. J. (1967). "Biochemistry of the Avian Embryo". Interscience Publishers, Wiley, London.
Rosenberg, H. and Ennor, A. H. (1966). *Biochim. biophys. Acta* **115**, 23–32.
Rossi, N. and Grazi, E. (1969). *Eur. J. Biochem.* **7**, 348–352.
Rowe, P. B., Coleman, M. D. and Wyngaarden, J. B. (1970). *Biochemistry* **9**, 1498–1505.
Roy, D. N. and Bird, H. R. (1959). *Poult. Sci.* **38**, 192–196.
Rudnick, D. and Waelsch, H. (1955). *J. exp. Zool.* **129**, 309–341.
Sallach, H. J. and Fahien, L. A. (1969). *In* "Metabolic Pathways" (D. M. Greenberg, ed.), Vol. 3, pp. 1–94. Academic Press, London.
Sanadi, D. R. and Bennett, M. J. (1960). *Biochim. biophys. Acta* **39**, 367–369.
Sanchez de Jimenez, E., Aurora Brunner, L. and Soberon, G. (1967). *Archs Biochem. Biophys.* **120**, 175–185.
Sanford, P. E., Wei, A. J. and Clegg, R. E. (1954). *Poult. Sci.* **33**, 585–589.
Scholz, R. W. (1970). *Comp. Biochem. Physiol.* **36**, 503–512.
Scholz, R. W. and Featherston, W. R. (1969). *J. Nutr.* **98**, 193–201.
Schulman, M. P. (1961). *In* "Metabolic Pathways" (D. M. Greenberg, ed.), Vol. 2, pp. 389–457. Academic Press, London.
Seely, J. H. and Benoiton, L. (1968). *Can. J. Biochem.* **46**, 387–389.
Sentenac, A. and Fromageot, P. (1964). *Biochim. biophys. Acta* **81**, 289–300.
Sentenac, A. Chapeville, F. and Fromageot, P. (1963). *Biochim. biophys. Acta* **67**, 672–673.
Sheid, B. and Hirschberg, E. (1967). *Am. J. Physiol.* **213**, 1173–1176.

Shiflett, J. M. and Haskell, B. E. (1969). *J. Nutr.* **98**, 420–426.
Shinwari, M. A. and Falconer, I. R. (1967). *Biochem. J.* **104**, 53–54*P*.
Sisken, B., Sano, K. and Roberts, E. (1961). *J. biol. Chem.* **236**, 503–507.
Skadhauge, E. and Schmidt-Nielsen, B. (1967). *Am. J. Physiol.* **212**, 1313–1318.
Smith, G. H. and Lewis, D. (1963). *Br. J. Nutr.* **17**, 433–444.
Snetsinger, D. C. and Scott, H. M. (1961). *Poult. Sci.* **40**, 1675–1681.
Snoke, J. E. (1956). *J. biol. Chem.* **223**, 271–276.
Solomon, J. B. (1959). *Devl Biol.* **1**, 182–198.
Speck, J. F. (1949). *J. biol. Chem.* **179**, 1405–1426.
Sperber, I. (1947). *K. LandtbrHögsk. Annlr* **15**, 108–112.
Spivey Fox, M. R., Briggs, G. M. and Ortiz, L. O. (1957). *J. Nutr.* **62**, 539–549.
Spivey Fox, M. R., Ludwig, W. J. and Baroody, M. C. (1961). *Proc. Soc. exp. Biol. Med.* **107**, 723–727.
Struck, J. and Sizer, I. W. (1960). *Archs Biochem. Biophys.* **90**, 22–30.
Sugahara, M. and Ariyoshi, S. (1967a). *Agric. biol. Chem.* **31**, 106–110.
Sugahara, M. and Ariyoshi, S. (1967b). *Agric. biol. Chem.* **31**, 1270–1275.
Sugahara, M., Morimoto, T., Kobayashi, T. and Ariyoshi, S. (1967). *Agric. biol. Chem.* **31**, 77–84.
Tamir, H. and Ratner, S. (1963a). *Archs Biochem. Biophys.* **102**, 249–258.
Tamir, H. and Ratner, S. (1963b). *Archs Biochem. Biophys.* **102**, 259–269.
Tasaki, I. and Takahashi, N. (1966). *J. Nutr.* **88**, 359–364.
Tatibana, M. and Ito, K. (1967). *Biochem. biophys. Res. Commun.* **26**, 221–227.
T'ing-Sen, H. (1959). *Biokhimiya* **24**, 528–534.
Tolosa, E. A., Chepurnova, N. K., Khomutov, R. M. and Severin, E. S. (1969). *Biochim. biophys. Acta* **171**, 369–371.
Trufanov, A. V. and Pavlova, Z. M. (1951). *Biokhimiya* **16**, 537–541.
Udenfriend, S. and Cooper, J. R. (1952). *J. biol. Chem.* **194**, 503–511.
Van Den Berg, C. J., Van Kempen, G. M. J., Schade, J. P. and Veldstra, H. (1965). *J. Neurochem.* **12**, 863–869.
Vecchio, D. A. and Kalman, S. M. (1968). *Archs Biochem. Biophys.* **127**, 376–383.
Vidaver, G. A., Romain, L. F. and Haurowitz, F. (1964). *Archs Biochem. Biophys.* **107**, 82–87.
Vorotnitskaya, N. E., Spivak, V. A. and Polyanovskii, O. L. (1968). *Biokhimiya* **33**, 375–382.
Voss, J. C. and Waisman, H. A. (1966). *Comp. Biochem. Physiol.* **17**, 49–58.
Wagner, C., Payne, N. A. and Briggs, W. (1969). *Expl Cell Res.* **55**, 330–338.
Walaas, E. and Walaas, D. (1956). *Acta chem. scand.* **10**, 122–133.
Walker, J. B. (1960). *J. biol. Chem.* **235**, 2357–2361.
Walker, J. B. (1961). *J. biol. Chem.* **236**, 493–498.
Walker, J. B. (1963). *Adv. Enzyme Reg.* **1**, 151–168.
Walsh, D. A. and Sallach, H. J. (1965). *Biochemistry* **4**, 1076–1085.
Walsh, D. A. and Sallach, H. J. (1966). *J. biol. Chem.* **241**, 4068–4076.
Walsh, D. A. and Sallach, H. J. (1967). *Biochim. biophys. Acta* **146**, 26–34.
Weber, R. (1965, 1967). "The Biochemistry of Animal Development". 2 vols. Academic Press, London.
Weissbach, H. and Dickerman, H. (1965). *Physiol. Rev.* **45**, 80–97.
Weissbach, H. and Sprinson, D. B. (1953). *J. biol. Chem.* **203**, 1023–1030.
Weissbach, H., Peterkofsky, A., Redfield, B. G. and Dickerman, H. (1963). *J. biol. Chem.* **238**, 3318–3324.
Wiggert, B. O. and Cohen, P. P. (1966). *J. biol. Chem.* **241**, 210–216.

Williams, J. N., Nichol, C. A. and Elvehjem, C. A. (1949). *J. biol. Chem.* **180**, 689–694.
Willis, J. E. and Sallach, H. J. (1964). *Biochim. biophys. Acta* **81**, 39–54.
Wilson, R. G. and Henderson, L. M. (1960). *J. biol. Chem.* **235**, 2099–2102.
Wu, C. (1963). *Comp. Biochem. Physiol.* **8**, 335–351.
Zimmerman, H. J., Dujovne, C. A. and Levy, R. (1968). *Comp. Biochem. Physiol.* **25**, 1081–1089.

14 The Role of Vitamins in Metabolic Processes

M. E. COATES

National Institute for Research in Dairying, University of Reading, Reading, England

I. Introduction

Vitamins were first recognized as dietary accessory factors needed in trace amounts to maintain normal health, well-being and productive capacity in animals and man. Lack of one or more of these factors led to certain distinctive "deficiency syndromes". Such minute amounts were required that it seemed reasonable to assign a catalytic role to the vitamins; in many instances this has since proved true. For convenience a classification has customarily been made into the fat-soluble vitamins A, D, E and K and the water-soluble vitamins B and C. Present indications are that the same classification may also cover a difference in mechanism of action. Whereas vitamins B and C have mostly been shown to play the part of co-factors in particular enzymic reactions, the function of fat-soluble vitamins is much less clear. Recent evidence suggests that several may have hormone-like properties, exerting their effects through an influence on genetic information.

The fowl is, in many respects, similar to other members of the animal kingdom in its dependence on, and utilization of, vitamins. In this chapter only a brief outline can be given of the role of the vitamins in metabolic processes in higher animals. Points of particular significance to the fowl have been emphasized and marked differences between avian and mammalian systems are noted. For more detailed information reference should be made to the specialist reviews mentioned in the text, and to standard textbooks on the subject (see, for instance, Moore, 1957; Robinson, 1966; Sebrell and Harris, 1968).

II. Vitamin A

A. FORMS

Vitamin A activity is shown by a number of naturally-occurring substances. All are isoprenoid compounds and include the alcohols retinol (vitamin A_1) and 3-dehydroretinol (vitamin A_2), their esters, their corresponding aldehydes, retinal (retinene or vitamin A_1 aldehyde), 3-dehydroretinal (retinene-2), and retinoic acid (vitamin A_1 acid). The preformed vitamins are found almost exclusively in animal tissues. Certain plant carotenoids convertible *in vivo* to vitamin A are termed provitamins A; they include the α-, β- and γ-carotenes and cryptoxanthin. In the vitamins and provitamins A the isoprenoid structure allows the occurrence of stereoisomers of which the all-trans isomers are, in general, the most biologically active.

The different forms of vitamin A are not all interchangeable in their physiological roles. In mammals the esters are hydrolysed in the intestine; the resulting alcohol passes into the mucosal cells where it is re-esterified and thence transported via the lymphatics to the blood. In the chicken the liver is an important site of storage of vitamin A. Retinol is the predominant form of the vitamin in blood of animals deprived of food, where it is maintained at a fairly constant level until the hepatic stores are exhausted. On mobilization, the stored esters undergo hydrolysis and the resulting alcohol is transported as a lipoprotein complex. Birds fed diets low in protein were reported by Nir and Ascarelli (1966, 1967) to exhibit a decreased rate of depletion of hepatic vitamin A. A concomitant fall in plasma proteins was concluded to interfere with the transport of vitamin A and hence with its mobilization from the liver.

1. *Utilization of Carotenoids*

The carotenoids are more stable than the preformed vitamin. In all animals so far examined, the major site of conversion of carotenoids to vitamin A is the intestinal wall. The chick is no exception (Thompson *et al.*, 1950). There is apparently also some extra-intestinal conversion since Bieri (1955) found that intravenously injected β-carotene and, to a lesser extent, cryptoxanthin were utilized by vitamin A-deficient chicks. It is still a matter for controversy whether the carotenoid precursor molecule is subjected to central fission or to terminal oxidation to yield the vitamin. It is, however, generally accepted that, weight for weight, β-carotene has half the biological activity of the performed vitamin. This conversion appears to be achieved at levels of intake near the physiological requirement but β-carotene is less well utilized at higher levels intended to induce hepatic storage (Ely, 1959; Marusich and Bauernfeind, 1963; Parrish *et al.*, 1963). Others found β-carotene to have less than its theoretical equivalence of vitamin A activity at all levels of supplementation. It may be that other dietary components interfere with its availability to the bird, which would account for the general lack of agreement between different investigators. Conversely, high dietary levels of vitamin A interfere with absorption of carotenoids, thus suppressing pigmentation of carcass or egg yolk (Dua and Day, 1964; Dua *et al.*, 1965; Gutzmann and Donovan, 1966).

B. METABOLIC ROLES

1. *Vision*

The function of vitamin A in vision has been fully reviewed by Wald (1960). Retinal is the prosthetic group of the visual pigments in all terrestrial vertebrates. The two pigments of the retina, rhodopsin in the rods and iodopsin in the cones, are formed by combination of vitamin A aldehyde with two different proteins (opsins). The synthesis can be represented as follows:

$$\text{retinol} \underset{\text{DPNH}}{\overset{\text{DPN}^+}{\rightleftharpoons}} \text{retinal} \begin{cases} + \text{rod opsin} \longrightarrow \text{rhodopsin} \\ + \text{cone opsin} \longrightarrow \text{iodopsin} \end{cases}$$

For *in vivo* combination with opsin the retinal must be in the 11-cis form. Isomerization from all-trans retinal is enzymic and by the action of light. On exposure to light the pigments are bleached, with release of all-trans retinal and the corresponding opsin. During this process nervous impulses are set up which, when transmitted to the brain, excite visual sensations. Although the first chemical isolation of a visual pigment, iodopsin, was made from the chicken eye (Wald and Zussman, 1938), comparatively few studies have been made of the visual cycle in the fowl. (See also Chapter 46.) However, from electroretinogram studies, Armington and Thiede (1956) concluded that the photochemistry of vision in the chicken is similar to that of other animals. Birds reared on a diet containing retinoic acid as a source of vitamin A became blind after about twenty weeks, indicating that there is no mechanism for the conversion of retinoic acid to retinal in the chicken eye (Thompson *et al.*, 1965).

2. Morphogenic Effects

a. Mucosal epithelia. In most animals atrophy of mucus-secreting epithelia is a first sign of lack of vitamin A. In chicks deprived of vitamin A, Aydelotte (1963) showed that the secretory cells of the conjunctiva, oesophagus and trachea were gradually replaced by stratified keratinizing epithelium. The corneal epithelium was less severely affected, so that frank xerophthalmia is rarely observed in birds. Howell and Thompson (1967a, b) confirmed Jungherr (1943) in that metaplasia of the nasal epithelium was the first lesion to appear. Excess vitamin A brings about the opposite effects. Fell and Mellanby (1953) observed that explants of chick ectoderm cultured in the presence of high levels of retinol developed a mucus-secreting epithelium in place of the normal squamous keratinizing cells. Further studies, summarized by Fell (1960), showed that the uptake of inorganic sulphate (indicative of formation of acidic mucopolysaccharides) paralleled the secretory activity of the cells, suggesting that the synthesis of acidic mucopolysaccharides is blocked by keratinization but proceeds normally when keratinization is prevented by vitamin A.

b. Cartilage and bone. The development of cartilage and bone in young animals, including the chick, is very sensitive to the vitamin A status. In deficiency, bone resorption is diminished but deposition continues, with consequent thickening and deformity of the skeletal elements. In the chick, all sequences of bone growth are retarded, and endochondrial bone growth is particularly affected, Wolbach and Hegsted (1952a). Howell and Thompson (1965, 1967a, b) also observed derangements in bone growth in chicks and adult fowl; they did not confirm the retardation of endochondrial bone-growth but instead reported an increased number of osteoblasts in the subperiosteal bone, with formation of subperiosteal cartilage. They considered this abnormality to be the more significant indication of vitamin A deficiency since it occurred also in adult birds in which growth of endochondrial bone had ceased. Havivi and Wolf (1967) also recorded biochemical changes compatible with increased metabolic activity in bones from chicks deprived of vitamin A.

In general, the effects of excess vitamin A on bone and cartilage are the reverse of those of deficiency (Wolbach and Hegsted, 1952b). In particular, there is suppression of osteoblastic activity (Baker *et al.*, 1967). Studies *in vitro* with explants of cartilaginous limb-bone rudiments from chick embryos have again indicated involvement of vitamin A in mucopolysaccharide metabolism (Fell *et al.*, 1956), but this mechanism is still under investigation (see, for instance, Dingle *et al.*, 1966).

c. Nervous tissue. Ataxia is a consistent sign of vitamin A inadequacy in young chicks. Controversy exists as to whether the cause is due to a primary effect of the deficiency on nervous tissue or is secondary to compression of the central nervous system by abnormal bone growth in the vertebral canal. After detailed histology, Wolbach and Hegsted (1952a) concluded that neurological disturbances in chicks deprived of vitamin A resulted entirely from compression of the CNS by deformed vertebrae and cranial bones. This conclusion is supported by Howell and Thompson (1967a, b) who found little evidence for a direct effect of vitamin A-deficiency on nervous tissue. They also considered that the nervous signs could be explained by compression of the CNS, although they differed from earlier authors as to the nature of the bone deformities responsible.

3. Resistance to Infection

Animals deprived of vitamin A are, in general, very susceptible to infection; there are a number of records of flocks on a low intake of vitamin A being more severely affected by common poultry diseases than controls with adequate vitamin A. Failure by deficient birds to resist infection is no doubt partly due to the changes in epithelial membranes, particularly in the case of intestinal parasitic infestation. There is, however, some evidence that vitamin A may be concerned in antibody production. Panda and Combs (1963) reported that chicks fed on diets low in vitamin A responded to *Salmonella pullorum* antigen with lower agglutinin titres than controls on an adequate diet. The bursa size was significantly reduced in the deficient birds. Leutskaja (1963, 1964) also concluded that vitamin A was required for antibody formation in chickens challenged with antigen from *Ascaridia galli.*

C. TOXICITY

Some morphogenic effects of excess vitamin A have already been mentioned. Biochemical studies on chick embryo tissues have led to the conclusion that vitamin A brings about release of proteolytic enzymes and other hydrolases from lysosomes (Dingle *et al.*, 1961; Fell and Dingle, 1963). Taylor *et al.* (1968) investigated the effects of high doses of vitamin A on the plasma enzymes of lysosomal origin in young chicks; on a diet supplemented with retinol at 1700 times the normal requirement, the plasma activities of acid phosphatase, β-glucuronidase and arylsulphatase were all reduced. There was also a marked reduction in packed cell volume, which the authors believed due to increased fragility of the erythrocytes. Increased red cell fragility in chicks given excess vitamin A was reported by March *et al.* (1966), who suggested that it was an indirect effect of thyroid suppression, leading to a greater proportion of older, fragile cells in the circulating blood.

III. Vitamin D

A. FORMS

Vitamins of the D group arise by ultraviolet irradiation of their corresponding provitamins, which are sterols characterized by a $\Delta 5,7$ double bond in ring B of the nucleus. The two most important are vitamin D_2 (ergocalciferol) formed on irradiation of the plant sterol, ergosterol, and vitamin D_3 (cholecalciferol) from irradiation of the animal sterol, 7-dehydrocholesterol. Although both vitamins are equally effective for mammals, birds utilize ergocalciferol very poorly. The potency of vitamin D_2 for the chick is of the order of a fortieth (Günther and Tekin, 1964) to a tenth (Chen and Bosmann, 1964) that of vitamin D_3. Apparently the two vitamins are equally well taken up by the organs of deficient chicks, but ergocalciferol is markedly less well retained (Imrie *et al.*, 1967). Birds reared in a sunny environment have a reduced need for dietary vitamin D since sunlight acts on its precursors in the superficial tissues.

B. METABOLIC ROLE

1. General Considerations

Rickets is the classic sign of lack of vitamin D in young birds and is characterized by failure of calcification in the cartilage zone of developing bone, with widening of the epiphyseal cartilage plate. In older birds osteomalacia develops, with loss of calcium from the skeleton. Egg production dwindles and shell quality deteriorates. These effects result from poorer absorption of calcium from the gut, disturbances in bone mineralization and are accompanied by a rise in plasma alkaline phosphatase. Abnormal blackening of the feathers has been reported in chicks deprived of vitamin D_3, and a possible interaction with thyroid hormone has been postulated (Glazener and Briggs, 1948).

Broadly speaking, vitamin D functions metabolically as a part of the mechanism of calcium homeostasis. Together with parathyroid hormone and calcitonin, it aids control of calcium and phosphorus levels in the body, although no direct involvement of vitamin D in phosphate metabolism has been substantiated. In rats parathyroid mobilization of calcium cannot take place in absence of vitamin D (Harrison *et al.*, 1958) but the reduction of plasma calcium by calcitonin is independent of the vitamin (Morii and DeLuca, 1967). The exact biochemistry of the action of vitamin D has long remained, and still is, unexplained but recent work on birds and mammals has made considerable progress towards its elucidation. (See reviews by DeLuca, 1967, and Norman, 1968.) Several workers have reported a lag of some hours between the administration of vitamin D and the induction of a physiological response; this might be due to a need for conversion of vitamin D into an active form in the tissues. Conversely, the suggestion has been made that vitamin D may have a hormone-like action, and the delay is caused by the time taken to activate genetic information which induces the physiological response.

2. Absorption of Calcium

In vivo and *in vitro* studies on chicks have shown that vitamin D_3 is directly concerned in active transport of calcium across the small intestinal mucosa and that bile salts facilitate the process (Coates and Holdsworth, 1961; Sallis and Holdsworth, 1962). Holdsworth (1965) proposed that vitamin D_3 inhibits return of calcium from the mucosal cell to the lumen by means of a metabolically operated pump. In contrast, the results of Wasserman *et al.* (1966) support an effect of vitamin D_3 on diffusional permeability, possibly by some alteration of cell membrane structure.

3. *Mineralization of Bone*

Evidence with chicks refutes the suggestion that vitamin D has a direct effect on the deposition of calcium in bone. Several workers found no difference between the incorporations of ^{45}Ca ions into rachitic and normal chick tibia (Migicovsky and Emslie, 1950; Migicovsky and Jamieson, 1955; Coates *et al.*, 1961). From histological and chemical studies by Migicovsky and his co-workers it seems more likely that vitamin D is necessary for the maturation of those cartilage cells concerned in the mechanism of bone growth. Belanger and Migicovsky (1958) observed a failure of maturation and enlargement of epiphyseal cartilage in rachitic chicks; a substance able to bind calcium *in vitro*, presumed to be chondroitin sulphate, accumulated in the cartilage matrix. After giving vitamin D_3 the cartilage rapidly matured. There was then a progressive loss of the calcium binder from the matrix and it was suggested that chondroitin sulphate acts as a localized trap for calcium. Degradation of chondroitin sulphate is then brought about, possibly enzymically, by the mature cartilage cells, with consequent release of calcium for mineralization of bone. In support of this hypothesis Cipera and Willmer (1963) found that the curative effect of vitamin D_3 was accompanied by increased activities of alkaline phosphatase, hexosamine synthetase and pyrophosphatase in epiphyseal cartilage. Increases in galactosamine, induced by vitamin D_3, indicate that the vitamin does influence the metabolism of chondroitin sulphates (sulphated glucurono-galactosaminoglycans) in the epiphyseal cartilage of the chick (Cipera, 1967).

C. TOXICITY

Although high levels of vitamin D are toxic, it appears that the chick can tolerate at least 100 times its normal requirement without adverse effect. Growth is depressed, but vitamin D_3 is ten times as toxic as vitamin D_2 in this respect (Chen and Bosmann, 1965). A hypercalcaemia, believed to be due to increased absorption of calcium and increased bone resorption, was observed in chicks given 50,000 i.u. vitamin D_3/kg diet by Taylor *et al.* (1968). There was at the same time a decrease in plasma acid phosphatase and inorganic phosphorus. No general antagonism between vitamins A and D was noted but, as excess vitamin A had the opposite effect on plasma calcium, phosphorus and acid phosphatase, the net result of excess of the two vitamins together was to retain the levels of all three near to normal.

IV. Vitamin E

A. FORMS

The vitamins E are tocopherols; at least seven occur in nature. They are methyl, dimethyl or trimethyl derivatives of tocol. The most important biologically is α-tocopherol (5,7,8-trimethyltocol). It occurs naturally in the L-form, which is about 1·35 times as potent for the chick as the synthetically produced DL-form (Pudelkiewicz *et al.*, 1960; Marusich *et al.*, 1967) and twice as potent as the L-form (Dam and Søndergaard, 1964). The esters are more stable than the free tocopherols, and DL-tocopheryl acetate is commonly used as a dietary supplement.

B. METABOLIC ROLE

Vitamin E plays several, apparently unrelated, roles in avian metabolism. The manifestations of deficiency differ in character and severity according to the composition of the diet. Chemically, it behaves as an antioxidant and there is controversy whether or not its antioxidant properties can entirely account for its biological activities. The importance of vitamin E in poultry nutrition has been reviewed by Ames (1956), and its role in metabolic processes in the bird by Dam (1962) and Scott (1962a, b).

1. Vitamin E as an Antioxidant

One effect of deprivation of vitamin E in young birds is encephalomalacia or "crazy chick disease", so-called because the birds exhibit ataxia, with violent spasms of unco-ordinated movement. In advanced cases, cerebellar lesions, notably oedema and haemorrhages, are seen. The condition occurs at the period of most rapid brain growth and is manifested on diets containing high levels of unsaturated fatty acids. It can be prevented by non-physiological antioxidants such as diphenyl-*p*-phenylene diamine (DPPD), methylene blue or 2,6-di-tertiary-butyl-*p*-cresol (DBC) (Machlin and Gordon, 1960; Scott and Stoewsand, 1961). Apparently, the protective effect of the antioxidant is exerted directly in the tissues and is not the indirect result of preventing the formation of toxic oxidation products in the food (Scott and Stoewsand, 1961). It might seem logical to suppose, therefore, that α-tocopherol also acts directly as a tissue antioxidant (see review by Tappel, 1962), but a number of experimental findings do not fit this hypothesis. For instance, although Glavind and Søndergaard (1964) found decreased levels of antioxidants in brains and cerebella of chicks with encephalomalacia, the decrease was in the aqueous, not the lipid, phase. No lipid peroxidases were detected in the brains of chicks with encephalomalacia and, furthermore, compared with other tissues, chick brain is very low in vitamin E (Dam, 1962). In extensive studies Green and collaborators could find no evidence for increased peroxidation in chicks deprived of vitamin E (Diplock *et al.*, 1967; Bunyan *et al.*, 1967a). After reviewing evidence from many sources, Green and Bunyan (1969) concluded that "Although there is a close relation between vitamin E and the metabolism of unsaturated fat, . . . it cannot be interpreted solely in terms of the peroxidisability of the fat and an antioxidant function for vitamin E."

2. Interrelationships of Vitamin E, Selenium and Sulphur-containing Amino Acids

Another consequence of vitamin E deficiency in chicks is exudative diathesis, in which plasma exudes from the capillaries and collects in the subcutaneous and connective tissues. It is accompanied by some haemorrhage, so that the birds present a bruised, discoloured appearance. The protein content of the exudate is sufficiently close to that of plasma to conclude that the condition is not the result of osmotic disturbance but is, more probably, due to local capillary damage. Exudative diathesis occurs on diets containing polyunsaturated fatty acids, and in this event some improvement can be brought about by other antioxidants, possibly by sparing vitamin E. It develops also on diets without excessive quantities of fat and, in these circumstances, antioxidants are without effect. In either instance the condition can be prevented or cured by dietary supplements containing inorganic selenium. Increased lysosomal hydrolase activity has been observed in the exudate, in the surrounding tissues and, to a lesser extent, in the kidney, liver and spleen of chicks with exudative diathesis (Bunyan *et al.*, 1967b). Although this suggests a connexion between lysosomal enzymes and causation of the disease, the authors point out that the increased hydrolase activity could be due to some other pathological change resulting from deficiency of vitamin E or selenium or both. Diplock *et al.* (1967) traced the metabolism of ^{3}H-α-tocopherol in birds with exudative diathesis. As it was unaffected by selenium deficiency it appears that vitamin E and selenium may function in two alternative biochemical pathways.

On diets low in vitamin E and in sulphur-containing amino acids, chicks develop muscular dystrophy, a degenerative condition characterized by the appearance of white striations in the skeletal muscle. It is more severe on diets containing unsaturated fatty acids, but occurs also on diets that are virtually fat-free. Non-physiological antioxidants are without effect in preventing or curing the disorder. High supplements (of the order of 0·5 p.p.m.) of selenium only partly ameliorate the dystrophy, but dietary methionine, cystine or vitamin E relieve it completely. From a time-sequence study, Desai *et al.* (1964b) concluded that the sulphur-containing amino acids play a direct role in preventing muscular dystrophy since the signs were more rapidly reversed by methionine than by vitamin E. The high degree of tissue peroxidation was markedly reduced by vitamin E but was much less affected by methionine; thus it appeared not to be directly associated with the dystrophy. Other sulphur-containing compounds have been investigated for their effect on muscular dystrophy, and there is evidence that cysteine is the functional compound involved in its prevention (Hathcock *et al.*,

1968). Deficiency of arginine protects chicks against the development of the condition (Nesheim *et al.*, 1960). Increased hydrolase activity in the muscle of dystrophic chicks has been reported (Desai *et al.*, 1964a, b; Bunyan *et al.*, 1967b). Such an effect could arise through breakdown of the dystrophic tissue and subsequent invasion by phagocytes, known to be rich in hydrolases.

Thus, there appears to be a metabolic interrelationship between the actions of vitamin E, selenium and the sulphur-containing amino acids in the chick. Selenium is mainly concerned in the prevention of exudative diathesis, and is effective at very low levels of supplementation—of the order of 0·1 p.p.m. diet (Nesheim and Scott, 1958). There is no evidence for specific involvement of the sulphur-containing amino acids in this condition, but they play a major role in the prevention of muscular dystrophy. Selenium also exerts some protective effect, but at much higher levels than are necessary to prevent diathesis. Muscular dystrophy appears to arise from an inability to make use of tissue vitamin E, since the tocopherol content of dystrophic muscle is much higher than normal (Diplock *et al.*, 1967), This concept is supported by Desai and Scott (1965), who found that administration of selenium to birds on a dystrophy-inducing diet increased the plasma levels of tocopherol. They suggest that selenium, possibly as a seleno-lipoprotein, may conserve vitamin E by controlling its retention or preventing its destruction in the tissues.

3. Reproduction

In spite of the well-known involvement of vitamin E—the "anti-sterility vitamin"—in mammalian reproduction, comparatively few studies have been made on its role in reproductive processes in poultry. In a long-term experiment Adamstone and Card (1934) noted no adverse effect on fertilizing capacity of cock semen after a year on a diet deficient in vitamin E. Testicular degeneration and reduction in fertilizing capacity was, however, observed in some birds after two years on the diet. More recently polyunsaturated fatty acids have been used to exacerbate the bird's need for vitamin E. Adult male chickens maintained for 25 weeks on a diet with a low content of linoleic acid showed no adverse effects of deprivation of vitamin E. Birds on a similar diet with 7·3% linoleic acid produced semen with a lower sperm count and fertilizing capacity (Arscott *et al.*, 1965). Fertility was rapidly restored when vitamin E supplementation (166 mg/kg diet) was begun at 28 weeks (Arscott and Parker, 1967). Comparable results were observed in hens (Machlin *et al.*, 1962). After eight weeks on a diet high in linoleic acid but without vitamin E, egg production, fertility and hatchability were all drastically reduced. Restoration to normal levels was attained by addition to the diet of either 100 i.u. vitamin E/220/kg or 0·3% ethoxyquin. No adverse effects on reproduction were observed on a diet having a low level of linoleic acid.

V. Vitamin K

A. FORMS

A number of natural and synthetic quinones have vitamin K activity. The phylloquinones (e.g. vitamin K_1, 2-methyl-3-phytyl-1,4-naphthoquinone) contain saturated isoprenoid side chains. They are of vegetable origin, and are particularly abundant in green leaves. The menaquinones (e.g. vitamin K_2, menaquinone-7,2-methyl-3-difarnesyl-1, 4-naphthoquinone) contain unsaturated isoprenoid side chains and are synthesized by micro-organisms. Several simpler synthetic naphthoquinones are active. Most commonly used are menaphthone (vitamin K_3, 2-methyl-1,4-naphthoquinone) and its water-soluble derivatives, e.g. menaphthone sodium bisulphite and menaphthone sodium diphosphate. The biological activities of various compounds of the vitamin K group have been investigated; the results are summarized by Griminger (1966). Weight for weight, 1 mg vitamin K_1 is equivalent to 0·38 mg menaphthone and 0·73 mg menaphthone sodium bisulphite but it is usual to compare the biological activities of the vitamins K on a molar basis; for the chicken vitamin K_1 and menaphthone sodium bisulphite are roughly equal but menaphthone is only about 40% as effective (Nelson and Norris, 1960).

B. METABOLIC ROLE

The essential need for vitamin K to maintain normal blood clotting has long been recognized. In its absence, clotting time is prolonged leading to haemorrhages throughout the muscles and in the abdominal cavity as results of very minor injury. Anaemia occurs, partly from loss of blood and partly from hypoplasia of the bone marrow (Scott, 1966). Hatchability is poor in eggs from hens given inadequate vitamin K; haemorrhagic patches occur in some of the dead embryos (Griminger, 1964). Vitamin K is synthesized by the intestinal flora, hence signs of deficiency are less severe in coprophagous species. Out of reach of their excreta, chicks appear to derive little of their vitamin K requirement from products of microbial origin; the chick has thus become the preferred subject for studies on vitamin K.

In broad outline, the mechanism of blood clotting may be represented as follows:

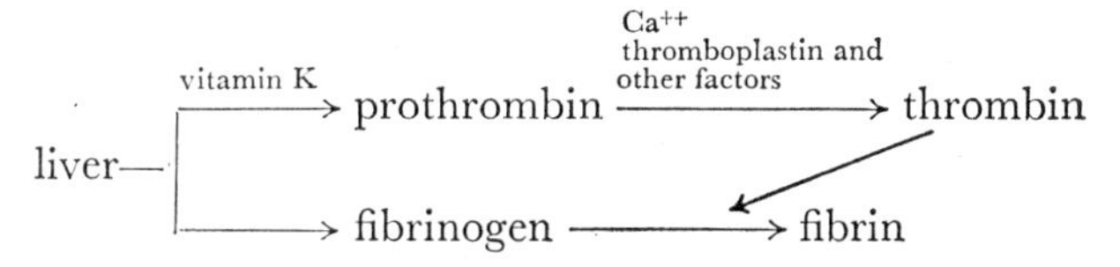

Clot formation is brought about by enzymic degradation of the soluble plasma protein, fibrinogen, to the insoluble protein, fibrin, by the protease, thrombin. The latter is formed from its precursor, prothrombin, which is produced in the liver. Vitamin K is essential for the synthesis of prothrombin but this mechanism is not yet elucidated. The chemically similar ubiquinones are known to play a vital part in tissue respiration. It has been proposed that vitamin K may likewise be involved, particularly in oxidative phosphorylation mechanisms specifically concerned in prothrombin synthesis (Martius, 1966). An alternative hypothesis is that vitamin K induces formation of prothrombin by way of a prothrombin-specific messenger RNA (Olson, 1964; Olson *et al.*, 1967).

C. ANTAGONISTS

Several types of compound increase the requirement of the chick for vitamin K, apparently through different mechanisms. By far the most active are substances resembling dicoumarol 3,3′-methylenebis(4-hydroxy-coumarin). Their anticoagulant effect can be reversed by vitamin K_1 but not by compounds of the K_3 group (Griminger and Donis, 1960; Griminger, 1965). This antagonism seems to depend on blockage of the vitamin K-dependent step in the synthesis of prothrombin. Prolonged clotting-times are induced in chicks by feeding sulphonamides or tetracycline and can be restored to normal with menaphthone compounds and with vitamin K_1. Such drugs are likely to interfere with intestinal production of vitamin K, although Griminger (1965) considers that reduction of microbial synthesis is not the complete explanation for their anticoagulant action.

VI. Vitamin B Complex

The vitamins loosely classed together as the "B complex" all take part as co-factors in one or more fundamental biochemical reactions. Nicotinic acid can be synthesized by animals from tryptophan, although not usually in sufficient quantity to satisfy the full requirement. There is no evidence of endogenous synthesis of other B vitamins, but all are produced by microbial action in the alimentary tract of the chick. However, birds kept out of contact with their excreta obtain little or no benefit from these (Coates *et al.*, 1968). With the exception of vitamin B_{12}, tissue storage of the B vitamins is low; a fairly continuous dietary supply is thus essential. B vitamins are, however, widely distributed in natural foods, particularly cereal grains, hence frank deficiencies are seldom seen under practical feeding conditions. Because of the vital role they play in the bird's metabolic processes, deprivation of any one of them is shown in poor growth, general unthriftiness and, frequently, poor reproductive performance.

A. THIAMIN

1. Metabolic Role

Thiamin (vitamin B_1) consists of a substituted pyrimidine ring joined through a methylene bridge to a substituted thiazole ring. As its pyrophosphate it takes part in at least two catabolic exergonic sequences by which glucose is oxidized to CO_2 and water. In glycolysis thiamine pyrophosphate is a co-factor ("cocarboxylase") in the decarboxylation of pyruvate. In the pentose phosphate pathway it functions, as a component of transketolase, in the reconversion of pentoses to hexoses. The level of transketolase activity in the erythrocytes has been used as an index of thiamin nutrition in the laying hen (Padhi and Combs, 1965).

Animals and birds deprived of thiamin accumulate pyruvate and lactate in their tissues. Pyruvate is the acetyl donor in the synthesis of acetylcholine, concerned in the chemical transmission of nerve impulses. Its catabolism is important in brain, for which glucose is the main source of energy. Thus, abnormalities of the nervous system are major consequences of a lack of thiamine. In the chicken they show as ataxia, frequently accompanied by retraction of the neck. Mortality is high and convulsions often precede death. In adult birds deficiency of thiamin is rarely encountered in practice but there are indications from experiments with thiamin antagonists that inadequacy might lead to poor reproductive performance. Adding high concentrations of amprolium (see below) to the diet of cocks and hens depresses fertility and hatchability possibly through induced deficiency of thiamin (Polin *et al.*, 1962). Injection of hatching eggs with anti-thiamin drugs resulted in a high embryonic mortality (Naber *et al.*, 1954; Polin *et al.*, 1962).

Because of the specific role in glucose metabolism of thiamine, its requirement is related to the dietary carbohydrate content. An increased need for thiamin was demonstrated when sucrose was substituted for maize oil in a broiler diet (Thornton and Schutze, 1960). In growing chicks Peacock and Combs (1969) noted a reduced requirement for thiamin when the major source of energy was fat instead of carbohydrate; the presence of fat did not reduce the need for thiamin when dietary carbohydrate was adequate both for energy production and conversion to tissue fat.

2. *Thiamin Antagonists*

Certain foods, not commonly used in poultry diets, contain enzymes (thiaminases) which destroy thiamin. For example raw fish, clams, shrimps and bracken are reported to contain thiaminases. Synthetic analogues of thiamin (e.g. neopyrithiamin, where a pyridine nucleus replaces the thiazole ring) are B_1-antagonists, presumably because of their closely related structure which blocks the chain of reactions normally involving thiamine. Another antagonist is the coccidiostat amprolium, 1-(4-amino-2-*n*-propyl-5-pyrimidylmethyl)-2-picolinium chloride hydrochloride. However, to counteract the effect of unit weight of thiamin, about 500 times the weight of amprolium is estimated to be necessary (Ott *et al.*, 1965).

B. RIBOFLAVIN

1. *Metabolic Role*

Riboflavin (vitamin B_2), 6,7-dimethyl-9-(1′-D-ribitol)isoalloxazine is a yellow substance containing a pentitol residue attached to an isoalloxazine ring. Riboflavin-5′-phosphate alone (flavin mononucleotide, FMN), or linked with adenosine-5′-phosphate (flavin adenine dinucleotide, FAD), are the prosthetic groups of the flavoproteins, which act as hydrogen carriers in biological oxidations. As such, one or both are concerned in the oxidation of amino acids, glucose, fatty acids and purines. In the livers of mammals deprived of riboflavin, reduced activity of flavoprotein enzymes (e.g. xanthine oxidase, D-amino acid oxidase) has been observed. A characteristic sign of deficiency in the chick is "curled-toe-paralysis" but the condition cannot be explained in terms of known biochemistry. Riboflavin is extremely important for normal embryonic development. In eggs from birds deprived of riboflavin three peaks of embryo mortality have been reported, one at three days, one between twelve and fourteen days and one towards the end of incubation. Micromelia, oedema and demyelination have been observed in the dead embryos and, in those surviving to the later stages, "clubbed down" is a characteristic feature.

The transfer of riboflavin to the egg has been studied in some detail (Bolton, 1953; Brown, 1957). Experiments with ^{14}C-riboflavin have shown that part of the riboflavin in the egg is derived directly from the food and part from tissue stores (Blum and Jacquot, 1964). At the onset of lay the free riboflavin content of the blood plasma rises (Boucher *et al.*, 1959). The vitamin is stored in the liver, largely as flavoproteins and, in the normal bird, a proportion of the hepatic store is readily mobilized for egg production. In hens depleted of riboflavin, hepatic storage decreases to a level at which it is tenaciously retained at the expense of egg formation. The egg's content (initially about equally distributed between yolk and white) thereafter dwindles rapidly and production ceases (Blum and Jacquot-Armand, 1965). The magnum can fix free riboflavin as a non-dialysable form, presumably a flavoprotein, thus providing storage for transfer to the egg (Blum and Jacquot, 1964). A flavoprotein and its corresponding apoprotein have been isolated from egg white by Rhodes *et al.*, (1959). It is remarkable in that the flavin moiety is riboflavin and not its 5′-phosphate.

An inherited riboflavin deficiency in embryos from a strain of White Leghorns (Maw, 1954) is due to a recessive gene that prevents adequate transfer to the egg (Buss *et al.*, 1959), by suppression of the renal reabsorption of the vitamin. The excessive losses in the urine thus reduce the blood level of free riboflavin and hence the amount available for the egg (Cowan *et al.*, 1966).

C. NICOTINIC ACID

Nicotinic acid (3-pyridine-carboxylic acid) and its corresponding amide have equal vitamin activity for higher animals. Nicotinamide forms part of two pyridine nucleotides which, in combination with a variety of apoenzymes, catalyse dehydrogenation of many important metabolites. Nicotinamide phosphate is combined with adenosine monophosphate to give nicotinamide-adenine dinucleotide (NAD, formerly coenzyme I or DPN). When this substance carries a third phosphoric acid residue it forms nicotinamide-adenine dinucleotide phosphate (NADP, formerly coenzyme II or TPN). In the dehydrogenations catalysed by NAD or NADP, two hydrogen atoms are transferred to the coenzyme; reoxidation of the reduced form is effected by flavoproteins (see above).

Birds deprived of nicotinic acid grow poorly and are generally unthrifty; perosis is a frequent sign. In laying and breeding birds egg production and hatchability are adversely affected. Manoukas (1967) established a linear relationship between dietary nicotinic acid and hatchability. Some nicotinic acid can be produced in avian tissues from tryptophan; over and above the bird's basic dietary tryptophan requirement of about 0·15% can be used for this synthesis, consequently sparing the amount needed in the diet. However, the extent of synthesis appears to be influenced by a variety of unknown factors since quantitative estimates of the equivalence of tryptophan to nicotinic acid have varied widely (see for instance Childs *et al.*, 1952; Patterson *et al.*, 1956; Manoukas *et al.*, 1968). There is no evidence that an excess of nicotinic acid spares the bird's requirement for tryptophan.

Nicotinic acid occurs in cereals in a bound form that is poorly utilized by higher animals (Chaudhuri and Kodicek, 1960). Using hatchability as criterion of adequacy, Manoukas *et al.* (1968) showed that only 30 and 36% of the nicotinic acid present in maize meal and wheat bran respectively was available to the hen.

D. PYRIDOXINE

Pyridoxine (vitamin B_6) is the general name for a group of pyridine derivatives consisting of pyridoxol, 2-methyl-3-hydroxy-4,5-di(hydroxymethyl)-pyridine, pyridoxal and pyridoxamine (its 4-formyl and 4-aminomethyl analogues respectively). All are equally active for the chick (Davies *et al.*, 1959). They usually occur in nature as their 5-phosphates. It is probable that the alcohol and amine are converted to the aldehyde in the tissues. Pyridoxal is a cofactor of considerable importance in the metabolism of amino acids. It is a coenzyme for some amino acid decarboxylases, transaminases, dehydrases and sulphydrases. It is concerned in the synthesis of tryptophan from indole and serine, and in the conversion of tryptophan to nicotinic acid. Suggestions that pyridoxine may function as a co-factor in the synthesis of arachidonic acid from linoleic acid have not been upheld by the results of Moore *et al.* (1968). Although they found increased linoleate and decreased arachidonate in the liver lipids of chicks on a low pyridoxine intake, the effect was associated with changes in the proportions of triglycerides and phospholipids and was apparently not due to a direct involvement of the vitamin in the biosynthesis of arachidonate from linoleate. Daghir and Porooshani (1968) were similarly unable to demonstrate involvement of pyridoxine in the interconversions of polyunsaturated fatty acids.

Severe deprivation of vitamin B is rapidly lethal to the young chick. Growth virtually ceases, the birds become hyperexcitable, frequently showing violent convulsions. Death follows within a few days. Anaemia has been observed by several investigators. Lowered serum β-globulin levels and reduced activity of some of the pyridoxine-dependent enzymes, particularly glutamic-oxalacetic transaminase, have been noted (Kirchgessner and Maier, 1968a, b). A reduced amine oxidase activity in the aorta of chicks deprived of pyridoxine was accompanied by reduced conversion of lysine to desmosine and a lower content of elastin (Hill and Kim 1967). In adults, egg production and hatchability are depressed. An increased requirement for pyridoxine on diets high in choline has been shown by Saville *et al.* (1967). They suggest that excess choline is catabolized to pyruvate through betaine, glycine and serine, a pathway that requires pyridoxine for the last two steps.

E. PANTOTHENIC ACID

Pantothenic acid *N*(2,4-dihydroxy-3,3-dimethylbutyryl)-3-aminopropionic acid is part of the complex nucleotide coenzyme A. In the tissues it combines with thioethanolamine to give pantetheine which, when phosphorylated and combined with an adenine nucleotide, forms coenzyme A. One of the several major metabolic roles of coenzyme A is in acetylations; through formation of the derivative acetylcoenzyme A, an acetyl residue can be transferred, usually reversibly, between two metabolites. The transfer is catalysed by a variety of enzymes, according to the particular reaction concerned. Coenzyme A is thus involved in the metabolism of carbohydrates and fatty acids, and in the synthesis of acetylcholine and of sterols. It can also form derivatives with other acyl groups; for example, succinyl-coenzyme A is an intermediate in the citric acid cycle and, in the biosynthesis of fatty acids, malonylcoenzyme A is the means by which two carbon atoms are added to the fatty acid chain.

As might be expected from the number of metabolic pathways in which coenzyme A plays a part, deprivation of pantothenic acid results in generalized unthriftiness and failure in reproduction. Young chicks develop a characteristic scaly dermatitis, appearing first at the corners of the mouth and later on the feet. There is frequently a sticky exudate from the eyes. In hens, hatchability of their eggs is seriously depressed, the peak of embryonic mortality becoming progressively earlier the more severe the depletion (Beer *et al.*, 1963).

On diets low in vitamin B_{12} the bird's need for pantothenic acid increases but the mechanism of the relationship between the two vitamins has not been explained.

F. BIOTIN

Biotin is a cyclic ureide, 2′-keto-3,4-imidazolido-2-tetrahydro-thiophene-valeric acid. Two forms have been isolated, α-biotin from hen's egg yolk and β-biotin from beef liver. Biotin occurs naturally as the D-isomer, frequently in a combined form. Several biotin-protein complexes have been isolated from animal tissues and biocytin, in which biotin is combined with lysine, occurs in yeast. Egg white contains a glycoprotein, avidin, with which biotin forms a complex that cannot be readily broken down by the digestive proteases. Thus biotin deficiency is produced in animals by the feeding of raw egg white but, as avidin is denatured by heat, cooked egg white has no adverse effect on biotin availability.

Biotin acts as a coenzyme in carboxylation reactions involving the fixation of CO_2. In this role it is concerned in the metabolism of carbohydrate, fatty acids and proteins. For instance, in pigeon and turkey livers the enzyme catalysing the reversible conversion of malate to pyruvate and CO_2 is reduced in biotin-deficiency (Ochoa *et al.*, 1947). In fatty acid synthesis biotin is a cofactor in the carboxylation of acetylcoenzyme A to malonylcoenzyme A. Decreased incorporation of acetate into liver lipids of chicks deprived of biotin has been reported by Balnave and Brown (1967). The reduced incorporation of amino acids into tissue proteins observed in biotin-deficient chicks by Dakshinamurti and Mistry (1963) was considered to be the indirect result of a reduced synthesis of dicarboxylic acids.

Gross signs of biotin deficiency are seldom seen in practice. They have been experimentally produced with biotin deficient diets, frequently accelerated by the inclusion of raw egg white. In young birds, growth is depressed, though less severely than by deficiency of the other B vitamins. There is a high incidence of perosis and a scaly dermatitis that appears first on the feet and later around the eyes and mouth (*cf.* pantothenic acid deficiency). Scott (1968) postulated that enlarged hocks and leg weakness in growing turkeys result from inadequate biotin intake. Biotin is of considerable importance in embryonic development. Mortality peaks occur at about the third day and again towards the end of incubation. Micromelia is a characteristic sign in affected embryos (Cravens *et al.*, 1944; Couch *et al.*, 1948).

G. FOLIC ACID

Folic acid is the group name used for the naturally occurring pteroylglutamic acids. The parent substance, pteroylmonoglutamic acid, consists of a pteridine nucleus linked through 4-aminobenzoic acid to a glutamic acid residue. Analogous compounds, with further molecules of glutamic acid attached in peptide linkage to the first, frequently occur in nature; the tri- and hepta-glutamates are notable examples.

Folic acid takes part in a number of different metabolic reactions, very many of which involve the incorporation of a single carbon unit. In the tissues, in the presence of ascorbic acid, pteroyglutamic acid is reduced to its tetrahydro derivative which is the functional coenzyme. This substance can take up a formyl or hydroxymethyl group and thereby facilitate transfer to another molecule. Among the many biochemical transformations in which incorporation of a one-carbon unit is catalysed by folic acid are the biosynthesis of purines, the interconversion of glycine and serine, and the formation of the labile methyl groups in methionine and choline.

Because of its involvement in purine, and hence nucleic acid, synthesis folic acid is essential for the formation of red blood cells, and birds deprived of it rapidly develop a macrocytic anaemia. Growth virtually ceases, there is a high incidence of perosis and feathering is extremely poor. In the absence of folic acid, the oviduct of immature female chicks fails to develop in response to hormone treatment (Hertz, 1950). In the hormone-stimulated oviducts of deficient birds production of water-soluble proteins, particularly the albumin fraction, is greatly reduced (Brown and Badman, 1965). There is ample evidence for the essential role of folic acid in hatchability but no generally characteristic feature of the deficient embryo has been reported.

The need for folic acid is increased on diets with inadequate choline, as might be expected in view of its role in methyl group synthesis (Young *et al.*, 1955). Diets rich in protein have been shown to increase the chick's requirement for folic acid. It is suggested that the consequently greater demand for folic acid for uric acid formation takes precedence over its other metabolic functions (Creek and Vasaitis, 1963).

H. VITAMIN B_{12}

1. *Occurrence*

The vitamins B_{12} (cobalamins) are a group of corrinoid compounds in which a corrin nucleus, with a cobalt atom at the centre, is linked to a nucleotide. The nucleotide consists of ribose phosphate and either a benzimidazole or a purine derivative, but only the benzimidazole-containing compounds have vitamin B_{12} activity for higher animals (Coates *et al.*, 1956). The most stable member of the group is cyanocobalamin [(5,6-dimethylbenzimidazolyl)cobamide cyanide], in which a cyanide group is linked to the cobalt atom. This compound is the one most usually produced commercially as a dietary supplement, but in the animal body vitamin B_{12} occurs in the hydroxy- form, with an OH group in place of the CN.

The cobalamins arise entirely as the result of microbial synthesis. Several members of the group have been isolated from faeces and gut contents of animals. Although not naturally present in vegetable material, they may be found in plant products (e.g. groundnuts) that have become contaminated with moulds capable of synthesizing cobalamins. Vitamin B_{12} is well stored by animal tissues and is therefore present in most proteins of animal origin; before its identity was established its activity was recognized in the so-called "animal protein factor".

2. *Metabolic Role*

Although vitamin B_{12} or its analogues has been implicated in several pathways of microbial metabolism, its role in the biochemistry of higher animals is far from clear. Many functions have been proposed for vitamin B_{12} but so far only two are at all well established. One physiologically active form of the vitamin contains a 5′-deoxyadenosine moiety in place of the CN or OH grouping on the cobalt atom. This substance (5,6-dimethyl-benzimidazolecobamide coenzyme) acts as coenzyme in the isomerization of methylmalonyl-coenzyme A to succinyl-coenzyme A, one of the steps in the pathway of conversion of propionate to succinate. In consequence, chicks deprived of vitamin B_{12} have a reduced content of the hepatic enzyme methylmalonyl-coenzyme A mutase (Erfle *et al.*, 1964). The other known function of vitamin B_{12} in animal tissues is in the biosynthesis of labile methyl groups, a series of reactions in which folic acid is also intimately involved. Both vitamins are required for normal growth of animals on a diet containing homocystine but without a source of methyl groups (Bennett, 1950). A derivative of folic acid, 5-methyltetrahydrofolate, is an intermediate in the biosynthesis of methionine from homocysteine, and an enzyme capable of effecting the transfer of the methyl group from 5-methyl tetrahydrofolic acid to homocysteine has been isolated from mammalian and avian livers. A marked lowering of the enzymatic methyl transfer activity was observed in the livers of chicks deprived of vitamin B_{12} (Dickerman *et al.*, 1964). It is assumed, therefore, that vitamin B_{12} is an essential co-factor in methionine biosynthesis, although the methyl transferase has not yet been fully characterized.

Vitamin B_{12} is required in minute amounts and frank deficiency is rarely seen in poultry practice, for two reasons. First, the eggs from hens fed on an adequate diet contain ample vitamin B_{12} and the chick hatches with sufficient reserves to last for many weeks. Second, the vitamin is produced in considerable quantity by micro-organisms in the alimentary tract and, although direct absorption does not take place, the ingestion of quite small amounts of excreta provides a significant contribution to the bird's requirement. In order to demonstrate severe deficiency in chicks, they must be hatched from the eggs of dams with a low vitamin B_{12} intake. Such chicks grow poorly and mortality is high, although the immediate cause of death is not known. There is an accumulation both of oxidized and reduced forms of coenzyme A in the liver (Boxer *et al.*, 1955). Adult birds show little obvious effect of deficiency but their eggs fail to hatch. Most embryo deaths occur during the third week of incubation. The greatest demand for vitamin B_{12} appears to occur after organogenesis is complete, since normal hatching can be restored if the vitamin is injected into the yolk sac of deficient eggs at any time up to the 14th day of incubation. In this respect the 5′-deoxyadenosyl coenzyme is more effective than cyanocobalamin (Coates *et al.*, 1964). Micromelia and internal haemorrhage are typical of vitamin B_{12}-deficient embryos (Ferguson *et al.*, 1955). Changes in liver lipid composition indicative of poorer utilization of triglyceride have been reported (Moore and Doran, 1962). In the phospholipid fraction a considerable excess of saturated over unsaturated fatty acids has been observed in deficient embryos (Noble and Moore, 1965). The derangement in lipid pattern in vitamin B_{12} deficiency seems peculiar to embryonic liver, and its mechanism remains a matter for speculation.

Because of its involvement in the biosynthesis of methyl groups, the need for vitamin B_{12} is less when the diet contains ample methyl donors such as methionine or choline. However, the fact that they cannot entirely replace the vitamin emphasizes its multiple role in animal metabolism.

VII. Ascorbic Acid

Ascorbic acid (vitamin C, 1-threo-2,3,4,5,6-pentahydroxy-2-hexenoic acid-4-lactone) is an optically active compound with strong reducing properties. Only the L-form is biologically active. In the tissues, by transfer of two hydrogen atoms, it forms a reversible oxidation-reduction system with dehydroascorbic acid. This property appears fundamental to its biological functions and, although its role in animal metabolism is not fully understood, vitamin C is believed to take part in reactions involving electron transfer.

In most higher animals ascorbic acid is synthesized in the tissues from glucose in three enzymically-catalysed stages, D-glucose ⟶ D-glucuronic acid ⟶ L-gulonic acid ⟶ L-ascorbic acid. A few exceptional races (notably primates, guinea pigs and an Indian bulbul, *Pycnonotus cafer*) lack the enzyme necessary for the final conversion of L-gulonic acid to L-ascorbic acid and are therefore dependent on a dietary source of vitamin C. All three enzymes are present in the chicken kidney (Roy and Guha, 1958) hence, for all normal purposes, dietary ascorbic acid is not required. There is some evidence, however, that in certain conditions of "stress" the fowl is unable to synthesize adequate amounts and in such circumstances a supplement can be beneficial. In particular, additional vitamin C has been shown to mitigate the adverse effects of high environmental temperature; improved semen production in cocks and egg production in hens maintained in hot climates have been reported (Perek and Snapir, 1963; Perek and Kendler, 1963). An association of ascorbic acid metabolism with stress was shown by Freeman (1967), who noted an immediate fall in the adrenal ascorbic acid level of birds exposed to handling stress. Reduced blood levels of ascorbic acid have been observed in birds suffering from infections of *Salmonella pullorum*, *S. gallinarum* and parasitic infestations (Satterfield *et al.*, 1940) but claims that high dietary supplements of ascorbic acid increase the bird's resistance to infection have yet to be fully substantiated.

References

Adamstone, F. B. and Card, L. E. (1934). *J. Morph.* **56**, 339–359.
Ames, S. R. (1956). *Poult. Sci.* **35**, 145–159.
Armington, J. C. and Thiede, F. C. (1956). *Am. J. Physiol.* **186**, 258–262.

Arscott, G. H. and Parker, J. E. (1967). *J. Nutr.* **91**, 219–222.
Arscott, G. H., Parker, J. E. and Dickinson, E. M. (1965). *J. Nutr.* **87**, 63–68.
Aydelotte, M. B. (1963). *Br. J. Nutr.* **17**, 205–210.
Baker, J. R., Howell, J. McC. and Thompson, J. N. (1967). *Br. J. exp. Path.* **48**, 507–512.
Balnave, D. and Brown, W. O. (1967). *Metabolism* **16**, 1164–1173.
Beer, A. E., Scott, M. L. and Nesheim, M. C. (1963). *Br. Poult. Sci.* **4**, 243–253.
Belanger, L. F. and Migicovsky, B. B. (1958). *J. exp. Med.* **107**, 821–828.
Bennett, M. A. (1950). *J. biol. Chem.* **187**, 751–756.
Bieri, J. G. (1955). *Archs Biochem. Biophys.* **56**, 90–96.
Blum, J.-C. and Jacquot, R. (1964). *C. r. hebd. Séanc. Acad. Sci., Paris* **259**, 4145–4147.
Blum, J.-C. and Jacquot-Armand, Y. (1965). *Annls Nutr. Aliment.* **19**, C599–C610.
Bolton, W. (1953). *J. agric. Sci., Camb.* **43**, 120–122.
Boucher, R. V., Buss, E. G. and Maw, A. J. G. (1959). *Poult. Sci.* **38**, 1190.
Boxer, G. E., Shonk, C. E., Gilfillan, E. W., Emerson, G. A. and Oginsky, E. L. (1955). *Archs Biochem. Biophys.* **59**, 24–32.
Brown, W. O. (1957). *J. agric. Sci., Camb.* **49**, 88–94.
Brown, W. O. and Badman, H. G. (1965). *Poult. Sci.* **44**, 206–210.
Bunyan, J., Diplock, A. T. and Green, J. (1967a). *Br. J. Nutr.* **21**, 217–224.
Bunyan, J., Green, J., Diplock, A. T. and Robinson, D. (1967b). *Br. J. Nutr.* **21**, 127–136.
Buss, E. G., Boucher, R. V. and Maw, A. J. G. (1959). *Poult. Sci.* **38**, 1192.
Chaudhuri, D. K. and Kodicek, E. (1960). *Br. J. Nutr.* **14**, 35–42.
Chen, P. S. and Bosmann, M. B. (1964). *J. Nutr.* **83**, 133–139.
Chen, P. S. and Bosmann, H. B. (1965). *J. Nutr.* **87**, 148–154.
Childs, G. R., Carrick, C. W. and Hauge, S. M. (1952). *Poult. Sci.* **31**, 551–558.
Cipera, J. D. (1967). *Can. J. Biochem. Physiol.* **45**, 729–734.
Cipera, J. D. and Willmer, W. S. (1963). *Can. J. Biochem. Physiol.* **41**, 1490–1493.
Coates, M. E., Davies, M. K., Dawson, R., Harrison, G. F., Holdsworth, E. S., Kon, S. K. and Porter, J. W. G. (1956). *Biochem. J.* **64**, 682–686.
Coates, M. E., Doran, B. M. and Harrison, G. F. (1964). *Ann. N.Y. Acad. Sci.* **112**, 837–843.
Coates, M. E., Ford, J. E. and Harrison, G. F. (1968). *Br. J. Nutr.* **22**, 493–500.
Coates, M. E., Harrison, G. F. and Holdsworth, E. S. (1961). *Br. J. Nutr.* **15**, 149–155.
Coates, M. E. and Holdsworth, E. S. (1961). *Br. J. Nutr.* **15**, 131–147.
Couch, J. R., Cravens, W. W., Elvehjem, C. A. and Halpin, J. G. (1948). *Anat. Rec.* **100**, 29–48.
Cowan, J. W., Boucher, R. V. and Buss, E. G. (1966). *Poult. Sci.* **45**, 538–541.
Cravens, W. W., McGibbon, W. H. and Sebesta, E. E. (1944). *Anat. Rec.* **90**, 55–64.
Creek, R. D. and Vasaitis, V. (1963). *Poult. Sci.* **42**, 1136–1141.
Daghir, N. J. and Porooshani, J. M. (1968). *Poult. Sci.* **47**, 1094–1098.
Dakshinamurti, K. and Mistry, S. P. (1963). *J. biol. Chem.* **238**, 297–301.
Dam, H. (1962). *Vitams Horm.* **20**, 527–540.
Dam, H. and Søndergaard, E. (1964). *Z. ErnährWiss.* **5**, 73–79.
Davies, M. K., Gregory, M. E. and Henry, K. M. (1959). *J. Dairy Res.* **26**, 215–220.
DeLuca, H. F. (1967). *Vitams Horm.* **25**, 315–367.
Desai, I. D., Calvert, C. C. and Scott, M. L. (1964a). *Archs Biochem. Biophys.* **108**, 60–64.

Desai, I. E., Calvert, C. C., Scott, M. L. and Tappel, A. L. (1964b). *Proc. Soc. exp. Biol. Med.* **115**, 462–466.
Desai, I. E. and Scott, M. L. (1965). *Archs Biochem. Biophys.* **110**, 309–315.
Dickerman, H. W., Redfield, B. G., Bieri, J. G. and Weissbach, H. (1964). *Ann. N.Y. Acad. Sci.* **112**, 791–798.
Dingle, J. T., Lucy, J. A. and Fell, H. B. (1961). *Biochem. J.* **79**, 497–508.
Dingle, J. T., Fell, H. B. and Lucy, J. A. (1966). *Biochem. J.* **98**, 173–181.
Diplock, A. T., Bunyan, J., McHale, D. and Green, J. (1967). *Br. J. Nutr.* **21**, 103–114.
Dua, P. N. and Day, E. J. (1964). *Poult. Sci.* **43**, 1511–1514.
Dua, P. N., Tipon, H. C. and Day, E. J. (1965). *Poult. Sci.* **44**, 1365–1366.
Ely, C. M. (1959). *Poult. Sci.* **38**, 1316–1324.
Erfle, J. D., Clark, J. M. and Johnson, B. C. (1964). *Ann. N.Y. Acad. Sci.* **112**, 684–694.
Fell, H. B. (1960). *Proc. Nutr. Soc.* **19**, 50–54.
Fell, H. B. and Dingle, J. T. (1963). *Biochem. J.* **87**, 403–408.
Fell, H. B. and Mellanby, E. (1953). *J. Physiol., Lond.* **119**, 470–488.
Fell, H. B., Mellanby, E. and Pelc, S. R. (1956). *J. Physiol., Lond.* **133**, 89–100.
Ferguson, T. M., Rigdon, R. H. and Couch, J. R. (1955). *Archs Path.* **60**, 393–400.
Freeman, B. M. (1967). *Comp. Biochem. Physiol.* **23**, 303–305.
Glavind, J. and Søndergaard, E. (1964). *Acta chem. scand.* **18**, 2173–2181.
Glazener, E. W. and Briggs, G. M. (1948). *Poult. Sci.* **27**, 462–465.
Green, J. and Bunyan, J. (1969). *Nutr. Abstr. Rev.* **39**, 321–345.
Griminger, P. (1964). *Poult. Sci.* **43**, 1289–1290.
Griminger, P. (1965). *J. Nutr.* **87**, 337–343.
Griminger, P. (1966). *Vitams Horm.* **24**, 605–618.
Griminger, P. and Donis, O. (1960). *J. Nutr.* **70**, 361–368.
Günther, K. and Tekin, C. (1964). *Arch. Tierernähr.* **14**, 431–449.
Gutzmann, W. C. and Donovan, G. A. (1966). *Poult. Sci.* **45**, 1088–1089.
Harrison, H. C., Harrison, H. E. and Park, E. A. (1958). *Am. J. Physiol.* **192**, 432–436.
Hathcock, J. N., Hull, S. J. and Scott, M. L. (1968). *J. Nutr.* **94**, 147–150.
Havivi, E. and Wolf, G. (1967). *J. Nutr.* **92**, 467–473.
Hertz, R. (1950). *Texas Rep. Biol. Med.* **8**, 154–158.
Hill, C. H. and Kim, C. S. (1967). *Biochem. biophys. Res. Commun.* **27**, 94–99.
Holdsworth, E. S. (1965). *Biochem. J.* **96**, 475–483.
Howell, J. McC. and Thompson, J. N. (1965). *Br. J. exp. Path.* **46**, 18–24.
Howell, J. McC. and Thompson, J. N. (1967a). *Br. J. Nutr.* **21**, 741–750.
Howell, J. McC. and Thompson, J. N. (1967b). *Br. J. exp. Path.* **48**, 450–454.
Imrie, M. H., Neville, P. F., Snellgrove, A. W. and DeLuca, H. F. (1967). *Archs Biochem. Biophys.* **120**, 525–532.
Jungherr, E. (1943). *Bull. Storrs agric. Exp. Stn* **250**.
Kirchgessner, M. and Maier, D. A. (1968a). *Arch. Tierernähr.* **18**, 300–308.
Kirchgessner, M. and Maier, D. A. (1968b). *Arch. Tierernähr.* **18**, 309–315.
Leutskaja, Z. K. (1963). *Dokl. Akad. Nauk SSSR* **153**, 243–245.
Leutskaja, Z. K. (1964). *Dokl. Akad. Nauk SSSR* **159**, 464–465.
Machlin, L. J. and Gordon, R. S. (1960). *Proc. Soc. exp. Biol. Med.* **103**, 659–663.
Machlin, L. J., Gordon, R. S., Marr, J. E. and Pope, C. W. (1962). *J. Nutr.* **76**, 284–290.
Manoukas, A. G. (1967). *Diss. Abstr.* **B28**, 2497B.
Manoukas, A. G., Ringrose, R. C. and Teeri, A. E. (1968). *Poult. Sci.* **47**, 1836.
March, B. E., Coates, V. and Biely, J. (1966). *Can. J. Physiol. Pharmacol.* **44**, 295–300.
Martius, C. (1966). *Vitams Horm.* **24**, 441–445.

Marusich, W. L., Ackerman, G. and Bauernfeind, J. C. (1967). *Poult. Sci.* **46**, 541–548.
Marusich, W. and Bauernfeind, J. C. (1963). *Poult. Sci.* **42**, 949–957.
Maw, A. J. G. (1954). *Poult. Sci.* **33**, 216–217.
Migicovsky, B. B. and Emslie, A. R. G. (1950). *Archs Biochem. Biophys.* **28**, 324–328.
Migicovsky, B. B. and Jamieson, J. W. S. (1955). *Can. J. Biochem. Physiol.* **33**, 202–208.
Moore, J. H. and Doran, B. M. (1962). *Biochem. J.* **84**, 506–513.
Moore, J. H., Coates, M. E. and Williams, D. L. (1968). *Proc. Nutr. Soc.* **27**, 32A.
Moore, T. (1957). "Vitamin A". Elsevier, Amsterdam.
Morii, H. and DeLuca, H. F. (1967). *Am. J. Physiol.* **213**, 358–362.
Naber, E. C., Cravens, W. W., Baumann, C. A. and Bird, H. R. (1954). *J. Nutr.* **54**, 579–591.
Nelson, T. S. and Norris, L. C. (1960). *J. Nutr.* **72**, 137–144.
Nesheim, M. C., Calvert, C. C. and Scott, M. L. (1960). *Proc. Soc. exp. Biol. Med.* **104**, 783–785.
Nesheim, M. C. and Scott, M. L. (1958). *J. Nutr.* **65**, 601–618.
Nir, I. and Ascarelli, I. (1966). *Br. J. Nutr.* **20**, 41–53.
Nir, I. and Ascarelli, I. (1967). *Br. J. Nutr.* **21**, 167–180.
Noble, R. C. and Moore, J. H. (1965). *Biochem. J.* **95**, 144–149.
Norman, A. W. (1968). *Biol. Rev.* **43**, 97–137.
Ochoa, S., Mehler, A., Blanchard, M. L., Jukes, T. H., Hoffmann, C. E. and Regan, M. (1947). *J. biol. Chem.* **170**, 413–414.
Olson, R. E. (1964). *Science, N.Y.* **145**, 926–928.
Olson, R. E., Li, L. F., Philipps, G., Berry, E. and Ryland, E. (1967). *Fedn Proc. Fedn Am. Socs exp. Biol.* **26**, 698.
Ott, W. H., Dickinson, A. M. and van Iderstine, A. (1965). *Poult Sci.* **44**, 920–925.
Padhi, P. and Combs, G. F. (1965). *Poult. Sci.* **44**, 1405.
Panda, B. and Combs, G. F. (1963). *Proc. Soc. exp. Biol. Med.* **113**, 530–534.
Parrish, D. B., Zimmerman, R. A., Sanford, P. E. and Hung, E. (1963). *J. Nutr.* **79**, 9–17.
Patterson, E. B., Hunt, J. R., Vohra, F., Blaylock, L. G. and McGinnis, J. (1956). *Poult. Sci.* **35**, 499–504.
Peacock, R. G. and Combs, G. F. (1969). *Poult. Sci.* **48**, 1857.
Perek, M. and Kendler, J. (1963). *Br. Poult. Sci.* **4**, 191–200.
Perek, M. and Snapir, N. (1963). *Br. Poult. Sci.* **4**, 19–26.
Polin, D. C., Porter, C. C., Wynosky, E. R. and Cobb, W. R. (1962). *Poult. Sci.* **41**, 372–380.
Pudelkiewicz, W. J., Matterson, L. D., Potter, L. M., Webster, L. and Singsen, E. P. (1960). *J. Nutr.* **71**, 115–121.
Rhodes, M. B., Bennett, N. and Feeney, R. E. (1959). *J. biol. Chem.* **234**, 2054–2060.
Robinson, F. A. (1966). "The Vitamin Co-factors of Enzyme Systems". Pergamon Press, Oxford.
Roy, R. N. and Guha, B. C. (1958). *Nature, Lond.* **182**, 319–320.
Sallis, J. D. and Holdsworth, E. S. (1962). *Am. J. Physiol.* **203**, 497–505.
Satterfield, G. H., Moseley, M. A., Gauger, H. C., Holmes, A. D. and Tripp, F. (1940). *Poult. Sci.* **19**, 337–344.
Saville, D. G., Solvyns, A. and Humphries, C. (1967). *Aust. vet. J.* **43**, 346–348.
Scott, M. L. (1962a). *Nutr. Abstr. Rev.* **32**, 1–8.
Scott, M. L. (1962b). *Vitams Horm.* **20**, 621–632.
Scott, M. L. (1966). *Vitams Horm.* **24**, 633–647.
Scott, M. L. (1968). *Feedstuffs, Minneap.* **40**, 24–25.

Scott, M. L. and Stoewsand, G. S. (1961). *Poult. Sci.* **40**, 1517–1523.
Sebrell, W. H. and Harris, R. S. (1968). "The Vitamins". (W. H. Sebrell and R. S. Harris eds.) Academic Press, New York.
Tappel, A. L. (1962). *Vitams Horm.* **20**, 493–510.
Taylor, T. G., Morris, K. M. L. and Kirkley, J. (1968). *Br. J. Nutr.* **22**, 713–721.
Thompson, J. N., Howell, J. McC., Pitt, G. A. J. and Houghton, C. I. (1965). *Nature, Lond.* **205**, 1006–1007.
Thompson, S. Y., Coates, M. E. and Kon, S. K. (1950). *Biochem. J.* **46**, xxx.
Thornton, P. A. and Schutze, J. V. (1960). *Poult. Sci.* **39**, 192–199.
Wald, G. (1960). *Vitams Horm.* **18**, 417–430.
Wald, G. and Zussman, J. (1938). *J. biol. Chem.* **122**, 449–460.
Wasserman, R. H., Taylor, A. N. and Kallfelz, F. A. (1966). *Am. J. Physiol.* **211**, 419–423.
Wolbach, S. B. and Hegsted, D. M. (1952a). *Archs Path.* **54**, 13–29.
Wolbach, S. B. and Hegsted, D. M. (1952b). *Archs Path.* **54**, 30–38.
Young, R. J., Norris, L. C. and Heuser, G. F. (1955). *J. Nutr.* **55**, 353–362.

15 The Role of Trace Elements in Metabolic Processes

E. J. BUTLER

Houghton Poultry Research Station, Houghton, Huntingdon, England

I. Introduction

A trace element is defined as one which usually occurs in concentrations be-below 100 μg/g. This figure represents the sensitivity limit of most classical gravimetric and volumetric analyses; when these were the only methods available, amounts too small to be thus measured were referred to as traces.

In biological materials the term "trace" is applied regardless of whether or not the element is known to be required for the metabolism or structure of the organism. Traces of many elements occur constantly in animal fluids and tissues but so far only 8 have been classified as essential constituents, *viz.*, iron, copper, zinc, manganese, cobalt, molybdenum, selenium and iodine, and are known as the micro-nutrient elements. When present in larger amounts they are toxic. These elements occur in a confined region of the Periodic Table between atomic numbers 25 to 53 and 5 of them, viz. manganese, iron, cobalt, copper and zinc occupy 6 consecutive places. The above definition serves to distinguish the micro-nutrient elements from the 11 macro-nutrient elements, viz. hydrogen, carbon, nitrogen, oxygen, phosphorus, sulphur, chlorine, sodium, potassium, calcium and magnesium which precede them in the Periodic Table and in general have different functions. However, this differentiation of function is not complete since metallic elements of both classes play similar roles as components and activators of enzyme systems. Although the iron content of fluids and tissues which contain appreciable amounts of haemoglobin or ferritin frequently exceeds 100 μg/g it is usually classed as a trace element because the concentration present in other materials rarely exceeds this figure. As will be seen later, its inclusion among the *bona fide* trace elements is also justified on functional grounds.

In this chapter present knowledge of the metabolic functions of the essential trace elements is summarized and an attempt is made to explain the pathogenesis in terms of biochemical lesions of those syndromes produced by dietary deficiencies of these elements. In this field, as in others, relatively little biochemical work has been carried out on birds compared with that on mammals and in order to present a coherent account it has been necessary to make frequent reference to information on mammalian systems. Although transposition of information of this kind can be, and has been, very misleading, it seems to be justified for a general discussion of the subject since the same trace elements appear to be required by both classes and the manifestations of a deficiency of a particular element are very similar. Greater care must be exercised in the interpretation of more fundamental data since many important differences have already been discovered, particularly in the composition and properties of enzymes and the regulation of enzymatic activity.

As in the case of the vitamins, the first indications that certain trace elements played an essential part in animal metabolism came from studies of deficiency diseases and since these studies still serve to guide and stimulate research on the biochemistry of these elements a summary of the main features of the deficiency syndrome is given for each element before its functions are discussed. Cobalt and iodine have not been included since they function almost entirely as constituents of vitamin B_{12} and the thyroid hormones respectively which are discussed elsewhere (see Chapters 14 and 17). Several nutritional and physiological aspects of the subject have also been omitted; these include dietary requirements, factors affecting availability, mechanisms of absorption and transport, and the distribution of the elements in the body. Information on these aspects has been reviewed elsewhere (Underwood 1962, 1966; Agricultural Research Council, 1963; Mills, 1964; National Academy of Sciences—National Research Council, 1966).

II. Metals in Enzyme Systems

The essential trace elements fulfill their roles in metabolic processes in combination with organic molecules as either components or activators of enzyme systems and as constituents of hormones, co-enzymes and oxygen carriers. Metallic elements have special and often specific functions in enzyme systems which merit separate consideration. The following discussion is intended to serve as a general introduction to the subject which has been expounded elsewhere (e.g. Bray and Harrap, 1959; Mahler, 1961; Proceedings of a Symposium on Biological Aspects of Metal Binding, 1961; Vallee and Coleman, 1964; Vallee and Williams, 1968a; Hamilton, 1969; Williams, 1970).

A. METALLO-ENZYMES

Enzymes which require the presence of a metal for maximal activity have been classified as metallo-enzymes or metal-activated enzymes according to the strength of the bonds between the metal and the enzyme protein (Vallee, 1955). In the former the metal is an integral component of the enzyme and there is virtually no dissociation under physiological conditions. Enzymes of this kind tend to contain those metals which form the most stable chelate complexes such as copper, iron and zinc. The metal is always present in stoichiometric amounts, usually 1 to 4 but sometimes 6 or 8 atoms per molecule of the enzyme. Iron is sometimes present in a porphyrin complex (haem) as in cytochrome oxidase and catalase.

A particular metal tends to occur in enzymes which catalyse a certain type of reaction; for example copper occurs in several oxidases which utilize molecular oxygen and zinc has been found in several dehydrogenases which require NAD as a co-enzyme. Some enzymes contain two metals; cytochrome oxidase contains both iron and copper and xanthine dehydrogenase both iron and molybdenum.

The metal requirement of these enzymes *in vivo* may not be as specific in all cases as was once believed. Zinc has been removed from carboxypeptidase A and carbonic anhydrase *in vitro* without denaturing the enzyme and has been replaced by other divalent metals; the cobalt derivatives particularly had appreciable activity. The firmly bound manganese in pyruvate carboxylase has been shown to be exchangeable with magnesium *in vivo* (Griminger and Scrutton, 1970).

B. METAL-ACTIVATED ENZYMES

In these enzymes the metal is only loosely held and usually can be removed by dialysis at pH 7. It is progressively lost during purification of the enzyme and little may remain in the final product.

Activation is non-specific and with a particular enzyme several cationic metals are effective though they often differ considerably in efficiency. Activation has been observed with at least 15 different metals including the macro-nutrient elements sodium, potassium, calcium and magnesium, and several such as cadmium, rubidium and caesium which at present are not regarded as essential elements. In general, metals which show this behaviour are those which form relatively weak complexes with organic molecules and activation effects by copper or metals in the trivalent state are rarely observed. Some essential metals are potent antagonists of activating metals; for example the activation of inorganic pyrophosphatase by manganese or magnesium is antagonized by calcium.

Enzymes which catalyse the same type of reaction are usually activated *in vitro* by the same metals. The phosphotransferases for instance, are activated by magnesium, manganese and zinc, the decarboxylases by manganese and to a lesser extent by zinc and magnesium, and the peptidases by cobalt, manganese and zinc.

The ionic radius of the metal ion and its stereochemistry appear to be the chief factors which determine its activating efficiency. With mitochondrial ATP-ase for example only those ions with radii between 0·7 and 1 Å and which can form tetrahedral or octahedral complexes, can function as activators (Selwyn, 1968).

A major problem presented by enzymes of this kind is to determine which metal or metals function as activators *in vivo*. It has often been assumed that the most efficient activator *in vitro* is the one which is physiologically active but there is no real justification for this conclusion without other evidence. The activity of the enzyme *in vivo* will clearly be determined by the concentrations of the various activating and inhibiting ions present and the regulation of these concentrations might well be the basis of control mechanisms.

C. CATALYTIC FUNCTIONS

In many enzyme systems the metallic participant may be considered to have a catalytic as distinct from a purely structural function. A mechanism for this action was first suggested by Hellerman (1937) who put forward what is now known as the bridge theory to explain the activation of arginase by manganese and other divalent metals. According to this theory the metal facilitates the reaction by binding the substrate to the active centre of the enzyme to form the transition state complex as shown diagrammatically in Fig. 1a. The theory has since been extended to other systems and more information obtained on the behaviour of the metal. In some systems, e.g. mammalian carboxypeptidase A, the formation of this complex brings the

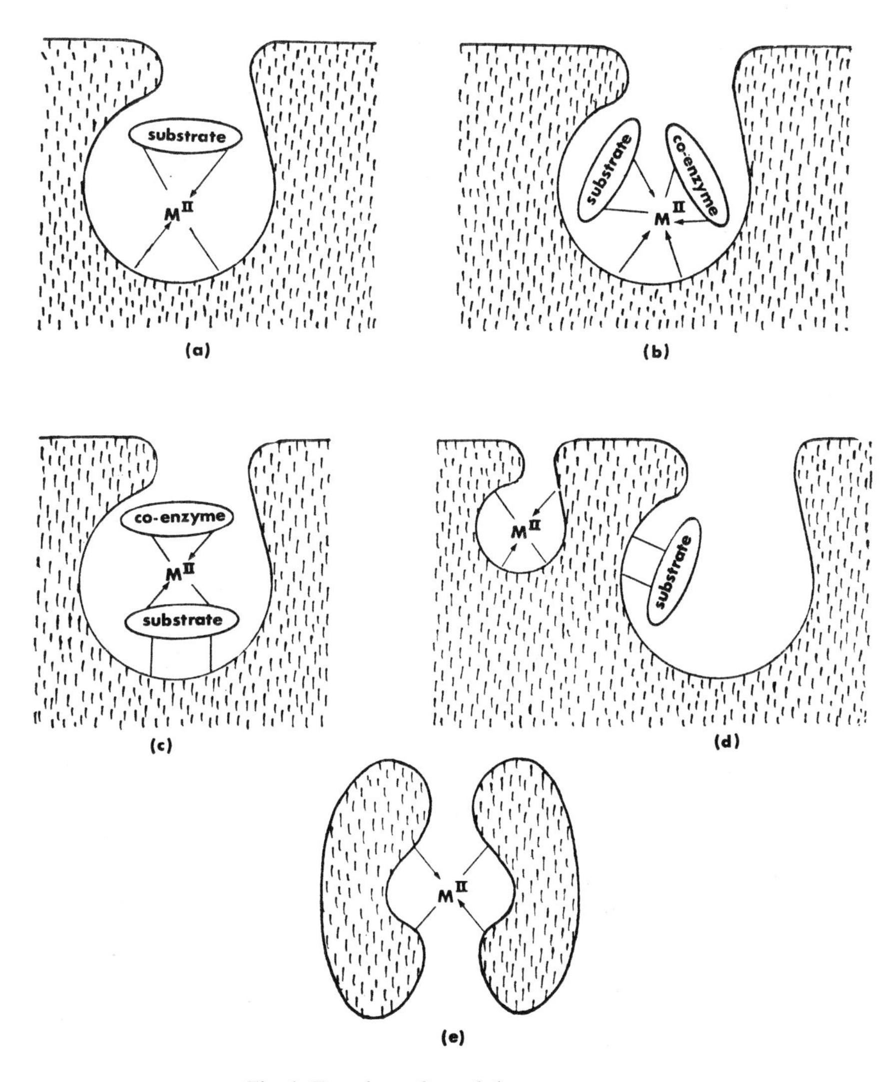

Fig. 1. Functions of metals in enzyme systems.

active groups in the substrate and enzyme or co-enzyme into suitable juxtaposition. In others the effect of the metal on the electronic configuration of the substrate may be more important and may result in stabilizing a reactive intermediate. For instance, in hydrolytic reactions the substrate may be activated by withdrawal of electrons towards attack by nucleophilic reagents such as OH^-. When a co-enzyme or prosthetic group is present the metal may also bind it to the active centre as indicated in Fig. 1b and enhance its reactivity. Pyridoxal phosphate, thiamine pyrophosphate and all the flavin and adenine nucleotides which function as co-enzymes or prosthetic groups are able to form complexes with trace metals; several reaction mechanisms involving the metal have been proposed (see Mahler, 1961). Metals which can undergo valency changes, such as iron and copper, are of particular importance in oxidation-reduction systems since they can participate in the transfer of electrons when bound to other components.

There is evidence from the absorption and EPR spectra of some copper and non-haem iron enzymes that the binding of the metal affects the sterochemical and electronic configuration of the active centre. Since the structure and bond characteristics of the metal–protein complex will, in many systems, represent a compromise between the preferred configurations of the metal and protein, the complex can be considered to be under strain and therefore more reactive. This has been termed an "entatic" state (Vallee and Williams, 1968b) and may be particularly important in facilitating electron transfer reactions which involve a change in the valency of the metal. The two valency states may prefer different site symmetries, for example mono- and divalent copper prefer tetrahedral and tetragonal structures respectively and neither may be consistent with the stereochemistry of the binding site on the enzyme protein. The normal configuration of the bonds from the metal will therefore be distorted, possibly more in one valency state than the other. An entatic situation will also arise when the sterochemical configuration of the metal does not change but the two valency states normally form different types of bond with different bond-lengths, as in the case of iron.

So far in this discussion it has been assumed that the metal is bound to the enzyme protein at the active centre. This may not always be so. It is possible that in some systems the metal may be bound at an allosteric site (Fig. 1c) and by inducing electronic displacements or conformational changes at the active centre may facilitate the bonding of the substrate directly to the centre. In some metal-activated systems the metal may not be bound to the enzyme protein at all and may function by complexing with the substrate or co-enzyme to facilitate its binding to the active centre. Several ATP-phosphotransferases are activated by magnesium and manganese in this way, an important function of the metal being to neutralize two of the negative charges on the tri-phosphate chain. Metals can also promote enzyme reactions by forming a stable complex with one of the reaction products so removing it from the equilibrium. The effect of divalent metals on the glucose-1-phosphate dismutase reaction is explained in this way, the complex being formed with glucose 1,6-diphosphate.

D. STRUCTURAL FUNCTIONS

Evidence that metals are responsible for maintaining the tertiary and quaternary structures of certain enzyme proteins has been recently obtained. Removal of the metal causes denaturation or dissociation into subunits and disintegration of the active centre. Zinc appears to perform both catalytic and structural functions in some dehydrogenases. Involvement of a metal in the quaternary structure of an enzyme protein is shown diagrammatically in Fig. 1d.

E. METALLO-COENZYMES

Two classes of co-enzymes contain a metal as a functional component, viz. the cytochromes which are haem proteins and vitamin B_{12} and its derivatives which contain cobalt. Other co-enzymes or prosthetic groups such as FAD and pyridoxal phosphate, may combine with metals in enzyme systems but usually do not contain metals when isolated from animal tissues.

A sequence of four cytochromes (b, c_1, c and a) forms a major part of the respiratory chain where they function as hydrogen or electron carriers (Fig. 2); cytochrome b_5 does this in microsomes. Transfer of each hydrogen atom or electron involves a change in the valency of the haem iron atom. In mammalian cytochrome *c* the process appears to be accompanied by a change in the groups which bind the protein to the fifth and sixth co-ordination positions of the iron atom (Harbury and Loach, 1960).

Vitamin B_{12} co-enzymes are required by methylmalonyl CoA mutase and methionine synthetase present in the liver (see Chapter 47). These co-enzymes contain trivalent cobalt chelated to four pyrrole rings which together form a corrin ring resembling the porphyrin ring except that one of the methine bridges is missing. 5,6-Dimethylbenzimidazole which is attached to the corrin ring by ribose phosphate is bound to the fifth co-ordination position. The remaining binding site can be occupied by various groups such as hydroxyl, cyanide, methyl or 5-deoxy-adenosine and is thought to be used for carrying groups during the course of enzyme reactions.

III. Iron

Deficiency of iron in the chick produces retardation of growth, a hypochromic microcytic anaemia with poikilocytosis, achromatrichia and eventually death (Elvehjem and Hart, 1929; Hill and Matrone, 1961; Davis *et al.*, 1962, 1968).

The text of this section and Fig. 2 are based on present knowledge of mammalian systems and although a detailed comparison of the composition and organization of the respiratory chain in birds and mammals has not been made the information available does not indicate the existence of gross differences.

A. RESPIRATION

To a large extent the changes accompanying iron deficiency reflect the importance of this metal in the respiratory systems. Besides being responsible for the binding of oxygen to haemoglobin and myoglobin, iron is a functional constituent of components of the respiratory chain. This is an integrated series of enzymes and co-enzymes located in the cristae of the mitochondria and is responsible for the final stage in the oxidation of fatty acids, carbohydrates and amino acids and produces most of the energy required by the animal in the form of ATP and heat.

Hydrogen atoms are removed from the various metabolites by specific dehydrogenases, such as those involved in the β-oxidation of fatty acids and the citric acid cycle, and are first transferred either to NAD or to co-enzyme Q (ubiquinone) and then transported through the system as protons and electrons by a series of carrier molecules arranged sequentially in order of increasing redox potential (shown diagrammatically in Fig. 2). The hydrogen atoms are finally combined with molecular oxygen, by the action of cytochrome oxidase (cyt. a_3), to form water. Three of these transfer reactions are energetically coupled to the synthesis of ATP from ADP and inorganic phosphate ("oxidative phosphorylation") which controls the activity of the respiratory chain.

Iron occurs in components of the respiratory chain as porphyrin complexes (haem) which are the functional groups of the five cytochromes present. It is also bound to other ligands when it is known as non-haem iron. This is frequently associated with stoichiometric amounts of labile sulphur and is probably bound to it. Iron in this form (indicated by Fe in Fig. 2) appears to participate in electron transport and it has been suggested that the interaction of the metal with the sulphur atoms results in strongly delocalized orbitals over the whole complex, as in the porphyrins, thereby enabling the complex to behave both as a donor and acceptor of electrons (Williams, 1965). Non-haem iron is also thought to be involved in oxidative phosphorylation (Butow and Racker, 1965, Boyer, 1968) possibly as a complex with co-enzyme Q (Moore and Folkers, 1964).

Succinic dehydrogenase [succinate oxidoreductase EC 1.3.99.1] which is integrated with the respiratory chain, also contains non-haem iron and labile sulphur.

B. MICROSOMAL OXIDATIONS

Microsomes from mammalian tissues contain two haem proteins, cytochromes b_5 and P-450, and non-haem iron. Cytochrome b_5 takes part in an extremely rapid oxidation of $NADH_2$ catalysed by a flavoprotein enzyme. Cytochrome P-450 is the terminal oxidase in the hydroxylation of steroids

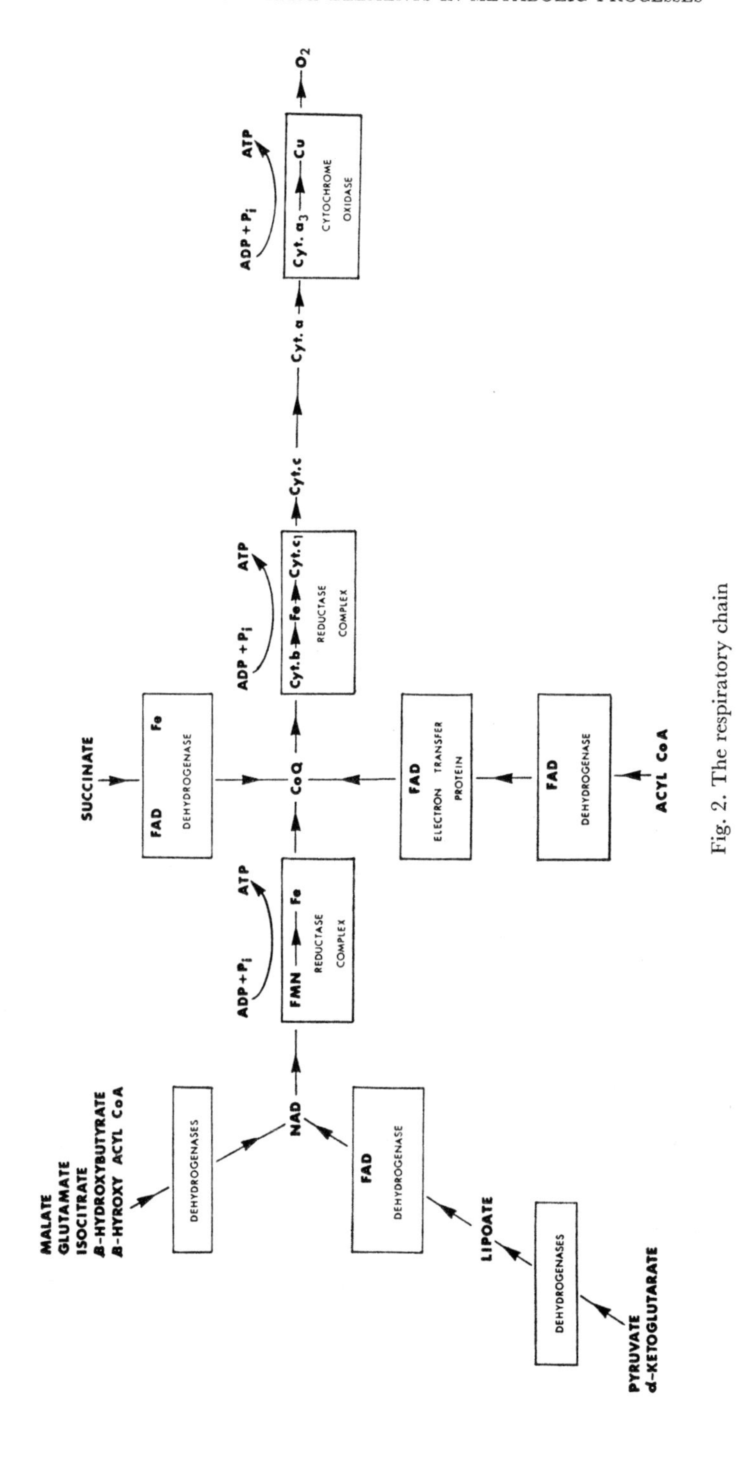

Fig. 2. The respiratory chain

and drugs (Omura and Sato, 1964) utilizing molecular oxygen and hydrogen atoms transferred from $NADPH_2$ by a system containing a flavoprotein dehydrogenase and possibly non-haem iron (Fig. 3). An enzyme containing non-haem iron and labile sulphur has been identified in the system in bovine adrenal cortex mitochondria which carries out C-11 hydroxylations of steroids (Omura *et al.*, 1965).

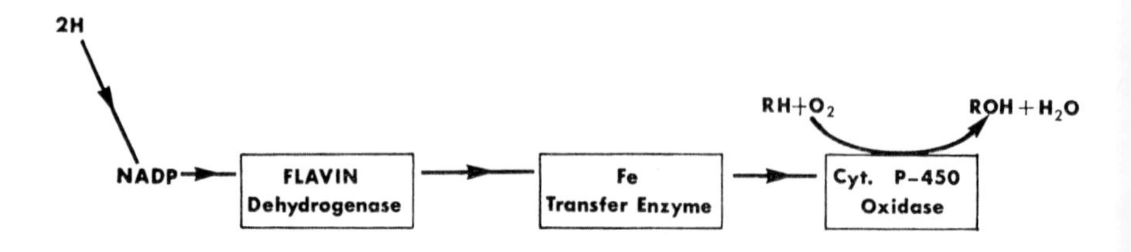

Fig. 3. Hydrogen transport system for hydroxylations.

C. DETOXIFICATION

Two mammalian hydroperoxidases, viz. peroxidase and catalase contain haem groups. Their function is assumed to be the destruction of hydrogen peroxide formed by aerobic dehydrogenases such as the amino acid and amine oxidases. Peroxidase [donor: H_2O_2 oxidoreductase, EC 1.11.1.7] contains one protohaem group per molecule and utilizes a variety of hydrogen donors:

$$H_2O_2 + RH_2 \longrightarrow 2H_2O + R$$

Catalase (H_2O_2:H_2O_2 oxidoreductase, EC 1.11.1.6) contains four haem groups per molecule and does not require a hydrogen donor:

$$2H_2O_2 \longrightarrow 2H_2O + O_2$$

D. PORPHYRIN SYNTHESIS

Iron appears to be involved in the synthesis of the porphyrin molecule. Experiments with chick erythrocytes have suggested that ferrous iron may be required for the synthesis of δ-aminolaevulic acid which is a precursor of the porphyrins (Brown, 1958).

E. FEATHER PIGMENTATION

A red iron-containing pigment has been isolated from feathers (Nickerson, 1946). An incomplete synthesis of this pigment is probably mainly responsible for the feather depigmentation observed in iron deficiency (Davis *et al.*, 1962).

IV. Copper

Copper deficiency in the chick produces retardation of growth, leg weakness, deformed and fragile bones, achromatosis of the feathers, mild anaemia and subcutaneous and internal haemorrhage (Elvehjem and Hart, 1929; Gallagher, 1957; Hill and Matrone, 1961; O'Dell *et al.*, 1961). Mortality is high and results mainly from rupture of the aorta or one of the other major blood vessels (O'Dell *et al.*, 1961).

When hens are fed a diet low in copper there is a slight fall in egg production but hatchability is markedly reduced (Bird *et al.*, 1963). The embryos usually die before the 10th day of incubation and show retarded tissue development and differentiation, abnormalities in the spinal cord, oedema, anaemia and haemorrhage (Bird, 1967; Savage, 1968; O'Dell, 1968).

Studies of these abnormalities at the biochemical level have shown that copper is involved in several synthetic pathways as well as in the terminal stage of mitochondrial respiration.

A. MITOCHONDRIAL RESPIRATION

As in the mammal, a deficiency of copper in the chick quickly produces a fall in the activity of cytochrome oxidase (Hill and Matrone, 1961; Hunt and Landesman, 1968; Rucker *et al.*, 1969). This enzyme [cytochrome c:O_2 oxido-reductase, EC 1.9.3.1] catalyses the terminal reaction in the respiratory chain (see Section IIIA above).

Cytochrome oxidase from mammalian heart muscle contains equal numbers of copper atoms and haem-α groups which take part in the transfer of electrons during the reaction (Wainio *et al.*, 1959; Griffiths and Wharton, 1961; Morrison *et al.*, 1963; Beinert and Palmer, 1964). At present there is no information in the literature on the composition of the avian enzyme.

The reduction in the activity of cytochrome oxidase brought about by copper deficiency probably has a profound effect on the metabolism of the animal, particularly on the synthetic pathways by restricting the production of ATP, and a considerable retardation of growth and development is to be expected.

B. SYNTHESIS OF CONNECTIVE TISSUE

The histological lesions in arteries and bones from copper-deficient chicks and embryos suggest that copper is required for the synthesis of certain structural components of connective tissues. The arterial defects occur in the elastic lamellae, weaken the walls of the vessels and lead to aneurysms and ruptures (O'Dell *et al.*, 1961; Carlton and Henderson, 1963; Simpson and Harms, 1964; Simpson *et al.*, 1967). In the skeleton there is a deficiency of osteoid tissue (osteoporosis) causing increased fragility and deformation of the long bones (Carlton and Henderson, 1964).

Synthesis of elastin and collagen is impaired and, in both cases, there is a defect in the mechanism for forming cross-links between peptide chains.

Elastin is deficient in the walls of the aorta, is much more soluble than normal elastin, contains a much higher percentage of lysine and fewer desmosine and isodesmosine groups (Starcher *et al.*, 1964; O'Dell *et al.*, 1966). These groups act as cross-links and each is formed from four lysine side chains on adjacent peptide chains after three of them have been oxidatively deaminated by monoamine oxidase (Partridge *et al.*, 1964, 1966; Miller *et al.*, 1964). The structure and formation of these links is shown in Fig. 4.

Monoamine oxidase therefore plays a vital part in the formation of cross-links in elastin and its activity in the aortas of copper-deficient chicks is considerably reduced (Bird *et al.*, 1966; Kim and Hill, 1966; Hill *et al.*, 1967). Purified monoamine oxidase (monoamine:O_2 oxidoreductase (deaminating) EC 1.4.3.4) obtained from mammalian blood plasma and tissues contained copper as an essential structural component (Yamada *et al.*, 1963; Nara *et al.*, 1966; Buffoni *et al.*, 1968). Although the composition of the avian enzyme has not yet been determined, the fact that the *in vitro* activity of preparations from aortas of copper-deficient chicks is increased considerably by the addition of copper indicates that it also incorporates that element (Bird *et al.*, 1966). Similar lesions are produced by lathyritic agents, such as β-aminopropionitrile, by inhibiting monoamine oxidase.

Evidence of defective cross-linking in the collagen present in connective tissues of the copper deficient chick, including aorta and bone, comes from solubility studies (O'Dell *et al.*, 1966; Chou *et al.*, 1968; Rucker *et al.*, 1969). Since the first stage in the formation of one type of cross-link in rat skin collagen appears to involve the oxidative deamination of lysine side chains by amine oxidase (Barnstein *et al.*, 1966; Piez, 1968) the failure to form these links in the tissues of the copper-deficient chick may again be attributed to the reduction in monoamine oxidase activity which has been found (Chou *et al.*, 1968; Rucker *et al.*, 1969). Although this is obviously an important biochemical lesion in connective tissue it is not the only manifestation of copper deficiency at this level. For instance, the activity of cytochrome oxidase is also reduced in the mitochondria of skeletal tissue (Hunt and Landesman, 1968; Rucker *et al.*, 1969) and this may be expected to limit the amount of ATP which is available for anabolic pathways.

C. HAEMATOPOIESIS

Anaemia develops in the later stages of copper deficiency and appears to be more difficult to produce in the fowl than in the mammal (Hill and Matrone, 1961). In the chick both the haemoglobin content of the blood and the number of erythrocytes are reduced but the mean corpuscular haemoglobin content is not significantly altered. This indicates that copper is required for erythropoiesis and suggests that it is also involved in the synthesis of haemoglobin.

The erythrocytopenia may be partly explained by a depressed phospholipid synthesis following deficiency in copper (Gallagher *et al.*, 1956a).

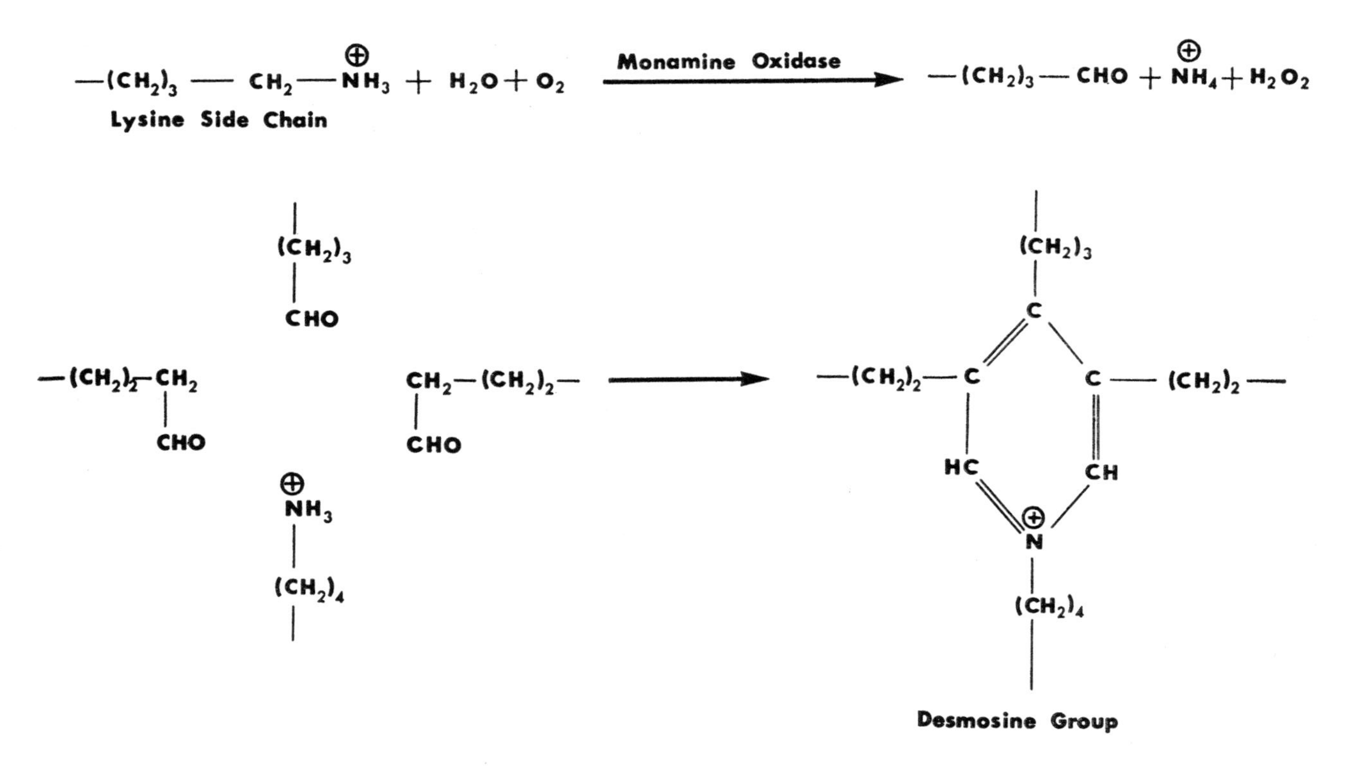

Fig. 4. Formation of cross-links in elastin.

Recent *in vitro* experiments with liver microsomes from copper-deficient chicks indicate that this may be due to a reduced activity of α-glycerophosphate acyl transferase [acyl-CoA:L-glycerol-3-phosphate O-acyl transferase, EC 2.3.15] which catalyses the synthesis of phosphatidic acids by acylation of α-glycerophosphate at the expense of acyl-coenzyme A derivatives (Di Paolo and Newberne, 1970).

The role of copper in haemoglobin synthesis has been debated since the classic demonstration of Hart *et al.* (1928) that copper as well as iron is needed to prevent anaemia in rats reared on milk diets. It has been reported that the synthesis of protoporphyrin from glycine by chick erythrocytes *in vitro* is reduced by copper deficiency and that the addition of copper increases the rate of synthesis (Anderson and Tove, 1958). However, recent experiments do not confirm this finding (Sardesai and Orten, 1969) and in pigs, copper deficiency produces an increase in protoporphyrin synthesis rather than a decrease (Lee *et al.*, 1968).

There remains the possibility that copper promotes the formation of haemoglobin by enhancing the synthesis of globin and/or the utilization of iron. Evidence of the latter function is conclusive in mammals where copper deficiency reduces the absorption of iron and its levels in the plasma and increases its deposition in the liver and spleen. The mobilization of iron as transferrin from the pig's liver is promoted by caeruloplasmin (Osaki *et al.*, 1970), a copper protein with oxidase activity present in the plasma of both birds and mammals and believed to facilitate the binding to transferrin of iron by oxidizing it to the trivalent state (Osaki *et al.*, 1966).

In the fowl plasma levels of caeruloplasmin are considerably lower than in mammals (Starcher and Hill, 1965) and each molecule contains fewer copper atoms, viz. 5 compared with 8 (Starcher and Hill, 1966). These differences are reflected in the whole blood copper content (Beck, 1956).

D. SYNTHESIS OF MELANIN PIGMENTS

Copper is a constituent of mammalian tyrosinase [*o*-diphenol:O_2 oxidoreductase, EC 1.10.3.1] which catalyses the formation of melanin pigments from tyrosine (Lerner *et al.*, 1950) and it is believed that copper deficiency produces achromatosis by reducing the activity of this enzyme. Although the presence of copper in fowl tyrosinase does not appear to have been demonstrated there is evidence of a requirement for copper. For instance, tyrosinase in epithelial melanocytes is inhibited by cyanide and by sodium diethyldithiocarbamate which form complexes with copper (Charles and Rawles, 1940; Misuraca *et al.*, 1969). It is therefore probable that the reduction in feather pigmentation which has been noted in copper-deficient chicks (Hill and Matrone, 1961) reflects this requirement.

E. OTHER POSSIBLE FUNCTIONS

The discovery that 3,4-dihydroxyphenylethylamine β-hydroxylase of the bovine adrenal medulla contains two copper atoms per molecule (Goldstein *et al.*, 1965; Friedman and Kaufman, 1965) demonstrated that copper is required for the synthesis of noradrenaline, and it has been confirmed that copper deficiency reduces the synthesis of this hormone in the rat (Missala *et al.*, 1967). At present it is not known whether copper also performs this function in the fowl.

The same question arises regarding 3-mercaptopyruvate sulphur transferase [EC 2.8.1.2] involved in the conversion of cysteine to pyruvate. A purified preparation of this enzyme from rat liver contained one atom of copper per molecule (Kun and Fanshier, 1959).

V. Zinc

The features of the zinc deficiency syndrome in the fowl show that this element is required primarily for the growth and development of the skeleton and the formation and maintenance of epithelial tissue.

In young chicks, a deficiency quickly causes a reduction in the consumption of food and water which results in dehydration and stunted growth and therefore tends to obscure the primary lesions. Skeletal abnormalities develop producing leg weakness and ataxia. These abnormalities are more severe in birds than they are in mammals. The long bones are shortened and thickened and may be crooked and the joints are enlarged and rigid. Feather development is impaired and a necrotic dermatitis appears, particularly on the legs and feet. Respiration is laboured (O'Dell and Savage, 1957; O'Dell *et al.*, 1958; Young *et al.*, 1958; Morrison and Sarett, 1958; Roberson and Schaible, 1958; Patrick, 1958; Rahman *et al.*, 1961). Histological examination reveals additional lesions in the epithelium of the oesophagus and in the pancreas (hyperplasia) (O'Dell *et al.*, 1958).

When zinc deficiency is induced in the laying hen, egg production and hatchability fall (Savage and O'Dell, 1959; Blamberg *et al.*, 1960; Kienholtz *et al.*, 1961; Savage, 1968). Hatched chicks are weak, unable to stand and show signs of respiratory distress. Skeletal development is obviously impaired and the feathers are poorly formed. Death usually follows within a few days (Turk *et al.*, 1959; Kienholtz *et al.*, 1961). When the deficiency is severe, the embryo shows gross faults in the development of the limbs and trunk. Entire limbs may be absent and sometimes the embryo consists only of skull and viscera (Blamberg *et al.*, 1960; Kienholtz *et al.*, 1961; O'Dell, 1968).

A. OSTEOGENESIS

The ash content of the skeleton of the zinc-deficient chick is not markedly reduced (O'Dell *et al.*, 1958; Rahman *et al.*, 1961) indicating that mineralization of that osteoid tissue which is formed is not greatly impaired. This is confirmed by histology which reveals changes in the epiphyseal cartilage. These include a considerable reduction in width, the almost complete disappearance of regular columns of cartilage cells and the presence of excessive amounts of extracellular matrix which does not stain normally (O'Dell *et al.*, 1958; Young *et al.*, 1958). In the degenerating region of the epiphyseal plate cells do not mature and degenerate normally unless they are near a blood vessel (Westmoreland and Hoekstra, 1969a).

These abnormalities are associated with a reduction in alkaline phosphatase activity in the bone (Britton, 1968; Westmoreland and Hoekstra, 1969b). This enzyme [orthophosphoric monoester phosphohydrolase EC 3.1.8.1] contains zinc (Britton, 1968) and is believed to be involved in several osteogenic processes other than calcification (Robison, 1923), including the differentiation and maturation of osteoblasts and chondroblasts (Westmoreland and Hoekstra, 1969b), the degeneration of cartilage cells (Henrichsen, 1958; Westmoreland and Hoekstra, 1969b) and the synthesis of mucopolysaccharides of the ground substance (Moog and Wenger, 1952). The hexosamine content of the epiphyseal plate does not appear to be altered by zinc deficiency but there is evidence that the uptake of sulphate is reduced (Nielsen and Ziporin, 1969).

An interesting but as yet unexplained finding is that dietary supplements of histidine, histamine and various anti-arthritic agents such as aspirin, indomethacin and cortisone will prevent the appearance of gross leg abnormalities in the zinc-deficient chick (Nielsen and Sunde, 1967; Nielsen *et al.*, 1968; Reimann, 1969; Hoekstra, 1969). However, these substances have no effect on the microscopic changes in the epiphyseal cartilage (Westmoreland and Hoekstra, 1969a, b) or on any of the other features of the zinc deficiency syndrome.

B. FORMATION OF EPITHELIAL TISSUE

Studies of the changes in the mammalian skin and oesophagus produced by zinc deficiency indicate that zinc is involved in the proliferation and differentiation of epithelial cells (Barney *et al.*, 1968; Pories and Strain, 1970). The primary biochemical lesion in these tissues has not yet been identified and, in view of the relatively large amounts of zinc present, this would appear to be a particularly promising field of work. The impairment of feather development has been attributed to atrophy of the follicles caused by hyperkeratinization of the epidermis (O'Dell *et al.*, 1958).

C. TRANSPORT AND UTILIZATION OF CARBON DIOXIDE

The transport of carbon dioxide, its utilization for controlling the pH of body fluids and secretions and for the production of carbonate ions are facilitated by carbonic anhydrase [carbonate hydro-lyase, EC 4.2.1.1.] which catalyses the formation of bicarbonate which then dissociates to give carbonate:

$$CO_2+OH^- \rightleftharpoons HCO_3^- \rightleftharpoons H^+ + CO_3^{2-}$$

Besides being concerned with respiration, kidney function and all secretory processes in the fowl, carbonic anhydrase in the shell gland mucosa is involved in the production of carbonate from metabolic carbon dioxide for egg shell formation (Benesch *et al.*, 1944; Simkiss, 1961; Hodges and Lorcher, 1967; Bernstein *et al.*, 1968).

Mammalian carbonic anhydrases contain one firmly bound atom of zinc per molecule and it has been assumed that the avian enzymes are also zinc enzymes. A partially purified preparation of carbonic anhydrase from duck erythrocytes contained 0·14% zinc (Leiner *et al.*, 1962) but the zinc content of the fowl enzyme has not been determined. Although a slight reduction in shell thickness has been associated with zinc deficiency in the fowl (Supplee *et al.*, 1958; Savage and O'Dell, 1959) this has not been a constant finding (Kienholtz *et al.*, 1961). This lack of response may be due to the presence of excessive amounts of carbonic anhydrase in the uterus, as in other tissues (Maren, 1967), and also to the strength of the bonds between zinc and the enzyme protein. The high affinity of the apoenzyme for zinc helps to explain why the activity of the enzyme in chick erythrocytes is not affected by dietary zinc deficiency (O'Dell *et al.*, 1958).

Carbonic anhydrase is present in bone but is confined to osteogenic cells in the epiphyseal areas which are destined for haematopoiesis; it does not appear to be involved in calcification (Ellison, 1965).

D. NUCLEIC ACID METABOLISM

The investigation of the effect of zinc deficiency on the activity of metabolic pathways is complicated by its influence on appetite and the feeding pattern and the results are difficult to interpret if the animals in the control group are not fed in exactly the same way as those receiving the zinc-deficient diet (Mills *et al.*, 1969). For this reason some of the conclusions reported in the literature must be regarded as tentative until more exact experiments have been carried out.

Zinc deficiency does not affect the concentrations of nucleic acids in the liver of the chick (Turk, 1966) and here the fowl appears to differ from the rat which shows a marked fall in the synthesis of DNA before its food intake has declined (Williams and Chesters, 1970). Furthermore, there are indications that the catabolism of RNA may be accelerated in the zinc-deficient rat (Macapinlac *et al.*, 1968; Somers and Underwood, 1969).

E. PROTEIN METABOLISM

The protein metabolism of the zinc-deficient fowl does not appear to have been studied. Experiments with rats have indicated an enhanced oxidation of several amino acids including leucine, lysine, glycine, cysteine and methionine (Theuer and Hoekstra, 1966; Hsu *et al.*, 1967; Buchanan and Hsu, 1969). The rate of protein synthesis does not appear to be markedly impaired.

Zinc is an essential constituent of mammalian pancreatic carboxypeptidases A and B [EC 3.4.2.1. and 3.4.2.2.] (Vallee and Neurath, 1954; Cox *et al.*,

1962) and also those of the spiny Pacific dogfish (Lacko and Neurath, 1967). The zinc content of the fowl enzymes is not known. Although the activity of pancreatic carboxypeptidase A is reduced in the zinc-deficient rat (Vallee *et al.*, 1963; Hsu *et al.*, 1966; Mills *et al.*, 1967) the digestion of proteins does not seem to be impaired (Mills *et al.*, 1967).

F. CARBOHYDRATE AND LIPID METABOLISM

The effect of zinc deficiency on the metabolism of carbohydrates and lipids by the fowl has not been investigated. In the rat, the catabolic pathways are not affected (Theuer and Hoekstra, 1966) although significant amounts of firmly bound zinc have been found in three of the mammalian dehydrogenases involved. These enzymes catalyse the phosphorylation of D-glyceraldehyde-3-phosphate, the oxidation of lactate and malate have been regarded as zinc enzymes (see review by Parisi and Vallee, 1969). However, there is now considerable doubt whether zinc is essential in the first two (Sund, 1968). Crystalline lactate dehydrogenase from chicken heart and breast muscle did not contain stoichiometric amounts of zinc (Pesce *et. al.*, 1964) and there have been no reports indicating its presence in the other two enzymes in the fowl.

G. FUNCTION AND STORAGE OF INSULIN

Insulin has a high affinity for zinc and since it was discovered that the binding of zinc enhances the crystallization of insulin and prolongs its hypoglycaemic action in the mammal (Scott and Fisher, 1936) there has been much speculation regarding the functional significance of this interaction. In mammals, with the possible exception of the guinea pig, insulin appears to be stored as its zinc complex in the β-cells of the islets of Langerhans (Maske, 1957; Logothetopoulos *et al.*, 1964; Mohnike and Montz, 1964). There is no such information for the fowl.

Zinc-deficient rats show reductions in their response to insulin, plasma insulin levels and tolerance to glucose (Quarterman *et al.*, 1966) but this may be due to an accelerated hepatic degradation of insulin when reductive cleavage of the molecule occurs under the influence of glutathione-insulin transhydrogenase. Zinc deficiency increases the synthesis of glutathione (Hsu *et al.*, 1968) which in turn may promote the degradation of insulin. No investigations along these lines have been carried out with the fowl.

VI. Manganese

Compared with mammals, the fowl has a high dietary requirement for manganese and consequently is a more sensitive animal for studying biochemical functions of this element. Their levels of blood and tissue manganese are not markedly different suggesting that the higher requirement of the fowl may be due to relatively poor absorption, a view supported by experiments with radioactive manganese (Underwood, 1962).

When chicks are reared on a diet low in manganese growth is retarded and they develop a crippling leg deformity known as perosis (Wilgus *et al.*, 1936, 1937; Gallup and Norris, 1937, 1939a; Lyons *et al.*, 1938). This is characterized by a gross enlargement and malformation of the tibiometatarsal joint which causes the gastrocnemius tendon to slip from its condyles. The bones are structurally weak and the shafts may be bent.

Manganese deficiency in the laying hen causes a marked fall in both egg production and hatchability and a reduction in the thickness and quality of the shell (Lyons and Insko, 1937a, b; Gallup and Norris, 1937; Caskey and Norris, 1938; Schaible *et al.*, 1938; Lyons, 1939; Gallup and Norris, 1939b; Hill and Mathers, 1968). Embryos frequently die during the last few days of incubation and may show chondrodystrophy (Lyons and Insko, 1937a, b). The chief features of this condition are micromelia, "parrot" beak, oedema and a protruding abdomen. In viable chicks a spastic condition may be evident soon after hatching. This is recognized by tetanic spasms, head retraction and extreme nervousness (Caskey and Norris, 1938; Caskey *et al.*, 1944). No histological lesions have been found in the brain and the condition may be caused by abnormal bone development near the vestibular apparatus of the inner ear as in rats and guinea pigs (Shrader and Everson, 1967). Micromelia produced in embryos persists through adult life (Caskey and Norris, 1940).

A. SYNTHESIS OF MUCOPOLYSACCHARIDES

Early studies of the skeletal abnormalities produced by manganese deficiency were concerned with the calcification process. Although the bone ash is slightly reduced, radiographic and histological examinations showed that calcification is not greatly impaired (Gallup and Norris, 1938; Caskey *et al.*, 1939). Furthermore, neither the amount nor the location of ^{45}Ca or ^{32}P deposited in the bone is influenced significantly by the manganese content of the diet (Parker *et al.*, 1955).

The discovery of distinctive histological changes involving the matrix of epiphyseal cartilage (Wolbach and Hegsted, 1953) stimulated an investigation of its chemical composition which showed that manganese deficiency results in an impaired ability of the chondrocytes to synthesize acid mucopolysaccharides, particularly chondroitin sulphate (Leach and Muenster, 1962; Leach, 1967, 1968). *In vitro*, manganese activates the enzymes involved in two stages in the synthetic pathway, viz. the polymeric reactions concerning UDP-N-acetyl-galactosamine and UDP-glucuronic acid which form the polysaccharide, and the incorporation of galactose into the trisaccharide which attaches the polysaccharide to protein (Leach *et al.*, 1969).

There is evidence that an impaired synthesis of mucopolysaccharides is also responsible for the reduction in shell thickness and quality which occurs when a deficiency of manganese is produced in the laying hen (Longstaff and Hill, 1970).

B. GLUCONEOGENESIS

Pyruvate carboxylase [pyruvate: CO_2—ligase (ADP), EC 6.4.1.1.] present in chicken liver mitochondria contains four firmly bound manganese atoms per molecule (Scrutton *et al.*, 1966; Mildvan and Scrutton, 1967). This enzyme catalyses the formation of oxaloacetate from pyruvate which is one of the rate-limiting reactions in gluconeogenesis:

$$\text{Pyruvate} + HCO_3^- \xrightarrow[\text{ATP} \;\; \text{ADP}+P_i]{\text{Acetyl-CoA, } Mg^{2+}} \text{Oxaloacetate}$$

Each molecule of the enzyme contains four biotin residues which become carboxylated during the first stage of the reaction when acetyl-coenzyme A and magnesium ions are required. It has been proposed (Mildvan and Scrutton, 1967) that, at each of the four active centres, manganese links both the pyruvate molecules and the biotin-carbon dioxide complex to the enzyme protein as shown in Fig. 5.

A deficiency of manganese does not, however, produce a reduction in the activity of this enzyme in the liver of the chick. Manganese is then replaced by magnesium which appears to be equally effective in facilitating the reaction (Griminger and Scrutton, 1970). The presence of manganese in pyruvate

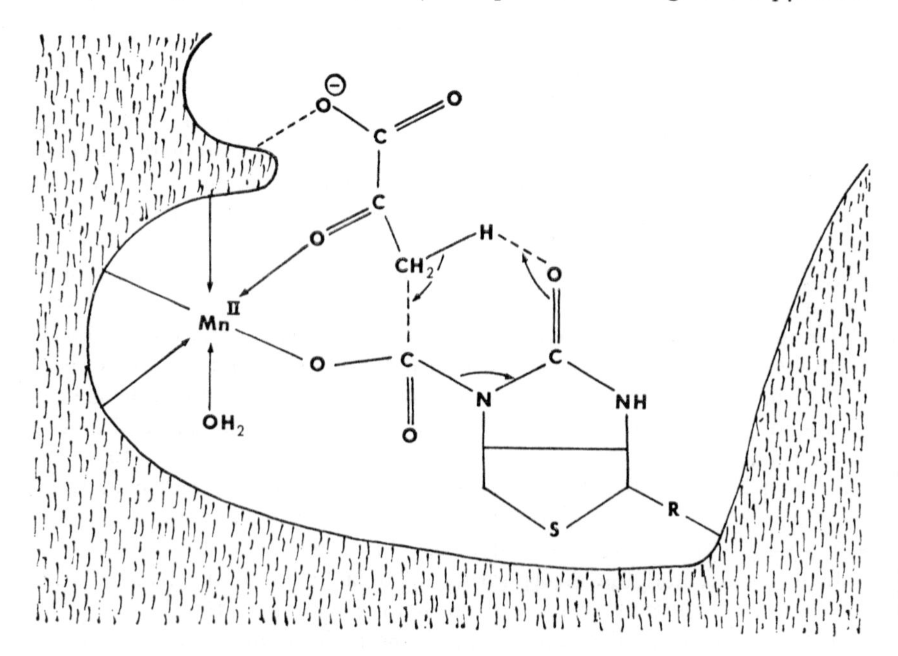

Fig. 5. Function of manganese in pyruvate carboxylase.

carboxylase under normal circumstances partly accounts for its occurrence in mitochondria (Maynard and Cotzias, 1955; Thiers and Vallee, 1957).

Evidence for the involvement of manganese in the remainder of the gluconeogenetic pathway has come from *in vitro* experiments with rat kidney cortex slices (Rutman *et al.*, 1965). Of the metals examined, only manganese and calcium stimulated the formation of glucose from oxaloacetate and other intermediates of the citric acid cycle. Some of the enzymes involved, such as enolase, phosphoglyceromutase and fructose-1,6-diphosphatase are activated *in vitro* by manganese. This property is shared by magnesium, which is usually a more efficient activator, yet no stimulation of gluconeogenesis was observed when magnesium was added to the rat kidney slices. The effect of manganese on this system might have been produced indirectly through a stimulation of oxidative phosphorylation (Lindberg and Ernster, 1954) and a consequent increase in the supplies of ATP.

C. REPRODUCTION

The fall in egg production caused by manganese deficiency is presumably due to a depression of ovulation which has been noted in mammals. Testicular degeneration is another feature of the mammalian deficiency syndrome and may also occur in the fowl. The functions of manganese in ovulation and spermatogenesis are unknown.

D. OTHER POSSIBLE FUNCTIONS

One feature of the manganese deficiency syndrome in mammals is an excessive deposition of fat (Amdur *et al.*, 1946; Plumlee *et al.*, 1956); this has not been observed in the manganese-deficient chick (Wolbach and Hegsted, 1953) but may have escaped notice in the laying hen. Since choline as well as manganese is capable of mobilizing fat in the manganese-deficient animal (Amdur *et al.*, 1946) it has been suspected that manganese may be involved in the function or metabolism of choline (Cotzias, 1958). This possibility is supported by the fact that perosis is also produced by a deficiency of choline. However, the perosis produced by manganese deficiency appears to be refractory to choline supplements alone (Coates *et al.*, 1947).

Manganese and other divalent metals activate several proteases and peptidases *in vitro* but there is little information on the metal or metals which perform this function *in vivo*. Manganese deficiency reduces the proteolytic activity of pancreatic homogenates in the rat but not in the chick (Johnson *et al.*, 1959).

Manganese deficiency lowers the activity of arginase [L-arginine urea hydrolase, EC 3.5.3.1] in the mammalian liver and it has been reported that crystalline preparations of this enzyme are activated only by this metal (Bach and Whitehouse, 1954). Involvement of manganese in the last stage of urea formation has therefore been established. However, this is not such an important function in avian species since they are unable to use the urea cycle for the excretion of nitrogen. Arginase is present in relatively small amounts in the liver of the fowl and is activated by manganese (Tamir and Ratner, 1963).

VII. Selenium

A. INTERACTIONS WITH VITAMIN E

Besides increasing the effectiveness of α-tocopherol in preventing signs of vitamin E deficiency in the chick (see Chapter 14) inorganic selenium appears to exert some protective action *per se*. Protection by selenium alone is complete in the case of exudative diathesis (Patterson *et al.*, 1957; Schwarz *et al.*, 1957), partial in the case of encephalomalacia (Century and Horwitt, 1964; Jenkins *et al.*, 1965) and muscular dystrophy (Ewen and Jenkins, 1967). Since some of the "vitamin E-free" diets used in experiments of this kind may have contained appreciable quantities of tocopherols (Green *et al.*, 1961) these observations cannot be taken as unequivocal evidence that selenium has functions independent of those of α-tocopherol. However, experiments with diets containing crystalline amino acids have shown that a dietary deficiency of selenium does produce a specific deficiency syndrome in the presence of high dietary levels of vitamin E (Thompson and Scott, 1969; Scott and Thompson, 1970). This syndrome is characterized by retarded growth, poor feather development and fibrotic degeneration of the pancreas. The latter feature may partly explain the potentiation of α-tocopherol activity by selenium since the reduction in the output of lipase caused by degeneration of the acinar cells impairs the digestion of fats and the formation of lipid-bile salt micelles which are required for the absorption of tocopherols (Scott and Thompson, 1970).

Besides facilitating the absorption of α-tocopherol in this way, selenium increases its retention by the chick. This effect is selective for the *d*-epimer and may be explained by the suggestion that a selenolipoprotein, which migrates with the γ-globulin fraction on electrophoresis, may be responsible for the transport of *d*-α-tocopherol (Desai and Scott, 1965; Scott, 1966). This protein does not contain selenium analogues of the sulphur amino acids and appears to bind selenium through the sulphur atoms of cysteine residues (Jenkins *et al.*, 1969).

B. ANTIOXIDANT PROPERTIES

Some of the inter-relationships observed between α-tocopherol and selenium may reflect similarities in their metabolic functions. For example, it has been suggested that, like the tocopherols, selenium functions as a biological antioxidant (Tappel, 1962) and may be particularly involved in the protection of unsaturated tissue lipids against peroxidation.

However, although selenium forms compounds with strong antioxidant properties *in vitro* it is now extremely doubtful whether this is an important aspect of its biochemistry (Green and Bunyan, 1969).

C. MITOCHONDRIAL METABOLISM

Both selenium and vitamin E appear to be involved, independently and not as antioxidants, in the synthesis of ubiquinones which function as hydrogen carriers in the respiratory chain and elsewhere (Green *et al.*, 1961). An impairment of mitochondrial respiration may therefore be expected in the selenium-deficient animal and reductions in the oxidation of α-ketoglutarate and pyruvate by rat liver preparations have in fact been observed (Connolly and Schwarz, 1963, 1965; Bull and Oldfield, 1967). However, it is unlikely that these effects are due to the presence of inadequate amounts of ubiquinone (as coenzyme Q) since the oxidation of succinate is not affected (Bull and Oldfield, 1967). They support the theory that selenium has specific biochemical functions in mitochondrial metabolism (Schwarz, 1965) one of which may be performed as a constituent of lipoate dehydrogenase (Schwarz 1962). Selenium occurs in at least three valency states in rat liver mitochondria and the amount of the acid-labile fraction present, presumably selenide, is dependent upon the vitamin E status of the animal. It has been suggested that this fraction may be present instead of sulphur in certain non-haem iron proteins and that one function of vitamin E is to protect it from oxidation to a higher valency state (Diplock *et al.*, 1968; Diplock, 1970). This would, of course, explain the potentiation of selenium activity by vitamin E (Calvert *et al.*, 1962).

VIII. Molybdenum

Dietary supplements of molybdate prevent the mortality and growth depression produced by tungstate in chicks (Higgins *et al.*, 1956) and stimulates their growth when they are maintained on purified diets (Reid *et al.*, 1956; Anders and Hill, 1970). No other clinical abnormalities attributable to molybdenum deficiency have been reported.

Both the simple and conditioned molybdenum deficiency states are accompanied by a decrease in the activity of tissue xanthine dehydrogenase (Higgins *et al.*, 1956; Anders and Hill, 1970). This enzyme catalyses the hydroxylation of both hypoxanthine and xanthine to form uric acid, the excretion of which by the chick is markedly reduced by molybdenum deficiency; unlike the mammalian enzyme (xanthine oxidase) it utilizes NAD (and not molecular oxygen) as the hydrogen acceptor and does not contain copper. Purified preparations of xanthine dehydrogenase, from chicken liver and kidney, contain molybdenum, non-haem iron and FAD. The most recent preparation of the liver enzyme contained FAD, molybdenum, iron and labile sulphide in the ratio of 1:1:4:4 (Rajagopalan and Handler, 1967). Earlier preparations (Remy *et al.*, 1955; Landon and Carter, 1960) contained larger amounts of iron, probably due to contamination with ferritin. Iron deficiency also reduces the activity of this enzyme in the chick (Davis *et al.*, 1968). This function of iron and molybdenum is far more important in birds than in mammals since they normally excrete a much greater proportion of uric acid (see Chapter 9).

The other metabolic functions of molybdenum in the fowl are still obscure. Xanthine dehydrogenase also oxidizes aldehydes and may perform the same functions as mammalian aldehyde oxidase which has a similar composition (Handler *et al.*, 1964). A role for molybdenum in the synthesis of haemoglobin is suggested by the effect of molybdate in preventing the "usual" fall in haemoglobin levels during the first week after hatching when administered with iron (Anders and Hill, 1970).

IX. Other elements

Traces of many elements in addition to those discussed above occur constantly in the fluids and tissues of animals and *in vitro* activate various enzymes non-specifically. With the increasing use of highly purified synthetic diets and the elimination of environmental contamination through the use of plastic cages, troughs and isolators (Smith and Schwarz, 1967) information is accumulating which suggests that several of these elements may perform essential metabolic functions. For instance, chicks reared in plastic isolators on a purified diet developed an abnormal gait and leg deformities and showed changes in the pigmentation of the skin of the legs (Nielsen, 1970). All these abnormalities were prevented by the addition of nickel to the diet. A small but statistically significant growth response to bromide has been obtained with chicks (Bosshardt *et al.*, 1956).

Mammalian experiments suggest several functions for chromium, including the control of carbohydrate metabolism in association with insulin (see review by Mertz, 1969). Vanadium, an essential element for tunicates, may be involved in the mineralization of the skeleton in mammals and a similar function has been suggested for strontium.

Practically nothing is known of the biological behaviour of other ubiquitous elements such as titanium and rubidium. Future research with the delicate and exact techniques now available may show whether they have any metabolic roles or whether their presence in the fowl's tissues is accidental.

References

Agricultural Research Council (1963). "The Nutrient Requirements of Farm Livestock, No. 1", London.
Amdur, M. O., Norris, L. C. and Heuser, G. F. (1946). *J. biol. Chem.* **164**, 783–784.
Anders, E. and Hill, C. (1970). *Fedn Proc. Fedn Am. Socs exp. Biol.* **29**, 766 (Abstr. 2937).
Anderson, R. L. and Tove, S. B. (1958). *Nature, Lond.* **182**, 315.
Bach, S. J. and Whitehouse, D. B. (1954). *Biochem. J.* **57**, xxxi.

Barney, G. H., Orgebin-Crist, M. G. and Macapinlac, M. B. (1968). *J. Nutr.* **95**, 526–534.

Barnstein, P., Kang, A. H. and Piez, K. A. (1966). *Proc. natn. Acad. Sci. U.S.A.* **55**, 417–424.

Beck, A. B. (1956). *Aust. J. Zool.* **4**, 1–18.

Beinert, H. and Palmer, G. (1964). *J. biol. Chem.* **239**, 1221–1227.

Benesch, R., Barron, N. W. and Newson, C. A. (1944). *Nature, Lond.* **153**, 138–139.

Bernstein, R. S., Nevalainen, J., Schraer, F. and Schraer H. (1968). *Biochim. biophys. Acta* **159**, 367–376.

Bird, D. W. (1967). *Diss. Abstr.* **27**, 2935.

Bird, D. W., O'Dell, B. L. and Savage, J. E. (1963). *Poult. Sci.* **42**, 1256.

Bird, D. W., Savage, J. E. and O'Dell, B. L. (1966). *Proc. Soc. exp. Biol. Med.* **123**, 250–254.

Blamberg, D. L., Blackwood, U. V., Supplee, W. C. and Combs, G. F. (1960). *Proc. Soc. exp. Biol. Med.* **104**, 217–220.

Bosshardt, D. K., Huff, J. W. and Barnes, R. H. (1956). *Proc. Soc. exp. Biol. Med.* **92**, 219–221.

Boyer, P. D. (1968). *In* "Biological Oxidations" (T. P. Singer, ed.), pp. 193–235, Interscience Publishers, New York, London and Sydney.

Bray, R. C. and Harrap, K. R. (1959). *Rep. Prog. Chem.* **55**, 343–353.

Britton, W. M. (1968). *Diss. Abstr.* **29B**, 256.

Brown, E. G. (1958). *Nature, Lond.* **182**, 313–315.

Buchanan, P. J. and Hsu, J. M. (1969). *Fedn Proc. Fedn Am. Socs exp. Biol.* **28**, 762 (Abstr. 2815).

Buffoni, F., Corte, L. D. and Knowles, P. F. (1968). *Biochem. J.* **106**, 575–576.

Bull, R. C. and Oldfield, J. E. (1967). *J. Nutr.* **91**, 237–246.

Butow, R. and Racker, E. (1965). *In* "Non-Heme Iron Proteins: Role in Energy Conversion" (A. San Pietro, ed.) pp. 383–392. Antioch Press, Yellow Springs, Ohio.

Calvert, C. C., Nesheim, M. C. and Scott, M. L. (1962). *Proc. Soc. exp. Biol. Med.* **109**, 16–18.

Carlton, W. W. and Henderson, W. (1963). *J. Nutr.* **81**, 200–208.

Carlton, W. W. and Henderson, W. (1964). *Avian Dis.* **8**, 48–55.

Caskey, C. D., Gallup, W. D. and Norris, L. C. (1939). *J. Nutr.* **17**, 407–417.

Caskey, C. D. and Norris, L. C. (1938). *Poult. Sci.* **17**, 433.

Caskey, C. D. and Norris, L. C. (1940). *Proc. Soc. exp. Biol. Med.* **44**, 332–335.

Caskey, C. D., Norris, L. C. and Heuser, G. F. (1944). *Poult. Sci.* **23**, 516–520.

Century, B. and Horwitt, M. K. (1964). *Proc. Soc. exp. Biol. Med.* **117**, 320–322.

Charles, D. R. and Rawles, M. E. (1940). *Proc. Soc. exp. Biol. Med.* **43**, 55–58.

Chou, W. S., Savage, J. E. and O'Dell, B. L. (1968). *Proc. Soc. exp. Biol. Med.* **128**, 948–952.

Coates, M. E., Kon, S. K., Shepherd, E. E. and White, E. G. (1947). *J. comp. Path. Ther.* **57**, 232–239.

Connolly, J. D. and Schwarz, K. (1963). *Fedn Proc. Fedn Am. Socs exp. Biol.* **22**, 652 (Abstr. 2941).

Connolly, J. D. and Schwarz, K. (1965). *Fedn Proc. Fedn Am. Socs exp. Biol.* **24**, 623 (Abstr. 2725).

Cotzias, G. C. (1958). *Physiol. Rev.* **38**, 503–532.

Cox, D. J., Wintersberger, E. and Neurath, H. (1962). *Biochemistry* **1**, 1078–1082.

Davis, P. N., Norris, L. C. and Kratzer, F. H. (1962). *J. Nutr.* **78**, 445–453.

Davis, P. N., Norris, L. C. and Kratzer, F. H. (1968). *J. Nutr.* **94**, 407–417.
Desai, I. D. and Scott, M. L. (1965). *Archs Biochem. Biophys.* **110**, 309–315.
Di Paolo, R. V. and Newberne, P. M. (1970). *Fedn Proc. Fedn Am. Socs exp. Biol.* **29**, 695 (Abstr. 2537).
Diplock, A. T. (1970). *In* "Trace Element Metabolism in Animals" (C. F. Mills, ed.), pp. 190–203. E. and S. Livingstone, Edinburgh.
Diplock, A. T., Baum, H. and Lucy, J. A. (1968). *5th Fedn europ. Biochem. Socs, Prague* **121** (Abstr. 484).
Ellison, A. C. (1965). *Proc. Soc. exp. Biol. Med.* **120**, 415–418.
Elvehjem, C. A. and Hart, E. B. (1929). *J. biol. Chem.* **84**, 131–141.
Ewen, L. M. and Jenkins, K. J. (1967). *J. Nutr.* **93**, 470–474.
Friedman, S. and Kaufman, S. (1965). *J. biol. Chem.* **240**, 4763–4773.
Gallagher, C. H. (1957). *Aust. vet. J.* **33**, 311–317.
Gallagher, C. H., Judah, J. D. and Rees, K. R. (1956). *Proc. R. Soc.* **B145**, 195–205.
Gallup, W. D. and Norris, L. C. (1937). *Poult. Sci.* **16**, 351–352.
Gallup, W. D. and Norris, L. C. (1938). *Science, N.Y.* **87**, 18–19.
Gallup, W. D. and Norris, L. C. (1939a). *Poult. Sci.* **18**, 76–82.
Gallup, W. D. and Norris, L. C. (1939b). *Poult. Sci.* **18**, 83–88.
Goldstein, M., Lauber, E. and McKereghan, M. B. (1965). *J. biol. Chem.* **240**, 2066–2072.
Green, J., Diplock, A. T., Bunyan, J., Edwin, E. F. and McHale, D. (1961). *Nature, Lond.* **190**, 318–335.
Green, J. and Bunyan, J. (1969). *Nutr. Abstr. Rev.* **39**, 321–345.
Griffiths, D. E. and Wharton, D. C. (1961). *J. biol. Chem.* **236**, 1857–1862.
Griminger, P. and Scrutton, M. C. (1970). *Fedn Proc. Fedn Am. Socs exp. Biol.* **29**, 765 (Abstr. 2934).
Hamilton, G. A. (1969). *Adv. Enzymol.* **32**, 55–96.
Handler, P., Rajagopalan, K. V. and Aleman, V. (1964). *Fedn Proc. Fedn Am. Socs exp. Biol.* **23**, 30–38.
Harbury, H. A. and Loach, P. A. (1960). *J. biol. Chem.* **235**, 3640–3645.
Hart, E. B., Steenbock, H., Waddell, J. and Elvehjem, C. A. (1928). *J. biol. Chem.* **77**, 797–812.
Hellerman, L. (1937). *Physiol. Rev.* **17**, 454–484.
Henrichsen, E. (1958). *Acta orthop. scand.* **27**, 173–191.
Higgins, E. S., Richert, D. A. and Westerfeld, W. W (1956). *J. Nutr.* **59**, 539–559.
Hill, C. H. and Matrone, G. (1961). *J. Nutr.* **73**, 425–431.
Hill, C. H., Starcher, B. and Kim, C. S. (1967). *Fedn Proc. Fedn Am. Socs exp. Biol.* **26**, 129–133.
Hill, R. and Mathers, J. W. (1968). *Br. J. Nutr.* **22**, 625–633.
Hodges, R. D. and Lorcher, K. (1967). *Nature, Lond.* **216**, 609–610.
Hoekstra, W. G. (1969). *Am. J. clin. Nutr.* **22**, 1268–1277.
Hsu, J. M., Anilane, J. K. and Scanlan, D. E. (1966). *Science, N.Y.* **153**, 882–883.
Hsu, J. M., Crowder, S. F. and Buchanan, P. J. (1967). *Fedn Proc. Fedn Am. Socs exp. Biol.* **26**, 524 (Abstr. 1491).
Hsu, J. M., Anthony, W. L. and Buchanan, P. J. (1968). *Proc. Soc. exp. Biol. Med.* **127**, 1048–1051.
Hunt, C. E. and Landesman, J. (1968). *Fedn. Proc. Fedn Am. Socs exp. Biol.* **27**, 476 (Abstr. 1461).
Jenkins, K. J., Ewan, L. M. and McConachie, J. D. (1965). *Poult. Sci.* **44**, 615–617.

Jenkins, K. J., Hidiroglou, M. and Ryan, J. F. (1969). *Can. J. Physiol. Pharm.* **47**, 459–467.
Johnson, R. R., Bentley, O. G. and Sutton, J. S. (1959). *J. Nutr.* **67**, 513–524.
Kienholtz, E. W., Turk, D. E., Sunde, M. L. and Hoekstra, W. G. (1961). *J. Nutr.* **75**, 211–221.
Kim, C. S. and Hill, C. H. (1966). *Biochem. biophys. Res. Commun.* **24**, 395–400.
Kun, E. and Fanshier, D. W. (1959). *Biochim. biophys. Acta* **32**, 338–348.
Lacko, A. and Neurath, H. (1967). *Fedn Proc. Fedn Am. Socs exp. Biol.* **26**, 279 (Abstr. 132).
Landon, E. J. and Carter, C. E. (1960). *J. biol. Chem.* **235**, 819–824.
Leach, R. M. (1967). *Fedn Proc. Fedn Am. Socs exp. Biol.* **26**, 118–120.
Leach, R. M. (1968). *Poult. Sci.* **47**, 828–830.
Leach, R. M. and Muenster, A. M. (1962). *J. Nutr.* **78**, 51–56.
Leach, R. M., Muenster, A. M. and Wien, E. M. (1969). *Archs Biochem. Biophys.* **133**, 22–28.
Lee, G. R., Cartwright, G. E. and Wintrobe, M. (1968). *Proc. Soc. Exp. Biol. Med.* **127**, 977–981.
Leiner, M., Beck, H. and Eckert, H. (1962). *Z. phys. Chem.* **327**, 144–165.
Lerner, A. B., Fitzpatrick, T. B., Calkins, E. and Summerson, W. H. (1950). *J. biol. Chem.* **187**, 793–802.
Lindberg, O. and Ernster, L. (1954). *Nature, Lond.* **173**, 1038–1039.
Logothetopoulos, J., Kaneko, M., Wrenshall, G. A. and Best, C. H. (1964). *In* "The Structure and Metabolism of the Pancreatic Islets" (S. E. Brolin, B. Hellman and H. Knutson, eds.), pp. 333–344. Pergamon Press, Oxford.
Longstaff, M. and Hill, R. (1970). *In* "Trace Element Metabolism in Animals" (C. F. Mills, ed.), pp. 137–139. E. and S. Livingstone, Edinburgh.
Lyons, M. (1939). *Bull. Ark. agric. Exp. Stn* **374**.
Lyons, M. and Insko, W. M. (1937a). *Poult. Sci.* **16**, 365–366.
Lyons, M. and Insko, W. M. (1937b). *Bull. Ky agric. Exp. Stn* **371**, 61–75.
Lyons, W., Insko, W. M. and Martin, J. H. (1938). *Poult. Sci.* **17**, 12–16.
Macapinlac, M. P., Pearson, W. N., Barney, G. H. and Darby, W. J. (1968). *J. Nutr.* **95**, 569–577.
Mahler, H. R. (1961). *In* "Mineral Metabolism" (E. L. Comar and F. Bronner, eds.), Vol. IB, pp. 743–897. Academic Press, London and New York.
Maren, T. H. (1967). *Physiol. Rev.* **47**, 595–781.
Maske, H. (1957). *Diabetes* **6**, 335–341.
Maynard, L. S. and Cotzias, G. C. (1955). *J. biol. Chem.* **214**, 489–495.
Mertz, W. (1969). *Physiol. Rev.* **49**, 163–239.
Mildvan, A. S. and Scrutton, M. C. (1967). *Biochemistry* **6**, 2978–2994.
Miller, E. J., Martin, G. R. and Piez, K. A. (1964). *Biochem. biophys. Res. Commun.* **17**, 248–253.
Mills, C. F. (1964). *Proc. Nutr. Soc.* **23**, 38–45.
Mills, C. F., Quarterman, J., Chesters, J. K., Williams, R. B. and Dalgarno, A. C. (1969). *Am. J. clin. Nutr.* **22**, 1240–1249.
Mills, C. F., Quarterman, J., Williams, R. B., Dalgarno, A. C. and Panic, B. (1967). *Biochem. J.* **102**, 712–718.
Missala, K., Lloyd, K., Gregoriads, G. and Sourkes, T. L. (1967). *Eurp. J. Pharm.* **1**, 6–10.
Misuraca, G., Nicolaus, R. A., Prota, G. and Ghiara, G. (1969). *Experientia* **25**, 920–922.

Mohnike, G. and Montz, V. (1964). *In* "The Structure and Metabolism of the Pancreatic Islets" (S. E. Brolin, B. Hellman and N. Knutson, eds.), pp. 75–81, Pergamon Press, Oxford.
Moog, F. and Wenger, E. L. (1952). *Am. J. Anat.* **90**, 339–371.
Moore, H. W. and Folkers, K. (1964). *J. Am. chem. Soc.* **86**, 3393–3394.
Morrison, A. B. and Sarett, H. P. (1958). *J. Nutr.* **65**, 267–280.
Morrison, M., Hori, S. and Mason, H. S. (1963). *J. biol. Chem.* **238**, 2220–2224.
Nara, S., Gomes, B. and Yasunobu, K. T. (1966). *J. biol. Chem.* **241**, 2774–2780.
National Academy of Sciences—National Research Council (1966). "Nutrient Requirements of Domestic Animals. No. 1 Nutrient Requirements of Poultry", 5th ed. Publ. 1345, Washington.
Nickerson, M. (1946). *Physiol. Zoöl.* **19**, 66–77.
Nielsen, F. H. (1970). *Fedn Proc. Fedn Am. Socs exp. Biol.* **29**, 696 (Abstr. 2541).
Nielsen, F. H. and Sunde, M. L. (1967). *Proc. Soc. exp. Biol. Med.* **124**, 1106–1110.
Nielsen, F. H., Sunde, M. L. and Hoekstra, W. G. (1968). *J. Nutr.* **94**, 527–533.
Nielsen, F. H. and Ziporin, Z. Z. (1969). *Fedn Proc. Fedn Am. Socs exp. Biol.* **28**, 762 (Abstr. 2814).
O'Dell, B. L. (1968). *Fedn Proc. Fedn Am. Socs exp. Biol.* **27**, 199–204.
O'Dell, B. L., Bird, D. W., Ruffles, D. L. and Savage, J. B. (1966). *J. Nutr.* **88**, 9–14.
O'Dell, B. L., Hardwick, B. C. and Reynolds, G. (1961). *Proc. Soc. exp. Biol. Med.* **108**, 402–405.
O'Dell, B. L., Newberne, P. M. and Savage, J. E. (1958). *J. Nutr.* **65**, 503–518.
O'Dell, B. L. and Savage, J. E. (1957). *Fedn Proc. Fedn Am. Socs exp. Biol.* **16**, 394 (Abstr. 1690).
Omura, T., Sanders, E., Cooper, D. Y., Rosenthal, O. and Estabrook, R. W. (1965). *In* "Non-Heme Iron Proteins: Role in Energy Conversion", (A. San Pietro, ed.), pp. 401–412. Antioch Press, Yellow Springs, Ohio.
Omura, T. and Sato, R. (1964). *J. biol. Chem.* **239**, 2370–2378.
Osaki, S., Johnson, D. A. and Frieden, E. (1966). *J. biol. Chem.* **241**, 2746–2751.
Osaki, S., Johnson, D. A., Tophem, R. W. and Frieden, E. (1970). *Fedn Proc. Fedn Am. Socs exp. Biol.* **29**, 695 (Abstr. 2538).
Parisi, A. F. and Vallee, B. L. (1969). *Am. J. clin. Nutr.* **22**, 1222–1239.
Parker, H. E., Andrews, F. N., Carrick, C. W., Creek, R. D. and Hauge, S. M. (1955). *Poult. Sci.* **34**, 1154–1158.
Partridge, S. M., Elsden, D. F., Thomas, J., Dorfman, A., Telser, A. and Ho, P. L. (1964). *Biochem. J.* **93**, 30C–33C.
Partridge, S. M., Elsden, D. F., Thomas, J., Dorfman, A., Telser, A. and Ho, P. L. (1966). *Nature, Lond.* **209**, 399–400.
Patrick, H. (1958). *Fedn Proc. Fedn Am. Socs exp. Biol.* **17**, 487 (Abstr. 1912).
Patterson, E. L., Milstrey, R. and Stokstad, E. L. R. (1957). *Proc. Soc. exp. Biol. Med.* **95**, 617–620.
Pesce, A., McKay, R. H., Stolzenbach, F., Cahn, R. D. and Kaplan, N. O. (1964). *J. biol. Chem.* **239**, 1753–1761.
Piez, K. A. (1968). *A. Rev. Biochem.* **37**, 547–570.
Plumlee, M. P., Thrasher, D. M., Beeson, W. M., Andrews, F. N. and Parker, H. E. (1956). *J. Anim. Sci.* **15**, 352–367.
Pories, W. J. and Strain, W. H. (1970). *In* "Trace Element Metabolism in Animals" (C. F. Mills, ed.), pp. 75–77. E. and S. Livingstone, Edinburgh.
Proceedings of a Symposium on Biological Aspects of Metal Binding (1961). *Fedn Proc. Fedn Am. Socs exp. Biol. Suppl.* **10**.

Quarterman, J., Mills, C. F. and Humphries, W. R. (1966). *Biochem. biophys. Res. Commun.* **25**, 354–358.
Rahman, M. M., Davies, R. E., Deyre, C. W., Reid, B. L. and Couch, J. R. (1961). *Poult. Sci.* **40**, 195–200.
Rajagopalan, K. V. and Handler, P. (1967). *J. biol. Chem.* **242**, 4097–4107.
Reid, B. L., Kurnich, A. A., Svacha, R. L. and Couch, J. R. (1956). *Proc. Soc. exp. Biol. Med.* **93**, 245–248.
Reimann, E. M. (1969). *Diss. Abstr.* **29B**, 4518–4519.
Remy, C. N., Richert, D. A., Doisy, R. J., Wells, I. C. and Westerfeld, W. W. (1955). *J. biol. Chem.* **217**, 293–305.
Roberson, R. H. and Schaible, P. J. (1958). *Poult. Sci.* **37**, 1321–1323.
Robison, R. (1923). *Biochem. J.* **17**, 286–293.
Rucker, R. B., Parker, H. E. and Rogler, J. C. (1969). *J. Nutr.* **98**, 57–63.
Rutman, J. Z., Meltzer, L. E., Kitchell, J. R., Rutman, R. J. and George, P. (1965). *Am. J. Physiol.* **208**, 841–846.
Sardesai, V. M. and Orten, J. M. (1969). *Fedn Proc. Fedn Am. Socs exp. Biol.* **28**, 300 (Abstr. 236).
Savage, J. E. (1968). *Fedn Proc. Fedn Am. Socs exp. Biol.* **27**, 927–931.
Savage, J. E. and O'Dell, B. L. (1959). *Wld's Poult. Sci. J.* **15**, 264–276.
Schaible, P. J., Bandemer, S. L. and Davidson, J. A. (1938). *Tech. Bull. Mich. (St. Coll.) agric. Exp. Stn* **159**.
Schwarz, K. (1962). *Vitams Horm.* **20**, 463–484.
Schwarz, K. (1965). *Fedn Proc. Fedn Am. Socs exp. Biol.* **24**, 58–67.
Schwarz, K., Bieri, J. G., Briggs, G. M. and Scott, M. L. (1957). *Proc. Soc. exp. Biol. Med.* **95**, 621–625.
Scott, D. A. and Fisher, A. M. (1936). *J. Pharmac. exp. Ther.* **58**, 78–92.
Scott, M. L. (1966). *Ann. N.Y. Acad. Sci.* **138**, 82–89.
Scott, M. L. and Thompson, J. N. (1970). *Fedn Proc. Fedn Am. Socs exp. Biol.* **29**, 499 (Abstr. 1433).
Scrutton, M. C., Utter, M. F. and Mildvan, A. S. (1966). *J. biol. Chem.* **241**, 3480–3487.
Selwyn, M. J. (1968). *Nature, Lond.* **219**, 490–493.
Shrader, R. E. and Everson, G. J. (1967). *J. Nutr.* **91**, 453–460.
Simkiss, K. (1961). *Biol. Rev.* **36**, 321–367.
Simpson, C. F. and Harms, R. H. (1964). *Expl molec. Path.* **3**, 390–400.
Simpson, C. F., Jones, J. E. and Harms, R. H. (1967). *J. Nutr.* **91**, 283–291.
Smith, J. C. and Schwarz, K. (1967). *J. Nutr.* **93**, 182–188.
Somers, M. and Underwood, E. J. (1969). *Aust. J. biol. Sci.* **22**, 1277–1282.
Starcher, B. and Hill, C. H. (1965). *Comp. Biochem. Physiol.* **15**, 429–434.
Starcher, B. and Hill, C. H. (1966). *Biochim. biophys. Acta* **127**, 400–406.
Starcher, B., Hill, C. H. and Matrone, G. (1964). *J. Nutr.* **82**, 318–322.
Sund, H. (1968). *In* "Biological Oxidations" (T. P. Singer, ed.), pp. 603–705. Interscience Publishers, New York, London and Sydney.
Supplee, W. C., Blamberg, D. L., Keene, O. D., Combs, C. F. and Romoser, G. L. (1958). *Poult. Sci.* **37**, 1245–1246.
Tamir, H. and Ratner, S. (1963). *Archs Biochem. Biophys.* **102**, 249–258.
Tappel, A. L. (1962). *Vitams Horm.* **20**, 493–510.
Theuer, R. C. and Hoekstra, W. G. (1966). *J. Nutr.* **89**, 448–454.
Thiers, R. E. and Vallee, B. L. (1957). *J. biol. Chem.* **226**, 911–920.
Thompson, J. N. and Scott, M. L. (1969). *J. Nutr.* **97**, 335–342.

Turk, D. E. (1966). *Poult. Sci.* **45**, 608–611.
Turk, D. E., Sunde, M. L. and Hoekstra, W. G. (1959). *Poult. Sci.* **38**, 1256.
Underwood, E. J. (1962). "Trace Elements in Human and Animal Nutrition", 2nd ed. Academic Press, New York and London.
Underwood, E. J. (1966). "The Mineral Nutrition of Livestock". Central Press Ltd., Aberdeen.
Vallee, B. L. (1955). *Adv. Protein Chem.* **10**, 317–384.
Vallee, B. L. and Coleman, J. E. (1964). *In* "Comprehensive Biochemistry" (M. Florkin and E. H. Stotz, eds.), Vol. 12, pp. 165–235. Elsevier Publishing Co., Amsterdam, London and New York.
Vallee, B. L. and Neurath, H. (1954). *J. Am. chem. Soc.* **76**, 5006–5007.
Vallee, B. L., Riordan, J. F. and Coleman, J. E. (1963). *Proc. natn Acad. Sci. U.S.A.* **49**, 109–116.
Vallee, B. L. and Williams, R. J. P. (1968a). *Chemy Britain* **4**, 397–402.
Vallee, B. L. and Williams, R. J. P. (1968b). *Proc. natn. Acad. Sci. U.S.A.* **59**, 498–505.
Wainio, W. W., Wende, C. V. and Shimp, N. F. (1959). *J. biol. Chem.* **234**, 2433–2436.
Westmoreland, N. and Hoekstra, W. G. (1969a). *J. Nutr.* **98**, 76–82.
Westmoreland, N. and Hoekstra, W. G. (1969b). *J. Nutr.* **98**, 83–89.
Wilgus, H. S., Norris, L. C. and Heuser, G. F. (1936). *Science, N.Y.* **84**, 252–253.
Wilgus, H. S., Norris, L. C. and Heuser, G. F. (1937). *J. Nutr.* **14**, 155–167.
Williams, R. B. and Chesters, J. K. (1970). *In* "Trace Element Metabolism in Animals" (C. F. Mills, ed.), pp. 164–166. E. and S. Livingstone, Edinburgh.
Williams, R. J. P. (1965). *In* "Non-Heme Iron Proteins: Role in Energy Conversion" (A. San Pietro, ed.), pp. 7–32, Antioch Press, Yellow Springs, Ohio.
Williams, R. J. P. (1970). *R. Inst. Chem Rev.* **1**, 13–38.
Wolbach, S. B. and Hegsted, D. M. (1953). *Archs Path.* **56**, 437–453.
Yamada, H., Yasunobu, K. T., Yamano, T. and Mason, H. S. (1963). *Nature, Lond.* **198**, 1092–1093.
Young, R. J., Edwards, H. M. and Gillis, M. B. (1958). *Poult. Sci.* **37**, 1100–1107.

16

The Pituitary Gland

A. STOCKELL HARTREE and F. J. CUNNINGHAM

Department of Biochemistry, University of Cambridge, Cambridge, England and Department of Physiology and Biochemistry, University of Reading, Reading, England

I. Structure of the Pituitary Gland

The hypophysis or pituitary gland of the domestic fowl, located just below the brain, secretes hormones which cause specific target organs to respond in a typical manner. Its secretions are essential to normal metabolism, and it also acts as a co-ordinator of other endocrine glands.

A. GENERAL ANATOMY AND TERMINOLOGY

The pituitary gland is a small lobed organ situated in the sella turcica, a depression of the sphenoid bone just posterior to the optic chiasma at the floor of the diencephalon. It consists of two distinct parts, the adenohypophysis and the neurohypophysis. The adenohypophysis is derived from the oral epithelium. It arises from Rathke's pouch, a hollow diverticulum which extends upwards from the stomadeum to the floor of the diencephalon in the region of the infundibulum. Meanwhile, an outgrowth of the infundibulum forms the neural lobe. During development these two structures become adjacent and together form part of the hypophysis.

To describe the orientation of the gland and its relations to adjacent structures the following terms are used: Rostral or anterior, which is the direction towards the beak, and caudal or posterior which is the opposite direction. Towards the brain is referred to as dorsal and towards the oral cavity is ventral. The terminology used to describe the subdivisions of the gland has been fully described by Wingstrand (1951) in his classic monograph on the structure and development of the avian pituitary. A slightly modified nomenclature of the different parts of the gland is shown in Table 1.

The pars distalis forms the bulk of the adenohypophysis and is situated ventrally or rostroventrally of the neurohypophysis. The most comprehensive investigation of the structure of the adenohypophysis is that of Rahn and Painter (1941) who examined the pituitaries of 18 species of birds including the domestic fowl. According to these workers the adenohypophysis in birds consists of two parts, the pars distalis and the pars tuberalis. Furthermore, they distinguished two histologically different divisions of the pars distalis which they referred to as the cephalic or rostral and caudal lobes. They denied the existence of a structural pars intermedia which is found in mammals and, although some diverging statements have appeared, this observation was definitely corroborated by the extensive investigation of Wingstrand (1951).

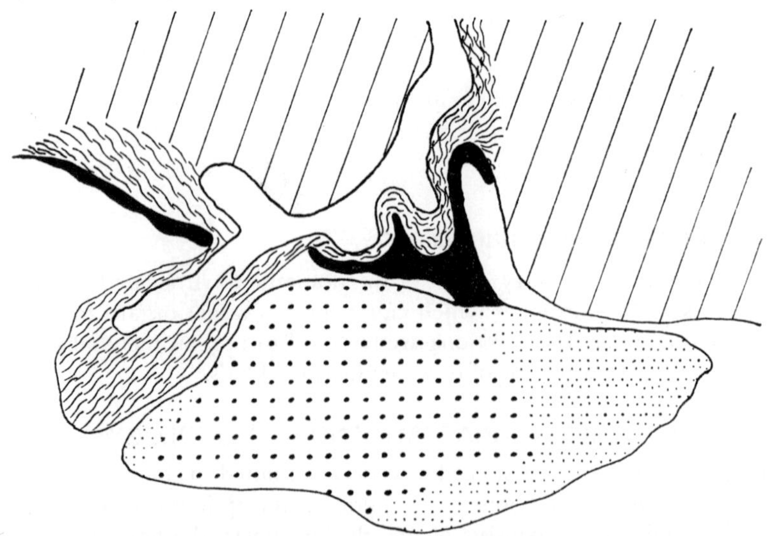

Fig. 1. Sagittal section of the pituitary gland. Reproduced with permission from Rahn and Painter (1941). Key: Large dots, caudal lobe of pars distalis; small dots, cephalic lobe of pars distalis; solid black, pars tuberalis; irregular lines, infundibular process; straight lines, brain tissue.

Table 1

Nomenclature and subdivisions of the pituitary gland in the domestic fowl

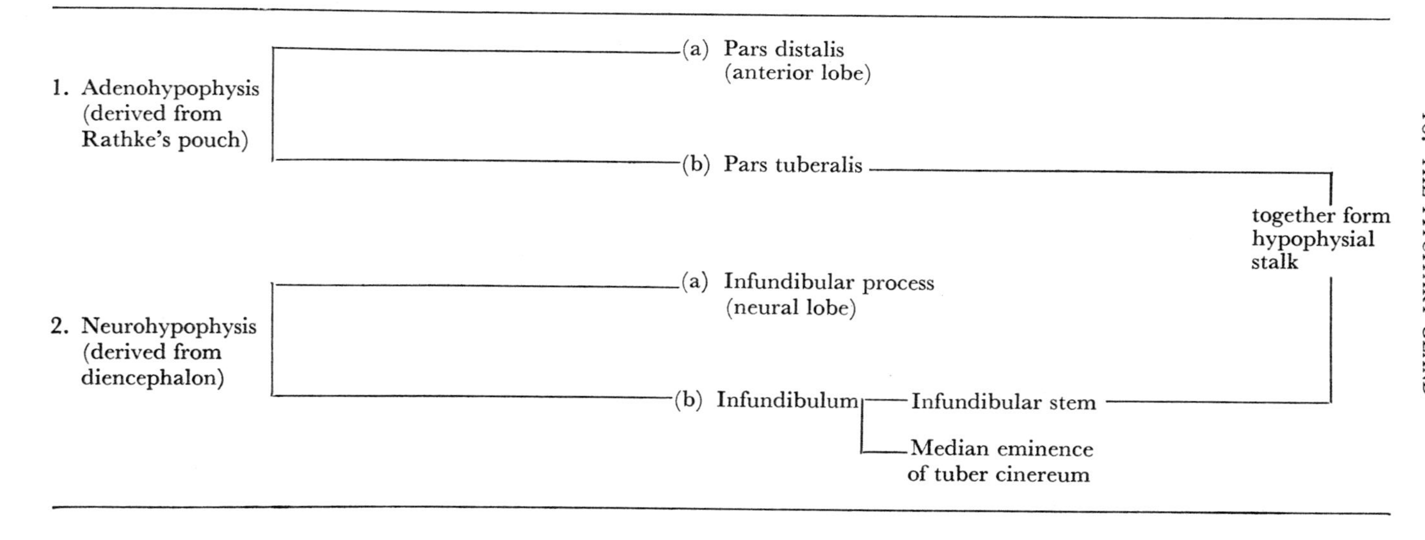

Division	Subdivision	Part	
1. Adenohypophysis (derived from Rathke's pouch)	(a) Pars distalis (anterior lobe)		
	(b) Pars tuberalis		together form hypophysial stalk
2. Neurohypophysis (derived from diencephalon)	(a) Infundibular process (neural lobe)		
	(b) Infundibulum	Infundibular stem	together form hypophysial stalk
		Median eminence of tuber cinereum	

The pars tuberalis consists of a strip of cells which covers parts of the ventral surface of the diencephalon and usually extends to the caudal side of the infundibular stem (see Fig. 1). Thus a collar of cells is formed around the infundibulum. The pars tuberalis is connected with the pars distalis by a string of cells which arise in the mid-region of the median eminence and pass to the pars distalis. This division of the pars tuberalis is associated with the portal blood vessels and runs independently of the infundibular stem to the pars distalis and not along it as in many animals.

The adenohypophysis and the neurohypophysis are separated by a connective tissue sheath. The neurohypophysis consists of a distinct neural lobe and the other two parts, the median eminence and the infundibular stem, are not directly separable. According to the terminology of Wingstrand (1951) the term posterior lobe is not acceptable in the domestic fowl since there is no pars intermedia.

The appearance of the neurohypophysis varies considerably from one avian species to another. In the domestic fowl there is a distinct neural lobe which is broadest near its base and which is clearly demarcated from the infundibular stem. The electron microscope studies of Duncan (1956) on the neural lobe indicate that the tissue consists of endings of nerve fibres from the neurosecretory supraoptico-hypophysial tract. These endings contain granules of diameter 1000 to 2000 Å and are surrounded by processes of glial pituicytes. The infundibular stem can be considered to be a tubular projection of the diencephalic wall which carried the neural lobe. In the fowl the stem is morphologically distinct from the median eminence which is the ventral or rostroventral wall of the diencephalon from the optic chiasma to the infundibular stem. The histological appearance of the median eminence and the infundibular stem are, however, closely related. The lumen of the neurohypophysis, referred to as the recessus infundibuli, extends throughout the infundibular stem and is continuous with the third ventricle. The recessus is wide and the entire organ appears as a thin-walled wrinkled sac.

The median eminence is intimately related to the dense primary capillary plexus of the hypophysial portal system. It develops as the horizontal floor of the third ventricle and extends from the optic chiasma rostrally to the infundibular stem caudally. It never reaches far up the side of the hypothalamus but is restricted to the most ventral parts of the wall.

In cross-section it is possible to distinguish three layers through the median eminence:

1. *The ependymal layer* consists of the cell bodies of the ependyma. There are a few scattered nerve fibres and sometimes a few nerve cells. The processes of the ependymal cells can be seen to run through the other two layers to the external surface.

2. *The fibre layer* consists of bundles of nerve fibres which run from rostral or rostro-lateral directions towards the infundibular stem. The bundles form a

distinct tract called the *tractus hypophyseus*. The bundles of fibres are separated from each other by the processes of the ependymal cells.

3. *The glandular layer* is situated in the superficial part of the wall and is characterized by the partial or complete absence of nerve fibres and cell nuclei. It is traversed by numerous fine processes arising from ependymal and glial cells.

The characteristic structure and innervation of the glandular layer indicates that it has a specialized function. The accumulation of a stainable colloid and the extremely rich blood supply support the idea that its function is glandular and that the secretions produced diffuse into the portal capillaries.

The structure of the infundibular stem is essentially the same as that of the median eminence. The ependymal lining of the ventricle is always present and the fibre layer is thicker in the stem. This is a reflection of the fact that the nerve fibres of the tractus hypophyseus are concentrated in this region. The glandular layer, however, is not well developed although it can be recognized.

B. INNERVATION OF THE HYPOPHYSIS

The nerve supply to the avian hypophysis was first studied by Drager (1945) who worked on the chick. He used pyridine-silver preparations and

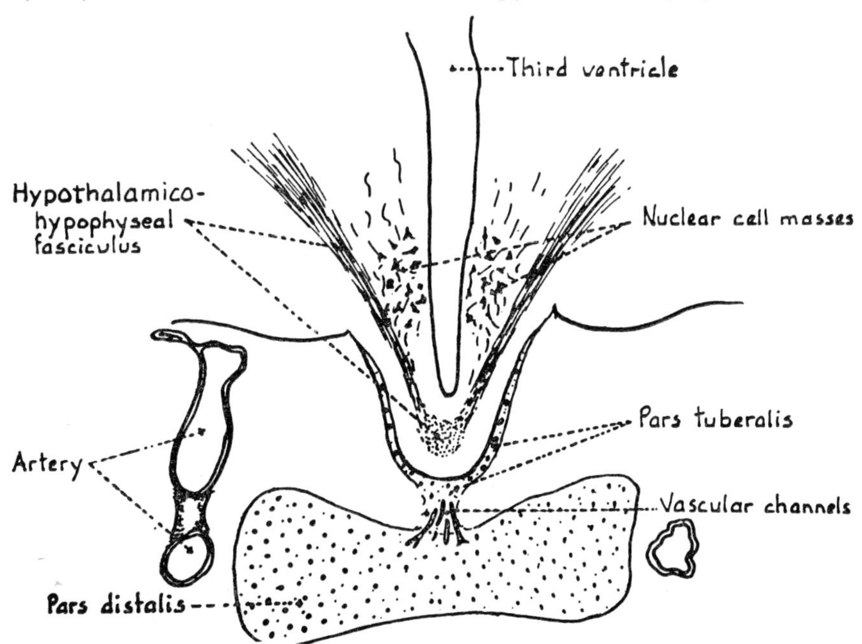

Fig. 2. A composite drawing constructed from frontal sections of the hypophysis. Reproduced with permission from Drager (1945).

revealed a conspicuous bundle of descending nerve fibres in the lower part of each lateral wall of the third ventricle (Fig. 2). The fibres from both sides converged in the floor of the third ventricle and passed posteriorly in the hypophysial stalk. Some of these fibres were derived from the large cells situated in the wall of the hypothalamic portion of the third ventricle. Another group of fibres could be followed from an anterior region in the neighbourhood of the upper portion of the optic chiasma but the exact origin of these fibres could not be ascertained from the pyridine-silver preparations.

The hypothalamico-hypophysial fasisculus, after its formation in the floor of the third ventricle, passes by way of the stalk to terminate in the neural lobe. From the hypophysial fasisculus a few small nerve fibres pass into the tissue of the pars tuberalis. These fibres, however, could not be followed far into the glandular substance. Other nerve fibres, autonomic in origin, accompany vascular channels into this part of the pituitary gland. In general there are relatively few fibres of either origin in this division of the chicken hypophysis.

In the region where the pars tuberalis contacts the pars distalis there are a considerable number of thin-walled vascular channels which are continuous with the sinusoids of the anterior lobe. There are no nerve fibres, however, associated with or accompanying these channels into the pars distalis.

The arteries adjacent to the hypophysis are accompanied by prominent nerve bundles which often exhibit small ganglionic masses along their course. However, nerve fibres entering the glandular tissue of the anterior lobe from the capsular region were not observed.

Okamoto and Ihara (1960) reported that neurosecretory fibres penetrated the pars distalis of the domestic fowl but these workers used an aldehyde-fuchsin stain, which will also stain fine bundles of connective tissue fibres. Confirmation of this observation is, therefore, required. After exhaustive studies on many species of birds, Wingstrand (1951) could conclude only that those few nerve fibres which can be demonstrated in the adenohypophysis are autonomic fibres associated with blood vessels and that they are not involved in the control of secretory activity of the pars distalis. The available evidence indicates that there is no direct nervous control of the pars distalis by the hypothalamus. Of special significance in the function of the median eminence are its vascular connections with the adenohypophysis.

C. HYPOTHALAMIC NUCLEI

Huber and Crosby (1929) recognized distinctive clusters of large ganglionic cells in the anterior hypothalamus. These observations were confirmed by Kurotsu (1935); Kuhlenbeck (1936) identified these groups of cells as the paraventricular and supraoptic nuclei. Following the introduction of the chromalum-haematoxylin staining technique (Bargmann *et al.*, 1950), neurosecretion was demonstrated in these cells. Wingstrand (1951) demonstrated neurosecretory material in the paraventricular and supraoptic nuclei of the pigeon and this system has been studied in the domestic fowl by Mikami (1960) and by Graber and Nalbandov (1965).

Two important efferent hypothalamic nerve tracts have been observed in birds (Wingstrand, 1951), the supraopticohypophysial tract and the tuberohypophysial tract. The former consists of neurosecretory axons from the supraoptic and paraventricular nuclei. The fibres of this tract pass partly to the neural lobe and partly to the median eminence. The tuberohypophysial tract is formed by fibres from the infundibular nucleus (Wingstrand, 1951) and possibly by fibres from the ventromedial nucleus.

D. HYPOPHYSIAL PORTAL BLOOD VESSELS

A detailed study of the blood vessels in the chicken pituitary gland (Fig. 3) was carried out by Green (1951). The superior hypophysial arteries supply the primary capillary plexus on the median eminence. The portal vessels collect the effluent blood from this plexus and pass downwards to the pars distalis penetrating a thick septum of dura mater on the way. In the pars distalis the vessels spread in all directions and drain into the venous sinuses surrounding the gland. The neural lobe receives a completely independent blood supply from a superficial plexus of blood vessels which is distinct from

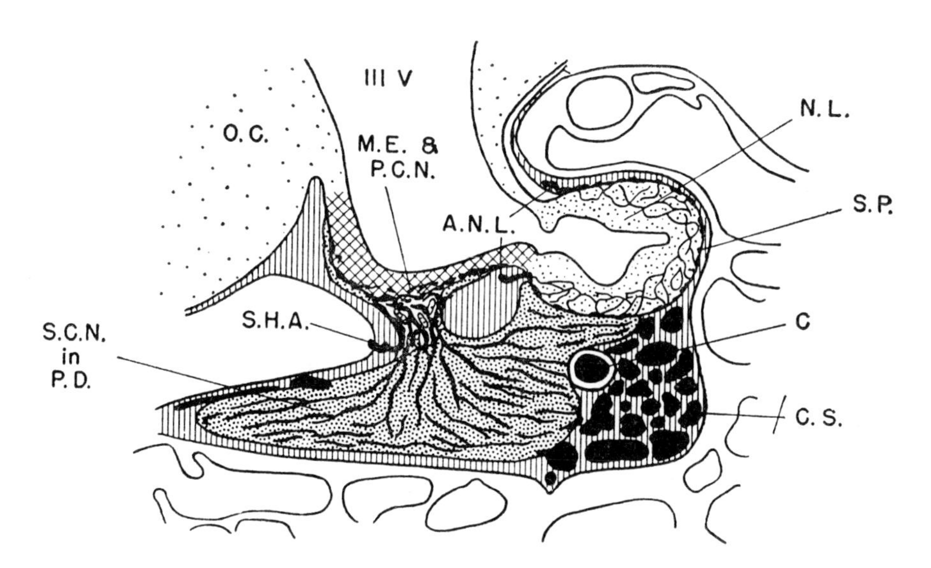

Fig. 3. The vascular relationships of the hypophysis. Reproduced with permission from Green (1951). Key: N.L., neural lobe; M.E., median eminence; P.C.N., primary capillary net; S.H.A., superior hypophysial artery; S.C.N., secondary capillary net; P.D., pars distalis; O.C., optic chiasma; III.V., third ventricle; A.N.L., artery of neural lobe; C.S., cavernous sinus; S.P., superficial plexus.

the primary capillary plexus. These vessels enter the gland at regular intervals on its surface.

There is a good deal of evidence to suggest that the portal vessels in mammals are concerned with controlling the secretions of the pars distalis (Harris, 1955). The concept involves (i) the production of substances in the median eminence, (ii) the release of this material through nervous impulses acting on the median eminence, (iii) the transport of the substances by the portal vessels to the pars distalis, (iv) the action of these substances on cells in the pars distalis to cause liberation of hormones. The studies of Assenmacher (1958) strongly support the idea that the portal vessels perform a similar role in birds.

E. CYTOLOGY OF THE ADENOHYPOPHYSIS

The secretory cells of the pars distalis of the vertebrate pituitary gland have been classified into three types, acidophils, basophils and chromophobes, according to differences in morphology and staining reactions. Rahn (1939), using differential staining techniques, first demonstrated two different types of acidophil in the pars distalis of the fowl. With azan stain the caudal lobe acidophils were stained intensely, while the acidophils, which were lightly stained with orange G, were found exclusively in the cephalic lobe. Since then there have been many investigations of the cytology of the pars distalis in the domestic fowl and other cell types have been distinguished (Payne, 1942, 1943; Matsuo, 1954; Legait and Legait, 1955; Mikami, 1958).

The different systems of nomenclature adopted for the cells of the pars distalis and the different staining techniques used by these workers have led to confusion and difficulty in comparing results. Recently Herlant (1964) and Purves (1966) reviewed the literature relating to the cytology of the adenohypophysis, in mainly mammalian species, and attempted to rationalize both staining procedures and nomenclature. It would be advantageous to have a similar system for the fowl. S. O. Amin and A. B. Gilbert (personal communication) point out that a modern appraisal of pituitary cytology in this species is essential since a significant amount of the early work was carried out in broody hens, a condition which is not normally found in modern breeds. They have studied laying hens and confirmed the observations of earlier workers that the adenohypophysis is divisible into two distinct lobes, namely the caudal and cephalic zones. The cells in both zones are arranged in follicles and each follicle has a central lumen which is often filled with stainable "colloid". Individual follicles are surrounded by a capsule of connective tissue within which are found the many blood vessels which make up the extensive blood supply of the gland. Six cell types were distinguished and these could be broadly separated into ones characteristically more basic in their staining affinities and others more acidic. In any given follicle, cells of both kinds were usually found. The basophils stained prominently with the so-called basic dyes and were generally large, ellipsoidal or spherical, with a large, often eccentric, nucleus. These cells could be further separated into two classes designated type I and type II on the basis of their staining reactions with PAS, aldehyde fuchsin and alcian blue. Both types of basophil were present throughout both zones in considerable numbers, although type I was more prominent.

Type I cells were PAS-positive and had intensely staining PAS-positive granules. In contrast to type II cells they remained pink with alcian blue stain and stained dark green with aldehyde-fuchsin-light green. Type II were the largest cells found in the pituitary gland of the hen and there appeared to be two distinct forms, each of which was considered to be related to its functional state. In normal laying hens, the aldehyde-fuchsin-positive cells were found in the peripheral part of the gland only, but cells within the central part, which were clearly similar, remained negative. These cells were almost chromophobic but they could be distinguished from chromophobes by their size, distribution and vacuolated cytoplasm.

Cells were observed which stained with most of the basic dyes but at the same time showed affinity for the acidic dyes. These were referred to as type III cells and their characteristic colour was a mixture of the two stains. They were small, slightly elongated or cuboidal, with a large nucleus and little cytoplasm, and were found exclusively in the central parts of the cephalic zone.

Two clear types of acidophil cells types IV and V were observed. Type IV cells, which were few in number, were extremely obvious as a result of their large granules and distinctly acidic cytoplasm. The granules were intensely PAS positive and stained with all the acidic dyes used in the investigation. They were medium-sized ellipsoidal cells and their distribution was highly characteristic. They were found between the follicles throughout both zones of the anterior pituitary. Other cells were also observed between the follicles; they stained similarly to type IV cells but did not contain granules and were not easily seen.

Type V cells were large spherical or ellipsoidal cells with a prominent large, centrally placed, nucleus and contained many granules. They were restricted almost entirely to the caudal zone where they formed the major part of the cell population. Follicles were frequently seen in which these cells were nearly the only type present.

The chromophobic or type VI cells were small and spherical with a centrally located nucleus. They were found in clusters throughout the gland and were not clearly organized into follicles.

Ideally, each cell type should be related to its function. Herlant (1964) and Purves (1966) discussed the histochemistry and histology of the pituitary in relation to cell function. In the case of mammals the relationship between the staining properties of the cells and their function is reasonably well understood. The results of S. O. Amin and A. B. Gilbert (personal communication) suggest that cells with staining properties similar to mammalian cells perform similar functions.

They concluded that type I cells, which were active during the laying cycle of the hen, are typical of those secreting follicle-stimulating hormone (FSH) as described for many other species. Type III cells appear to be identical to the luteinizing hormone (LH) secreting cells. Both types are, therefore, gonadotrophs. This LH secreting cell has not previously been identified in the fowl, although it has been in the duck (Herlant *et al.*, 1960; Tixier-Vidal, 1963) and the quail (Tixier-Vidal *et al.*, 1968).

The type II cells which gave a positive staining reaction with aldehyde-fuchsin and alcian blue, were considered to be thyrotrophs, since the same staining reaction is given by the cells which secrete thyroid-stimulating hormone (TSH) in other species.

The identification of two types of acidophil agrees well with studies on other avian and mammalian species. The cells were distinguished by their staining reactions and different distributions. Type IV cells stained bright yellow with orange G and distinct granules were observed after treatment with erythrosine and acid fuchsin. It was proposed that these cells are identical to the luteotrophs (prolactin secreting cells) of mammals which are active only during pregnancy. They were not readily visible in the laying hen, which might be expected if they secrete prolactin since this hormone, in the hen, is associated with a cessation of laying. In non-laying hens, however, type IV cells were extensively distributed in great numbers throughout the pituitary.

Type V acidophils tended to be restricted to the caudal zone and were markedly organophilic. Their appearance and general staining reactions are similar to the somatotrophs (growth hormone secreting cells) of mammals.

It was thought that the small chromophobes present throughout the gland were non-secretory but that they might be undifferentiated cells which can give rise to further active cells when required.

Amin and Gilbert have been unable positively to identify any cell type which could have been regarded as secreting adrenocorticotrophic hormone (ACTH). Tixier-Vidal *et al.* (1968) suggest that such cells are rare and can be seen only after treatment with metapirone, a drug which interferes with the synthesis of adrenocortical steroids, thereby causing an increased production of ACTH.

In a further series of experiments Amin and Gilbert (1970) considered that, during growth, the functional activity of the pituitary would vary in relation to the physiological state of the birds, and that it should be possible to relate the functional activity of the various cell types to the effect that their hormones are producing. From the age of ten weeks onwards it was found that all of the cell types in the pars distalis varied in number and there was an associated change in functional activity. The cells most affected were the gonadotrophs (secreting both FSH and LH) and the luteotrophs. The changes in these cells were clearly related to the onset of sexual maturity. The FSH cells were the first to increase in number and activity and this probably correlated with a general increase in steroid secretion by the ovary as measured by the increase in weight of the oviduct. LH cells became obvious at the same time as ovarian follicles matured prior to ovulation. As age increased, the acidophilic somatrophs decreased and, at sexual maturity, the thyrotrophs became more obvious. This latter observation is compatible with an increase in metabolic rate which occurs at this age.

There are few studies of the ultrastructure of the anterior pituitary gland of the fowl. Payne (1965) and Gilbert and Amin (1969) reported the results of preliminary studies. The latter authors (1969 and personal communication) recognize six distinctive cell types and observe a remarkable similarity to cell types present in the mammalian pituitary (Barnes, 1962; Herlant, 1964; Green, 1966). On the basis of a comparative study with the results of their observations with the light microscope they have been able to assign a probable function to each type of cell. It is encouraging that a good correlation was achieved between the two sets of observations and it should be noted that no cells secreting either ACTH or MSH (melanocyte-stimulating hormone) were identified. The lack of ACTH secreting cells is compatible with the proposal of Nalbandov (1966) that there is another source of ACTH in the fowl (see Section II D).

II. Secretions of the Pars Distalis

A. GONADOTROPHINS (FSH AND LH)

The fowl, like the mammal, is dependent on pituitary hormones for normal gonadal development and function. (See also Chapters 50, 52, 57, 61.) Hypophysectomy prevents the onset of sexual maturity in immature birds and in adult birds causes atrophy of the gonads, regression of the accessory reproductive organs and characteristic changes in plumage (Hill and Parkes, 1934). Similar effects are observed when the connexions between the anterior lobe and the median eminence are severed (Shirley and Nalbandov, 1956a) or when the adenohypophysis is transplanted to the kidney capsule (Ma and Nalbandov, 1963). These effects in hypophysectomized fowls can be reversed by daily injection of extracts of chicken anterior pituitary powder. If gonadotrophin preparations of mammalian origin are used only partial restoration of gonadal function is obtained (Nalbandov *et al.*, 1951; Opel and Nalbandov, 1961; Mitchell, 1967a).

Starvation in the laying hen causes atresia of the ovarian follicles which can be prevented by injection of chicken pituitary powder or purified mammalian LH and FSH (Morris and Nalbandov, 1961). Calcium deficiency also results in ovarian atresia which can be prevented by injection of unfractionated avian pituitary powder (Taylor *et al.*, 1962). Precocious gonadal development in immature chickens is stimulated by daily implantation of a chicken hypophysis (Domm, 1931) or by daily injection of desiccated avian pituitaries but not by prolonged injection of mammalian gonadotrophins (Das and Nalbandov, 1955; Taber *et al.*, 1958). In the newly-hatched chick, however, the testes are sensitive to stimulation by both mammalian and avian gonadotrophins (Breneman *et al.*, 1962) injected over a period of 3 d. Moreover, mammalian FSH, followed by LH, stimulates follicle production and ovulation in chickens one to three weeks earlier than egg laying would normally commence (Nalbandov and Card, 1946). Because of the differences in response to avian pituitary material as compared with mammalian gonadotrophins, Nalbandov and his co-workers have suggested that avian pituitary LH is qualitatively different from mammalian LH or that the avian pituitary secretes two types of LH with two different functions (Nalbandov *et al.*, 1951; Nalbandov, 1966). However, Knobil and Sandler (1963) have suggested that the differences in response may be explained by other factors such as immunological inactivation, the ratio of FSH to LH administered, or the necessity of a full complement of pituitary hormones rather than purified fractions. Further experimentation with purified chicken pituitary fractions will be required to determine the reasons for the observed response differences, but it seems likely that prolonged injection of mammalian gonadotrophins would induce formation of antibodies in the fowl which would inhibit the activity of the mammalian hormones.

1. Methods for Bioassay

The methods employed for assay of chicken gonadotrophins have generally been the same ones used to measure the corresponding mammalian hormones. Certain assays measure total gonadotrophic activity while others are considered specific for determination of FSH or LH. Injection of gonadotrophin into newly-hatched chicks results in a significant increase in testis weight (Byerly and Burrows, 1938) which is frequently used as an assay for gonadotrophic activity (Breneman *et al.*, 1959). A more sensitive response parameter is the uptake of radioactive phosphorus by the chick testis (Breneman *et al.*, 1962). Purified preparations of mammalian LH and FSH are active in both assays, and there is little or no synergism when LH and FSH are administered together (Breneman *et al.*, 1962). FSH activity can be determined specifically by measuring the increase of ovarian weight in the immature rat treated simultaneously with the test substance and human chorionic gonadotrophin (Steelman and Pohley, 1953). A similar assay using mice has been developed by Brown (1955). Measurement of ovarian ascorbic acid depletion (OAAD) in pseudopregnant rats (Parlow, 1961) is used as an assay for LH. Frankel *et al.* (1965) reported that a factor containing OAAD activity was present in the plasma of adenohypophysectomized cockerels and Jackson and Nalbandov (1969a) reported the presence of a heat stable component with OAAD activity in the anterior pituitary gland of the cockerel. In both cases evidence was presented that the factor was not LH but that it might be arginine vasotocin which was shown to have OAAD activity (Frankel *et al.*, 1965). Therefore the OAAD activity of unfractionated chicken pituitaries or plasma may not necessarily be a measure of LH content. Burns (1970a) has developed an assay which he considers specific for LH activity in which the uptake of ^{32}P by the testes of newly-hatched chicks is determined after injection of chicken pituitary material treated with neuraminidase. This treatment results in destruction of FSH activity in chicken pituitary preparations as measured by the essay of Steelman and Pohley (1953) and the residual activity is assumed to be a measure of the LH content of the material, since other anterior lobe hormones were inactive in the system.

2. Changes in Gonadotrophin Content with Physiological State

Fluctuations in OAAD activity in the plasma and pituitary glands of laying hens have been related to the time of ovulation (Nelson *et al.*, 1965a; Tanaka *et al.*, 1966; Tanaka and Yoshioka, 1967; Bullock and Nalbandov, 1967; Heald *et al.*, 1967; Heald *et al.*, 1968a) but the lack of specificity of this assay method when applied to chicken pituitary glands makes interpretation of such data extremely difficult.

The OAAD activity in the pituitary and plasma of cockerels increases with age (Nelson *et al.*, 1965b) as does the pituitary weight and its gonadotrophin content when assayed by the testis weight response (Breneman, 1955). Herrick *et al.* (1962) confirmed that pituitary gonadotrophin content per unit weight increased with age in males but not in females. Serum gonadotrophin was higher in young chickens and non-laying hens than in laying hens and breeding males (Bailey and Phillips, 1952). After castration there is hypertrophy of the pituitary in male chickens but the total gonadotrophin content of the gland remains constant (Herrick *et al.*, 1962). Injection of oestradiol or progesterone, but not testosterone resulted in increased pituitary OAAD activity in the mature domestic fowl (Heald *et al.*, 1968b). The pituitary OAAD activity in non-laying hens is higher than the levels normally found in laying hens (Tanaka *et al.*, 1966) as is the total gonadotrophin potency (Nakajo and Imai, 1961). These differences are primarily in the gonadotrophin of the cephalic lobe whereas the caudal lobe gonadotrophin is roughly the same for both groups (Nakajo and Imai, 1961).

3. Purification and Properties of FSH and LH

Purified preparations of FSH and LH from mammalian pituitary glands are glycoproteins with molecular weights in the range 25,000 to 30,000 (Butt, 1967). Chicken anterior pituitary powder fractionated by an ammonium sulphate precipitation method originally used for sheep pituitary glands (Fraps *et al.*, 1947) yielded a presumed LH fraction which induced premature ovulation in hens. Alternative methods effective in purification of mammalian FSH and LH were used by Stockell Hartree and Cunningham (1969) and the resulting fractions assayed for FSH and LH activities by the methods of Brown (1955) and Parlow (1961). Extraction of acetone-dried chicken pituitary powder with 6% ammonium acetate, pH 5·1, in 40% ethanol resulted in a 20-fold concentration of the FSH and LH activities and their partial separation was achieved by chromatography on CM-cellulose at pH 5·5 in 4 mM ammonium acetate. The FSH fraction was further purified by chromatography on DEAE-cellulose and hydroxylapatite to a potency 100-fold greater than in crude pituitary powder. After chromatography of the LH fraction on DEAE-cellulose and Amberlite IRC-50, LH activity was concentrated 300-fold compared with the original pituitary powder. However, neither of these purified preparations was homogeneous on acrylamide gel electrophoresis or gel-filtration (A. Stockell Hartree, unpublished). The biological potencies of chicken FSH and LH were very much lower than the corresponding mammalian hormones (Stockell Hartree and Cunningham, 1969) and it is likely that the assay methods using mice and rats as test animals are less sensitive to the chicken hormones than to mammalian gonadotrophins. The ^{32}P uptake by chick testis (Breneman *et al.*, 1962) is a more sensitive assay for chicken gonadotrophin, but purified chicken FSH and LH preparations are less active than the crude gonadotrophin extract (Furr and Cunningham, 1970). It is likely, therefore, that both hormones are required for a maximum response in this assay. Burns (1970b) observed that chicken pituitary powder after inactivation of FSH with neuraminidase stimulated the uptake of ^{32}P by the ovaries of 1- to 3-day-old pullets. This effect, presumably due to LH, was obtained with only 1 μg of chicken pituitary powder, but bovine LH was very much less active. The maximum response of the ovary occurred 30 min. following injection of the hormone, but the ribonucleic acid fraction of the ovary showed a high response in only 15 min.

Tests of crude gonadotrophin extract in the hypophysectomized lizard (*Anolis carolinensis*) suggest that in this species the chicken pituitary fraction is more active in gonadotrophic activity than is the corresponding sheep fraction. Injection of partially purified FSH from chicken, sheep or human pituitaries into intact female lizards results in development of ovarian follicles, ovulation and enlargement of the oviduct. A comparison of the dosages required for this response also suggests that *Anolis* is relatively more sensitive to chicken FSH than is the rodent (Licht and Pearson, 1969; Licht and Stockell Hartree, 1971).

4. *Immunological Properties of Chicken Gonadotrophins*

By immunological techniques, only relatively weak cross-reactions between chicken and mammalian gonadotrophins have been demonstrated. Parkes and Rowlands (1937) produced antisera to ox and horse pituitary extracts and to human chorionic gonadotrophin and pregnant mare's serum gonadotrophin. These antisera failed to inhibit the activity of endogenous gonadotrophins when injected into laying hens although they were shown to produce a passive immunity against the endogenous hormones in the rabbit.

Moudgal and Li (1961) showed that there was no cross-reaction in gel diffusion tests between chicken pituitary powder and an antiserum to sheep LH. Desjardins and Hafs (1964) confirmed this observation and, although they found no cross-reaction between a chicken pituitary extract and an antiserum to equine LH in a gel diffusion test, a cross-reaction was observed by the more sensitive complement fixation technique. Bullock *et al.* (1967) raised an antiserum to chicken pituitary powder, and this reacted with both chicken pituitary powder and with chicken plasma in gel diffusion studies. None of the mammalian gonadotrophin preparations tested cross-reacted in this system. Furthermore, antiserum to sheep LH did not cross-react with the chicken material although it did cross-react with some mammalian pituitary gonadotrophins. By the technique of haemagglutination-inhibition, a weak immunological cross-reaction was demonstrated between chicken LH fractions and antiserum to human pituitary LH (Stockell Hartree and Cunningham, 1969) or antiserum to human chorionic gonadotrophin (A. Stockell Hartree, unpublished). In addition cross-reactions between these human hormones and chicken LH have been demonstrated by complement fixation (F. J. Cunningham, unpublished), and a cross-reaction between chicken plasma and human chorionic gonadotrophin was detected using a radioimmunological method (Bagshawe *et al.*, 1968).

Cunningham *et al.* (1970) have described preliminary attempts to measure gonadotrophin in fowl's plasma by a radioimmunoassay system. They used an antiserum raised in rabbits to a purified preparation of chicken LH (Stockell Hartree and Cunningham, 1969) and the same antigen labelled with ^{125}I. In the solid phase radioimmunoassay system of McNeilly (1970) a linear inhibition curve was obtained, using the unlabelled LH fraction as standard. When the method was applied to the plasma of a laying hen, a line of slope similar to the standard was obtained, using a starting concentration of 0·2 ml of plasma and doubling dilutions from this. The immunological activity of a chicken pituitary FSH preparation, in this system, was low and no activity was detected in plasma obtained from hypophysectomized birds. Unpublished observations indicate that the labelled preparation of LH contains TSH activity and before the method is suitable for specifically measuring LH levels in plasma it will be necessary to overcome this problem.

5. *The Control of Gonadotrophin Secretion from the Pars Distalis*

The functional activity of the pars distalis, at least, as far as gonadotrophin secretion is concerned depends upon the integrity of its vascular connections with the median eminence.

Shirley and Nalbandov (1956a) found that if the hypophysial portal vessels are sectioned and regeneration prevented, there is a complete and permanent regression of the gonads. In addition Ma and Nalbandov (1963) observed a complete and permanent regression of testicular function if the pars distalis is transplanted to the kidney capsule.

The results of a variety of experiments suggest that the release of ovulation-inducing hormone in the hen is controlled by a neural mechanism. Fraps (1953) showed that ovulation is induced in a significant proportion of hens following treatment with several barbiturates including pentobarbital sodium (Nembutal). The same barbiturates enhanced the ovulation-inducing activity of progesterone administered at low dosages. Phenobarbital sodium, on the other hand, fails to induce ovulation and suppresses the effect of progesterone when this steroid is administered at a normally effective dosage. Spontaneous and progesterone-induced ovulation can also be inhibited by adrenergic blocking drugs such as dibenamine and the more potent dibenzyline and by the cholinergic blocker atropine (Zarrow and Bastian, 1953; van Tienhoven *et al.*, 1954). Ferrando and Nalbandov (1969), however, have shown that the large dosages of adrenergic blocking drugs which have to be used to completely blockade ovulation in birds can exert this effect locally at the level of the ovary. It is possible, therefore, that the blocking effect of these drugs might not be mediated through a central mechanism since Gilbert (1965) observed that follicles in the ovaries of chickens are well innervated with both adrenergic and cholinergic neurones.

The induction of ovulation with progesterone involves the activation of a neural mechanism since lesions of the anterior hypothalamus made within $2\frac{1}{2}$ h after the subcutaneous injection of progesterone prevent ovulation (Ralph and Fraps, 1959a). Furthermore, the intrahypothalamic but not the intrapituitary injection of progesterone does result in ovulation (Ralph and Fraps, 1960).

Huston and Nalbandov (1953) claimed that the presence of an irritant in the oviducal magnum suppresses ovulation for as long as 3 weeks. It was interesting that in these anovulatory birds neither the combs, ovaries nor oviducts regressed. The ovarian follicles were not atretic and could be ovulated by treatment with progesterone or mammalian luteinizing hormone. Since transection of the hypophysial portal vessels leads rapidly to ovarian regression, follicular atresia and a decreased size of the comb, it appears that basal amounts of gonadotrophin are released throughout the anovulatory period. The maintenance of the comb and oviduct implies continued secretion of androgen and oestrogen from the ovary, functions dependent upon a supply of gonadotrophin. Huston and Nalbandov (1953) and Nalbandov (1959) considered that these results were best explained by the assumption that there was a neurogenic inhibition, arising from the magnum, which prevented the periodic release in sufficient quantities of the necessary ovulation-inducing hormone.

Further work, however, on the effect on ovulation of foreign bodies in the oviduct has not borne out these conclusions (Sykes, 1962; Lake and Gilbert, 1964; Bullock and Nalbandov, 1967). Their findings and the report of Opel (1965a) that ovulation continues normally even in the absence of the oviduct, has led Bullock and Nalbandov (1967) to disclaim an obligatory role for the oviduct in the mechanism controlling the release of ovulation-inducing hormone.

The hypothalamic control of anterior pituitary function has been frequently investigated by means of electrolytic stimulation or lesioning experiments. Ralph (1959) placed electrolytic lesions in a ventral part of the preoptic hypothalamus which caused an immediate cessation and prolonged interruption of ovulation in adult laying hens. This neural region might function in maintenance of follicles, since some of the hens which were effectively lesioned and did not resume ovulation during the 1-month experimental period were found, at the time of killing, to have atrophic ovaries and oviducts.

A series of observations by Ralph and Fraps (1959b) has implicated other diencephalic sites in the mechanisms controlling the secretion of gonadotrophin by the pars distalis. Lesions scattered throughout an essentially dorsocaudal thalamic region are effective in preventing ovulation and induce regression of the ovaries and oviducts. Also effective are lesions throughout an area of the hypothalamus which includes most of the preoptic paraventricular nuclei, part of the magnocellular paraventricular and lateral hypothalamic nuclei, the anterior and supraoptic hypophysial tracts, the tuberal and mammillary regions and the median eminence. Despite the fact that reproductive dysfunction could not be correlated consistently with specific loci, the failure to detect regeneration of the ovaries or oviducts when the birds were killed following lesioning suggests that some specific neural mechanism was permanently interrupted.

Ovulation was delayed following electrical stimulation of the anterior median eminence but this was believed to be due to electrolytic deposition of iron rather than to the effects of the stimulation (Opel and Fraps, 1961). When deposition of iron was prevented, electrical stimulation of the preoptic area as well as sham operation still delayed ovulation (Opel, 1963). It might be expected, in view of the results of experiments involving lesions of the hypothalamus, that electrical stimulation of selected areas would induce ovulation. The significance of the lack of a positive effect is unclear.

A further approach in the study of the hypothalamic mechanism controlling the secretory activity of the pars distalis is to investigate the capacity of extracts of hypothalamic tissue to cause the release of pituitary hormones. Recently Opel and Lepore (1967) have found that direct intrapituitary infusion of extracts of fowl hypothalamus produced premature ovulation in laying hens. The infusion of arginine vasotocin or oxytocin, alone or in combination, was ineffective. This would suggest that there is present in the hypothalamus of the fowl a gonadotrophin releasing factor. Jackson and Nalbandov (1969b) showed acid extracts of chicken hypothalamic tissue to contain a substance which stimulates the *in vitro* release of luteinizing hormone from rat anterior pituitary glands maintained in an incubation medium. Arginine vasotocin in concentrations equal to that in the extracts had no effect on LH release. This neurohypophysial principle was suspected of having gonadotrophin releasing activity, since a substance resembling it has been detected in extracts of chicken hypothalamus (Jackson and Nalbandov, 1969c).

These findings support the conclusion that gonadotrophin secretion in the domestic fowl is controlled by a neurohumoral mechanism. Definitive proof of this concept depends on the purification of specific gonadotrophin releasing factors and their demonstration in hypophysial portal blood.

B. PROLACTIN

The protein hormone, prolactin, secreted by the anterior pituitary gland of many species has been studied and purified most extensively from beef and sheep sources. The functioning of the crop-sac gland in pigeons and of milk secretion in mammals were shown by Riddle and co-workers in the 1930's to be controlled by prolactin (Riddle, 1963).

Injection of prolactin into fowls results in broody (nesting) behaviour in cocks, and in both laying and non-laying hens. In hens it leads to a decrease in both ovarian tissue and oestrogen production and usually results in cessation of egg-laying (Riddle, 1963). Similar findings had been reported by Nalbandov (1945) who showed that males injected with prolactin would care for normal chicks rather than kill them or allow them to die of neglect. The treated cocks also exhibited a decrease in size of both testis and comb, cessation of crowing and lack of sex drive. These effects were nullified if FSH or androgen were injected with prolactin.

The most generally used assay methods for prolactin are based on stimulation of the pigeon crop-sac gland. The assay is discussed in detail by Lyons and Dixon (1966).

Burrows and Byerly (1936) showed that pituitary glands of broody hens had higher crop-sac stimulating activity than pituitaries from laying hens or males. These findings were confirmed by Saeki and Tanabe (1955) who reported that pituitaries of broody hens contained approximately 0·20 i.u. per gland as compared with 0·07 i.u. per gland in laying hens. Nakajo and Tanaka (1956) concluded that the higher prolactin content in broody hens was due chiefly to an increase in the prolactin of the caudal lobe, the cephalic lobe content being nearly equal to that of laying hens.

As yet there have been no reports of purification of prolactin from pituitaries of domestic fowl. Highly purified sheep and beef prolactin have each been shown to be proteins with molecular weights of approximately 23,000 (Lyons and Dixon, 1966), and the complete amino-acid sequence of sheep prolactin has recently been reported (Li *et al.*, 1969b).

It is now widely accepted that in mammals the secretion of prolactin is chronically inhibited by the hypothalamus. Evidence for this view is based on experiments *in vivo* utilizing hypothalamic lesions, section of the pituitary stalk and transplantation of the pituitary gland, as well as upon work *in vitro* with pituitary tissue cultures or short-term incubations (Schally *et al.*, 1968). There is no evidence that prolactin secretion is increased in the chicken following transplantation of the pituitary (Ma and Nalbandov, 1963). The observation of Meites and Nicoll (1966) that chicken hypothalamic extracts increase prolactin release from pigeon pituitary glands *in vitro* suggests that there is hypothalamic stimulation of prolactin secretion in the domestic fowl.

C. THYROTROPHIN (TSH)

The thyroid gland of the fowl, as with other birds and mammals, is under hypophysial control. Hypophysectomy results in decreases of both thyroid weight (King, 1969) and secretory epithelium (Nalbandov and Card, 1943; Mitchell, 1967b). (See also Chapter 17.)

A standard method for the bioassay of TSH depends on the depletion of ^{131}I from the thyroid of fasting 1-day-old chicks pre-treated with thyroxine and propylthiouracil (Bates and Condliffe, 1960). Injection of mammalian TSH into the chick also causes an increase in the water content of the thyroid gland (Solomon, 1961). Very little information is available, however, about the TSH content of chicken pituitary glands.

Purified mammalian TSH has been shown to be a glycoprotein of molecular weight approximately 25,000 (Shome *et al.*, 1968). Although purification of chicken TSH to this degree has not been reported, a gonadotrophin extract representing 3% by weight of chicken pituitary powder also contained most of the thyroid stimulating activity of this powder (Mitchell, 1967b).

The factors which control TSH secretion are as yet incompletely understood. Although transection of the hypophysial portal vessels resulted in atrophy of the gonads, the weight and histological appearance of the thyroids was not altered (Shirley and Nalbandov, 1956a). Following transplantation of the pituitary, the thyroid soon continues to function normally and has a healthy histological appearance (Ma and Nalbandov, 1963). However, Egge and Chiasson (1963) have reported that lesions in the preoptic division of the nucleus supraopticus resulted in a change in thyroid histology indicative of some reduction in activity of the gland. Kanematsu and Mikami (1970) have recently obtained evidence that secretion of TSH in the chicken is regulated by a region of the anterior hypothalamus extending from the area ventrocaudal to the anterior commissure to the area dorsocaudal to the optic chiasma. Destruction of this area resulted in a decreased uptake of ^{131}I by the thyroid and its atrophy.

D. ADRENOCORTICOTROPHIN (ACTH)

Hypophysectomy of young chickens results in rapid degeneration of adrenal cortical and medullary tissues which after a time may regain their normal structures, to some extent (Nalbandov and Card, 1943). (See also

Chapter 20.) However, Shirley and Nalbandov (1956a) reported that after complete isolation of the adenohypophysis, the histology and weight of the adrenals in laying hens remained normal. Resko *et al.* (1964) measured corticosterone levels in adrenal venous plasma of chickens and observed an increase after injection of mammalian ACTH or crude chicken anterior pituitary powder and a decrease after hypophysectomy. Corticosterone continued to be produced for some time after hypophysectomy but was greatly decreased if the median eminence was also damaged, thus implicating that site as a possible additional source of ACTH.

A rapid and sensitive bioassay for ACTH depends on its effectiveness in depletion of adrenal ascorbic acid in the hypophysectomized rat (Sayers *et al.*, 1948). Salem *et al.* (1970a) have shown that ACTH in chicken pituitary extracts can be assayed in this system and that the concentration of ACTH is much higher in the chicken pituitary gland than in that of the rat. Siegel (1962) showed that the hormone could be assayed in 3-week-old chicks if either increase in adrenal weight, decrease in adrenal cholesterol or increase in serum cholesterol were the response measured. Both mammalian ACTH and acid extracts of chicken anterior pituitaries stimulate the secretion *in vitro* of corticosterone and aldosterone from chicken adrenal tissue (de Roos and de Roos, 1964). More detailed information on methods of measuring the effects of ACTH on the chicken adrenal has been given by Frankel (1970).

Although chicken ACTH has not been purified and characterized, the complete amino acid sequence of this hormone from several mammalian species has been determined. It consists of a single polypeptide chain containing 39 amino acid residues. Full activity is exhibited by the polypeptide containing the first 24 amino acids and this portion of the molecule is identical in all species studied (Evans *et al.*, 1966).

Injection of mammalian ACTH results in an increase in weight of the adrenal of cockerels (Zarrow *et al.*, 1962). In laying hens administration of ACTH produces an increase in both free fatty acids and glucose levels in plasma (Heald *et al.*, 1965).

The adrenals of the domestic fowl are unaffected by transection of the hypophysial portal vessels (Shirley and Nalbandov, 1956a) and Egge and Chiasson (1963) found no correlation between hypothalamic lesions in a variety of sites and the level of corticoids in the blood. Thus it appears that the adrenal is able to function independently of the connexions between the hypothalamus and the pars distalis. It has also been reported (Frankel *et al.*, 1967) that lesions in many hypothalamic areas do not affect adrenal function in intact or hypophysectomized cockerels. However, electrolytic lesions in the ventral tuberal area of the hypothalamus result in decreased corticosterone levels in adrenal vein plasma of intact cockerels whilst in adenohypophysectomized cockerels the levels increased following lesioning. These data suggest a complex hypothalamic-hypophysial control of the adrenals in cockerels and that the role of the ventral hypothalamus is modified following removal of the adenohypophysis. More recently Salem *et al.* (1970b) demonstrated the presence of both an ACTH-releasing factor and a factor with ACTH-like activity in the chicken hypothalamus. They suggest that the latter might be an α-MSH, a substance known to possess slight ACTH activity.

E. GROWTH HORMONE (GH)

Hypophysectomy of growing chickens at 8 to 9 weeks of age results in dwarfism although body weight continues to increase (Nalbandov and Card, 1942). When cockerels were hypophysectomized at 3 to 4 weeks both food intake and bone growth were decreased (King, 1969). These findings suggest that the chicken, like the mammal, secretes a pituitary hormone which stimulates growth.

The usual assay animal for measurement of growth-promoting activity is the hypophysectomized rat and the response measured can be either weight gain or increase in width of epiphysial cartilage (Russell, 1955). Attempts to measure growth-promoting activity of chicken pituitaries by these methods have not been very successful. Hazelwood and Hazelwood (1961) reported that chicken pituitary extract was approximately one eighth as active as rat pituitary extract in increasing tibial cartilage width of hypophysectomized rats. Low doses of the chicken extract were effective in the weight gain assay, but high doses were not. Therefore it appears that the hypophysectomized rat is not a suitable species for assay of avian growth hormone. Similar findings were described by Nalbandov (1966) who quoted the work of Hirsch (1961). In addition, large daily doses of beef or pig GH were not effective in increasing body weight or nitrogen retention in hypophysectomized chickens. Crude chicken pituitary powder, however, significantly increased growth-rate, nitrogen-retention and bone growth in both hypophysectomized and normal chicks (Hirsch, 1961).

The chemistry of mammalian growth hormone from several species has been studied extensively and the most recent estimates of molecular weight are in the range 20,000 to 26,000 (Andrews, 1966; Wilhelmi and Mills, 1969). For each species GH is a protein composed of a single polypeptide chain which in the human contains 188 amino acid residues (Li *et al.*, 1969a). Chemical similarities between chicken and mammalian growth hormones seem likely since there is an immunological cross-reaction between rat GH and chicken pituitary extracts (Hayashida, 1969).

A hypothalamic factor which stimulates secretion of GH by the adenohypophysis has been demonstrated in a number of mammals and also in the pigeon and the frog (Schally *et al.*, 1968). Although studies have not yet been performed on the domestic fowl, it is likely that release of pituitary GH in this species is mediated by a similar factor.

F. OTHER FACTORS

In addition to the anterior lobe hormones already described Kleinholz and Rahn (1940) have reported the presence of intermedin (or MSH) in the pars distalis of the chicken. This hormone, normally found in the intermediate lobe of species that possess this structure, causes darkening of the skin particularly in amphibia and fish. The cephalic portion of the chicken anterior lobe is 20 times more potent in this activity than the caudal portion and is also more active on a weight basis than the intermediate lobes of cattle (Kleinholz and Rahn, 1940).

A crude fraction prepared from chicken anterior lobes was shown to stimulate lipolysis in chicken adipose tissue *in vitro* (Langslow and Hales, 1969; Stockell Hartree *et al.*, 1969). It is not known whether this activity is present in one of the hormones already discussed or whether there is a specific lipolytic hormone in fowls.

A substance resembling the neurohypophysial hormone, arginine vasotocin, has been found in large amounts in the anterior pituitary of the cockerel. This factor is heat stable, readily dialysable, inactivated by thioglycollate and has potent vasodepressor activity (Jackson and Nalbandov, 1969a).

III. Neurohypophysial Hormones

The neurohypophysis of the fowl has been reported (Sawyer, 1966b) to contain the polypeptide hormone oxytocin, which occurs in mammals, birds, amphibia and reptiles; and the related hormone, vasotocin, which has been found in most non-mammalian vertebrates thus far studied. More recently Acher *et al.* (1970) reported isolation of mesotocin, a hormone with pharmacological properties similar to oxytocin, from chicken neural lobes as well as from lungfish, amphibia and reptiles. The chemical structures of these hormones are given in Table 2. A brief description of assay methods for oxytocic activity and vasotocin appears here, but a more detailed discussion can be found in Sawyer (1966a).

Table 2

Amino acid sequences of chicken neurohypophysial hormones

CyS-Tyr-Ile-Glu(NH_2)-Asp(NH_2)-CyS-Pro-Leu-Gly(NH_2)
1 2 3 4 5 6 7 8 9
(disulfide bridge between CyS 1 and CyS 6)

Oxytocin

CyS-Tyr-Ile-Glu(NH_2)-Asp(NH_2)-CyS-Pro-Ile-Gly(NH_2)

Mesotocin (8-isoleucine oxytocin)

CyS-Tyr-Ile-Glu(NH_2)-Asp(NH_2)-CyS-Pro-Arg-Gly(NH_2)

Vasotocin (8-arginine oxytocin)

A. OXYTOCIC HORMONES

The detailed pharmacological studies of Munsick *et al.* (1960) showed that fowl neurohypophysial extract contains significant amounts of milk-ejection and rat oxytocic activity after digestion with the enzyme trypsin. The active principle was assumed to be oxytocin and, on the basis of the activities measured, its concentration in fowl neural lobes was determined as 1·2 μg of oxytocin per mg. This concentration is considerably less than in ox posterior pituitary powder (5 μg/mg). Acher *et al.* (1960) isolated oxytocin from chicken neural lobes and showed that it was identical to mammalian oxytocin in its pharmacological activity, behaviour on paper chromatography, electrophoretic mobility and amino acid analysis. Subsequently Acher *et al.* (1970) reported isolation of mesotocin from chicken neural lobes and suggested that the oxytocin identified in their previous studies had been introduced with the mammalian neurophysin added to facilitate precipitation of the chicken hormones. However, it is not yet clear whether fowl pituitaries contain both oxytocin and mesotocin or whether only one oxytocic hormone is present. One of the standard assays for oxytocin is its effect on lowering the blood pressure of the anaesthetized fowl (Coon, 1939; Thompson, 1944), and mesotocin is about 10% more active than oxytocin in this assay (Berde and Konzett, 1960; Jacquenoud and Boissonas, 1961). However, the concentration of oxytocic activity in the chicken neurohypophysis is too low to be of physiological importance (Munsick *et al.*, 1960). It is of interest that removal of the N-terminal amino group of oxytocin to yield desamino oxytocin results in a 60% increase in the fowl vasodepressor activity (Chan and du Vigneaud, 1962), and substitution of threonine for glutamine in position 4 increases this activity three-fold compared to oxytocin (Manning and Sawyer, 1970). Kook *et al.* (1964) showed that oxytocin causes a transient but marked increase of both glucose and free fatty acids in cocks and hens, but the small amounts of oxytocic activity in the pituitary make it unlikely that this effect is of physiological significance. The half-life of oxytocin following an intravenous injection in cocks and hens is approximately 10 min (Hasan, 1967). However, the physiological role of the oxytocic hormone of the chicken pituitary is thus far unknown.

The studies of Munsick *et al.* (1960) showed that determination of oxytocic activity on the rat uterus in the absence of magnesium was one of the most specific assays for oxytocin, its activity being more than ten-fold greater than any other naturally-occurring neurohypophysial hormone known at that time. The activity of mesotocin in this assay (Acher *et al.*, 1970) is slightly lower than that of oxytocin (Munsick *et al.*, 1960).

B. VASOTOCIN

The structure of vasotocin (8-arginine oxytocin) is midway between those of the mammalian hormones oxytocin and arginine-vasopressin (3-phenylalanine, 8-arginine oxytocin). In order to study the pharmacological properties of this "hybrid" molecule, it was synthesized by Katsoyannis and de Vigneaud (1958) before it was known to occur naturally. Munsick *et al.* (1960) showed that the chicken antidiuretic activity of vasotocin was significantly higher than that of arginine vasopressin. Moreover vasotocin had an *in vitro* oxytocic activity on the shell gland more than twice that of arginine vasopressin and twenty times that of oxytocin. Their detailed studies of the pharmacological activities of chicken neurohypophysial extracts suggested that oxytocin and vasotocin were responsible for all of the activities observed, and that the concentration of vasotocin (12 μg/mg of posterior pituitary powder) was ten-fold greater than that of oxytocin. Chauvet *et al.* (1960) isolated vasotocin and also arginine vasopressin from fowl neurohypophyses, but in order to facilitate precipitation of the peptide hormones they added equine neurophysin to the fowl extract. From the pharmacological studies of Munsick *et al.* (1960) and chromatographic studies of Heller and Pickering (1961), it seemed unlikely that appreciable amounts of arginine vasopressin were present in fowl neural lobes, and it is possible that vasopressin was introduced with the equine neurophysin. This interpretation is supported by more recent work of Munsick (1964) who performed careful chromatographic fractionations of neurohypophysial extracts from chickens and turkeys and found oxytocin and vasotocin, but no vasopressin. Acher *et al.* (1970) have now confirmed these findings.

Sawyer (1960) demonstrated that arginine vasotocin was far more active in increasing the water permeability of the frog bladder than were other naturally-occurring neurohypophysial hormones, and this method is now widely used to assay vasotocin.

Munsick *et al.* (1960) concluded that the antidiuretic and oxytocic activities of vasotocin in the domestic fowl are sufficiently high to be of physiological importance. Bentley (1966) confirmed the observations of Kook *et al.* (1964) on the hyperglycaemic action of oxytocin in the chicken and showed that vasotocin has the same effect except that the increase in blood glucose persists longer than with oxytocin. Because of the relatively high concentration of vasotocin in the chicken neural lobe it was concluded that this effect is probably of physiological significance.

Tanaka and Nakajo (1962) showed that there is an abrupt decrease in the vasotocin content of the chicken neurohypophysis at oviposition while oxytocin decreases only slightly, and they suggested that oviposition in the hen may be caused by the prompt discharge of vasotocin from the posterior pituitary. Their interpretation has received strong support from subsequent experimental work. Douglas and Sturkie (1964) and Sturkie and Lin (1966) reported that the concentration of vasotocin in the blood of laying hens increases markedly a few minutes before oviposition and returns to resting levels 10 to 20 min after laying. Sturkie and Lin (1966) also showed that stimuli which usually cause release of vasopressin (i.e. intracarotid injection of hypertonic saline, injection of 10% glucose or haemorrhage) in mammals were ineffective in causing release of vasotocin in hens. Their experiments suggested that vasotocin was most likely a stimulus for oviposition and that oviposition was not the inducer for release of vasotocin. Contraction of the oviduct and premature oviposition induced by other methods did not cause release of vasotocin. This interpretation is supported by the work of Opel (1964, 1966) who showed that stimulation of the preoptic area of the brain of chickens caused premature oviposition and release of vasotocin. Moreover the duration of the elevated level of vasotocin in the blood (Sturkie and Lin, 1966) is approximately equal to the half-life of an intravenous injection of vasotocin in the hen determined by Hasan (1967) as 20 min. Munsick *et al.* (1960) found that intravenous injection of approximately 0·5 μg of vasotocin induced oviposition within 90 s in an intact hen. Ewy and Rzasa (1967) reported that intravenous injection of vasotocin (0·1 μg/kg) caused contraction of the oviduct in the anaesthetized hen. A larger dose (0·5 μg/kg) caused a strong single contraction of the uterus, but oxytocin at 1 μg/kg caused only slight contractions. The sensitivity of the shell gland to either hormone was greatly increased if injection occurred near the time of expected oviposition (Rzasa and Ewy, 1970).

C. EFFECTS OF NEUROHYPOPHYSECTOMY AND EXTRAHYPOPHYSIAL SOURCES OF HORMONES

Shirley and Nalbandov (1956b) reported that after neurohypophysectomy in the fowl ovulation and oviposition occurred at the same rate as in unoperated controls, but there was an increase in the red blood cell count with marked polydipsia and polyuria. Their operation would leave more or less intact, the median eminence proximal to the site of stalk section as a possible alternative source of neurohypophysial hormones. More recent studies by Hirano (1964) have shown that these hormones are found in both the median eminence and pars nervosa of chickens, but their concentration is about 30-fold higher in the pars nervosa.

Information on secretion of neurohypophysial hormones in birds has been summarized by Farner and Oksche (1962) and a more recent general discussion of neurosecretion has been published by Sachs (1969). Neurosecretory material is demonstrable by staining with aldehyde-fuchsin and chromalum-hematoxylin dyes; more recently it has been revealed as electron-dense granules by electron microscopy. There is now a large body of evidence obtained mainly from histological studies that the hormones are formed in hypothalamic neurosecretory cells and are transported via neurosecretory axons to the neuropophysis in association with a stainable carrier protein. Graber and Nalbandov (1965) reported that in White Leghorn cockerels dehydration effects a depletion of aldehyde-fuchsin positive neurosecretory material from the posterior pituitary, but not from the median eminence. The neurosecretory material of the median eminence was also unaffected by castration, adenohypophysectomy or exposure to light. Opel (1965b) confirmed the conclusion of Shirley and Nalbandov (1956b) that the posterior lobe of the chicken pituitary is not essential to oviposition and pointed out that the neurohypophysis may serve as a reservoir for the hormones which are in fact produced by the hypothalamus. Thus the machinery necessary for accumulation and release of enough hormone to effect oviposition may be in components of the hypothalamo-neurohypophysial system proximal to the neural lobe. Further studies of Opel (1966) demonstrated that after removal of the pituitary neural lobe, oviposition was associated with a significant loss of vasotocin activity from the stalk median eminence region and increase of this activity in plasma to more than 30 times that of the preoviposition level. These experiments suggest that oviposition in neural lobectomized hens results from an abrupt release of vasotocin from the stalk-median eminence region.

D. CONTROL OF SECRETION OF NEUROHYPOPHYSIAL HORMONES

It is now generally accepted on the basis of evidence from many species that secretion of neurohypophysial hormones is under hypothalamic control. In the domestic fowl, the effects of lesions in certain specific areas of the hypothalamus can be attributed to decreased secretion of neurohypophysial hormones. Ralph and Fraps (1959b) and Egge and Chiasson (1963) reported cessation of egg laying in the hen after electrolytic production of lesions in the diencephalon. Ralph and Fraps (1959b) found that lesions in the preoptic hypothalamus, supraoptico-hypophysial tract, or dorsocaudal thalamus resulted in a longer period of interruption to ovulation than when lesions were in the central diencephalon. McFarland (1959) demonstrated that an electrolytic lesion involving the hypothalamus and the hypothalamo-hypophysial tract of cocks resulted in diabetes insipidus. Ralph (1960) reported increased water intake (polydipsia) in some hens sustaining lesions of the supraoptic region, but not from lesions elsewhere in the anteroventral hypothalamus. Permanent polyuria and in some cases polydipsia was produced in chickens by lesions in the nerve tracts from the supraoptic and paraventricular nuclei to the neural lobe (Koike and Lepkovsky, 1967). There was a low level of vasotocin in the posterior pituitary, but the chickens appeared normal in other respects.

Dehydration in the domestic fowl was found to cause hypothalamic neurosecretory cells to display increased activity which was accompanied by a decrease in both stainable neurosecretion and antidiuretic activity in the neural lobe but not in the median eminence if the osmotic stress was sufficiently severe (Legait, 1959; Graber and Nalbandov, 1965). Release of neurohypophysial hormones can also result from stimulation of specific areas of the hypothalamus. Opel (1964) reported that electrical stimulation of the preoptic hypothalamus induced premature oviposition of the terminal egg of a 2-egg sequence. Even insertion of electrodes without passage of current was effective and further studies (Opel, 1966) showed that there was release of vasotocin but not oxytocin, associated with induced oviposition.

Studies on other species of birds have suggested that the median eminence and the neural lobe are controlled separately, the median eminence being activated by long photoperiods and the neural lobe by dehydration (Dodd *et al.*, 1966). For the domestic fowl, however, there was no correlation between presence of neurosecretory material in the median eminence and dehydration or exposure to light (Graber and Nalbandov, 1965).

References

Acher, R., Chauvet, J. and Lenci, M. T. (1960). *Biochim. biophys. Acta* **38**, 344–345.

Acher, R., Chauvet, J. and Chauvet, M.-T. (1970). *Eur. J. Biochem.* **17**, 509–513.

Amin, S. O. and Gilbert, A. B. (1970). *Br. Poult. Sci.* **11**, 451–458.

Andrews, P. (1966). *Nature, Lond.* **209**, 155–157.

Assenmacher, I. (1958). *Archs Anat. microsc. Morph. exp.* **47**, 447–572.

Bagshawe, K. D., Orr, A. H. and Godden, J. (1968). *J. Endocr.* **42**, 513–518.

Bailey, R. L. and Phillips, R. E. (1952). *Poult. Sci.* **31**, 68–71.

Bargmann, W., Hild, W., Ortmann, R. and Schiebler, T. H. (1950). *Acta neuroveg.* **1**, 233–275.

Barnes, B. G. (1962). *Endocrinology* **71**, 618–628.

Bates, R. W. and Condliffe, P. G. (1960). *Recent Prog. Horm. Res.* **16**, 309–352.

Bentley, P. J. (1966). *J. Endocr.* **34**, 527–528.

Berde, B. and Konzett, H. (1960). *Medna Exp.* **2**, 317–322.

Breneman, W. R. (1955). *Mem. Soc. Endocr.* **4**, 94–113.

Breneman, W. R., Zeller, F. J. and Beekman, B. E. (1959). *Poult. Sci.* **38**, 152–158.

Breneman, W. R., Zeller, F. J. and Creek, R. O. (1962). *Endocrinology* **71**, 790–798.

Brown, P. S. (1955). *J. Endocr.* **13**, 59–64.

Bullock, D. W., Mittal, K. K. and Nalbandov, A. V. (1967). *Endocrinology* **80**, 1182–1184.

Bullock, D. W. and Nalbandov, A. V. (1967). *J. Endocr.* **38**, 407–415.

Burns, J. M. (1970a). *Comp. Biochem. Physiol.* **34**, 727–731.

Burns, J. M. (1970b). *Comp. Biochem. Physiol.* **35**, 867–872.

Burrows, W. H. and Byerly, T. C. (1936). *Proc. Soc. exp. Biol. Med.* **34**, 841–844.

Butt, W. R. (1967). *In* "The Chemistry of the Gonadotrophins", pp. 58–74, C. C. Thomas, Springfield, Illinois.

Byerly, T. C. and Burrows, W. H. (1938). *Endocrinology* **22**, 366–369.

Chan, W. Y. and du Vigneaud, V. (1962). *Endocrinology* **71**, 977–982.

Chauvet, J., Lenci, M. T. and Acher, R. (1960). *Biochim. biophys. Acta* **38**, 571–573.

Coon, J. M. (1939). *Archs int. pharmacodyn. Thér.* **62**, 79–99.

Cunningham, F. J., Myres, R. P. and McNeilly, J. R. (1970). *J. Reprod. Fert.* **23**, 538–539.

Das, B. C. and Nalbandov, A. V. (1955). *Endocrinology* **57**, 705–710.

Desjardins, C. and Hafs, H. D. (1964). *J. anim. Sci.* **23**, 903–904.

Dodd, J. M., Perks, A. M. and Dodd, M. H. I. (1966). *In* "The Pituitary Gland" (G. W. Harris and B. T. Donovan, eds.), Vol. 3, pp. 578–623. Butterworths, London.

Domm, L. V. (1931). *Proc. Soc. exp. Biol. Med.* **29**, 308–312.
Douglas, D. S. and Sturkie, P. D. (1964). *Fedn Proc. Fedn Am. Socs exp. Biol.* **23**, 150.
Drager, G. A. (1945). *Endocrinology* **36**, 124–129.
Duncan, D. (1956). *Anat. Rec.* **125**, 457–471.
Egge, A. S. and Chiasson, R. B. (1963). *Gen. comp. Endocr.* **3**, 346–361.
Evans, H. M., Sparks, L. L. and Dixon, J. S. (1966). *In* "The Pituitary Gland" (G. W. Harris and B. T. Donovan, eds.), Vol. 1, pp. 317–373. Butterworths, London.
Ewy, Z. and Rzasa, J. (1967). *Gen. comp. Endocr.* **9**, 449.
Farner, D. S. and Oksche, A. (1962). *Gen. comp. Endocr.* **2**, 113–147.
Ferrando, G. and Nalbandov, A. V. (1969). *Endocrinology* **85**, 38–42.
Frankel, A. I. (1970). *Poult. Sci.* **49**, 869–921.
Frankel, A. I., Gibson, W. R., Graber, J. W., Nelson, D. M., Reichert, L. E. and Nalbandov, A. V. (1965). *Endocrinology* **77**, 651–657.
Frankel, A. I., Graber, J. W. and Nalbandov, A. V. (1967). *Gen. comp. Endocr.* **8**, 387–396.
Fraps, R. M. (1953). *Poult. Sci.* **32**, 899.
Fraps, R. M., Fevold, H. L. and Neher, B. H. (1947). *Anat. Rec.* **99**, 571–572.
Furr, B. J. A. and Cunningham, F. J. (1970). *Br. Poult. Sci.* **11**, 7–13.
Gilbert, A. B. (1965). *Q. Jl exp. Physiol.* **50**, 437–445.
Gilbert, A. B. and Amin, S. O. (1969). *J. Reprod. Fert.* **20**, 363–364.
Graber, J. W. and Nalbandov, A. V. (1965). *Gen. comp. Endocr.* **5**, 485–492.
Green, J. D. (1951). *Am. J. Anat.* **88**, 225–311.
Green, J. D. (1966). *In* "The Pituitary Gland" (G. W. Harris and B. T. Donovan, eds.), Vol. I, pp. 127–146 Butterworths, London.
Harris, G. W. (1955). *In* "Neural Control of the Pituitary Gland" Arnold, London.
Hasan, S. H. (1967). *Gen. comp. Endocr.* **9**, 456–457.
Hayashida, T. (1969). *Nature, Lond.* **222**, 294–295.
Hazelwood, R. L. and Hazelwood, B. S. (1961). *Proc. Soc. exp. Biol. Med.* **108**, 10–12.
Heald, P. J., McLachlan, P. M. and Rookledge, K. A. (1965). *J. Endocr.* **33**, 83–95.
Heald, P. J., Furnival, B. E. and Rookledge, K. A. (1967). *J. Endocr.* **37**, 73–81.
Heald, P. J., Rookledge, K. A., Furnival, B. E. and Watts, G. D. (1968a). *J. Endocr.* **41**, 197–201.
Heald, P. J., Rookledge, K. A., Furnival, B. E. and Watts, G. D. (1968b). *J. Endocr.* **41**, 313–318.
Heller, H. and Pickering, B. T. (1961). *J. Physiol., Lond.* **155**, 98–114.
Herlant, M. (1964). *Int. Rev. Cytol.* **17**, 299–382.
Herlant, M., Benoit, J., Tixier-Vidal, A. and Assenmacher, I. (1960). *C. r. hebd. Séanc. Acad. Sci., Paris* **250**, 2936–2938.
Herrick, R. B., McGibbon, W. H. and McShan, W. H. (1962). *Endocrinology* **71**, 487–491.
Hill, R. T. and Parkes, A. S. (1934). *Proc. R. Soc.* **B115**, 402–409.
Hirano, I. (1964). *Endocr. jap.* **11**, 87–95.
Hirsch, L. J. (1961). *Ph.D. Thesis, University of Illinois.*
Huber, G. E. and Crosby, E. C. (1929). *J. comp. Neurol.* **48**, 1–225.
Huston, T. M. and Nalbandov, A. V. (1953). *Endocrinology*, **52**, 149–156.
Jackson, G. L. and Nalbandov, A. V. (1969a). *Endocrinology* **84**, 1218–1223.
Jackson, G. and Nalbandov, A. V. (1969b). *Endocrinology* **84**, 1262–1265.
Jackson, G. and Nalbandov, A. V. (1969c). *Endocrinology*, **85**, 113–120.
Jacquenoud, P. A. and Boissonas, R. A. (1961). *Helv. chim. Acta* **44**, 113–122.

Kanematsu, S. and Mikami, S. (1970). *Gen. comp. Endocr.* **14**, 25–34.
Katsoyannis, P. G. and du Vigneaud, V. (1958). *J. biol. Chem.* **233**, 1352–1354.
King, D. B. (1969). *Gen. comp. Endocr.* **12**, 242–255.
Kleinholz, L. H. and Rahn, H. (1940). *Anat. Rec.* **76**, 157.
Knobil, E. and Sandler, R. (1963). *In* "Comparative Endocrinology" (U. S. von Euler and H. Heller, eds.), Vol. 1, pp. 447–491, Academic Press, New York.
Koike, T. and Lepkovsky, S. (1967). *Gen. comp. Endocr.* **8**, 397–402.
Kook, Y., Cho., K. B. and Yun, K. O. (1964). *Nature, Lond.* **204**, 385–386.
Kuhlenbeck, H. (1936). *Morph. Jb.* **77**, 61–109.
Kurotsu, T. (1935). *Proc. Sect. Sci. K. ned. Akad. Wet.* **38**, 784–797.
Lake, P. E. and Gilbert, A. B. (1964). *Res. vet. Sci.* **5**, 39–45.
Langslow, D. R. and Hales, C. N. (1969). *J. Endocr.* **43**, 285–294.
Legait, H. (1959). *Thèse, Louvain Nancy.*
Legait, E. and Legait, H. (1955). *C. r. Ass. Anat.* **84**, 188–199.
Li, C. H., Dixon, J. S. and Liu, W. K. (1969a). *Archs Biochem. Biophys.* **133**, 70–91.
Li, C. H., Dixon, J. S., Lo, T. B., Pankov, Y. A. and Schmidt, K. D. (1969b), *Nature, Lond.* **224**, 695–696.
Licht, P. and Pearson, A. K. (1969). *Gen. comp. Endocr.* **13**, 367–381.
Licht, P. and Stockell Hartree, A. (1971). *J. Endocr.* **49**, 113–124.
Lyons, W. R. and Dixon, J. S. (1966). *In* "The Pituitary Gland" (G. W. Harris and B. T. Donovan, eds.), Vol. 1, pp. 527–581. Butterworths, London.
Ma, R. C. S. and Nalbandov, A. V. (1963). *In* "Advances in Neuroendocrinology" (A. V. Nalbandov, ed.), pp. 306–311. University of Illinois Press, Urbana, Illinois.
Manning, M. and Sawyer, W. H. (1970). *Nature, Lond.* **227**, 715–716.
Matsuo, S. (1954). *Jap. J. zootech. Sci.* **25**, 63–69.
McFarland, L. Z. (1959). *Anat. Rec.* **133**, 411.
McNeilly, J. R. (1970). *J. Endocr.* **48**, xxxiii.
Meites, J. and Nicoll, C. S. (1966). *A. Rev. Physiol.* **28**, 57–88.
Mikami, S. (1958). *J. Fac. Agric. Iwate Univ.* **3**, 473–545.
Mikami, S. (1960). *J. Fac. Agric. Iwate Univ.* **4**, 359–379.
Mitchell, M. E. (1967a). *J. Reprod. Fert.* **14**, 249–256.
Mitchell, M. E. (1967b). *J. Reprod. Fert.* **14**, 257–263.
Morris, T. R. and Nalbandov, A. V. (1961). *Endocrinology* **68**, 687–697.
Moudgal, N. R. and Li, C. H. (1961). *Archs Biochem. Biophys.* **95**, 93–98.
Munsick, R. A. (1964). *Endocrinology* **75**, 104–112.
Munsick, R. A., Sawyer, W. H. and van Dyke, H. B. (1960). *Endocrinology* **66**, 860–871.
Nakajo, S. and Imai, K. (1961). *Poult. Sci.* **40**, 739–744.
Nakajo, S. and Tanaka, K. (1956). *Poult. Sci.* **35**, 990–994.
Nalbandov, A. V. (1945). *Endocrinology* **36**, 251–258.
Nalbandov, A. V. (1959). *In* "Comparative Endocrinology" (A. Gorbman, ed.), pp. 161–173, Wiley, New York.
Nalbandov, A. V. (1966). *In* "The Pituitary Gland" (G. W. Harris and B. T. Donovan, eds.), Vol. 1, pp. 295–316, Butterworths, London.
Nalbandov, A. V. and Card, L. E. (1942). *Poult. Sci.* **21**, 474.
Nalbandov, A. V. and Card, L. E. (1943). *J. exp. Zool.* **94**, 387–413.
Nalbandov, A. V. and Card, L. E. (1946). *Endocrinology* **38**, 71–78.
Nalbandov, A. V., Meyer, R. K. and McShan, W. H. (1951). *Anat. Rec.* **110**, 475.
Nelson, D. M., Norton, H. W. and Nalbandov, A. V. (1965a). *Endocrinology* **77**, 889–896.

Nelson, D. M., Norton, H. W. and Nalbandov, A. V. (1965b). *Endocrinology* **77**, 731–734.

Okamoto, S. and Ihara, Y. (1960). *Anat. Rec.* **137**, 485–499.

Opel, H. (1963). *Proc. Soc. exp. Biol. Med.* **113**, 488–492.

Opel, H. (1964). *Endocrinology* **74**, 193–200.

Opel, H. (1965a). *Anat. Rec.* **151**, 394.

Opel, H. (1965b). *Endocrinology* **76**, 673–677.

Opel, H. (1966). *Anat. Rec.* **154**, 396.

Opel, H. and Fraps, R. M. (1961). *Proc. Soc. exp. Biol. Med.* **108**, 291–296.

Opel, H. and Lepore, P. D. (1967). *Poult. Sci.* **46**, 1302.

Opel, H. and Nalbandov, A. V. (1961). *Endocrinology* **69**, 1016–1028.

Parkes, A. S. and Rowlands, I. W. (1937). *J. Physiol., Lond.* **90**, 100–103.

Parlow, A. F. (1961). *In* "Human Pituitary Gonadotrophins" (A. Albert, ed.) pp. 300–312, Thomas, Springfield.

Payne, F. (1942). *Biol. Bull. mar. biol. Lab., Wood's Hole* **82**, 79–111.

Payne, F. (1943). *Anat. Rec.* **86**, 1–13.

Payne, F. (1965). *J. Morph.* **117**, 185–199.

Purves, H. D. (1966). *In* "The Pituitary Gland" (G. W. Harris and B. T. Donovan, eds.), Vol. 1, pp. 147–232, Butterworths, London.

Rahn, H. (1939). *J. Morph.* **64**, 483–517.

Rahn, H. and Painter, B. T. (1941). *Anat. Rec.* **79**, 297–311.

Ralph, C. L. (1959). *Anat. Rec.* **134**, 411–431.

Ralph, C. L. (1960). *Am. J. Physiol.* **198**, 528–530.

Ralph, C. L. and Fraps, R. M. (1959a). *Endocrinology* **65**, 819–824.

Ralph, C. L. and Fraps, R. M. (1959b). *Am. J. Physiol.* **197**, 1279–1283.

Ralph, C. L. and Fraps, R. M. (1960). *Endocrinology* **66**, 269–272.

Resko, J. A., Norton, H. W. and Nalbandov, A. V. (1964). *Endocrinology* **75**, 192–200.

Riddle, O. (1963). *Anim. Behav.* **11**, 419–432.

de Roos, R. and de Roos, C. C. (1964). *Gen. comp. Endocr.* **4**, 602–607.

Russell, J. (1955). *In* "The Hypophyseal Growth Hormone, Nature and Actions" (R. W. Smith, Jr., O. H. Gaebler and C. N. H. Long, eds.), pp. 17–27. Blakiston and McGraw-Hill, New York.

Rzasa, J. and Ewy, Z. (1970). *J. Reprod. Fert.* **21**, 549–550.

Sachs, H. (1969). *Adv. Enzymol.* **32**, 327–372.

Saeki, Y. and Tanabe, Y. (1955). *Poult. Sci.* **34**, 909–919.

Salem, M. H. M., Norton, H. W. and Nalbandov, A. V. (1970a). *Gen. comp. Endocr.* **14**, 270–280.

Salem, M. H. M., Norton, H. W. and Nalbandov, A. V. (1970b). *Gen. comp. Endocr.* **14**, 281–289.

Sawyer, W. H. (1960). *Endocrinology* **66**, 112–120.

Sawyer, W. H. (1966a). *In* "The Pituitary Gland" (G. W. Harris and B. J. Donovan, eds.) Vol. 3, pp. 307–329. Butterworths, London.

Sawyer, W. H. (1966b). *In* "The Pituitary Gland" (G. W. Harris and B. T. Donovan, eds.), Vol. 3, pp. 288–306. Butterworths, London.

Sayers, M. A., Sayers, G. and Woodbury, L. A. (1948). *Endocrinology* **42**, 379–393.

Schally, A. V., Arimura, A., Bowers, C. Y., Kastin, A. J., Sawano, S. and Redding, T. W. (1968). *Recent Prog. Horm. Res.* **24**, 497–588.

Shirley, H. V. and Nalbandov, A. V. (1956a). *Endocrinology* **58**, 694–700.

Shirley, H. V. and Nalbandov, A. V. (1956b). *Endocrinology* **58**, 477–483.

Shome, B., Parlow, A. F., Ramirèz, V. D., Elrick, H. and Pierce, J. G. (1968) *Archs Biochem. Biophys.* **126**, 444–455.

Siegel, H. S. (1962). *Gen. comp. Endocr.* **2**, 385–388.
Solomon, D. H. (1961). *Endocrinology* **69**, 939–957.
Steelman, S. L. and Pohley, F. M. (1953). *Endocrinology* **53**, 604–616.
Stockell Hartree, A. and Cunningham, F. J. (1969). *J. Endocr.* **43**, 609–616.
Stockell Hartree, A., Langslow, D. R. and Hales, C. N. (1969). *J. Endocr.* **44**, 287–288.
Sturkie, P. D. and Lin, Y. C. (1966). *J. Endocr.* **35**, 325–326.
Sykes, A. H. (1962). *J. Reprod. Fert.* **4**, 214.
Taber, E., Claytor, M., Knight, J., Gambrell, D., Flowers, J. and Ayers, C. (1958). *Endocrinology* **62**, 84–89.
Tanaka, K., Fujisawa, Y. and Yoshioka, S. (1966). *Poult. Sci.* **45**, 970–973.
Tanaka, K. and Nakajo, S. (1962). *Endocrinology* **70**, 453–458.
Tanaka, K. and Yoshioka, S. (1967). *Gen. comp. Endocr.* **9**, 374–379.
Taylor, T. G., Morris, T. R. and Hertelendy, F. (1962). *Vet. Rec.* **74**, 123–125.
Thompson, R. E. (1944). *J. Pharmac. exp. Ther.* **80**, 373–382.
Tienhoven, A. van, Nalbandov, A. V. and Norton, H. W. (1954). *Endocrinology*, **54**, 605–611.
Tixier-Vidal, A. (1963). *In* "Cytologie de l'Adenohypophyse" (J. Benoit and C. Da Lage, eds.) Editions du C.N.R.S., Paris.
Tixier-Vidal, A., Follet, B. K. and Farner, D. A. (1968). *Z. Zellforsch. mikrosk. Anat.* **92**, 610–635.
Wilhelmi, A. E. and Mills, J. B. (1969). *Colloq. Int. Centre natn. Rech. Sci., Paris*, no. 177, 165–173.
Wingstrand, K. G. (1951). "The Structure and Development of the Avian Pituitary", Gleerup, Lund.
Zarrow, M. X. and Bastian, J. W. (1953). *Proc. Soc. exp. Biol. Med.* **84**, 457–459.
Zarrow, M. X., Greenman, D. L., Kollias, J. and Dalrymple, D. (1962). *Gen. comp. Endocr.* **2**, 177–180.

17 The Thyroid Glands

I. R. FALCONER

Department of Applied Biochemistry and Nutrition,
School of Agriculture, University of Nottingham
Sutton Bonington, Loughborough, Leicestershire, England

I. Structure and Development of the Thyroid Glands

A. ANATOMY AND HISTOLOGY

In the chicken and other avian species the thyroid glands are small paired, dark red oval structures. They are situated low in the neck, each close to a common carotid artery about 5 mm cranial to the angle formed by the union of the subclavian and common carotid arteries. The thyroid glands in all species are characterized by an abundant blood supply originating from the carotid arteries. In the chicken there are both cranial and caudal thyroid arteries, arising from a branch of the common carotid. Venous drainage is into the jugular veins which also lie close to the thyroid glands (Fig. 1).

In an adult bird of 2 kg body weight the two thyroid lobes each weigh about 50 mg, but the weights will vary widely depending on the iodine content of the diet, the season and the strain of bird. A connective tissue capsule encloses the secretory cells of the thyroid, which are arranged as a single layer in roughly spherical follicles (Fig. 2). Within the follicles is a protein gel (colloid) consisting largely of an iodinated protein, thyroglobulin, of molecular weight about 625,000 daltons. Between the thyroid follicles lie arterioles, capillaries, venous sinuses, venules and lymphatics. There is a relatively sparse nerve supply, mainly associated with smooth muscle fibres in the walls of the arterial vessels.

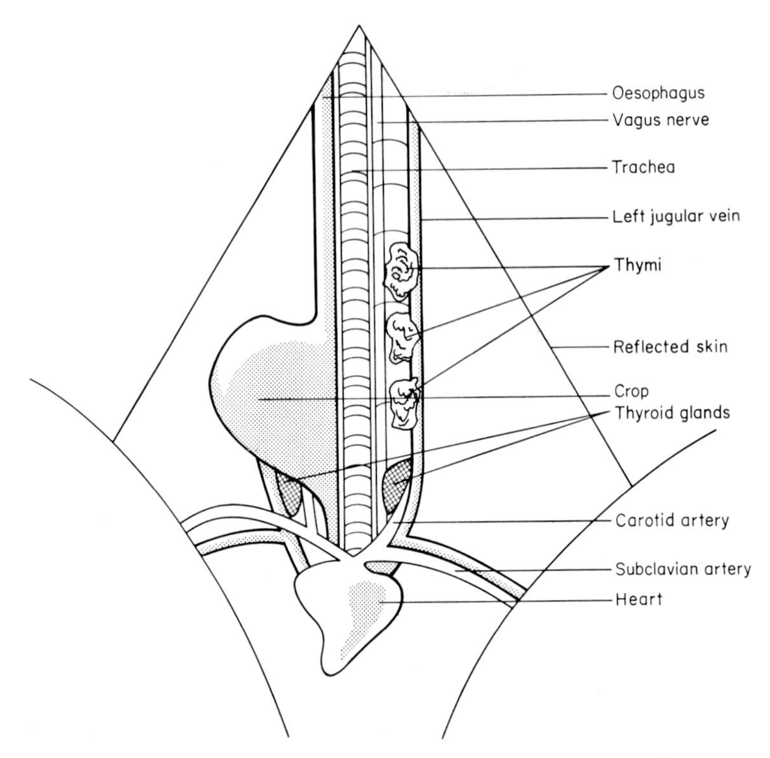

Fig. 1. Dissection of chicken neck to show the position of the thyroid glands.

The thyroid secretory cells show considerable polarity, with the outer (basal) portion of the cell in close contact with capillary sinuses. The endothelial lining of the capillaries is perforated allowing free diffusion of plasma to the thyroid cell surface; numerous channels entering the secretory cells from their base have been observed. The secretory cells have large well-defined nuclei containing one or more nucleoli and many dark staining areas of nucleic acid (Fig. 3). The basal portions of the cells contain mitochondria and abundant endoplasmic reticulum. Golgi zones and droplets of secretory protein are observed in the apical region of the cells. The surface of the cells facing the interior of the follicle is extended into microvilli, characteristic of an absorptive membrane.

A quiescent thyroid gland shows an accumulation of intrafollicular protein, which stretches and flattens the secretory cells lining the follicles. When such glands become active the volume of intrafollicular protein decreases, and the secretory cells become columnar. A rapid increase in protein droplets within the cells is observed, indicating that these may be engulfed follicular protein as well as newly-synthesized protein. The hydrolysis of thyroglobulin yields free amino acids, including the two thyroid hormones, thyroxine and tri-

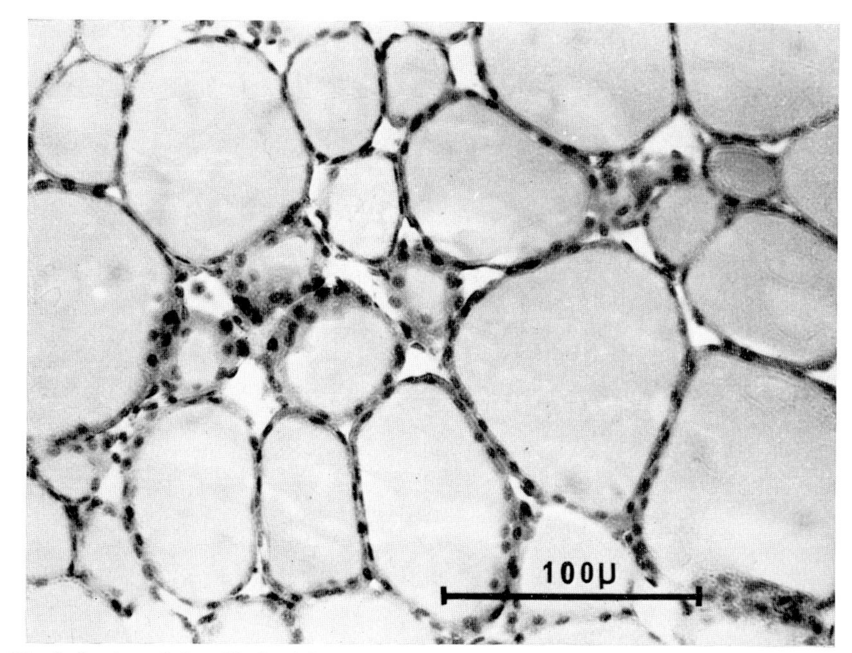

Fig. 2. Section of thyroid gland from a 6-week-old chick, showing spherical follicles of cells filled with colloid.

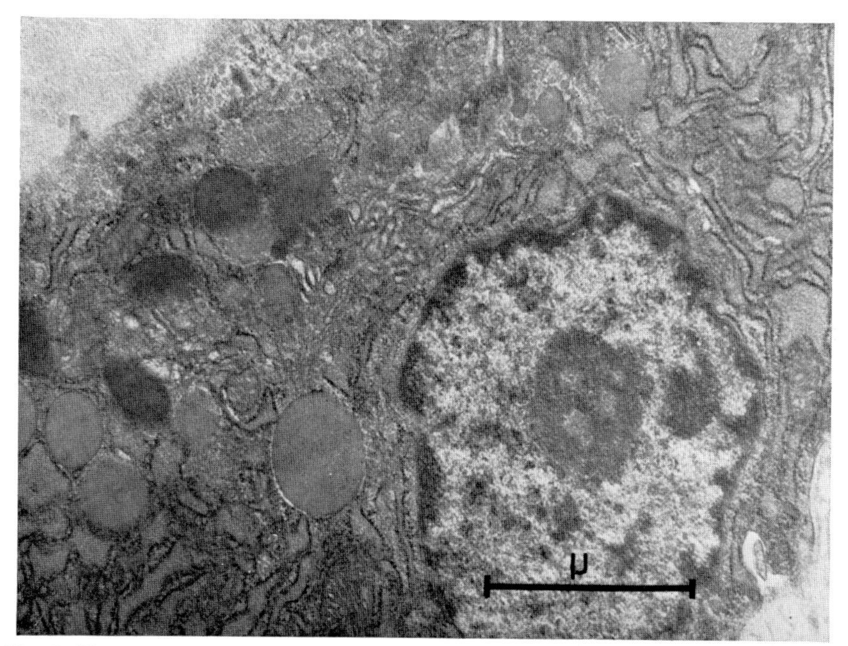

Fig. 3. Electron micrograph of a thyroid cell from a 6-week-old chick, with a prominent nucleus containing a nucleolus, large protein droplets in the cytoplasm and abundant rough endoplasmic reticulum.

iodothyronine. These hormones diffuse from the thyroid cells into the blood, and the other amino acids are re-utilized within the gland.

B. EMBRYOLOGY

The embryonic origins of the avian thyroid gland have been well described by Romanoff (1960). The thyroid is one of the earliest endocrine glands to develop in the chick embyro, appearing first as a median outgrowth of the pharynx on the 2nd d of incubation. By the 5th d the thyroid anlage has formed a two-lobed structure, and the cord of cells joining the pharynx to the thyroid lobes has become a vestigial strand of cells. The thyroid tissue, by this stage, has migrated to almost its final location, just anterior to the third aortic arches in the embryo.

By the 8th d of incubation mesenchyme has invaded the thyroid tissue and the glandular cells have become separated into plates and cords of cells. The next major change is the accumulation of colloid droplets and organization of the secretory cells into spherical follicles which occurs from the 10th to 13th d of incubation. During the latter half of incubation the diameter of the follicles progressively increases due to the accumulation of colloid and cell division.

It appears likely that thyroid hormone is being synthesized from the 11th d of incubation, as autoradiography has identified ^{131}I-labelled colloid in thyroids from embryos of that age (Hansborough and Khan, 1951). However, the iodine content of the thyroid does not show any dramatic increase until the 16th d which indicates that little secretion occurs until this time (Sun, 1932). The tissues of the chick embryo are sensitive to thyroid hormone, as the length of incubation can be affected by injecting either thyroid hormone or antithyroid drugs into the egg. Romanoff and Laufer (1956) showed a marked increase in the incubation period resulting from thiourea injection and Beyer (1952) showed a decrease in the incubation period and an increase in chick weight after thyroxine treatment of the egg. The thyroid gland of mammalian foetuses is similarly active before birth, and a congenital deficiency of thyroid hormone in a species such as the sheep results in high postnatal mortality (Falconer, 1966).

II. Thyroid Hormone Biosynthesis

Little direct investigation of thyroid hormone biosynthesis in avian glands has been attempted. However, analyses of avian thyroid tissue for iodinated tyrosines and thyroid hormones has shown quantities comparable with mammalian glands (Shellabarger and Pitt-Rivers, 1958; Mellen and Wentworth, 1959), and there is at present no evidence for a diversity of mechanisms for thyroid hormone biosynthesis in different classes of animal.

The thyroid gland possesses an ability to concentrate iodide from the bloodstream and therefore will accumulate a variable proportion of the dietary iodide, sufficient to supply the requirement for thyroid hormone biothesis. The thyroid hormones, thyroxine and triiodothyronine (Fig. 4), both contain iodine, and in the process of thyroid hormone biosynthesis monoiodotyrosine and diiodotyrosine are also formed. The effective supply and

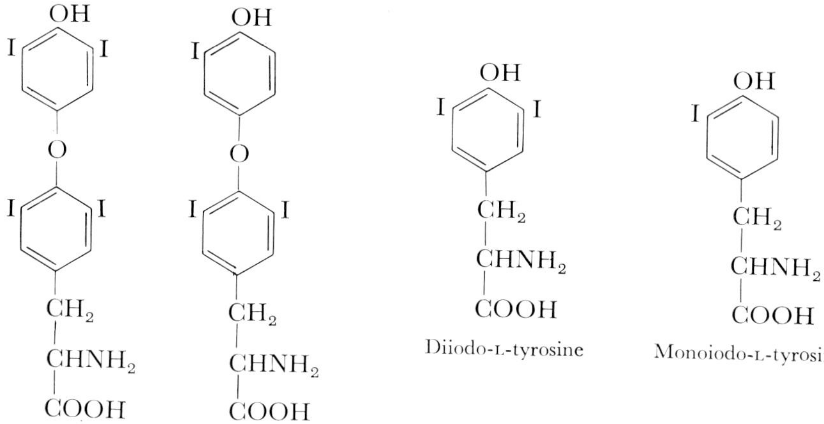

L-Thyroxine L-Triiodothyronine Diiodo-L-tyrosine Monoiodo-L-tyrosine

Fig. 4. Formulae of the thyroid hormones thyroxine and triiodothyronine, and of the iodotyrosines.

metabolism of iodine in the thyroid is therefore of considerable importance, as iodine deficiency results in malfunction of the thyroid gland and inadequate hormone production.

A peroxidase system in the thyroid cells is responsible for the oxidation of iodide ions to iodine (for review see Stanbury, 1967). Iodine reacts with the phenolic ring of free tyrosine and of tyrosine in proteins to form iodotyrosines. Whether this process requires an enzyme in the cell has not been established, but the absence of general iodination of proteins in the thyroid indicates a high degree of specificity in thyroglobulin iodination.

Thyroglobulin is biosynthesized as polypeptide sub-units containing normal un-iodinated amino acids which are modified and aggregated to form thyroglobulin (Thompson and Goldberg, 1968). After polypeptide biosynthesis, the thyroglobulin sub-units are modified by the addition of substituted hexoses such as D-glucosamine, to form a glycoprotein (Cheftel and Bouchilloux, 1968). Iodination probably occurs from the time of completion of the glycoprotein sub-unit until after the final molecule of thyroglobulin has left the cell and passed into the follicular colloid.

The combination of iodinated tyrosyl groups to form thyroxine probably occurs through an enzymic process using one free iodotyrosine molecule, and one iodotyrosine bound into the thyroglobulin molecule. The pathway is considered to go via an enol-pyruvate derivative of iodotyrosine, which is then eliminated on linking with the phenolic hydroxyl of a second iodotyrosine (Blasi *et al.*, 1969). Thus thyroglobulin molecules contain both thyroxine and triiodothyronine, as well as mono- and di-iodotyrosines.

For the liberation of thyroxine and triiodothyronine into the blood, follicular thyroglobulin is hydrolysed to free amino acids. These, with the exception of the thyroid hormones, are re-utilized within the cells.

III. Thyroid Hormone Transport in the Blood

Thyroxine and triiodothyroxine both occur in the blood of chickens, in a ratio of 6 to 4 (Wentworth and Mellen, 1961). The quantity of circulating thyroid hormones, expressed as protein-bound iodine per 100 ml plasma, varies between 1 and 2 μg in untreated adult birds, which is under half of the quantity found in the plasma of domestic mammals or man (Mellen and Hardy, 1957). This marked reduction is attributed to a difference in the protein binding of thyroid hormones in avian blood. Mammalian plasma contains an α_2-globulin which selectively binds thyroxine and to a lesser extent triiodothyronine, and normally carries the major proportion of circulating thyroid hormone. This is absent in avian blood, which transports thyroid hormones in free solution and loosely bound to albumin and pre-albumin.

A second consequence of the reduced thyroid hormone binding in avian blood is the relatively short half-life of thyroxine (8·3 h) and triiodothyronine (7·2 h) in chicken plasma during the first 24 h after administration (Heninger and Newcomer, 1964) compared with mammalian plasma. For example, in guinea-pig plasma the half-life of thyroxine is 31·3 h and triiodothyronine 30·2 h, in sheep plasma the half-life of thyroxine is 37 h (Frienkel and Lewis, 1957; Ray and Premachandra, 1964).

The potencies of thyroxine and triiodothyronine are similar in the chicken, unlike the mammal where the latter is from 5 to 7 times more potent than the former (see Heninger and Newcomer, 1964). This difference in potency in mammals is again related to protein-binding of the two hormones—thyroxine being more strongly bound than triiodothyronine and therefore less able to diffuse into the tissues (Tata and Shellabarger, 1959).

IV. The Measurement of Thyroid Hormone Secretion

During the last ten years considerable effort has been employed in developing effective methods of measuring thyroid hormone secretion in the fowl. The establishment of these methods is a necessary preliminary to measuring thyroid activity under different physiological conditions, and with differing treatments of the bird.

A comparative study of four of the most frequently used techniques of thyroid investigation was carried out by Singh *et al.* (1968a). The two methods of determination of thyroid secretion rate considered most accurate by these authors were based upon (i) measurement of thyroid radioactive iodine output, and (ii) measurement of the thyroxine degradation rate.

(i) The first of these methods employed the administration of a single dose of radioactive [^{131}I]iodide to the chicks by intramuscular injection, followed by measurement of thyroid radioactivity over a period of nine days. From a semi-logarithmic plot of radioactivity as the percentage dose against time, the half-life ($T\frac{1}{2}$, in days) can be found for thyroid-^{131}I. The rate constant for decrease in radioactivity (K_4^*) is equal to $0{\cdot}693/T\frac{1}{2}$, as

$$\text{Radioactivity (\% dose at time } t) = (\text{Max. \% dose at } t = 0)\ e^{-K_4^* t}$$

The rate constant K_4^- includes a component of recirculated ^{131}I from the degradation of labelled hormone. To eliminate this error a correction is made based upon the proportion of the original dose of ^{131}I taken up by the thyroids, and the corrected rate constant (K_4) calculated.

$$K_4 \text{ (corrected rate constant)} = \frac{1-[(\text{Max. \% dose at } t = 0) \div 100]}{K_4^*}$$

At the end of the thyroid radioactivity measurements the glands are removed from the birds and their iodine content determined by chemical analysis.

Using this method:

Thyroid secretion rate (as thyroxine, μg/d)

$$= \text{Thyroid iodine content } (\mu g) \times K_4 \times 1{\cdot}592$$

where 1·592 corrects the thyroid iodine released into its equivalent weight of thyroxine.

This technique determined an average secretion rate of 1·1 μg/d 100 g body weight of thyroxine in 39 chicks of 9 weeks of age (Singh *et al.*, 1968a).

An alternative method of calculation based on the direct measurement of ^{131}I released from the thyroid used, instead of thyroid iodine analysis, the determination of plasma protein-bound ^{131}I and protein-bound ^{127}I, to find the specific activity of secreted hormone (Robertson and Falconer, 1961).

(ii) The second method employed measurement of the plasma ^{131}I for a period of 12 h after a single intravenous injection of [^{131}I]thyroxine together with chemical analysis of plasma protein-bound ^{127}I. The data obtained for plasma ^{131}I were plotted on a semi-logarithmic graph as a percentage of the injected dose/ml against time (in hours), and the half-life determined. From this graph the rate constant for decrease in plasma radioactivity (K) is equal to $0{\cdot}693/T\frac{1}{2}$.

Using this method:

Thyroid secretion rate (as thyroxine, μg/d)

$$= \text{Extra-thyroidal thyroxine} \times K \times 24 \times 1{\cdot}527$$

Where K = degradation rate of plasma [^{131}I]thyroxine, %/h.

24 corrects to a daily secretion rate.

1·529 corrects the weight of iodine to its equivalent weight of thyroxine.

To find the quantity of extra-thyrodial thyroxine, the volume within the bird in which thyroxine distributes must also be determined. This is done by extrapolating the graph of plasma radioactivity to zero time, when the thyroxine distribution space (ml)

$$= 100 \div \% \text{ dose of } ^{131}\text{I/ml plasma at } t = 0.$$

Having determined the thyroxine distribution space, then extra-thyroidal thyroxine = plasma protein-bound ^{127}I × thyroxine distribution space.

The average secretion rate by this method was 2·03 μg/d 100 g body weight in chicks 6 to 7 weeks old (Singh *et al.*, 1968a). No clear explanation of the difference between the results of these two methods was given, but one possibility of reaching an excessive value by this method lies in the measurement of total plasma radioactivity. If, instead, only plasma thyroxine radioactivity had been measured, any rapid clearance of [^{131}I]iodide present in the thyroxine sample would not be measured. The values obtained by both of these methods are in the same range as thyroid secretion rates calculated for other species of bird and for mammals.

V. Control of Thyroid Secretion

The direct control of the activity of the thyroid gland in homeothermic animals is exerted by the pituitary secretion of thyroid stimulating hormone (TSH). This hormone has been purified from the anterior pituitary gland of a number of species, and is a peptide of molecular weight of about 30,000 daltons. TSH circulates in the plasma in very low concentrations, in man 1–10 mU/100 ml appear to be normal. The highest potency TSH isolated has 50 mU/mg, so that 10 mU represents approximately 0·2 μg of hormone.

The effects of TSH on the thyroid are diverse, and can be detected within 1 to 30 min of administration of the hormone. Blood flow through the gland increases within one minute, and an increased output of thyroid hormone within 20 min. Ultrastructural studies have shown the engulfing of follicular colloid into cell cytoplasm within 10 min of TSH administration and the hydrolysis of thyroglobulin contained in this colloid provides free thyroid hormones for secretion.

The longer-term effects of TSH on thyroid function include an increased iodide uptake, increased hormone biosynthesis and an increased gland size.

Secretion of TSH is inversely related to the thyroid hormone content of the blood, forming a negative feedback system. Thus a high concentration of circulating thyroid hormone acts on the pituitary to suppress the release of TSH and hence indirectly to reduce the secretion of thyroid hormones. In addition to this control by negative feedback, the central nervous system can directly regulate TSH release through a hypothalamic neurosecretion called thyrotrophin releasing factor. This, and other releasing factors, are liberated into the blood vessels of the hypophysial portal system, at or near their initiation in the hypothalamus, and pass in the portal blood to the anterior pituitary. A rapid release of TSH occurs after the infusion of the releasing factor into the pituitary. Burgus *et al.* (1970) have recently isolated and identified the ovine TSH-releasing factor as 2-pyrrolidone-5-carboxy-histidyl-proline amide (Fig. 5) and it appears likely that this tripeptide is also the releasing factor in other mammalian species.

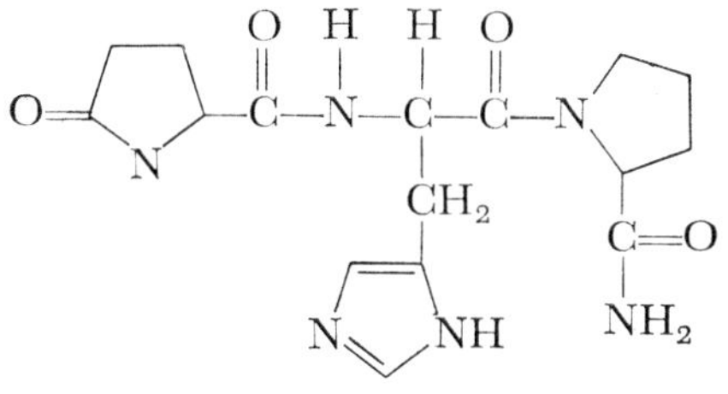

O H H O
O C—N—C—C—N
N
CH_2
C=O
N NH NH_2

Fig. 5. The structure of ovine TSH-releasing factor, as established by Burgus *et al.* (1970). The molecule is a tripeptide, 2-pyrrolidone-5-carboxy-histidyl-proline amide; the pyrrolidone group may have been formed by cyclization of a terminal glutamate residue during extraction.

The area of the hypothalamus which appears to control the secretion of TSH-releasing factor is in the region above and behind the optic chiasma as lesions in the area between the anterior commissure, posterior commissure and optic chiasma suppress thyroid activity in fowls, and lesions in the supraoptico-hypophysial tract reduce thyroid activity in mammals (see Brown-Grant, 1966; Kanematsu and Mikami, 1970).

This higher control of TSH secretion appears to be important in response to a variety of stresses, such as cold, which affect thyroid activity. It is also likely that thyroid changes which are associated with reproduction are mediated through the hypothalamic regulation of pituitary TSH release (for a general discussion of thyroid control by TSH secretion see Brown-Grant, 1966).

VI. Thyroid Function and Growth

An extensive study of the relationship between age and thyroid secretion rate in White Leghorn and crossbred cockerels under a controlled environment was carried out by Tanabe (1965). His experiments showed a linear decrease in secretion rate with age over the period of 14 to 100 d (Fig. 6). This decline is similar to that seen in post-pubertal mammals, and is reflected by a decrease in the metabolic rate of both birds and mammals.

The relationship between the rate of growth of birds and thyroid secretion rate was also calculated from the data obtained by Tanabe (1965), and shows a linear increase in rate of body weight gain with increasing thyroxine secretion rate. However, this correlation may be entirely fortuitious, as the

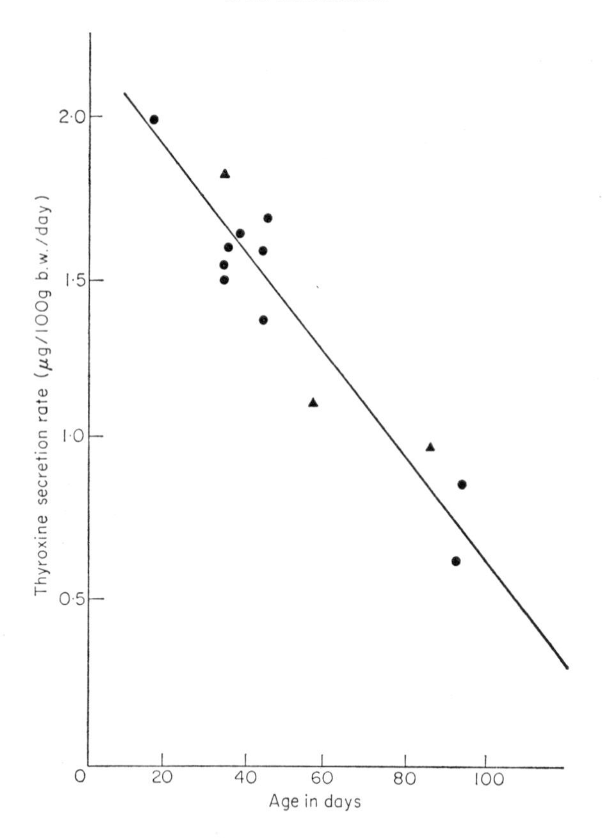

Fig. 6. Relationship between thyroxine secretion rate and age of cockerels. ● represents the rate of thyroxine secretion of White Leghorns, and ▲ represents the rate of crossbreds. Reproduced with permission from Tanabe (1965).

correlation of rate of gain with age shows a linear decrease in gain with increasing age. For validity all the birds used in the measurement need to be of the same age, not, as was the case, over the whole range from 14 to 100 d of age. In an investigation of the relationship between thyroid secretion and growth rate in sheep under carefully controlled conditions a curvilinear relationship was demonstrated, with marked decreases in growth rate in animals which were hyper- or hypo-thyroid (Draper *et al.*, 1968).

This type of relationship is also probable in the growing chick, as the results of Singh *et al.* (1968b) indicate increases in growth rate with thyroxine administration at 2 or 3 μg/d 100 g body weight in birds given an antithyroid drug, whereas a dose of 4 μg/d resulted in a lower growth rate, similar to that shown by chicks receiving 1 μg/d 100 g body weight. Goitrogen-treated chicks, with no hormone supplementation, had a very poor growth rate as would be predicted from studies on cretinism in mammals.

It would therefore appear that moderate increases in thyroid hormone availability in chicks will stimulate accelerated growth, provided that the additional hormone does not exceed about 1·5 times the normal secretion rate.

VII. Thyroid Function and the Environment

The variation in thyroid secretion with season of the year was first investigated in the chick by Reineke and Turner (1945). They showed maximum secretion during October and November, with the lowest levels from April to August. Since these early investigations, the elevation of thyroid activity in birds during the winter months has been recorded by many other workers.

Thyroid secretion rate in adult birds has been shown to be increased by exposure to 4·4°C for short periods, and cold has also been shown to cause increased TSH release in birds (Hendrich and Turner, 1965). Similar responses to cold have been widely demonstrated in mammals.

High environmental temperatures (30 to 35°C) have a depressing effect on thyroid secretion, in almost all experiments reported. Only under extreme conditions of heat (45 to 48°C) has an activation of the thyroid in birds been observed (Chaudhuri and Sadhu, 1961). The speed of response of the mammalian thyroid to elevations of body temperature is almost immediate, indicating that a mechanism other than the normal negative feedback regulation of the thyroid is involved.

The elegant experiments of Andersson *et al.* (1962, 1963) have shown that cooling mammals (goats) results initially in a fall in body temperature, followed by a rise in temperature with a parallel rise in circulating thyroid hormone. By warming the pre-optic area of the brain during cooling of the body, the increase in thyroid hormone secretion was completely prevented. It was therefore clear that a temperature-regulating centre in the hypothalamus was initiating the response, presumably through the secretion of hypothalamic TSH-releasing factor.

The physiological importance of the regulation of thyroid output by environmental temperature is presumably through the influence of thyroid hormones on metabolic rate. In mammals, the direct relationship between thyroid activity and basal metabolic rate (BMR) has been well established and in human clinical practice measurement of BMR may be used as an indirect assessment of thyroid activity. Increasing environmental temperatures have been shown to result in a reduction of metabolic rate in mammals, which closely corresponds to the observed reduction in thyroid activity (Collins and Weiner, 1968). Since the maintenance of body temperature in homeothermic species is of considerable importance, the role of the thyroid gland in the metabolic regulation of heat production under both hot and cold conditions will be of value for survival.

Very little data exist for the effects of thyroid activity on metabolic rate in birds. The injection of thyroxine into chickens resulted in a rise in metabolic rate of only short duration, probably because of the rapid rate of destruction of thyroid hormones in the bird (Singh *et al.*, 1968b). Until further evidence is available, it would appear likely that birds and mammals are similar in their thyroid-response to environmental temperature.

VIII. Thyroid Function and Reproduction

If chicks are thyroidectomized, the gonads remain juvenile and comb and wattle size are reduced. Even in adult hens and cockerels, removal of the thyroid glands leads to a marked reduction in egg production or to testicular atrophy. These observations were made early in the study of avian thyroid function (Taylor and Burmester, 1940; Blivaiss, 1947) and resulted in extensive studies of the effect of thyroid hormone on egg production, with the aim of increasing the performance of laying hens. Since both egg laying and thyroid activity in hens decline during the summer months, the feeding of diets containing thyroid hormones has been employed to maintain a raised level of hormone availability during the summer in laying hens. At the optimum rate of thyroid hormone intake the seasonal fall in egg production has been partially prevented (Turner *et al.*, 1945) but at higher rates of hormone feeding, egg production and body weight decrease and mortality increases.

Studies of naturally occurring differences in thyroid activity and egg production have proved interesting, as hens laying four-egg sequences were found to have a higher thyroxine secretion rate than similar hens laying two-egg sequences (Booker and Sturkie, 1950). The patterns of reproductive response to thyroid hormone therefore resemble that shown by growth, as a deficiency of hormone induced by thyroidectomy or the feeding of goitrogens reduce both growth and reproductive performance. Moderate rates of hormone administration or naturally high rates of secretion will increase egg production, particularly during the warmer months when thyroid activity is normally depressed by elevated temperatures. Similarly, a moderate rate of thyroid hormone treatment stimulates an increased growth rate. However, an excessive dosage of thyroid hormone has progressively more deleterious consequences on both growth and reprodution, with increasing doses.

The influence of antithyroid drugs and thyroid hormone administration on testicular development and spermatogenesis in the male bird have also been studied. Hypothyroidism results in reduced testicular development and fertility, depending on the severity of the hormone deficiency. The effect of thyroid hormone administration is more complex; as would be expected high dosages were shown to impair fertility, but low dosages increased the sperm concentration in seminal fluid (Wilwerth *et al.*, 1954).

IX. Thyroid Function, Moulting and Feather Growth

In a number of avian species an increased thyroid secretion has been observed just preceding the annual moult, during the summer when thyroid activity is otherwise low. This is particularly well defined in the mallard duck, where the increased thyroid activity occurs one month earlier in males than in females, and the males also moult one month earlier than the females (Höhn, 1961).

Thyroidectomy of birds abolishes moulting and also has marked effects on feather growth. The feathers grow at a reduced rate and are abnormally elongated and narrowed; they are also loose in texture due to a reduction in the formation of interlocking barbules. In dark-coloured breeds of domestic fowl there is a reddening of feather pigmentation (Höhn, 1961).

Thyroxine treatment reverses the effects of thyroidectomy, and in intact birds can cause the formation of darker coloured, more rounded feathers. The administration of a large dose of thyroxine causes a sudden moult, with a reduction in egg laying and atrophy of the comb and wattles. Hypophysectomy or treatment with gonadotrophin-inhibiting drugs also causes moulting. Höhn (1961) concluded that moulting is caused by a reduction in gonadotrophin secretion which can be artificially induced by a high dose of thyroxine. The role of thyroid hormones in the process of moulting may be in their requirement for the stimulation of growth of new feathers, which is a normal preliminary to moulting.

It is apparent that in many areas our knowledge of thyroid physiology in the fowl is inadequate, and that almost no experimental data exists on the biochemistry of the avian thyroid or on the mode of action of thyroid hormones in birds. With the recent improvements in methodology for studying avian thyroid function, and in biochemical techniques for studying hormone action, the opportunity exists for much valuable research in this field.

Acknowledgements

The author wishes to express his gratitude to Dr. Margaret Birkinshaw for her skilled electron microscopy and anatomical drawing, and to Dr. P. W. Murphy and Mr. J. Hall for their assistance with thyroid histology.

References

Andersson, B., Ekman, L., Gale, C. C. and Sundsten, J. W. (1962). *Acta physiol. scand.* **56**, 94–96.

Andersson, B., Ekman, L., Gale, C. C. and Sundsten, J. W. (1963). *Acta physiol. scand.* **59**, 12–33.

Beyer, R. E. (1952). *Endocrinology* **50**, 497–503.

Blasi, F., Fragomele, F. and Covelli, I. (1969). *Endocrinology* **85**, 542–551.

Blivaiss, B. B. (1947). *J. exp. Zool.* **104**, 267–310.

Booker, E. E. and Sturkie, P. D. (1950). *Poult. Sci.* **29**, 240–243.

Brown-Grant, K. (1966). *In* "The Pituitary Gland" (G. W. Harris and B. T. Donovan, eds.), Vol. 2, pp. 235–269. Butterworths, London.

Burgus, R., Dunn, T. F., Desiderio, D., Ward, D. N., Vale, W. and Guillemin, R. (1970). *Nature, Lond.* **226**, 321–325.

Chaudhuri, S. and Sadhu, D. P. (1961). *Nature, Lond.* **192**, 560–561.
Cheftel, C. and Bouchilloux, S. (1968). *Biochim. biophys. Acta* **170**, 15–28.
Collins, K. J. and Weiner, J. S. (1968). *Physiol. Rev.* **48**, 785–839.
Draper, S. A., Falconer, I. R. and Lamming, G. E. (1968). *J. Physiol., Lond.* **197**, 659–665.
Falconer, I. R. (1966). *Biochem. J.* **100**, 190–196.
Frienkel, H. and Lewis, D. (1957). *J. Physiol., Lond.* **135**, 288–300.
Hansborough, L. A. and Khan, M. (1951). *J. exp. Zool.* **116**, 447–454.
Hendrich, C. E. and Turner, C. W. (1964). *Proc. Soc. exp. Biol. Med.* **117**, 218–222.
Heninger, R. W. and Newcomer, W. S. (1964). *Proc. Soc. exp. Biol. Med.* **116**, 624–628.
Höhn, E. O. (1961). *In* "Biology and Comparative Physiology of Birds" (A. J. Marshall, ed.), Vol. II, pp. 97–102. Academic Press, New York.
Kanetmatsu, S. and Mikami, S.-I. (1970). *Gen. comp. Endocr.* **14**, 25–34.
Mellen, W. J. and Hardy, L. B. (1957). *Endocrinology* **60**, 547–551.
Mellen, W. J. and Wentworth, B. C. (1959). *U.S. Atomic Energy Commission Pub.* TID-7578:77.
Ray, A. K. and Premachandra, B. N. (1964). *Endocrinology* **74**, 800–802.
Reineke, E. P. and Turner, C. W. (1945). *Poult. Sci.* **24**, 499–503.
Robertson, H. A. and Falconer, I. R. (1961). *J. Endocr.* **21**, 411–420.
Romanoff, A. L. (1960). "The Avian Embryo" pp. 866–878. McMillan, New York.
Romanoff, A. L. and Laufer, H. (1956). *Endocrinology* **59**, 611–619.
Shellabarger, C. J. and Pitt-Rivers, R. (1958). *Nature, Lond.* **181**, 546.
Singh, A., Reineke, E. P. and Ringer, R. K. (1968a). *Poult. Sci.* **47**, 205–211.
Singh, A., Reineke, E. P. and Ringer, R. K. (1968b). *Poult. Sci.* **47**, 212–219.
Stanbury, J. B. (1967). *In* "Endocrine Genetics", (S. G. Spickett ed.), pp. 107–136. University Press, Cambridge.
Sun, T. B. (1932). *Physiol. Zoöl.* **5**, 384–396.
Tanabe, Y. (1965). *Poult. Sci.* **44**, 591–595.
Tata, J. R. and Shellabarger, C. J. (1959). *Biochem. J.* **72**, 608–613.
Taylor, L. W. and Burmester, B. R. (1940). *Poult. Sci.* **19**, 326–331.
Thompson, J. A. and Goldberg, I. H. (1968). *Endocrinology* **82**, 805–817.
Turner, C. W., Kempster, H. L., Hall, N. M. and Reineke, E. P. (1945). *Poult. Sci.* **24**, 522–533.
Wentworth, B. C. and Mellen, W. J. (1961). *Poult. Sci.* **40**, 1275–1276.
Wilwerth, A. M., Martinez-Campos, C. and Reineke, E. P. (1954). *Poult. Sci.* **33**, 729–735.

18 The Parathyroid Glands

T. G. TAYLOR

Department of Physiology and Biochemistry,
University of Southampton,
Southampton, England

I. Structure

There are initially four parathyroid glands in the embryo which arise from the third and fourth visceral pouches. The pair on each side usually fuse to give one glandular mass. This is light brown and is usually close, or even attached, to the posterior poles of the thyroids but considerable variation occurs in its precise location. Each parathyroid consists of chief (parenchyma) cells arranged in cords separated by connective tissue, as in mammals, but no oxyphil cells are present (Benoit, 1950). The fine structure of the parathyroid gland of the laying hen has been described by Nevalainen (1969); in the chief cells the Golgi apparatus is well developed and large numbers of small secretory granules are present in this region, but there are few large storage granules. Nevalainen suggests that most of the secretory product is liberated in the form of small granules and that relatively few of these coalesce to form mature storage granules.

Accessory parathyroid tissue occurs in the thymus and in the ultimobranchial bodies (Nonidez and Goodale, 1927; Dudley, 1942).

In females, the parathyroids enlarge with the onset of reproductive activity (MacOwan, 1931–2). They undergo hypertrophy and hyperplasia when diets deficient in either calcium (Hurwitz and Griminger, 1961) or vitamin D (Nonidez and Goodale, 1927) are given. Hertelendy (1962) fed chicks diets deficient in either calcium, or vitamin D_3 or both and Table 1

Table 1

The effects of diets low in calcium and/or vitamin D_3 on the mean body weights and mean parathyroid weights of chicks 36 to 39 d of age (Hertelendy, 1962)

	Diet			
	Normal calcium		Low calcium	
	Normal D_3	Low D_3	Normal D_3	Low D_3
Body weight (g)	668	349	553	261
Parathyroid weight (mg)	14·0	51·5	58·0	30·5
Parathyroid weight as percentage body weight (× 1000)	2·1	14·8	10·5	11·7

shows the effects of these diets on the weight of the parathyroids in relation to body size. Diets relatively deficient in available phosphorus (0·4%) and containing excessive amounts of calcium (3·0%) cause atrophy of the parathyroids of growing pullets and a depression in their secretory function as judged by their histological appearance (Shane *et al.*, 1969). Increasing the available phosphorus in the diet to 1·2% prevents the degeneration of the glands (Shane *et al.*, 1968).

II. General Function

A. PARATHYROIDECTOMY

The classical method of investigating the physiological function of an endocrine gland consists of studying the effects of its removal and the administration, to intact animals, of an extract of the gland. As in mammals, the parathyroid glands are concerned in the regulation of the level of ionic calcium in the plasma and hence in the tissue fluids; parathyroidectomy results in a fall, and injections of parathyroid extract in a rise, in the plasma calcium level.

Parathyroidectomy has been carried out in mature chickens by a number of different workers, some of whom reported a high mortality (Urist *et al.*, 1960) while others experienced few losses (Polin *et al.*, 1957; Polin and Sturkie, 1958). The response to parathyroidectomy may be quite variable; some birds show a marked reduction in the level of plasma calcium while others show no response at all (Polin and Sturkie, 1958). Birds that survive for 48 h after the operation recover completely, usually within a week and sometimes within as short a time as 2 d, due to the hyperplasia and hypertrophy of accessory parathyroid tissue. It seems reasonable to suggest that birds that do not exhibit a fall in plasma calcium after parathyroidectomy and those that recover particularly rapidly after the operation, possess unusually large amounts of actively secreting accessory tissue. Complete removal of all parathyroid tissue does not appear to have been achieved in chickens, but Benoit *et al.* (1941) were probably successful in this operation in ducks, judging by the extremely low levels of plasma calcium reached and by the short time that their birds survived (20 to 24 h).

B. EFFECTS OF PARATHYROID EXTRACTS ON PLASMA CALCIUM LEVELS

Chickens respond to bovine parathyroid preparations in much the same way as mammals; however, the maximum hypercalcaemic response is reached earlier and the effect is relatively short-lived. Polin *et al.* (1957) observed maximum responses in laying hens and in cocks 3·5 h after subcutaneous injections of parathyroid extracts (PTE); the plasma levels had returned to normal 7 h after the injections. Similar experiments were carried out by Hertelendy (1962) who observed maximum elevations in the total plasma calcium after 2 h and a return to pre-injection levels after 5 h. After intravenous injections, Hertelendy (1962) found that the response was even faster and more transient. A significant elevation in total plasma calcium occurred after 0·5 h, maximum levels were observed after 1·5 h and normal values were restored in 2·5 h.

A recent report by Candlish and Taylor (1970) indicates that the response of the plasma diffusible calcium to PTE is even more rapid than that observed by Hertelendy (1962) for the total calcium level. In the former work the wing veins of heparin-treated hens were connected directly to the sample line of an Auto Analyser. The blood was passed through a dialyser so that only diffusible calcium was measured and a hypercalcaemic response to PTE was observed in less than 10 min (Table 2). The response lasted for at least 15 min and took the form of an elevated plateau. It is possible that this rapidly-induced hypercalcaemia reflects the immediate response of the bone

Table 2

The effect of intravenous injections of parathyroid extract (PTE) on changes in the blood calcium of laying fowls (Candlish and Taylor, 1970)

Treatment	No. of birds	Mean response time (min) ± S.E.	Increase in blood calcium (mg/100 ml) ± S.E.
PTE 20 USP units/kg	4	8·1 ± 2·1	2·9 ± 0·7
PTE 100 USP units/kg	4	7·2 ± 1·9	4·4 ± 1·1
Solvent for PTE (control)	3	—	Nil

cells to parathyroid hormone (PTH) postulated by Nichols (1970) and discussed in Section III. On the other hand, it may have been due to a redistribution of calcium between the diffusible and non-diffusible fractions under the influence of the hormone.

Immature birds also respond to bovine PTE and Polin *et al.* (1957) have proposed a biological assay based on the hypercalcaemic response in 5- to 7-week-old chicks.

Polin *et al.* (1957), Polin and Sturkie (1957, 1958) and Hertelendy (1962) have all reported that the hypercalcaemic response to a standard dose of PTE is far greater in the laying hen than in the cock and Urist *et al.* (1960) and Hertelendy (1962) also showed that oestrogen-treated cocks gave a response in total calcium similar in magnitude to that in laying hens, but that the rise in diffusible calcium was of the same order in hens as in normal and oestrogen-treated cocks. The greater response in laying hens compared with cocks in the level of total calcium is presumably due to the plasma lipophosphoproteins binding additional calcium as the ionic calcium rises, since both the major calcium fractions of the blood, the diffusible and non-difusible, are in equilibrium with one another.

There is some disagreement among different workers as to whether or not PTH is necessary for the hypercalcaemic action of oestrogen. Those who claim that it is essential (e.g. Benoit, 1950) worked with parathyroidectomized ducks which suffered very severe falls in plasma calcium levels, while those who were successful in obtaining a response to oestrogen (Riddle *et al.*, 1945) maintained the plasma calcium of their birds (pigeons) after parathyroidectomy at a considerably higher level before oestrogen was administered by feeding either aluminium hydroxide (to prevent absorption of phosphate) or calcium gluconate together with dihydrotachysterol (a sterol which causes bone resorption in much the same way as an excess of vitamin D). The increase in plasma calcium following oestrogen treatment is due to the release into the plasma of a specific calcium-binding phosphoprotein synthesized by the liver (see Chapter 41) and there is no reason to believe that PTH is necessary for the synthesis of this protein. However, a certain minimum concentration of calcium may be necessary for it to be transported from the liver. If this is so—and it seems to offer the best explanation for resolving what appears to be conflicting evidence—oestrogen will not induce a hypercalcaemic response when, for any reason, plasma calcium is exceptionally low; PTH cannot therefore be considered essential for the action of the sex hormone. According to this theory, then, it is the level of ionic calcium in the plasma that is important in this respect, not PTH itself.

C. HALF-LIFE OF THE HORMONE

The half-life of PTH in the blood of birds is not known but in view of the short period during which a hypercalcaemic response to exogeneous hormone is observed, it is unlikely to be long. In the rat it has been estimated to be of the order of 20 min (Melick *et al.*, 1965).

D. GENERAL MODE OF ACTION

It is now generally accepted that there are two major effects of parathyroid hormone in mammals; it increases both the rate of bone resorption and the urinary excretion of phosphate. It also exerts a small increase in the intestinal absorption of calcium (Cramer, 1963; Care and Keynes, 1964). The actions of the hormone in birds appear to be similar, although relatively little work has been carried out, and no experiments seem to have been conducted with avian PTH.

III. Physiological Action

A. ACTION ON BONE

The effect on bone resorption is brought about by an increase in the activity and numbers of the osteoclasts, which attack the surfaces of the bone (see Hancox, 1956 for a review) and also by a process known as osteolysis, whereby the large osteocytes embedded in the bone resorb the tissue from the inside (Belanger *et al.*, 1963). Osteoclasis has been described in the medullary bone of pigeons during the egg cycle by Bloom *et al.* (1941) and in hens by Bloom *et al.* (1958) and osteolysis has been shown to occur in the medullary bone of laying hens, particularly after treatment with PTE, by Taylor and Belanger (1969).

There are no *a priori* reasons for supposing that the physiological role of the parathyroids in birds is any different from that in mammals, i.e. the maintenance of the plasma level of ionic calcium within the narrow limits necessary for the normal functioning of muscles, nerves and, indeed, of all tissues and cells. In the laying hen this task is particularly exacting because of the sudden and enormous drains on the blood calcium that are periodically made during egg shell calcification. It seems probable that soon after the period of rapid shell calcification begins there is a fall in the plasma ionic calcium (see also Chapter 55) (Taylor and Hertelendy, 1961), which causes increased secretion of PTH into the blood thus stimulating skeletal resorption. The medullary bone, the fine spicules of which present a huge surface area to the blood sinuses that permeate the marrow cavities, is thought to respond most rapidly to the hormone (Taylor and Belanger, 1969). When shell calcification is completed, the plasma calcium level rises, secretion of PTH is reduced and with it the rate of bone resorption. At present, evidence in favour of this suggested train of events is purely circumstantial and cannot be confirmed or denied until an assay is developed for PTH, sensitive enough to detect changes in plasma concentration during the egg cycle.

The control of medullary bone resorption during the egg cycle and the possible role of oestrogen in this control are discussed further in Chapters 26 and 55.

B. ACTION ON THE KIDNEY

An increased urinary excretion of phosphate in chickens injected with PTE has been reported by Levinsky and Davidson (1957). This was confirmed by Martindale (1969) using laying hens and shown to be due to a decrease in tubular resorption. Other effects of PTE have been studied by Candlish (1970); these include a water diuresis which was almost instantaneous and an increase in the excretion of calcium, hydroxyproline and uronic acids within 12 min after 125 USP units PTE/kg were injected intravenously. Buchanan (1961) reported that intramuscular injections of PTE, in amounts which did not exert a hypercalcaemic effect (50 USP units/kg) in cocks, markedly increased the urinary calcium and phosphorus within 30 min. Smaller doses (down to 5 USP units/kg) gave only the phosphaturic response and the minimum dose required to give a clear-cut calcium response was 37·5 units/kg. Glomerular filtration rates were not affected by the PTE in these experiments.

IV. Biochemical Action

The mode of action of PTH at a biochemical level is not fully understood. Bone destruction clearly involves both the solubilization of bone mineral and the breakdown of the organic matrix. The former can most readily be achieved by means of acids, while destruction of the organic matrix must presumably require the intervention of enzymes; a number of the lysosomal acid hydrolases possess the specificity necessary for breaking down the complex polysaccharides of the ground substance and the collagen of the fibrous matrix (Vaes, 1965: Woods and Nichols, 1965). Lactic acid appears to be the most important of the organic acids involved in bone resorption (Borle *et al.*, 1960a, b).

Vaes (1965) has demonstrated that PTE stimulates the formation of acid and the release and synthesis of a number of acid hydrolases of lysosomal origin in explants of murine bone growing in tissue culture. The enzymes studied included β-glucuronidase, β-*N*-acetyl-glucosaminidase, acid phosphatase, hyaluronidase, β-galactosidase and cathepsin. The skull bones of new-born rats also showed increased activity of the same acid hydrolases after treatment with PTE for three days (Vaes, 1966).

Acid phosphatase is commonly used as an indicator-enzyme for lysosomal acid hydrolases in general and Taylor *et al.* (1965) have shown that the plasma level of this enzyme is elevated during the main period of shell formation in the hen and that the level falls precipitously at about the time of oviposition. Morris and Taylor (1970) observed that the urinary excretion of acid phosphatase by laying hens was more than three times as high on shell-forming days than on days when egg formation did not occur. Urinary hydroxyproline was also elevated on shell-forming days and there was a high correlation between the concentrations of hydroxyproline and acid phosphatase, which is consistent with the view that this enzyme is released from bone during resorption.

Both osteoclasts and mature osteocytes are rich in lysosomes but the mechanism by which PTH stimulates the release of the acid hydrolases from these cells is not clear. One theory, put forward by Wells and Lloyd (1968), suggests that this is achieved by cyclic 3′, 5′-adenosine monophosphate (cyclic AMP), formation of which from ATP by the enzyme adenyl cyclase is stimulated by PTH in both kidney (Chase and Aurbach, 1968a) and bone cells (Chase and Aurbach, 1968b). It has been shown (Rasmussen *et al.*, 1964) that actinomycin D, which inhibits DNA-mediated protein synthesis at the transcription level, prevents the long-term effect of PTH on the blood calcium, but not the short-term effect (up to 6 h). This is taken to mean that actinomycin D inhibits the synthesis of new lysosomal enzymes but not the release of those already present in the lysosomes.

Nichols (1970) has postulated that the immediate response of bone cells, particularly osteocytes, to PTH is to increase their rate of calcium turnover, under the stimulus of an increased level of cyclic AMP. According to this theory the rate at which the bone cells pump calcium from the fluid bathing the bone tissue into the channels in the bones that communicate with the vascular system is greatly increased by PTH. This effect on the bone cells does not itself involve bone resorption, i.e. the destruction of both organic and mineral components, but, if the PTH-stimulus continues, bone resorption ensues. According to Nichols' theory, the accumulation of calcium by the bone cells that accompanies the increase in the rate of calcium turnover, sets in train the events discussed above, i.e. the release of acids and acid hydrolases, that result in bone resorption.

References

Belanger, L. F., Robichon, J., Migicovsky, B. B., Copp, D. H. and Vincent, J. (1963). *In* "Mechanisms of Hard Tissue Destruction" (R. F. Sognnaes, ed.), pp. 531–556. American Association for the Advancement of Science, Washington, D.C.

Benoit, J. (1950). *In* "Traité de Zoologie" (P. P. Grassé, ed.), Vol. 15, pp. 290–334. Masson, Paris.

Benoit, J., Stricker, P. and Fabiani, G. (1941). *C. r. hebd. Séanc Soc. biol. Paris* **135**, 1600–1602.

Bloom, M. A., Domm, L. V., Nalbandov, A. V. and Bloom, W. (1958). *Am. J. Anat.* **102**, 411–453.

Bloom, W., Bloom, M. A. and McLean, F. C. (1941). *Anat. Rec.* **81**, 443–475.

Borle, A. B., Nichols, N. and Nichols, G. (1960a). *J. biol. Chem.* **235**, 1206–1210.

Borle, A. B., Nichols, N. and Nichols, G. (1960b). *J. biol. Chem.* **235**, 1211–1214.

Buchanan, J. D. (1961). *In* "The Parathyroids" (R. O. Green and R. V. Talmage, eds.), pp. 334–352. Charles C. Thomas, Springfield, Illinois.

Candlish, J. K. (1970). *Comp. Biochem. Physiol.* **32**, 703–707.

Candlish, J. K. and Taylor, T. G. (1970). *J. Endocr.* **48**, 143–144.

Care, A. D. and Keynes, W. M. (1964). *Proc. R. Soc. Med.* **57**, 867–870.
Chase, L. R. and Aurbach, G. D. (1968a). *In* "Parathyroid Hormone and Thyrocalcitonin (Calcitonin)" (R. V. Talmage and L. F. Belanger, eds.), pp. 247–257. Excerpta Medica Foundation, Amsterdam.
Chase, L. R. and Aurbach, G. D. (1968b). *In* "Abstracts of brief communications, 3rd int. Endocrine Congress", p. 87. Excerpta Medica Foundation, Amsterdam.
Cranmer, C. F. (1963). *Endocrinology* **72**, 192–196.
Dudley, J. (1942). *Am. J. Anat.* **71**, 65–97.
Hancox, N. M. (1956). *In* "The Biochemistry and Physiology of Bone" (G. H. Bourne, ed.), pp. 213–250, Academic Press, New York.
Hertelendy, F. (1962). *Ph.D. thesis, University of Reading.*
Hurwitz, S. and Griminger, P. (1961). *J. Nutr.* **23**, 177–185.
Levinsky, N. G. and Davidson, D. G. (1957). *Am. J. Physiol.* **191**, 530–536.
MacOwan, M. M. (1931–2). *Q. Jl exp. Physiol.* **21**, 383–392.
Martindale, L. (1969). *J. Physiol., Lond.* **203**, 82*P*–83*P*.
Melick, R. A., Aurbach, G. D. and Potts, J. T. (1965). *Endocrinology* **77**, 198–202.
Morris, K. M. L. and Taylor, T. G. (1970). *Annls Biol. anim. Biochim. Biophys.* **10**, 185–190.
Nevalainen, T. (1969). *Gen. comp. Endocr.* **12**, 561–567.
Nichols, G. (1970). *Calc. Tiss. Res.* **4** (Supplement) 61–63.
Nonidez, F. J. and Goodale, H. D. (1927). *Am. J. Anat.* **38**, 319–341.
Polin, D. and Sturkie, P. D. (1957). *Endocrinology* **60**, 778–784.
Polin, D. and Sturkie, P. D. (1958). *Endocrinology* **63**, 177–182.
Polin, D., Sturkie, P. D. and Hunsaker, W. (1957). *Endocrinology* **60**, 1–5.
Rasmussen, H., Arnaud, C. and Hawker, C. (1964). *Science, N.Y.* **144**, 1019–1020.
Riddle, O., Rauch, V. M. and Smith, G. C. (1945). *Endocrinology* **36**, 41–47.
Shane, S. M., Young, R. J. and Krook, L. (1968). *Proc. Cornell Nutr. Conf.* 126–131.
Shane, S. M., Young, R. J. and Krook, L. (1969). *Avian Dis.* **13**, 558–567.
Taylor, T. G. and Belanger, L. F. (1969). *Calc. Tiss. Res.* **4**, 162–173.
Taylor, T. G. and Hertelendy, F. (1961). *Poult. Sci.* **40**, 115–123.
Taylor, T. G., Williams, A. and Kirkley, J. (1965). *Can. J. Physiol.* **43**, 451–457.
Urist, M. R., Deutsch, N. M., Pomerantz, G. and McLean, F. C. (1960). *Am. J. Physiol.* **199**, 851–855.
Vaes, G. (1965). *Expl Cell Res.* **39**, 470–474.
Vaes, G. (1966). *In* "Calcified Tissues, 1965" (H. Fleisch, H. J. J. Blackwood and M. Owen, eds.), pp. 56–59. Springer-Verlag, Berlin.
Wells, H. and Lloyd, W. (1968). *In* "Parathyroid Hormone and Thyrocalcitonin (Calcitonin)" (R. V. Talmage and L. F. Belanger, eds.), pp. 332–333. Excerpta Medica Foundation, Amsterdam.
Woods, J. F. and Nichols, G. (1965). *J. Cell Biol.* **26**, 747–757.

ADDENDUM

The most important biochemical reactions leading to the formation of increased amounts of citric and lactic acids by the bone cells have been reviewed by Tenenhouse *et al.* (1970). An increased intracellular concentration of calcium inhibits the enzyme isocitrate dehydrogenase which results in an accumulation of citric acid and of intermediates of the glycolysis pathway, while an increase in cyclic AMP activates the phosphorylase system and with it the rate of glycolysis, leading to an increase in the intracellular concentration of lactic acid.

Tenenhouse, A., Rasmussen, H. and Nagata, N. (1970). *In* "Calcitonin 1969" (S. Taylor and G. Foster, eds.), pp. 418–426. Heinemann, London.

19 Ultimobranchial Glands and Calcitonin

K. SIMKISS and C. G. DACKE[1]

Department of Zoology and Comparative Physiology, Queen Mary College, University of London, and Department of Physiology and Biochemistry, University of Reading, Reading, England

I. Introduction

The concentration of calcium ions in the plasma of vertebrates is regulated and maintained within narrow limits by the activities of certain endocrine systems. Of these the most important is the parathyroid glands, the hormone from which affects the skeletal, renal and intestinal systems so as to increase the level of plasma calcium and decrease that of phosphate. It was once thought that a decrease in the rate of secretion of parathyroid hormone would be adequate to lower the level of plasma Ca^{2+} should there be an excess in the blood (McLean, 1958). The experiments of Copp *et al.* (1961) questioned this, for they found a very rapid fall in plasma calcium when the thyroid-parathyroid system of dogs was perfused with a calcium-rich solution. The fall in plasma calcium was greater in these perfused animals than in thyroid-parathyroidectomized animals. This was difficult to explain on a simple feedback hypothesis since it implied that the abolition of parathyroid secretion (by surgical removal) produced a slower response than an inhibition of hormone secretion (by perfusion with calcium-rich solutions). (See also Chapter 18.) They therefore suggested that there must be some active compound which increased the rate of calcium removal from the blood of the intact animal, i.e. a hormone with the opposite effect of parathyroid hormone and named it calcitonin.

II. Source of Calcitonin: the Ultimobranchial Body

The cells responsible for the secretion of calcitonin arise from the endoderm of the 6th branchial pouch of the embryo. In mammals these cells migrate

[1] Present address: Space Sciences Research Center, University of Missouri, Columbia, Missouri, 65201, U.S.A.

into the neck region during development and may thus become incorporated to some extent in the parathyroids and thymus but they are found mainly in the thyroid gland where they can be identified in the adult as the parafollicular C cells (Pearse and Carvalheira, 1967). Hormone extracted from the mammalian thyroid was originally called thyrocalcitonin (Hirsch *et al.*, 1964) and it may differ slightly from avian calcitonin. In most other vertebrates, including the birds, the cells from the 6th branchial pouch are found in the adult mainly as a small gland in the neck called the ultimobranchial body (Copp *et al.*, 1968) although there may also be a few cells in other glands, such as the thyroid (Pearse and Carvalheira, 1967). The ultimobranchial body is a rich source of calcitonin.

The ultimobranchial glands of the fowl are not encapsulated but are small (2 to 3 mm) structures lying in the neck posterior to the thyroid and parathyroid glands. In the freshly-killed bird they are pink due to their rich blood supply from the carotid arteries. Blood from the glands drains into the jugular veins. The glands are found just posterior to the parathyroids close to the origin of the carotid and subclavian arteries (Fig. 1). A detailed review

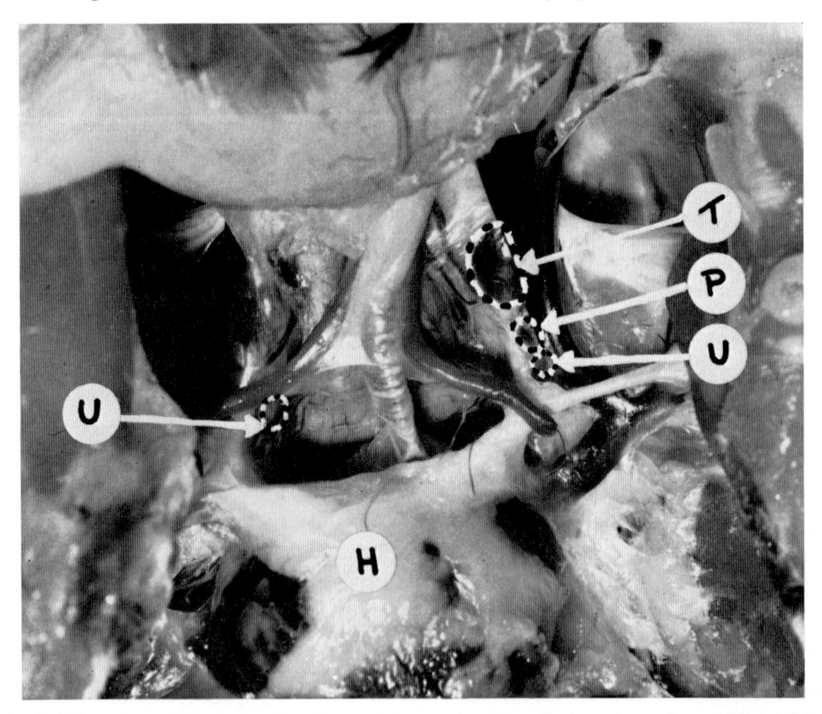

Fig. 1. Dissection of the ventral surface of the neck of the fowl to show the position of the thyroids (T), parathyroids (P), and ultimobranchial glands (U) on the left-hand side of the body. A black pointer has been inserted along the side of these glands and indicates the position of the small ultimobranchials. On the right-hand side of the body the ultimobranchial gland is often slightly deeper and more posterior in position. Heart (H).

of the anatomy and structure of the ultimobranchial glands of birds has been given by Hodges (1970) but there appears to be considerable variation in size and position between the two sides of the body and between different individuals and strains. The glands are reported to have a complex and well-developed nerve supply from the vagus (Nonidez, 1935), recurrent laryngeal and sympathetic systems (Dudley, 1942). It has been suggested that stimulating the vagus increases calcitonin secretion (Hodges, 1970) although futher work is necessary to investigate this possibility.

The histology of the development of the ultimobranchial glands has been described by a number of authors. The C cells are abundant and frequently arranged in cords. From about the 11th day of embryonic development onwards, it is often possible to identify encapsulated parathyroid nodules within the ultimobranchial tissue. It is also possible to identify vesicular cells from embryos of about 15 d of incubation onwards (Dudley, 1942). The functions of all these cells are not clear but it is generally accepted that the C cells are the source of calcitonin and they have a well-developed capillary bed associated with them whilst the vesicular cells are only poorly vascularized (Hodges, 1970). The ultrastructures of the ultimobranchial glands of the hatched fowl, pigeon and turtle dove and the chick embryo have been studied in detail by Stoeckel and Porte (1967, 1969, 1970).

III. Calcitonin

Mammalian calcitonin (thyrocalcitonin) is a polypeptide containing 32 amino acids and with a molecular weight of between 3,500 and 4,500 depending upon the origin and method of extraction (O'Dor *et al.*, 1969). The polypeptide contains no lysine or isoleucine but has a 1 to 7 intrachain disulphide bridge followed by a 23 amino acid chain ending with an N-terminal prolinamide. Avian calcitonin appears to be similar in that it contains about 32 amino acids and has a molecular weight of about 4,500, which is similar to that of the rat and human hormone extracted by the same technique (Copp, 1969).

Calcitonin has been isolated from the fowl, the turkey (Copp *et al.*, 1967) the goose (Bates *et al.*, 1969) the pigeon (Moseley *et al.*, 1968) and the quail (see Hirsch and Munson, 1969). According to Tauber (1967), Urist (1967), Kraintz and Puil (1967) and Copp (1969) the hormone is found only in the ultimobranchial glands of the bird but MacIntyre (1967) and Moseley *et al.* (1968) consider that there are traces of the hormone in the thyroids of the fowl and that the thyroid and ultimobranchial glands of the pigeon contain almost equal amounts when compared as MRC units/mg dry fat free gland. There are traces of calcitonin in the thyroid of the goose (Bates *et al.*, 1969).

The ultimobranchial glands of birds contain very high concentrations of the hormone when compared with the mammalian thyroids. This is presumably because they are much richer in C cells for, when the concentration of the hormone is expressed in terms of MRC units/kg body weight, it is apparent that there is relatively little difference between birds and mammals (Table 1).

Table 1

Calcitonin (thyrocalcitonin) content of various glands of different vertebrates (after Copp, 1969)

Species	MRC units/g fresh weight gland			MRC units/kg body weight
	Thyroid	Ultimobranchial	Parathyroid	
Dog	1–4	—	1·5–3·3	0·25–0·50
Rat	5–15	—	—	0·20–0·60
Fowl	—	30–120	—	0·50–0·80
Turkey	—	60–100	—	0·50–0·90

The weight of the ultimobranchial glands and the amount of calcitonin which they contain both increase as the birds mature. The quantity of hormone/kg body weight tends to decrease in older birds although the values of different workers show considerable variation (Table 2). The decrease in hormone content of the birds has been variously interpreted as indicating that the hormone exerts its maximal effect in the young animal (Dent *et al.*, 1969) or as showing that the hormone plays an important part in the egg laying cycle of the female (Wittermann *et al.*, 1969).

Table 2

Levels of plasma calcium, weight of ultimobranchial bodies and calcitonin content of domestic fowls of different ages

Age (d)	Plasma calcium (mEq/litre)	Wet weight of ultimobranchial glands (mg)	MRC mU calcitonin/bird	MRC mU calcitonin/kg body weight
(Data from Wittermann *et al.*, 1969)				
21	6·01 ± 0·08	17·1	2324	2800
84	6·34 ± 0·15	25·2	4508	3450
252♀	9·14 ± 0·60	31·6	3969	2060
(Data from Dent *et al.*, 1969)				
Embryo: 18	—	0·5	42	2500
Hatched: 3	4·50	0·4	163	5200
35	4·95	3·4	595	2000
77	5·00	5·0	1085	1400
105	4·95	6·4	1351	1000
490♂	5·40	5·6	1495	600
490♀	10·50	5·3	1060	600

IV. Assay and Effects of Calcitonin

Calcitonin is assayed by injecting the hormone into rats and measuring the fall in plasma calcium within the next 1 to 2 h. The logarithmic dose-response curve is compared with that produced by an MRC standard preparation. There are a large number of individual variations of the assay method (Hirsch and Munson, 1969) but two particular features are important. First, the assay is very variable depending upon the age and strain of the rats used and second it has not yet been possible to assay the hormone in a non-mammalian animal. Thus all the bioassays of avian calcitonin have been performed upon either rats or mice (Parsons and Reynolds, 1968) and attempts to devise a bioassay using birds have all been unsuccessful. In fact, although avian calcitonin is extremely potent when injected into mammals, it is very difficult to obtain a measurable response in either adult fowl or young chicks (Urist, 1967; Low and Brown, 1968; Kraintz and Intscher, 1969). This is an unfortunate situation since it is very probable that there are species-specific effects of the hormone due to variations in chemical composition. The bird may also respond in a number of unique ways but it is not yet possible to investigate them because of the difficulty in detecting a measurable phenomenon.

All the evidence from the study of the mammalian system suggests that the hormone has a rapid turnover (half-life of 5 to 15 min in the rabbit), that its primary effect is upon the bones, and that it causes a lowering of plasma calcium and phosphate levels by inhibiting bone resorption. Young and growing animals normally have a much more active turnover of their skeletal system and since the hormone appears to act by inhibiting bone resorption it is not surprising that calcitonin has a greater effect upon young animals than upon fully grown adults (Copp, 1969).

V. Rates of Secretion

It is possible to extract calcitonin from the plasma of the fowl and to demonstrate that the quantity of hormone present fluctuates with the calcium content of the diet of the bird (Low and Brown, 1968). This suggests that the rate of hormone secretion varies with the concentration of plasma calcium present in the bird and two attempts have been made to demonstrate this by measuring secretion rates *in vivo*.

In one set of experiments the ultimobranchial glands of the fowl were perfused with plasma containing 5 mEq/litre calcium which produces a basal secretion rate of about 200 μU calcitonin/gland min. This rises to a maximal secretion rate of about 1200 μU on perfusion with plasma containing 10 mEq/litre. It is therefore possible to demonstrate a 5 to 6-fold increase in secretion rates of calcitonin in the hypercalcaemic fowl. This means that during 1 h of maximal stimulation the bird would secrete about 80 mU of hormone or 10 to 20% of its total calcitonin content (Tables 1, 2) which suggests that birds probably possess only minimal stores of the hormone but have the capacity for its rapid synthesis (Ziegler *et al.*, 1970). Certainly inhibitors of protein synthesis diminish the secretion rate of calcitonin in the bird while substances which influence the formation of cyclic 3′,5′-adenosine monophosphate (cyclic AMP) increase the secretion rate of calcitonin.

A second set of experiments comparing the secretion rates of calcitonin in the bird and mammal is of further interest since Bates *et al.* (1969) have found that, in the goose, the ultimobranchial gland secretes hormone at a rate of about 2280 ± 940 μU/min/kg body weight. This is about 25 times greater than the rate in the pig or sheep. The goose appeared, however, to be rather unresponsive to hypercalcaemic stimulation and showed an increase in secretion rate of only about 15 to 47% per mEq Ca as compared with 230 to 960% per mEq Ca in a pig of the same "biological age".

VI. Possible Functions of Calcitonin

The evidence suggesting that calcitonin plays an active role in the normal calcium metabolism of birds is largely circumstantial. According to Urist (1967) the ultimobranchial glands hypertrophy in the laying hen and Copp *et al.* (1968) report that there is a discharge of secretory granules from the cells during hypercalcaemia in young cockerels. Similar results have been obtained by feeding young chicks a diet rich in calcium (Cipera *et al.*, 1970).

The ultimobranchial glands increase in size as the fowl matures (Table 2). During egg laying there is an increase in the total plasma calcium level and a fall in the calcitonin content of the hen (Table 2). This could indicate either a decreased synthesis or an increased secretion of the hormone and as these two explanations are entirely different in their physiological implications it is difficult to comment upon the significance of these results (Witterman *et al.*, 1969). Removal of the ultimobranchial glands has no effect upon the plasma calcium, phosphorus or alkaline phosphatase levels of the fowl and the growth of the skeleton and its opacity to X-rays are not influenced by the loss of these glands even after three months. There is, however, a decline in the ability of the operated birds to compensate for injections of parathyroid hormone and plasma calcium levels remain elevated for much longer periods of time in these birds as compared with sham-operated controls (Brown *et al.*, 1969). This would suggest that the effect of calcitonin may normally be to prevent overshooting in the parathyroid regulation of the bird's plasma calcium level. The converse is presumably also true since parathyroid hormone appears to produce a faster response in the bird than in the mammal and it has been suggested that this may be why it is difficult to demonstrate a hypocalcaemic effect in the bird following calcitonin injections. Thus, Kraintz and Intscher (1969) claim that in partially parathyroidectomized cockerels it is possible to demonstrate a hypercalcaemia following injections of mammalian calcitonin.

It has also been suggested that porcine calcitonin has a significant effect within 30 min of injection into vitamin D-deficient chicks (Gulat-Marnay, 1968).

It will be apparent, therefore, that the available experimental evidence is not completely convincing in showing that calcitonin is physiologically important in controlling the level of plasma calcium. Injections of the hormone produce little effect in birds and ablation of the ultimobranchial glands produces only minor side-effects. Furthermore, the secretion of the hormone in response to hypercalcaemia is under much less sensitive control in birds than in mammals (Bates *et al.*, 1969). There is now a growing body of opinion that an important function of calcitonin may be to protect the skeleton from excessive resorption. The experimental evidence relating to this concept has been reviewed by Hirsch and Munson (1969) and the only experiment relating to it in birds is that of Brown *et al.* (1969) already mentioned.

The possible importance of calcitonin during the reproductive cycle of the fowl is at the moment an unexplored field. It is likely, however, that in the next few years the concentration of calcitonin in the blood will be determined at various stages of the egg-laying cycle. It should also be possible to determine the effect of the hormone upon the medullary bone of the bird and upon the general homeostasis of its calcium metabolism.

There is an increasing indication that the embryo is sensitive to parathyroid hormone (Ranly and Runnels, 1969) and Taylor (1970) has suggested that calcitonin may have an important endocrine role in the embryo. The mean level of plasma calcium in the 14-day-old embryo is only 5·1 mg/100 ml and it increases steadily thereafter to reach a mean of 8·8 mg/100 ml on the 19th day. However, in 20-day-old embryos, in which pulmonary respiration has not yet been initiated, there is a fall to 7·3 mg/100 ml (Taylor, 1963), which suggests that the calcium level is being actively lowered at this time. Support for this suggestion comes from the recent observations of P. Lewis and T. G. Taylor (unpublished) who have shown that there is a striking increase in the plasma calcitonin between the 17th and 20th d of incubation, while the hormone could not be detected in the plasma of chicks the day after hatching.

References

Bates, R. F. L., Bruce, J. and Care, A. D. (1969). *J. Endocr.* **45**, xiv–xv.

Brown, D. M., Perey, D. Y. E., Dent, P. B. and Good, R. A. (1969). *Proc. Soc. exp. Biol. Med.* **130**, 1001–1004.

Cipera, J. D., Chan, A. S. and Belanger, L. F. (1970). "Calcitonin 1969" pp. 320–326. Heinemann, London.

Copp, D. H. (1969). *J. Endocr.* **43**, 137–161.

Copp, D. H., Cockcroft, D. W. and Kueh, Y. (1967). *Can. J. Physiol. Pharmacol.* **45**, 1095–1099.

Copp, D. H., Cockcroft, D. W., Kueh, Y. and Melville, M. (1968a). "Calcitonin." Proceedings of Symposium on Thyrocalcitonin and C cells (S. Taylor, ed.), pp. 306–321. Heinemann, London.

Copp, D. H., Davidson, A. G. F. and Cheney, B. (1961). *Proc. Can. Fedn biol. Socs* **4**, 17.

Copp, D. H., Webber, W. A., Low, B. S., Kueh, Y. and Biely, J. (1968b). *Proc. Can. Fedn biol. Socs* **11**, 34.

Dent, P. B., Brown, D. M. and Good, R. A. (1969). *Endocrinology* **85**, 582–585.

Dudley, J. (1942). *Am. J. Anat.* **71**, 65–89.

Gulat-Marnay, Ch. (1968). *Annls Nutr. Aliment.* **22**, 57–58.

Hirsch, P. F. and Munson, P. L. (1969). *Physiol. Rev.* **49**, 548–622.

Hirsch, P. F., Voelkel, E. F. and Munson, P. L. (1964). *Science, N.Y.* **146**, 412–413.

Hodges, R. D. (1970). *Annls Biol. anim. Biochim. Biophys* **10**, 255–275.

Kraintz, L. and Intscher, K. (1969). *Can. J. Physiol. Pharmacol.* **47**, 313–315.

Kraintz, L. and Puil, E. A. (1967). *Can. J. Physiol. Pharmacol.* **45**, 1099–1103.

Low, B. S. and Brown, J. C. (1968). *Proc. Can. Fedn biol. Socs* **11**, 34–35.

MacIntyre, I. (1967). *Calc. Tissue Res.* **1**, 173–182.

McLean, F. C. (1958). *Science, N.Y.* **127**, 451–457.

Moseley, J. M., Matthews, E. W., Breed, R. H., Galante, L., Tse, A. and Macintyre, I. (1968). *Lancet* **i**, 108–110.

Nonidez, J. F. (1935). *Anat. Rec.* **62**, 47–73.

O'Dor, R. K., Parkes, C. O. and Copp, D. H. (1969). *Endocrinology* **85** 582–585.

Parsons, J. A. and Reynolds, J. J. (1968). *Lancet* **i**, 1067–1070.

Pearse, A. G. E. and Carvalheira, A. F. (1967). *Nature, Lond.* **214**, 929–930.

Ranly, D. M. and Runnels, P. R. (1969). *Texas Rep. Biol. Med.* **27**, 795.

Stoeckel, M. E. and Porte, E. (1967). *C. r. Séanc. Soc. Biol.* **161**, 2040–2043.

Stoeckel, M. E. and Porte, E. (1969). *Z. Zellforsch. mikrosk. Anat.* **94**, 495–512.

Stoeckel, M. E. and Porte, E. (1970). *In* "Calcitonin 1969" pp. 327–338. Heinemann, London.

Tauber, S. D. (1967). *Proc. natn. Acad. Sci. U.S.A.* **28**, 1684–1687.

Taylor, T. G. (1963). *Biochem. J.* **87**, 7*P*.

Taylor, T. G. (1970). *Annls Biol. anim. Biochim. Biophys.* **10**, 83–91.

Urist, M. R. (1967). *Am. Zoöl.* **7**, 883–895.

Witterman, E. R., Cherian, A. G. and Radde, I. C. (1969). *Can. J. Physiol. Pharmacol.* **47**, 175–180.

Ziegler, R., Delling, G. and Pfeiffer, E. F. (1970). *In* "Calcitonin 1969" pp. 301–310. Heinemann, London.

20 The Adrenal Glands

J. W. WELLS and P. A. L. WIGHT

Agricultural Research Council's Poultry Research Centre, King's Buildings, Edinburgh, Scotland

I. Structure

A. INTRODUCTION

The two adrenal glands of the fowl are situated one on each side of the median line just anterior to the bifurcation of the caudal vena cava and close to the gonads and the anterior division of the kidneys. The glands are yellow, irregular and oval and are about the same size. In adults, they weigh from 100 to 200 mg each, considerable variation being due to breed, age, health and environmental factors (Crile and Quiring, 1940; Kar, 1947; Arvy and Gabe, 1951; Payne, 1955; Garren and Shaffner, 1956). The adrenals of young chicks are, relative to body weight, heavier than those of adults, but reported differences in weight associated with sex (Breneman, 1954) are probably insignificant if calculated as a percentage of the total body weight (Juhn and Mitchell, 1927; Sauer and Latimer, 1931).

Abbreviations used in this chapter: AAAD = adrenal ascorbic acid depletion; ACTH = corticotrophin; CNS = central nervous system; CRF = corticotrophin releasing factor; DNA = deoxyribonucleic acid; PAS = periodic acid-Schiff reaction; RNA = ribonucleic acid.

The adrenal is invested by a thin capsule of connective tissue from which fine septa ramify between cords and islands of parenchymal cells. This interparenchymal network contains only collagen and reticular fibres but, according to Günther (1906), a few fine elastic fibres are also present in the capsule. The latter contains many blood vessels, bundles of nerve fibres, plexuses of autonomic ganglia and even occasional Herbst corpuscles.

B. VASCULAR AND NERVE SUPPLY

The major blood supply comes from the anterior renal arteries but, in addition, a small artery from the aorta supplies the posterior part of the left gland and a secondary trunk from the anterior renal artery supplies the posterior part of the right gland. These vessels form extensive capillaries which drain into a network of venous sinuses. Although there may be a large venous sinus in the middle of the gland, there is no true central vein such as there is in man. Ultimately, the sinuses lead to an adrenal vein which joins the caudal vena cava. Goodchild (1969) has described a lateral adrenal vein which constitutes an adrenal portal (afferent) system coming mainly from the muscles of the flank.

Many non-myelinated fibres penetrate the gland from a sympathetic plexus on the dorso-medial surface of the anterior pole.

C. HISTOGENESIS

The adrenal has two main component tissues, cortex and medulla (Fig. 1): they are derived from different germ layers. The cortical cells are mesodermal and arise from the embryonic peritoneal epithelium, whereas the cells of the medulla are ectodermal and originate from the sympathetic trunks. The morphological relationship of cortex and medulla varies in the different classes of vertebrates; in many fishes the two tissues are quite separate, in marsupials and eutherian mammals the medulla lies in the centre of the cortex and in amphibia, in reptiles and birds they usually intermingle. Because of this intermingling Hartman *et al.* (1947) suggested that the cortex should be termed interrenal tissue and the medulla chromaffin tissue. Their distribution throughout the gland is fairly uniform (W. G. Siller and G. M. MacKenzie, personal communication) but the ratio of one type of tissue to the other is variable. Thus Payne (1955) found that in some young birds cortex comprised 90% of the gland, an observation not incompatible with the prolonged sympathetic migration and slow histogenesis of the medullary tissue (Venzke, 1953); by maturity there were about equal amounts of cortex and medulla. Sauer and Latimer (1931) calculated that the adult female has approximately 30% more cortical tissue than the male, in proportion to body weight. According to Sivaram (1965) the medulla has a consistently higher per cent volume than the cortex in the adult. The proportion of each is undoubtedly influenced by a variety of factors including age, sex, health and environment (Elliot and Tuckett, 1906; Kar, 1947; Sauer and Latimer,

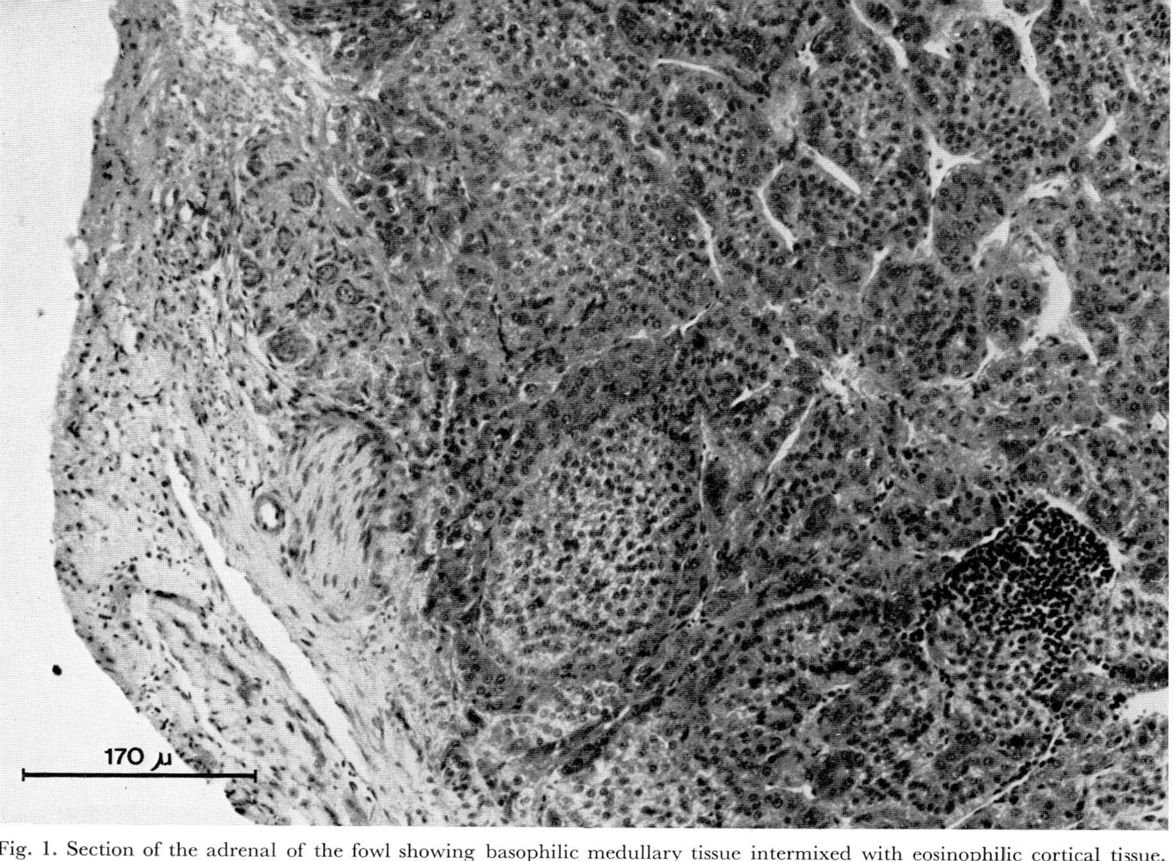

Fig. 1. Section of the adrenal of the fowl showing basophilic medullary tissue intermixed with eosinophilic cortical tissue. Adjacent to the capsule is a small ganglion. A focus of lymphocytes can be seen among the parenchymal cells. Haematoxylin and eosin.

1931; Oakberg, 1951). Nevertheless, determination of the norm, laborious as it may prove, would be of considerable value in physiological and pathological studies.

D. STRUCTURE OF THE CORTEX

The cortex of the fowl's adrenal gland cannot be divided into the three distinct zones of its human counterpart although several observers have thought that there are some regional differences. The cortical cells are aligned in strands which form loops against the capsule and it has been suggested (Chester Jones *et al.*, 1962) that this indicates zoning. Sauer and Latimer (1931) thought that, in addition to the glomerulosa-like appearance of these peripheral strands, the architecture of the cortical cells in the centre of the gland resembled that of the mammalian reticular zone. Sivaram (1964, 1965) considered that mitochondria were more abundant in the peripheral than the central region and also that the activity of certain enzymes was higher in the former. Arvy (1962) has recorded differences in the amount of Δ^5-3β-hydroxysteroid hydrogenase in the two regions. Kondics and Kjaerheim (1966) demonstrated ultrastructural differences in the mitochondria at the two sites. On the other hand, Fujita (1961) was not convinced that there were significant ultrastructural variations. Kar (1947) suggested that cortical cells differentiate from certain mesenchymal capsular cells and migrate progressively towards the centre of the gland and Sivaram (1964, 1965) confirmed the presence of a higher peripheral metabolism, mitoses and nuclear pyknosis, which might be accounted for by such a progression. In summary, it seems that although the fowl's adrenal cortex is relatively undifferentiated, compared with that of mammals, the peripheral strands show slight morphological differences from the remainder.

Histological sections along the length of the cortical strands show a double row of cells; transversely the strand is composed of stacks of about six radially arranged cells. There is no lumen. In paraffin-embedded sections these columnar cells are eosinophilic, granular and vacuolated; the vacuoles represent sites of lipid removed in histological processing. These abundant lipid globules accumulate in the outer part of the cell adjacent to the blood sinuses which run between the strands, with the result that the nucleus tends to lie in the part of the cell which is in the middle of the strand. The lipids are acetone-soluble, birefringent, fluorescent, sudanophilic and give positive Lieberman-Burchardt, Schiff (plasmal) and dinitrophenylhydrazine reactions. It has been claimed that this battery of reactions signifies the presence of ketosteroids, but this is questionable (*cf.* Pearse, 1960). Carotenoids are present in these lipids (Findlay, 1920) and impart a yellow colour to the fresh gland. The lipids are uniformly distributed throughout all the cortical tissue.

The nucleus is spherical and usually has one or two nucleoli. Mitochondria are present in some cells and in the "fuchsinophil cells" of Uotila (1939) are extremely abundant. A Golgi apparatus can be demonstrated (Sivaram,

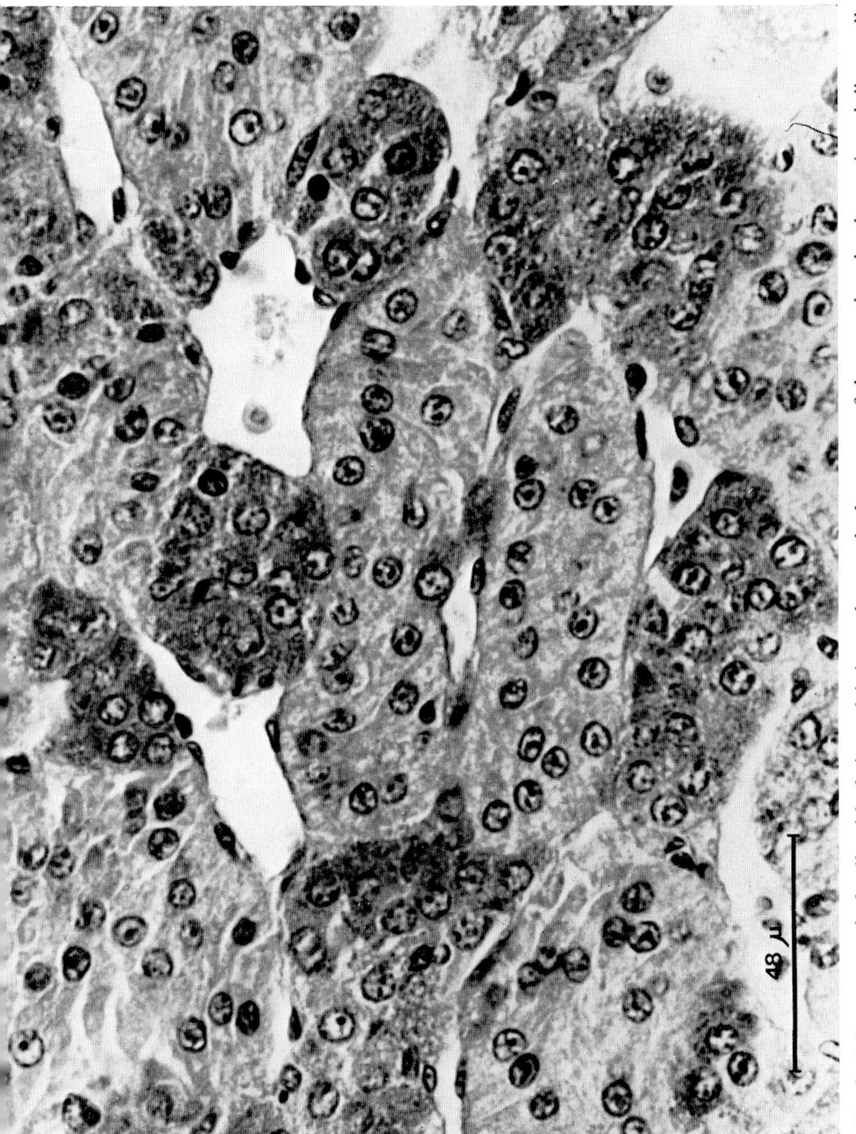

Fig. 2. Columnar cortical cells with their nuclei situated towards the centre of the strand and polygonal medullary cells with basophilic granules in their cytoplasm and a central nucleus. Endothelial cell nuclei can be identified in the wall of the sinus. Haematoxylin and eosin.

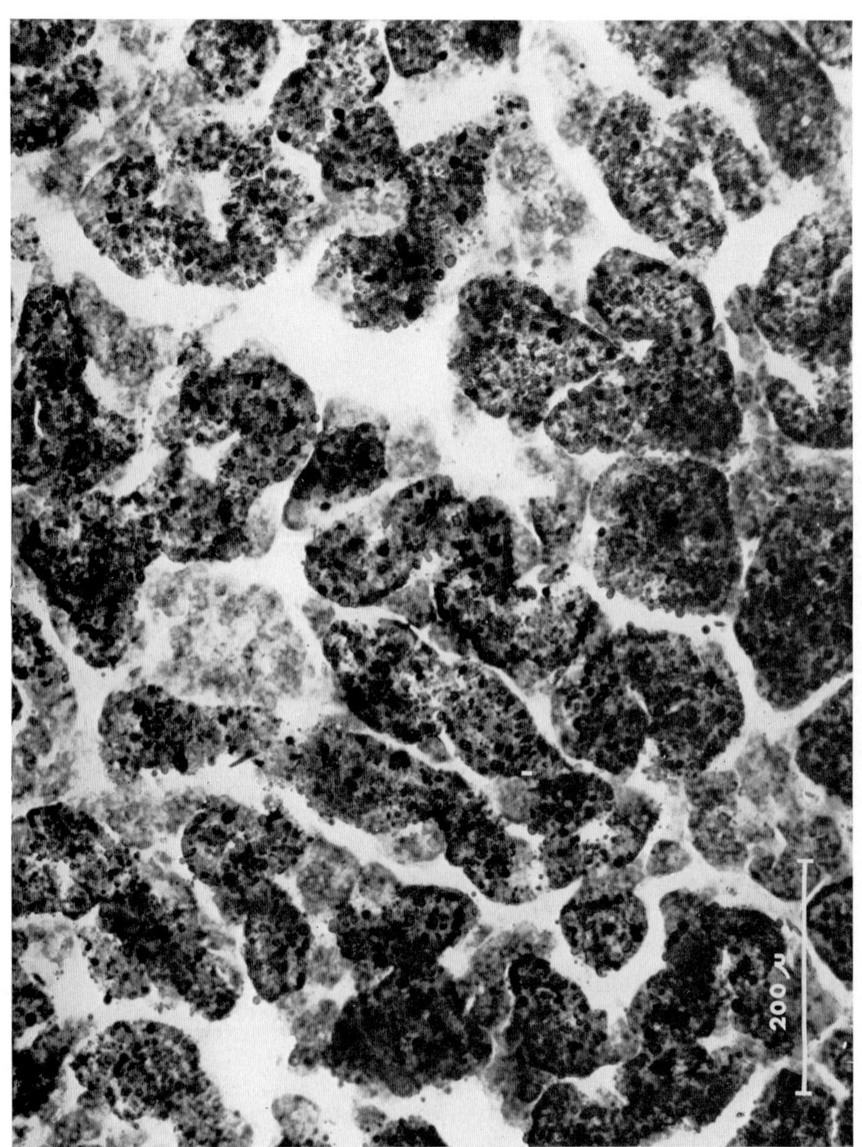

Fig. 3. Lipid in the cortical cells. Frozen section; Sudan IV.

1965). The cortex also gives positive histochemical reactions for acid and alkaline phosphatases, Δ^5-3β-hydroxysteroid hydrogenase (Arvy, 1962), aliphatic esterase, cholinesterase and succinate dehydrogenase (Sivaram, 1964, 1965). Much ascorbic acid is present (Morita *et al.*, 1961).

E. ULTRASTRUCTURE OF THE CORTEX

Despite differences between the structures of the avian and mammalian adrenals as seen by light microscopy, the ultrastructure of the cortical cells in the two classes is basically similar. Kjacrheim (1968) records numerous mitochondria, mostly spherical (diameter 0·5 to 1·0 μm) and having tubular inner structures. The mitochondria in a peripheral band of cortical cells about 200 or 300 μm deep are said to be rod-shaped and about 2 μm long (Kondics and Kjaerheim, 1966). Membrane-limited vacuoles containing homogeneous, electron-translucent material are numerous, particularly in the basal region; these correspond to the lipid droplets seen by light microscopy. This lipid is probably associated with the abundant smooth endoplasmic reticulum. Dense bodies surrounded by a triple membrane are present and there are also bodies having both electron-translucent and electron-dense components which Kjaerheim (1968) considered to be intermediary forms between lipid droplets and dense bodies. There is a Golgi apparatus of moderate size and a few cytoplasmic filaments, micro-tubules and cilia. The cells rest on a basal lamina adjacent to the blood sinuses.

F. STRUCTURE OF THE MEDULLA

The intensely basophilic medullary cells are easily distinguished from the eosinophilic cortical cells in routine histological sections stained with haematoxylin and eosin. There is a thin zone of medullary cells about 90 μm wide (Sivaram, 1965) just beneath the capsule but the bulk of the cells occurs as irregular masses among the cortical cords. These medullary cells are polygonal, larger than cortical cells and have a spherical, centrally-situated nucleus with a chromatin network and one or two nucleoli. Their cytoplasm contains numerous, small basophilic granules. Physiologically inactive cells are reported (Payne, 1955) to have a darker, more condensed nucleus and more numerous basophilic granules than the actively secreting cells. Mitochondria are present and there is a small Golgi apparatus although this is infrequently seen (Sivaram, 1965).

The cytoplasm of the medullary cells stains dark brown with osmic acid and brown with chrome salts, the latter being due to oxidation of the catecholamine-containing, basophilic granules and constitutes the "chromaffin reaction" after which the cells and the granules are sometimes named. The medullary cells are coloured green by dilute neutral ferric chloride (Vulpian reaction) and give positive ferric ferricyanide and argentaffin reactions. They fluoresce intensely after the formaldehyde condensation method for biogenic monoamines. Adrenaline and noradrenaline are present, the latter

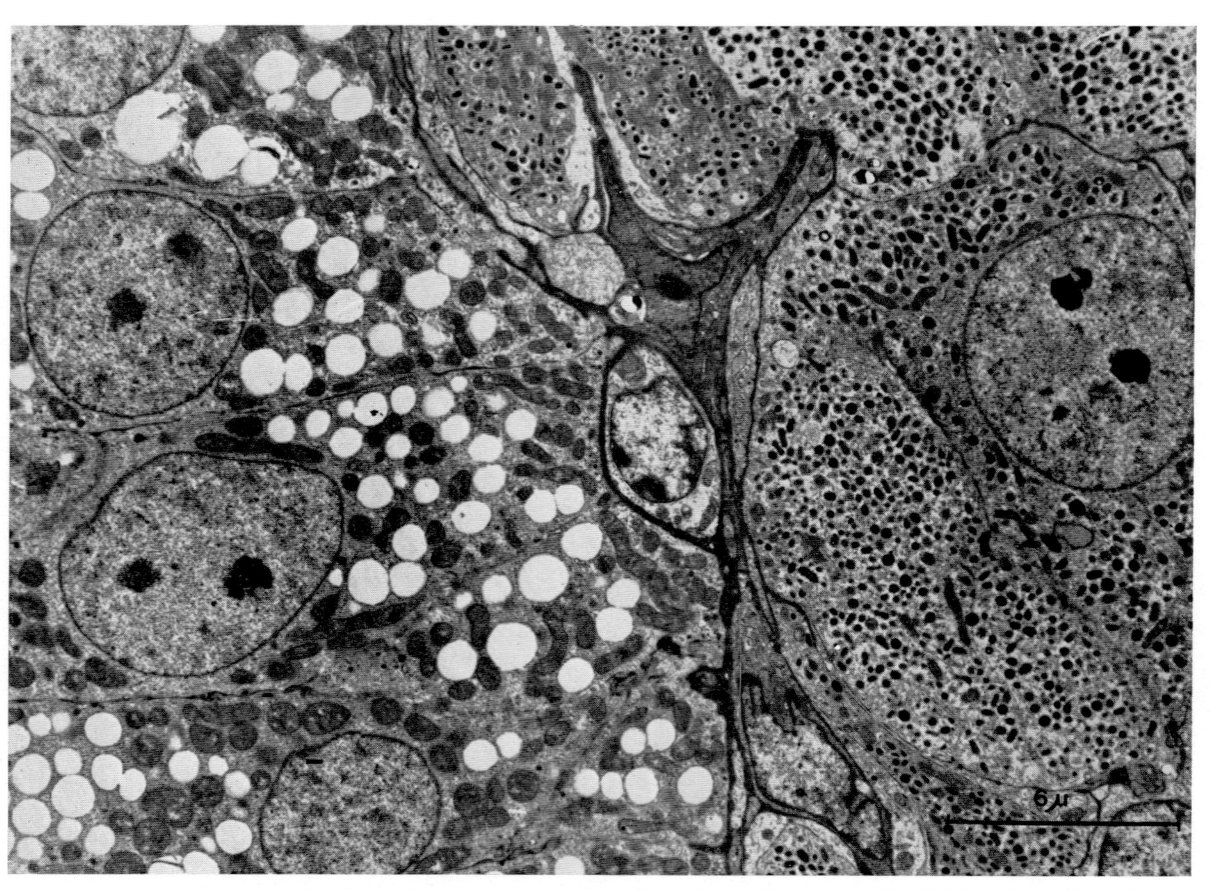

Fig. 4. On the left are cortical cells with numerous mitochondria and vacuoles representing lipid, and on the right are medullary cells packed with electron-dense catecholamine granules.

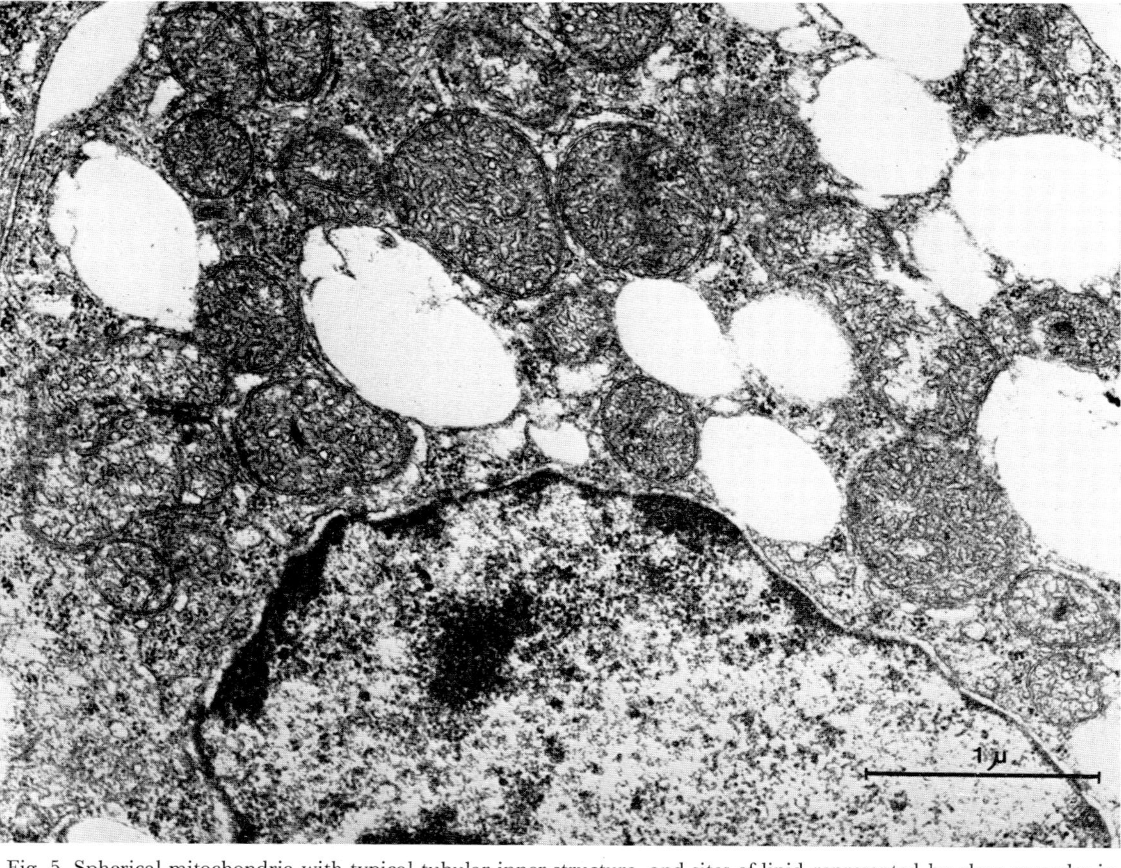

Fig. 5. Spherical mitochondria with typical tubular inner structure, and sites of lipid represented by clear vacuoles in a cortical cell. Part of the nucleus is visible.

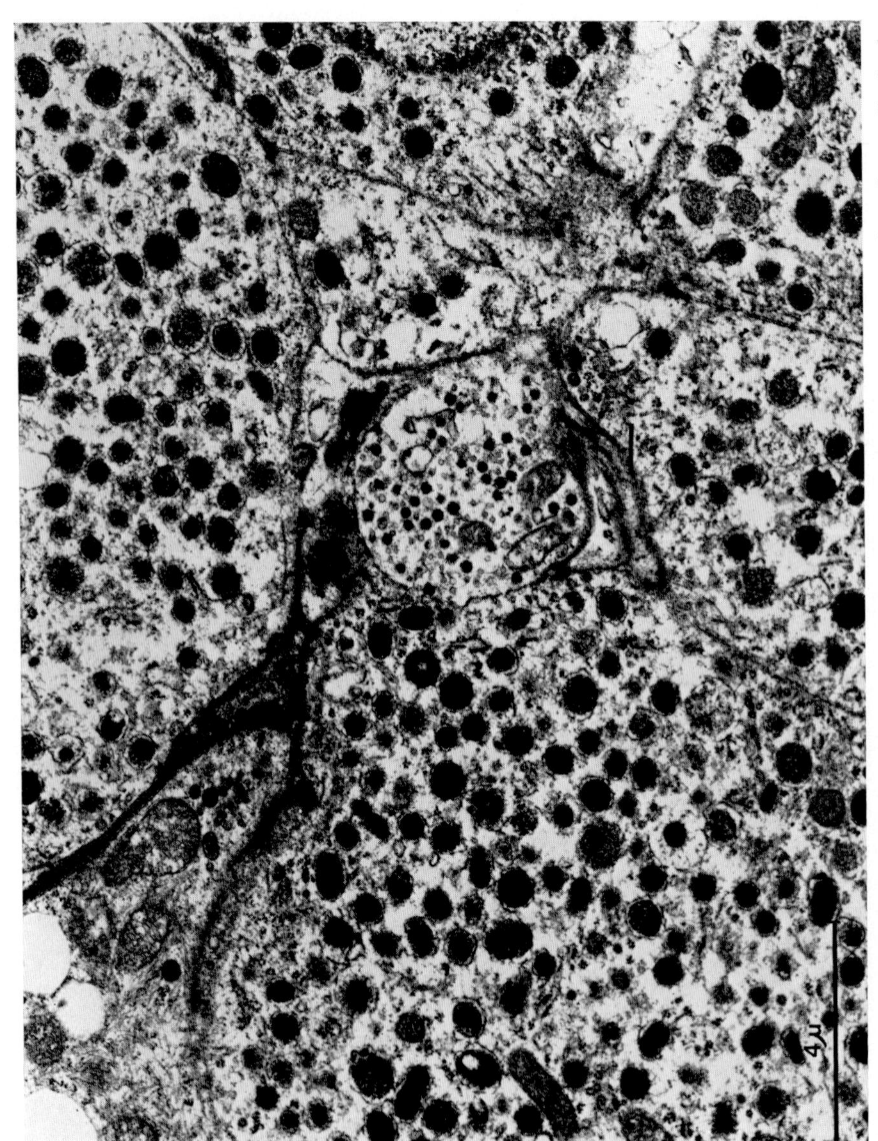

Fig. 6. Membrane-bound catecholamine granules in the cytoplasm of medullary cells. A non-myelinated nerve lies in the intercellular space.

being histochemically predominant (Ghosh, 1962). The medullary cells which contain noradrenaline may be distinguished from those containing adrenaline by several histochemical techniques (Eränkö, 1957; Coupland *et al.*, 1964). The cells are PAS negative but stain metachromatically with buffered toluidine blue (Sivaram and Sharma, 1964) and this is thought to be due to phosphate groups of adenosine triphosphate utilized in the synthesis and storage of catecholamines. Histochemically the cells show high concentrations of ascorbic acid, adenosine triphosphatase and acid phosphomonoesterase (Sivaram, 1964).

G. ULTRASTRUCTURE OF THE MEDULLA

Electron microscopy reveals that the cytoplasm of the medullary cells is packed with electron-dense granules bounded by a three-layered membrane. The majority are spherical and range up to about 0·45 μm in diameter, but some are distinctly elongated and these may be up to about 0·45 μm × 0·12 μm in outline. Mitochondria are elongated ovals with a rather dense matrix and lamina cristae. They occur in only moderate numbers, often in a perinuclear location. A few short strands of granular endoplasmic reticulum are present. Kano (1959) did not observe a Golgi apparatus, but this does occur although it is small and rather insignificant; in mammals some importance has been attached to the relationship between the chromaffin granules and the Golgi apparatus (*cf.* Elfvin, 1967). There is a basal lamina at the free border but between adjoining medullary cells is an intercellular space which contains non-myelinated neurites having dense-cored vesicles and synapsing with the parenchymal cells. Elfvin (1965) has suggested that, in rats, discharged medullary material may reach the vascular system along these canaliculi-like intercellular spaces.

Several kinds of cells which are either undifferentiated or of uncertain function may be found in the fowl's adrenal. Associated with the chromaffin tissue are small cells with a round nucleus and scant cytoplasm which Payne (1955) called medullary satellite cells. Groups of ganglion cells may invade the parenchyma at the periphery. Small groups of medium sized "ganglion-like" cells, intermediate in appearance between ganglion cells and giving a positive chromaffin reaction, may be found at the edge of the gland (Rabl, 1891; Payne, 1955). They are probably an intermediate state in the development of the medullary cells. Small foci of lymphocytes may be observed amongst the parenchymal cells; Wheeler (1950) found these foci in 51% of apparently normal birds.

II. Adrenal Function

A. FACTORS USED TO ASSESS ADRENAL FUNCTION

1. Adrenal Weight

In a large-scale study Breneman (1954) observed that absolute adrenal weight increased at a constant rate from hatching until 100 d of age, after

Table 1

Some recorded adrenal measurements

Breed or strain	Age (weeks)	Sex		Source
		Mass		
			Relative adrenal weight mg/100 g b.w.[a]	
New Hampshire	3	M and F	14·3	Garren and Shaffner (1956)
New Hampshire	4	M and F	12·6	Garren and Shaffner (1956)
New Hampshire	5	M and F	10·0	Garren and Shaffner (1956)
New Hampshire	6	M and F	9·4	Garren and Shaffner (1956)
New Hampshire	8	M and F	10·2	Garren and Shaffner (1956)
New Hampshire	10	M and F	9·0	Garren and Shaffner (1956)
Brown Leghorn	3	M and F	18·3	Freeman *et al.* (1966)[b]
"Broiler" stock	10	M and F	7·6	W. Bolton and W. Dewar (personal communication)
White Leghorn (40 weeks in production)	—	F; layers	6·4	Wolford and Ringer (1962)
White Leghorn (40 weeks in production)	—	F; non-layers	8·3	Wolford and Ringer (1962)
White Leghorn	72	F; layers	2·7–3·1	R. G. Wells (personal communication)
		Ascorbic Acid		
			Concentration μg/100 mg adrenal	
Brown Leghorn	3	M and F intact	187	Freeman (1970a)[b]
Brown Leghorn	3	Bursectomized	150	Freeman (1970a)[b]
White Leghorn	6	M	186	Perek and Eckstein (1959)

		Ascorbic Acid		
			Concentration µg/100 mg adrenal	
White Leghorn	12	F	335	Perek and Eckstein (1959)
White Leghorn	52	F	161	Perek and Eckstein (1959)
White Leghorn (40 weeks in production)	—	F; layers	215	Wolford and Ringer (1962)
White Leghorn (40 weeks in production)	—	Non-layers	167	Wolford and Ringer (1962)
		Cholesterol		
			Concentration µg/100 mg adrenal	
Brown Leghorn	3	M and F	1·61 (total)	Freeman *et al.* (1966)[b]
Brown Leghorn	3	M and F	1·28 (ester)	Freeman *et al.* (1966)[b]
Various strains	8	M and F	2·2–2·5 (total)	Siegel and Siegel (1969)
White Leghorn (40 weeks in production)	—	F; layers	3·2 (total)	Wolford and Ringer (1962)

[a] Grams body weight.
[b] Calculated from data.

which the growth curve flattened out, whereas the gonads and thyroids exhibited different types of growth curves. The relative adrenal mass, expressed as a percentage of the body weight, falls with increasing age, although in some strains this was not apparent with younger birds (Breneman, 1941). Table 1 presents an arbitary selection of mean relative adrenal weights based on studies using relatively large numbers of birds. The standard errors of the means are often large and may reflect in part experimental difficulties in removal of the glands.

2. *Ascorbic Acid*

Compared with that in the rat, the ascorbic acid concentration in the fowl's adrenal gland, about 200 μg per 100 mg tissue (Table 1), would be considered "depleted". Although the decrease of ascorbate in mammalian adrenals is used as a quantitative measure of adrenal stimulation by corticotrophin, at the molecular level the function of ascorbic acid has not been fully defined. It appears to act as an electron donor in the hydroxylation reactions used in the biosynthesis of corticosteroids (Kernstein *et al.*, 1958) and catecholamines (Friedman and Kaufman, 1965). Since ascorbic acid has been detected histochemically in both cell types (see Section I F), its depletion may reflect activity of cortical and medullary cells.

3. *Adrenal Cholesterol*

There is not enough information to enable one to decide whether the total cholesterol content, or the proportion esterified, changes with age (Table 1). A recent study of six stocks consisting of three closed and three random-bred populations revealed no significant sexual dimorphism for adrenal cholesterol concentration (Siegel and Siegel, 1969).

4. *Catecholamines*

The results of two investigations into the catecholamine content of adrenals from birds of different age groups are shown in Table 2. The apparent discrepancies in the biogenic amine content and in the proportion of constituents may arise from the variations existing between breeds or strains or perhaps may reflect the use of different analytical techniques.

The biosynthesis of noradrenaline appeared to be similar to that observed for mammalian adrenal preparations (Hagen, 1956). Apart from a brief account on the recovery of radioactivity in excreta after injecting [*N*-methyl-^{14}C]adrenaline into young birds (Scott, 1962), the metabolism of these amines in the fowl has not been studied.

5. *The Adrenocorticoid Steroid Hormones*

The first definitive study of the steroid secretion of the avian adrenal was made by Phillips and Chester Jones (1957) who showed that adrenal effluent plasma from capons contained predominantly corticosterone with

Table 2

Catecholamines in adrenal

Age (days)	Total amines μg/100 mg adrenal	% noradrenaline	Source
1	825	60	Shepherd and West (1951)
7	950	65	
77	1000	60	
110	1010	80	
200	1125	70	
1–7	530	50	Callingham and Cass (1966)[a]
8–15	530	50	
16–23	660	53	
24–56	700	46	

[a] Calculated from figure.

much smaller amounts of cortisol, cortisone and aldosterone; the last was detected only after prolonged corticotrophin (ACTH) treatment. Subsequently, de Roos (1961) by *in vitro* incubation, confirmed the identity of corticosterone as the major adrenal corticoid. Aldosterone was also characterized in addition to a tentative identification of 11-dehydro-corticosterone. Contrary to previous reports (Phillips and Chester Jones, 1957; Urist and Deutsch, 1960) cortisol and cortisone were not detected, in agreement with Nagra *et al.* (1960) who examined the adrenal venous plasma of three species of gallinaceous birds including the fowl. The corticosteroids from similar incubations of adrenals from three other birds belonging to different orders, viz. the white king pigeon, the western gull and the white Pekin duck, showed a similar pattern of products. Similarly, Brown (1961) reported that peripheral plasma of the turkey contained corticosterone and cortisol in the ratio of 20 to 1. The above evidence implying that 17α-hydroxylation is a minor biosynthetic pathway in birds, was confirmed recently by analysis of adrenal venous plasma from cockerels using the sensitive double isotope derivative technique (Kliman and Peterson, 1960) when cortisol was not detected (Taylor *et al.*, 1970).

Various workers (Newcomer, 1959a, b; Nagra *et al.*, 1960, 1963; Brown, 1961) have attempted to assess adrenal function in the fowl using procedures developed for human clinical studies. However Frankel *et al.* (1967d) demonstrated decisively that these procedures, when applied to adrenal venous plasma of the cockerel, gave gross errors in estimates of corticosterone, arising out of incomplete purification of extracts. Stenlake *et al.* (1970) have now shown that in human plasma, by a typical fluorimetric procedure (Mattingly, 1962) the major interfering fluorogens are cholesterol and its esters which are not removed by preliminary extraction of the plasma with hydrocarbon solvents. Frankel *et al.* (1967d) found that their extended method, using double solvent partition prior to paper chromatography gave results in close agreement with a double isotope derivative dilution procedure. Four pools of adrenal plasma from cockerels had a range of values from 8·0 to 15·0 μg corticosterone per 100 ml compared with the concentrations of 2·1 to 3·3 μg for adenohypophysectomized males.

Evidently a counter-current purification process is required for analysis of peripheral plasma. Using silica-gel column chromatography Brown (1968) obtained values of 1·05 μg corticosterone per 100 ml plasma for unstressed 4-week-old chickens. Using florisil adsorption chromatography as the purification stage before the acid fluorescent reaction, J. W. Wells and J. Culbert (unpublished) have found corticosterone levels of less than 6 μg per 100 ml in peripheral plasma of pullets and laying hens of various strains. It should be mentioned that fluorimetric procedures are very sensitive to impurities and the occasional high values may reflect incomplete separation of lipids which are high in plasma of the laying hen. Using 2 ml plasma for analysis, the amount of corticosterone actually measured is usually near the limit of the fluorimetric determination; thus Frankel *et al.* (1967d) reported that the sensitivity for a single estimation with a maximum error of 100% was 1·2 μg corticosterone per 100 ml adrenal venous plasma.

In certain studies analysis of adrenal tissue for changes in corticosterone concentration may be more suitable for the measurement of adrenal response. Siegel and Siegel (1969) reported that chickens 57 d of age had 2·63 to 3·92 μg corticosterone per 100 mg adrenal tissue while J. Culbert and J. W. Wells (unpublished) found in a group of laying hens 17 months of age, that the adrenals contained 18·7 μg per 100 mg tissue.

Although no studies have appeared on the transport and metabolism of corticosterone, Seal and Doe (1963) found that the cortisol-binding capacity of fowl serum increased as the temperature decreased, implying that the plasma may contain a globulin similar to transcortin. Kinetic studies of corticosteroid metabolism have yet to be made in the fowl; the short half-life of corticosterone in the duck of 10 to 11 min (Donaldson and Holmes, 1965) compared with that of approximately 60 min in man (Peterson, 1959), suggests that the turnover rate may be quite rapid in the fowl.

The production of steroid sex hormones by the fowl adrenal has not been reported. J. Culbert and J. W. Wells (unpublished) obtained evidence for the presence of progesterone and testosterone in adrenal tissue. In laying hens, the pool of these steroids per pair of adrenals was approximately half that found in the ovary. Measurement of these hormones in adrenal venous effluent plasma will be necessary before it can be decided that the adrenal glands are a source of the sex hormones.

B. BIOGENESIS OF ADRENAL STEROIDS

Most of our knowledge on the biosynthetic pathways of the adrenal steroid hormones is based on experiments using mammalian preparations. The synthesis of corticosteroids in adrenal tissue involves various hydroxylating enzymes located in different cellular organelles. Thus cholesterol, the immediate precursor, is transformed into pregnenolone within the mitochondrion by an enzyme complex which hydroxylates first C-20, then C-22 before cleavage of the side chain by 20,22 desmolase. Pregnenolone is converted into progesterone by a microsomal 3β-hydroxysteroid dehydrogenase. Hydroxylation of C-21, utilizing atmospheric oxygen, is accomplished by a microsomal enzyme complex involving cytochrome P450. The resulting deoxycoticosterone re-enters the mitochondrion for the final 11β-hydroxylation step.

To arrive at a picture of the probable biogenesis of corticosteroids in the fowl adrenal, it will be necessary to discuss results of experiments using adrenal tissue of other avian species, bearing in mind that de Roos (1961) demonstrated that the secretory pattern was similar. Sandor *et al.* (1965) established that the isotopic label of $[1-{}^{14}C]$-acetate was incorporated into cholesterol, corticosterone, 18-hydroxycorticosterone and aldosterone by incubation with duck and goose adrenal slices. Similar enzyme systems appeared to be present also in embryonic fowl adrenal tissue (Bonhommet and Weniger, 1967). Although Sandor *et al.* (1965) found that only duck adrenal tissue homogenates, but not slices, transformed $[4-{}^{14}C]$-cholesterol into corticosterone, Hall and Koritz (1966) showed that $[7-{}^{3}H]$-cholesterol of high specific activity was converted to corticosterone by quartered fowl adrenals. Other intermediates, e.g. pregnenolone, progesterone and deoxycorticosterone, have been shown *in vitro* to be precursors of corticosterone. Finally, the conversion of $[4-{}^{14}C]$-corticosterone into radioactive 18-hydroxycorticosterone and aldosterone (Donaldson *et al.*, 1965) by duck adrenal slices completes the biosynthetic pathway shown in Fig. 7.

Two groups (Whitehouse and Vinson, 1967; Sandor and Lanthier, 1970) have attempted to establish the hydroxylation sequence from pregnenolone to aldosterone in the avian adrenal by kinetic studies on the competitive incorporation of various substrates with tritium and ${}^{14}C$ labels into the corticosteroids using duck adrenal preparations. Their results supported the main route outlined in Fig. 7 but there is disagreement about the role of 11β-hydroxyprogesterone, first obtained as a minor product from duck and chicken adrenal incubation experiments (Sandor *et al.*, 1963). This conflict may have arisen from experimental variations such as conditions of incubation or analysis of media alone instead of extraction of the whole reaction mixture. Often, the mass of added precursors may have been comparable with the adrenal pool of these intermediates, thus altering the kinetics of the reactions involved. Perhaps the use of pulse-labelling with high specific activity precursors or the superfusion technique (Hainsworth and Grant,

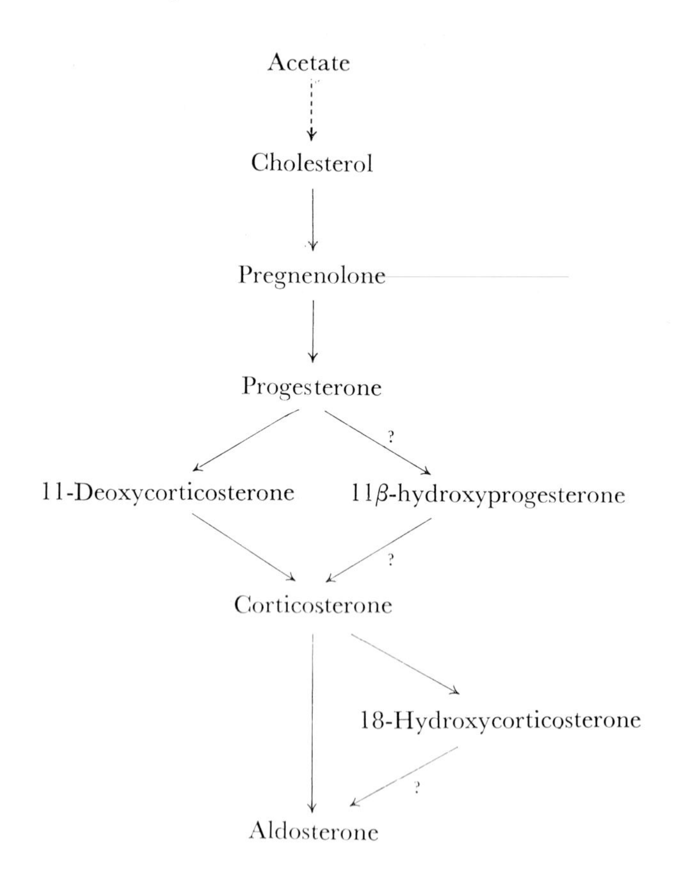

Fig. 7. Biosynthetic pathways to the corticosteroids.

1970) will solve these problems. It should be noted that there is no evidence that 18-hydroxycorticosterone is the direct precursor of aldosterone.

In vitro, mammalian ACTH promotes steroidogenesis by chicken adrenal preparations (de Roos, 1961; Hall and Koritz, 1966). On the basis of the enhanced production of corticosterone and aldosterone *in vitro*, after adding ACTH and progesterone, de Roos (1966) suggested that ACTH exerts its influence beyond progesterone in the biosynthetic chain. However, this concept appears untenable since, with isotopically labelled precursors, the presence of ACTH in incubation mixtures caused a decline in the specific activities of corticosterone and aldosterone, thus reflecting an increase in the synthesis of endogenous precursors between cholesterol and progesterone (Donaldson *et al.*, 1965; Frankel *et al.*, 1967e; Macchi, 1967). Furthermore, addition to duck adrenal tissue of 3′,5′-adenosine cyclic monophosphate which mediates ACTH-stimulated steroidogenesis, was followed by an increase in corticosterone production without any significant effect on that of aldosterone (Macchi, 1967). The pressor octapeptide, angiotensin II, did not appear to enhance the biosynthesis of aldosterone by the fowl adrenal *in vitro* (de Roos and de Roos, 1963).

In the duck adrenal, the 11β- and 18-hydroxylases have been located predominantly in the mitochondrial fraction. As Sandor (1969) has noted, the avian adrenal resembles that of reptiles and amphibia in possessing biosynthetic equipment similar to that postulated for the zona glomerulosa in the mammal.

C. REGULATION OF ADRENAL CORTICAL ACTIVITY

It is generally accepted, on the basis of experiments with mammals, that adrenal function is controlled by the central nervous system via a hypothalamic-hypophysial axis as diagramatically illustrated in Fig. 8. (See also Chapter 16.)

In birds, anatomical connections between the hypothalamus and the anterior pituitary have been described (Benoit, 1962) and adrenal dependence on the integrity of the hypothalamus was demonstrated when lesions in the ventral tuberal area resulted in decreased steroidogenesis (Frankel *et al.*, 1967b). A corticotrophin releasing factor (CRF) has now been characterized in crude extracts of the hypothalamus (Salem *et al.*, 1970b).

The available evidence fully supports the belief that the fowl's anterior pituitary has a regulating role in steroid output; its removal was followed by a decreased adrenal weight (Brown *et al.*, 1958a; Resko *et al.*, 1964) and caused a decline in corticosterone levels in adrenal venous plasma (Frankel *et al.*, 1967a, b, c; Taylor *et al.*, 1970). Although Resko *et al.* (1964) did not obtain evidence for ACTH release from hypophysial autotransplants into the kidney

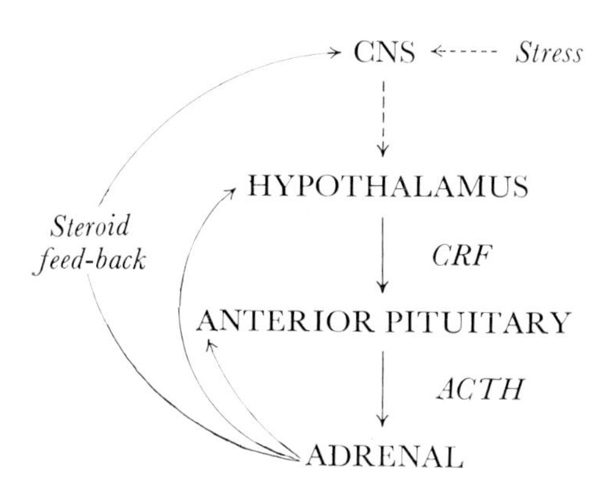

Fig. 8. Diagrammatic representation of the relationship between the CNS and the hypothalamo-pituitary-adrenal axis.

of cockerels, Salem *et al.* (1970a) reported that fowl pituitaries contained relatively more ACTH than the rat.

Early studies suggested that the fowl adrenal was relatively autonomous (Brown *et al.*, 1958a; Newcomer, 1959) but this was disputed by Frankel *et al.* (1967b). These authors showed that corticosterone levels in the adrenal venous plasma of hypophysectomized cockerels rose during short-term surgical stress, implying continued control of corticoid secretion by the CNS. Further, dexamethasone, a synthetic cortisol analogue, completely suppressed corticosterone release in hypophysectomized cockerels. Dexamethasone does not inhibit steroid biosynthesis by adrenal slices *in vitro*, but probably acts at the pituitary, the hypothalamus, or higher centres (Dallman and Yates, 1968; Russel *et al.*, 1969). The presence of an ACTH-like substance in hypothalamic extracts, but distinct from CRF appears to have been confirmed recently by Salem *et al.* (1970b) who speculated that it may be an α-melanocyte-stimulating hormone. The regulation of the avian adrenal seems more complex than was originally supposed (Frankel, 1970).

Despite *in vitro* studies on adrenal tissue, where addition of ACTH increased biosynthesis of aldosterone *inter alia* (de Roos, 1969), analysis of adrenal venous plasma did not reveal an increase in its aldosterone concentration following ACTH infusion over 30 min, although corticosterone levels rose by 250% over that observed before stimulation (Taylor *et al.*, 1970). The latter authors found that the renal-pressor system in the fowl was similar to the renin-angiotensin system described for mammals, but neither kidney extracts nor sodium depletion had any apparent effect on steroidogenesis.

D. RESPONSE TO ACTH

Selye's speculations (1950) on the nature of stress and the pituitary-adrenal response has stimulated many to study the effects of exogenous ACTH in attempts to obtain a reliable index of stress in fowls. Since pure preparations of avian ACTH are not available, most of the work has used materials from mammalian sources. It is of interest to consider the various effects used to assess the action of ACTH.

1. Steroid Secretion

Although Nagra *et al.* (1960) obtained a rise in adrenal venous plasma corticosteroid after ACTH treatment, other studies using similar non-specific assays on peripheral plasma reported either a decrease (Newcomer, 1959a) or no change (Siegel and Siegel, 1966) in corticoid levels following long-term treatment. On the other hand, Breitenbach (1962) found a significant response, within an hour, to 10 U/kg in 4- to 8-week-old birds.

Using a specific procedure, Frankel *et al.* (1967b) found increases of 650% and 300% in corticosterone concentrations in adrenal venous plasma of intact and hypophysectomized cockerels within 1 h after intravenous ACTH (16 U/kg). Similar responses were obtained on intact birds using a double isotope-dilution-derivative method by an independent group (Taylor *et al.*, 1970) who infused ACTH for 30 min. Brown (1968) found that peripheral corticosterone levels of 4-week-old chicks were significantly higher 6 h after injection of ACTH (8U). In contrast, using an adult hen, J. W. Wells and J. Culbert (unpublished) observed that the maximum response to intravenous injection of a synthetic corticotrophin (Synacthen, approximately 10 U/kg) occurred within 30 min and that the steroid level returned to normal within 60 min. With intramuscular administration of a long-acting depot preparation, the peak appeared between 1 and 2 h. Greenman *et al.* (1967) found that a 200% rise in corticosterone production *in vitro* by adrenal slices from adult hens occurred 15 min after intravenous injection of ACTH (5 to 6 U/kg).

The evidence shows that the fowl adrenal does respond to mammalian ACTH and that, probably, the varying results reported arise from a combination of different dose levels and sampling times after treatment.

2. *Adrenal Weight*

The literature on the response to ACTH measured in terms of change in adrenal mass is conflicting. Often, no response has been observed, even with massive doses, e.g. up to 30 U/kg (Siegel and Beane, 1961), in short term experiments (Perek *et al.*, 1959; Wolford and Ringer, 1962; Freeman *et al.*, 1966) or after prolonged treatment of immature and adult birds (Jailer and Boas, 1950; Wolford and Ringer, 1962; Siegel and Siegel, 1966). On the other hand, Howard and Constable (1958) and Buzickovic *et al.* (1968) found evidence for hypertrophy. The use of adrenal mass as a criterion of function evidently requires caution, especially as Flickinger (1966) observed surprising differences in adrenal weight (as well as cortical-to-medullary-tissue ratios) in response to varying dose levels of ACTH in males compared with females of the same age group. Frankel *et al.* (1967c) found no correlation between adrenal weight and corticosterone production. However, Freeman (1970a) reported that both intact and bursectomized chickens when injected with ACTH (5 U) showed increased adrenal weights within 10 and 15 min respectively. The increase in both groups was maintained at least up to 50 min. It seems that dose level and time of sampling are important factors when adrenal weight changes are used to assess function.

3. *Adrenal Ascorbic Acid Depletion* (*AAAD*)

Although depletion of adrenal ascorbic acid in the rat is the basis of a sensitive bio-assay for ACTH (Sayers *et al.*, 1948), the effect of the hormone on this component in the fowl adrenal is variable (Freeman *et al.*, 1966) and most authors have reported no depletion (Jailer and Boas, 1950; Howard and Constable, 1958; Elton *et al.*, 1959; Breitenbach, 1962; Mazini, 1966; Greenman *et al.*, 1967). Perek and Eckstein (1959), Perek *et al.*, 1959 and Perek and Eilat (1960a, b) reported that bursectomized chickens showed

AAAD after ACTH but there was disagreement on this effect in the laying hen (Perek and Eckstein, 1959; *cf.* Wolford and Ringer, 1962). In embryonic adrenals Freeman (1968) observed that depletion occurred within 5 min but ascorbic acid levels returned to normal 30 min after injection. Subsequently Freeman (1970a) reported that in bursectomized 3-week-old birds the mean adrenal *content* fell from 30·5 to 17·4 μg by 25 min after injection of ACTH (5 U), while the adrenals of intact chicks of the same age increased their their ascorbic acid content between 35 to 40 min.

The bursa of Fabricius may thus affect the lability of ascorbic acid in avian adrenal tissue (see Chapter 23).

4. *Adrenal Cholesterol*

On the whole, the depletion of adrenal cholesterol levels appears to be readily observed following both short-term (Howard and Constable, 1958; Siegel and Beane, 1961) and long-term treatment with ACTH (Wolford and Ringer, 1962; Siegel, 1962a; Siegel and Siegel, 1966), although some failures to detect a change have been recorded (Elton *et al.*, 1959; Buzickovic *et al.*, 1968). In a kinetic study of cholesterol depletion, Freeman *et al.* (1966) showed that utilization of esterified cholesterol accounted for the fall in the sterol content up to 6 h after ACTH injection.

5. *Metabolic Endpoints*

Other secondary changes caused by increased steroid output following ACTH administration have been examined including regression of the spleen and the bursa of Fabricius (Siegel and Beane, 1961), changes in blood cell components (Hublé, 1955; Newcomer, 1957), hyperglycaemia (Bell, 1961; Siegel and Beane, 1961; Siegel, 1962b) and hypercholesteraemia (Siegel and Siegel, 1966), but many of these phenomena must be regarded only as useful qualitative indicators of increased adrenal activity (Siegel, 1962b) since they could be sensitive to dietary changes or reproductive activity in the bird.

E. FACTORS EVOKING ADRENAL RESPONSE

From the preceding discussion it is evident that the fowl's adrenal cortical tissue is under control of a specific "trophic" hormone, ACTH, the secretion of which is regulated by the central nervous system. By contrast, the medullary tissue is a specialized part of the sympathetic nervous system and may be expected to respond rapidly to stimulation through that system.

In common with other endocrine glands, the adrenal functions as part of the normal homeostatic mechanism and, by virtue of its relationship with the CNS, responds to stress. In this context, any event acting as a threat, real or apparent, to the biological integrity of the organism may constitute a stress. It is implied that a stressor may be an unconditional stimulus acting directly on the organism, e.g. the physical environment or attack by micro-organisms, or a conditional stimulus which acts indirectly and is a product of individual past experience and may perhaps have a genetic background (Wolff, 1960).

This stereotyped response of the organism to a variety of stressors was postulated by Selye (1950) as the general adaption syndrome (GAS). In mammals GAS is manifested by adrenal enlargement with increased secretion of corticosteroids which in turn cause involution of the thymus and lymphatic organs *inter alia* (see Chapter 45). Although much attention has been devoted to the response of cortical tissue under stress (e.g. by Brown, 1968) the function of the medullary secretion has not been fully explored in the fowl (Draper and Lake, 1968), mainly because of the difficulties of assessing whether the catecholamines have been involved.

1. Effects of Environment

Although Garren and Shaffner (1956) reported that exposure to low temperatures (6·7°C) for 1 to 4 weeks was followed by an increase in adrenal weights in chicks, Buzickovic *et al.* (1968), using a lower temperature (−5°C) for a shorter period, observed only ascorbic acid depletion in adrenals of bursectomized and intact chicks. In adult hens kept at temperatures of 1·7° and 17·8°C for varying periods, none of the usual indicators of adrenal function changed, but the differential leucocyte counts appeared to reflect some reaction (Wolford and Ringer, 1962). Turkeys responded to cold stress by increased corticosterone levels in the peripheral plasma (Brown, 1968).

Relative humidity may be an important feature neglected in most work since both Newcomer (1958) and Perek and Bedrak (1962) reported that cold and dampness caused an increase in "acidophils" and a depletion of adrenal ascorbic acid and cholesterol.

Very little research has been done on heat stress except in relation to increases in plasma citric acid (Hill *et al.*, 1961). The level of protein in the diet or vitamin supplementation appeared to modify this response to low (4°C) and high (40°C) temperatures (Hill, 1961a, b).

High altitudes (above 3,600 m) caused activation of the pituitary-adrenal axis within 5 h as indicated by increased adrenal weights and regression of the spleen and bursa of Fabricius (Garren and Shaffner, 1956). Acute anoxia increased "acidophils" measured 6 to 7 h later (Newcomer, 1958).

2. Other Physical Stimuli

That the corticosterone concentration in adrenal efferent venous plasma rose following surgery was demonstrated conclusively by Frankel *et al.* (1967a, b). Following debeaking, marked AAAD was observed in chicks, lasting up to 12 h (Perek and Bedrak, 1962).

Handling of chicks resulted in AAAD without gland hypertrophy within 10 to 15 min (Freeman, 1967, 1969), but in hens only blood cell components altered (Wolford and Ringer, 1962). Restraint for varying periods increased the "acidophil" count (Newcomer, 1958, 1959a).

Vigorous exercise caused marked adrenal enlargement involving both medullary and cortical tissues. Adrenal hypertrophy appeared to be indicative of, but not necessarily allied with, successful resistance to exercise stress. Gonadal hormones have some interaction since males resisted stress better than females, whose resistance was increased by androgen treatment (Conner and Shaffner, 1954).

3. *Noxious Stimuli*

AAAD occurred within 10 min of histamine injection (3 mg/kg) into chick embryos but the ascorbic acid content was replete within 1 h (Freeman, 1968). This rapid response probably accounts for a previous failure to observe reaction to histamine (15 mg/kg) although aspirin caused AAAD (Howard and Constable, 1958). "Acidophils", on the other hand, increased 4 to 6 h after either histamine or formaldehyde injections (Newcomer, 1958).

4. *Disease*

Inoculation of birds with the fowl typhoid organism *Salmonella gallinarum* resulted in marked hypertrophy of the pituitary and adrenal cortical tissue in those animals which became infected (Garren and Barber, 1955).

5. *Other Hormones*

a. Adrenaline. Although adrenal weights increased as a result of adrenaline injection (Jailer and Boas, 1950), Zarrow and Baldwin (1952) did not observe AAAD, perhaps because of their time of sampling.

b. Insulin. It has been suggested that the resistance of birds to insulin is due to differences in the primary structure of insulins of various origins and to the presence in avian plasma of a "factor" which inactivates mammalian insulin preparations (Hazelwood *et al.*, 1968) (see Chapter 21). As a result of the non-availability of fowl insulin, most workers have used large doses of mammalian preparations to effect hypoglycaemia.

Catecholamines in fowl adrenals underwent a reversible decrease paralleling the fall in plasma glucose after insulin injection. There is disagreement in the relevant publications (West, 1951; Schams and Karg, 1967) as to whether adrenaline or noradrenaline were the major constituents depleted by this treatment.

The role of the adrenal cortical tissue of young birds and hens following insulin hypoglycaemia was not apparent using the criteria of adrenal mass or ascorbic acid concentration (Howard and Constable, 1958; Lepkovsky *et al.*, 1965). However, Mazini (1966) noted that after the 15th d, the embryonic adrenal reacted to hypoglycaemic shock by a 35 to 60% increase in mass.

Hypoglycaemia due to insulin (1 U/kg) or tolbutamide (50 mg/kg) did not alter the corticosterone concentration in peripheral plasma up to 3 h later. With protamine zinc insulin (8 U/kg/d) injected intramuscularly over a 6-d period into one commercial strain of adult hens, a rise of corticosterone levels was observed between days 5 and 8 (J. Culbert and J. W. Wells, unpublished).

This slow response by the fowl cortical tissue reflects a major difference between avian and mammalian physiologies. The rapid rise in plasma corticoid levels following insulin-induced hypoglycaemia is used to test the integrity of the hypothalamic-pituitary-adrenal axis in mammals (Landon *et al.*, 1963; Bassett and Hinks, 1969).

c. Thyroid hormones. Hypothyroidism, induced by thiouracil, had no effect on the adrenal weight of 4-week-old chicks over a 3- to 17-d period, although both bursa and spleen regressed indicating a possible increase in steroid secretion. In birds of the same age group, addition of thyroprotein to the food to promote hyperthyroidism did not produce an increased adrenal mass until 12 d later (Garren and Shaffner, 1956).

6. *Food and Water Deprivation*

Adrenal responses in young birds appeared to be more sensitive to food and water restrictions (Conner, 1959; Buzickovic *et al.*, 1968) than in adult hens (Wolford and Ringer, 1962). Corticosterone levels have not been measured under such conditions.

7. *Mineral Deficiencies*

Despite hypertrophy and histological changes in adrenals of sodium-deficient cockerels, the concentrations of both corticosterone and aldosterone in the adrenal venous plasma did not differ from that of control birds (Taylor *et al.*, 1970). These results confirm observations that birds on diets deficient in calcium and sodium did not have enhanced corticosterone levels in the peripheral plasma (B. Hughes, J. Culbert and J. W. Wells, unpublished).

8. *Social Interaction*

a. Social rank. There are two conflicting publications on social competition; one (Flickinger, 1961) claims that, after sexual maturity, adrenal weights of cockerels are correlated inversely with social rank but the other (Siegel and Siegel, 1961) reported no significant effect.

b. Population density. The effect on the pituitary-adrenal axis of variation in floor areas (ranging from 372 to 929 cm^2 per bird), indicated by increased adrenal weight and depletion of adrenal cholesterol, was not manifest until after 11 weeks of age (Siegel, 1960; Siegel and Siegel, 1969). Broilers reared in pens from hatch to 10 weeks with floor areas between 464 and 929 cm^2 per bird did not differ in their relative adrenal sizes, although live-weight gains and food conversion were affected (W. Bolton and W. A. Dewar, personal communication).

In a recent large-scale study on the performance of laying hens in cages at stocking densities from 1 to 16 birds per cage (i.e. areas of 1391 to 432 cm^2 per bird) R. G. Wells (personal communication) was unable to detect differences in adrenal weights or cholesterol content after caging from 20 to 72 weeks, despite a higher mortality rate under the more crowded conditions.

9. Frustration

Fowls which responded to frustrating situations by a stereotyped behaviour pattern (Duncan, 1970) failed to show any change in corticosterone concentrations in their peripheral plasma (J. Culbert, I. J. H. Duncan and J. W. Wells, unpublished).

In conclusion, most of the physical stressors or unconditional stimuli studied appear to activate the fowl's adrenal although, by contrast with mammals, enlargement does not necessarily occur. Some of the biological changes used to assess adrenal response in the fowl, e.g. plasma citric acid concentration or numbers of "acidophils", may be due to enhanced catecholamine secretion. It would be desirable to confirm that corticosteroid output is increased before many of the findings can be accepted as conclusive. The evidence for increased adrenal secretion due to social interaction, e.g. increased population density, appears inconclusive when compared with the well-documented effects observed in small mammals (Christian *et al.*, 1965).

Further, the response of the sympathoadrenomedullary system to stresses induced by conditional stimuli requires study since, in humans, both the adrenal medulla and cortex respond with increased hormone secretion during anxiety states (Levi, 1969).

F. THE ROLES OF THE ADRENAL HORMONES

1. Catecholamines

In the young bird, exogenous adrenaline increased oxygen consumption more markedly than did noradrenaline. Both amines influenced carbohydrate metabolism (see Chapter 11), e.g. by causing hyperglycaemia and depletion of glycogen stores in muscles and liver, but there were quantitative differences in the effects of the hormones, depending upon the age of the bird (Freeman 1966). Neither hormone proved as effective as triiodothyronine or thyroxine as thermogenic stimulants and it was concluded that the catecholamines had a minor role in thermoregulation (Freeman, 1970b). (See Chapter 48.) The depletion of adrenal catecholamines in insulin-induced hypoglycaemia has already been discussed. (Section II E 5b.) However these hormones are not primary regulators of plasma glucose, since feeding an adrenolytic agent, dibenyline, did not affect normal glucose levels but prevented the characteristic rebound after insulin injection (Hazelwood and Lorenz, 1959).

Catecholamines were ineffective in altering the lipid metabolism of the bird; *in vivo*, free fatty acid levels were unchanged (Heald, 1966; Langslow *et al.*, 1969) and, *in vitro*, neither amine stimulated free fatty acid release from avian adipose tissue (Carlson *et al.*, 1964).

On infusion into fowls both catecholamines have vasoconstrictor properties;

they increase systolic and diastolic pressures with but little or no response on heart rate (Akers and Peiss, 1963; Draper and Lake, 1968). Although the duration of the response of the cardiovascular system with noradrenaline was half that with adrenaline, on the basis of the greater noradrenaline content of the adrenal gland, Akers and Peiss (1963) concluded that this amine was the more active adrenergic agent in the fowl. However, catecholamine secretion by the adrenal has not been measured.

Draper and Lake (1968) pointed out that the blood pressure response to catecholamines would probably result in the diversion of blood away from peripheral organs. If this occurred in the ovarian and oviducal tissues the defence reaction, i.e. adrenal medullary response, may account for sudden drops in egg production. By injecting adrenaline, albeit in large quantities, a drop or cessation in egg laying followed (see also Sykes, 1955). Alternatively, adrenaline may activate the pituitary-adrenal axis.

2. Corticosteroids

The essential role of the adrenal cortical tissue was demonstrated when birds, subjected to bilateral adrenalectomy without steroid replacement therapy, died within 60 h. Somewhat unexpectedly, when compared with mammals, the excretion of urinary sodium decreased while that of potassium remained unchanged. Again, plasma sodium did not alter while the plasma potassium concentration rose to twice its normal level *ante mortem* (Brown *et al.*, 1958b). Sodium depletion did not enhance outputs of aldosterone or corticosterone (Taylor *et al.*, 1970). These facts underline the need for further examination of the role of adrenal steroids in the electrolyte balance.

The involvement of corticosteroids in carbohydrate metabolism has been well explored. Adrenalectomy resulted in a large decrease in blood glucose which was restored by steroid treatment (Brown *et al.*, 1958b). Various corticoids have been used to effect hyperglycaemia and increased liver glycogen deposition (Brown *et al.*, 1958a; Greenman and Zarrow, 1961), presumably through gluconeogenesis since steroid-treated birds had less carcase protein, particularly skeletal muscle (Bellamy and Leonard, 1965) and showed an increased urinary nitrogen excretion (Brown *et al.*, 1958a, b).

Chronic steroid administration depressed body weight (Bellamy and Leonard, 1965; Adams, 1968), despite hyperphagia (Nagra and Meyer, 1963; Nagra *et al.*, 1963); the liver weight was unaffected. The principal effect appeared to be an enhanced lipogenesis, shown by increased plasma lipid and fat deposition. Of the lipid constituents, the specific activity of fatty acids derived from administered [U$-^{14}$C]glucose increased with steroid administration; analysis of the fatty acids revealed a rise in the stearic to oleic acid ratio (Nagra and Meyer, 1963).

Massive doses of ACTH, i.v., (100 U) or corticosterone (20 mg) either inhibited or caused premature ovulation (van Tienhoven, 1961); such vast quantities of ACTH would seem unphysiological. Nevertheless, the increased steroid output stimulated by ACTH caused a linear reduction in size of both oviduct and testes. The rate of decline in egg production appeared to be dependent on the dose injected (Flickinger, 1966).

Chronic steroid treatment was also followed by involution of the bursa of Fabricius (Zarrow *et al.*, 1961) and of the thymus (Bellamy and Leonard, 1965). Destructive osteoporosis in the hen was observed on treatment with massive cortisone dosage i.e. 50–300 mg/week for 2 to 8 weeks (Urist and Deutsch, 1960), but cortisone is not known to occur in the fowl.

In young birds cortisol had an anabolic effect on liver growth, in contrast to its catabolic action on lymphoid tissue (Bellamy and Leonard, 1965). In this respect the effect on the duck of prednisolone, a synthetic glucocorticoid, deserves mention: a single injection increased the incorporation of isotopically labelled uridine into the RNA of liver, thymus and spleen without effect on DNA synthesis; repeated doses, however decreased nucleotide incorporation into both RNA and DNA of all tissues (Bottoms *et al.*, 1969).

Obviously, the actions of adrenal steroids are manifold, with different effects on various tissues.

Experiments based on corticosteroid insufficiency or excess do not necessarily reveal the nature of their normal function. It should be emphasized that the corticosteroids play a "permissive" role in the metabolic response of the organism to stress; thus activation of adrenal cortical tissue tends to preserve homeostasis rather than to cause a state of hypercorticalism (Ingle, 1952). The interpretation of their role is complicated by the possibility that the effect of the corticosteroids may be caused by direct interaction with translation and transcription of genetic information within the target organ and so affecting enzyme synthesis, by intervening in the transport of metabolites through alteration of membrane permeability, or by feed-back on the CNS, hypothalamus or even the anterior pituitary. The output of the "trophic" hormones which regulate the production of other steroid and thyroid hormones would thus be influenced. (See further Chapter 16.)

References

Adams, B. M. (1968). *J. Endocr.* **40**, 145–151.

Akers, T. K. and Peiss, C. N. (1963). *Proc. Soc. exp. Biol. Med.* **112**, 396–399.

Arvy, L. (1962). *C. r. hebd. Séanc. Acad. Sci., Paris* **255**, 1803–1804.

Arvy, L. and Gabe, M. (1951). *C. r. hebd. Séanc. Acad. Sci., Paris* **232**, 260–262.

Bassett, J. M. and Hinks, N. T. (1969). *J. Endocr.* **44**, 387–403.

Bell, D. J. (1961). *Nature, Lond.* **190**, 913.

Bellamy, D. and Leonard, R. (1965). *Gen. comp. Endocr.* **5**, 402–410.

Benoit, J. (1962). *Gen. comp. Endocr.* (Suppl.) **1**, 254–274.

Bonhommet, M. and Weniger, J. P. (1967). *C. r. Séanc. Soc. Biol.* **161**, 2052–2055.

Bottoms, G. D., McCracken, M. D. and Carlton, W. W. (1969). *Poult. Sci.* **48**, 1420–1425.

Breitenbach, R. P. (1962). *Poult. Sci.* **41**, 1318–1324.

Breneman, W. R. (1941). *Endocrinology* **28**, 946–954.

Breneman, W. R. (1954). *Endocrinology* **55**, 54–64.
Brown, K. I. (1961). *Proc. Soc. exp. Biol. Med.* **107**, 538–542.
Brown, K. I. (1968). *In* "Environmental Control in Poultry Production" (T. C. Carter, ed.), pp. 101–113. Oliver and Boyd Ltd., Edinburgh.
Brown, K. I., Brown, D. J. and Meyer, R. K. (1958a). *Am. J. Physiol.* **192**, 43–50.
Brown, K. I., Meyer, R. K. and Brown, D. J. (1958b). *Poult. Sci.* **37**, 680–684.
Buzickovic, P., Brmalj, V. and Srebocan, V. (1968). *Veterinarski Arhiv* **38**, 373–376.
Callingham, B. A. and Cass, R. (1966). *In* "Physiology of the Domestic Fowl" (C. Horton-Smith and E. C. Amoroso, eds.), pp. 279–285. Oliver and Boyd, Ltd., Edinburgh.
Carlson, L. A., Liljedahl, S. O., Verdy, M. and Wirsen, C. (1964). *Metabolism* **13**, 227–231.
Chester Jones, I., Phillips, J. G. and Bellamy, D. (1962). *Br. med. Bull.* **18**, 110–114.
Christian, J. J., Lloyd, J. A. and Davis, D. E. (1965). *Recent Prog. Horm. Res.* **21**, 501–571.
Conner, M. H. (1959). *Poult. Sci.* **38**, 1340–1343.
Conner, M. H. and Shaffner, C. S. (1954). *Endocrinology* **55**, 45–53.
Coupland, R. E., Pyper, A. S. and Hopwood, D. (1964). *Nature, Lond.* **201**, 1240–1242.
Crile, G. W. and Quiring, D. P. (1940). *Ohio J. Sci.* **40**, 219–259.
Dallman, M. F. and Yates, F. E. (1968). *In* "The Investigation of Hypothalamic-Pituitary Adrenal Function". Memoirs of the Society for Endocrinology (V. H. T. James and J. Landon, eds.), No. 17, pp. 39–72. University Press, Cambridge.
Donaldson, E. M. and Holmes, W. N. (1965). *J. Endocr.* **32**, 329–336.
Donaldson, E. M., Holmes, W. N. and Stachenko, J. (1965). *Gen. comp. Endocr.* **5**, 542–551.
Draper, M. H. and Lake, P. E. (1968). *In* "Environmental Control in Poultry Production" (T. C. Carter, ed.), pp. 87–100. Oliver and Boyd Ltd., Edinburgh.
Duncan, I. J. H. (1970). *In* "Aspects of Poultry Behaviour" (B. M. Freeman and R. F. Gordon, eds.), pp. 15–31. Oliver and Boyd Ltd., Edinburgh.
Elfvin, L.-G. (1965). *J. Ultrastruct. Res.* **12**, 263–286.
Elfvin, L.-G. (1967). *J. Ultrastruct. Res.* **17**, 45–62.
Elliot, T. R. and Tuckett, I. (1906). *J. Physiol., Lond.* **34**, 322–369.
Elton, R. L., Zarrow, I. G. and Zarrow, M. X. (1959). *Endocrinology* **65**, 152–157.
Eränkö, O. (1957). *Nature, Lond.* **179**, 417–418.
Findlay, G. M. (1920). *J. Path. Bact.* **23**, 482–489.
Flickinger, G. L. (1961). *Gen. comp. Endocr.* **1**, 332–340.
Flickinger, G. L. (1966). *Poult. Sci.* **45**, 753–761.
Frankel, A. I. (1970). *Poult. Sci.* **49**, 869–921.
Frankel, A. I., Graber, J. W. and Nalbandov, A. V. (1967a). Proceedings of the Second International Congress on Hormonal Steroids, Excerpta Medica, Intern. Congr. Ser. No. 132, pp. 1104–1113, Amsterdam.
Frankel, A. I., Graber, J. W. and Nalbandov, A. V. (1967b). *Endocrinology* **80**, 1013–1019.
Frankel, A. I., Graber, J. W. and Nalbandov, A. V. (1967c). *Gen. comp. Endocr.* **8**, 387–396.
Frankel, A. I., Cook, B., Graber, J. W. and Nalbandov, A. V. (1967d). *Endocrinology* **80**, 181–194.
Frankel, A. I., Graber, J. W., Cook, B. and Nalbandov, A. V. (1967e). *Steroids* **10**, 699–707.
Freeman, B. M. (1966). *Comp. Biochem. Physiol.* **18**, 369–382.

Freeman, B. M. (1967). *Comp. Biochem. Physiol.* **23**, 303–305.
Freeman, B. M. (1968). *Comp. Biochem. Physiol.* **24**, 905–914.
Freeman, B. M. (1969). *Comp. Biochem. Physiol.* **29**, 639–646.
Freeman, B. M. (1970a). *Comp. Biochem. Physiol.* **32**, 755–761.
Freeman, B. M. (1970b). *Comp. Biochem. Physiol.* **33**, 219–230.
Freeman, B. M., Chubb, L. G. and Pearson, A. W. (1966). *In* "Physiology of the Domestic Fowl" (C. Horton-Smith and E. C. Amoroso, eds.), pp. 103–112. Oliver and Boyd Ltd., Edinburgh.
Friedman, S. and Kaufman, S. (1965). *J. biol. Chem.* **240**, 4763–4773.
Fujita, H. (1961). *Z. Zellforsch. mikrosk. Anat.* **55**, 80–88.
Garren, H. W. and Barber, C. W. (1955). *Poult. Sci.* **34**, 1250–1258.
Garren, H. W. and Shaffner, C. S. (1956). *Poult. Sci.* **35**, 266–272.
Ghosh, A. (1962). *Gen. comp. Endocr. Suppl.* **I**, 75.
Goodchild, W. M. (1969). *Br. Poult. Sci.* **10**, 183–185.
Greenman, D. L. and Zarrow, M. X. (1961). *Proc. Soc. exp. Biol. Med.* **106**, 459–462.
Greenman, D. L., Whitley, L. S. and Zarrow, M. X. (1967). *Gen. comp. Endocr.* **9**, 422–427.
Günther, G. (1906). *In* "Handbuch der vergleichenden mik. Anat. der Haustiere" (W. Ellenbergen, ed.), Vol. 1, pp. 264–266. Paul Parry, Berlin.
Hagen, P. (1956). *J. Pharmac. exp. Ther.* **116**, 26.
Hainsworth, I. R. and Grant, J. K. (1970). *Biochem. J.* **117**, 25*P*.
Hall, P. F. and Koritz, S. B. (1966). *Endocrinology* **79**, 652–654.
Hartman, F. A., Knouff, R. A., McNutt, A. W. and Carver, J. E. (1947). *Anat. Rec.* **97**, 211–221.
Hazelwood, R. L., Kimmel, J. R. and Pollock, H. G. (1968). *Endocrinology* **83**, 1331–1336.
Hazelwood, R. L. and Lorenz, F. W. (1959). *Am. J. Physiol.* **197**, 47–51.
Heald, P. J. (1966). *In* "Physiology of the Domestic Fowl" (C. Horton Smith and E. C. Amoroso, eds.), pp. 113–124. Oliver and Boyd Ltd., Edinburgh.
Hill, C. H. (1961a). *Poult. Sci.* **40**, 762–765.
Hill, C. H. (1961b). *Poult. Sci.* **40**, 1311–1315.
Hill, C. H., Warren, M. K. and Garren, H. W. (1961). *Poult. Sci.* **40**, 422–424.
Howard, A. N. and Constable, B. J. (1958). *Biochem. J.* **69**, 501–505.
Hublé, J. (1955). *Poult. Sci.* **34**, 1357–1359.
Ingle, D. J. (1952). *Proc. Soc. Endocr.* **8**, xxiii–xxxvii.
Jailer, J. W. and Boas, N. F. (1950). *Endocrinology* **46**, 314–318.
Juhn, M. and Mitchell, J. B. (1927). *Am. J. Physiol.* **88**, 177–182.
Kano, M. (1959). *Archvm. histol. jap.* **18**, 25–56.
Kar, A. B. (1947). *Anat. Rec.* **99**, 177–198.
Kerstein, H. S., Leonhäuer, S. and Staudinger, Hj. (1958). *Biochim. biophys. Acta* **29**, 350–357.
Kjaerheim, Å., (1968). *Z. Zellforsch. mikrosk. Anat.* **91**, 429–455.
Kliman, B. and Peterson, R. E. (1950). *J. biol. Chem.* **235**, 1639–1648.
Kondics, L. and Kjaerheim, Å. (1966). *Z. Zellforsch. mikrosk. Anat.* **70**, 81–90.
Landon, J., Wynn, V. and James, V. H. T. (1963). *J. Endocr.* **27**, 183–192.
Langslow, D. R., Butler, E. J., Hales, C. N. and Pearson, A. W. (1969). *J. Endocr.* **46**, 243–260.
Lepkovsky, S., Len, R., Koike, T. and Bouthilet, R. (1965). *Am. J. Physiol.* **208**, 589–592.
Levi, L. (1969). *In* "Studies of Anxiety" (M. H. Lader, ed.), pp. 40–52. Headley Bros., Ltd., Kent.

Macchi, I. A. (1967). Proceedings of the Second International Congress on Hormonal Steroids, Excerpta Medica Intern. Congr. Ser. No. 132, pp. 1094–1103, Amsterdam.

Mattingly, D. (1962). *J. clin. Path.* **15**, 374–379.

Mazini, T. I. (1966). *Stanovlenie Endokr. Funkts. Zarodyshevom Razv Akad. Nauk SSSR Inst. Morfol Zhivotn, 97–107*; *Chem. Abstr.* **66**, 83510c.

Morita, S., Masayuki, D. and Ogami, E. (1961). *Jap. J. vet. Sci.* **23**, 353–355.

Nagra, C. L., Baum, G. J. and Meyer, R. K. (1960). *Proc. Soc. exp. Biol. Med.* **105**, 68–70.

Nagra, C. L., Breitenbach, R. P. and Meyer, R. K. (1963). *Poult. Sci.* **43**, 770–775.

Nagra, C. L. and Meyer, R. K. (1963). *Gen. comp. Endocr.* **3**, 131–138.

Newcomer, W. S. (1957). *Proc. Soc. exp. Biol. Med.* **96**, 613–616.

Newcomer, W. S. (1958). *Am. J. Physiol.* **194**, 251–254.

Newcomer, W. S. (1959a). *Am. J. Physiol.* **196**, 276–278.

Newcomer, W. S. (1959b). *Endocrinology* **65**, 133–135.

Oakberg, E. F. (1951). *Growth* **15**, 57–78.

Payne, F. (1955). *J. exp. Zool.* **128**, 259.

Pearse, A. G. E. (1960). "Histochemistry Theoretical and Applied". I. and A. Churchill Ltd., London.

Perek, M. and Bedrak, E. (1962). *Poult. Sci.* **41**, 1149–1156.

Perek, M. and Eckstein, B. (1959). *Poult. Sci.* **38**, 996–999.

Perek, M., Eckstein, B. and Eshkol, Z. (1959). *Endocrinology* **64**, 831–832.

Perek, M. and Eilat, A. (1960a). *J. Endocr.* **20**, 251–255.

Perek, M. and Eilat, A. (1960b). *Endocrinology* **66**, 304–305.

Peterson, R. E. (1959). *Recent Prog. Horm. Res.* **15**, 231–261.

Phillips, J. G. and Chester Jones, I. (1957). *J. Endocr.* **16**, iii.

Rabl, H. (1891). *Arch. mikrosk. Anat.* **38**, 492–523.

Resko, J. A., Norton, H. W. and Nalbandov, A. V. (1964). *Endocrinology* **75**, 192–200.

de Roos, R. (1961). *Gen. comp. Endocr.* **1**, 494–512.

de Roos, R. (1969). *Gen. comp. Endocr.* **13**, 455–459.

de Roos, R. and de Roos, C. C. (1963). *Science, N.Y.* **141**, 1284.

Russel, S. M., Dhariwal, A. P. S., McCann, S. M. and Yates, F. E. (1969). *Endocrinology* **85**, 512–521.

Salem, M. H. M., Norton, H. W. and Nalbandov, A. V. (1970a). *Gen. comp. Endocr.* **14**, 270–280.

Salem, M. H. M., Norton, H. W. and Nalbandov, A. V. (1970b). *Gen. comp. Endocr.* **14**, 281–289.

Sandor, T. (1969). *Gen. comp. Endocr.* Suppl. **2**. 284–298.

Sandor, T., Lamoureux, J. and Lanthier, A. (1963). *Endocrinology* **73**, 629–636.

Sandor, T., Lamoureux, J. and Lanthier, A. (1965). *Steroids* **6**, 143–157.

Sandor, T. and Lanthier, A. (1970). *Endocrinology* **86**, 552–559.

Sauer, F. C. and Latimer, H. B. (1931). *Anat. Rec.* **50**, 289–298.

Sayers, M. A., Sayers, G. and Woodbury, L. A. (1948). *Endocrinology* **42**, 379–393.

Schams, D. and Karg, H. (1967). *Berl. Münch. Tierärztl. Wschr.* **80**, 28–31.

Scott, J. L. (1962). *Poult. Sci.* **41**, 1009–1011.

Seal, U. S. and Doe, R. P. (1963). *Endocrinology* **73**, 371–376.

Selye, H. (1950). "The Physiology and Pathology of Exposure to Stress. A Treatise based on the Concept of the General Adaption Syndrome and the Diseases of Adaption" *Acta Inc.*, Montreal.

Shepherd, D. M. and West, G. B. (1951). *Br. J. Pharm. Chemother.* **6**, 665–674.
Siegel, H. S. (1960). *Poult. Sci.* **39**, 500–510.
Siegel, H. S. (1962a). *Poult. Sci.* **41**, 321–334.
Siegel, H. S. (1962b). *Gen. comp. Endocr.* **2**, 385–388.
Siegel, H. S. and Beane, W. L. (1961). *Poult. Sci.* **40**, 216–219.
Siegel, H. S. and Siegel, P. B. (1961). *Anim. Behav.* **9**, 151–158.
Siegel, H. S. and Siegel, P. B. (1966). *Poult. Sci.* **45**, 901–912.
Siegel, P. B. and Siegel, H. S. (1969). *Poult. Sci.* **48**, 1425–1433.
Sivaram, S. (1964). *Anat. Rec.* **148**, 336.
Sivaram, S. (1965). *Can. J. Zool.* **43**, 1021–1031.
Sivaram, S. and Sharma, D. R. (1964). *J. Histochem. Cytochem.* **12**, 852.
Stenlake, J. B., Davidson, A. G., Williams, W. D. and Downie, W. W. (1970). *J. Endocr.* **46**, 209–220.
Sykes, A. H. (1955). *Poult. Sci.* **34**, 622–628.
Taylor, A. A., Davis, J. O., Breitenbach, R. P. and Hartroft, P. M. (1970). *Gen. comp. Endocr.* **14**, 321–333.
Tienhoven, A. van (1961). *Acta endocr. Copenh.* **38**, 407–412.
Uotila, U. U. (1939). *Anat. Rec.* **75**, 439–451.
Urist, M. R. and Deutsch, N. M. (1960). *Endocrinology* **66**, 805–818.
Venzke, W. G. (1953). *Am. J. vet. Res.* **14**, 219–229.
West, G. B. (1951). *J. Pharm. Pharmac.* **3**, 400–408.
Wheeler, R. S. (1950). *Poult. Sci.* **29**, 784.
Whitehouse, B. and Vinson, G. P. (1967). *Gen. comp. Endocr.* **9**, 161–171.
Wolff, H. G. (1960). *In* "Stress and Psychiatric Disorder" (J. M. Tanner, ed.), pp. 17–30. Blackwell, Oxford.
Wolford, J. H. and Ringer, R. K. (1962). *Poult. Sci.* **41**, 1521–1529.
Zarrow, M. X. and Baldwin, J. T. (1952). *Endocrinology* **50**, 555–561.
Zarrow, M. X. Greenman, D. L. and Peters, L. E. (1961). *Poult. Sci.* **40**, 87–93.

21 The Role of the Endocrine Pancreas and Catecholamines in the Control of Carbohydrate and Lipid Metabolism

D. R. LANGSLOW and C. N. HALES[1]

Houghton Poultry Research Station, Houghton, Huntingdon, England
and Department of Biochemistry, University of Cambridge, Cambridge, England

I. Introduction

The role of the endocrine pancreas in the metabolism of birds has fascinated investigators for many years. Minkowski (1893), following his classic observations with von Mehring (1889) on the dog after pancreatectomy, discovered that removal of the pancreas in ducks and geese failed to produce signs of *diabetes mellitus*. These observations, later confirmed by many workers (for review see Hazelwood, 1965), suggested that birds and mammals differed considerably in their regulation of carbohydrate and lipid metabolism. In addition, the normal plasma glucose concentration of fowls is 2 to 3 times

[1] Present address: The Welsh National School of Medicine, Department of Chemical Pathology, Royal Infirmary, Cardiff, Wales.

higher than that of most mammals. However, there appear to be no major differences in the pathways by which glucose and free (non-esterified) fatty acids (FFA) can be utilized by both birds and mammals (see Chapters 11 and 12).

The pancreas has both exocrine and endocrine functions. The exocrine secretions are considered in Chapter 2. This chapter is confined mainly to the fowl but is extended to cover relevant comparative aspects of avian and mammalian metabolisms. The functions of glucagon and insulin in mammalian metabolism have been considered in recent reviews by Taylor (1967), Unger and Eisentraut (1967), Foa (1968), Hales (1968), Rodbell *et al.* (1968), Villar-Palasi (1968), Sutherland and Robison (1969), and Mayhew *et al.* (1969).

II. Endocrine Pancreas

A. STRUCTURE

The fowls' pancreas is an elongated gland, located in the duodenal loop and comprises three main lobes, splenic, dorsal and ventral (Calhoun, 1933). Part of the ventral lobe has been designated the third lobe on account of its islet distribution (Mikami and Ono, 1962). The splenic lobe has no secretory duct, is rich in islets and may have no exocrine function. Oakberg (1949) studied the distribution of alpha and beta islets in the fowl in relation to age, weight and sex. The absolute number of alpha islets increased up to 100 d of age, and beta islets, up to 300 d of age whilst the number of alpha and beta islets per unit body weight increased during the first 10 d after hatching, but thereafter declined. Beta islets are found throughout the pancreas whereas alpha islets predominate in its proximal portion. The non-random distribution of alpha and beta islets was confirmed by Mikami and Ono (1962).

Unlike human and other mammalian pancreas, both fowl and duck pancreas have two distinct types of islets of Langerhans. The light (beta) islets consist of beta-cells and alpha$_1$-cells. The dark (alpha) islets, consist predominantly of alpha$_2$-cells and alpha$_1$-cells; a few alpha islets contain clusters of beta cells (Hellman and Hellerstrom, 1960; Hellman, 1961; Hellerstrom *et al.*, 1964). Some agranular cells are also found. The difference between the two types of alpha-cells has been characterized in ducks on the basis of (a) the position within the islets, (b) cytoplasmic granulation, (c) reactions to various staining procedures, and (d) cell size and nuclear volume (Hellerstrom *et al.*, (1964). Histochemical studies of the three main cell types have shown important differences in enzyme content (Hellerstrom, 1963).

The ultrastructure of the alpha$_1$-cells suggests a secretory activity (Machino and Sakuma, 1968) but the product has not been identified. An extract of pigeon alpha$_1$-cells inhibited insulin secretion *in vitro* by islets of obese-hyperglycaemic mice (Hellman and Lernmark, 1969). Immunofluorescent studies have suggested that human alpha$_1$-cells may contain gastrin (Lomsky *et al.*, 1969). The alpha$_2$-cells contain glucagon (see review by Lundquist *et al.*, 1970) and the beta-cells contain insulin. Electron microscopy of fowl beta-cells shows granules containing needle- or bar-shaped crystalloid structures, similar to those found in the dog (Sato *et al.*, 1966). Generic variations in the ultrastructure of the granules which are observed are unlikely to be due only to differences in amino-acid sequence of the insulin molecules (Bjorkman and Hellman, 1967). The results of X-ray studies of pig insulin crystals have recently been published (Adams *et al.*, 1969). The rhombohedral unit cell of the crystals contained two zinc ions and six insulin molecules and each zinc ion was in contact with three B10 histidine residues. The amount and manner of zinc binding by insulin is likely to be important in determining the structure of the stored hormone. Other factors which may be important include the presence of insulin derivatives related to its synthesis. Steiner *et al.* (1969) have shown that insulin is synthesized initially as a single-chain precursor molecule "proinsulin", which is later cleaved to insulin, free amino acids and the "C-peptide". The latter appears to be present in pancreas in amounts equimolar with insulin. During the stimulation of insulin secretion, C-peptide is also released in amounts equimolar to that of released insulin suggesting that the former is also stored in the insulin-containing granules. This peptide, and other compounds present in the granule, may therefore also modify the structure of stored insulin. The alpha$_2$-cell granules are densely packed and are spherical in fowls and ducks (Bjorkman and Hellman, 1964; Sato *et al.*, 1966; Machino, 1966; Machino and Sakuma, 1967).

B. HORMONAL CONTENT

1. *Insulin*

Shortly after the first extraction of insulin by Banting and Best (1922), crude fowl pancreatic extracts were prepared and bioassayed (Redenbaugh *et al.*, 1926; Jephcott, 1932 quoted by Haist, 1944); the insulin content of chicken pancreas is low, containing only 1 to 2 mg/100 g wet weight tissue compared with 10 to 15 mg/100 g wet weight in mammalian pancreas (Haist, 1944; Kimmel *et al.*, 1968; Langslow, 1969).

Table 1

The variation in amino-acid sequence between ox, pig and fowl insulins (from Smith, 1966)

	A chain			B chain		
	8	9	10	1	2	27
Ox	Ala	Ser	Val	Phe	Val	Thr
Pig	Thr	Ser	Ileu	Phe	Val	Thr
Fowl	His	Asn	Thr	Ala	Ala	Ser

Crystalline fowl insulin was first prepared by Mirsky *et al.* (1963) and by Smith (1964). There were six different amino acids in the sequence of fowl insulin as compared to porcine and ox insulins (Table 1, Smith, 1966); the significance of this difference is not clear. Smith (1966) tried to relate variations in the primary structure of different insulins to their evolutionary relationships but without success. The possible importance of these sequence differences to the sensitivity of fowls to insulin is discussed later (Section III B). Fowl insulin was heterogeneous on polyacrylamide gel electrophoresis. The minor bands were not identified but may result from contamination by proinsulin and other derivatives of insulin and proinsulin ox insulin is similarly heterogeneous (Mirsky and Kawamura, 1966; Kimmel *et al.*, 1968 (Steiner *et al.*, 1969).

2. Glucagon

Bioassay on the hyperglycaemic factor extractable from fowl pancreas showed it to contain about 10 times more glucagon than an equivalent weight of mammalian pancreas (Vuylsteke and DeDuve, 1953). Assan *et al.* (1968) found glucagon in extracts of fowl pancreas with cross reactivity identical to glucagon from a number of mammals. The glucagon content of duck pancreas has been measured immunologically and this has confirmed the earlier observations of a high glucagon content (Samols *et al.*, 1969a).

3. Other Factors

Two crystalline polypeptides have been isolated from fowl pancreas. Ryan and Tominatsu (1965) detected no biological activity to be associated with their polypeptide of minimum molecular weight about 6,900.

The polypeptide "APP" (Kimmel *et al.*, 1968) has a minimum molecular weight of about 4,200 and contains 36 amino-acid residues in a sequence partially determined by Kimmel *et al.* (1971). They found that APP was present in the plasma of both 4-week-old chickens (range 4 to 5 ng/ml) and laying hens (range 2 to 12 ng/ml). Acid–alcohol extracts of fowl pancreas contained 8·6 mg APP/100 g of tissue but no APP activity was obtained in similar extracts of fowl intestine. APP (0·3 mg/kg body weight) injected i.v. into fowls did not affect the blood glucose concentration. On the other hand, the blood glycerol concentration fell sharply and immediately and the liver glycogen was depleted. The effect on blood glycerol concentration certainly suggested an anti-lipolytic role for APP. Hazelwood *et al.* (1971) have also shown a gastrin-like action of APP; both acid and pepsin secretion were promoted although it was not as potent as pentagastrin. This gastrin-like action of APP may be related to the extract of alpha$_1$-cells of pigeon pancreas isolated by Hellman and Lernmark (1969) who prepared a water extract from the alpha$_1$-cells of pigeon pancreas which *in vitro* inhibited insulin secretion from murine pancreatic islets. This inhibitor may be identical with gastrin which inhibited the glucose-stimulated insulin release from mouse islets (Hellman and Lernmark, 1970). Alpha$_1$-cells have been reported to contain gastrin in some mammals (Lomsky *et al.*, 1969).

C. PANCREATECTOMY

Numerous conflicting reports exist on the effects of pancreatectomy in birds. This is not surprising in view of the surgical problems and the diffuse nature of the avian pancreas; it is supplied by blood vessels which also supply adjacent parts of the intestine. Destruction of these vessels led to gangrene and death in ducks (Sprague and Ivy, 1936). Thus, ischaemic damage to surrounding tissues is likely if pancreatectomy is to be complete. Pancreatectomy has been performed on fowls by Giaja (1912), Koppanyi *et al.* (1926), Batt (1940), Stamler *et al.* (1950), Hazelwood (1958), Lepkovsky *et al.* (1964) and Koike *et al.* (1964). Mild hyperglycaemia, glycosuria and ketosis sometimes follow but are not permanent and most birds remain in a fair state of health for several days. Despite normal feeding, the birds lose weight and the gonads atrophy unless fed raw pancreas. Death was generally attributed to impaired digestion. Thus, it appears that pancreatectomized fowls have normal glucose metabolism although some pancreatic tissue may have remained intact. No measurements were made of plasma insulin concentrations. Depancreatized fowls develop fasting ketosis as do normals and respond in the same way to exogeneous insulin (Koike *et al.*, 1964). However, Stamler *et al.* (1950) reported that pancreatectomy in fowls altered their lipid metabolism with resultant ketonaemia, lipaemia and spontaneous atherosclerosis when a fat-supplemented diet was fed but that their glucose tolerance was still normal.

In contrast to the effect of total pancreatectomy in fowls, Mikami and Ono (1962) found that partial pancreatectomy, when the third and splenic lobes were removed, produced a brief hyperglycaemic phase followed by severe hypoglycaemia together with convulsions and death within 12 to 36 h. The hypoglycaemia could be alleviated by glucose or glucagon. The birds recovered under this therapy and began to feed again but severe hypoglycaemia returned within a few hours.

The beta-cells of the pancreatic islets of fowls and ducks seem to resist damage by alloxan (Scott *et al.*, 1945; Lukens, 1948; Mirsky and Gitelson, 1957; Langslow *et al.*, 1970). Streptozotocin, another diabetogenic agent (Dulin *et al.*, 1967; Brosky and Logothetopoulos, 1969; Pitkin and Reynolds, 1970), did not destroy fowl pancreatic beta-cells although the dose administered was 10 times that found effective on the rat (Langslow *et al.*, 1970 and unpublished). Neither alloxan nor streptozotocin produced any long-term alterations in plasma glucose, insulin or FFA concentrations.

Administration of synthalin A (believed to destroy glucagon-secreting cells), produced initially a brief hyperglycaemic phase, followed by intense hypoglycaemia, convulsions and death (Beekman, 1956). In rabbits, synthalin A produced the same biphasic change in blood sugar but both phases were much more prolonged (Davis, 1952). Cobaltous chloride produced a persistent hyperglycaemia in fowls, an effect produced neither by nickel chloride nor ferrous chloride (Hazelwood and Lorenz, 1957). However, they could not rule out a direct effect on the liver by cobalt, as well as a possible effect on the $alpha_2$-cells, as suggested by Volk *et al.* (1953). Pretreatment of rabbits with cobalt chloride subsequently increased insulin secretion from their pancreas *in vitro* (Telib and Pfeiffer, 1969).

Hence, results from both partial pancreatectomy and alpha cytotoxins suggest that $alpha_2$-cell secretion may play a homeostatic role in avian carbohydrate metabolism.

Most investigators who have pancreatectomized ducks have found results similar to those obtained with fowls. Occasionally, mild hyperglycaemia and glycosuria were seen but the ducks generally recovered rapidly (Minkowski, 1893; Weintraud, 1894; Paton, 1905; Fleming, 1919; Seitz and Ivy, 1929; Sprague and Ivy, 1936). However, Mialhe (1958) concluded pancreatectomy to be incomplete in these earlier experiments. This is certainly indicated in the pancreatectomy by Mirsky *et al.* (1964). Mialhe (1958) found that complete removal of the duck's pancreas produced hypoglycaemia and death from convulsions 2 to 24 h later. Glucose kept the duck alive during this period after which it survived with a low blood sugar concentration. The pancreatectomized duck was much less able to control its blood sugar level. After feeding, blood sugar levels were either normal or elevated, while starvation produced severe hypoglycaemia. This condition was ascribed to the absence of both insulin and glucagon, the latter apparently being necessary for the avoidance of starvation hypoglycaemia (Mialhe, 1960). Treatment of the pancreatectomized duck with both insulin and glucagon can normalize both the fasting blood sugar concentration and the sugar tolerance curve (Mialhe, 1957). Partial pancreatectomy in ducks led to a variety of syndromes. Transient post-operative diabetes was followed by a period with normal metabolism. Subsequently, three main syndromes were observed and were correlated with the nature of the remaining islet cells; normal metabolism (functioning $alpha_2$-cells and beta-cells), diabetes (normal $alpha_2$-cells only) and a new syndrome involving degranulated $alpha_2$-cells and vacuolated beta-cells (Mialhe, 1969).

Samols *et al.* (1969a) have recently measured the immunoreactive insulin and glucagon concentrations, following pancreatectomy of ducks by Mialhe's method. While the plasma insulin concentration fell below detectable concentrations, the plasma glucagon fell sharply but did not reach zero. This may have been due to cross-reacting material from the gut. The plasma glucagon concentration was higher than that reported in man and dog. Mirsky *et al.* (1964) have induced hyperglycaemia in ducks with anti-insulin serum and the blood glucose concentration of normal ducks returned to the initial value much more rapidly than that of pancreatectomized ducks.

D. REGULATION OF HORMONE SECRETION

1. Insulin Secretion

Studies on the control of insulin secretion in fowls *in vivo* have been carried out by Langslow *et al.* (1970). Contrary to what is observed in starved mammals, the plasma insulin concentrations in both male and female immature starved fowls exhibited only minor fluctuations. The changes appeared to be unrelated to the length of starvation (at least up to 72 h), and to the changes in plasma glucose and FFA concentrations. Glucose administered either orally or via the heart (Fig. 1) caused the plasma insulin concentration to rise although the intracardiac route proved the more potent stimulus of

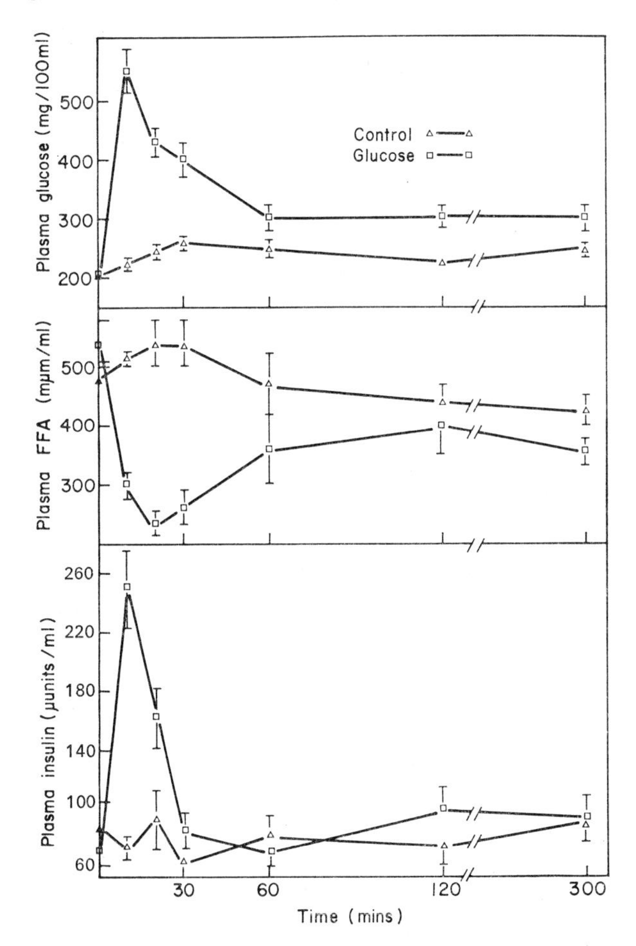

Fig. 1. Effect of intracardiac glucose (1 g) on the plasma glucose, FFA and insulin concentrations of 7-week-old chickens. The control group was given 0·9% saline. Reproduced with permission from Langslow *et al.* (1970).

insulin secretion. However, this difference may be due to the very different rate at which glucose reached the plasma, and to the fact that the change in plasma glucose concentration was greater with intracardiac glucose (Langslow *et al.*, 1970). Similar results have been obtained with ducks (Samols *et al.*, 1969a) thus contrasting with the effect in man where glucose *per os* is a much more potent stimulus of insulin secretion than intravenous administration possibly because of the release of insulin-stimulating factors from the gut (Elrick *et al.*, 1964; McIntyre *et al.*, 1964). This factor could be the glucagon-like substance from the gut. Extracts from canine jejunum and from porcine ileum and jejunum, containing glucagon-like activity, stimulated insulin secretion (Valverde *et al.*, 1968; Moody *et al.*, 1970) although there was no exact correlation between the degree of stimulation and the amount of glucagon-like immunoreactivity. Buchanan *et al.* (1969) found no effect of crude extracts of rat jejunum on insulin secretion from rat islets of Langerhans. Marco *et al.* (1970) have stated that the increase in gut glucagon-like activity in the plasma of dogs follows the increase in the insulin secretion induced by intraduodenal administration of several monosaccharides and have suggested that the stimulation of insulin secretion produced by impure gut extracts was not due to their glucagon-like activity.

Oral administration of amino acids to young fowls resulted in a small but significant increase in both plasma glucose and insulin concentrations. These were accompanied by a modest increase in plasma α-amino nitrogen concentration (Langslow *et al.*, 1970). Similar changes in plasma insulin concentration have been reported for mammals (Floyd *et al.*, 1966; Hertelendy *et al.*, 1969; Unger and Eisentraut, 1970).

A rapid increase in the fowl's plasma glucose concentration was not always accompanied by a rise in plasma insulin concentration; both adrenaline and glucagon induced hyperglycaemia although both failed to increase the plasma insulin concentration. Adrenaline caused a small fall in plasma insulin concentration (Langslow *et al.*, 1970) as previously observed in man (Porte *et al.*, 1965). The failure of glucagon to affect the plasma insulin concentration of fowls (Fig. 2) was in marked contrast to its action in man, dogs and ducks where glucagon increased the plasma insulin concentration (Samols *et al.*, 1965; Ketterer *et al.*, 1967; Mialhe, 1969). The difference between chickens and ducks, in this respect, is interesting.

Tolbutamide given to fed chickens produced hypoglycaemia which persisted for 2 h (Mirsky and Gitelson, 1957); this was accompanied by increased insulin secretion (Langslow, 1969). Samols *et al.* (1969b) reported that tolbutamide, as well as increasing insulin secretion, reduced the plasma glucagon concentration in ducks; this may be partly responsible for their hypoglycaemia. However, hepatectomized and pancreatectomized fowls respond to tolbutamide as do normals. This suggests a peripheral action of the drug (Hazelwood, 1958).

Attempts to produce diabetic chickens by the injection of alloxan or

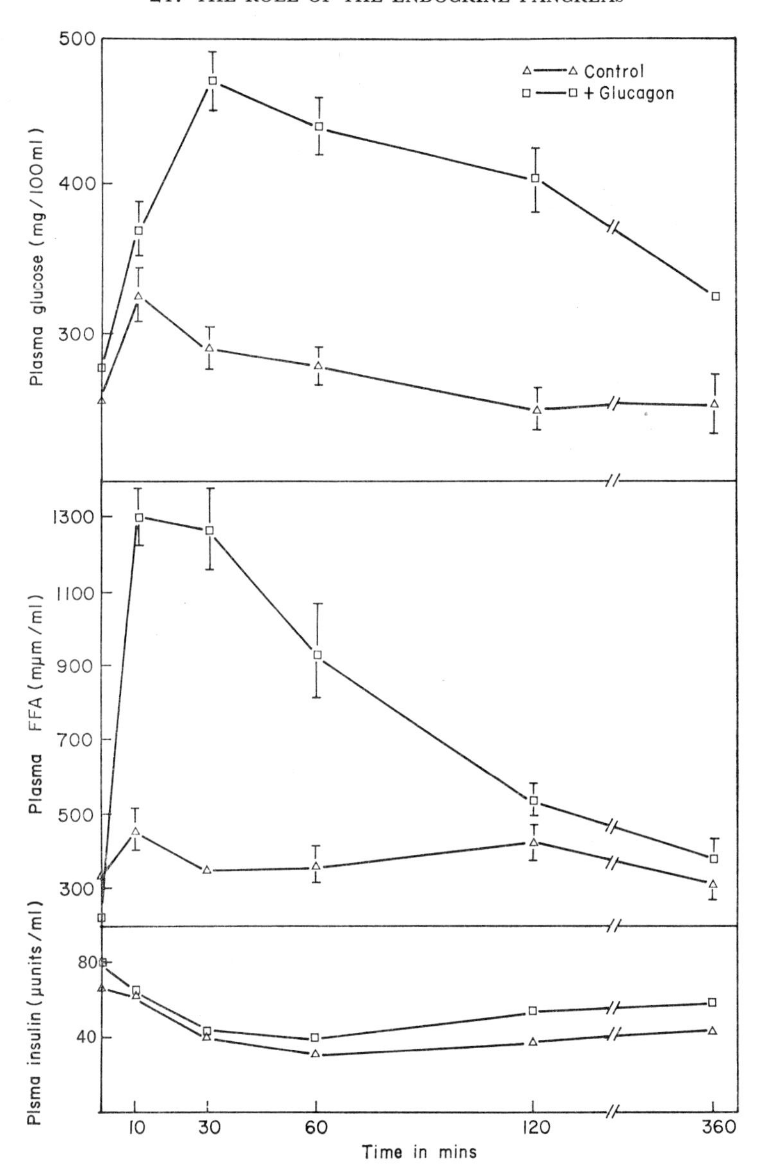

Fig. 2. Effect of intracardiac glucagon (0·2 mg) on the plasma glucose, FFA and insulin concentrations of 8-week-old chickens. The control group of chickens was given 0·9% saline.

streptozotocin have failed; the beta-cells being unaffected (Lukens, 1948; Mirsky and Gitelson, 1957). Langslow *et al.* (1970) showed this to be probably due to failure of the substances to depress the plasma insulin concentration.

Diazoxide, which inhibits insulin secretion in man, and from rabbit pancreas *in vitro* (Graber *et al.*, 1966; Howell and Taylor, 1966), produced hyperglycaemia and a modest fall in the plasma insulin concentration of fowls; no concomitant increase in plasma FFA concentration due to diazoxide was found (Langslow *et al.*, 1970).

2. *Glucagon Secretion*

Ducks are the only birds in which the plasma glucagon concentration has been studied. Samols *et al.* (1969a) demonstrated a fall in peripheral plasma glucagon concentration after pancreatectomy. The plasma glucagon concentration has been reported to be much higher in ducks starved overnight (0·9 ng/ml) than in dogs or humans (0·2 to 0·4 ng/ml) (Leclercq-Meyer *et al.*, 1970). Both intrajejunal and intravenous glucose suppressed the peripheral plasma glucagon concentration of ducks but it is not known what proportion of the glucagon immunological activity of the plasma is contributed by pancreatic glucagon. Insulin, pancreozymin and an increase in plasma FFA concentration produced a significant increase in peripheral plasma glucagon concentration (Samols *et al.*, 1969a). In ducks, tolbutamide produced a significant increase in peripheral plasma glucagon after an initial fall (Samols *et al.*, 1969b).

III. Regulation of Plasma Glucose and FFA Concentrations

A. INTRODUCTION

Most of the total stored energy of chickens is contained in fat depots. A little glycogen is stored in the liver and muscle. The food intake is largely carbohydrate (about 70% in chickens up to 6 weeks of age and about 80% in older birds with the remainder being mostly protein). However, normal growth can occur when neutral fat supplies virtually all the non-protein dietary calories (Brambila and Hill, 1966; Renner and Elcombe, 1967). Ketonaemia can result from diets rich in fat. Fatty acid can be synthesized in the liver and stored in adipose tissue as triglyceride. (See also Chapters 11 and 12.) The normal blood glucose and lactate levels in fowls are much higher than those of most mammals, although the plasma FFA concentrations are similar (Hazelwood, 1965; Bell and Culbert, 1968). Many investigators have measured the blood glucose rather than the plasma glucose concentration in chickens. This is not ideal—see Chapter 39—since virtually all the blood glucose is confined to the plasma (Riis and Herstad, 1967).

B. NUTRITIONAL EFFECTS

1. *Deprivation of Food*

In most mammals, lack of food reduces both the plasma glucose and insulin concentrations and increases the plasma FFA concentration. In fowls, age, sex and temperature modify the responses. Immature chickens 6 to 8 weeks old) responded to starvation with hypoglycaemia, a raised plasma FFA concentration and no fall in the plasma insulin concentration (Langslow *et al.*, 1970). However, the laying hen, which normally has an elevated plasma FFA concentration, showed a fall in plasma FFA concentration after starving (Heald and Rookledge, 1964). At an environmental temperature of 22°C, chickens can maintain their whole blood glucose concentration during starvation for 7 d (Hazelwood and Lorenz, 1959). Langslow *et al.* (1970) found that, at 15°C, fowls (6 weeks old) did not maintain their plasma glucose concentration. Initially, a mild hypoglycaemia lasted for 48 h, but then the plasma glucose concentration began to increase; a 72-h starvation was accompanied by a 20% loss in body weight (Langslow *et al.*, 1970).

Liver glycogen of starved fowls (5 to 11 weeks old), was depleted during the first day of starvation, was partially replenished, while the cardiac glycogen increased for 48 h before decreasing. The increase in blood non-protein nitrogen suggested that gluconeogenesis was responsible both for the maintenance of blood glucose and the restoration of liver glycogen (Hazelwood and Lorenz, 1959). (See also Chapter 11.)

Both normal and pancreatectomized fowls responded similarly to a 12-h deprivation of food but hypophysectomized chickens became more rapidly hypoglycaemic (Koike *et al.*, 1964). Hypophysectomy terminates true growth but the bird continues to gain weight and becomes obese. Hypophysectomy also retards lipid mobilization in cockerels, although they have normal plasma glucose and FFA concentrations (Gibson and Nalbandov, 1966a).

2. *Administration of Glucose and Amino Acids*

The mammalian plasma FFA concentration is closely related to nutritional status and an inverse relationship exists between the plasma glucose and FFA concentrations under physiological conditions (Randle *et al.*, 1963). An intravenous glucose load to female chickens (aged 28 or 18 weeks) failed to produce a marked drop in the plasma FFA concentration (Heald *et al.*, 1965). This contrasts with the effect in younger birds (7 to 8 weeks old) where both oral and intracardiac glucose (Fig. 1) produced rapid and prolonged falls in the plasma FFA concentration (Langslow *et al.*, 1970). In these experiments the dose of injected glucose was almost twice that used by Heald *et al.* (1965). In the young chickens both routes of glucose administration stimulated insulin secretion (see Section II D); their plasma glucose concentrations remained high for at least 5 h after both oral and intracardiac loads. (Langslow *et al.*, 1970.) This contrasts with man where both return to normal after 2 or 3 h.

In the above experiments it was found that both oral and intracardiac glucose would produce marked and persistent decreases in plasma α-amino nitrogen concentrations. Amino acids *per os* produced mild hyperglycaemia together with a modest increase in plasma insulin concentration (Langslow *et al.*, 1970). The hyperglycaemia may have been due to an increased secretion of glucagon.

C. HORMONES

1. *Insulin*

One observed difference between avian and mammalian carbohydrate and lipid metabolism is the often reported "avian resistance" to exogenous insulin. This is reflected both in the lack of hypoglycaemic coma and convulsions produced by large doses of mammalian insulins and in the less severe hypoglycaemia produced by insulin in fowls, ducks and pigeons (Chen *et al.*, 1945). In most mammals, insulin produces severe hypoglycaemia and a fall in the plasma FFA concentration. In birds, in contrast to mammals, the hypoglycaemia is less intense and the plasma FFA concentration rises (Hazelwood and Lorenz, 1959; Heald *et al.*, 1965; Lepkovsky *et al.*, 1967; Langslow *et al.*, 1970).

Lepkovsky *et al.* (1967) observed that fowls could tolerate large doses of mammalian insulin (50 i.u./kg body weight) without convulsing. This cannot be explained by an increased release of pancreatic glucagon since both pancreatectomized and normal chickens reacted similarly to insulin. Pretreatment with reserpine, to deplete adrenal catecholamines, did not affect the response of the fowl to insulin (Heald *et al.*, 1965).

Differences in the primary sequences of chicken, bovine and porcine insulins (see Section II B) have been suggested as a reason for the resistance and insensitivity of fowls to mammalian insulins. Structural differences might either affect binding between hormone and its tissue receptor or mammalian insulins may be inactivated much more rapidly by chicken plasma. The half-life of both bovine and chicken insulin in fowls (measured by the fall of plasma insulin concentration after giving 120 milli-units/kg to young birds) is 4·5 min (D. R. Langslow, unpublished) and is similar to that in man (Cerasi and Luft, 1969; Orskov and Christensen, 1969).

Data on rat adipose tissue and diaphragm muscle and on chicken adipose tissue and muscle—all *in vitro*—suggest no differences in response or sensitivity to bovine and chicken insulins (Goodridge, 1968a; Hazelwood *et al.*, 1968; Langslow and Hales, 1969; Langslow, 1969). The only *in vivo* data suggested that young chickens are more sensitive to fowl insulin than to ox insulin as judged by the duration of the hypoglycaemia produced. Both bovine and fowl insulins (0·02 units/kg body weight) gave the same effect on blood glucose concentration 20 min after the injection. The difference in potency was only apparent later (Hazelwood *et al.*, 1968).

In geese and owls, insulin induced hypoglycaemia, but did not reduce the plasma FFA (Grande, 1969a, 1970); in ducks insulin produced hypoglycaemia and a rise in plasma FFA concentration. There was also a rise in plasma glucagon concentration (Samols *et al.*, 1969a). There was no change in the plasma FFA concentration of pancreatectomized ducks given insulin (Mialhe, 1969). This observation is consistent with a role of glucagon in raising plasma FFA concentration during insulin induced hypoglycaemia in the duck.

2. Glucagon

Injection of fowls with mammalian pancreatic glucagon raises both the plasma glucose and FFA concentrations (Fig. 2); the latter change is both large and rapid. The maximum hyperglycaemic effect followed the peak of plasma FFA concentration. Glucagon increased the glycogen content of both heart and liver in fed, but not fasted fowls but did not increase the plasma insulin concentration, despite the hyperglycaemia (Beekman, 1958; Hazelwood and Lorenz, 1959; Heald *et al.*, 1965; Langslow *et al.*, 1970). This contrasts with the mammalian response where the hyperglycaemia is accompanied by a fall in the plasma FFA concentration. This is probably due to the increase in insulin secretion (Drieling *et al.*, 1962; Samols *et al.*, 1965) since no fall in plasma FFA concentration was seen in pancreatectomized dogs given glucagon (Sokal *et al.*, 1966). Ox insulin did not inhibit the increase in plasma FFA concentration due to glucagon in geese, ducks and roosters (Grande, 1968, 1969a, b; Mialhe, 1969).

In hypophysectomized ducks, glucagon did not produce hyperglycaemia although the plasma FFA concentration increased. The hyperglycaemic response to glucagon can be restored with pituitary hormone therapy (Desbals *et al.*, 1968). Glucagon increased the plasma insulin concentration of ducks (Mialhe, 1969).

Discussion on the relative functions of mammalian pancreatic glucagon and gut glucagon-like activity has been reviewed by Anon. (1968) and Unger (1968); a similar problem appears to exist in the fowl and duck. Extracts containing immunological glucagon-like activity prepared from the duodenum and jejunum of both birds did not give a dilution curve parallel to that of pancreatic glucagon on radioimmunoassay. There is no immunological difference between glucagons obtained from pancreatic extracts of man, pig, chicken and duck (Assan *et al.*, 1968). Fowl gut-extracts are thought to contain glucagon-like activity from their lipolytic action on isolated chicken fat cells by their activity (Langslow and Hales, 1970). The significance of these results is uncertain with regard to the physiological role of glucagon in chickens.

3. Adrenaline and Noradrenaline

Little has been done on the effects of adrenal catecholamines in fowls in which noradrenaline is the major adrenal catecholamine (Shepherd and West, 1951). Increased systolic blood pressure (but no rise in plasma FFA concentration) accompanied the infusion of noradrenaline to 8-week-old male birds (Carlson *et al.*, 1964). Intravenous noradrenaline in chickens (12–16 weeks old), while producing hyperglycaemia, gave no increase in plasma FFA concentration (Heald, 1966); age appeared to play some part in this response. Intraperitoneal noradrenaline produced an increasing rise in plasma FFA concentration, up to 4 weeks of age but at 8 weeks of age the response was much reduced. The hyperglycaemic response to noradrenaline increased with age as the adipokinetic response declined (Freeman, 1969) suggesting an interaction between these processes. The foregoing differences probably reflect the differences in routes of administration of noradrenaline, in the ages of the birds and in dosage. Adrenaline proved a more potent hyperglycaemic agent than noradrenaline in young birds. Both adrenaline and noradrenaline decreased muscle glycogen (Freeman, 1966).

Intracardiac adrenaline produced intense hyperglycaemia but no change in plasma FFA concentration of young chickens (Langslow *et al.*, 1970). Consistent with this is the finding that fowl adipose tissue cells *in vitro* are insensitive to the lipolytic action of adrenaline and noradrenaline (Langslow and Hales, 1969); in man and dog both hormones are powerfully adipokinetic (Fiegelson *et al.*, 1961; Porte *et al.*, 1965).

In mammals both glucagon and insulin induce the release of catecholamines (Bethune *et al.*, 1957; Sarcione *et al.*, 1963). The fowls' response to glucagon and insulin was not affected by previous depletion of adrenal catecholamines (Heald *et al.*, 1965).

4. Corticosteroids

Both hydrocortisone and corticosterone produce hyperglycaemia and an increase in liver glycogen levels when daily intramuscular injections are given to adult chickens although neither cortisone nor desoxycorticosterone evoked a similar response (Greenman and Zarrow, 1961).

5. Other Hormones and Drugs

Hypophysectomy terminated the true growth of chickens but they continued to gain weight with resulting obesity (Nalbandov and Card, 1943). Despite this obesity, plasma FFA and glucose concentrations remained unaltered. Hypophysectomy impaired the normal response to lack of food with a slower rise in plasma FFA concentration. Subcutaneous injection of powdered fowl pituitary produced a rise in plasma FFA concentration (Gibson and Nalbandov, 1966a). The effects of several hormones on plasma FFA and glucose concentrations of fowls are summarized in Table 2.

The hypoglycaemic drug tolbutamide produced hypoglycaemia, persisting for up to 2 h in fed fowls. Despite the low blood glucose of some birds, no convulsions were observed (Hazelwood and Lorenz, 1957; Mirsky and Gitelson, 1957). Tolbutamide has been observed to produce a small but significant increase in the plasma insulin concentration which returned to zero time level long before the end of the hypoglycaemic phase (Langslow, 1969). Both pancreatectomized and functionally hepatectomized chickens responded to tolbutamide with hypoglycaemia. The hepatectomized birds were hypersensitive to insulin and insulin with tolbutamide was synergistic (Hazelwood, 1958).

The plasma FFA concentration of fowls responded biphasically to intracardiac tolbutamide. Initially there was a depression of the plasma FFA concentration (up to 30 min after injection) and this was followed by a rise to supranormal levels after 2 h (Langslow, 1969); changes in the plasma FFA concentration may be due to effects on glucagon secretion. In ducks, Samols *et al.* (1969b) have found that tolbutamide reduced the plasma glucagon concentration as well as increasing the plasma insulin concentration. Mirsky and Gitelson (1957) found tolbutamide to have the same effect on normal, pancreatectomized and enterectomized chickens. The results with pancreatectomized chickens, if the pancreatectomy was complete, suggested no involvement of pancreatic glucagon in the response to tolbutamide.

Table 2

The plasma FFA and glucose concentrations of fowls in response to various hormones

	Plasma FFA of laying hens	Plasma FFA of immature chickens	Plasma glucose
Oestrogens	—	↑	NE
Testosterone	↓	NE	—
Thyroxin	NE	NE	—
Gonadotrophins	↑	NE	—
Hog pituitary extract	—	↑	—
Chick anterior pituitary extract	↑	NE	—
ACTH (mammalian)	↑	↑	↑
Long-acting ACTH	↓	—	↑
Oxytocin	—	—	↑
Vasotocin	—	—	↑

NE = No effect; ↑ = an increase; ↓ = a decrease; — = no information

(From Bell, 1961; Rudman *et al.*, 1962a; Heald and Rookledge, 1964; Heald *et al.*, 1965; Bentley, 1966; Gibson and Nalbandov, 1966a.)

IV. Effects of Hormones on Tissue Carbohydrate and Lipid Metabolism

A. MUSCLE

Little has been described on chicken muscle tissues *in vitro* and most available data relate to embryo heart. Insulin (at high concentrations) stimulated glucose uptake and the synthesis of DNA, RNA and protein in cultures of 13-day-old chick embryo hearts; Leslie and Paul (1954) thus concluded that insulin had a growth-promoting action. Equal concentrations of bovine and chicken insulins gave equal stimulation to the rate of accumulation of 2-amino isobutyric acid and glycine into embryo heart (Guidotti *et al.*, 1968); further evidence for the identical potency of action of the two insulins *in vitro*.

Langslow (1969) described a preparation of fowl skeletal muscle which was suitable for *in vitro* studies. Muscles chosen were the obliquus abdominis externus and the rectus abdominis. These muscles can be rapidly dissected out from the chicken, are extremely thin and may be removed by cutting a minimum number of muscle fibres. Muscles from both left and right sides were used and were cut in half, along the direction of the fibres, to provide paired pieces of tissue.

High concentrations of both ox and chicken insulin were required to stimulate glucose uptake into these muscles. No effect was detected below 1 μg insulin/ml and 10 μg insulin/ml was maximal. Insulin also stimulated the incorporation of [U-^{14}C]-glucose into glycogen although the effect was much smaller than that found in rat (Langslow, 1969; Beloff-Chain *et al.*, 1955).

B. ADIPOSE TISSUE

1. Lipolysis

There is considerable species variation with respect to both the minimum dose and the actual hormones which elicit a lipolytic response from adipose tissue of various animals. Table 3 illustrates some of these aspects.

Early experiments on lipolysis had shown that fowl adipose tissue readily responded to glucagon but either weakly or not at all to ACTH, adrenaline and noradrenaline (Rudman *et al.*, 1962b; Carlson *et al.*, 1964; Gibson and Nalbandov, 1966b). Development of a preparation of isolated chicken fat cells allowed a more specific study of their lipolytic sensitivity (Langslow and Hales, 1969). Adipose tissue fat cells of the fowl are extremely sensitive to glucagon; as little as 0·1 ng glucagon/ml stimulated release of glycerol and FFA (Fig. 3). This sensitivity allows the bio-assay of glucagon in human serum (Langslow and Hales, 1970). Glucose added to the incubation medium did not significantly alter the response of the fat cells to glucagon. Adipose tissue from 7-day-old chickens was 10 times more sensitive to lipolytic stimulation by glucagon than was embryonic adipose tissue (Goodridge, 1968b). Adrenaline, noradrenaline, porcine ACTH and long-acting ACTH did stimulate lipolysis in chicken fat cells but the effects were small and needed high concentrations (> 1 μg/ml) for a significant effect (Langslow and Hales, 1969).

The role of the pituitary has been investigated. Adipose tissue from hypophysectomized cockerels had a basal rate of lipolysis *in vitro* lower than in normals, as well as a reduced lipase activity (Gibson and Nalbandov, 1966b). Crude fowl pituitary fractions were not themselves lipolytic but a combination of hydrocortisone sodium succinate and crude chicken "growth

Table 3

Effect of several hormones on lipolysis in adipose tissue from rat, man, rabbit, guinea-pig, pig, fowl, hamster and dog

	Adrena-line	Nor-adrenaline	Gluca-gon	ACTH	TSH	GH + steroid	Vaso-pressin	α-MSH	Insulin	Prosta-glandin E_1
at	[a] 0·01	[a] 0·01	[a] 0·001	[a] 0·01	[a] 0·1	[a] Highly sensitive	N	N	[a] 0·00001	[a] < 0·001
Man	[a] 0·001	< 0·1	[a] N	[a] N	[a] 0·1	[a] Variable small response	—	—	[a] 0·000004	[a] < 0·01
Rabbit	[a] N	< 10	[a] N	0·1	N	—	0·1	0·1	N	< 0·01
Guinea-pig	0·2	0·2	—	1	1	—	3	10	N	—
Pig	N	N	—	N	N	—	N	N	—	—
Fowl	[a] 1	[a] 1	[a] 0·0001	1	—	Small response	—	—	[a] N	[a] < 0·0001
Hamster	[a] 0·01	0·1	[a] N	[a] 0·01	[a] N	—	N	N	< 0·01	—
Dog	< 0·1	< 0·1	—	N	< 10	—	N	< 10	—	—

The concentration given (in μg/ml) is the minimum reported dose required for an effect. [a] refers to results with fat cells; < indicates the lowest concentration that was tested; N is no effect detectable; — is no information. (Compiled from Rudman *et al.* (1962a, b); Rodbell (1964); Fain *et al.* (1965); Rodbell (1965); Burns and Hales (1966a, b); Rudman and Shank (1966); Arnold and McAuliff (1968); Bergstrom *et al.* (1968); Burns and Langley (1968); Perry (1968); Langslow and Hales (1969); Moscovitz and Fain (1969); Rudman and Del Rio (1969); Langslow (1970); P. L. Storring (1970) personal communication.

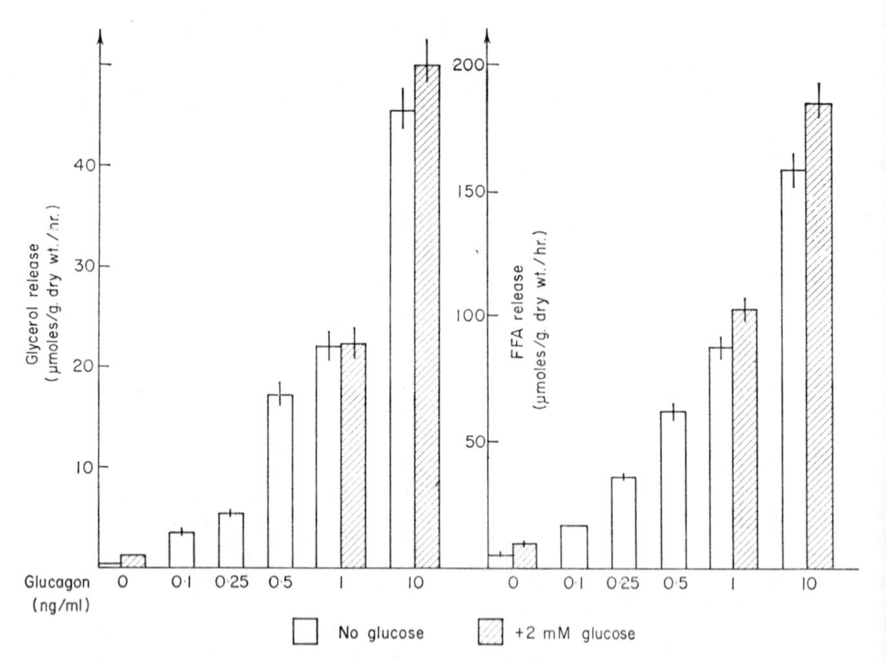

Fig. 3. Effect of various concentrations of glucagon on the rate of glycerol and FFA release by isolated fat cells prepared from chicken adipose tissue in the presence and absence of glucose.

hormone" (so designated by analogy with mammalian pituitary fractionation) were lipolytic over a four-hour incubation period (Langslow and Hales, 1969; Stockell-Hartree *et al.*, 1969). This resembles the effect with rat adipose tissue (Fain *et al.*, 1965).

The mediation of lipolytic activity in fowls' adipose tissue is probably similar to that in other animals': both theophylline and dibutyryl 3′,5′-(cyclic)-AMP are potent stimulators of lipolysis (Langslow and Hales, 1969).

Inhibition studies of lipolysis *in vitro* have shown that neither ox nor fowl insulins antagonized glucagon-stimulated lipolysis in fowl adipose tissue pieces or in the isolated fat cells (Goodridge, 1968a; Langslow and Hales, 1969). At high concentrations, insulin potentiated the effect of glucagon; this was surprising since insulin alone was not lipolytic and, thus, contamination of the insulin by glucagon was unlikely. Goodridge (1968a) suggested that this effect might have physiological significance in regulating lipolysis but his effects required relatively high concentrations of insulin. The site within the bird from which the adipose tissue was taken did not influence the response to glucagon and insulin. Prostaglandin E_1 was potently anti-lipolytic to glucagon and adrenaline stimulated lipolysis (Fig. 4). As little as 0·5 ng prostaglandin E_1/ml inhibited the lipolytic stimulation produced by 1 ng glucagon/ml by 50% (Langslow, 1970). Prostaglandin E_1 also inhibited theophylline-stimulated lipolysis but not that potentiated dibutyryl 3',5'-(cyclic)-AMP, suggesting an effect on cyclic AMP production (Langslow, 1971) similar to that of prostaglandin E_1 on rat fat cells (Butcher and Baird, 1968).

Glucagon also appears to be important in the regulation of lipolysis in other birds (Goodridge, 1964; Goodridge and Ball, 1965; Evans, 1969).

2. Lipogenesis

Fowl adipose tissue has a low lipogenic capacity compared with the rat (Winegrad and Renold, 1958; Goodridge and Ball, 1966). In both fowls and pigeons, the major site of fatty acid synthesis from glucose is the liver rather than adipose tissue (Goodridge and Ball, 1967; O'Hea and Leveille, 1969). (See also Chapters 11 and 12.) In adipose tissue from young chickens, glucagon slightly inhibited glucose oxidation and markedly inhibited synthesis of glycogen, glyceride-glycerol and fatty-acid from [U-^{14}C]-glucose. Insulin together with glucagon inhibited all these parameters much more than did

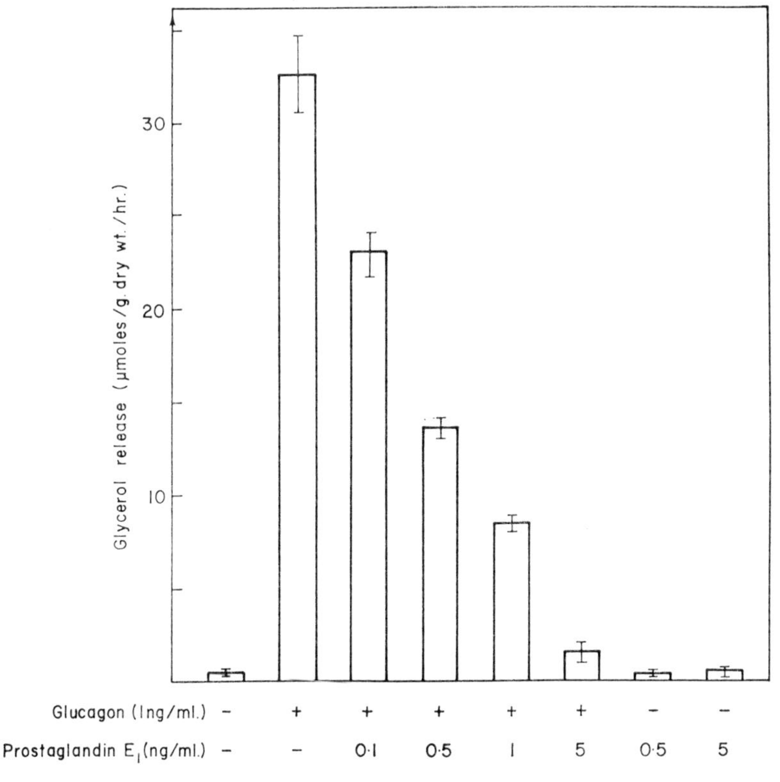

Fig. 4. Effect of prostaglandin E_1 on glucagon-stimulated lipolysis by isolated fat cells prepared from chicken adipose tissue.

glucagon alone, whereas insulin alone slightly increased them. Ox and fowl insulins had identical effects on these parameters (Goodridge, 1968b). It is possible that the inhibitory effects of glucagon and glucagon plus insulin were secondary to the high intracellular concentrations of FFA produced. Insulin did not stimulate the incorporation of acetate and pyruvate into fatty acids. High concentrations of insulin were required to stimulate glucose uptake and glycogen synthesis from [U-^{14}C]-glucose. The tissue was equally insensitive to both ox and chicken insulins although rat fat cells were extremely sensitive to both insulins (Langslow, 1969). The very low activity of the pentose phosphate cycle dehydrogenases and the malic enzyme in chicken adipose tissue is consistent with its low lipogenic capacity (Goodridge, 1968d; O'Hea and Leveille, 1968) although chicken and pigeon adipose tissue both readily converted glucose and pyruvate to glyceride-glycerol to very similar extents (Goodridge and Ball, 1966).

In contrast to the liver, fowl adipose tissue showed no marked changes in the enzymes concerned with lipogenesis, after hatching (Goodridge, 1968d).

The weak effects of insulin on lipogenesis and lipolysis appeared to be typical of birds in general. Insulin had no effect on glucose uptake and lipid synthesis from glucose and acetate in the adipose tissue of migratory finches. In spite of this these birds deposited a weight of fat up to half their body weight prior to migration (Goodridge, 1964).

The source of adipose tissue triglyceride appears to be lipid synthesized in the liver which is transported to the depots as low-density or β-lipoprotein (O'Hea and Leveille, 1969). Adipose tissue of both chicken and pigeon have an active lipoprotein lipase system capable of mediating uptake of fatty acid from plasma triglyceride (Korn and Quigley, 1957; Goodridge and Ball, 1967). The fowl's adipose tissue thus appears to play a minor role in fatty acid synthesis and to act largely as a depository for material synthesized by the liver.

C. LIVER

Virtually nothing is known of the role of the fowl's endocrine pancreas in hepatic regulation including the regulation of glycogen synthesis, glycolysis and gluconeogenesis. Most biochemical investigations have been concerned with lipogenesis. The effects of nutrition on this process have been studied; it is likely that these are, at least partially, hormonally mediated. This section is a brief summary of what is known of these effects.

In vivo, at least 90% of the total lipid synthesized from glucose or acetate is of hepatic origin in the fowl; in rats, adipose tissue accounted for most of the lipid synthesis (Leveille *et al.*, 1968; O'Hea and Leveille, 1969). Pigeons resemble chickens in this respect (Goodridge and Ball, 1967) and hepatic lipogenesis was as high in mature females as in young chicks (Husbands and Brown, 1965).

Hepatic lipogenesis responds to dietary variations. In rats, starvation or a fat-rich diet decreased lipogenesis while a high carbohydrate diet increased it (Lyon *et al.*, 1952; Hill *et al.*, 1960). Goodridge (1968c) studied the incorporation of [U-^{14}C]-glucose into carbon dioxide, glycogen and fatty acid, in chicken liver slices. During the embryonic period, the rates of incorporation were low and stable. Fatty acid synthesis and glucose oxidation increased rapidly when the chicks were fed after hatching and reached a plateau after 6 d. Glycogen synthesis increased rapidly for about 11 d after hatching. The increased rate of lipogenesis did not develop if food was withheld from the chicks after hatching. Thus the increase in glucose metabolism after hatching was probably due to the change from a fat-rich diet (yolk) to a carbohydrate-rich one (post-hatching).

As early as 30 min after the withdrawal of food from 6-week-old chickens, the rate of hepatic lipogenesis declined (as measured by [U-^{14}C]-glucose and [1-^{14}C]-acetate incorporation into liver lipid), both in intact chickens and in *in vitro* studies. In the intact chickens, aphagia was accompanied by a fall in the plasma triglyceride and an increase in the plasma FFA concentration (Yeh *et al.*, 1970).

A hepatic metabolic abnormality is associated with a disease of chickens, the "fatty-liver and kidney syndrome". Young chickens (5 to 7 weeks old) are susceptible to the disease which is sometimes fatal. Obese laying hens are susceptible to the "fatty-liver syndrome" which is usually detected after a fall in egg production. Its cause is not known but may be related to high-energy diets, husbandry or to genetic constitution. Fatty livers can be produced in ducks by the infusion of glucagon, but not by adrenaline or noradrenaline; triglyceride deposited in the liver probably originates from the raised plasma FFA concentration (Grande, 1969c). Fatty liver observed in mammals (Carlson, 1968) is often associated with the lipaemia of *diabetes mellitus* (Carlson, 1969).

Goodridge (1968e) studied lipogenesis in liver slices from 3-week-old chickens. Starvation reduced glucose utilization and the activities of both the hepatic ATP-citrate lyose and the NADP-linked malate dehydrogenase; re-feeding increased these to supra-normal levels. Isocitrate dehydrogenase activity responded to starvation by increasing and to re-feeding by decreasing. Thus increased substrate delivery to the liver appeared to be the main stimulus to the increased rate of glucose metabolism and to the changes in enzymic activities.

Activities of some NADP-dependent enzymes have been studied. The pentose phosphate pathway dehydrogenases [E.C. 1.1.1.49 and E.C. 1.1.1.44] were inactive in chicken liver both before and after hatching. The ratio of the conversion of C_1 of glucose to C_b of glucose into $^{14}CO_2$ was unity, suggesting an inactive pathway (Duncan, 1968; Goodridge, 1968d). Liver isocitrate dehydrogenase [E.C. 1.1.1.42] increased to a maximum on the day of hatching and thereafter fell. Both the NADP-linked malate dehydrogenase [E.C. 1.1.1.40] and the ATP-citrate lyase [E.C. 4.1.3.8] increased dramatically after the chick hatched and was fed, and this coincided with the increase in lipogenesis (Goodridge, 1968d). These enzyme changes were probably secondary to an increased flow of glucose to fatty acids since neither the NADP-linked malate dehydrogenase nor the citrate cleavage enzyme increased if food was withheld from newly hatched chicks.

An important rate limiting enzyme in fatty acid synthesis is acetyl CoA carboxylase [E.C. 6.4.1.2]. Mature female chickens responded rapidly to a change from a normal to a fat-rich diet. The activities of both the citrate cleavage enzyme and acetyl CoA carboxylase fell within 2 h; the acetyl CoA carboxylase activity was only 3% of its original value after 48 h on a fat-rich diet. Isocitrate dehydrogenase activity was constant for 24 h, and thereafter declined slowly, suggesting a reduced requirement for NADPH (Pearce, 1968). This confirms the observations of Weiss *et al.* (1967) who found that laying birds maintained on a fat-rich diet incorporated less [1-C^{14}]-acetate into liver lipid than control birds. However, Goodridge (1969) found that an increase in dietary fat did not suppress hepatic lipogenesis in young chickens and that the NADP-linked malate dehydrogenase activity was unaffected. This difference may be due to the different dietary regimes employed; in Goodridge's (1969) diet, the increase in percentage of fat was accompanied by only a small fall in percentage of carbohydrate but this was not true of all the other diets. Hepatic lipogenesis was also reduced when the protein content of the diet of growing chickens was increased at the expense of carbohydrate (Yeh and Leveille, 1969).

In 1-day-old chicks, kept on a carbohydrate-free diet, the rate of hepatic gluconeogenesis increased compared with those fed a normal high-carbohydrate diet. Thus there was an increase in net glucose production by the liver and this may have been controlled by glucose-6-phosphatase activity (Allred and Roehrig, 1970). The rate of glucose uptake by the perfused liver of fed chickens was proportional to the glucose concentration of the perfusate (Bickerstaffe *et al.*, 1970).

V. Conclusions

The role of the fowl's endocrine pancreas in the control of carbohydrate and lipid metabolism is far from resolved. The sensitivity of chickens to the hyperglycaemic and lipolytic actions of glucagon has suggested that this hormone may play a key role in glucose and lipid homeostases. The importance of insulin is less certain. The insulin secretory system is responsive to both glucose and amino acids but the failure of starvation to alter the plasma insulin concentration is not consistent with day-to-day changes in insulin concentration due to normal nutrition. It is possible that the regulation of the plasma insulin concentration involves factors which modify the sensitivity of the secretory mechanism to glucose and amino acids. In addition, both chicken muscle and adipose tissue preparations are insensitive to the actions of insulin on glucose uptake, oxidation and conversion to glycogen.

Acknowledgements

The authors would like to thank the British Egg Marketing Board for research grants and Professor F. G. Young and Dr R. F. Gordon for their interest and encouragement.

References

Adams, M. J., Blundell, T. L., Dodson, E. J., Dodson, G. G., Vijayan, M., Baker, E. N., Harding, M. M., Hodgkin, D. C., Rimmer, B. and Skeats, S. (1969). *Nature, Lond.* **224**, 491–495.

Allred, J. B. and Roehrig, K. L. (1970). *J. Nutr.* **100**, 615–622.

Anon (1968). *Lancet* **ii**, 1021–1022.

Arnold, A. and McAuliff, M. (1968). *Experientia* **24**, 436.

Assan, R., Rosselin, G. and Tchobroutsky, G. (1968). *In* "Proceedings of International Symposium on Protein and Polypeptide Hormones (Ed. M. Margoulies) Part I, pp. 87–90. Excerpta Medica Foundation, Amsterdam.

Banting, F. G. and Best, C. H. (1922). *J. Lab. clin. Med.* **7**, 251–266.

Batt, H. T. (1940). *Thesis, University of Toronto.*

Beekman, B. E. (1956). *Endocrinology* **59**, 708–712.

Beekman, B. E. (1958). *Poult. Sci.* **37**, 595–599.

Bell, D. J. (1961). *Nature, Lond.* **190**, 913.

Bell, D. J. and Culbert, J. (1968). *Comp. Biochem. Physiol.* **25**, 627–637.

Beloff-Chain, A., Catanzaro, R., Chain, E. B., Masi, I., Pocchiari, F. and Rossi, C. (1955). *Proc. R. Soc.* **B143**, 481–503.

Bentley, P. J. (1966). *J. Endocr.* **34**, 527–528.

Bergstrom, S., Carlson, L. A. and Weeks, J. R. (1968). *Pharmac. Rev.* **20**, 1–48.

Bethune, J. E., Goldfien, A., Zileli, M. and Despointes, R. M. (1957). *Fedn Proc. Fedn Am. Socs exp. Biol.* **16**, 11.

Bickerstaffe, R., West, C. E. and Annison, E. F. (1970). *Biochem. J.* **118**, 427–432.

Bjorkman, N. and Hellman, B. (1964). *In* "The Structure and Metabolism of the Pancreatic Islets" (S. E. Brolin, B. Hellman and H. Knutson eds.), pp. 131–142. Pergamon Press, London.

Bjorkman, N. and Hellman, B. (1967). *Experientia* **23**, 721–722.

Brambila, S. and Hill, F. W. (1966). *J. Nutr.* **88**, 84–92.

Brosky, G. and Logothetopoulos, J. (1969). *Diabetes* **18**, 606–611.

Buchanan, K. D., Vance, J. E. and Williams, R. H. (1969). *Diabetes* **18**, 381–386.

Burns, T. W. and Hales, C. N. (1966a) *Lancet* **i**, 796–798.

Burns, T. W. and Hales, C. N. (1966b). *Lancet* **ii**, 1111–1113.

Burns, T. W. and Langley, P. (1968). *J. Lab. clin. Med.* **72**, 813–823.

Butcher, R. W. and Baird, C. E. (1968). *J. biol. Chem.* **243**, 1713–1717.

Calhoun, M. L. (1933). *Iowa State Coll. J. Sci.* **7**, 261.

Carlson, L. A. (1968). *In* "Proceedings of International Symposium on Protein and Polypeptide Hormones". (M. Margoulies, ed.), Part I, pp. 143–149. Excerpta Medica Foundation, Amsterdam.

Carlson, L. A. (1969). *Diabetologia* **5**, 361–365.

Carlson, L. A., Liljedahl, S. O., Verdy, M. and Wirsen, C. (1964). *Metabolism* **13**, 227–231.

Cerasi, E. and Luft, R. (1969). *Horm. Metab. Res.* **1**, 221–223.

Chen, K. K., Anderson, R. C. and Maze, N. (1945). *J. Pharm. exp. Therap.* **84**, 74–77.

Davis, J. C. (1952). *J. Path. Bact.* **64**, 575–584.

Desbals, P., Desbals, B. and Mialhe, P. (1968). *J. Physiol., Paris* **60**, Suppl. 1, 240.
Drieling, D. A., Bierman, E. L., Debons, A. F., Elsbach, P. and Schwartz, I. L. (1962). *Metabolism* **11**, 572–578.
Dulin, W. E., Lund, G. H. and Gerritson, G. C. (1967). *Diabetes* **16**, 512–513.
Duncan, M. J. (1968). *Can. J. Biochem.* **46**, 1321–1326.
Elrick, H., Stimmler, L., Hlad, C. J. Jr. and Arai, Y. (1964). *J. clin. Endocr. Metab.* **24**, 1076–1082.
Evans, A. J. (1969). *Ph. D. thesis, University of Edinburgh.*
Fain, J. N., Kovacev, V. P. and Scow, R. O. (1965). *J. biol. Chem.* **240**, 3522–3529.
Feigelson, E. B., Pfaff, W. W., Karmen, A. and Steinberg, D. (1961). *J. clin. Invest.* **40**, 2171–2179.
Fleming, G. B. (1919). *J. Physiol., Lond.* **53**, 236–246.
Floyd, J. C., Fajans, S. S., Conn, J. W., Knopf, R. F. and Rull, J. (1966). *J. clin. Invest.* **45**, 1487–1502.
Foa, P. P. (1968). *Ergebn. Physiol.* **60**, 142–219.
Freeman, B. M. (1966). *Comp. Biochem. Physiol.* **18**, 369–382.
Freeman, B. M. (1969). *Comp. Biochem. Physiol.* **30**, 993–996.
Giaja, J. (1912). *C. r. Soc. biol.* **73**, 102–104.
Gibson, W. R. and Nalbandov, A. V. (1966a). *Am. J. Physiol.* **211**, 1345–1351.
Gibson, W. R. and Nalbandov, A. V. (1966b). *Am. J. Physiol.* **211**, 1352–1356.
Goodridge, A. G. (1964). *Comp. Biochem. Physiol.* **13**, 1–26.
Goodridge, A. G. (1968a). *Am. J. Physiol.* **214**, 897–901.
Goodridge, A. G. (1968b). *Am. J. Physiol.* **214**, 902–907.
Goodridge, A. G. (1968c). *Biochem. J.* **108**, 655–661.
Goodridge, A. G. (1968d). *Biochem. J.* **108**, 663–666.
Goodridge, A. G. (1968e). *Biochem. J.* **108**, 667–673.
Goodridge, A. G. (1969). *Can. J. Biochem.* **47**, 743–746.
Goodridge, A. G. and Ball, E. G. (1965). *Comp. Biochem. Physiol.* **16**, 367–381.
Goodridge, A. G. and Ball, E. G. (1966). *Am. J. Physiol.* **211**, 803–808.
Goodridge, A. G. and Ball, E. G. (1967). *Am. J. Physiol.* **213**, 245–249.
Graber, A. L., Porte, D. Jr. and Williams, R. H. (1966). *Diabetes* **15**, 143–148.
Grande, F. (1968). *Proc. Soc. exp. Biol. Med.* **128**, 532–536.
Grande, F. (1969a). *Proc. Soc. exp. Biol. Med.* **130**, 711–713.
Grande, F. (1969b). *Proc. Soc. exp. Biol. Med.* **131**, 740–744.
Grande, F. (1969c). *Abstract to 6th F.E.B.S. Meeting, Madrid*, p. 198.
Grande, F. (1970). *Proc. Soc. exp. Biol. Med.* **133**, 540–543.
Greenman, D. L. and Zarrow, M. X. (1961). *Proc. Soc. exp. Biol. Med.* **106**, 459–462.
Guidotti, G. G., Borghetti, A. F., Gaja, G., Loreti, L., Ragnotti, G. and Foa, P. P. (1968). *Biochem. J.* **107**, 565–574.
Haist, R. E. (1944). *Physiol. Rev.* **24**, 409–444.
Hales, C. N. (1968). *In* "Carbohydrate Metabolism and its Disorders". (F. Dickens, P. J. Randle and W. J. Whelan, eds.), Vol. 2, pp. 25–50. Academic Press, London.
Hazelwood, R. L. (1958). *Endocrinology* **63**, 611–618.
Hazelwood, R. L. (1965). *In* "Avian Physiology". (P. D. Sturkie, ed.), pp. 313–371. Baillière, Tindall & Cassell, London.
Hazelwood, R. L., Kimmel, J. R. and Pollock, H. G. (1968). *Endocrinology* **83**, 1331–1336.
Hazelwood, R. L. and Lorenz, F. W. (1957). *Endocrinology* **61**, 520–527.
Hazelwood, R. L. and Lorenz, F. W. (1959). *Am. J. Physiol.* **197**, 47–51.

Hazelwood, R. L., Turner, D. S. and Kimmel, J. R. (1971). *Endocrinology*, **88**, A55.
Heald, P. J. (1966). *In* "Physiology of the Domestic Fowl" (C. Horton-Smith and E. C. Amoroso, eds.), pp. 113–124. Oliver and Boyd, Edinburgh and London.
Heald, P. J., McLachlan, P. M. and Rookledge, K. A. (1965). *J. Endocr.* **33**, 83–95.
Heald, P. J. and Rookledge, K. A. (1964). *J. Endocr.* **30**, 115–130.
Hellerstrom, C. (1963). *Z. Zellforsch. mikrosk. Anat.* **60**, 688–710.
Hellerstrom, C., Hellman, B., Petersson, B. and Alm, G. (1964). *In* "The Structure and Metabolism of the Pancreatic Islets", (S. E. Brolin, B. Hellman and H. Knutson, eds.), pp. 117–130. Pergamon Press, London.
Hellman, B. (1961). *Acta endocr. Copenh.* **36**, 603–608.
Hellman, B. and Hellerstrom, C. (1960). *Z. Zellforsch. mikrosk Anat.* **52**, 278–290.
Hellman, B. and Lernmark, A. (1969). *Endocrinology* **84**, 1484–1488.
Hellman, B. and Lernmark, A. (1970). *In* "The Structure and Metabolism of Pancreatic Islets". (S. Falkmer, B. Hellman and I. B. Taljedaho, eds.), pp. 453–462. Pergamon Press, London.
Hertelendy, F., Machlin, L. and Kipnis, D. M. (1969). *Endocrinology* **84**, 192–199.
Hill, R., Webster, W. W., Linazoro, J. M. and Chaikoff, I. L. (1960). *J. Lipid Res.* **1**, 150–158.
Howell, S. L. and Taylor, K. W. (1966). *Lancet* **i**, 128–129.
Husbands, D. R. and Brown, W. O. (1965). *Comp. Biochem. Physiol.* **14**, 445–451.
Ketterer, H., Eisentraut, A. M. and Unger, R. H. (1967). *Diabetes* **16**, 283–288.
Kimmel, J. R., Pollock, G. H. and Hazelwood, R. L. (1968). *Endocrinology* **83**, 1323–1330.
Kimmel, J. R., Pollock, H. G. and Hazelwood, R. L. (1971). *Fedn Povc. Fedn Am. Socs. exp. Biol.* **30**, Abstr. 1318.
Koike, T., Nalbandov, A., Dimick, M., Matsumura, Y. and Lepkovsky, S. (1964). *Endocrinology* **74**, 944–948.
Koppanyi, T., Ivy, A. C., Tatum, A. L. and Jung, F. T. (1926). *Am. J. Physiol.* **76**, 212–213.
Korn, E. D. and Quigley, T. W. (1957). *J. biol. Chem.* **226**, 833–839.
Langslow, D. R. (1969). *Ph.D. thesis, University of Cambridge.*
Langslow, D. R. (1971). *Biochem. biophys. Acta.* (In press).
Langslow, D. R., Butler, E. J., Hales, C. N. and Pearson, A. W. (1970). *J. Endocr.* **46**, 243–260.
Langslow, D. R. and Hales, C. N. (1969). *J. Endocr.* **43**, 285–294.
Langslow, D. R. and Hales, C. N. (1970). *Lancet* **i**, 1151–1152.
Leclercq-Meyer, V., Mialhe, P. and Malaisse, W. J. (1970). *Diabetologia* **6**, 121–129.
Lepkovsky, S., Dimick, M. K., Furuta, F., Snapir, N., Park, R., Narita, N. and Komatsu, K. (1967). *Endocrinology* **81**, 1001–1006.
Lepkovsky, S., Nalbandov, A. V., Dimick, M. K., McFarland, L. Z. and Pencharz, R. (1964). *Endocrinology* **74**, 207–211.
Leslie, I. and Paul, J. (1954). *J. Endocr.* **11**, 110–124.
Leveille, G. A., O'Hea, E. K. and Chakrabarty, K. (1968). *Proc. Soc. exp. Biol. Med.* **128**, 398–401.
Lomsky, R., Langr, F. and Vortel, V. (1969). *Nature, Lond.* **223**, 618–619.
Lukens, F. D. W. (1948). *Physiol. Rev.* **28**, 304–330.
Lundquist, G., Brolin, S. E., Unger, R. H. and Eisentraut, A. M. (1970). *In* "The Structure and Metabolism of Pancreatic Islets" (S. Falkmer, B. Hellman and I. B. Taljedahl, eds.), pp. 115–121. Pergamon Press, London.

Lyon, I., Masri, M. S. and Chaikoff, I. L. (1952). *J. biol. Chem.* **196**, 25–32.
Machino, M. (1966). *Nature, Lond.* **210**, 853–854.
Machino, M. and Sakuma, H. (1967). *Nature, Lond.* **214**, 808–809.
Machino, M. and Sakuma, H. (1968). *J. Endocr.* **40**, 129–130.
Marco, J., Faloona, G. R. and Unger, R. H. (1970). *Diabetes* **19**, Suppl. 1, 366–367.
Mayhew, D. A., Wright, P. H. and Ashmore, J. (1969). *Pharmac. Rev.* **21**, 183–212.
McIntyre, N., Holdsworth, C. D. and Turner, D. S. (1964). *Lancet* **ii**, 20–21.
Mialhe, P. (1957). *C. r. hebd. Séanc. Acad. Sci. Paris* **244**, 385.
Mialhe, P. (1958). *Acta. endocr. Copenh.* Suppl. **36**.
Mialhe, P. (1960). *C. r. hebd. Séanc. Acad. Sci. Paris* **154**, 1867–1868.
Mialhe, P. (1969). Proceedings of the 3rd International Conference on Endocrinology, Mexico, pp. 158–163. Excerpta Medica, Amsterdam.
Mikami, S. and Ono, K. (1962). *Endocrinology* **71**, 464–473.
Minkowski, O. (1893). *Arch. exp. Path. Pharm.* **31**, 85–189.
Minkowski, O. and von Mehring, J. (1889). *Arch. exp. Path. Pharm.* **26**, 371–387.
Mirsky, I. A. and Gitelson, S. (1957). *Endocrinology* **61**, 148–152.
Mirsky, I. A., Jinks, R. and Perisutti, G. (1963). *J. clin. Invest.* **42**, 1869–1872.
Mirsky, I. A., Jinks, R. and Perisutti, G. (1964). *Am. J. Physiol.* **206**, 133–135.
Mirsky, I. A. and Kawamura, K. (1966). *Endocrinology* **78**, 1115–1119.
Moody, A. J., Markussen, J., Schaich Fries, A., Steenstrup, C., Sundby, F., Malaisse, W. and Malaisse-Lagae, F. (1970). *Diabetologia* **6**, 135–140.
Moscovitz, J. and Fain, J. N. (1969). *J. clin. Invest.* **48**, 1802–1808.
Nalbandov, A. V. and Card, L. E. (1943). *J. exp. Zool.* **94**, 387–413.
Oakberg, E. F. (1949). *Am. J. Anat.* **84**, 279–310.
O'Hea, E. K. and Leveille, G. A. (1968). *Comp. Biochem. Physiol.* **26**, 111–120.
O'Hea, E. K. and Leveille, G. A. (1969). *Comp. Biochem. Physiol.* **30**, 149–159.
Orskov, H. and Christensen, N. J. (1969). *Diabetes* **18**, 653–659.
Paton, W. (1905). *J. Physiol., Lond.* **32**, 59–64.
Pearce, J. (1968). *Biochem. J.* **109**, 702–704.
Perry, M. C. (1968). *Ph.D. thesis, University of Cambridge.*
Pitkin, R. M. and Reynolds, W. A. (1970). *Diabetes* **19**, 85–90.
Porte, D., Graber, A., Kuzuya, T. and Williams, R. H. (1965). *J. clin. Invest.* **44**, 1087.
Randle, P. J., Garland, P. B., Hales, C. N. and Newsholme, E. A. (1963). *Lancet* **i**, 785–789.
Redenbaugh, H. E., Ivy, A. C. and Koppanyi, T. (1926). *Proc. Soc. exp. Biol. Med.* **23**, 756–757.
Renner, R. and Elcombe, A. M. (1967). *J. Nutr.* **93**, 31–36.
Riis, P. M. and Herstad, O. (1967). *Acta agric. scand.* **17**, 3–12.
Rodbell, M. (1964). *J. biol. Chem.* **239**, 375–380.
Rodbell, M. (1965). *In* "Handbook of Physiology" (A. E. Renold and G. F. Cahill, eds.), Section 5, pp. 471–482. American Physiological Society, Washington, D.C.
Rodbell, M., Jones, A. B., Chiappe de Cinigolani, G. E. and Birnbaumer, L. (1968). *Recent Prog. Horm. Res.* **24**, 215–254.
Rudman, D., Brown, S. J. and Malkin, M. F. (1962a). *Endocrinology* **72**, 527–543.
Rudman, D. and Del Rio, A. E. (1969). *Endocrinology* **85**, 209–213.
Rudman, D., Hirsch, R. L., Kendall, F. E., Seidman, F. and Brown, S. J. (1962b). *Recent Prog. Horm. Res.* **18**, 89–123.

Rudman, D. and Shank, P. W. (1966). *Endocrinology* **79**, 565–571.
Ryan, C. A. and Tominatsu, Y. (1965). *Archs Biochem. Biophys.* **111**, 461–466.
Samols, E., Marri, G. and Marks, V. (1965). *Lancet* **ii**, 415–416.
Samols, E., Tyler, J. M., Marks, V. and Mialhe, P. (1969a). Proceedings of the 3rd International Conference on Endocrinology, Mexico, pp. 206–219. Excerpta Medica, Amsterdam.
Samols, E., Tyler, J. and Mialhe, P. (1969b). *Lancet* **i**, 174–176.
Sarcione, E. J., Back, N., Sokal, J. E., Mehlmann, B. and Knoblock, E. (1963). *Endocrinology* **72**, 523–526.
Sato, T., Herman, L. and Fitzgerald, P. J. (1966). *Gen. comp. Endocr.* **7**, 132–157.
Scott, C. C., Harris, P. N. and Chen, K. K. (1945). *Endocrinology* **37**, 201–207.
Seitz, I. J. and Ivy, A. C. (1929). *Proc. Soc. exp. Biol. Med.* **26**, 463–464.
Shepherd, D. B. and West, G. B. (1951). *Br. J. Pharmac. Chemother.* **6**, 665–674.
Smith, L. F. (1964). *Biochim. biophys. Acta* **82**, 231–236.
Smith, L. F. (1966). *Am. J. Med.* **40**, 662–666.
Sokal, J. E., Aydin, A. and Kraus, G. (1966). *Am. J. Physiol.* **211**, 1334–1338.
Sprague, R. and Ivy, A. C. (1936). *Am. J. Physiol.* **115**, 389–394.
Stamler, J., Bolene, C., Katz, L. N., Harris, R. and Pick, R. (1950). *Fedn Proc. Fedn Am. Socs exp. Biol.* **9**, 121.
Steiner, D. F., Clark, J. L., Nolan, C., Rubenstein, A. H., Margoliash, E., Aten, B. and Oyer, P. E. (1969). *Recent Prog. Horm. Res.* **25**, 207–268.
Stockell-Hartree, A., Langslow, D. R. and Hales, C. N. (1969). *J. Endocr.* **44**, 287–288.
Sutherland, E. W. and Robison, G. A. (1969). *Diabetes* **18**, 797–819.
Taylor, K. W. (1967). *In* "Hormones in Blood". (C. H. Gray and A. L. Bacharach, eds.), Vol. I, pp. 47–81. Academic Press, London.
Telib, M. and Pfeiffer, E. F. (1969). *Horm. Metab. Res.* **1**, 44.
Unger, R. H. (1968). *Am. J. med. Sci.* **255**, 273–276.
Unger, R. H. and Eisentraut, A. M. (1967). *In* "Hormones in Blood" (C. H. Gray, and A. L. Bacharach, eds.), Vol. I, pp. 83–128. Academic Press, London.
Unger, R. H. and Eisentraut, A. M. (1970). *In* "The Structure and Metabolism of Pancreatic Islets". (S. Falkmer, B. Hellman and I. B. Taljedahl, eds.), pp. 141–155. Pergamon Press, London.
Valverde, I., Rigopoulou, D., Exton, J., Ohneda, A., Eisentraut, A. and Unger, R. H. (1968). *Am. J. med. Sci.* **255**, 415–420.
Villar-Palasi, C. (1968). *Vitams Horm.* **26**, 65–118.
Volk, B. W., Lazarus, S. S. and Goldner, M. G. (1953). *Proc. Soc. exp. Biol. Med.* **82**, 406–411.
Vuylsteke, C. A. and DeDuve, C. (1953). *Arch. Intern. Physiol.* **61**, 273–274.
Weintraud, W. (1894). *Arch. exp. Path. Pharm.* **34**, 303–312.
Weiss, J. F., Naber, E. C. and Johnson, R. M. (1967). *J. Nutr.* **93**, 142–152.
Winegrad, A. I. and Renold, A. E. (1958). *J. biol. Chem.* **233**, 267–272.
Yeh, Y. and Leveille, G. A. (1969). *J. Nutr.* **98**, 356–366.
Yeh, Y., Leveille, G. A. and Forbes, R. M. (1970). *Fedn Proc. Fedn Am. Socs exp. Biol.* **29**, 425.

22 The Pineal Gland

P. A. L. WIGHT

Agricultural Research Council's Poultry Research Centre, King's Buildings, Edinburgh, Scotland

I. Introduction

The pineal gland, or epiphysis cerebri, of the fowl is a bluntly conical, pink structure. It is situated on the dorsum of the brain in the triangular space between the cerebellum and the cerebral hemispheres (Fig. 1) and is protected from the exterior by the overlying meninges, skull and cutaneous tissues. It may be easily seen by exposing the transverse depression and the external sagital groove of the calvarium. Beneath the point of intersection of these grooves lies the pineal; if a disc of skull is removed in this region and the meninges carefully turned back, the gland will be revealed. It often remains attached to the deflected meninges.

In an egg-laying strain of fowl the pineals of adults were found to be about 5 mg in weight and about $3{\cdot}5 \times 2{\cdot}0$ mm in length and breadth respectively. The mean weights of the glands of birds of this strain at 7 different ages are given in Table 1. In this survey the weight of the pineal was found to increase with age after hatching up to maturity. During the first two weeks, the rate of relative increase in weight of the pineal was about the same as that of the body; subsequently the rate of increase became considerably less. There was no consistent evidence of a relationship between sex and pineal weight. Pineals are present in nearly all vertebrates but their gross and microscopic morphologies vary considerably in different classes. The *complete* pineal system is in two parts, the epiphysis or pineal proper, and the parietal or parapineal organ. The latter generally occurs in the lower vertebrates and is

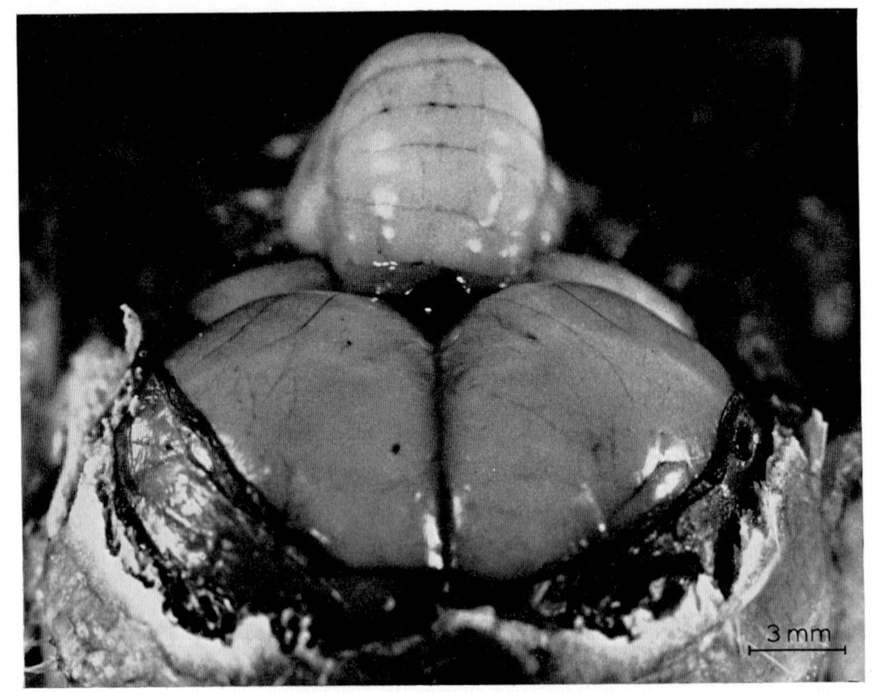

Fig. 1. The location of the pineal in the triangular space between the cerebral hemispheres and the cerebellum. In this specimen a systemic injection of trypan blue has stained the pineal but not the brain, which is protected by the blood-brain barrier.

most remarkable in certain lizards where it constitutes a lens-containing pineal "eye" situated just beneath the scales on the dorsum of the head. Although some avian embryos may have a parapineal organ (Krabbe, 1955), there is none in the fowl, which has the epiphysis alone.

II. Vascular and Nerve Supply

The pineal is surrounded by a very vascular extension of the meninges. A rich network of blood vessels, the vete pinealis, accompanies the trabeculae. The major blood supply is by the posterior meningeal artery, a branch of the cranial vamus of the left internal carotid artery, which ascends the stalk and forms a circus vasculosus at the base of the gland (Beattie and Glenny, 1966). Large venous sinuses overly the anterior surface. The fowl's pineal is a circumventricular organ and, like that of mammals, is unprotected by the blood-brain barrier. Consequently, vital dyes injected systematically diffuse into the gland (Fig. 1) but not into the brain, despite the common origin of the two structures. Blood flow through the rat pineal is rapid, the rate per gram of tissue being surpassed only by that through the kidney (Goldman and Wurtman, 1964). While there is no information on the volume or velocity

Table 1

Weight of pineal glands of domestic fowls at different ages

Age (weeks)	Male			Female		
	No. of birds	Body weight Mean ± S.E. (g)	Pineal weight Mean ± S.E. (mg)	No. of birds	Body weight Mean ± S.E. (g)	Pineal weight Mean ± S.E. (mg)
1 day	10	40 ± 1·7	0·97 ± 0·14	10	40 ± 1·0	1·39 ± 0·12
1	10	59 ± 1·9	1·82 ± 0·13	10	60 ± 2·3	1·90 ± 0·12
2	10	128 ± 4·9	3·06 ± 0·11	10	108 ± 6·5	2·55 ± 0·17
5	10	539 ± 16·4	3·53 ± 0·19	10	526 ± 35·4	3·25 ± 0·17
10	12	1522 ± 50·0	4·62 ± 0·22	9	1207 ± 41·0	4·10 ± 0·29
14	10	1750 ± 83·0	4·21 ± 0·24	10	1720 ± 113·0	4·25 ± 0·24
25	10	2831 ± 195·0	4·87 ± 0·33	13	2339 ± 72·0	4·78 ± 0·34

All birds were a commercial egg-laying strain (Thornber's '404') kept under standard conditions.

of flow in birds, the organ is, as we shall see, metabolically very active and probably has a correspondingly copious vascular supply.

Because of its reputed direct or indirect response to illumination, the innervation of the pineal has important physiological implications. Both Stieda (1869) and Studnička (1905) found no nerve fibres in the pineal of fowls, but Stammer (1961), Quay (1965a) and Ariëns Kappers (1965) have shown the organ to be well innervated. Silver preparations reveal abundant fine neurites in the trabeculae surrounding the follicles. Electron microscopy shows most neurites to be unmyelinated (Figs. 9, 10), (Fujie, 1968) but a few are myelinated (Bischoff, 1969). Many of these have the ultrastructural (Fujie, 1968; Bischoff, 1969) and histochemical (Wight and MacKenzie, 1970) characteristics of adrenergic, sympathetic nerve fibres (Fig. 14), but Wight and MacKenzie (1970) found acetylcholine in the trabeculae and suggested that this may originate from nerves. Cholinergic innervation occurs in lower vertebrates; Wurtman *et al.* (1968), suggest that a change from these to the exclusively adrenergic innervation of some mammals takes place in the more advanced phyla.

Although no anatomical studies of the course of these autonomic nerves have been reported in the fowl, physiological observations (Lauber *et al.*, 1968) indicate that they are post-ganglionic fibres coming from the superior cervical ganglia. In mammals, they probably enter the pineal via the tentorium cerebelli and accompany the pial vessels. In mammals some fibres also pass up the pineal stalk from the habenula and posterior commissural regions but these appear to return without terminating in the pineal. The present author agrees with Ariëns Kappers (1965) that there are neurites in the stalk in the fowl, but the origins and terminations of these have not been determined. Oksche *et al.* (1969) saw nerve bundles in the pineal stalk of the pigeon.

Quay and Renzoni (1963) have described a neurosecretory nucleus in *Passer domesticus*. There are no neurons in the fowl's pineal nor is "Gomori positive" (Bern and Knowles, 1966) neurosecretory material demonstrable by Bargman's chrome alum-haematoxylin-phloxine technique.

III. Histogenesis

The pineal is of ectodermal origin. It develops in the 3 to 4-day-old chick embryo from a thickening in the diencephalon, dorsal and posterior to the velum transverum. This thickening gives rise to a diverticulum, almost always single, from which ependyma-lined vesicles bud-off. Rarely, two diverticuli are formed, thus indicating the primary bilateral origin of the pineal. It is possible that at first the vesicles communicate with the third ventricle via the lumen of the stalk of the pineal although Krabbe (1955) does not think they ever communicate with the lumen and certainly they are closed sacs at the time of hatching. Even by the 8th d (Fig. 2) the majority of

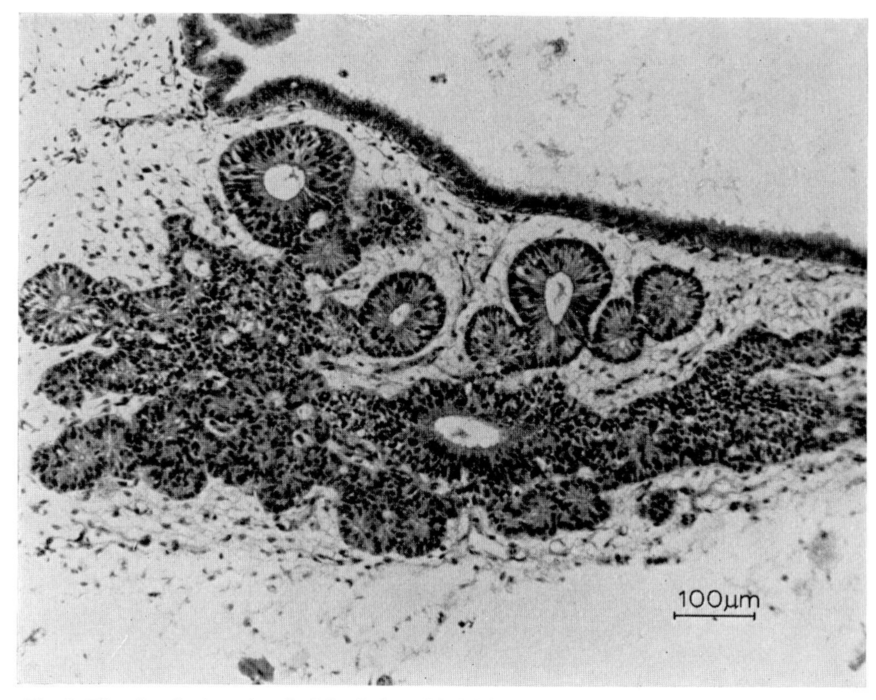

Fig. 2. The developing pineal of the 8-day-old chick embryo. Vesicles have budded off from the stalk, but even at this early stage they are blind sacs cut off from the stalk's lumen. Haematoxylin and eosin.

vesicles appear to be closed off and loose mesenchyme has proliferated between the vesicles separating them from each other as do the trabeculae in the adult. The stalk itself becomes little more than a vasculo-fibrous cord, which means that humoral mediation of pineal secretion via the lumen of the stalk, and thence the cerebrospinal fluid, can be ruled out (Quay and Renzoni, 1967). Small clumps of pineal tissue are sometimes present along the length of the stalk. By about 3 months of age the vesicles have diminished in size and come closer together and the organ has assumed the appearance it will have for the usual life-span of the adult (Figs. 3, 4).

IV. Structure

Studnička (1905) grouped avian pineals in three classes according to the occurrence and patency of the vesicles and central lumen of the organ and its stalk; simple tubular, follicular and solid lobular. That of the adult is usually said to be solid lobular although a few follicles are nearly always present (Fig. 4); the gland is more correctly described as intermediate between the solid lobular and follicular types. It is composed of vesicles or rosettes of cells

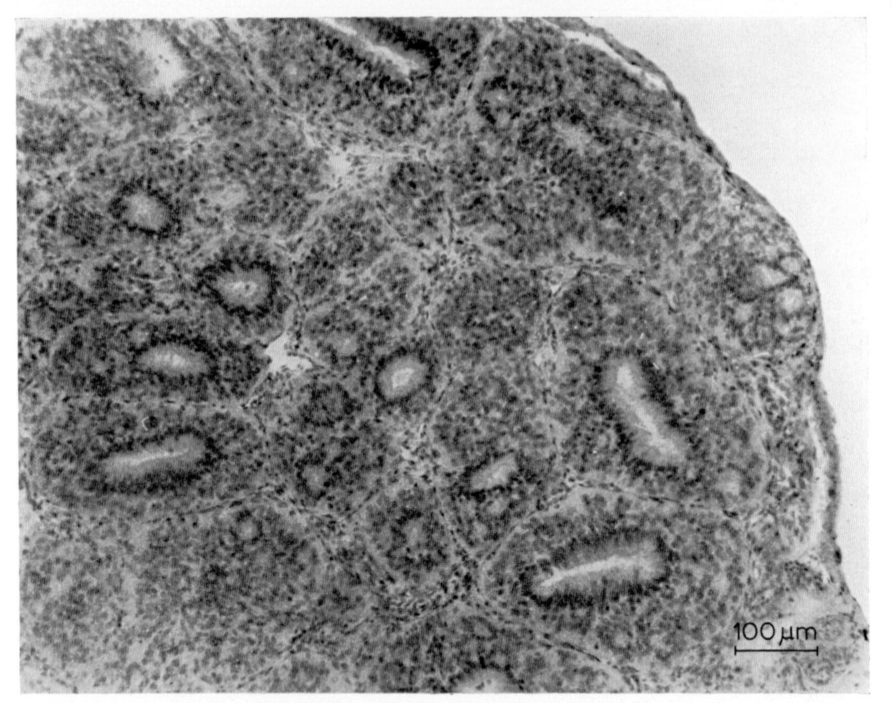

Fig. 3. The pineal of a 1-month-old chick. The organ is becoming more condensed although several vesicles have lumena. Haematoxylin and eosin.

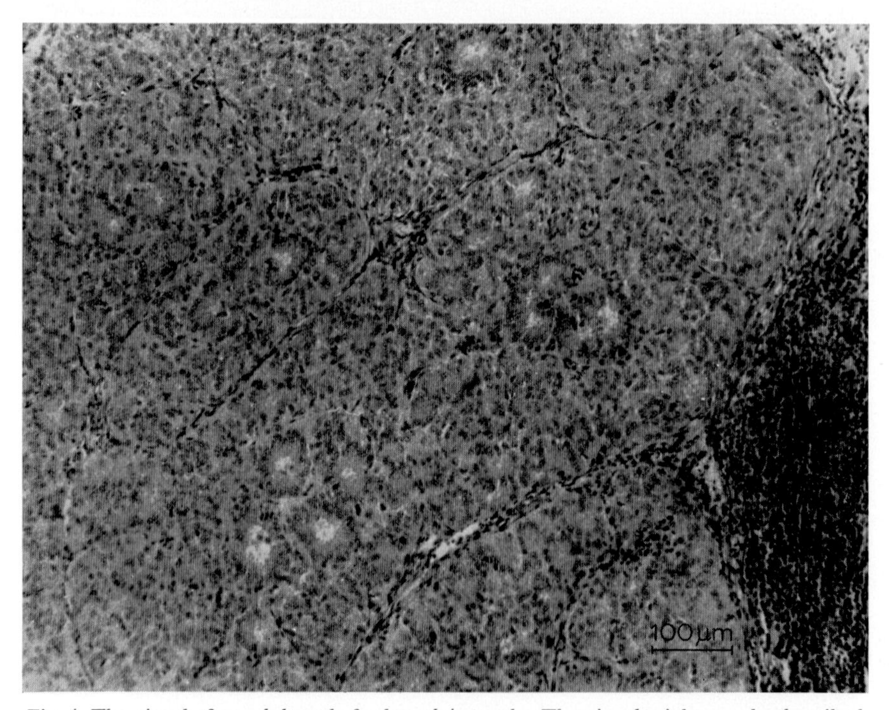

Fig. 4. The pineal of an adult male fowl aged 4 months. The pineal might now be described as of the solid lobular type although some follicles are present. A focus of lymphocytes can be seen. Haematoxylin and eosin.

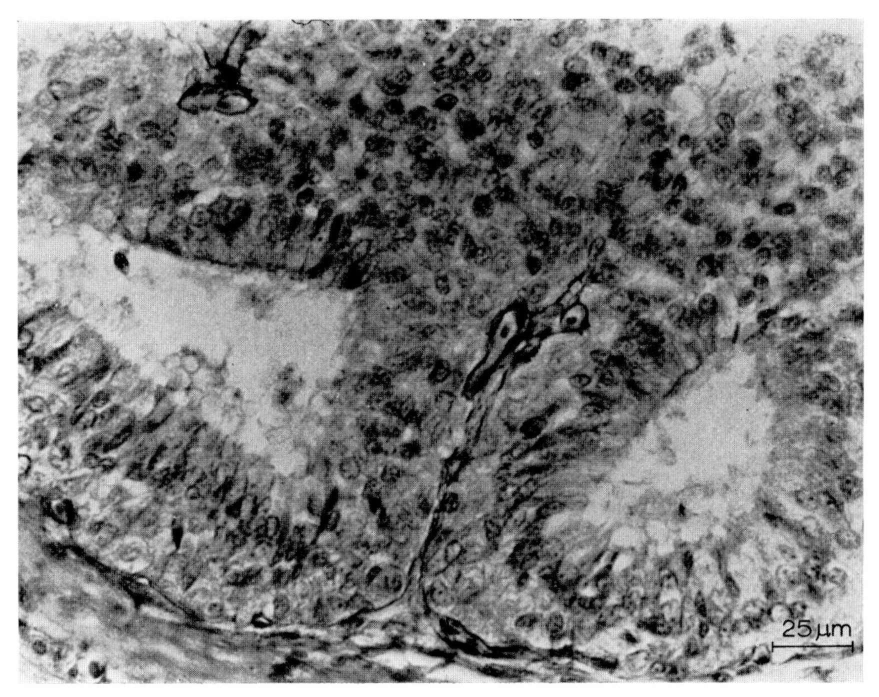

Fig. 5. Lumena of the vesicles containing fine threads which are probably the cilia and lamellated processes of pinealocytes. Periodic acid-Schiff reactions.

derived therefrom, supported by thin intervesicular septa and lobulated by trabeculae arising from a rather thin surrounding capsule. Each lobule incorporates several vesicles or groups of cells. In the young fowl, two kinds of cells have been described with the light microscope. The ependymocytes are tall, columnar cells of ependymal type with a basally-situated nucleus; these line the vesicles. Fine eosinophilic threads are often attached to their apical border (Fig. 5). The hypendymocytes are polygonal and are concentrically layered external to the former. In mature birds it is difficult to differentiate the two types; all the parenchymal cells are then usually termed pineocytes or pinealocytes. In haematoxylin- eosin stained sections, pinealocytes have finely granular, acidophilic cytoplasm with a spherical, rather vesicular nucleus which often shows a nucleolus. Under the light microscope these cells bear no resemblance to vertebrate photoreceptors nor do routine stains indicate any form of secretory activity. If the vesicles have a lumen (Fig. 5), it may contain the apical threads mentioned above, often in a reticulum (Fig. 4), and it may also contain what appears to be some coagulum, globules and some cell debris. Although foci of pinealocytes sometimes project into the lumen, there is really no evidence of holocrine secretion.

In addition to the pineal parenchymal cells and the trabeculae which contain blood vessels, nerve fibres, connective tissue, fibroblasts and fat-laden macrophages, large foci of mature lymphocytes are often seen. The pineal was once classed as a lymphatic organ (Henle, 1871) and Romieu and Jullien (1942) thought it an important source of lymphocytes. Foci of lymphocytes are widespread in avian tissues and it seems likely (*cf.* Spiroff, 1953) that no more significance can be attached to their presence in the pineal than at any other site.

V. Ultrastructure

The pineal has been recognized from antiquity but only about the beginning of the present century did studies effect an advance in knowledge leading to a general concept of the photic associations of the organ. Except in the case of certain lower vertebrates, the light microscope is inadequate to identify photoreceptors in the pineal but the electron microscope has not only confirmed the retinal-like structure of the cells of certain reptiles and amphibia but shown that, besides several mammals (*cf.* Wurtman *et al.*, 1968), birds also have pineal components reminiscent of rudimentary photoreceptors. Oksche and Vaupel-von Harnack (1965a) first observed in chicken embryos lamellated projections, cilia and basal centrioles which were slightly reminiscent of photoreceptor-type outer segments and this has been confirmed in young fowls (Oksche and Vaupel-von Harnack, 1966; Fujie, 1968; Bischoff, 1969). In fact, "photoreceptor" processes in the fowl bear only a distorted resemblance to the neatly stacked membranes of the retinal rods and cones, while it should perhaps be mentioned that neither Morita (1966) nor Ralph and Dawson (1968) found evidence of electrical responses to light in avian pineals. Further, although some pinealocytes show neuromorphological features, secretory properties predominate in most. This is indicated by the presence of abundant cytoplasmic organelles and by certain subsidiary features such as fenestration of the endothelium of the pineal capillaries. Histochemistry of the cells also points to the prominance of protein synthesis and secretion (Wight and Mackenzie, 1971).

Investigators of the ultrastructure of the fowl's pineal have classified the parenchymal cells into three types. It may be hypothesized, however, that two of these types are modifications of a basic pinealocyte varying between a cell which is essentially secretory and one which is essentially nervous and shows degrees of development and complexity of its photoreceptor process and synaptic contacts. Ultrastructural examination of pineals from sparrows, ducks, pigeons, magpies and quails shows variations around the essentially similar structure of the pinealocytes (Oksche and Vaupel-von Harnack, 1965b, 1966; Collin, 1966, 1969; Bischoff and Richter, 1966; Bischoff, 1969; Oksche *et al.*, 1969).

First, the nervous type of cell is referred to as a pinealocyte by Fujie (1968)

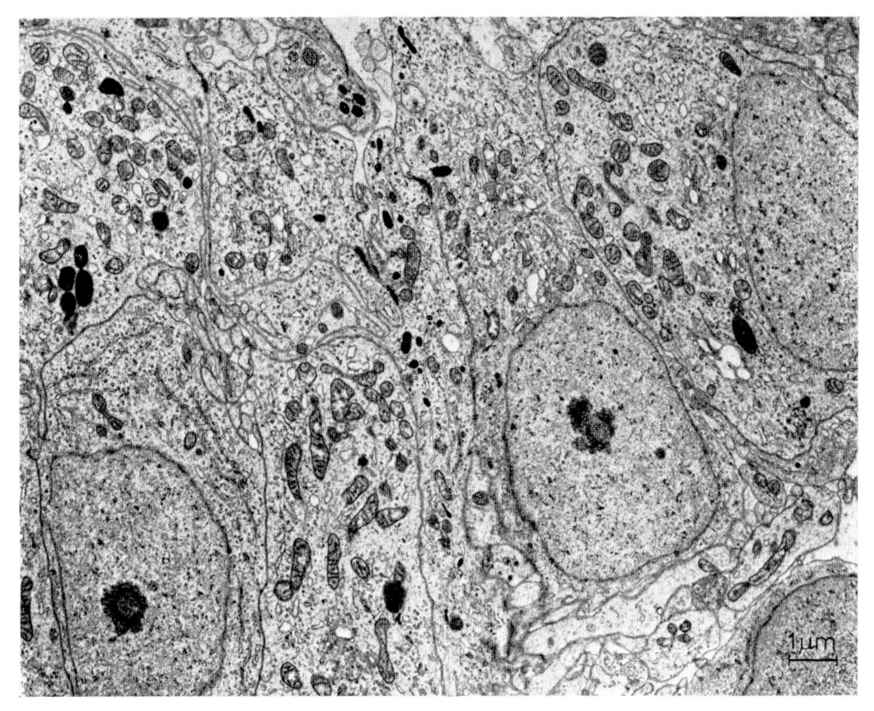

Fig. 6. Pinealocytes adjacent to the lumen of a vesicle. Most of these cells appear to be of the secretory type and their apices, some with microvilli, project into the lumen. The cells are extremely rich in organelles including numerous mitochondria, lysosomes, smooth and rough endoplasmic reticulum, ribosomes, smooth vesicles and a few dense-cored vesicles.

and as a photoreceptor by Bischoff (1969). The body of this cell has three distinct regions; base, narrow neck or colliculus and apex. Its cytoplasm contains a wealth of organelles (Figs. 7, 8). In the base are the nucleus, many mitochondria, a prominent Golgi complex, numerous smooth vesicles of smooth endoplasmic reticulum, some rough endoplasmic reticulum and free ribosomes. There are also varying numbers of lipid droplets, lysosomes, small dense-cored vesicles and medium-sized electron-dense granules surrounded by a triple membrane and having a total diameter of 100 nm to 300 nm. Along the periphery of the base are some synaptic ribbons surrounded by synaptic vesicles. The base may show lobulated end-feet. The most notable features of the colliculum are its microtubule content and the prominent maculae adherentes which join adjacent cells (Figs. 7, 8). The apex contains a great number of mitochondria, a few small, dense-cored vesicles and small numbers of smooth vesicles, rough endoplasmic reticulum and ribosomes. Its most distinctive feature is a cilium (Figs. 8, 9), which usually has a $9+0$ axial filament complex, and which arises from a basal corpuscle and associated ciliary rootlets and projects into the follicular lumen. The cilium may

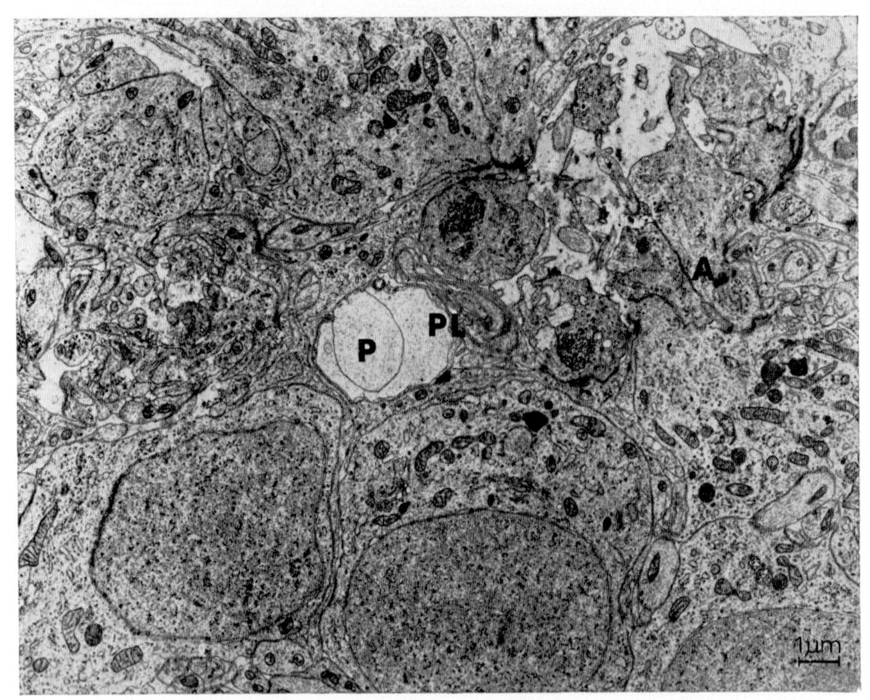

Fig. 7. Pinealocytes adjacent to a lumen. Some of these cells are of the "photoreceptor" type and they show both swollen (P) and lamellated (PL) apical processes. Maculae adherentes (A) are prominent in the region of the colliculi and cytoplasmic organelles are abundant.

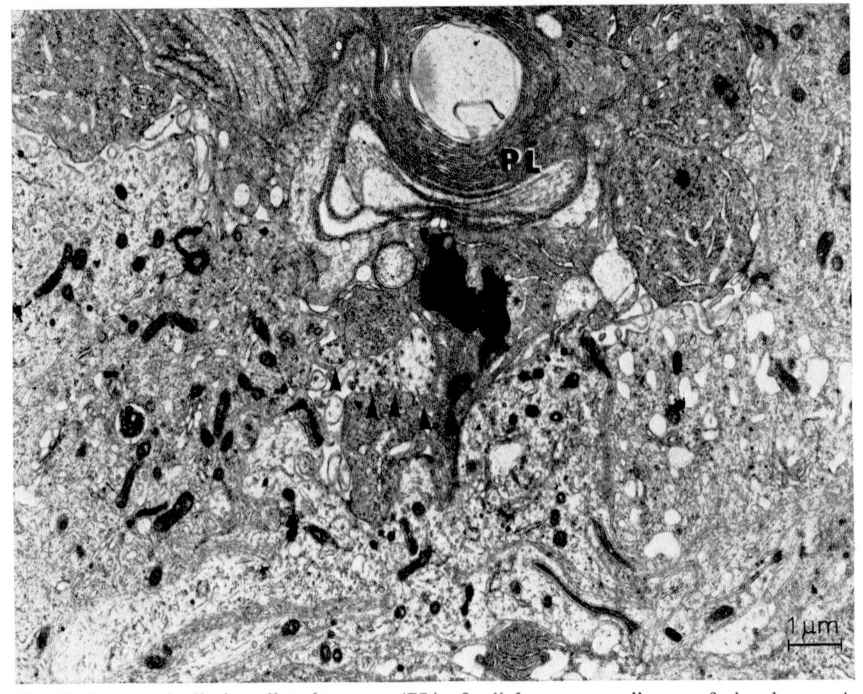

Fig. 8. Concentrically lamellated process (PL) of a "photoreceptor" type of pinealocyte. A large accumulation of osmiophilic lipid is present in the centre of the field and close to it are four cilia (arrows) showing the 9+0 axial configuration.

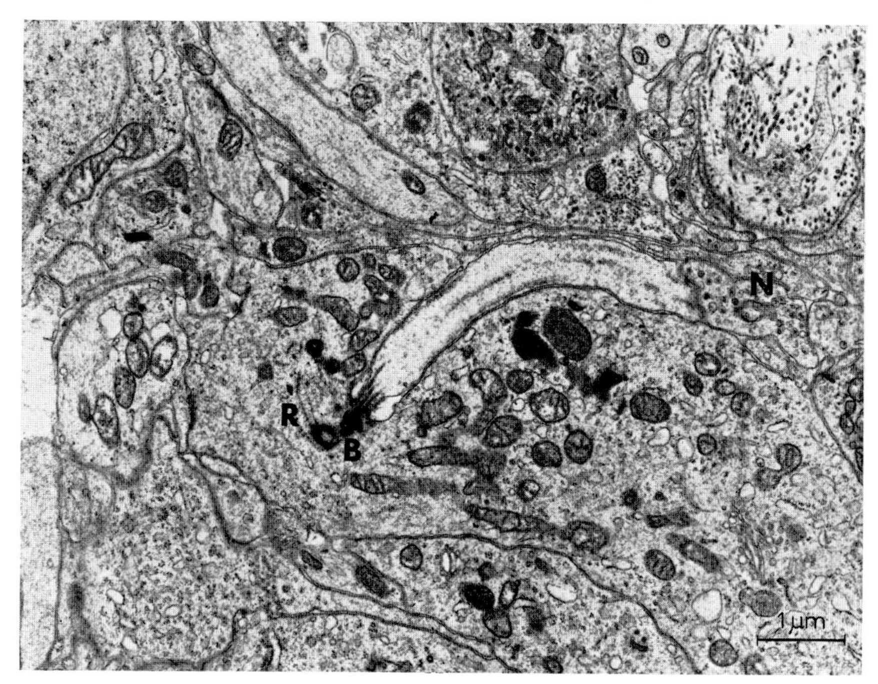

Fig. 9. Cilium arising from the apex of a "photoreceptor" type cell. Rootlets (R) are associated with the basal corpuscle (B). Above the cilium is a non-myelinated nerve fibre (N) containing dense-cored vesicles.

have a swollen, electron-transparent, saccular termination or show the concentric, membranous lamellations (Figs. 7, 8) which resemble the outer segment rudimentary retinal rod or cone.

The second type of cell is termed "secretory" by Bischoff (1969) and is inserted between the photoreceptor cells and like them has an apex, colliculum and base containing a nucleus and many organelles (Fig. 6). However, this cell lacks both cilia and synapses and its most distinctive feature is the presence of numerous medium-sized, membrane-limited electron-dense granules, between 100 nm and 200 nm in diameter, throughout the cytoplasm and particularly in the apex. The apex may project into the lumen and show many vesicular profiles. This cell seems to correspond to the "supporting" cells of Fujie (1968) which also lack cilia and have abundant medium-sized granules but are said to have numerous apical microvilli. Bischoff (1969) has described, in the fowl, another kind of pineal cell intermediate in appearance between the photoreceptor and the secretory types, thus adding some weight to the view that all these cells are varieties of a basic pinealocyte. These he called "ependymal" cells; they have three regions as do the others but have both cilia and microvilli although synaptic ribbons are absent. The cilia are said to have a 9+2 axial configuration. Bischoff (1969) has suggested that the ependymal cells may function as a supporting cell or perhaps as a precursor of the other types.

The third type of pineal cell is the glial cell, situated in the peripheral part of the lobule. It has a dense nucleus and numerous cytoplasmic filaments. Fujie (1968) estimated that these cells comprise only 3 to 5% of the total in the pineal lobule; few are present in mammals (Del Rio-Hortega, 1932). These cells are sometimes described as astrocytes but, unlike their counterpart in the human pineal, they do not stain by Cajal's gold chloride.

The electron microscope also reveals non-myelinated axons, many containing small dense-cored vesicles (Figs. 9, 10), associated with the pineal cells. The pinealocytes rest on a basal lamina; in the trabeculae can be seen collagen fibrils, capillaries, neurites and fat-laden macrophages (Fig. 11).

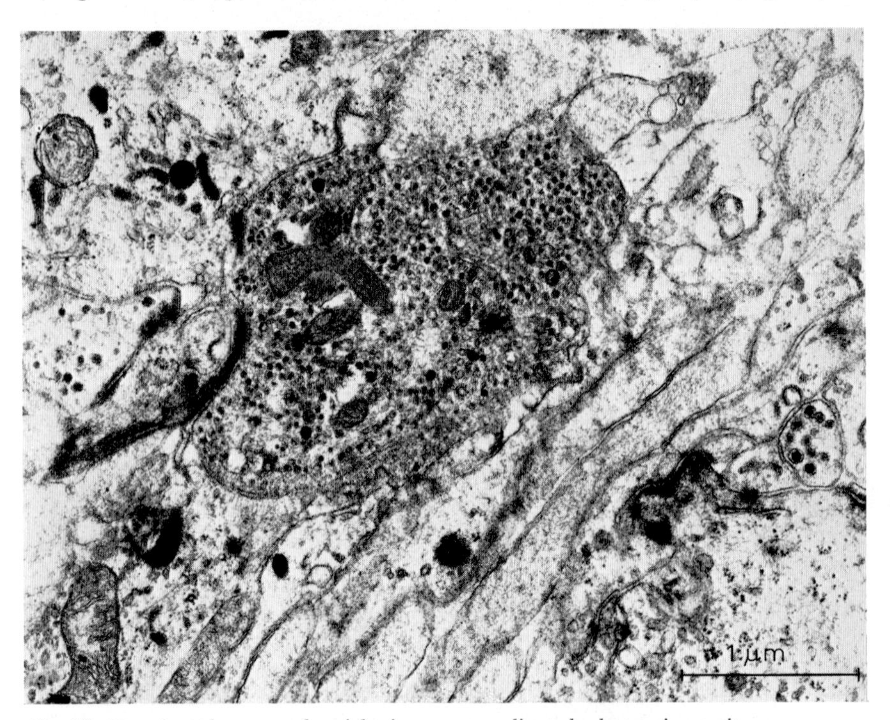

Fig. 10. Abundant dense-cored vesicles in a non-myelinated adrenergic neurite.

VI. Histochemistry

The great number of organelles in the pinealocytes indicates considerable metabolic activity and this is substantiated by histochemistry. Lipid is abundant (Desogus, 1928; Thillard, 1968) and triglycerides, phospholipids and cholesterol and its esters have been identified (Wight and MacKenzie, 1971). The lipid occurs in the pinealocytes and macrophages of the trabe-

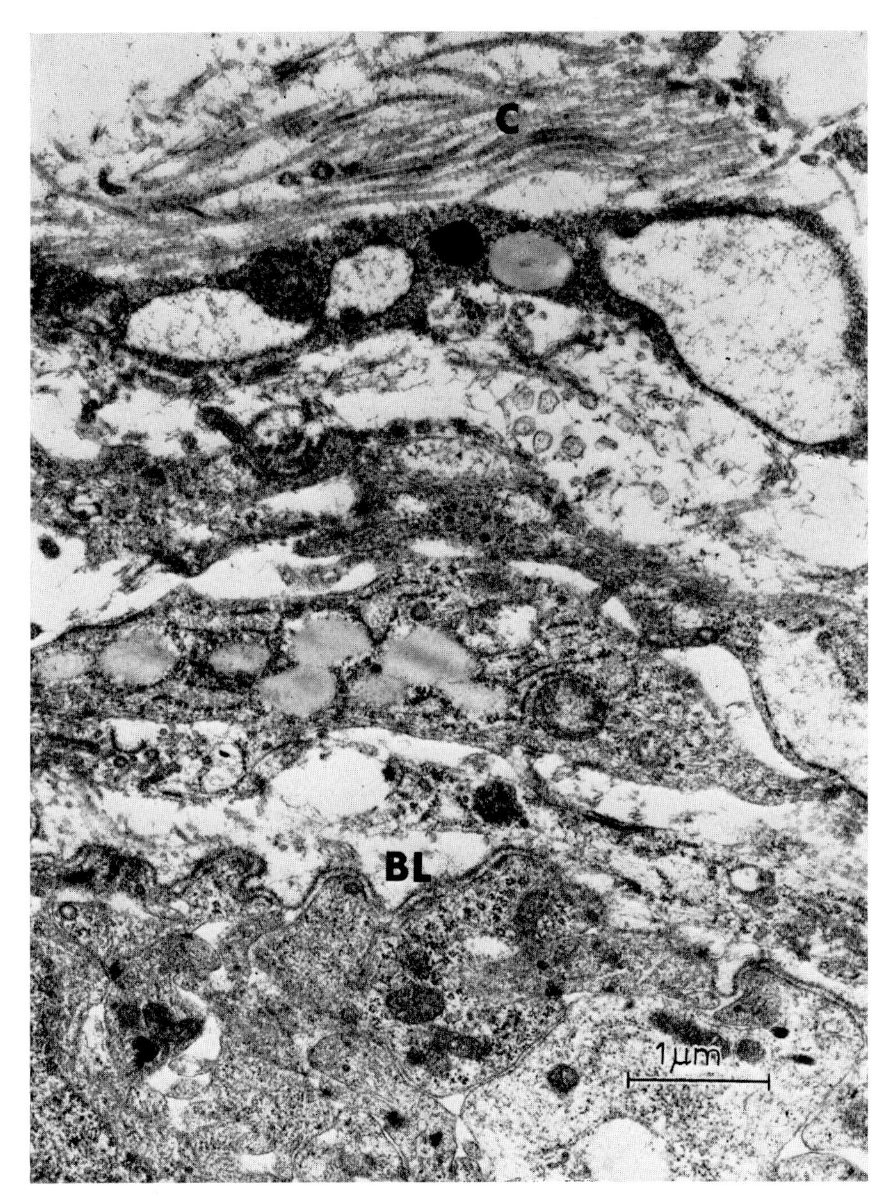

Fig. 11. Lipid-laden macrophages in the trabeculae. Collagen fibres (C) are present at the top of the illustration and the bases of pinealocytes with a basal lamina (BL) can be seen at the bottom.

culae (Fig. 12). It has been suggested that lipid moves from the pinealocytes to the trabecular blood vessels and, in consequence, is associated with a secretion. On the other hand Thillard (1968) noted that phospholipid of the

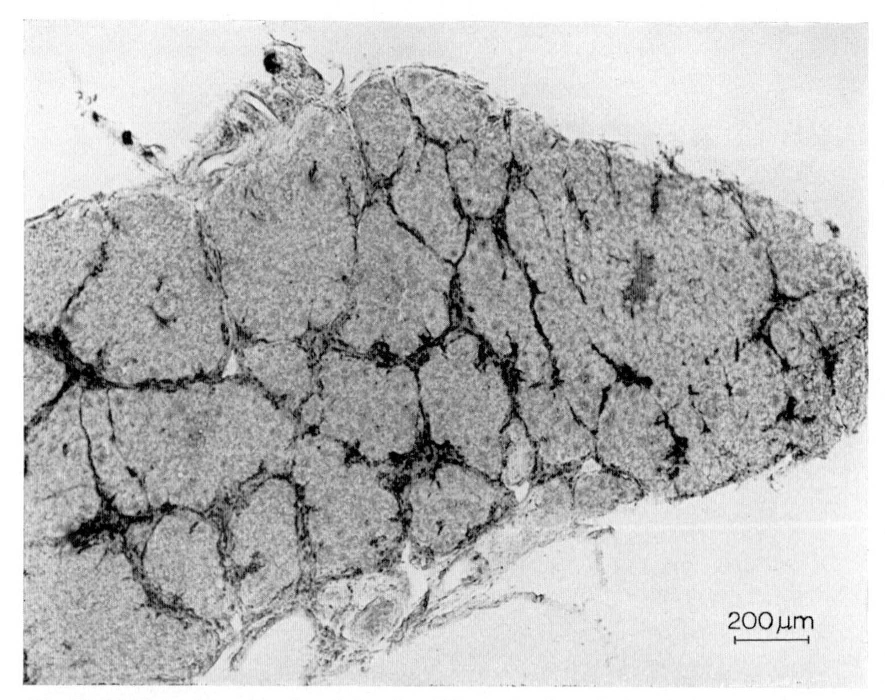

Fig. 12. Lipids in the trabeculae of the pineal of an adult hen. Sudan black B after Elftman's chromation.

pinealocyte apex increased when chickens were reared in constant illumination and suggested that lipid moved in the reverse direction, ultimately into the cerebrospinal fluid. Perhaps the most probable hypothesis (Prop, 1965) is that energy for synthetic purposes is provided by lipid, particularly as lipase activity is high in the fowl's pineal (Wight and MacKenzie, 1971). Desogus (1928) suggested that pineal lipid increased when the ovary was active; this observation has never been confirmed.

The gland has considerable enzymic activities (Fig. 13). Alkaline and acid phosphatases, adenosine triphosphatase, lipase, non-specific esterase, NAD-diaphorase, cytochrome oxidase, β-glucuronidase and aminopeptidase activities are all present (Wight and MacKenzie, 1971). Thus the pineal is not vestigial but actively functional. In particular, energy transfer through "high-energy" phosphates is an important part of cell transport and metabolism and the histochemical reactions for acid phosphatase and adenosine triphosphatase (ATP-ase) were particularly intense. Further, adenosine

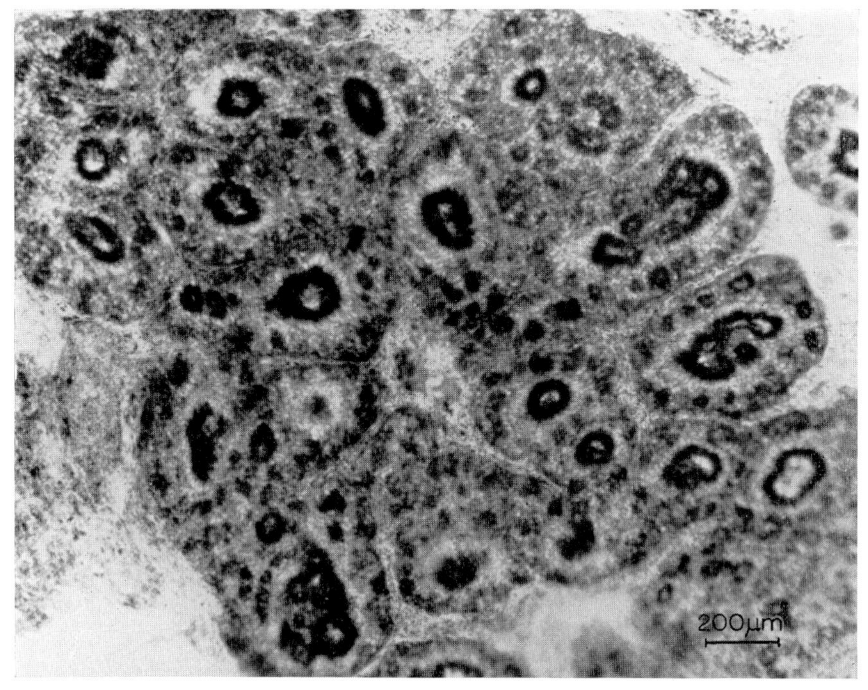

Fig. 13. Intense reaction for acid phosphatase in the pineal of an 8-week-old female. Azo-coupling method using Fast Garnet GBC.

triphosphate is essential for protein synthesis and the relatively strong activity of its hydrolase, together with the large amounts of RNA which histochemical tests reveal, are important indications of active protein synthesis.

ATP-ase is closely involved in the function of mitochondrial membranes and structure; there are large numbers of mitochondria in the avian pinealocytes. Succinic dehydrogenase, NAD-diaphorase and cytochrome oxidase are intramitochondrial. Succinate is one of the best indicators of Krebs' cycle activity from the histochemical aspect (Pearse, 1960), and the strong reactions found indicate a highly active tricarboxylic acid cycle and cytochrome system in the fowl's pineal. Acid phosphatase and β-glucuronidase are associated with lysosomes and the former enzyme has also been localized in the Golgi apparatus, in small vesicles and in cisternae of the endoplasmic reticulum. In fowl pinealocytes all these structures are present and some are abundant.

The fluorescent formaldehyde-condensation method shows that monoamines (Fig. 14), probably serotonin, are present (Wight and MacKenzie, 1971). In the pineal of lizards serotonin is associated with both granulated vesicles between 60 nm and 300 nm diameter and an extravesicular pool

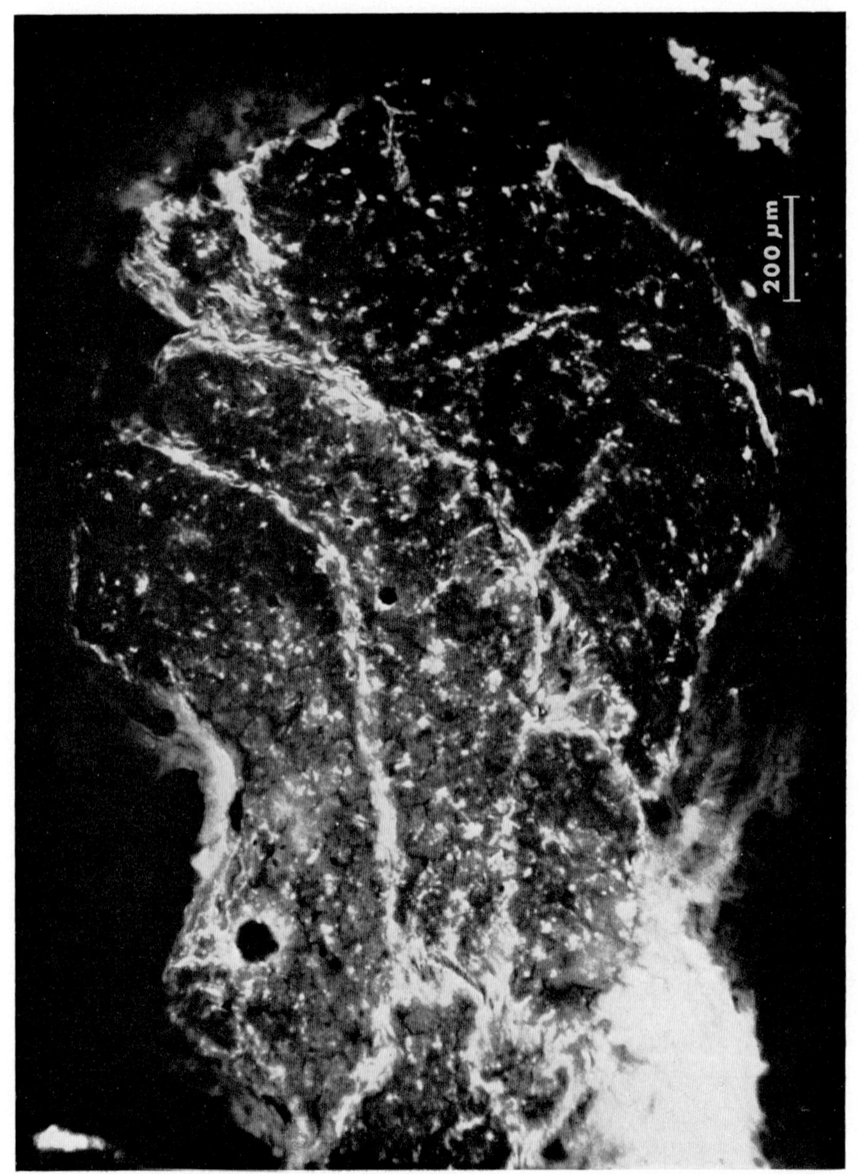

Fig. 14. Adrenergic nerves and biogenic amines in the pineal of an adult female. Fluorescent formaldehyde-condensation method.

spread over the cytoplasm of the pinealocyte (Wartenberg and Baumgarten, 1969). As noted above such granulated vesicles occur in the fowl's pineal. Histamine occurs in some mammalian pineals, perhaps associated with mast cells (Machado *et al.*, 1965), but it has not been identified in that of the fowl (Wight and MacKenzie, 1971).

Fowl pineal cells do not contain histochemically detectable amounts of glycogen, intracytoplasmic mucopolysaccharides, Gomori-positive neurosecretory substance, iron or calcium. Acervuli, which are common in man and have been found in several other mammals, are absent from the pineals of fowls although calcareous concretions have been observed in aged herons and geese (Stammer, 1961). PAS-positive material, probably associated with cilia, is present in some lumina.

VII. Biochemistry

Biochemical studies of the pineal gland have been mostly concerned with the indole product melatonin, biogenic amines and associated enzymes. Some interest has been taken in the organ's lipids.

It has been long known (McCord and Allen, 1917) that the bovine pineal contained a substance which could reduce the skin pigmentation of amphibians (Fig. 15). The active principle, *N*-acetyl-5-methoxy-tryptamine, was

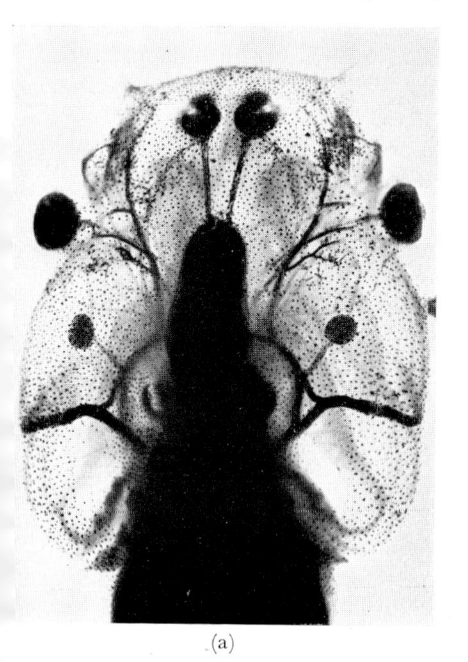

(a)

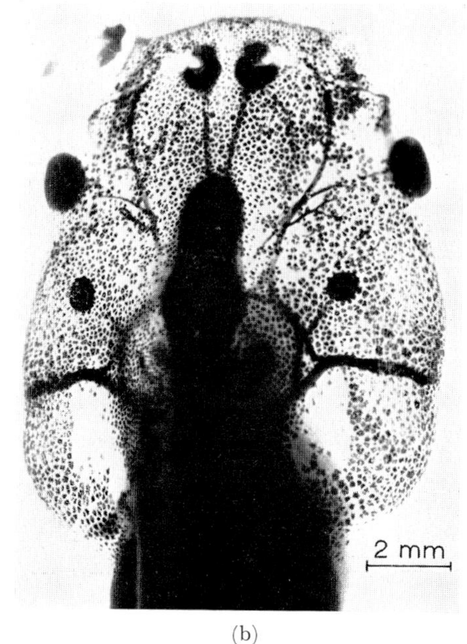

(b)

Fig. 15. *Xenopus laevis* larvae. An aqueous homogenate of avian pineals has been added to the water of the larva on the left causing a marked contraction of the melanophores.

isolated and synthesized (Lerner *et al.*, 1960) and named melatonin from its effect on frog skin melanophores. Melatonin is produced mainly in the pineal —in mammals perhaps only in the pineal—but in the fowl a certain amount may be formed in the retina of the lateral eyes (Quay, 1965b). Circulating melatonin appears to be widely distributed in the tissues. In some mammals experimentally injected [acetyl-^{3}H] melatonin is rapidly taken up by the pineal, eye, ovary, nervous tissue and some endocrine glands as well as other tissues (Kopin *et al.*, 1961; Wurtman *et al.*, 1964); the highest concentrations were found in the pineal.

Melatonin is an *O*-methylated indole synthesized in the pineal from *N*-acetylserotinin from the methyl donor *S*-adenosylmethionine and the methylating enzyme hydroxyindole-*O*-methyl transferase (HIOMT), which can *O*-methylate several hydroxyindoles. In the mammalian pineal, *N*-acetylserotonin appears to be the major source of melatonin. HIOMT is present in the fowl's pineal (Axelrod *et al.*, 1964; Quay, 1965c) in much greater amounts than in mammals and has been assayed by incubating aqueous homogenates with [methyl-^{14}C]-*S*-adenosyl-L-methionine and *N*-acetylserotonin and measuring the resultant radioactive melatonin. If birds are kept in darkness the enzymic activity is decreased whereas in light it is increased (Axelrod *et al.*, 1964; Lauber *et al.*, 1968). This is opposite to the effect which darkness has on the pineal HIOMT of the rat. HIOMT has also been identified in the retina of the fowl (Quay, 1965b). Although there are no records that serotonin occurs in the retinae of birds, it has been found in the retina of four major classes of vertebrates (Welsh, 1964); it is therefore probably present at this locus in the fowl. As Quay (1965b) has pointed out, this means that retinae have the melatonin precursors and, although it is not known whether retinal melatonin is secreted as a hormone, a compensatory increase following pinealectomy is a possibility which should not be overlooked in experimental work.

Pinealocytes take up tryptophan from the blood and can synthesize melatonin. Tryptophan is hydroxylated to 5-hydroxytryptophan, decarboxylated to serotonin, acetylated to *N*-acetylserotonin and finally methylated to melatonin as mentioned above. Necessary catalysts, co-factors and donors have been shown to be present in the pineal. Details of the reactions have been recorded and reviewed elsewhere (*cf.* Wurtman *et al.*, 1968). Although this biosynthetic procedure has been determined with mammalian pineals, particularly in organ culture, evidence of the presence in the fowls' pineals of biogenic amines (Wight and MacKenzie, 1971), HIOMT (Axelrod *et al.*, 1964; Quay, 1965c) and melatonin (van de Veerdonk, 1965) suggests that a similar pathway may be followed in this genus.

Melatonin can be assayed fluorometrically in 3 N-HCl after extraction into chloroform (Axelrod and Weissbach, 1961), but for the detection of small amounts in biological material a bioassay procedure using either the skin of *Rana pipiens* (Lerner and Wright, 1960) or *Xenopus laevis* larvae (Quay and Bagnara, 1964) is much more sensitive. Using such a *Xenopus* bioassay technique, van de Veerdonk (1965) fractionated an aqueous extract of pineals of domestic fowls using Sephadex G25 and estimated that considerably more melatonin was present than in the gland of most mammals.

As well as melatonin, several biogenic amines and their metabolites have been identified in mammalian pineals including serotonin, noradrenaline, dopamine and histamine. In the fowl, despite the scientific interest and economic advantage which might result from increased knowledge of a light-associated gland, reliable biochemical information is very limited. Melatonin and HIOMT have been identified and histochemical studies have shown that biogenic amines are present in the pinealocytes and catecholamines in the nerves. Serotonin has been identified in the pineal of the pigeon (Quay, 1966). From this one may assume that amines similar to those in mammals are present but this has not been proved. In mammals, some of these compounds show diurnal rhythms and are quantitatively changed under the influence of different lighting regimes. Lack of similar information for the fowl is a handicap to a full understanding of the function of the pineal gland. Hydroxytryptophol and 5-methoxy-tryptophol occur in the fowl's pineal (Wurtman and Axelrod, 1966); these may be formed by deamination of serotonin by monoamine oxidase and subsequently be *O*-methylated by HIOMT (Axelrod and Weissbach, 1960). Monoamine oxidase activity appears to be unaffected when hens are kept in either continuous light or darkness (Axelrod *et al.*, 1964).

VIII. Physiology

There are few studies on the physiology of the fowl's pineal. Yet information obtained from mammals has opened up new concepts of its activity which will certainly prove broadly applicable to the fowl. No attempt has been made here to cover all the numerous publications in the field, but the major advances cannot be ignored even though they were not obtained by studying the fowl. The few lines on the subject in most textbooks of avian physiology are quite inadequate (Höhn, 1961; Sturkie, 1965). The pineal can no longer be regarded as vestigial. The metabolism of radioactive phosphorus in the rat pineal is two or three times as vigorous as in the anterior lobe of the hypophysis and it shows considerably higher activity than the rest of the brain (Börell and Orström, 1945).

Historical theories of pineal function are sufficiently entertaining to have stimulated several reviews (Gladstone and Wakeley, 1940; Kitay and Altschule, 1954), but the earliest works which need be mentioned here are those of Gutzeit (1896) and other clinicians (*cf.* Del Rio-Hortega, 1932) who observed sexual precocity in patients with pineal tumours and thus drew attention to its gonodal associations, and the anatomical studies (*cf.* Studnička, 1905) which led to the concept of pineal photoreception.

Although there now seems no doubt of the involvement of the pineal in the growth and function of the fowl's reproductive system, experiments designed to study this association have not always given unequivocal results. (See Chapter 57 for further discussion.) This may be attributed partly to technical imperfections, including damage to other parts of the central nervous system during pinealectomy, and partly to the multiplicity of factors normally affecting the reproductive system and of which the pineal is only one. Shellabarger (1953), and Kitay and Altschule (1954), list the relevant publications, the majority of which claim that hypertrophy of the fowl's gonads follows pinealectomy and that gonadal atrophy follows the injection of pineal extracts. Badertscher (1924), who gives a good critical review of pineal experiments in the fowl up to that date, and more recently Stalsberg (1965), have observed no change after pinealectomizing fowls. The most significant of these studies are perhaps those of Shellabarger (1952, 1953), who found that if chicks were autopsied 20 d after pinealectomy at one day of age, the testes were atrophied but by 40 to 65 d after the operation, the testes were hypertrophied. This suggests that this aspect of the pineal's function may be age dependent. When pineal extract was administered to the pinealectomized fowls, the testes' weights were not significantly different from the sham-operated controls. The injection of pineal extracts to intact birds reduced the weight of the testes. The pineals of capons were hypertrophied. More recently Singh and Turner (1967) have demonstrated that injected melatonin decreases the weight of the testes and ovaries of developing chickens. This may not apply to other avian genera for while Homma *et al.* (1967) found that melatonin inhibited the growth of the quail's gonads, Sayler and Wolfson (1967, 1968) found that the pineal stimulates gonadal maturation in quail. However, the sum of the evidence seems to be that in the fowl, the pineal produces a substance which delays the development of the gonads.

Urechia and Grigoriu (1922) found that pinealectomized fowls had hypertrophied pituitaries; later, Shellabarger (1953) also found that the gonadotropic potency of the anterior pituitary was increased in pinealectomized chicks and reduced in intact chicks which had received additional pineal material. This observation that pineal action on the reproductive system is mediated through the pituitary was a most important advance and predated current theories of the mode of action of the pineal.

In mammals it now seems certain that the pineal's effect on the gonads is mediated through brain centres acting on the pituitary; Moszkowska (1963) suggested that the route may be through the hypothalamo-pituitary axis. Motta *et al.* (1967) thought that alteration in the weight of the reproductive organs of pinealectomized male rats may be caused by an influence of the pineal on the release of luteinizing (LH) and follicle-stimulating (FSH) hormones. Anton-Tay *et al.* (1968) have shown that intraperitoneal injection of melatonin modifies the level of activity of serotonin-containing neurones and Wurtman and Anton-Tay (1969) suggest that these neurones are the "endocrine site of action" of the pineal. Nerve cells in the midbrain are particularly affected and many of these project to the hypothalamus. The main areas in the mammalian brain which are sensitive to these pituitary-influencing pineal compounds are located in the median eminence of the hypothalamus, although there are also sensitive neurones in the reticular substance of the midbrain. Several pineal compounds are involved (*cf.* Martini, 1969). Melatonin and 5-hydroxytryptophol reduce pituitary LH, while 5-methoxytryptophol and serotonin reduce FSH but not LH. These indoles are indeed present in the fowl but it has not been discovered whether they have similar effects and modes of action.

Although the hypothalamo-hypophyseal site of action of pineal hormones has been investigated most often, additional sites of action have been proposed (Quay, 1969). Thus, if the pineal acts on the hepatic metabolism of steroids, this could affect the feed-back control of certain hypophyseal functions. Alternatively, the pineal may control the homeostasis of cerebral metabolism and transport of potassium and sodium and thus affect the hypothalamo-hypophyseal systems.

Light has a most important effect on pineal function. Neglect of this factor has probably confused the interpretation of many pinealectomy experiments and, indeed, it may be necessary in both mammals and birds, as Sayler and Wolfson (1967) have shown in the quail, to keep the animals under either a gonadal stimulatory or depressive lighting regimen in order to bring out effects of pinealectomy. Unlike the frog, in which Dodt and Heerd (1962) demonstrated electrical response to direct photic stimulus of the pineal, it is improbable for both anatomical and physiological reasons that light acts directly on the mammalian or avian pineal (Morita, 1966; Ralph and Dawson, 1968). The effect of light of these higher vertebrates is via the eyes, the brain and the superior sympathetic ganglia from which nerves run to the pineal. In the fowl there may be alternative modes of action of light on the pineal (Lauber *et al.*, 1968) and cholinergic nerves may be involved (Wight and MacKenzie, 1970) but no more details of the nervous pathways are known. In mammals, however, it has been demonstrated (Moore, 1969) that impulses from the retina leave the optic nerves immediately after the chiasma and travel via the inferior accessory optic tract and the medial forebrain bundle into the lateral hypothalamus. They terminate in the medial terminal nucleus of the inferior accessory optic tract, near the junction of the hypothalamus and mesencephalon, from where they are presumed to pass through the brain to the spinal cord and the sympathetic nerves.

The sympathetic nerves probably act on the pineal through release of noradrenaline. If noradrenaline is added to the medium in cultured rat pineals, melatonin production is increased (Axelrod *et al.*, 1968). Noradrenaline may act by stimulating increased HIOMT production (Axelrod *et al.*, 1968) or by increasing the formation of the enzyme which acetylates serotonin (Klein *et al.*, 1970).

In the fowl, there is both morphological and biochemical evidence of the effect of light. The amount of cytoplasm of the pinealocytes increases in chickens reared during long photoperiods; the cells contain increased numbers of small granules (70 nm diameter), lysosomes and lipid droplets (Fujie, 1968). Cored synaptic vesicles in the unmyelinated pineal nerves were thought to decrease in number. Similarly, when fowls were kept in darkness, McFarland *et al.* (1969) found the volume of the pinealocytes and the amount of lipid were both decreased. Light increases the amount of pineal HIOMT in the fowl (Axelrod *et al.*, 1964; Lauber *et al.*, 1968). In the pigeon, the daily rhythm of pineal 5-hydroxytryptophan can be influenced by altering the lighting regimen (Quay, 1966).

Some of the effects of light may be mediated neither through the retina nor directly on the pineal. Lauber *et al.* (1968) have shown that the pineal still responds to photo alterations in sympathectomized, enucleated fowls. Light can pass through the wall of the orbital cavities of enucleated ducks to stimulate the hypothalamus (Benoit and Assenmacher, 1959) and the latter in turn could affect the pineal through hormonal feed-back mechanisms. The lipid content of rodent pineals can be modified by the experimental administration of oestrodiol and gonadotropins (Zweens, 1965).

Morphological changes have been observed in the pineal at different stages of the sexual cycle of rodents (Zweens, 1965) and certain birds including the fowl (Desogus, 1928) although in the latter the changes are not clear cut (Milcu *et al.*, 1964a). In male ducks, radioactive phosphorus uptake is greatest when the testes are least active (Milcu *et al.*, 1964b). In mammals it seems probable that the pineal may affect oestrus (Wurtman *et al.*, 1968). On the other hand, it apparently does not participate in the "Zeitgeber" for ovulation in the rat (Alleva *et al.*, 1970), nor is it required for the photocontrol of egg laying in the fowl (Harrison and Becker, 1969).

In addition to its association with the reproductive system, the pineal may be involved in the function of other endocrine glands. Farrell (1959) claimed to have identified an aldosterone-releasing hormone in bovine pineal extracts which he called adrenoglomerulotropin. Several aspects of this work have been adversely criticized but other evidence from mammals suggests that the possibility of a relationship between the pineal, adrenal and electrolyte balance should not be ignored (*cf.* Wurtman *et al.*, 1968). In the fowl Singh and Turner (1967) found that injections of melatonin decreased adrenal weight.

Several mammalian studies (*cf.* Wurtman *et al.*, 1968) have shown that melatonin treatment decreases thyroid cell-size, ^{131}I uptake, and secretion of thyroid hormone. Information about connexions between the pineal and thyroid of thc fowl is lacking.

The pineal shows both truly endogenous (circadian) and exogenous 24-h rhythms in the bird. For example, a rhythmic production of serotonin has been identified in pigeons (Quay, 1966) and in quail melatonin shows a rhythmic increase during the dark part of the 24-h cycle (Ralph *et al.*, 1968). Despite this, there is little evidence that the pineal participates in other 24-h rhythms in birds. It is essential for the persistence of circadian locomotor rhythms in the sparrow in which it is a crucial component of the endogenous time-measuring system (Gaston and Menaker, 1968). Intravenous or intraperitoneal injections of melatonin into chicks induces sleep (Barchas *et al.*, 1967) but there is no evidence that it is associated with the normal rhythm of sleep.

The mammalian pineal has also been associated with several less thoroughly investigated functions including hypertrophy of the parathyroid (Miline and Krstic, 1966), immunological phenomena (Csaba *et al.*, 1966), mast cells (Csaba *et al.*, 1968), and stress (Miline *et al.*, 1968).

References

Alleva, J. J., Waleski, Mary V. and Alleva, F. R. (1970). *Life Sci.* Pt. I **9**, 241–246.
Anton-Tay, F., Chou, C., Anton, S. and Wurtman, R. J. (1968). *Science, N.Y.* **162**, 277–278.
Ariëns Kappers, J. (1965). *Prog. Brain Res.* **10**, 87–153.
Axelrod, J. and Weissbach, H. (1960). *Science, N.Y.* **131**, 1312.
Axelrod, J. and Weissbach, H. (1961). *J. biol. Chem.* **236**, 211–213.
Axelrod, J., Wurtman, R. J. and Winget, C. M. (1964). *Nature, Lond.* **201**, 1134.
Axelrod, J., Shein, H. and Wurtman, R. J. (1968). Cited by Wurtman, R. J., Axelrod, J. and Kelly, D. E. (1969). "The Pineal" p. 59. Academic Press, New York.
Barchas, J., Da Costa, F. and Spector, S. (1967). *Nature, Lond.* **214**, 919–920.
Badertscher, J. A. (1924). *Anat. Rec.* **28**, 177–197.
Beattie, C. W. and Glenny, F. H. (1966). *Anat. Anz.* **118**, 396–404.
Benoit, J. and Assenmacher, I. (1959). *Recent Prog. Horm. Res.* **15**, 143–164.
Bern, H. A. and Knowles, F. G. W. (1966). In "Neuroendocrinology" (L. Martini and W. F. Ganong, eds.), 1, 140–142, Academic Press, London.
Bischoff, M. B. and Richter, W. R. (1966). Proc. VIth Int. Congr. Electron Microscopy, vol. 2, pp. 523–534. Maruzan Co., Ltd., Tokyo.
Bischoff, M. B. (1969). *J. Ultrastruct. Res.* **28**, 16–20.
Börell, V. and Orström, A. (1945). *Acta physiol. scand.* **10**, 231–242.
Collin, J.-P. (1966). *C. r. hebd. Séanc. Soc. Biol.* **160**, 1876–1880.
Collin, J-P. (1969). *C. r. hebd. Séanc. Soc. Biol.* **163**, 1137–1142.
Csaba, G., Bodoky, M., Fischer, J. and Acs, T. (1966). *Experientia* **22**, 168–169.
Csaba, G., Dunay, C., Fischer, J. and Bodoky, M. (1968). *Acta anat.* **71**, 565–580.
Del Rio-Hortega, P. (1932). *In* "Cytology and Cellular Pathology of the Nervous System". (W. Penfield, ed.) Vol. 2, p. 668. Hoeber, New York.
Desogus, V. (1928). *Monitore zool. ital.* **39**, 58–71.
Dodt, E. and Heerd, E. (1962). *J. Neurophysiol.* **25**, 405–429.
Farrell, G. (1959). *Endocrinology* **65**, 239–241.
Fujie, E. (1968). *Archvm histol. jap.* **29**, 271–303.
Gaston, S. and Menaker, M. (1968). *Science, N.Y.* **160**, 1125–1127.
Gladstone, R. J. and Wakeley, C. P. G. (1940). "The Pineal Organ". Baillière, Tindall and Cox, London.
Goldman, H. and Wurtman, R. J. (1964). *Nature, Lond.* **203**, 87–88.
Gutzeit, R. (1896). Cited by Del Rio-Hortega, P. in "Cytology and Cellular Pathology of the Nervous System". (W. Penfield, ed.), Vol. 2, 638. Hoeber, New York.

Harrison, P. C. and Becker, W. C. (1969). *Proc. Soc. exp. Biol. Med.* **132**, 161–164.
Henle, F. G. J. (1871). Cited by Del Rio-Hortega, P. (1932) *In* "Cytology and Cellular Pathology of the Nervous System" (W. Penfield, ed.), Vol. 2, p. 642. Hoeber, New York.
Höhn, E. O. (1961). *In* "Biology and Comparative Physiology of Birds" (A. J. Marshall ed.), Vol. 2, p. 112. Academic Press, New York.
Homma, K., McFarland, L. Z. and Wilson, W. D. (1967). *Poult. Sci.* **46**, 314–319.
Kitay, J. I. and Altschule, M. D. (1954). "The Pineal Gland: A Review of the Physiologic Literature". University Press, Harvard.
Klein, D. C., Berg, G. R. and Weller, J. (1970). *Science, N.Y.* **168**, 979–980.
Kopin, I. J., Pare, C. M. B., Axelrod, J. and Weissbach, H. (1961). *J. biol. Chem.* **236**, 3072–3075.
Krabbe, K. H. (1955). *J. comp. Neurol.* **103**, 139–149.
Lauber, J. K., Boyd, J. E. and Axelrod, J. (1968). *Science, N.Y.* **161**, 489–490.
Lerner, A. B., Case, J. D. and Takahashi, Y. (1960). *J. biol. Chem.* **235**, 1992–1997.
Lerner, A. B. and Wright, R. M. (1960). *Meth. biochem. Analysis* **8**, 295–307.
Machado, A. B. M., Faleiro, L. C. and Da Silva, W. D. (1965). *Z. Zellforsch. mikrosk. Anat.* **65**, 521–529.
Martini, L. (1969). *Gen. comp. Endocr.* Suppl. **2**, 214–226.
McCord, C. P. and Allen, F. P. (1917). *J. exp. Zool.* **23**, 207–224.
McFarland, L. Z., Wilson, W. O. and Winget, C. M. (1969). *Poult. Sci.* **48**, 903–907.
Milcu, St. M., Postelnicu, D. and Georgeta Ionescu. (1964a). *St. Cercet. Endocr.* **15**, 485–486.
Milcu, St. M., Postelnicu, D. and Juvina, E. (1964b). *C. r. hebd. Séanc. Soc. Biol.* **158**, 436–437.
Miline, R. and Krstic, R. (1966). *Z. Zellforsch. mikrosk. Anat.* **69**, 428–437.
Miline, R., Krstic, R. and Devečerski, V. (1968). *Acta anat.* **71**, 352–402.
Moore, R. Y. (1969). *Nature, Lond.* **222**, 781–782.
Morita, Y. (1966). *Experientia* **22**, 402.
Moszkowksa, A. (1963). *Annls Endocr.* **24**, 215–225.
Motta, M., Fraschini, F. and Martini, L. (1967). *Proc. Soc. exp. Biol. Med.* **126**, 431–435.
Oksche, A. and Vaupel-von Harnack, M. (1965a). *Naturwissenschaften* **52**, 662–663.
Oksche, A. and Vaupel-von Harnack, M. (1965b). *Prog. Brain Res.* **10**, 237–258.
Oksche, A. and Vaupel-von Harnack, M. (1966). *Z. Zellforsch. mikrosk. Anat.* **69**, 41–60.
Oksche, A., Morita, Y. and Vaupel-von Harnack, M. (1969). *Z. Zellforsch. mikrosk. Anat.* **102**, 1–30.
Pearse, A. G. E. (1960). "Histochemistry: Theoretical and Applied". 2nd Ed. Churchill, London.
Prop, N. (1965). *Prog. Brain Res.* **10**, 454–464.
Quay, W. B. (1965a). *Prog. Brain Res.* **10**, 49–86.
Quay, W. B. (1965b). *Life Sci.* **4**, 983–991.
Quay, W. B. (1965c). *Pharmac. Rev.* **17**, 321–345.
Quay, W. B. (1966). *Gen. comp. Endocr.* **6**, 371–377.
Quay, W. B. (1969). *Gen. comp. Endocr.* suppl. **2**, 214–226.
Quay, W. B. and Bagnara, J. T. (1964). *Archs int. Pharmacodyn. Thèr.* **150**, 137–143.
Quay, W. B. and Renzoni, A. (1963). *Riv. Biol.* **56**, 363–407.
Quay, W. B. and Renzoni, A. (1967). *Riv. Biol.* **60**, 9–75.
Ralph, C. L. and Dawson, D. C. (1968). *Experientia* **24**, 147.

Ralph, C. L., Hedlund, L. and Murphy, W. A. (1968). *Comp. Biochem. Physiol.* **22**, 591–599.
Romieu, M. and Jullien, G. (1942). *C. r. hebd. Séanc. Soc. Biol.* **136**, 626–628.
Sayler, A. and Wolfson, A. (1967). *Science, N.Y.* **158**, 1478–1479.
Sayler, A. and Wolfson, A. (1968). *Endocrinology* **83**, 1237–1246.
Shellabarger, C. J. (1952). *Endocrinology* **51**, 152–154.
Shellabarger, C. J. (1953). *Poult. Sci.*, **32**, 189–197.
Singh, D. V. and Turner, C. W. (1967). *Proc. Soc. exp. Biol. Med.* **125**, 407–411.
Spiroff, B. E. N. (1953). *Ph.D. thesis, Northwestern University.*
Stalsberg, H. (1965). *Acta endocr., Copenh.*, suppl. **97**, 1–119.
Stammer, A. (1961). *Acta biol., Szeged* **7**, 65–75.
Stieda, L. (1869). *Z. wiss Zool.* **19**, 1–94.
Studnička, F. K. (1905). *In* "Lehrbuch der Vergleichenden Mikroskopischen Anatomie der Wirbeltiere" (A. Oppel, ed.), Vol. 5, pp. 210–220. Fischer, Jena.
Sturkie, P. D. (1965). "Avian Physiology" 2nd Ed. p. 657. Baillière, Tindall and Cassel, London.
Thillard, M. J. (1968). *C. r. hebd. Séanc. Soc. Biol.* **162**, 1074–1080.
Urechia, C.-I. and Grigoriu, C. (1922). *C. r. hebd. Séanc. Soc. Biol.* **87**, 815–816.
van de Veerdonk, F. C. G. (1965). *Nature, Lond.* **208**, 1324–1325.
Wartenberg, H. and Baumgarten, H. G. (1969). *Z. Anat. EntwGesch.* **128**, 185–210.
Welsh, J. H. (1964). In "Comparative Neurochemistry" (D. Richter, ed.), p. 355. Pergamon Press, Oxford.
Wight, P. A. L. and MacKenzie, G. M. (1971). *J. Anat.* **108**, 261–273
Wight, P. A. L. and MacKenzie, G. M. (1970). *Nature, Lond.* **228**, 474–475.
Wurtman, R. J., Axelrod, J. and Potter, L. T. (1964). *J. Pharmac. exp. Ther.* **143**, 314–318.
Wurtman, R. J. and Axelrod, J. (1966). Cited by Wurtman, R. J., Axelrod, J. and Kelly, D. E. (1968). "The Pineal." Academic Press, New York.
Wurtman, R. J., Axelrod, J. and Kelly, D. E. (1968). "The Pineal." Academic Press, New York.
Wurtman, R. J. and Anton-Tay, F. (1969). *Recent Prog. Horm. Res.* **25**, 493–522.
Zweens, J. (1965). *Prog. Brain Res.* **10**, 540–551.

23 The Endocrine Status of the Bursa of Fabricius and the Thymus Gland

B. M. FREEMAN

Houghton Poultry Research Station,
Houghton,
Huntingdon, England

I. Introduction

The functions of the bursa of Fabricius, named after Hieronymus Fabricius who first described the gland in 1621, and of the thymus gland have intrigued generations of research workers. Whilst their anatomy and structures have been known for many years it has only been in the last decade or so that some of their functions have been elucidated. Yet the question of their being endocrine glands still remains something of an enigma. In the following the evidence for considering them to be endocrine glands is assessed.

II. The Bursa of Fabricius

A. INTRODUCTION

The bursa of Fabricius is a lympho-epithelial gland unique to the class Aves (see also Chapter 45). It has been suggested, however, that the mammalian vermiform appendix or Peyer's patches may be its homologue

(Archer *et al.*, 1963; Cooper *et al.*, 1966a; Cooper *et al.*, 1967; Fichtelius, 1967; Fichtelius *et al.*, 1968). The importance of the bursa in conferring immunological competence on the fowl has been recognized for some time (Chang *et al.*, 1955; Glick *et al.*, 1956) but only recently has it been found that the immunological reactions of the bird can be classified as being either bursa- or thymus-dependent (Szenberg and Warner, 1962; 1964; Cooper *et al.*, 1965; Cooper *et al.*, 1966b).

The concept of the bursa of Fabricius as an endocrine gland has thus developed during the last decade although even now the evidence is not conclusive. At present it seems likely that the bursa produces secretions concerned with immunological competence, the activity of the adrenal "cortex" and possibly the thyroid.

B. HORMONAL SUPPRESSION AND SURGICAL REMOVAL OF THE BURSA

1. *Hormonal Suppression*

The bursa is sensitive to steroid hormones. Cortisone, deoxycorticosterone, 19-nortestosterone and 17-ethyl-19-testosterone all cause a marked involution of the bursa (Meyer *et al.*, 1959; Glick, 1960; Aspinall *et al.*, 1961, 1963; Mueller *et al.*, 1962; Raos *et al.*, 1962). This reaction has been applied to the embryo *in ovo* where the development of the bursa is suppressed or severely curtailed. Complete suppression may be obtained either by injecting a sterile suspension of 0·63 mg of 19-nortestosterone into the allantoic cavity of the embryo on the 5th day of incubation (Meyer *et al.*, 1959) or by dipping the fertile egg in a solution of the same (Glick and Sadler, 1961). The main disadvantage of this technique is that there is often an impairment in the chick's ability to void faecal material from the cloaca resulting in a very poor growth rate and often in death.

In an attempt to overcome this objection to the technique some workers have delayed treatment until the 12th or 13th day of incubation. Certainly the quality of the chick is superior but the possibility that some development of the bursa, or that bursal cell migration occurs, cannot be ruled out.

2. *Surgical Bursectomy*

This technique is simple and has been used with much success in the 1-day-old chick and even in the mature embryo (Cain *et al.*, 1968; Alten *et al.*, 1968). The technical details are given by Mueller *et al.* (1960). Mortality from surgical bursectomy is usually low and efficient functioning of the cloaca is unimpaired. The major disadvantage of the technique is that limited regeneration may occur (Ortega and Der, 1964) and cautery of the wound's surface is recommended. Furthermore, by delaying surgery until after hatching, there is a very real possibility of the transfer of bursal cells to other sites such as the thymus or spleen. It is therefore usual to supplement this treatment with whole-body X-irradiation; the usual dose is in the order of 600–

700 R for the 1-day-old chick (Cooper *et al.*, 1965; Cooper *et al.*, 1966a, b; Greaves *et al.*, 1968).

C. THE BURSA AND IMMUNOLOGICAL COMPETENCE

It has been shown many times that bursectomy leads to a greatly impaired ability to produce circulating antibody to a primary antigenic stimulus although some immunoglobulin can still be synthesized (Janković and Isaković, 1966; Claflin *et al.*, 1966). In its most extreme form, the bird may be agammaglobulinaemic (Meter *et al.*, 1969; Warner *et al.*, 1969; Perey *et al.*, 1970). At present at least three theories have been advanced to explain the role of the bursa in the immune response: (1) the bursa provides stem cells which migrate to peripheral sites and develop into immunologically competent cells; (2) the bursa provides a possibly unique environment for the maturation of immunologically competent cells; (3) the bursa secretes a hormone which stimulates the maturation of such cells in peripheral lymphatic sites. Certainly, a migration of bursal cells to the thymus does occur (Woods and Linna, 1965) and the finding that impairment to immunoglobulin synthesis by bursectomy is progressively reduced with the age of the bird (Chang *et al.*, 1957; Mueller *et al.*, 1962; Graetzer *et al.*, 1963a) would support the first hypothesis.

Jaffé and Fechheimer (1966) noted that, during embryonic development, cells of unknown origin do migrate into the bursa but further support for the second hypothesis is lacking. It was Glick (1958, 1960) who provided the first evidence that extracts of bursae might restore immunological competence to bursectomized chickens.

Recently, more substantive evidence seems to strengthen the view that the bursa of Fabricius is an endocrine gland. St. Pierre and Ackerman (1965) and Janković and Leskowitz (1965) reported independently that implantation of fragments of bursa, contained in diffusion chambers, into bursectomized birds led to a significant restoration of their immunological capabilities.

Three objections remain to this experimental technique. One is that antigen can diffuse into the chamber and so stimulate synthesis and release of immunoglobulins. Whilst it had been suggested that antibody synthesis does not occur within the bursa it now seems likely that it does occur (Binet *et al.*, 1964; Thorbecke *et al.*, 1968). The second objection is that, unless the system is maintained completely sterile during implantation, bacterial products may act as an adjuvant (Dent and Peterson, 1967; Dent *et al.*, 1968). Thus, when contaminated pieces of bursae from hatched chicks were used, the antibody response of bursectomized birds was enhanced but when the bursal samples were sterile (19-day-old embryos providing the material) no enhancement was noted. The third objection concerns the possibility of cell migration from the diffusion chamber itself. This aspect has been reviewed by Davies (1969).

Currently, attempts are being made to meet objections to the use of diffusion chambers including use of cell-free bursal extracts. Janković *et al.* (1967) used lipid and protein fractions of the bursa and found a significant enhancement of antibody production in bursectomized birds. They do not consider that the material is acting as an adjuvant but rather that their results are consistent with release of a bursal hormone. Similar results have been reported by May *et al.* (1967).

Whilst the evidence for a bursal hormone concerned with humoral immunological competence is still inconclusive, it must be allowed that it is very circumstantial. Indeed, one group of workers has proposed a quite detailed hypothesis—

> "... the bursal epithelial cells begin to synthesize a hormone around 16 to 18 d of embryonic life, which has a function analogous to erythropoietin. This hormone is initially released locally in small amounts, perhaps by cell–cell contact, and under its influence the immigrant stem-cells differentiate into early plasma blasts.... Alternatively, hormone-synthesizing cells (or even free hormone) leave the bursa and induce differentiation of the stem-cells in the spleen and other peripheral lumphoid tissues." (Thorbecke *et al.*, 1968.)

Whether the hormone postulated by these workers is at least analogous to the "lymphokines" recently described by Dumonde *et al.* (1969) requires further work.

D. THE BURSA AND THE ADRENAL GLAND

Modification of the response of the adrenal gland to stress following bursectomy was first described by Perek and Eilat (1960). Attempts to deplete the adrenal stores of ascorbic acid by treating the intact immature chicken with adrenocorticotrophic hormone (ACTH) has failed (Jailer and Boas, 1950; Elton and Zarrow, 1955; Howard and Constable, 1958; Elton *et al.*, 1959; Perek and Eckstein, 1959; Perek *et al.*, 1959; Perek and Eilat, 1960; Wolford and Ringer, 1962; Mazina, 1964; Freeman *et al.*, 1966), presumably because of the influence of the bursa.

The response of the adrenal to stressful stimuli of varying intensity has been shown to be somewhat more complex than hitherto thought. In Fig. 1 the changes in adrenal ascorbic acid following relatively small stimuli (handling) or relatively large stimuli (handling plus exogenous ACTH) are shown together with the modifying influence of the bursa. From his results Freeman (1967, 1969, 1970) Freeman (1970) has advanced the hypothesis that the bursa of Fabricius produces a factor (hormone) which facilitates the maturation or perhaps the functioning of the adrenal's ascorbic acid repletion mechanism. Thus Freeman (1970) postulates that, when the stimulus is large, repletion and depletion go on side by side in the bird, in contrast to the mammal (Sharma *et al.*, 1964). In the intact mature bird this results in no net reduction in adrenal ascorbic acid concentration but in the bursectomized bird, with the repletion mechanism impaired, the adrenal ascorbic acid becomes depleted. The existence of a hormone must remain hypothetical.

Radoiu *et al.* (1969) have shown, in rabbits, that sterile, cell-free extracts of lymph nodes impair the production of adrenal cortical hormones. At the same time, Freeman *et al.* (1966) and Freeman (1969) have shown that the

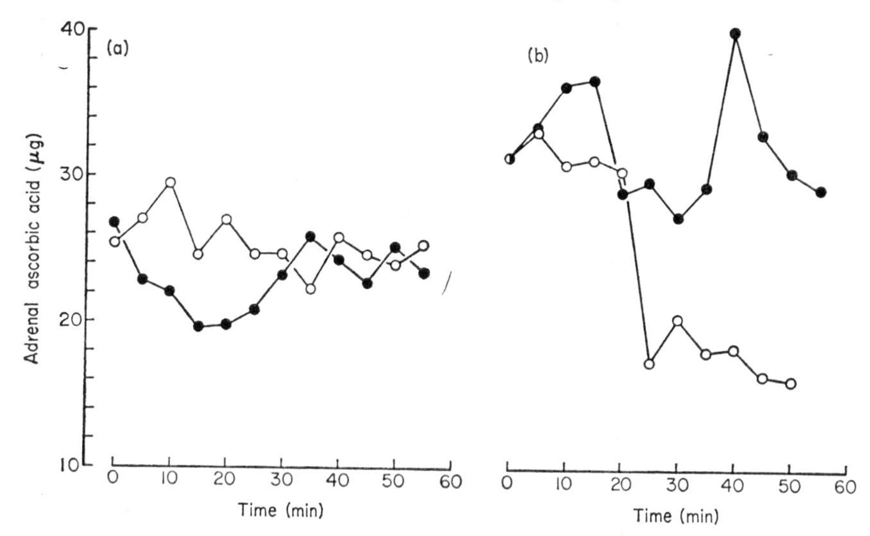

Fig. 1. Changes in the ascorbic acid content of the adrenal glands of the normal and bursectomized chicken in response to stress. For clarity the standard errors have been omitted. ● = Normal; ○ = bursectomized.

(a) Response to handling. The fall in the ascorbic acid content of the adrenals of normal birds was significant after 15 min. No significant changes in bursectomized birds were noted. (b) Response to handling and an interperitoneal injection of fast-acting ACTH (5 i.u./bird). There was a significant rise in normal birds between the 35th and 40th min. In the bursectomized birds there was a highly significant fall in adrenal ascorbic acid. Reproduced with permission from Freeman (1969, 1970).

bursa apparently has a similar action on the adrenal cortical tissue. Hyperglycaemia follows treatment by ACTH; the action of the ACTH is to stimulate the production of adrenal glucocorticoids. In the bursectomized bird the hyperglycaemia is significantly delayed when ACTH is given intraperitoneally, although the ultimate plasma glucose concentration in both normal and bursectomized chicks is similar—see Fig. 2. The response to an intramuscular dose of ACTH is somewhat different in that the immediate response is similar in both intact and bursectomized chickens; subsequently, the degree of hyperglycaemia increases only in the intact birds (Freeman *et al.*, 1966).

Recently, Freeman (1971) has shown that the action of the bursa on metabolism may be more fundamental than merely altering the rate of glucocorticoid secretion. The response to the hyperglycaemic action of adrenaline has been found to be approximately halved by bursectomy—see Fig. 3. Again, these experiments on carbohydrate mobilization suggest that the bursa is secreting a hormone.

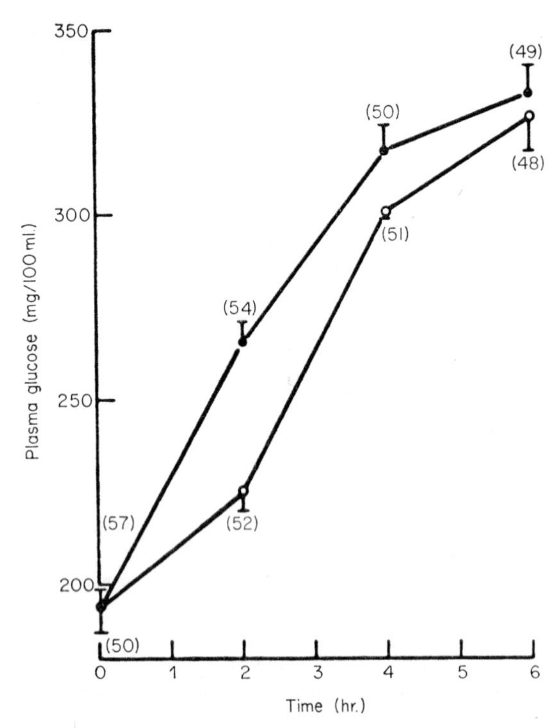

Fig. 2. Changes in the concentration of plasma glucose of normal and bursectomized chicks following treatment with ACTH. The differences 2 and 4 h after injection are significant. ● = Normal; ○ = bursectomized. Reproduced with permission from Freeman (1969).

E. THE BURSA AND THE THYROID

Pintea and Pethes (1967) have noted that the rate of ^{131}I uptake is depressed by bursectomy. The rate of thyroid hormone synthesis is also lower. Unfortunately no measurements have yet been made to determine whether any differences exist in the metabolic rate of the bursectomized chick as compared with its intact control. It is possible that a lowered metabolic rate might explain the differences noted in the glycaemic activity of adrenaline. No differences in body temperature have been noted (B. M. Freeman, unpublished observations).

F. THE BURSA AND ERYTHROPOIESIS

Alm and Peterson (1971) have reported that bursectomized-X-irradiated chicks develop a severe anaemia compared with X-irradiated intact controls. Mean corpuscular haemoglobin concentration is also depressed. These authors suggested that the bursa may produce a hormonal factor which influences erythropoiesis.

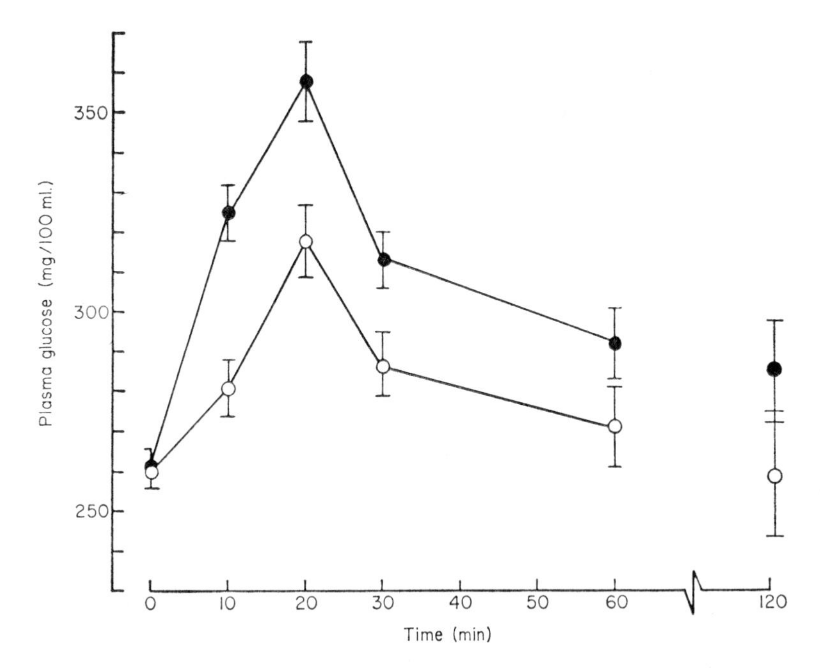

Fig. 3. The glycaemic response of normal and bursectomized chicks to adrenaline (300 μg/kg). The difference in response is highly significant. ● = Normal; ○ = bursectomized. Reproduced with permission from Freeman (1971).

G. THE ENDOCRINE STATUS OF THE BURSA OF FABRICIUS

Although the existence of a bursal hormone or hormones has yet to be conclusively demonstrated, the balance of evidence is now such that the bursa of Fabricius should be considered as an endocrine gland.

At present it would seem most reasonable to consider that a hormone is produced and of a type that may be classified as developmental, that is, concerned with the maturation of various tissues. Thus, whilst the existence of the bursa is not essential to life, it does seem to accelerate the maturation of tissues. In a chromatographic study of the bursa, Keméný *et al.* (1968) have found a substance (designated bF_3) present only in newly-hatched chicks. Its structure and function remain to be elucidated but it is nevertheless interesting to speculate that this substance might be the hormone itself.

III. The Thymus

A. INTRODUCTION

The unique nature of the bursa of Fabricius has ensured that a large literature should amass around it, once some indication of its function was found. The avian thymus might have been expected to attract a similar amount of attention once it was demonstrated that the immunological reactions of the animal could be classified as being bursa- or thymus-dependent. However, this has not been the case and, as a consequence, the status of the avian thymus as an endocrine gland is still equivocal.

B. TECHNIQUES OF THYMECTOMY

Although there is a significant reduction in the size of the thymus following treatment of the embryo either by injection or dipping the egg with testosterone or related steroids (Papermaster *et al.*, 1962; Pierce and Long, 1965) the thymus is many times less sensitive than the bursa. Complete atrophy of the thymic cortex is normally found in no more than 10% of hormone-treated eggs (Warner and Szenberg, 1962) and, in consequence suppression of the thymus *in ovo*, whilst desirable, cannot be effected by these means.

It is generally agreed that surgery offers the best chance of complete extirpation but there are several complicating factors which must be considered. Surgical thymectomy should be carried out as soon as possible after hatching in order that "seeding" can be either prevented or reduced to a minimum. Furthermore, it should be appreciated that removal of the caudal lobes of the thymus can be difficult owing to their proximity to the thyroid/parathyroid complex. For this reason a thorough examination of the bird should be carried out *post mortem* in order that its exact status can be determined.

The surgical technique has been described by Peterson *et al.* (1964) and Aspinall *et al.* (1963). Many now supplement surgery with whole body irradiation. The dose usually given to 1-day-old chicks is in the region of 600 to 700 R/bird (Cooper *et al.*, 1965, 1966b; Greaves *et al.*, 1968).

A final complication should be mentioned. There seems a distinct possibility that other lymphoid tissue in the bird may be able to function in lieu of the thymus (White and Goldstein, 1968). The use of irradiation may help to reduce the activity of such tissue.

C. THE THYMUS AND IMMUNOLOGICAL REACTIONS

1. *Effects of Neonatal Thymectomy on Immological Function*

The wasting syndrome often found in thymectomized mammals (see Miller and Osoba, 1967) has not been encountered in thymectomized chickens although a marked depression of growth normally occurs (Tettenborn, 1965; Cooper *et al.*, 1966b).

Following neonatal thymectomy, there is a significant fall in the number of circulating small lymphocytes (Warner and Szenberg, 1962; Isaković and Janković, 1964; Janković and Isaković, 1964; Cooper *et al.*, 1965, 1966b). However, there is no change in the number of plasma cells in the spleen (Warner and Szenberg, 1962; Isaković and Janković, 1964; Janković and Isaković, 1964). Where thymectomy is accompanied by whole-body irradiation, plasma cell depletion is particularly marked.

2. *Effect of Thymectomy on Cellular Immunity*

Following thymectomy there is a marked decrease in the ability of the chick to reject foreign tissue (the graft versus host reaction) Warner and Szenberg, 1962; Janković *et al.*, 1963; Aspinall *et al.*, 1963; Vojtiskova *et al.*, 1963; Janković and Isaković, 1964; Vojtiskova and Nouza, 1965; Cooper *et al.*, 1966). There is also an impairment in the ability to develop delayed hypersensitivity reactions as measured by the tuberculin reaction (Janković *et al.*, 1965; Cooper *et al.*, 1966b; Szenberg and Warner, 1967).

At the same time, in those diseases thought to induce production of cellular antibodies, the thymectomized chicken has been found to be less resistant. Pierce and Long (1965) found that oocyst production from birds infected with *Eimeria tenella* was increased while Radzichovskaja (1967) found that the growth of Rous sarcoma virus was enhanced.

3. *Effect of Thymectomy on Humoral Immunity*

Although there are many reports that neonatal thymectomy does not have any consistent effect on the ability of the chick to produce immunoglobulins (Warner and Szenberg, 1962; Graetzer *et al.*, 1963b) there is some evidence that humoral immune responsiveness to some antigens is reduced (Cooper *et al.*, 1965, 1966b; Chaperon, 1966), particularly in later life.

4. *The Endocrine Status of the Thymus in Immunological Reactions*

The immunological consequences of neonatal thymectomy are very similar in both mammals and birds (see Miller and Osoba, 1967). This fact is crucial in discussing the possible endocrine nature of the avian thymus since work on this aspect of the bird *per se* is extremely limited and to assume that reactions in the mammal probably occur in the bird is dangerous.

However, the first stage in assessing the function of the thymus has been completed. Its extirpation leads to considerable changes in the immunological reactivity of the bird although the changes may not be as quantitatively marked as they are in the mammal. "Seeding" remains a very real problem and it would perhaps be worthwhile to carry out work of this kind on the altricial (immature) rather than the precocial or mature avian species like the domestic fowl.

Replacement therapy, either as whole tissue grafts or as tissue extracts, has not yet been critically examined in the fowl. There is, however, a large literature concerning the mammal. This has been recently reviewed by Miller and Osoba (1967) and by White and Goldstein (1968). Only one paper has been located on the effect of thymic extracts on the bird. Rehn (1940) reported the successful induction of lymphocytosis in the pigeon using an oily extract of thymus. It is perhaps therefore only a matter of time before the rather more sophisticated extraction procedures as used in mammalian studies by Goldstein *et al.* (1966) Comsa (1966), Trainin *et al.* (1966, 1967) and Hand *et al.* (1967) will be applied to the fowl.

Again, the recent work describing cell-free factors such as the lymphokines (Dumonde *et al.*, 1969) adds weight to the concept of a hormone being secreted.

D. EFFECT OF THYMECTOMY ON METABOLIC PROCESSES

Thymectomy has been reported to influence at least one reaction other than the immunological reaction of the bird. This is the metabolism of calcium (Georgievsky, 1961) although the work has not been confirmed. Indeed, earlier work by Maughan (1938) suggests that calcium metabolism is not affected. However, it is interesting to find that calcium metabolism may be affected by the thymus, for such a relationship was noted by Bracci (1905) working on the rabbit, and more recently by Schwarz *et al.* (1953) with man and dog, although the finding that accessory parathyroid tissue can be present in the caudal lobes of the thymus (Nonidez and Goodale, 1927) suggests that the results of Georgievsky (1961) should be interpreted with care.

The thymus has also been claimed to modify carbohydrate metabolism (Baggio *et al.*, 1965; Bonifaci *et al.*, 1965; Fassetto *et al.*, 1965; Gaburro *et al.*, 1965a, b, c) and the metabolism of the thyroid (Rattini *et al.*, 1966). At the same time it has been claimed that differing thymic extracts affect differing metabolic pathways. Thus lipid fractions obtained by the use of non-polar solvents have been shown to have lymphocytopoietic properties in mammals (Bomskov and Sladovic, 1940; Nakamoto, 1957; Janković *et al.*, 1965); histone fractions have also been shown to have lymphocytopoietic properties (Hand *et al.*, 1967) as have glycoprotein and polypeptide fractions (Cosma, 1966; Goldstein *et al.*, 1966).

E. THE ENDOCRINE STATUS OF THE THYMUS GLAND

As has already been pointed out the avian thymus has received considerably less attention than the bursa of Fabricius. However, from the limited data available, it can now be stated that ablation of the thymus does bring about a profound change in the immunological activity of the fowl, and may also influence calcium metabolism. That its influence is expressed through a humoral factor or factors is less certain although comparisons with mammals suggest that this is now a distinct possibility.

References

Alm, G. V. and Peterson, R. D. A. (1971). *Nature, Lond.* **229**, 201–202.

Alten, J. van, Cain, W. A., Good, R. A. and Cooper, M. D. (1968). *Nature, Lond.* **217**, 358–360.

Archer, O. K., Sutherland, D. E. R. and Good, R. A. (1963). *Nature, Lond.* **200**, 337–339.

Aspinall, R. L., Meyer, R. K., Graetzer, M. A. and Wolfe, H. E. (1963). *J. Immun.* **90**, 872–877.

Aspinall, R. L., Meyer, R. K. and Rao, M. A. (1961). *Endocrinology* **68**, 944–949.

Baggio, P., Fassetto, G., Volpato, S. and Gravina, E. (1965). *Boll. Soc. ital. Biol. sper.* **41**, 76–79.
Binet, J. L., Pouliquen, Y., Meshaka, G. and Hurez, D. (1964). *Ann. N.Y. Acad. Sci.* **120**, 162–170.
Bomskov, C. and Sladovic, L. (1940). *Deut. med. Wschr.* **66**, 589–594.
Bonifaci, E., Baggio, P., Rattini, F. M. and Volpato, S. (1965). *Boll. Soc. ital. Biol. sper.* **41**, 210–212.
Bracci, C. (1905). *Riv. clin. pediat.* **3**, 572.
Cain, W. A., Alten, P. J. van, Good, R. A. and Cooper, M. D. (1968). *Fedn Proc. Fedn Am. Socs exp. Biol.* **27**, 493.
Chang, T. S., Glick, B. and Winter, A. R. (1955). *Poult. Sci.* **34**, 1187.
Chang, T. S., Rheims, M. S. and Winter, A. R. (1957). *Poult. Sci.* **36**, 735–738.
Chaperon, E. C. (1966). *Diss. Abstr.* **26**, 5603.
Claflin, A. J., Smithies, O. and Meyer, R. K. (1966). *J. Immun.* **97**, 693–699.
Cooper, M. D., Perey, D. Y., McKneally, M. F., Gabrielsen, A. E., Sutherland, D. E. R. and Good, R. A. (1966a). *Lancet* **i**, 1388–1391.
Cooper, M. D., Peterson, R. D. A. and Good, R. A. (1965). *Nature, Lond.* **205**, 143–146.
Cooper, M. D., Peterson, R. D. A., South, M. A. and Good, R. A. (1966b). *J. exp. Med.* **123**, 75–102.
Cooper, M. D., Perey, D. Y. E., Gabrielsen, A. E., Dent, P. B., Cain, W. A. and Good, R. A. (1967). *Fedn Proc. Fedn Am. Socs exp. Biol.* **26**, 752.
Comsa, J. (1966). *Arzneimittel-Forsch.* **16**, 18–22.
Davies, A. J. S. (1969). *Agents and Actions* **1** (2), 1–7.
Dent, P. B., Perey, D. Y. E., Cooper, M. D. and Good, R. A. (1968). *J. Immun.* **101**, 799–805.
Dent, P. B. and Peterson, R. D. A. (1967). *Fedn Proc. Fedn Am. Socs exp. Biol.* **26**, 621.
Dumonde, D. C., Wolstencroft, R. A., Panay, G. S., Matthew, M., Morley, J. and Howson, W. T. (1969). *Nature, Lond.* **224**, 38–42.
Elton, R. L., Zarrow, I. G. and Zarrow, M. X. (1959). *Endocrinology* **65**, 152–157.
Elton, R. L. and Zarrow, M. X. (1955). *Anat. Rec.* **122**, 473–474.
Fassetto, G., Bonifaci, E., Rattini, F. M. and Baggio, P. (1965). *Boll. Soc. ital. Biol. sper.* **41**, 207–210.
Fichtelius, K.-E. (1967). *Expl Cell Res.* **46**, 231–234.
Fichtelius, K.-E., Finstad, J. and Good, R. A. (1968). *Lab. Invest.* **19**, 339–351.
Freeman, B. M. (1967). *Comp. Biochem. Physiol.* **23**, 303–305.
Freeman, B. M. (1969). *Comp. Biochem. Physiol.* **29**, 639–646.
Freeman, B. M. (1970). *Comp. Biochem. Physiol.* **32**, 755–761.
Freeman, B. M. (1971) *J. Physiol., Lond.* **214**, 22*P*–23*P*.
Freeman, B. M., Chubb, L. G. and Pearson, A. W. (1966). *In* "Physiology of the Domestic Fowl" (C. Horton-Smith and E. C. Amoroso, eds.), pp. 103–112. Oliver and Boyd, Edinburgh.
Gaburro, D., Volpato, S., Baggio, P. and Bonifaci, E. (1965a). *Boll. Soc. ital. Biol. sper.* **41**, 79–82.
Gaburro, D., Volpato, S., Baggio, P. and Rattini, F. (1965b). *Boll. Soc. ital. Biol. sper.* **41**, 73–76.
Gaburro, D., Volpato, S., Bonifaci, E. and Rattini, F. M. (1965c). *Boll. Soc. ital. Biol. sper.* **41**, 204–207.
Georgievsky, V. I. (1961). *Izv. timiryazer. sel.'-khoz. Akad.* **4**, 113–126.
Glick, B. (1958). *Poult. Sci.* **37**, 240–241.
Glick, B. (1960). *Poult. Sci.* **39**, 1527–1533.

Glick, B., Chang, T. S. and Jaap, R. G. (1956). *Poult. Sci.* **35**, 224–225.
Glick, B. and Sadler, C. R. (1961). *Poult. Sci.* **40**, 185–189.
Goldstein, A. L., Slater, F. D. and White A. (1966). *Proc. natn. Acad. Sci. U.S.A.* **56**, 101.
Graetzer, M. A., Cote, W. P. and Wolfe, H. R. (1963a). *J. Immun.* **91**, 576–581.
Graetzer, M. A., Wolfe, H. R., Aspinall, R. L. and Meyer, R. K. (1963b). *J. Immun.* **90**, 878–887.
Greaves, M. F., Roitt, I. M. and Rose, M. E. (1968). *Nature, Lond.* **220**. 293–295.
Hand, T., Caster, P. and Luckey, T. D. (1967). *Biochem. biophys. Res. Commun.* **26**, 18–23.
Howard, A. N. and Constable, B. J. (1958). *Biochem. J.* **69**, 501–505.
Isaković, K. and Janković, B. D. (1964). *Int. Archs Allergy appl. Immun.* **24**, 296–310.
Jaffé, W. P. and Fechheimer, N. S. (1966). *Nature, Lond.* **212**, 92.
Jailer, J. W. and Boas, N. F. (1950). *Endocrinology* **46**, 314–318.
Janković, B. D. and Isaković, K. (1964). *Int. Archs Allergy appl. Immun.* **24**, 278–295.
Janković, B. D. and Isaković, K. (1966). *Nature, Lond.* **211**, 202–203.
Janković, B. D., Isaković, K. and Horvát, J. (1965). *Nature, Lond.* **208**, 356–357.
Janković, B. D., Isaković, K. and Horvát, J. (1967). *Experientia* **23**, 1062–1063.
Janković, B. D., Išvaneski, M., Milosević, D. and Popesković, L. (1963). *Nature, Lond.* **198**, 298–299.
Janković, B. D. and Leskowitz, S. (1965). *Proc. Soc. exp. Biol. Med.* **118**, 1164–1166.
Kemény, V., Péthes, G. and Kozma, M. (1968). *Comp. Biochem. Physiol.* **26**, 757–759.
Maughan, G. H. (1938). *Am. J. Physiol.* **123**, 319–325.
May, J. D., Hill, C. H. and Garren, H. W. (1967). *Fedn Proc. Fedn Am. Socs exp. Biol.* **26**, 769.
Mazina, T. I. (1964). *Byull. exsp. Biol. Med.* **57**, 54.
Meter, R. van, Good, R. A. and Cooper, M. D. (1969). *J. Immun.* **102**, 370–374.
Meyer, R. K., Rao, M. A. and Aspinall, R. L. (1959). *Endocrinology* **64**, 890–897.
Miller, J. F. A. P. and Osoba, D. (1967). *Physiol. Rev.* **47**, 437–520.
Mueller, A. P., Wolfe, H. R. and Meyer, R. K. (1960). *J. Immun.* **85**, 172–179.
Mueller, A. P., Wolfe, H. R., Meyer, R. K. and Aspinall, R. L. (1962). *J. Immun.* **88**, 354–360.
Nakamoto, A. (1957). *Acta haemat. jap.* **20**, 187–199.
Nonidez, J. F. and Goodale, H. D. (1927). *Am. J. Anat.* **38**, 319–341.
Ortega, L. G. and Der, B. K. (1964). *Fedn Proc. Fedn Am. Socs exp. Biol.* **23**, 546.
Papermaster, B. W., Griedman, D. I. and Good, R. A. (1962). *Proc. Soc. exp. Biol. Med.* **110**, 62–64.
Perek, M. and Eckstein, B. (1959). *Poult. Sci.* **38**, 996–999.
Perek, M., Eckstein, B. and Eshkol, Z. (1959). *Endocrinology* **64**, 831–832.
Perek, M. and Eilat, A. (1960). *J. Endocr.* **20**, 251–255.
Perey, D. Y. E., Dupuy, J. M. and Good, R. A. (1970). *Transplantation* **9**, 8–17.
Peterson, R. D. A., Burmester, B. R., Frederickson, T. N., Purchase, H. G. and Good, R. A. (1964). *J. natn. Cancer Inst.* **32**, 1343–1354.
Pierce, A. E. and Long, P. L. (1965). *Immunology* **9**, 427–439.
Pintea, V. and Pethes, G. (1967). *Sb. vys. Sk. zemed. les Fac. Brne.* **B36**, 449–452.
Radoiu, N., Zydeck, F. A. and Wolf, P. L. (1969). *Experientia* **25**, 643–645.
Radzichovskaja, R. (1967). *Proc. Soc. exp. Biol. Med.* **126**, 13–15.
Rao, M. A., Aspinall, R. L. and Meyer, R. K. (1962). *Endocrinology* **70**, 159–166.
Rattini, F. M., Fassetta, G., Bonifaci, E. and Baggio, P. (1966). *Boll. Soc. ital. Biol. sper.* **42**, 180–182.

Rehn, E. (1940). *Deut. med. Wschr.* **66**, 594–597.
Schwarz, H., Price, M. and Odell, C. A. (1953). *Metabolism* **2**, 261–267.
Sharma, S. K., Johnstone, R. M. and Quastel, J. H. (1964). *Biochem. J.* **92**, 564–573.
St. Pierre, R. L. and Ackerman, G. A. (1965). *Science, N.Y.* **147**, 1307–1308.
Szenberg, A. and Warner, N. L. (1962). *Nature, Lond.* **194**, 146–147.
Szenberg, A. and Warner, N. L. (1964). *Ann. N.Y. Acad. Sci.* **120**, 150–161.
Szenberg, A. and Warner, N. L. (1967). *Br. med. Bull.* **23**, 30–34.
Tettenborn, D. (1965). *Arch. Geflügelk.* **29**, 323–339.
Thorbecke, G. I., Warner, N. L., Hochwald, G. M. and Ohanian, S. H. (1968). *Immunology* **15**, 123–134.
Trainin, N., Bejerano, A., Strahilevitch, M. Goldring, D. and Small, M. (1966). *Israel J. med. Sci.* **2**, 549–559.
Trainin, N., Burger, M. and Kaye, A. (1967). *Biochem. Pharmac.* **16**, 711–720.
Vojtiskova, M., Masnerova, M. and Viklicky, J. (1963). *Folia Biol., Praha* **9**, 424–432.
Vojtiskova, M. and Nouza, K. (1965). *Folia Biol., Praha* **11**, 406–410.
Warner, N. L. and Szenberg, A. (1962). *Nature, Lond.* **196**, 784–785.
Warner, N. L., Uhr, J. W., Thorbecke, G. J. and Ovary, Z. (1969). *J. Immun.* **103**, 1317–1330.
White, A. and Goldstein, A. L. (1968). *Perspect. Biol. Med.* **11**, 475–489.
Wolford, J. H. and Ringer, R. K. (1962). *Poult. Sci.* **41**, 1521–1529.
Woods, R. and Linna, J. (1965). *Acta path. microbiol. scand.* **64**, 470–476.

24 Prostaglandins

E. W. HORTON

Department of Pharmacology,
The University Medical School,
Edinburgh, Scotland

I. Prostaglandins in Non-avian Species

A. INTRODUCTION

Fourteen prostaglandins and numerous metabolites have been isolated from animal sources. These acidic lipids are all derivatives of prostanoic acid (Fig. 1). Three principal groups of pharmacologically-active prostaglandins have

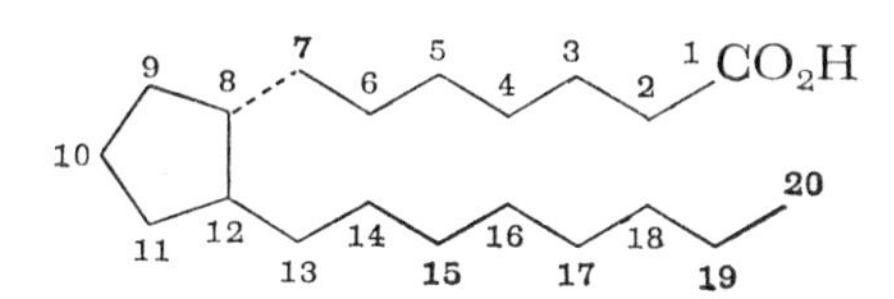

Fig. 1. Prostanoic acid.

been described, referred to as prostaglandins E, F and A respectively. The members of each group possess one or more double bonds, the compounds being designated E_1, E_2, E_3, etc. (Fig. 2).

Prostaglandins are synthesized in mammalian tissues from dihomo-γ-linolenic acid and arachidonic acid (Van Dorp *et al.*, 1964; Bergström *et al.*, 1964b; van Dorp, 1966). These acids in turn are derived from linoleic acid which is an essential constituent of the diet. Prostaglandins are metabolized in lungs, liver and kidneys; several metabolites have been isolated from mammalian urine. Details of these complex pathways and other information

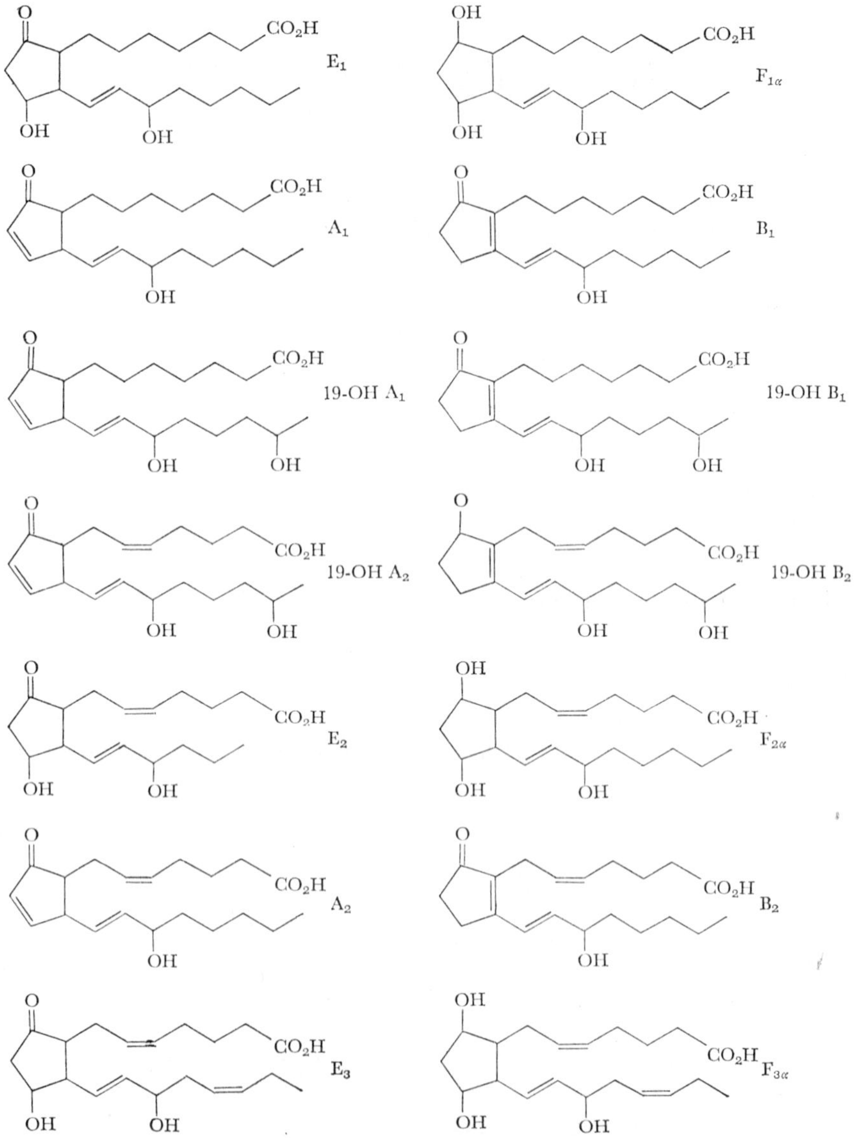

Fig. 2. Structural formulae of 14 naturally occurring prostaglandins.

concerning prostaglandins in non-avian species may be found in recent monographs and reviews (Bergström and Samuelsson, 1965, 1967; Bergström *et al.*, 1968; Pickles, 1967; von Euler and Eliasson, 1968; Ramwell and Shaw, 1968; Horton, 1968, 1969).

Prostaglandins are widely distributed in mammalian tissues, concentrations in human seminal plasma being exceptionally high (Table 1). Tissues from which one or more prostaglandins have been isolated include lung, brain, kidney, iris, pancreas, thymus and endometrium (see Horton, 1969 for references).

Table 1

Tissues from which prostaglandins have been isolated and identified by full structural elucidation

	E_1	E_2	E_3	$F_{1\alpha}$	$F_{2\alpha}$	$F_{3\alpha}$
Seminal plasma (human)	+	+	+	+	+	
Seminal plasma (sheep)	+	+	+	+	+	
Vesicular gland (sheep)	+	+	+	+		
Menstrual fluid (human)		+			+	
Lung (sheep)		+			+	
Lung (ox)					+	+
Lung (pig, guinea pig, monkey, human)					+	
Iris (sheep)					+	
Brain (ox)					+	
Thymus (calf)	+					
Renal medulla (rabbit)		+			+	

Prostaglandins are released from mammalian tissues in response to humoral, drug or nerve stimulation. There is spontaneous release from various parts of the central nervous system which may be increased by sensory nerve stimulation or by centrally-active drugs such as 5-hydroxytryptamine. Stimulation of the nerves to the rat diaphragm (Ramwell *et al.*, 1965), rat epididymal fat pad (Ramwell and Shaw, 1967), dog spleen (Davies *et al.*, 1968) and rat stomach (Bennett *et al.*, 1967; Shaw and Ramwell, 1968) all result in the increased output of one or more prostaglandins. Prostaglandins are released from lungs in response to the enzyme phospholipase A (Babilli and Vogt, 1964), from the rabbit eye on mechanical stimulation (Ambache *et al.*, 1965) and spontaneously from human medullary carcinoma of the thyroid (Williams *et al.*, 1968). A release into human uterine venous blood during the contractions of labour has also been reported (Karim, 1968). The significance of these observations has been discussed but remains obscure. Furthermore, the identification of the nanogram amounts of prostaglandins in these experiments has depended upon chromatographic and biological evidence; this is inconclusive. The observations require confirmation using techniques such as combined gas chromatography-mass spectrometry which provide more conclusive data (Thompson *et al.*, 1970).

B. PHARMACOLOGICAL ACTIONS

1. *Reproductive Tract*

There is evidence from both human and animal experiments that prostaglandins are absorbed from the vagina into the systemic circulation (Horton *et al.*, 1965; Eliasson and Posse, 1965; Sandberg *et al.*, 1967). The amounts reaching the circulation in this way after coitus may be sufficient to affect female reproductive smooth muscle since pregnant human myometrium *in vivo* responds to very small amounts of E_1 (Bygdeman *et al.*, 1967) and of $F_{2\alpha}$ (Karim *et al.*, 1968). Whether such a myometrial action in the non-pregnant woman is of importance in maintaining fertility is not known. There is evidence that some infertile men have low seminal prostaglandin levels especially in respect of the E group (Hawkins and Labrum, 1961; Bygdeman *et al.*, 1966; Samuelsson, 1968).

2. *Cardiovascular System*

In certain animals prostaglandins E_1 and $F_{1\alpha}$ have positive inotropic effects on the isolated heart. The predominant cardiovascular action of prostaglandins is however upon peripheral vessels. Prostaglandins E and A are potent dilators of arterioles thus causing a lowering of systemic arterial blood pressure on intravenous injection or infusion. Prostaglandin $F_{2\alpha}$ has the interesting pharmacological property of constricting veins and venules in rats and dogs (DuCharme and Weeks, 1967). This increases venous return and so increases cardiac output; as there is no change in peripheral resistance the systemic arterial blood pressure is raised by prostaglandin $F_{2\alpha}$ in these animals. A similar pressor effect of $F_{2\alpha}$ has been reported in the spinal chick (see below).

3. *Adipose Tissue*

Prostaglandin E_1 is a potent inhibitor of lipolysis induced by hormones or nerve stimulation (Steinberg *et al.*, 1963, 1964; Bergström *et al.*, 1964a; Shaw, 1966). The lipolytic mechanism is dependent upon the adenyl cyclase–cyclic 3′,5′-adenosine monophosphate system and there is some evidence that it is this common pathway which is blocked by prostaglandin. Furthermore since lipolytic hormones (and nerve stimulation) themselves release prostaglandins from adipose tissue, the possibility of a local negative feed-back mechanism can be envisaged. (See also Chapter 21.)

4. *Nervous System*

Prostaglandins, especially E_1 and $F_{2\alpha}$, have many sites of action on the mammalian central nervous system. On injection into the ventricular system of the brain prostaglandin E_1, but *not* $F_{2\alpha}$, causes sedation, stupor and catatonia. On intravenous injection in the decerebrate cat both prostaglandins potentiate decerebrate rigidity and cause contractions of extensor muscles by a central action (Horton, 1964; Horton and Main, 1965a, 1967a). Their discrete effect upon neurones in the medulla oblongata has been described by Avanzino *et al.* (1966). Both prostaglandins E_1 and $F_{2\alpha}$ either inhibit or facilitate spinal reflexes depending upon the conditions (Horton and Main, 1967a; Duda *et al.*, 1968), and prostaglandin E_1 on intravenous injection abolishes pentobarbitone tremor in the cat (Horton and Main, 1967b).

Numerous effects of prostaglandins E_1 and $F_{2\alpha}$ have been observed in mice. Both potentiate hexobarbitone sleeping time while E_1 causes ptosis and antagonizes the convulsant action of leptazol and maximal electroshock (Holmes and Horton, 1968a).

5. *Respiratory Smooth Muscle*

The bronchoconstrictor action of vagal stimulation or of histamine is reduced after an intravenous injection of prostaglandin E_1 in guinea-pigs and rabbits (Main, 1964). Respiratory smooth muscle *in vitro* is inhibited by prostaglandin E_1 and to a lesser extent by prostaglandin $F_{2\alpha}$ (Main, 1964; Horton and Main, 1965b). The observations have been made on tracheal muscle of several mammalian species, of which the cat is the most sensitive.

6. *Gastro-intestinal Tract*

Gastric-acid secretion in rats and dogs following pentagastrin or histamine is inhibited by prostaglandin E_1 (Shaw and Ramwell, 1968; Robert *et al.*, 1967). In the dog prostaglandins E_2 and A_1 are also very active in this respect. Although this inhibition has been observed in rats when prostaglandin E_1 is administered into the lumen of the stomach, no such inhibition of gastric secretion could be detected in experiments on human volunteers given prostaglandin E_1 orally (Horton *et al.*, 1968). The doses used (10 to 40 μg/kg) had a purgative action, an effect which was subsequently confirmed by quantitative measurements of human intestinal activity (Misiewicz *et al.*, 1969). References to the numerous *in vitro* studies of prostaglandins on gastro-intestinal smooth muscle can be found in the various reviews cited above.

7. *Platelet Aggregation*

Prostaglandins E_1 and E_2 have remarkable activity on the aggregation of platelets. Prostaglandin E_1 invariably inhibits aggregation induced by ADP, whereas in certain animals prostaglandin E_2 actually enhances platelet aggregation (Kloeze, 1967). No such action has yet been detected on platelets when prostaglandins are administered intravenously in man, although the aggregation of human platelets *in vitro* is inhibited by prostaglandin E_1.

8. *Kidney*

Both prostaglandins E_1 and A_1 increase urine flow and the excretion of sodium by the kidney in dogs (Herzog *et al.*, 1967). Both these prostaglandins are potent renal vasodilators and it is not clear to what extent the urinary changes may be secondary to these vascular effects. A role for prostaglandin in controlling either renal blood flow or sodium and water excretion has been postulated. The relationship of prostaglandins to the unidentified natriuretic hormone has not yet been elucidated (Lee and Ferguson, 1969).

II. Prostaglandins in the Domestic Fowl

A. OCCURRENCE

The natural distribution of prostaglandins in tissues of the domestic fowl has not been studied systematically. Substances with pharmacological and chromatographic properties identical to prostaglandins E_2 and $F_{2\alpha}$ have been isolated from both brain and spinal cord of adult fowls (Horton and Main, 1967c). It was estimated that both tissues contained not less than 100 ng/g of prostaglandin E_2 and approximately 10 ng/g prostaglandin $F_{2\alpha}$. These observations require confirmation with more recently developed methods which provide conclusive evidence of identification (Thompson *et al.*, 1970).

B. PHARMACOLOGICAL ACTIONS

1. *Central Nervous Actions*

Prostaglandin E_1, E_2 and E_3 in doses of 10 to 400 μg/kg injected intravenously cause sedation of rapid onset in the 3- to 5-d-old chick (Horton, 1964). There is a cessation of spontaneous movement, closure of the eyes and loss of the righting reflex (Fig. 3a and b). During this period chirping never occurs even when pressure is applied to a toe, although this produces immediate withdrawal. With prostaglandin E_1 the effects of 10 μg/kg last about 5 min and those of 200 μg/kg last about 75 min. There is always complete recovery, with resumption of normal posture, activity and chirping. When chicks were killed during maximum sedation, 2 min after an intravenous injection 2 μg of tritiated prostaglandin E_1 only 0·1 to 1·7% of the radioactivity could be recovered from the brain; the liver contained 5% and 85% of the injected activity was recovered from the remainder of the body (Holmes and Horton, 1968b). Even when the prostaglandin was injected towards the head via the common carotid artery, only 2% of the radioactivity could be extracted from the brain. Prostaglandins A_1, A_2 and B_1 have qualitatively similar actions but have less than 1% of the potency of prostaglandin E_1 (Jones, 1970).

In contrast, prostaglandin $F_{2\alpha}$ produces not sedation but extension and abduction of the legs and dorsiflexion of the neck (Fig. 3c and d) (Horton and Main, 1965a). The effect on the extensor muscles of the leg can be abolished by previous sectioning of the sciatic nerve (Fig. 4). Furthermore this action of $F_{2\alpha}$ can be elicited in the chick after mid-cervical cord transection and

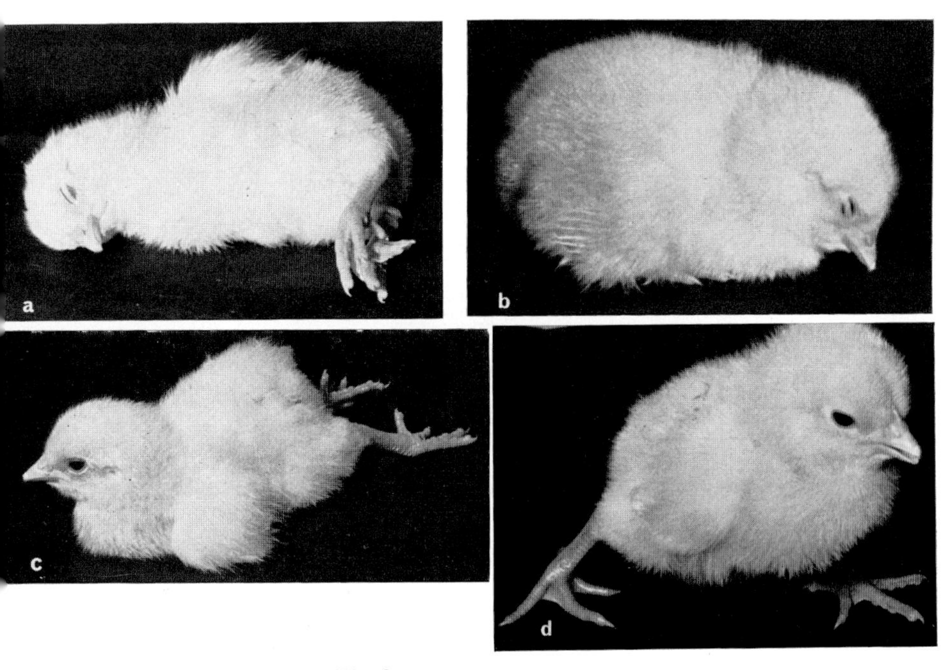

Fig. 3

(a) Chick, 39 g, photographed 1 min after an intravenous injection of 2 μg of prostaglandin E_1.
(b) Same chick 30 min later.
(c) Chick, 40 g, photographed 1 min after an intravenous injection of 4 μg of prostaglandin $F_{2\alpha}$.
(d) Same chick 22 min later.

Reproduced with permission from Horton and Main (1965a).

decapitation (the chick being ventilated artificially). It thus appears that $F_{2\alpha}$ acts directly upon the spinal cord in the chick though additional actions of $F_{2\alpha}$ on higher centres cannot be excluded (Horton and Main, 1967a). A spinal site of action has also been reported in experiments on non-avian preparations (Horton and Main, 1967a; Horton and Main, 1969; Duda *et al.*, 1968; Phillis and Tebecis, 1968).

In the unanaesthetized, decapitated chick with a mid-cervical transection, both prostaglandin $F_{2\alpha}$ and E_1 potentiate twitches of the gastrocnemius muscle elicited reflexly by electrical stimulation of the cut central stump of the contralateral sciatic nerve (Fig. 5). This effect is observed with $F_{2\alpha}$ in the chloralose-anaesthetized chick with intact brain but in this preparation prostaglandin E_1 *inhibits* the crossed extensor reflex, suggesting that E_1 may inhibit supraspinal centres (or excite descending inhibitory pathways) which mask the facilitatory action of E_1 observed in the spinal chick (Horton and Main, 1967a).

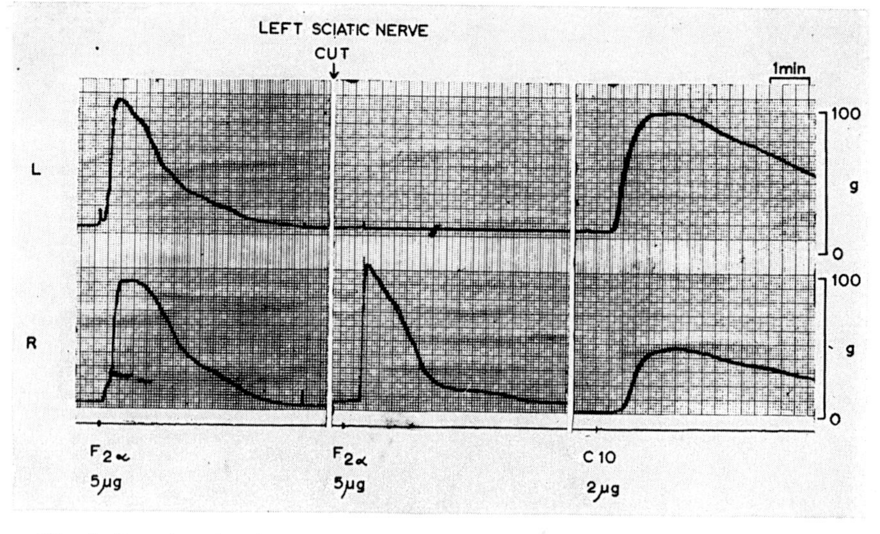

Fig. 4. Decapitated chick (32 g) spinalized in the mid-cervical region. Gastrocnemius muscle tension recorded isometrically (upper tracing left leg, lower tracing right leg). Responses to intravenous injections of prostaglandin $F_{2\alpha}$ (5 μg) and decamethonium iodide (2 μg). Between the first and second panel the left sciatic nerve was cut. There was an interval of 60 min before each injection. Reproduced with permission from Horton and Main (1967a).

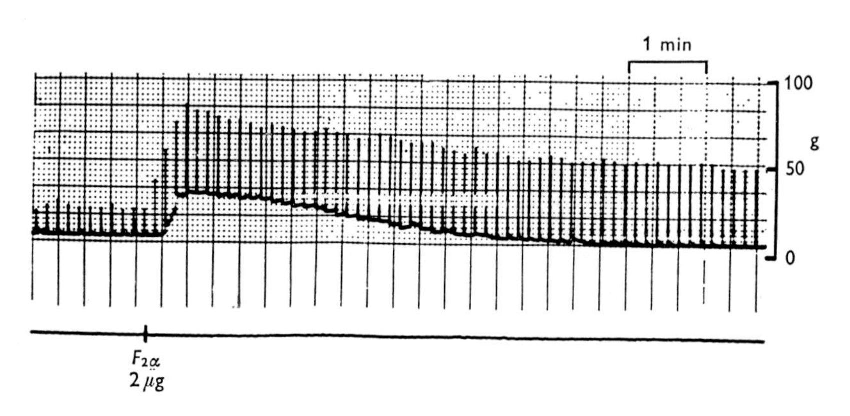

Fig. 5. Spinal chick (decapitated weight 45 g). Gastrocnemius muscle tension recorded isometrically. Reflex contractions of gastrocnemius muscle were elicited by 5 V pulses of 25 m duration. $F_{2\alpha}$=prostaglandin $F_{2\alpha}$ injected into the right external jugular vein. Reproduced with permission from Horton and Main (1965c).

The crossed extensor reflex is readily abolished in the chick by anaesthesia with urethane. However with this anaesthetic, tremor or shivering is very frequently observed. When tremor occurs it can be abolished by intravenous E_1 or $F_{2\alpha}$ (Horton and Main, 1967b).

Although tremor can also be abolished by direct application of E_1 to the cerebral hemispheres it is not certain that the mechanism of action is really a central nervous one. Changes in skin blood flow may account in part for this effect (Main and Wright, 1969).

The central nervous actions of prostaglandins on the chick are summarized in Table 2.

Table 2

Central nervous actions of prostaglandins in the chick

Anaesthetic	State of CNS	Route of injection	Prostaglandin	Response
None	Intact	Intravenous	E_1, E_2, E_3	Sedation, loss of righting reflex
			$F_{1\alpha}$, $F_{2\alpha}$	Contraction of extensor muscles
None	Spinal section at C4, decapitated	Intravenous	$F_{2\alpha}$	1. Contraction of gastrocnemius muscle 2. Potentiation of crossed extensor reflex
			E_1	Potentiation of crossed extensor reflex
Chloralose	Intact	Intravenous	$F_{2\alpha}$	1. Contraction of gastrocnemius muscle 2. Potentiation of crossed extensor reflex
			E_1	Inhibition of crossed extensor reflex
Urethane	Intact	Intravenous	$F_{2\alpha}$	1. Contraction of gastrocnemius muscle 2. Abolition of tremor
			E_1	Abolition of tremor
		Cerebral cortex	E_1	Abolition of tremor

2. *Cardiovascular Actions*

The effects of prostaglandins E_1 and $F_{2\alpha}$ have been observed on systemic arterial blood pressure in anaesthetized and spinal chicks. The results of one series of experiments (Horton and Main, 1967a) are summarized in Table 3. These prostaglandins exert little or no effect on the Locke-perfused chicken

Table 3

Actions of prostaglandins on the arterial blood pressure of anaesthetized and spinal chicks

Preparation	Prostaglandin	Number of experiments			
		Arterial blood pressure response			Total
		Pressor	Pressor-depressor	Depressor	
Spinal	$F_{2\alpha}$	14	0	0	14
	E_1	0	0	9	9
Chloralose anaesthesia	$F_{2\alpha}$	8	5	1	14
	E_1	0	0	8	8
Urethane anaesthesia	$F_{2\alpha}$	3	4	3	10
	E_1	0	0	15	15
All anaesthetized chicks	$F_{2\alpha}$	11	9	4	24
	E_1	0	0	23	23

heart and so the changes in blood pressure are probably due to changes in the peripheral vasculature. By analogy with mammals it is likely that prostaglandin E_1 is an arteriolar vasodilator whereas in the spinal chick the pressor action of $F_{2\alpha}$ may be attributable to venoconstriction and increased venous return thus increasing cardiac output as in the dog and the rat (DuCharme and Weeks, 1967). In contrast, in the anaesthetized chick with brain intact the response to $F_{2\alpha}$ was either depressor or had a strong depressor component (Horton and Main, 1967a). Whether this represents a supraspinal site of action or an inhibition of sympathetic tone (not present in the spinal chick) is unknown. The depressor action of prostaglandins $F_{1\alpha}$ and $F_{2\alpha}$ was confirmed in the chick anaesthetized with either chloralose, urethane or barbitone (de Boer *et al.*, unpublished).

Prostaglandin A_1 is a very weak depressor substance in the chick in contrast to its high vasodilator potency in mammals such as the cat (Jones, 1970). This represents a very significant and interesting class difference. As in mammals prostaglandin A_1 is not removed on passage through the lungs whereas approximately 90% of prostaglandin E_1 intravenously infused in the fowl does not appear in the aortic blood (Jones, 1970).

3. *Isolated Gastro-intestinal Smooth Muscle*

The isolated crop strip is contracted by prostaglandins E_1, $F_{2\alpha}$ and A_1. The mean threshold dose for E_1 was 2 ng/ml and prostaglandin A_1 had approximately 1% of the activity of E_1 (Horton and Jones, 1969).

Prostaglandin E_1 initiates, or if already present, enhances the rhythmic pendular contractions of the isolated chick oesophagus (Fig. 6). It also causes some potentiation of the contractile response of chick oesophagus longitudinal muscle produced by preganglionic parasympathetic nerve stimulation (Jones, 1970).

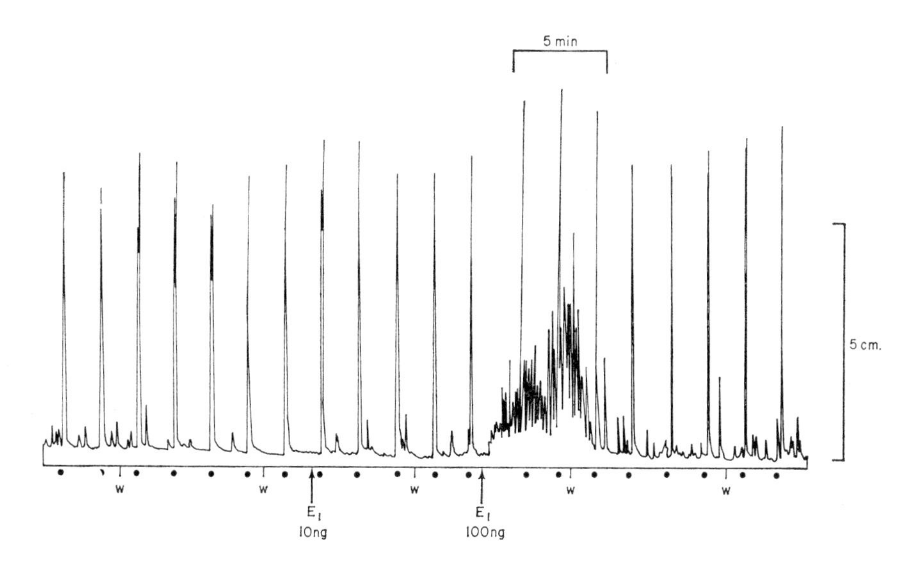

Fig. 6. Chick isolated oesophagus in 10 ml bath containing Krebs solution at 32°C gassed with 5% CO_2 in oxygen. Electrical stimulation of the right vagus at 2-min intervals with pulses of 10 V 0·5 ms at 5/s for 10 s. Contractions were recorded with an isotonic lever (10 × magnification and 2 g load). W = Wash; E_1 = prostaglandin E_1; Time marker 5 min. (Experiment by R. L. Jones.)

4. *Miscellaneous Actions*

Prostaglandins interfere with embryonic skin and feather development in the chick (Kischer, 1967). The mechanism of the effect and its significance await further investigation.

References

Ambache, N., Kavanagh, L. and Whiting, J. (1965). *J. Physiol., Lond.* **176**, 378–408.

Avanzino, G. L., Bradley, P. B. and Wolstencroft, J. H. (1966). *Br. J. Pharmac. Chemother.* **27**, 157–163.

Babilli, S. and Vogt, W. (1964). *J. Physiol., Lond.* **177**, 31*P*–32*P*.

Bennett, A., Friedmann, C. A. and Vane, J. R. (1967). *Nature, Lond.* **216**, 873–876.

Bergström, S. and Samuelsson, B. (1965). *A. Rev. Biochem.* **34**, 101–108.

Bergström, S. and Samuelsson, B. (eds.) (1967). "Prostaglandins". *Proceedings of the Second Nobel Symposium.* Almqvist and Wiksell, Stockholm.

Bergström, S., Carlson, L. A. and Orö, L. (1964a). *Acta physiol. scand.* **60**, 170–180.

Bergström, S., Danielsson, H. and Samuelsson, B. (1964b). *Biochim. biophys. Acta* **90**, 207–210.
Bergström, S., Carlson, L. A. and Weeks, J. R. (1968). *Pharmac. Rev.* **20**, 1–48.
Bygdeman, M., Hamberg, M. and Samuelsson, B. (1966). *Mem. Soc. Endocr.* **14**, 49–64.
Bygdeman, M., Kwon, S. and Wiquist, N. (1967). *In* "Proceedings of the Second Nobel Symposium" (S. Bergström and B. Samuelsson, eds.), pp. 93–96. Almqvist and Wiksell, Stockholm.
Davies, B. N., Horton, E. W. and Withrington, P. G. (1968). *Br. J. Pharmac. Chemother.* **32**, 127–135.
DuCharme, D. W. and Weeks, J. R. (1967). *In* "Proceedings of the Second Nobel Symposium" (S. Bergström and B. Samuelsson, eds.), pp. 173–181. Almqvist and Wiksell, Stockholm.
Duda, P., Horton, E. W. and McPherson, A. (1968). *J. Physiol., Lond.* **196**, 151–162.
Eliasson, R. and Posse, N. (1965). *Int. J. Fert.* **10**, 373–377.
Euler, U. S. von and Eliasson, R. (1968). "Prostaglandins". *Med. Chem. Monographs*, Vol. 8. Academic Press, New York.
Hawkins, D. F. and Labrum, A. H. (1961). *J. Reprod. Fert.* **2**, 1–10.
Herzog, J., Johnston, H. and Lauler, D. (1967). *Clin. Res.* **15**, 360.
Holmes, S. W. and Horton, E. W. (1968a). *In* "Prostaglandin Symposium of the Worcester Foundation for Experimental Biology", pp. 21–36. Interscience.
Holmes, S. W. and Horton, E. W. (1968b). *Br. J. Pharmac. Chemother.* **34**, 32–37.
Horton, E. W. (1964). *Br. J. Pharmac. Chemother.* **22**, 189–192.
Horton, E. W. (1968). *In* "Recent Advances in Pharmacology" (R. S. Stacey and J. M. Robson, eds.), Chapter 7. Churchill, London.
Horton, E. W. (1969). *Physiol. Rev.* **49**, 122–161.
Horton, E. W. and Jones, R. L. (1969). *Br. J. Pharmac. Chemother.* **37**, 705–722.
Horton, E. W. and Main, I. H. M. (1965a). *Int. J. Neuropharmac.* **4**, 65–69.
Horton, E. W. and Main, I. H. M. (1965b). *Br. J. Pharmac. Chemother.* **24**, 470–476.
Horton, E. W. and Main, I. H. M. (1965c). *J. Physiol., Lond.* **179**, 18*P*–20*P*.
Horton, E. W. and Main, I. H. M. (1967a). *Br. J. Pharmac. Chemother.* **30**, 568–581.
Horton, E. W. and Main, I. H. M. (1967b). *In* "Proceedings of the Second Nobel Symposium" (S. Bergström and B. Samuelsson, eds.), pp. 253–260. Almqvist and Wiksell, Stockholm.
Horton, E. W. and Main, I. H. M. (1967c). *Br. J. Pharmac. Chemother.* **30**, 582–602.
Horton, E. W. and Main, I. H. M. (1969). *In* "Prostaglandins, Peptides and Amines" (P. Mantegazza and E. W. Horton, eds.), pp. 121–122. Academic Press, London and New York.
Horton, E. W., Main, I. H. M. and Thompson, C. J. (1965). *J. Physiol., Lond.* **180**, 514–528.
Horton, E. W., Main, I. H. M., Thompson, C. J. and Wright, P. M. (1968). *Gut* **9**, 655–658.
Jones, R. L. (1970). *Ph.D. thesis, University of London.*
Karim, S. M. M. (1968). *Br. med. J.* **4**, 618–621.
Karim, S. M. M., Trussell, R. R., Patel, R. C. and Hillier, K. (1968). *Br. med. J.* **4**, 621–623.

Kischer, C. W. (1967). *Devl Biol.* **16**, 203–215.
Kloeze, J. (1967). *In* "Proceedings of the Second Nobel Symposium" (S. Bergström and B. Samuelsson, eds.), pp. 241–252. Almqvist and Wiksell, Stockholm.
Lee, J. B. and Ferguson, J. F. (1969). *Nature, Lond.* **222**, 1185–1186.
Main, I. H. M. (1964). *Br. J. Pharmac. Chemother.* **22**, 511–519.
Main, I. H. M. and Wright, P. M. (1969). *In* "Prostaglandins, Peptides and Amines" (P. Mantegazza and E. W. Horton, eds.), pp. 125–127. Academic Press, London and New York.
Misiewicz, J. J., Waller, S. L., Kiley, N. and Horton, E. W. (1969). *Lancet* **i**, 648–651.
Phillis, J. W. and Tebecis, A. K. (1968). *Nature, Lond.* **217**, 1076–1077.
Pickles, V. R. (1967). *Biol. Rev.* **42**, 614–652.
Ramwell, P. W. and Shaw, J. E. (1967). *In* "Proceedings of the Second Nobel Symposium" (S. Bergström and B. Samuelsson, eds.), pp. 283–292. Almqvist and Wiksell, Stockholm.
Ramwell, P. W. and Shaw, J. E. (eds.) (1968). "Prostaglandin Symposium of the Worcester Foundation for Experimental Biology". Interscience, New York.
Ramwell, P. W., Shaw, J. E. and Kucharski, J. (1965). *Science, N.Y.* **149**, 1390–1391.
Robert, A., Nezamis, J. E. and Phillips, J. P. (1967). *Am. J. dig. Dis.* **12**, 1073–1076.
Samuelsson, B. (1968). *In* "Prostaglandin Symposium of the Worcester Foundation for Experimental Biology" (P. W. Ramwell and J. E. Shaw, eds.), pp. 8–9. Interscience, New York.
Sandberg, F., Ingelman-Sandberg, A., Joelsson, I. and Rydén, G. (1967). *In* "Proceedings of the Second Nobel Symposium" (S. Bergström and B. Samuelsson, eds.), pp. 91–92. Almqvist and Wiksell, Stockholm.
Shaw, J. E. (1966). *Fedn Proc. Fedn Am. Socs exp. Biol.* **25**, 770.
Shaw, J. E. and Ramwell, P. W. (eds.) (1968). *In* "Prostaglandin Symposium of the Worcester Foundation for Experimental Biology", pp. 55–64. Interscience, New York.
Steinberg, D., Vaughan, M., Nestel, P. J. and Bergström, S. (1963). *Biochem. Pharmac.* **12**, 764–766.
Steinberg, D., Vaughan, M., Nestel, P. J., Strand, O. and Bergström, S. (1964). *J. clin. Invest.* **43**, 1533–1540.
Thompson, C. J., Los, M. and Horton, E. W. (1970). *Life Sci.* **9**, 983–988.
Van Dorp, D. A. (1966). *Mem. Soc. Endocr.* **14**, 39–46.
Van Dorp, D. A., Beerthuis, R. K., Nugteren, D. H. and Vonkeman, H. (1964). *Biochim. biophys. Acta* **90**, 204–207.
Williams, E. D., Karim, S. M. M. and Sandler, M. (1968). *Lancet* **i**, 22–23.

Author Index

Italic numbers are those pages where references are listed in the text.

B

C

D

F

H

K

M

N

O

P

Q

R

S

T

U

V

W

Y

Z

Subject Index

Numbers in italic type indicate pages on which figures occur.

C

F

J

K

L

N

P

S

Abbreviations used in the Text

AAAD, adrenal ascorbic acid depletion
ACTH, corticotrophin
ADP, adenosine diphosphate
ALD, anterior latissimus dorsi
AMP, adenosine monophosphate
"APP", (*see* polypeptides)
ATPase, adenosine triphosphatase
ATP, adenosine triphosphate
A-V, atrio-ventricular

BMR, basal metabolic rate
BSP, bromsulphonethalein

CBP, calcium binding protein
CoA, coenzyme A
CoE I, coenzyme I, nicotinamide adenine dinucleotide
CoE II, coenzyme II, nicotinamide adenine dinucleotide phosphate
CM, chylomicron
CNS, central nervous system
CRF, corticotrophin releasing factor

DBC, 2,6-di-tertiarybutyl-*p*-cresol
3,6 DBSP, 3,6-dibromsulphonphthalein
DCT, distal convoluted tubules
DHEA, dehydroepiandrosterone
DHT, 5α-dihydrotestosterone
DNA, deoxyribonucleic acid
DPN, nicotinamide-adenine dinucleotide (NAD)
DPNH, reduced DPN
DPPD, diphenyl-*p*-phenylenediamine

ECG, electrocardiogram
ECPD, electrochemical potential difference
EDTA, ethylene diamine tetra acetic acid
EEG, electroencephalogram, electroencephalography
EFA, essential fatty acids
epp, end plate potential
ERG, electroretinogram
ESR, erythrocyte sedimentation rate
EV, erythrocyte volume

FAD, flavine adenine dinucleotide
FFA, free fatty acids
FH_4, tetrahydrofolic acid
FMN, flavin mononucleotide
FSH, follicle-stimulating hormone

GABA, γ-amino butyric acid
GAS, general adaptation syndrome
GLC, gas-liquid chromatography
GOT, glutamic oxaloacetic transaminase
GPC, glyceryl phosphoryl choline
GPT, glutamic pyruvic transaminase
GTP, guanosine triphosphate
GVHR, graft-versus-host reaction

HDL, high density lipoproteins
HIOMT, hydroxyindole-*O*-methyl transferase
20α-HO-P, 20β-HO-P, 20α,20β hydroxy-preg-4-ene-3-one
Δ5-3β-HSDH, Δ5-3β-hydroxysteroid dehydrogenase
5-HT, 5-hydroxytryptamine

ICG, indocyanine green
ICI, Imperial Chemical Industries
ICSH, interstitial cell stimulating hormone
Ig, immunoglobulin

LDF, low density fraction
LDH, lactic dehydrogenase
LDL, low density lipoproteins
LH, luteinizing hormone
LSD, lysergic acid diethylamide

MCHC, mean corpuscular haemoglobin concentration
MDH, malic dehydrogenase

ME, metabolizable energy
mepp, miniature end plate potential
MAOI, monoamine oxidase inhibitors
MSH, melanocyte-stimulating hormone

NAD, nicotinamide adenine dinucleotide
NADH, reduced nicotinamide adenine dinucleotide
NADP, nicotinamide adenine dinucleotide phosphate
NADPH, reduced nicotinamide adenine dinucleotide phosphate
NDV, Newcastle disease virus
NEFA, non-essential fatty acids
NPN, non-protein nitrogen

OAAD, ovarian ascorbic acid depletion
5-OHDA, 5-hydroxydopamine
6-OHDA, 6-hydroxydopamine
OIH, ovulation-inducing hormone

PAPS, 3′-phosphoadenosine-5′-phosphosulphate
PAS, periodic acid-Schiff reaction (in this publication, *not* paraamino salicylic acid)
PBI, protein bound iodine
PC, phosphoryl choline
PCT, proximal convoluted tubules
PCV, packed cell volume
PLD, posterior latissimus dorsi
PLP, phospholipoprotein
PMS, pregnant mare's serum
PPP, pentose phosphate pathway
PTE, parathyroid extract
PTH, parathyroid hormone

RNA, ribonucleic acid
RQ, respiratory quotient
RRF, responsive receptive field

S-A, sino-atrial
SDH, sorbitol (D-glucitol) dehydrogenase

TBW, total body water
TLC, thin layer chromatography
TSH, thyroid-stimulating hormone

UDP, uridine diphosphate

VFA, volatile fatty acids
VLDL, very low density lipoproteins

WSF, water-soluble fraction
WSOP, water-soluble oviduct proteins